International Microwave Handbook

Edited by Andy Barter, G8ATD

Published jointly by The Radio Society of Great Britain, Cranborne Road, Potters Bar, Herts, EN6 3JE, UK and The American Radio Relay League, Inc, 225 Main Street, Newington, CT, 06111-1494 USA.

First published 2002

ISBN 1-872309-83-6

Publisher's note

The opinions expressed in this book are those of the authors and not necessarily those of the RSGB or ARRL. While the information presented is believed to be correct, the authors, the publisher and their agents cannot accept responsibility for the consequences arising for any inaccuracies or omissions.

Cover design: Anne McVicar

Production: Mark Allgar

Typography: Andy Barter, K M Publications, Luton

Printed in Great Britain by Nuffield Press Ltd of Oxford

Acknowledgements

In this edition of the International Microwave Handbook contributions have been assembled from around the world. The editor would like to thank the following radio amateurs for their contributions:

William A Parmley	KR8L
Curtis W Preuss	WB2V
G Sabbadini	I2SG
Gunthard Kraus	DG8GB
Gregor Storz	ZL1GSG
Bernd Kaa	DG4RBF
Wolfgang Schneider	DJ8ES
Matjaz Vidmar	S53MV
Mihael Kuhne	DB6NT
Angel Vilaseca	HB9SLV
Dieter Briggmann	DC6GC
Noel Hunkeler	F5JIO
Rainer Bertelsmeier	DJ9BV
Leo Lorentzen	OZ3TZ
Harald Fleckner	DC8UG
Phillipp Prinz	DL2AM
Werner Rahe	DC8NR
Paul Wade	W1GHZ
Peter Vogl	DL1RQ
Luis Cupido	CT1DMK
John Stephenson	KD6OZH
Hans Hellmuth	DL2CH

The following organisations, publications and companies have also provided material for this edition.

RSGB
ARRL
QST
QEX
CQ DL
CQ ZRS
OZ
DUBUS
VHF Communications
Agilent
Mini Kits

Finally the book would not be what it is without the help of two proof readers who have, hopefully, found all of the major mistakes. They were Pat Brambley and Don Ross, G4LOO.

There are also other contributors to the book since it has its roots in the three volume Microwave Handbook published by The RSGB in the late eighties and early nineties. Those three books were compiled by a group of dedicated radio amateurs who are acknowledged below.

M Dixon	G3PFR
J N Gannaway	G3YGF
B Chambers	G8AGN
M H Walters	G3JVL
S J Davies	G4KNZ/ZL2AZQ
S T Jewell	G4DDK
D J Robinson	WR3E/G4FRE
C W Suckling	G3WDG
H W Rees	G3HWR
S Page	GW3XYW

G4CNV	G4KGC	G4FSG
G3RPE	G3PHO	G4MBS
G8MWR	G3SEK	OE9PMJ
DC8UG	G3BNL	G4LQR
G3JHM	OZ9CR	G4PMK
G4HUP	G8TIR	G4COM
G8DEK	G4DDN	G6WWM
HB9MIN	N W Kent Beacon Group	

The original three book set also contained contributions from other publications and companies who are acknowledged below.

Radio Communication
VHF Communications
Dubus
CQ TV
QST
Microwave Journal
Microwave System News
Wireless World
Reprints by permission of Prentice-Hall
Andrew Corporation
Agilent (Hewlett-Packard)
Mitsubishi Electric UK
Avantek
NEC
Plessey Semiconductors
Microwave Associates

Contents

This is the first International Microwave Handbook to be published jointly by The Radio Society of Great Britain (RSGB) and The American Radio Relay League (ARRL). The microwave bands are an area for the amateurs who want to experiment and construct their own equipment, so there is a need for an source of good reference information plus good designs to fire their imagination.

This handbook owes a great deal to the three volume Microwave Handbook published by the RSGB in the late eighties and early nineties. In fact the theory sections are mainly taken from books one and two of that set because they are still a very good theoretical reference. The content of these books was very good and is a tribute to the efforts of the Microwave Committee who put the effort into the contents. Unfortunately the material used was not available in modern electronic format, so it has been reconstructed. Some of the diagrams are not of the best quality but it was felt more important to include them rather than loose this important reference work.

The techniques and devices available to the microwave experimenter are better now than they have ever been. The explosion in consumer electronics using microwave frequencies has given us semiconductor devices with impressive parameters at reasonable prices. The internet has given us an almost endless supply of data on new devices and access to much modern microwave design software. It is invaluable to see how other amateurs have used this information to produce working equipment, these ideas can be taken and used as they are or be used to improve the design.

The band chapters of this book contain a selection of designs using the latest technology that can reasonably be used by amateurs. There is a variation of designs ranging from ones that can be reproduced by most amateurs to those that require a high level of skill to make. One problem is that the designs often require very accurately produced printed circuit boards and it is not possible to reproduce them in a book like this in sufficient detail for reproduction. Fortunately they are usually available from the authors of the articles either as just PCBs or as complete kits of parts. The designs are from many different countries, this gives a good cross section of different techniques.

Andy Barter, G8ATD

Chapter 1

Operating Techniques

In this chapter :

- Operating Procedures
- Fixed Station Operation
- Portable Operation
- Microwave Calculations

Microwave operation should not be regarded simply as an extension of the operating techniques practised on the lower-frequency bands. Although some similarities do exist, there are many more ways in which practical microwave operation differs from the norm and it is the purpose of this chapter to outline these differences.

In general it is possible for the amateur to easily generate high powers only on the two lowest frequency bands, viz 1.3 and 2.3GHz. On 3.4GHz and above, the amateur is often confined to power levels which these days are regarded as QRP; that is, output levels of between a few milliwatts and say 10W. Feeder losses increase rapidly with frequency (hence the use of waveguide at the higher frequencies). These limitations are offset to some extent by using high-gain antenna systems, short lengths of high-quality feeder and low-noise receive preamplifiers, usually built around gallium arsenide fets and mounted at or very close to the antenna. Transmit multipliers may also be sited at or close to the antenna.

These facts, coupled with financial and other considerations, may determine whether the majority of the operation is from a fixed-station location or portable from hill-top sites. The advantages of a better site can often outweigh the disadvantages of low powers and other equipment limitations.

One very significant facet of microwave operation which is different to practices on other bands is that under normal conditions involving line-of-sight or troposcatter propagation, not only should the equipment capability be quite closely known but also the path losses can be quite accurately assessed (see Chapter 2, Systems analysis, and later in this chapter).

However, unknown factors such as obstructions or enhanced modes of propagation often provide the amateur with the thrill and excitement of working other stations at unexpected signal strengths and distances.

The contents of this chapter are divided into four main sections which deal with procedures, fixed-station operating, portable operating and the use of calculations as an aid to operating.

Operating Procedures

Comparison with HF/VHF

The equipment and some of the operating techniques needed to establish a contact on the microwave bands can be very different to those used at hf and vhf. The antennas used are highly directional and the general level of activity is much lower.

High-gain (and hence highly directional) antennas are an essential feature of most microwave stations because they are necessary to compensate for the much weaker signals which would be received by a dipole, which is physically much smaller at these frequencies. Feeder losses are higher, and transmitters and receivers often less potent than on the lower bands.

Although, on some of the shorter line-of-sight paths, signals may be strong enough to be found easily using the techniques common on the HF/VHF bands, many paths will be either obstructed or of such a length that the signals will not be much above noise even when all the antennas and equipment are correctly aligned, and in these cases a systematic approach is almost essential to guarantee success. Operating protocols are discussed later in the Procedures section, but it is probably wise at this point to summarise the main conditions to be met to guarantee a contact;

- The equipment must be capable of overcoming the path loss, so it must be reliable and its performance known, this should be established before use. The path losses are often fairly predictable and can be calculated beforehand to give the operator some idea of the likely levels of signal to be expected.

- Both stations must have their antennas pointed at each other to within the beamwidths. Using higher-gain antennas or working at higher frequencies both reduce the antenna beamwidths, so that they have to be positioned more accurately. This requires knowledge of the locations of both stations, calculation of the beam headings and a means of setting them. The beamwidths of some typical antennas are given in Table 1.1.

- The receiver must be set to within its bandwidth of the transmit frequency. Improvements in both receiver and transmitter stability have resulted in the use of narrower bandwidths at higher frequencies, so that frequencies need to be measured more accurately. Both stations should have a reference available with which frequencies can be set to a fraction of the receiver bandwidth. Table 1.2 gives the accuracies required for some of the typical modes and frequencies and a means of achieving them.

- The transmitter must be on when the receiving station is listening for it! Thus a strict protocol should be adopted to ensure this state of affairs.

If all these conditions are followed, then a test should just consist of listening: when the path loss falls to a value which the equipment can overcome, signals will be heard.

Table 1.1 : Comparison of antenna gains and beamwidths for various bands

Antenna	Band (MHz)	Gain (dB)	Beamwidth (degrees)
16-ele Yagi	144	16	24
17ft dish	144	16	26
25-ele loop Yagi	432	19	18
20ft dish	432	26	8
30ft dish	432	30	5
4 x loop Yagi	1,296	24	10
6ft dish	1,296	25	9
1ft dish	10GHz	28	7
4ft dish	10GHz	40	1.7
1ft dish	24GHz	35	3
4ft dish	24GHz	47	0.7

Table 1.2 : Marker frequency accuracies and marker crystal characteristics

Accuracy	Frequency	Fractional accuracy
100Hz	432MHz	2 parts in 10^7
1KHz	1,296MHz	8 parts in 10^7
1KHz	10GHz	1 part in 10^7
100kHz	10GHz	1 part in 10^5
1kHz	24GHz	4 parts in 10^8
100kHz	24GHz	4 parts in 10^6

Typical tolerances of crystals	
As built in circuit ± 30ppm	3 parts in 10^5
Variation with temp (0-60°) ± 30 ppm	3 parts in 10^5
Best case (10-30°)	2 parts in 10^6
Ovened crystal	1 part in 10^8
Off-air standard	1 part in 10^{10} to 10^{11}

This may all seem very obvious, but many of the developments which improve the performance of amateur equipment, for instance higher-gain antennas and narrower bandwidths, also make these requirements harder to meet.

If the operator must do something while listening, then tuning the receiver over a few bandwidths is acceptable and can help concentration. In practice, most people seem to relax one or two of these conditions considerably, usually by tuning a considerable distance on the receiver, and searching with the antenna heading. One can get away with this if the signals are fairly strong and consistent, but the chances of finding a weak or fading signal are greatly reduced. Consider the number of possible combinations of frequency and beam direction in the following example (on 10GHz):

Receiver bandwidth:	300kHz
Tuning range:	30MHz, giving 100 possible channels
Antenna beamwidth:	3°
Uncertainty in direction:	15°, giving five different headings

Thus there are 5 x 100 = 500 possible combinations to try. This searching considerably reduces the time spent listening in the correct place and so greatly reduces the chance of finding the signal, particularly if the equipment is also unreliable!

This has admittedly been a description of an ideal approach, but the closer one can get to it, the more successful the operator will be in the tests. Familiarity with, and successful experience in using, the equipment is the best way to become convinced of the effectiveness of this type of approach. When the equipment can be set up to these standards it will be found that signals which are only just above noise level can be detected, and the operator will then begin to notice the benefits obtainable by squeezing every decibel of performance out of the equipment.

With regard to frequency measurement, most absorption wavemeters are far too inaccurate to set a receiver frequency on their own since the 3dB bandwidth of the wavemeter (for instance on 10GHz) may be between 3 and 100MHz, depending on the Q of the wavemeter. A very useful target to aim for would be a receiver scale with 100kHz divisions which, if combined with a crystal calibrator generating pips every 10 to 100MHz in the band, would be accurate enough for most simple wideband equipment. An accuracy of 10ppm in the crystal will give 100kHz at 10GHz. Typical accuracies are given in Table 1.2.

For narrow band work, an accuracy of several orders of magnitude better is required. This could be obtained with an ovened crystal but this still has to be set on frequency in the first place using some more accurate reference.

The real solution the problem of accurate frequency standards is to compare or phase-lock the crystal calibrator to an off-air frequency standard such as Droitwich or Rugby. Many other standard frequency transmissions exist as part of a world-wide service; the two stations quoted are of most use to the UK operator. If both stations do this or even lock their transmitter or receiver local oscillators to this standard, then accuracy is guaranteed. Some of the standard transmissions also carry other useful information such as time signals (Rugby) and weather forecasts (Droitwich). An accurate reference such as this is also useful on the lower-frequency bands, and details of a suitable receiver and harmonic generator are given in [1].

Talkback

Regardless of the technical standard of the equipment, in very few instances (excepting perhaps, 1.3GHz) is it possible to call CQ on a microwave band and expect a reply. It is much more common to arrange a sked, a series of tests between respective stations, or to use a lower-frequency band (for instance 144 or 432MHz) to establish initial contact and possible to act as a talkback channel.. In the UK it has been common practice for some years to use 144.175MHz +/- 25kHz, (ssb) (or 144.33MHz +/- 10kHz) as microwave calling windows, especially during contests. Due to the very crowded nature of the 144-146 MHz band in continental Europe, it has become established practice there to call for microwave contacts on either 432.35MHz +/- 25kHz (ssb) or on 1,296.2MHz +/- 25kHz (ssb).

It is therefore desirable that the operator should have multimode facilities available for both 144 and 432MHz, to set up contacts on the higher bands for instances using 144MHz to set up contacts on 432 and 1,296MHz, or 432MHz to set up contacts on 1.3GHz and higher. Whichever method is used it may pay dividends to put out a call such as CQ from xxxx, looking for contacts on a,b,c band and listening on this frequency for any replies.

Procedures

Whether a sked or a series of tests is being undertaken it is good and sensible practice to adopt a disciplined, timed sequence of transmitting and receiving on the microwave band similar to that adopted for eme or meteor scatter QSOs. Although the equipment parameters and the basic path loss may have been estimated, there are usually some unknowns remaining, and the adoption of such a procedure is more likely to lead to successful contacts than the rather slacker procedures often used on the lower frequencies.

With these points in mind it is suggested that such a procedure and timetable be agreed with the other station before moving up in frequency to the band where the attempted contact is to take place. The requirements of such a protocol are these:

• A clock or wristwatch accurate to within a few seconds of standard time.

• Regular transmit/receive periods, for example 1min, 5min or any other mutually agreed period, changing on the minute.

• Agreement as to when to start, who starts, and for how long the sked should continue.

• If the sked fails, a return to talkback (or telephone!) to fix another sked for some time later.

• Agreement on a locator system to enable both stations to calculate the bearings and distances, for example latitude/longitude, National Grid References, or derived from them, for example the Maidenhead locator.

• The use, at least initially, of cw or ssb, but preferably the former.

• Frequency setting must be sufficiently accurate at both ends of the path to allow each operator to find the others signals.

Forecasting conditions

With regard to lifts or openings, much can be learned and anticipated by a combination of tactics, many of which should already be familiar to the more experienced users of vhf and uhf. Regular viewing and interpretation of the synoptic (weather) charts either on television or in daily newspapers, looking for co-channel interference on uhf tv, abnormal reception of distant uhf repeaters or, more recently, by the direct acquisition of weather satellite data, should enable the operator to predict potential lifts. Much confirmatory (or otherwise) evidence can be gleaned from regular monitoring of beacons or repeaters, particularly on 432MHz and higher.

Start to look for lifts when there is stable high-pressure weather over the country, and particularly when this drifts slowly with a slow decline of pressure. In the UK an easterly drift will usually give good openings to Western Europe. These conditions are most common in September and November, but are also common in the late spring. Over the sea, ducts may form almost daily during stable weather in the summer months which are particularly noticeable on the higher microwave bands; they are particularly good around dawn and dusk.

As should be evident from Fig 2.34, Chapter 2, the effectiveness of a duct or layer enhancement of a given thickness is frequency dependent, and the probability of the formation of shallow layers is much higher than that of the formation of deep ones. It follows since there is a relationship between the layer thickness and its ability to propagate a particular frequency, that there is a higher probability of enhanced conditions occurring at microwave frequencies than at vhf or uhf.

Depending on the height and thickness of the layers, regular signals such as beacons may either exhibit greatly enhanced signal strengths when the transmitting and receiving antennas lie in or close to the layer(s), or greatly diminished strengths if either or both antennas lie significantly above or below the layer(s). This is frequently observed with the UK beacons GB3MLY (432MHz) and GB3MLE (1,296MHz), European dx may be clearly audible on both bands while the beacons are completely inaudible! However poor conditions on 144 or 432MHz do not necessarily imply poor conditions on the higher bands for the reasons given above.

In contrast, very bad weather can also provide enhanced signals. Very heavy rain or thunder cells can scatter enough energy to bring the signal levels well above the normal troposcatter level. These signals sound similar to auroral signals on vhf.

Nevertheless, regular listening, logging of signals received and the accumulation of meteorological observations should, in the course of time, not only improve the operators chances of detecting and exploiting enhancements but could also make a significant contribution to the further understanding of microwave propagation. Above all, therefore, the serious microwave operator should be methodical.

Fixed-Station Operation

Aims of fixed operation

Bearing in mind the constraints of limited power, feeder losses and size/gain of the antenna installation, the microwave operator should consider a number of factors when setting up a fixed station. It is usually not sufficient to approach microwave operation on as casual a basis as might be possible on the lower vhf or the hf bands, where a few watts to a piece of wire will always give some sort of results.

The first decision the operator must make is What are my objectives? Are they short-to-medium distance general communication, troposcatter, use of tropo-enhancement modes, eme or what? These questions should be answered because, for terrestrial communications, site and antenna/feeder considerations are far more important in determining station effectiveness than at lower frequencies. The answer to such questions will, to a great extent, determine the band(s), power, mode(s) and antenna/feeder system aspirations of the operator. These might vary, for example, from full legal power and a large, fully steerable dish on 1.3 or 2.3GHz for eme down to a few milliwatts of wideband fm to a small dish or horn for semi-local communication on 10GHz.

The next consideration, having decided the objectives, is to see how these will fit into what may be broadly considered to be the 'domestic scene'. Is the site reasonably in the clear? Will there be planning consent problems for a large antenna installation? Is the feeder-run of minimum practicable length? And so on.

Finally must come the question of resourcing is, the time, skill and finance available or does the operator need to start in a modest way and develop the system as fast as these constraints allow? In general it is not too costly, time-consuming or technically demanding to produce a few hundred milliwatts to a few watts on the lower microwave bands, or a few tens of milliwatts on the higher bands. What will be costly, time-consuming and technically demanding, is the establishment of a first class antenna/feeder system, high power and multi-mode capability which will provide the user with the means to push terrestrial communications potential to its limits (when operated well!) and possibly, fulfil the stringent requirements of eme work. If the operator is prepared to compromise and sacrifice this ultimate performance for a usable, practical and less-demanding system which can be developed as resources permit, then communications will be assured over limited distances (perhaps a few tens to a few hundred km) and with limited modes under normal conditions; under lift conditions these distances could extend enormously and even the relatively poorly equipped station will give its operator the chance to secure dx contacts.

Antenna systems

On the lower bands (1.3, 2.3 and to some extent 3.4GHz) the choice of antenna will lie in the range from a single conventional or quad-loop Yagi through stacked/bayed multiple Yagis to, typically, a skeleton dish of perhaps 1 to 3m in diameter mounted on the station mast and fully rotatable. Larger dishes used for eme are usually mounted at ground level and need to be fully steerable. Details of suitable antennas for these bands are given in Chapter 3. With regard to the feeder, which is effectively an integral part of the design of a microwave antenna system, the principle to follow is the shortest length of the best-quality cable the operator can afford. Figures for feeder losses are given in Appendix I, and from this and the manufacturer's price list a choice can be made. It is, as far as the amateur is concerned, invariably a cost/effectiveness compromise.

For the higher bands (5.7GHz and up), the choice of antenna is relatively limited; a skeleton dish of perhaps 0.5 to 2m diameter or preferably a fly-swatter or periscope antenna will serve for all bands with the advantage that, with proper planning and installation, it can minimise or eliminate the need for a feeder and its associated losses. Again, some details of this type of antenna and feed assemblies are given in Chapter 3. There seems to be a growing practice (especially in continental Europe) to employ a skeleton dish with a single feeder and multiband feed covering typically 1.3 to 5.7GHz, and this approach should be considered a serious option. Suitable multiband feeds are described in (3), (4) and (5). It should be realised that such feeds represent a compromise solution and they are not particularly efficient compared with a dedicated, single-band feed. However, they may be acceptable to operators who are unable to accommodate more than single dish antenna on their mast, or afford to duplicate the feeder to more than one antenna.

At 10 and 24GHz a small solid dish of 0.3 to 0.6m diameter or a 20dB horn becomes a practical proposition, and it is desirable for the transceiver to be mounted near the dish to minimise the high feeder losses associated with such frequencies such a system might even be considered for 3.4 and 5.7GHz where the transmit multipliers and at least the receive preamp could be mast-mounted.

For all bands, receive performance is always greatly improved by the use of a low-noise mast-head receive preamplifier connected as close to the antenna as possible. The reader is referred to the appropriate band chapters for ideas regarding preamplifiers, switching relays, feeders and their interconnection.

Summary

To summarise, it is felt that the operator must first decide what he or she wants to achieve on the microwave band(s). Second, the operator must then decide how these objectives are to be attained, and then finally select the designs for equipment and antennas best suited to these needs. There is relatively little commercially produced equipment for microwave amateur use compared with the hf and vhf bands. This mean that the operator will often have to build from scratch or adapt existing (often surplus) equipment for these bands. This may be another very good reason for starting with relatively simple equipment and developing its potential as the need arises.

Portable Operation

Advantages of portable operation

If, as outlined above, the fixed station operator is often restricted to using low power, limited modes, small antennas and possibly high feeder losses from an indifferent site, effectiveness of operation can be maximised by adopting the techniques already outlined. Such an operator may still be able to obtain similar results to better-equipped fixed stations by the relatively simple expedient of taking his or her microwave equipment out portable to a better site at quite short notice if preparations are made in advance. By using the band and weather observations previously discussed and coupling these with snap portable operation it may prove possible to catch many of the quite spectacular enhancements that occur from time to time.

When very-low-loss propagation modes are involved, it matters little whether the signal generated is 100mW, 1W, 10W or more or whether the antenna/feeder and transmitter/receiver is highly sophisticated. The main advantage of a clear portable site is that local site obstructions and long feeders with their attendant losses can be largely avoided. Under these circumstances the operator can obtain good results from barefoot transverters or transmitter-receiver combinations. The more ambitious portable operator may choose to increase both power and antenna size as experience or needs increase, and in these respects portable operation may follow the same course of development as that outlined for fixed-station operation.

The same advice as for fixed-station operation should be followed with regard to the use of talkback and orderly procedures; again the use of cw, at least initially, is advised as the most potent means of communication, with ssb a close second best.

Whether snap or more regular portable operation is undertaken, then certain aspects of this type of operation should be carefully considered beforehand. It is possible to do much effective preparative work before undertaking such operation and this is as desirable as when planning a full-sized expedition. Although the logistics may be very much simpler, the principles remain the same.

First a packing list should be compiled to check that all equipment needed is available, known to be functional and can be loaded systematically (and safely) into the vehicle. Such a list is, to an extent, one of individual preference, but might look something like the one given below, depending on the band(s) involved and the type of site to be used.

Path list	Spare car keys
Licence copy	Microwave transceiver
Logbook	Microwave dish/antenna
Compass	Dish feed/dipole etc
Maps	Fully charged batteries
Spirit level	Binoculars
Notepad and pen	Cassette recorder
Talkback transceiver	Cables (all)
Talkback antenna	Spare clothes
Mast and guys	Thermos flask (full)
Tripod	Tools
Plastic bags	Spares

Choice of site

In choosing a site, reference may be made to detailed maps, first at the 1:625,000 scale to give some indication of the general topography of the terrain out to, say, 100km, and then at the 1:50,000 scale (or larger) which should indicate the local path obstructions which may play a significant part in the success or failure of operation from the site, particularly at the higher frequencies and under normal propagation conditions. The quality of the site can be quantified by calculating the extra path losses involved due to the obstructions that form the horizon in various directions. Details of troposcatter losses are given in (2).

Having decided on a site, check the access; if it is a roadside site served by a public road there will usually be no difficulty in operating. However, the site may be privately owned or be part of an estate or a national park, in which case there may be restrictions imposed on the operator. It is thus a good idea to check these points beforehand. Permission will seldom be refused if the landowner is approached in the right manner beforehand and the operator shows a sensible and responsible attitude while on the site. This will usually mean the adoption of some form of country code, most of which is common sense:

- Avoid climbing over walls and fences, and close gates.

- Do not leave litter.

- Keep to recognised footpaths as far as possible.

- Avoid disturbing livestock.

- Avoid the risk of fire or other damage

- Generally behave as one would expect visitors to behave when in ones own home or garden.

At the end of operation a word of thanks to the site owner is common courtesy and will often assist in the granting of permission to operate on future occasions.

When surveying the potential of a site the operator should work out most of the distances, bearings and path profiles in advance of operation simply by reference to maps. This is particularly appropriate for operation on the higher bands such as 10 and 24GHz or above, and a few hours spent in doing such work will free the valuable time on site for actually operating the station. By maintaining an up to date file of path details, quick reference can be made in the field and the chances of successful contacts greatly improved. Once such a file has been constructed it does not need to be done again, only modified to extended.

In the UK the microwave operator is singularly well provided with detailed maps in the Ordnance Survey (OS) series and has the choice of using the National Grid Reference (NGR) system, latitude and longitude, or the Maidenhead system which is based on latitude and longitude when calculating bearings and distances. These maps contain the information for both systems. NGRs are widely used within the UK, while the latter are more commonly used for longer-distance, international contacts. Several simple computer programs are available from the Internet (7) that should greatly simplify such calculations. Path profiles can be manually plotted using earth-profile plotting paper, an example of which is reproduced in fig 1.1. Alternatively a simple path-plotting program may be found on the Internet (8).

Armed with this information, orderly operating techniques as earlier discussed (and perhaps an element of skill and luck) the chances of successful operation will be controlled mainly by conditions. One additional item of equipment might be a special log sheet such as that shown in fig 1.2. This is just one of the possibilities designed to assist in the accumulation of useful data in an easily accessible form and which could be used to supplement the main station log or the portable log. No doubt the keen and methodical operator will devise his or her own sheet to suit the particular needs of the station.

Antennas and masts

Both the talkback and the microwave antennas can, unless very large, be supported on a single, well-guyed mast 3.5 to 8m (10 to 25ft) long, constructed of tubing which plugs or screws together, each section of which is short enough to fit easily within the available car space. Alternatively on the higher bands, for instance 10 or 24GHz, the most likely type of antenna will be a dish of typically 46cm (18in) to 92cm (36in) diameter. In this case, a lightweight sectional mast might be used to support the talkback antenna (such as an HB9CV or small Yagi), while the microwave antenna and transceiver might be mounted on a tripod or on the car roof. While mounting the equipment on the car roof does restrict the sites that can be accessed, it does provide a simple, stable platform to operate from, with power and shelter conveniently at hand. Both mast and tripod should be sited close together so that both sets of equipment can be operated simultaneously and the tripod should be guyed as well as the mast. The dish will present considerable windage and unless adequate guyed is likely to blow over, causing damage to the equipment, perhaps even before the operating session has commenced. The mast should be free enough to rotate easily, but not so free that even a light wind will cause it to blow round like a weather vane.

The microwave antenna mount should be capable of being clamped in place once it is aligned in the right direction. One advantage of using a single mast is that the ground-plate can be fitted with a compass scale which can be aligned while erecting the mast. A pointer can be fitted to the base of the mast, adjacent to this scale, and a reasonably accurate beam bearing obtained from the scale without the need constantly to take compass bearings for each successive contact which might lie in different directions. For very lightweight operation the

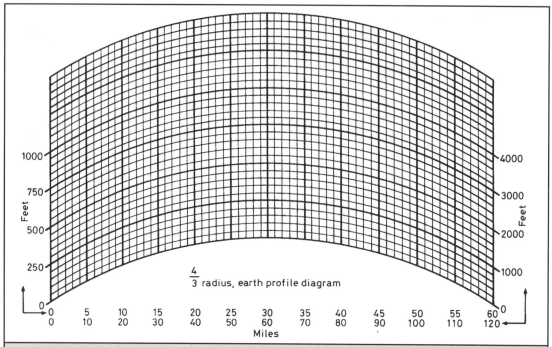

Fig 1.1 : Profile plotting sheet (altitude in feet, distance in statute miles - metric measures may, of course, be used if preferred). The home station altitude is entered at the 0 miles point and the distant station altitude at the approximate distance along either distance scale. Other spot heights at known distances are plotted in a similar manner. Joining the points together will produce a path profile. If a straight line joining the two stations together passes above the highest obstruction, then the path is line-of-sight. If it passes through either an obstruction or through the zero altitude line, then the path contains obstructions.

user might dispense with a mast altogether and rely on a wall, rock, fence post, cairn or survey trig-point to support the gear and antenna.

When using high-gain dishes having beamwidths of 5° or less, it is essential that not only does it point in the right direction but that it is also aligned in the vertical plane, that is the mast is truly vertical or the table of the tripod is truly horizontal. A spirit level will allow such alignment with a minimum of complication.

With regard to talk back, the operator should preferably try to use at least 25W on the talkback band. Very often, and for a great variety of reasons, the talkback signals may be considerably weaker than those received on the microwave band. Around 5 to 10dB antenna gain on the talkback band should be acceptable. On either the talkback band or the microwave band the modes used should be, in order of preference (effectiveness): cw, ssb, am, nbfm (with limiter) and wbfm. The aim of talkback is to provide a reliable communication link to assist the microwave operation. In some cases, the 432MHz repeater network may prove useful, particularly since it is not very heavily used. The telephone is, if available, also a very useful fallback if all else fails.

Bearings

When taking compass readings the operator should be well away from ferrous metal objects which may affect the accuracy of the readings cars, metal fences, power lines and even certain types of rock should be avoided. It is often advisable to take a series of readings from different places near the equipment in order to rule out the possibility of such errors. A suitable sighting compass can

be constructed from a map-reading compass as shown in fig 1.3(a). Sightings may be taken along the boom of a Yagi, along the axis of waveguide or across the face of a dish, not forgetting to add (or subtract) 90° to the compass sighting if this method is used. Sighting across the face of the dish (fig 1.3(b)) has the advantage that the operator is using an integral part of the antenna structure and that the sighting line can be at any angle of elevation, depending where the operator stands to take the sighting. This could be horizontally at landmarks or up to the sun, moon or stars without having to tilt the dish up at an angle. Sighting is made easier by painting the front and back of the dish in contrasting colours so that a sharp dividing line is seen from the side. At night the edges can be very clearly seen if a torch is shone to illuminate the front of the dish, leaving the back dark. To get good accuracy the operator should stand about 3m (10ft) from the dish, and and farther if possible.

Sighting on landmarks can be quite accurate provided a well-defined, visible object can be found. This is more difficult than might be expected, especially in the country as most natural features of the landscape tend to be poorly defined and hard to locate at a distance. Bear in mind that 1° corresponds to a distance of 150m (about 500ft) at a range of 8km (5 miles), in turn corresponding to 0.1in on a 1in map. The feature has to be pinpointed to this degree of accuracy so that good objects to sight on are water towers, cooling towers, church towers or steeples, radio masts and similar objects. Roads and other linear features are not very useful from a distance since it is very difficult to pinpoint any position along their length.

Fig 1.2 : Special microwave logging sheet. This is one suggested format which has proved useful in the field. No doubt the operator will devise his or her own layout and fill in useful information before going out portable, as explained in the text

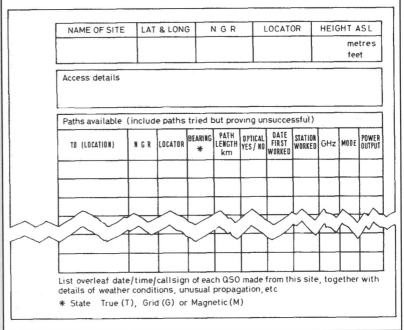

NAME OF SITE	LAT & LONG	N G R	LOCATOR	HEIGHT ASL
				metres feet

Access details

Paths available (include paths tried but proving unsuccessful)

TO (LOCATION)	N G R	LOCATOR	BEARING *	PATH LENGTH km	OPTICAL YES / NO	DATE FIRST WORKED	STATION WORKED	GHz	MODE	POWER OUTPUT

List overleaf date/time/callsign of each QSO made from this site, together with details of weather conditions, unusual propagation, etc
* State True (T), Grid (G) or Magnetic (M)

At night (in the northern hemisphere) the Pole Star gives a bearing of true north. By sighting up across the front of the dish at the star, the dish can be set due east or west. Other bearings can then be taken off the calibrated scale previously mentioned.

Microwave beacons or other amateur stations, the position of which is accurately known, might be used for beam heading, although the operator should beware of reflections off hills or large man made objects or the effects of other unusual propagation.

Using sun position is a very accurate method of beam setting and can serve several purposes. The position of the sun can be calculated from the equations given in an astronomical reference book (such as (6)), using a pocket calculator with reference to an almanac for some of the data. Briefly, the suns position is defined by two astronomical co-ordinates, its Greenwich hour angle (Gha) and declination (Dec). These are its longitude and latitude in the sky. They cannot be used directly for determining the azimuth of the sun and have to be converted to useful co-ordinates using two equations:

$E1 + \arcsin (\sin(Lat)\sin(Dec) + \cos(lat)\cos(Dec)\cos(Lha))$

$Az = \arccos (\cos(Lat)\sin(Dec)/\cos(E1) \tan(lat)\tan(E1))$

Where :

E1	=	Elevation of the sun
Az	=	Azimuth of the sun (its bearing relative to true north)
Lat	=	Latitude of station (+ve N, -ve S)
Dec	=	Declination of sun (+ve N, -ve S)
Gha	=	Greenwich hour angle of sun

Lha	=	Local hour angle
	=	Gha + long of station (east) - long of station (west)

These equations are easily solved using a scientific calculator, but a programmable calculator is to be preferred if a large number of calculations are envisaged.

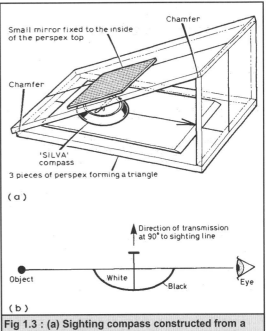

Small mirror fixed to the inside of the perspex top

Chamfer

Chamfer

'SILVA' compass

3 pieces of perspex forming a triangle

(a)

Direction of transmission at 90° to sighting line

Object   White   Black   Eye

(b)

Fig 1.3 : (a) Sighting compass constructed from a map reading compass. (b) Sighting across the face of a dish antenna

Clearly it is necessary to know the Gha and Dec of the sun and these are tabulated in various almanacs. One recommended almanac is the Nautical Almanac (obtained from HMSO in the UK) which tabulates these co-ordinates for both sun and moon (and the equations apply equally to the moon) for every hour of every day of the year. For those who do not wish to invest in such an almanac, it may be possible to obtain the data via a local reference library.

On a bright day, with the sun casting strong shadows, it is easy to see the shadow of the dish feed and simply adjust the dish until the shadow is central in the dish. On a cloudy day just point the dish at the approximate position of the sun if the receiver is sensitive enough to detect sun noise, optimise the position by maximising the noise by ear or watching the S-meter while adjusting the dish.

If you really must observe the sun with the unaided eye, then the very least eye protection is to look through are welders goggles or through a sheet of didymium glass intended to fit an are-welders mask. This glass is the only commonly available material that possesses enough optical density to prevent eye damage, since it absorbs light of all wavelengths from ultraviolet to infrared. Under no circumstances be tempted to use the naked eye, sunglasses, binoculars or a telescope. Another caution, especially on a clear sunny day and with a highly polished solid dish, there may be enough concentration of heat at the dish focus to damage the feed! However that would be a less consequence than eyes permanently damaged through a moments rashness.

Several kinds of north have been referred to indirectly and these are summarised below:

Magnetic north is used mainly when setting bearings with a compass and is about 6.5° west of true north in the UK at the present (2001) decreasing by about 0.5 in six years. The figure of 6.5° also varies with location by about 3°, being about 8° in Wales and 5° near the east coast of the UK. The exact figure should be obtained from a recent map of the area. Aeronautical maps are particularly useful for this, as they have an overprint of lat/long and lines of constant magnetic variation. The heights of ground are clearly marked and they are available in 1:250,000 and 1:500,000 scales (2.5km/cm, 5km/cm) from airport shops or the Civil Aviation Authority.

Grid north also differs from true north. This is the north with respect to which NGRs are calculated, and again the deviation is marked on each map.

True north is that which is obtained from lat/long and any astronomical measurements are referred to it. Ideally all bearings should be converted to true north before use to avoid confusion, in all cases the type of north to which a bearing refers should always be stated.

Powering portable equipment

With relatively simple and low-powered equipment there is little problem in powering it from batteries. If in good condition, the car battery might be used or a spare car battery carried. Alternatively, the smaller, non-spillable motorcycle-sized batteries or gel-acid batteries such as those used as power back up in alarm systems will suffice to run the equipment for quite long periods, especially when most of the operators time may be spent in receiving rather than transmitting. Since a lead-acid battery may only provide 11.6V when nearly discharged or as much as 13.8V when fully charged, it is a good idea

to be certain that the equipment to be used can function at say 11.6V.

For higher powers it may be necessary to carry a petrol or diesel driven generator. The operator should not forget to put batteries on charge as soon as the portable operating session is over, for there is nothing more frustrating than finding the batteries flat when next needed. Similarly, if a generator is used, it should be refuelled and serviced on a regular basis.

Microwave calculations

Many of the day-to-day calculations involved in microwave operation centre on the use of trigonometry to determine the distances and beam bearings to distant stations. The calculations can be carried out manually, with the aid of a scientific calculator, a programmable calculator or, more easily with a pocket portable or home microcomputer.

The mathematics of such calculations are given here which exploit latitude, longitude and Maidenhead locator systems as well as the UK based National Grid Reference (NGR) system to calculate distance and bearing between stations.

Location systems

There are several methods of varying accuracy which can be used, but all depend on some form of map reference system to specify the location of the two stations in what follows the terms home and distant refer to operator location and the location of the station to be worked, respectively. Some of the more common methods are:

- Latitude and longitude. This is the most fundamental and universally understood system, consisting of a network of imaginary circles drawn on the earths surface. It is capable of very great accuracy and is unambiguous. Maps for this system are available worldwide.

  Source: Most maps

  Accuracy: As great as is required

- QRA locator. This is based on latitude and longitude but has several disadvantages for calculation purposes. The references produced are not unique and repeat themselves over large distances. It is not capable of great accuracy and has to be converted back to latitude and longitude for bearing calculations. This conversion is tedious, and the sub-divisions of the squares are non-decimal and a very strange arrangement is used for the last letter of the reference. However, it has been widely used in Europe on the vhf bands.

  Source: Same as latitude and longitude but needs a look up table for conversion to QRA.

  Accuracy: Worse error +/-3km; the smallest square is about 4.4km square in Europe. See fig 1.4

- National Grid Reference (NGR). This method uses a rectangular grid of lines on a mercater projection map of the UK. It is very simple to use but is limited to the UK though its use has been extended to nearby coasts, for example France.

  Source: Ordnance Survey (OS) etc

  Accuracy: Fractions of a km, as great as required

- Universal Transverse Mercator (UTM). This method is

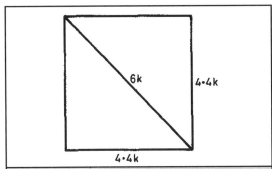

Fig 1.4 : The smallest QRA locator square, showing the greatest inaccuracy of ± 3km

similar to but not compatible with NGR. Maps are available world-wide. It has the accuracy and ease of use of the NGR over short ranges but becomes more complicated if the path crosses the boundary between two adjacent zones that occur about every 1,000km.

- Bearing and distance from a large town. This is a very quick but rather inaccurate method since towns are usually quite large and poorly defined in outline.

Source: any map

Accuracy: approximately 5 to 10km.

- Maidenhead locator. This is a newer system than the QRA locator, and one that overcomes some of its disadvantages. It is based on lat/long and has world-wide coverage. The standard version is accurate to about 3.5km but his can be improved by adding further characters to the reference.

Source: maps containing lat/long

What follows is a summary of an article by John Morris, GM4ANB which appeared in Radio Communications October 1984, Entitled The new locator system it explains how the then-new system works and provides much useful and practical information.

Vhf, uhf and microwave operators used the new locator system from 1 January 1985. For the uninitiated, a locator system is a way of giving the location of your station to within a few kilometres as quickly and efficiently as possible.

During the previous decade or so it has become common-place for stations making dx contacts on the vhf bands and above to exchange their locators, partly to allow the distance to be calculated, but also because collecting locator squares has become a popular sport. In most contests it is part of the rules that locators much be exchanged, again to allow the distance, and hence the score, for the contact to be calculated.

Until recently it was only in Europe that a locator system was in widespread use. The system, commonly called the QRA but in recent years more correctly termed QTH locator, was invented somewhere in middle Europe, and gradually spread to the whole continent. Unfortunately it was never really designed for more than local use, and was stretched to serve even Europe. Now other parts of the world, notably North America, are becoming inter-ested in the idea of a locator system, but there is just no way that the QRA can be persuaded to reach across the Atlantic. It is also a rather complicated system, and not

too easy to calculate.

The solution was to change to a new system. A single locator system, designed by amateurs to suit the special requirements of amateur radio, has now been adopted by all three regions in the IARU the international organisation of national radio societies. The adoption means that amateurs all over the world have a single well-defined way of telling each other their locations with a minimum of fuss and bother.

Some questions often asked:

Q: What difference will this make to operating?

A: Once the change over is complete there would be little difference on the surface. Instead of sending one string of characters such as YN27e you will send a different string such as IO83QP. After a short while the new system should become just as familiar as the old one, and you will turn your beam south (or whatever) just as automatically on hearing IO83PQ as you did on hearing YN27e.

Q Is my hard-earned collection of squares totally useless, so that I have to start collecting all over again?

A; No. The new system was deliberately designed to have a certain amount of compatibility with the old one, so that square collections, lists, awards and so on can continue as before. The big square are all in the same places but they have must been given different names. Conversion from old to new squares is quite simple, and will be explained.

Q: How do I find my new locator?

A: There are three ways of doing this. The first is to read and understand the description of the system, and work it out for yourself. The second is to follow the step-by-step procedure given in this article, using nothing more than pencil, paper, a calculator, and a few minutes time. The third is to use a computer, either your own or perhaps one brought along to the local club.
Q: What about maps?

A: Maps are available from the RSGB.

How the system works.

The world is divided along the lines of latitude and longitude into fields. Each field cover 20° of longitude from west to each and 10° of latitude from south to north, and it takes 324 of them to cover the world. The fields are labelled by two letters, each in the range A to R. The first letter gives the longitude of the field, starting from 108°W and working eastwards. The second letter gives the latitude of the field, starting from 90°S and working northwards.

For example, take field AA. The first A shows that the field covers 180°W to 160°W. The second A gives the latitude covered by the field as 90°S to 80°S. At the opposite end of the earth is field RR, which runs from 160°E to 180°E and 80°N to 90°N.

Most of Britain and Ireland is in field IO, which covers 20°W to 0°W and 50°N to 60°N. The Channel Islands, Isles of Scilly and the Lizard Point in Cornwall are in field IN, which covers 20°W to 0°W and 40°N to 50°N. Most of the Shetlands (roughly from Sandwick northwards) are in field IP, which covers 20°W to 0°W and 60°N to 70°N. That part of England east of Greenwich, including Norfolk, Suffolk, Essex, Kent and most of Cambridgeshire and parts of other counties is in field JO which covers 0°E to

20°E and 50°N to 60°N.

Each of the fields is divided, again along lines of latitude and longitude, into 100 squares. These are arranged as a 10 by 10 grid, so that each one covers 1° of latitude and 2° of longitude. Two digits, each 0 to 9, are used to label the squares. The first gives the longitude within the field, starting from the west, and the second the latitude, starting from the south.

For example, take square IO00. The field letters, IO show that is somewhere in the range 20°W to 0°W, and 50°N to 60°N. The first 0 says that the square is at the western end of the field, so that it covers 20°W to 18°W. The second 0 puts the square at the south of the field, giving its latitude coverage as 50°N to 51°N.

It may be noted that the locator squares coincide with the old QRA squares. Thus square IO83 is exactly the same as the old YN square; new JO01 is the same as old AL square; new IN79 is the same as old XJ; new IO86 is the same as old YQ, and so on.

Each of the locator squares is finally divided into a 24 by 24 grid of sub squares. Each of these covers 5' of longitude and 2.5' of latitude. They are labelled using two letters, each in the range 'A' to 'X', starting from the south west corner of the square. The first of the sub-square letters gives the longitude, and the second the latitude.

A full locator reference consists of the two field letters, two square digits and two sub-square letters. To give a full example, take locator IO83QP. The 'IO' part means that the field is 'IO', i.e. 20° W to 0°W, 50° N to 60° N. Th '83' part means that the square is 'IO83' which has its south west corner at 4° W and 53° N. The 'Q' says that the longitude is between 1° 20' and 1° 25' east of this. The 'P' gives the latitude as somewhere between 37.5' to 40' north of the southern edge of the square. Thus the area covered by the locator 'IO83QP' is from 2° 40' W to 2° 25' W and 53° 37.5' N to 53° 40' N. So, if you happen to live in Chorley in Lancashire, UK, your locator is IO83QP.

Fig 1.5 shows how the locator is built up from fields, squares and sub-squares. It may be noted that all of the longitude defining characters - the first, third and fifth - run from west to east, while the latitude defining characters - the second, fourth and sixth - go from south to north, In addition, at all levels the east-west size in degrees, is twice the north-south size.

In distance terms, each sub-square is about 4.6km from south to north. The east-west size varies with latitude, but in the middle of Britain (55° N) it is about 5.3km. If you take the accuracy of the system as being the farthest you can get from the middle of a sub-square without actually leaving it, this gives a maximum error of about 3.5km (at latitude 55°), which is quite adequate for most normal operation. Accuracy can be further improved by subdividing the smallest square in a decimal fashion and adding two more digits after the last character of the usual format, although this degree of accuracy (around 300m) is seldom required.

Computer Programs

The new locator system is very well suited indeed to computers, rather more so than the old system. There are many programs available that will do various things with the locator, such as contest scoring, distance and bearing calculations, and so on. An example can be found on the Internet (7)

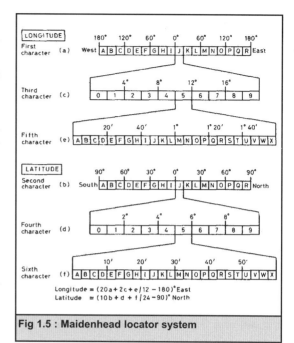

Fig 1.5 : Maidenhead locator system

Bearing calculations

Once the location of each station is know the bearing can be calculated using one or other of the following methods. Classical manual methods only are described here.

Spherical trigonometry

This can be used with the latitude/longitude (lat/long) system and gives exact values for bearings and distances. The calculations can be carried out with a scientific calculator and the bearings are given relative to true (geographic) north.

First the latitude and longitude of both stations are converted into the form of degrees and decimal fractions of degrees. The distant station's latitude and longitude are referred to as DN and DW (distant north, distant west), and the home station's latitude and longitude are HN and HW (home north, home west). Degrees north and west are positive, south and east negative.

DIFF represents the difference in longitude between the two stations (DW HW) and A represents the angle in degrees subtended at the center of the earth by the two stations. The bearings and distance are then calculated as follows:

	Lat	Long	
Distant DN =	** ***	DW = ** ***	(decimal degrees)
Home HN =	** ***	HW = ** ***	
		** ***	

DIFF = DW - HW

A = arccos(cos(DIFF)cos(DN)cos(HN) + sin(HN) sin(DN)

Path length = 111.15A km

$$X = \arccos\left(\frac{\sin(DN) - \cos(A)\sin(HN)}{\sin(A)\cos(HN)}\right)$$

$$Y = \arccos\left(\frac{\sin(HN) - \cos(A)\sin(DN)}{\sin(A)\cos(DN)}\right)$$

If DW > HW then: Bearing TO   = 360 - X

                             Bearing FROM = Y

Otherwise: Bearing TO   = X

                             Bearing FROM = 360 - Y

This method gives the exact great circle bearings relative to true north. It will be found that the bearings TO and FROM are not exactly 180 apart. This is quite correct and other methods which only give one bearing are only approximate.

To illustrate this method the calculation will be done for a path between Snowdon and Peterhead, with Peterhead as the home station.

Snowdon: 53°04′ N   04°05′ W

          DN =   53.067   DW =   +4.083

Peterhead: 57°24′ N   01°57′ W

          HN =   57.4   HW =   +1.95

          DW =   +4.083

          HW =   +1.950

          DIFF = 2.133

A=arccos(cos(2.133)cos(53.067)cos(57.4)
      sin(57.4)sin(53.067))   = 6.722

$$X = \arccos\left(\frac{\sin(53.067) - \cos(6.722)\sin(57.4)}{\sin(6.722)\cos(57.4)}\right) = 163.4°$$

$$Y = \arccos\left(\frac{\sin(57.4) - \cos(6.722)\sin(53.067)}{\sin(6.722)\cos(53.067)}\right) = 14.8°$$

DW (4.083) is greater than HW (1.95) so:

Bearing TO Snowdon = 360-163.4 = 196.6 °true

Bearing FROM Snowdon = 14.8° true

Note that 14.9 + 180 = 194.8 compared with 196.6

Simple trigonometry and NGR

This can be used with any rectangular grid system such as NGR or UTM. The National Grid is a rectangular grid drawn on a transverse Mercator projection map of the UK, with the prime meridian at 2 °W. This method of calculating the bearings assumes that the paths are straight lines on this map (which is not strictly correct as the map is a flat representation of a near-spherical earth). In practice this method is very accurate over paths up to about 1,000km.

In the NGR system, a point on the earth is usually specified by a two-letter reference for the 100km square and the number of kilometres east and north of the bottom left-hand corner of that square. Thus SZ710992 represents a point 471.0km east and 99.2km north of the origin of the grid.

An alternative way of describing the point is to specify the 100km square not by its letters, but by its number of kilometres east and north of the grid origin, giving an eight-digit reference. Thus SZ710992 is equivalent to

47100992. The latter method is far more useful for amateur purposes as the eastings and northings can be obtained directly by subtracting the references of the two stations, rather than having to work out where SZ square is, relative to the Home station square. It is far less prone to errors during calculations which is an important factor if the calculation is being carried out under pressure or in a hurry, such as during portable operation.

The numbers of hundreds of kilometres for each square can be found at each corner of an OS map in rather small print!

The bearings and distance are calculated using the simple geometry of a right-angled triangle. The only numbers required are the difference between the easterly and northerly co-ordinates DE, DN (distant east, distant north) and HE, HN (home east, home north), then the differences are DIFFE = DE - HE and DIFFN = DN - HN, and the angle X is as shown in Fig 2.11.

The calculations are as follows:

	km E	km N
Distant NGR	DE	DN
Home NGR	HE	HN
DIFF	DIFFE	DIFFN

Path length (km) = √(DIFFE² + DIFFN²)

If DIFFE is zero then:

                  (a) If DIFFN is positive, then
                     Bearing TO = 0
                     Bearing FROM = 180

                  (B) If DIFFN is negative, then
                     Bearing TO = 180
                     Bearing FROM = 0

Otherwise:

$$X = \arctan\left(\frac{|DIFFN|}{|DIFFE|}\right)$$

(The | | signs mean that the sign of DIFF taken to be positive, even if it is negative.)

If DIFFE is positive then: Bearing TO = 90 - X

                                Bearing FROM = 270 - X

Otherwise: Bearing TO = 270 + X

                                Bearing FROM = 90 + X

These bearings are relative to GRID north which differs from true north by a few degrees, the amount depending on the longitude of each station. The exact value is given on each OS map. The difference can be found approximately (in the UK) by taking the difference between the longitude of the station and that of the 2° W line (degrees west are positive, degrees east are negative), but this can be up to about 0.5° in error. Thus:

TRUE bearing TO   = GRID bearing TO
                            - (Home long - 2)

TRUE bearing FROM= GRID bearing FROM
                            -(Distant long - 2)

For example, take a path from Portsmouth to South Wales:

Portsmouth   SZ10992 =   47100992

South Wales ST260977 =   32601977

DE = 471.0   DN = 99.2

HE = 326.0   HN = 197.7

DIFFE = 145.0   DIFFN = -98.5

Path length = √(145² + 98.5²) = 175km

DIFFE is not zero, so

$$X = \arctan\left(\frac{|-98.5|}{|145.0|}\right) = 34.2$$

DIFFE is positive, so

Bearing TO Portsmouth = 90 + 34.2=124.2° GRID

Bearing FROM Portsmouth = 270+34.2=304.2° GRID

Now convert to TRUE bearings:

TRUE GRID at Portsmouth from map = 0.8°

TRUE bearing from Portsmouth = 304.2 + 0.8 = 305°

Longitude of Portsmouth = +0.994°

TRUE bearing FROM Portsmouth = 304.2-(0.994-2)
                              = 305.2°

TRUE GRID at South Wales from map = -0.7°

TRUE bearing to Portsmouth = 124.2-0.7 = 123.5°

Longitude of South Wales = 3.1°

TRUE bearing TO Portsmouth = 12.42 (3.1-2) = 123.1°

Drawing a line on a map

This is used mainly when the location is given as a bearing and distance from a town. It requires that both stations are shown on the same sheet of the map which, if it is small enough to be convenient to use, will probably not be accurate enough to give precise details of the location. This method is only suitable for very short paths, or for those stations using quite large beam widths, since it does not take account of the earths curvature.

Converting NGR to lat/long

If the NGR is known, the following relationships can be used to calculate the lat and long of the station, so that the more accurate spherical trigonometry method can be used.

If NG(n) is the number of kilometres north and NG(e) the number of kilometres east from the grid origin, then:

$$X = \frac{NG(n)+100}{111.166}$$

L = 49 + X - 0.000079X²

$$B = \frac{NG(e)-400}{111.55}$$

LAT = arcsin(cos(B) sin(L))

Y = arccos(tan(LAT)/tan(L))

If NG(e) is greater than 400km then longitude = 2 - Y

If NG (e) is less than 400km then longitude = 2 + Y

(Degrees west are positive)

In some cases the NGR co-ordinates may become negative if they are extended further south. Provided that the sign is used throughout the calculations, there will be no problems.

To summarise, the method of calculation chosen depends on the accuracy required. For vhf/uhf, QRA or line-drawing are acceptable methods. At the lower microwave frequencies comparatively wide beamwidths are still in use and quite large distances may be involved. Here QRA, Maidenhead, lat/long has many advantages and is highly accurate, although slightly more involved.

Conclusions

This chapter has attempted to point out the differences in techniques between operation on frequencies above 1GHz and those which are generally practiced on the hf, vhf and uhf bands.

The main technical difference is that the path losses are greater, and so the equipment has to have a much higher performance to achieve similar ranges. Much of the improvement is a result of using higher-gain antennas, and this inevitably means that the beamwidths are much narrower.

Thus, much importance is attached to calculating directions, accurately pointing antennas, and optimising equipment performance.

The narrow beamwidths have a major effect on operating techniques, calling CQ is much less common, and replies will usually only come from stations that are in the beam. Consequently, on the higher bands most of the contacts are the result of pre-arranged skeds, and a separate talkback link is used. However, enhanced propagation modes are more likely on these bands.

A scientific approach is therefore much more common, and the operator will often estimate the potential of the equipment and path parameters, under normal conditions, in advance of operation. Methods of achieving these ends are described and the effects of power, antenna gain and feeder losses on operating habits (fixed versus portable) discussed.

Finally, it is stressed that *method* is considered an important ingredient in the recipe for success.

References

[1] *Test Equipment for the Radio Amateur,* 2nd edn, H L Gibson, G2BUP, RSGB.

[2] Tropospheric scatter propagation, J N Gannaway, G3YGF, *Radio Communication,* Bol 57, p710.

[3] Wideband horn feed for 1.2 to 2.4GHz, Peter Riml, OE9PMJ, *Dubas* 2/86, pp 110-111.

[4] Log periodic antenna feed for 1.0 to 3.5GHz, Hans Schinnerling, DC8DE, *Dubas* 2/83, pp99-101.

[5] Multiband feed for 1 to 12 GHz, Klaus Neie, DL7QY, *Dubas* 2/80, pp66-76.

[6] *Practical Astronomy with Your Calculator,* Peter Duffett-Smith, Cambridge University Press. ISBN 0 521 29636 6.

[7] http://users.skynet.be/on1dht/
                    download/dhtlocat11.exe

[8] www.cix.co.uk/~g8dhe/
                    amateurradio/g4jnt_suite.zip

Chapter 2

System analysis and propagation

In this chapter :

- Analysis of communication system performance

- Microwave propagation

A basic communication system is shown in Fig 2.1. It consists of a transmitter and a receiver, each with its associated antenna, the two being separated by the path to be covered. In order to generate an intelligible output, the receiver requires a certain minimum signal to be collected by its antenna and presented to its input socket. Whether or not signals can be passed between any particular transmitter and receiver will thus depend primarily on the power of the transmitter, the sensitivity of the receiver and the loss associated with the path between them the path loss.

In practice, the size of transmitters and antennas and the sensitivity of receivers can be varied over a wide range. For example, the difference at a particular frequency between a small transmitter and a large one can be a factor of 1,000:1 or 10,000:1. The overall antenna gain may vary over a similar range, and there is potentially a fair range of choice of basic receiver sensitivity that is controlled by the amount of information that needs to be transmitted. This means that it is possible to design equipment having the capacity to cope with paths for which the path loss may differ by a factor of billions.

Usually the transmitter will generate a much stronger signal than the minimum required by the receiver, hence a large loss of signal between the two antennas can be tolerated. Despite this, for many paths of interest, the actual losses will be much greater and communication therefore impossible.

An added complication is that, even over a given path, the variation in path loss over a period of time can be large, and this may lead to problems. For example, if communication is to be maintained for a high proportion of the time, and 99.5 per cent is a common requirement for a professional link, then relatively powerful equipment will be required in order to cope with poor propagation conditions. Under favourable conditions, however, severe overloading may occur, and the equipment must be designed to also cope with this. On the other hand, if it is acceptable to be able to communicate only during periods of very favourable propagation conditions, then relatively simple equipment can be surprisingly effective, as the results obtained by amateurs over the years can testify.

It is probably true to say that communicating at microwave frequencies rather than at hf or vhf demands a more critical attitude to factors that affect propagation. The reasons are a mixture of practical and theoretical:

- At hf and vhf some degree of success can be virtually guaranteed from most sites when using relatively low power, inexpensive equipment with simple antennas success meaning communication over paths tens, hun-

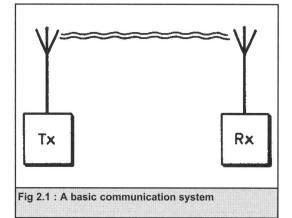

Fig 2.1 : A basic communication system

dreds or thousands of kilometres long according to the norm currently accepted. In this sense, operating at these frequencies can be regarded as being uncritical.

- By contrast, communication via microwaves over paths even a few kilometres in length may often prove impossible even though relatively powerful equipment is being used. This can be due to the effect of obstructions that, in terms of wavelength, appear larger as the frequency of transmission increases. Consequently, marginal changes in the path, produced by moving the transmitter or receiver a short distance, may have a large effect on the propagation loss and consequently determine whether or not signals can be passed.

- In more general terms, while the propagation of microwaves obeys the same laws as other forms of electromagnetic radiation, there is a tendency for one particular propagation mode to dominate to a greater extent than at vhf, for example, where signals beyond the horizon are usually received by a number of modes such as diffraction, refraction and reflection. Microwave propagation therefore tends to be more of a go/no-go phenomenon.

- Except in certain specialised areas, it is rare for amateurs operating at hf or vhf to attempt to correlate equipment parameters with performance. This is often done at microwave frequencies, however, and has proved invaluable as a creative tool in directing attention away from impractical objectives and towards more useful possibilities, sometimes with must surprising results.

For these reasons, the relationship between propagation

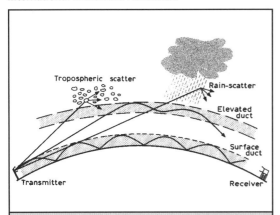

Fig 2.2 : Trans-horizon propagation mechanisms in the lower atmosphere

characteristics and equipment parameters can be regarded as being of particular significance at microwave frequencies.

Communication system analysis

As we have already seen, whether or not communication is possible between two sites will depend on the characteristics of the transmitter, the receiver and the path between them. The objective of system analysis is to quantify these factors so that the performance of a communication link can be predicted. It can be used to determine the initial design of equipment for a link, to determine the suitability of existing equipment to communicate over a given path and, if the way in which the path loss varies with time is also known, it enables one to predict the percentage of the total time that communication is practical or whether it is possible at all. In the latter case, the analysis will also indicate by how much the potency of the equipment needs to be increased to produce an acceptable level of communication.

The accuracy of the predicted performance will obviously depend on the reliability of the basic data. In general terms, equipment parameters such as transmitter power and antenna gain can be measured under the best conditions to within fractions of a decibel. Consequently, errors from this source are likely to be small compared with the uncertainty in the path loss. For this reason, even guesstimates of equipment parameters can provide a useful guide to the performance of the equipment and can prevent gross mistakes being made in either underestimating or overestimating its potential.

It can thus be seen that system analysis is of great practical value. Amateurs traditionally exploit the trial-and-error approach to their activities, including radio communication, and, while this philosophy has produced notably successes, it is suggested that system analysis deserves more attention for the following reasons:

- It may help to prevent the waste of effort on impossible tasks.

- It can direct attention towards areas of investigation which do not have obvious attractions.

- By providing a quantitative framework, it helps to isolate the factors which are controlling the performance of an equipment from those which are connected with propa-

gation phenomena or operating techniques.

- As will be seen later, it can be used to provide a rough check on the overall performance of a piece of equipment. This is of course especially valuable for those without elaborate test equipment.

Microwave propagation

Many of the propagation modes encountered at vhf and uhf are also applicable at microwave frequencies, with the exception of those that rely on ionised media such as aurora and sporadic-E. Along unobstructed paths, propagation will be via line-of-site, whereas trans-horizon propagation will usually rely on mechanisms such as diffraction, ducting or tropospheric, rain, or aircraft scatter, as shown in Fig.2.2 [1]. EME is also a possibility, especially on the lower microwave bands.

Analysis Of Communication System Performance

Consider the communication link shown schematically in Fig 2.3 As was noted in the introduction to this chapter, a receiver requires a certain minimum signal power at is input in order to generate a useful intelligible audio output. The transmitter power potentially available vastly exceeds this minimum signal, and consequently communication can take place between the transmitter and the receiver despite a huge loss of signal the path loss between them.

In simple terms, the ratio of the power generated by the transmitter to that required by the receiver is a measure of the maximum path loss that a particular equipment can tolerate. This maximum path loss may be referred to as the path-loss capability (plc) of the particular system.

Thus $\quad plc = \dfrac{\text{effective power of transmitter}}{\text{minimum power to operate receiver}}$

If the actual path loss exceeds the plc of the equipment, signals cannot be heard. If it is less, however, then signals will be heard, the signal-to-noise ratio of which will improve as the difference between the two increases. Thus the plc simply reflects the potency of a given transmitter-receiver combination.

Path-loss capability calculations

These require the specification of the transmitter, receiver and antenna characteristics as discussed below.

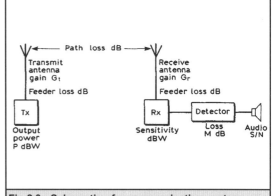

Fig 2.3 : Schematic of a communication system

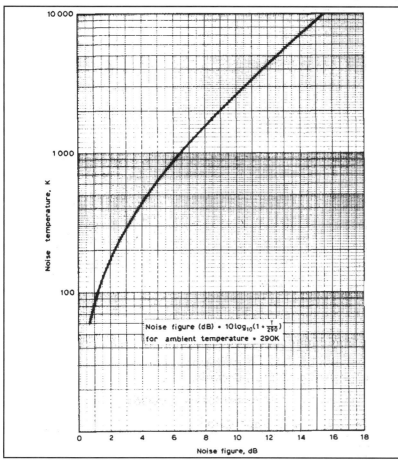

Fig 2.4 : Relationship between noise temperature and noise figure

Noise figure (dB) = $10\log_{10}(1 + \frac{T}{290})$ for ambient temperature = 290K

Transmitter

The figure of merit of a transmitter and its associated antenna is given by its effective isotropic radiated power (eirp), which is defined as that power which would be required to be radiated from an isotropic antenna to produce the same signal at the receiver as does the actual transmitter-antenna combination. Thus:

$$\text{eirp} = P \times G_t \text{ watts}$$

Where P is the transmitter output power in watts and G_t is the gain of the transmitting antenna in absolute terms.

In practice, it is convenient to convert all units into decibels with the reference power as 1W, ie 1W = 0dBW. An alternative reference level of 1mW is also sometimes used and this is indicated by the use of the term dBm. Values given in dBW and dBm thus differ by a factor of 1,000, i.e. 30dB. It is important to recognise that the value of the antenna gain is with reference to an isotropic radiator (as opposed, for example, to a dipole), and is the gain in the direction of the receiver. Feeder losses also need to be taken into account and can be included at this stage. Hence in terms of decibels:

eirp (dBWE) = power generated (dBW)
+ antenna gain (dB)
- feeder losses (dB)

As an example, consider a transmitter operating at 1.3GHz

Transmitter power of 20W	= + 13dBW
34-element antenna, gain	= + 19dB
Loss of 10m of RG9 cable	= - 2dB
Thus eirp	= + 30dBW

Receiver

The figure of merit of a receiver can be defined in terms of the minimum power its antenna must collect in order that the receiver may generate an intelligible output. This power is conveniently referred to as the effective receiver sensitivity (ers) and is defined as:

ers (dBW) = minimum detectable signal at receiver input (as a positive number)
+ antenna gain (dB)
- feeder losses (dB)

This relationship is seen to have a form similar to that corresponding to the eirp of a transmitter as described previously.

The effective receiver sensitivity can be quantified as follows:-

• The noise power generated internally by a receiver is given by the value of kTB, where:

K = Boltzmanns constant, 1.38×10^{-23} W/K/Hz

B = receiver noise bandwith (Hz). This can be taken to

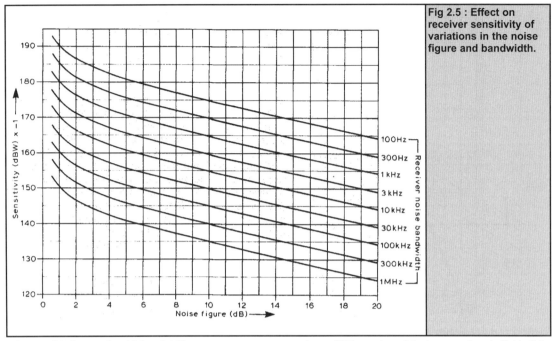

Fig 2.5 : Effect on receiver sensitivity of variations in the noise figure and bandwidth.

be equal to the receiver i.f. bandwidth.

T = noise temperature of the receiver (K). This is related to the noise figure of the receiver by the

equation:

Noise figure (dB) = 10 log(1 + T/290)

This relationship is shown in graphical form in Fig. 2.4. In terms of decibels the noise power generated is given by:

$$\text{noise power (dBW)} = -228.6 + 10\log(B)_{(Hz)} + 10\log(T)_{(K)}$$

As an example, the noise power generated by a receiver having an i.f bandwidth of 3kHz and a noise figure of 5dB, i.e. a noise temperature of 627K, is given by:

$$\begin{aligned}\text{noise power (dBW)} &= -228.6 + 10\log(3,000) \\ &\quad + 10\log(627) \\ &= 228.6 + 34.8 + 28.0 \\ &= 165.8 \text{ dBW}\end{aligned}$$

Values for kTB as a function of receiver bandwidth and noise figure obtained in this way are summarised in Fig.2.5

• The audio quality of a signal is directly related to the ratio of the audio power of the signal to that of the noise power generated by the receiver, and this obviously needs to be specified. As a guide, skilled operators can in some cases understand signals that are several decibels below the audio noise. For many practical purposes, therefore, the minimum acceptable signal-to-noise ratio can be taken as 0dB.

• During the detection process, the signal-to-noise ratio of the pre-detector (i.e. the i.f) signal may suffer significant degradation on conversion into an audio signal. Therefore, in order to maintain audio quality, it is necessary to increase the pre-detector signal by an amount M decibels equivalent to this loss. Approximate

values of M for various detectors are given in Table 2.1.

• Antenna gain: Gr decibels

• Losses such as feeder losses: L decibels.

Putting these factors together:

$$\begin{aligned}\text{ers(dBW)} &= -228.6 + 10\log(B) + 10\log(T) \\ &\quad + \text{snr} + M - Gr + L\end{aligned}$$

where snr is the audio signal-to noise ratio. As an example, consider the following receiver system:

Boltzmanns constant k	=	-228.6dBW
Bandwidth B = 3kHz, 10log(3000)	=	34.8dB
NF 5dB;T = 627K; 10log(T)	=	28.0dB
Audio snr	=	10.0dB
Product detector M	=	0dB
Antenna gain Gr	=	19.0dB
Cable losses L	=	3.0dB

Hence, the effective receiver sensitivity (ers) = -171.8dBW. Note that, for many practical cases, the value corresponding to the first three factors can be taken directly from Fig. 2.5.

Although in most cases the value for the ers as calculated

Table 3.1 : Approximate values of M for various detectors

Type of detector	Relative loss (dB)
Linear	0
(product, diode quadrature detector, diode with high bfo injection)	
Envelope	2.6
(diode detector)	
FM discriminator	
(with limiter)	10
(without limiter)	16
Slope detector	22

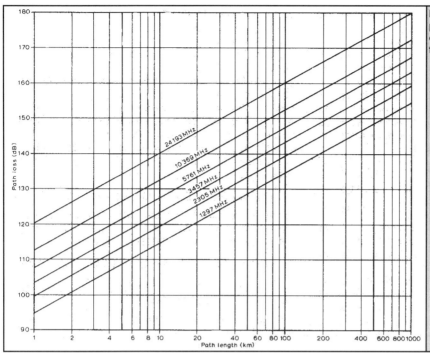

Fig 2.6 : Free-space path loss as a function of path length and frequency

Calculation of path loss

by the above method will be accurate enough, this may not be so when considering the influence of feeder loss on the effective noise performance of a receiver having a very low intrinsic noise figure. In such a case a more accurate estimate of the receiver ers is obtained by adding the feeder loss directly to the receiver noise figure before the conversion to effective noise temperature. Hence for the previous example:

noise figure + cable loss = 5 + 3 = 8.0dB

T = 1,540K (previously 627K); 10log(1,540) = 31.9dB

Hence ers (previously 171..8dBW) = -170.9dBW

Although the difference in the values of the ers predicted using the two methods is small (0.9dB), this may be very important in the context of, say, the implementation of an eme system.

In all cases, the effective receiver sensitivity will correspond to a very small power which will be represented by a large negative number in terms of decibels. As the overall sensitivity of the receiver system increases, the ers becomes more negative in value. Any additional losses in the receiver are added to this number to make it less negative.

Path-loss capability

The path-loss capability (plc) of a system is obtained by dividing the effective radiated power by the effective receiver sensitivity. When these values are expressed in decibels:

plc(dB) = eirp(dBW) - ers (dBW)

Using the previously worked examples:

plc(dB) = 13 + 19 2 - (-170.9)
        = 200.9dB

Calculation of path loss

The final step in the overall system analysis is that of determining the actual transmission loss of the path separating the transmitter and the receiver. If this loss is less than the plc of the system then communication is possible. The actual loss of a given path will depend critically upon whether it is unobstructed (corresponding to line-of-sight propagation) or obstructed, and whether or not anomalous propagation conditions are present. These cases will be considered in turn.

Line-of-sight propagation

This is represented by a transmitter and a receiver operated in free space. As there are no effects due to obstacles (including the atmosphere), this is a simple form of propagation for which the path loss can be calculated precisely from basic physical principles. Due to this fundamental nature, it is a reference mode against which path losses associated with other propagation modes are usually judged.

As will be familiar to many readers, the radiation emitted from an isotropic antenna spreads out equally in all directions and this means that the radiation intensity decreases as the square of the distance from the antenna. The path loss corresponding to this behaviour obeys the same law. Hence, a convenient form of the equation relating path loss, frequency and distance is:

path loss (dB) = 32.45 + 20log(f) + 20log(d)

where f is the frequency in megahertz and d is the path length in kilometres. As an example, the loss associated with a 100km path at 1,300MHz is given by:

path loss (dB) = 32.45 + 20log(1,300) + 20log(100)

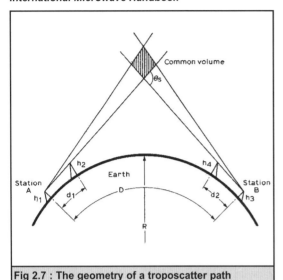

Fig 2.7 : The geometry of a troposcatter path

$$= \quad 32.45 + 62.28 + 40.0$$

$$= \quad 134.73\text{dB}$$

For many purposes this can be rounded to the nearest decibel, i.e. 135dB.

At higher frequencies, it is often more convenient to describe frequencies in terms of gigahertz rather than megahertz. The equivalent equation is then:

$$\text{path loss(dB)} = 92.45 + 20\log(f) + 20\log(d)$$

Values for the free-space path losses over terrestrial distances at some specific frequencies of amateur interest are given in Fig.2.6.

As noted earlier, the path loss associated with this propagation mode is relatively low compared with most other modes. This can be illustrated by calculating the transmitter power required to send signals corresponding to a single ssb speech channel over a 100km path. Using the following parameters:

Frequency	=	1.3GHz
Combined antenna gains	=	38dB
Receiver i.f bandwidth	=	3kHz
Receiver noise figure	=	5dB
Audio snr required	=	10dB
Product detector, M	=	0dB
Total cable losses	=	6dB

Then the path loss, as calculated previously, is 134.73dB. the plc of the equipment specified, as calculated by the method given earlier, is:

$$\text{plc(dB)} \quad = \quad P + 186.9$$

where P (dBW) is the transmitter power.

Since the plc of the equipment must at least equal the path loss, then

$$P + 186.9 \quad = \quad 134.73$$

From which P = -52.2dBW. This value corresponds to a power level of roughly 6µW. Similar equipment for 10GHz would require a transmitter power of roughly 0.4mW, or the combined gain of the antennas to be increased by

about 18dB. In practice, transmitter powers of this level are easily achieved with the minimum of facilities. This implies that normally there will be a large reserve of power available to cope with additional losses such as those incurred when perhaps the path is obstructed and communication can only be achieved via diffraction or indirectly by reflection. Such cases will be considered later.

Free-space path loss during anomalous propagation conditions

Under normal propagation conditions, then, the free-space path loss increases as the square of the distance between the transmitter and the receiver due to the spreading out of the transmitted energy in both the horizontal and vertical directions. Under certain conditions of anomalous propagation, however, the transmitted energy may be partially confined in the vertical direction and energy spreading then occurs mainly in the horizontal direction. This confining process leads to a reduction in the path loss such that in the ideal case of perfect confinement in the vertical direction it increases with distance rather than with the square of the distance, hence:

$$\text{path loss (dB)} \quad = \quad 92.45 + 20\log(f) + 10\log(d)$$

where f is in gigahertz and d is in kilometres.

For the 100km path at a frequency of 1.3GHz discussed earlier, the path loss under ideal anomalous propagation condition is given by:

$$\text{path loss (dB)} \quad = \quad 92.45 + 20\log(1.3) + 10\log(100)$$

$$= \quad 92.45 + 2.28 + 20.0$$

$$= \quad 114.73\text{dB}$$

This should be compared with the value of 134.73dB obtained under normal free-space or line-of-sight propagation conditions. The difference, 20dB, indicates why anomalous propagation conditions are so desirable, from an amateur point of view, since they enable contacts over much longer distances or over slightly obstructed paths to be made with no improvement in equipment specification.

The conditions under which anomalous propagation can occur are discussed in detail later in this chapter.

Path loss for propagation via troposcatter

This propagation mode requires that both antennas be pointed along the great-circle path between the two stations at as low an angle of elevation as possible. The two beams will then intersect in a common volume of the atmosphere near the centre of the path, as shown in Fig. 2.7. Propagation will be line-of-sight to the common volume from the transmitter. A very small fraction of the power passing through this volume will then be scattered in all directions by the irregularities in the atmosphere, and some of it will be in the direction of the receiver. This power then propagates by line-of-sight to the receiver. The height of the bottom of this scattering volume will depend on the path length, and to some extent on the horizons of the sites used at each end, but will be typically 600m on a 100km path and 10km on a 500km path. The loss in the scattering process is usually so large that the equipment is unlikely to have enough spare capacity to overcome the extra losses introduced by any additional obstructions in the path. The path loss increases by about

10dB for every degree of horizon angle at each station, and on paths of over 100km by about 9dB for every extra 100km of path length, so the choice of a site with a good horizon is vitally important since it can make a difference of several hundred kilometres to the range obtained.

The angle through which the signal is scattered is an important characteristic of a troposcatter path, as the loss involved increases with angle; the angle involved being usually only a few degrees. The relevant details of a troposcatter path are shown in Fig 2.7. The heights of each station are h1 and h3, and h2 and h4 are the heights of the obstructions forming the horizon at each station, at distances of d1 and d2 respectively. All heights are radius of the earth, 8,497 km. θs is the scattering angle which is determined by the path geometry, and consists of three terms: one depending on the overall path length, and two being characteristic of the sites at each end.

$$\theta_s = \frac{180D}{\pi R} + \frac{180}{\pi}\left(\frac{h_2 - h_1}{d_1} - \frac{d_2}{2R}\right) + \frac{180}{\pi}\left(\frac{h_4 - h_3}{d_2} - \frac{d_2}{2R}\right)$$

The path loss can now be expressed as the sum of several components:

• The free-space loss, as already defined above

• The loss in the scattering process:

$\quad$ L$_s$ $\quad$ = $\quad$ 21 + 10θs + 10log(f) (dB, deg)

This is an empirical expression derived from observed signal levels, and shows the variation of scattering efficiency with frequency and scattering angle. The loss increases by 10dB per degree of scattering angle.

These expressions are plotted in Fig 2.8, which shows the free-space loss and the sum of the free-space and scatter losses for comparison, indicating that much greater losses are involved in troposcatter, and that they increase very rapidly with distance.

• The aperture-to-medium coupling loss:

$\quad$ L$_{am}$ $\quad\quad$ = 2 + 2θs/α $\quad\quad\quad$ (dB,deg)

Where α is $\sqrt{(\theta 1 \times \theta 2)}$, the geometric mean of the two antenna beamwidths. This takes into account the size of the two beams and the way in which they intersect in the atmosphere, which affects the efficiency of coupling between them. It implies that there is no point in increasing the size of the antennas above a certain gain on a given path, as the expected increases in gain will not be realised when very-high-gain antennas are used. This condition occurs when the antenna beamwidths approach the scattering angle, i.e. a few degrees.

• Loss due to the variation of the mean radio refractive index of the atmosphere:

$\quad$ Ln = 0.2 (N-310) $\quad\quad\quad$ (dB)

were N is the refractive index expressed in millionths above unity. If N varied by 30 units, this would affect the path loss by 6dB, so it has a significant effect and probably accounts for the seasonal variations mentioned later.

The total troposcatter loss is the sum of all these terms. It is convenient to split it into two parts; one being the basic path loss and the other being the variable losses due to the nature of the sites used and the climatic conditions.

The first part, the troposcatter loss between two stations

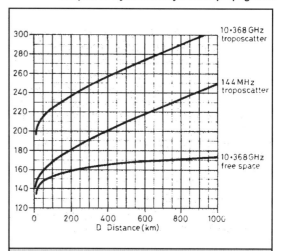

Fig 2.8 : Comparison of troposcatter and free-space path loss

on a smooth earth, is obtained by taking the terms which are either constant or depend on path length or frequency.

$$L = 55.5 + 20\log(D) + 30\log(f) + \frac{1,800D}{\pi R}(dB, km, MHz)$$

This loss is plotted against distance in Fig 2.9 for a frequency of 10,368MHz. The graph can be used at other frequencies by adding 30log(F/10,368) to the value read from the curve. Values of this term for various amateur bands are given in Table 2.2.

The remaining terms are the variable ones which depend on the sites or propagation conditions and weather, so these should then be added to the loss obtained from the graph.

$$L_V = 10\theta_a + 10\theta_b + \frac{2\theta_s}{\alpha} - 0.2(N - 310)(deg, Db)$$

$\quad$ (Site A) $\quad$ (Site B)

For most purposes the total loss can be taken as the loss from the graph plus the contribution from each site. The other two terms will have little effect, and the value of N is not likely to be known accurately.

Once the details of the sites are known, the values of θ_a

Table 2.2 : Correction to path loss given in Fig 2.9 for different frequencies

Band (GHz)	Correction (dB)
1.3	-27
2.3	-20
3.4	-15
5.6	-8
10	0
24*	11

* An additional allowance must be made for water vapour absorbtion on this band

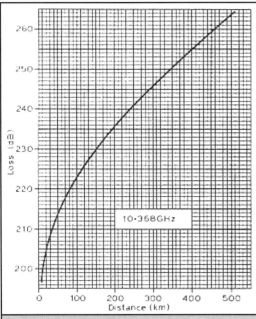

Fig 2.9 : Troposcatter path loss variation with distance

and θ_b can either be calculated using the expression for θ_s given earlier, or the loss ($10\theta_a$) can be found directly using Fig 2.10. In this, d is the distance to the first obstruction, and dh is the height of the obstruction above the site, see Fig 2.7.

$$Dh = h_{obstruction} - h_{site}$$

The actual height of the site does not appear explicitly in the expressions, only in as much as it determines where the first obstruction is and its height relative to a site. It

can be seen from the original expression for θs that both the elevation angle that the obstacle presents and its distance from the site are the important parameters in determining the path loss. A distant horizon is the key feature of a good site which, in simple terms, might be described as a place having a good view.

It is also very useful to calculate the loss from the site-dependent terms separately, as it provides a means of comparing the merits of various sites and is independent of frequency. Path profile plots should be performed for each direction of interest at each site to find the object causing the horizon and thus the values of d and h. A very good site can give negative values of this loss and so reduce the overall path loss. This loss is typically in the range -5 to +10dB.

As an example of a troposcatter path loss calculation, consider the Oxford-Hayling Island path, a distance of 110km. Taking the distances from the path profile shown in Fig 2.11, the site losses are:

Oxford

$$\theta_a = 57.3\left(\frac{184-77}{26,000} - \frac{26}{17,000}\right) = 0.23 - 0.09 = 0.14°$$

Loss = $10\theta a$ = 1.4dB

Hayling Island

$$\theta_b = 57.3\left(\frac{199-18}{18,000} - \frac{18}{17,000}\right) = 0.57 - 0.06 = 0.51°$$

Loss = $10\theta b$ = 5.1dB

The total loss due to the site is 6.5dB. The same result can be obtained by using the following values of dh and d in Fig 2.10.

Site A: d = 26km,   dh = 107m

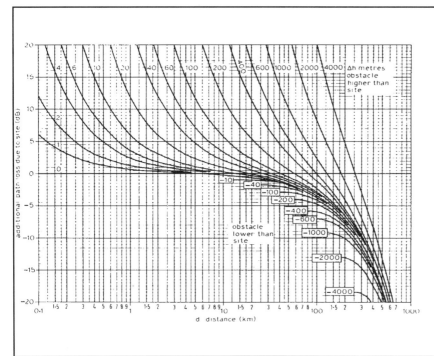

Fig 2.10 : The effect of site geography on path loss

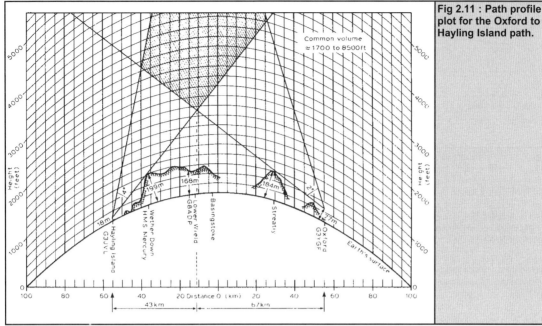

Fig 2.11 : Path profile plot for the Oxford to Hayling Island path.

Site B: d = 18km,   dh = 181m

Next, θ_s is needed to calculate the coupling loss. θ_s is the sum of the horizon angles at each site, plus the term in the total path length, 57.3d/R

$$\theta_s = \theta_a + \theta_b + 57.3d/R$$

$$= 0.14 + 0.51 + 57.3 \times 110/8,497$$

Now the coupling loss can be found. The antennas, 0.6m and 1.3m dishes, have beamwidths of 4° and 2° at 10GHz, so the mean is 2.5°. The coupling loss is thus 2.8/2.5, approximately 1dB. The path loss from Fig 2.9 is 224dB; hence the total loss at 10,368MHz is:

224 + 6.5 + 1 = 232dB

This value is the mean value of the loss averaged over a year. There are many factors that will affect this value slightly, and these are discussed later.

Knowing the path loss, the next step is to calculate the plc of the equipment at each end of the path. Hence:

G3JVL (transmit)

Transmitter output	=	7dBW
Feeder loss	=	-2dB
Antenna gain, 0.6m dish	=	34dBi
EIRP	=	39dBW

G3YGF (receive)

Receiver, nf 8dB,		
bandwidth 500Hz	=	169dBW
Antenna gain, 1.3m dish	=	39dBi
Feeder loss	=	-2dB
Threshold (cw)	=	OdB
ERS	=	206dBW
Hence plc	=	245dB

Since the path loss has been calculated as 232dB, the predicted signal-to-noise ratio is 13dB. This should be compared with the mean observed value of 10dB. The discrepancy is probably due to the seasonal and climatic variations, for which no allowance has been made, although there will always be a few decibels of uncertainly in the equipment parameters.

The potential of troposcatter communication

Details of various systems and the range that can be expected between two stations using them are given in Table 2.3 to illustrate the performance that should be expected under flat conditions from good sites. The

Table 2.3 : Range obtainable by troposcatter propagation on various amateur bands

Band (GHz)	Path loss (dB)	Range (km)	Equipment	Antenna gain (dBi)
1.3	258	760	100W, 3dB nf, 4x25ele Yagi	24
2.3	262	720	50W, 3dB nf, 1.8m dish	31
10.3	234	240	100mW, 10dB nf, 1.3m dish	39
10.3	254	440	1W, 3dB nf, 1.3m dish	39

Table 2.4 : Troposcatter range of equipment capable of eme operation

Band (GHz)	eme path loss (dB)	Tropo range (km)	Equipment	Antenna (dBi)
1.3	271	890	500W, 3dB nf, 500Hz,5.2m dish	34
2.3	276	860	100W 3dB nf, 500Hz, 5.2m dish	40
10.3	289	790	50W, 3db nf, 1kHz, 3.9m dish	50

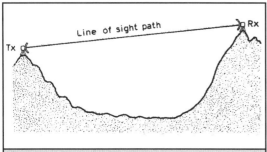

Fig 2.12 : A line of sight path

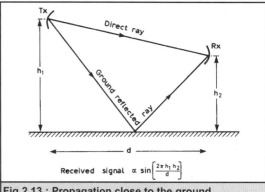

Received signal $\alpha \sin\left[\dfrac{2\pi h_1 h_2}{d}\right]$

Fig 2.13 : Propagation close to the ground

site-loss contributions are assumed to be zero, and the figures are given for a signal-to-noise ratio of 0dB in a 100Hz bandwidth, representing a weak cw signal. The range expected when using ssb in a 2kHz bandwidth is about 130km less on each band.

Table 2.4 gives the estimated troposcatter range between two systems which are capable of eme communication to illustrate the relative magnitudes of the problems involved. For distances approaching 1,000km, the challenge represented by the two modes of communication can be seen to be comparable. These tables also show that in theory the range attainable by troposcatter need not vary much with frequency. In practice, however, physically smaller antennas tend to be used on the higher frequencies, and it is also harder to generate comparable powers on the higher bands.

Microwave propagation

In the following sections, the various modes of propagation applicable at microwave frequencies will be considered. An appreciation of these will enable the amateur to estimate, as part of the system analysis procedure already discussed, how the transmission loss over a given path is likely to be affected, not only by the frequency of operations and the nature of the terrain along the path, but also by weather-related phenomena.

Free-space propagation

This is represented by a transmitter and a receiver operated in space. Since there are no effects due to obstructions to be considered, the path loss can be calculated precisely from basic physical principles, as discussed earlier in this chapter. It can be seen from Fig 2.6 that the transmission losses corresponding to path lengths of practical interest are very large. Nevertheless, this propagation mode is one of the least lossy.

Practical examples of this mode arise in space-to-space and earth-to-space communication at frequencies at which atmospheric absorption losses can be ignored.

Line-of-sight paths

As shown in Fig 2.12, a so-called line-of-sight path between two sites exists when the transmitting and the receiving antennas are so elevated that there is no obvious obstacle between them. That is, the transmitter can see the receiver and vice versa. This does not necessarily imply, however, that a person standing next to the receiving antenna literally can see the transmitter, assuming that his eyesight is good enough and the atmosphere is clear and still, since a path which is obstructed at optical frequencies may be line-of-sight at

microwave frequencies.

Provided that atmospheric effects are small, and that all potential obstacles are well removed from the propagation path, the propagation conditions approximate closely to the free-space mode discussed above. Consequently, even if the transmitting and receiving antennas have only modest gain, the transmitter output power required to pass a signal over a line-of-sight path which may be several hundreds of kilometres long is generally only a few milliwatts. Also, it is a fairly simple matter to calculate the system power budget for a given line-of-sight path and hence to estimate the probable signal-to-noise ratio at the receiver. A measurement of the actual received signal-to-noise ratio will then provide a check on the overall performance of the system. This approach is most useful since it does not rely on the amateur having access to sophisticated test equipment, only a calibrated attenuator.

In practice, effects due to path obstructions and the atmosphere can be of major significance; these will be discussed in the following sections.

Propagation over flat terrain

A simplified view of a communication link operating over a short terrestrial path is shown in Fig 2.13. The surface of the terrain is assumed to be smooth and the curvature of the earth has been neglected. Due to the proximity of the ground to the line-of-sight path, the received signal now has a contribution from the ground-reflected wave in addition to that from the direct wave. Hence for a fixed transmitting antenna height h1 the strength of the received signal will be found to vary periodically through a series of interference maxima and minima as the height h2 of the receiving antenna is varied. Such maxima are

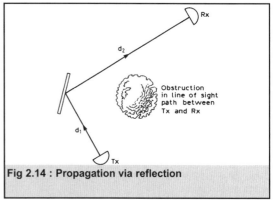

Fig 2.14 : Propagation via reflection

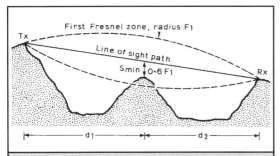

Fig 2.15 : Fresnel zone asurrounding line of sight ray on a microwave path

characteristically broad and the minima are sharp. The vertical separation s between adjacent receiving antenna positions for maximum and minimum signal strength is approximately given by:

$$s(m) = \frac{7.5 \times 10^4 d}{h_1 f} \quad \begin{array}{l}(km)\\ \\ (m,MHz)\end{array}$$

Hence, for a 10GHz transmitter at a height of 200m above ground level and operating over a 50km path, s = 3.75m.

In practice this simplified behaviour will be modified due to the effects of ground roughness at the reflection point, wave polarisation and earth curvature. Nevertheless, it does show that, especially when operating over marginal paths, the height of the receiving antenna and/or the transmitting antenna should be adjusted for best results whenever possible.

Propagation via reflection

Even if the direct path between the transmitter and the receiver is obstructed, it is sometimes still possible to achieve contact by reflecting signals from an object that is in view from both ends of the path. As can be seen from Fig 2.14, the transmission loss between the transmitter and the receiver will be proportional to:

$$\frac{1}{(d_1 d_2)^2}$$

Hence for minimum transmission loss the reflecting object should be situated close to one end of the path.

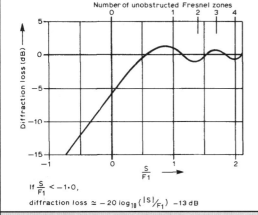

Fig 2.16 : Diffraction loss due to an ideal knife edge obstacle

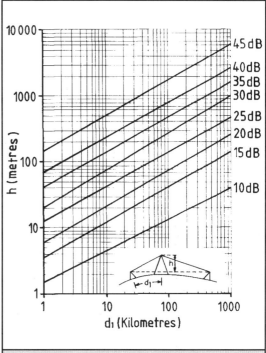

Fig 2.17 : Minimum obstacle loss at 10GHz

Knife-edge diffraction

If the transmitting and receiving antennas are not sufficiently elevated for the line-of-sight ray to propagate well away from any intervening obstruction, then additional transmission loss over the path will be incurred due to diffraction around the obstruction. In practice the microwave signal will not propagate from the transmitter to the receiver as a single ray but as a collection of rays or a wave front. The received signal is thus made up of contributions from a number of these rays, the final combination of which will be determined by the amplitudes and phases of individual components. Hence an obstacle, such as a hill, situated near to the line-of-sight ray may block some of the contributing rays or change their phases relative to others, thus normally leading to a decrease but possibly even an increase in the received signal level compared to its value in the absence of the obstacle. It is usually assumed that the obstacle will have a negligible effect on the received signal if it is situated more than a certain minimum distance away from the line-of-sight ray, as shown in Fig 2.15. Normally, this minimum distance is taken to be 0.6 of the first Fresnel

Table 2.5 : Correction factors

Band (GHz)	Diffraction loss correction factor for Fig 2.17
1.3	-9
2.3	-6
3.4	-5
5.7	-3
10	0
24	4

zone (often denoted as F$_1$) or:

$$S_{\min}(m) > 0.6F_1$$

$$S_{\min} > 10.4 \sqrt{\frac{d_1 d_2}{(d_1 + d_2)f}}$$

where d1 and d2 are in kilometres and are the distances from each end of the path to the potential obstruction, and f is in gigahertz.

If an obstacle is closer to the line-of-sight ray than the distance Smin, or the line-of-sight ray is actually obstructed, then normally the strength of the received signal will be diminished. To obtain an estimate of the additional path loss due to the obstacle it is often assumed that the latter can be replaced by a simple knife-edge, the diffraction loss properties of which are well known. Fig 2.16 shows the ideal diffraction loss for a line-of-sight ray, expressed in wavelengths. From this it can be seen why the value of Smin is often chosen to be 0.6F1 since this corresponds to the case of no diffraction loss. Note also that when the line-of-sight ray just touches the obstacle (S = O), the diffraction loss is 6dB. A more useful representation of Fig 2.16 is shown in Fig 2.17. This enables the minimum diffraction loss of a single knife-edge obstacle to be determined easily. Although drawn for 10GHz, values read from Fig 2.17 can be adjusted for other amateur bands by adding the correction factors given in Table 2.5.

It should be noted that the values for the diffraction loss obtained from Fig 2.17 have been found, in practice, to be optimistic by 6 to 15 dB at 10GHz. This discrepancy is due almost certainly to the fact that real obstacles are usually rounded rather than sharp-edged, and have a rough surface due to the presence of vegetation, trees or buildings. It is difficult to quote definitive values for the additional losses due to such roughness effects since they will depend on frequency and, more importantly, on the water content of the object in question. As a rough guide, however, measurements made at a frequency of 3.3GHz [3] suggest that the following should normally be regarded as opaque to microwaves:

• rows of trees in leaf, if more than two in depth

• trunks of trees, whether leafless or in leaf

• walls of masonry, if more than 20cm in thickness

• any but the lightest of wooden buildings, particularly if containing partitions.

In spite of these rather gloomy conclusions and their implications for propagation at microwave frequencies through an urban environment, test at 3.4GHz have shown that signals can be received through hedges and individual trees in leaf. Also in one case over a 1km path through a housing estate, the transmitting and receiving antennas being inside houses on opposite sides of a hill some 30m high. The lesson to be learned from this latter test is clear; even if on paper a particular path is not line-of-sight, it is still possible on occasion to transmit signals over it, as other propagation mechanisms may come to the rescue.

Techniques exist for dealing with the calculation of diffraction losses due to several obstacles lying along a single path, but these are outside the scope of this review. Further details may be found elsewhere [4].

Atmospheric refraction

Terrestrial paths inevitably involve the propagation of signals through the lower part of the atmosphere, the troposphere. This is a medium whose properties vary from place to place with factors such as pressure, temperature and humidity. The effect of these variations is to alter the refractive index of the medium that in turn affects how signals are bent or refracted in passing from the transmitter to the receiver.

Normally the earth's atmosphere has a refractive index n, the value of which is only slightly greater than unity, typically 1.0003. It is more convenient, therefore, to work in terms of the refractivity N, which is defined as:

N = (N - 1) x 10^6

Thus the refractivity corresponding to a refractive index of 1.000320 is 320.

In temperate climates at frequencies up to about 30GHz, the dependence of N on air pressure, temperature and humidity is given by:

$$N = \frac{77.6 \times p}{T} + \frac{3.733 \times 10^5 e}{T^2}$$

Where

p	=	the atmospheric pressure (mb)
E	=	the water vapour pressure (mb)
T	=	the air temperature (K)

The first term on the right-hand side of this equation corresponds approximately to the optical value of the refractivity; the second term, which must be included at microwave frequencies, accounts for the presence of water vapour in the atmosphere. Hence when propagating through the latter, light waves and microwaves are affected differently.

The normal variation of air pressure, temperature and humidity with height above the ground results in the refractivity decreasing in an approximately linear fashion with increasing height up to about 1km. The effect of this is to bend signals down towards the earth rather than allowing them to propagate in a straight line, as assumed earlier. If the refractivity remains substantially constant along the length of a path, the usual assumption except along mixed land-water paths, then the signals will follow a smooth curve whose radius of curvature normally is somewhat greater than that of the earth.

Over a greater range of heights, the mean value of the refractive index may often be well approximated by:

$$n(h) = 1 + N_s \times e^{(-b \cdot h)} \times 10^{-6}$$

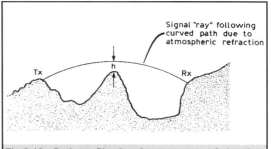

Fig 2.18 : Path profile showing curvature of signal ray due to atmospheric refraction

where

Ns   =   refractivity at the surface

h   =   height (km)

and

b is a constant determined from

$e^{-b} = 1 + dN/N_s$

where dN is the difference in N values at a height of 1km above the surface and at the surface.

It hs been shown from long-term studies of meteorological data that dN is In general correlated with the surface value Ns, but in a way that differs with the climate or region of the world, for example:

dN = -3.95xe$^{(0.0072Ns)}$	in the UK
dN = -9.30 xe$^{(0.004565Ns)}$	in the FRG
dN = - 7.32xe$^{(0.005577Ns)}$	*in the USA*

Although these expressions can be used to estimate dN when only surface meteorological data are available, this approach must be used with some caution, since the presence of anomalous propagation conditions, especially at the lower microwave frequencies, cannot usually be inferred only from a knowledge of conditions at the surface [5].

K factor

Under normal propagation conditions, signals follow a curved path, its radius of curvature being related to the refractivity of the intermediate atmosphere. To determine the proximity of potential obstacles to the line-of-sight ray, a path profile is plotted, on which the surface of the terrain immediately below the signal path is compared with the curved path of the signal ray itself. This is illustrated in Fig 2.18.

A more convenient technique avoids the need to plot the curved path of the signal ray. In this, the apparent radius of curvature of the earth is increased by such an amount that the relative positions of the terrain and the signal ray are maintained when the latter is taken to follow a straight path. This is shown in Fig 2.19. The ratio of the apparent radius of curvature of the earth to the actual value is known as the K factor, i.e.:

K =   apparent radius of earth
            real radius of earth

Since the curvature of a signal ray is related directly to both K and the refractivity N, then obviously K and N are

Table 2.6 : Estimate of K values

K value	Proportion of total time for which this value will be exceeded (%)
1	99
1.33	65
1.55	50
3.85	10
5	8
infinity	1.8

related also. This relationship is given by:

$$K = \frac{1}{1 + \left(\dfrac{\alpha}{n_s} \times \dfrac{d_n}{d_h} \right)}$$

where

a   =   real radius of the earth

n_s =   surface value of n

and

$\dfrac{dn}{dh}$ =   gradient of n at heights below 1km

Under average conditions near to the earths surface, the refractivity N decreases with height at the rate of about 40 N units per kilometre. Then taking n as typically 1.0003 results in a value for K of 1.33 (i.e. 4/3) which is the most commonly adopted starting value used in checking paths for line-of-sight conditions.

In practice the mean value of K can vary over wide limits. In a temperate maritime climate such as that of the UK, K can vary between about 1.32 in NE Scotland in February to about 1.45 in Cornwall in July or August [6]. Over the continental USA the variation in the mean value of K is even greater, ranging from about 1.25 in winter up to about 2 in summer [7]. Charts showing the world-wide

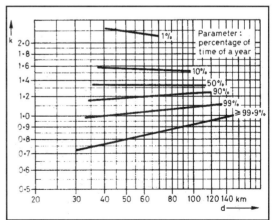

Fig 2.20 : Frequency distribution of the modified earth's radius factor K as a function of the path length (from Planning and Engineering of Radio Relay Links, Brodhage and Hormuth, Siemens AG, Berlin and Munich, 1977

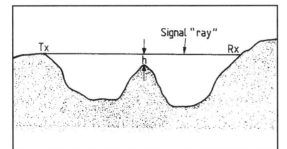

Fig 2.19 : Path profile drawn using the modified earth's radius for factor K to give a straight signal ray

distribution of Ns and N, from which mean values of K can be determined, are also available [8].

Since the troposphere is not homogeneous, the value of K will not only vary with time at a given location but also along a given path at a fixed time. Table 2.6 gives an estimate [9] of the percentage of the total time that the mean value of K at a particular location on land in a temperate climate will exceed a certain value. It is clear from this that enhanced propagation conditions are likely to prevail for a surprisingly large fraction of the total time.

When checking to see whether a given path is likely to be line-of-sight it is prudent to use a value of K averaged along the path rather than the value at any particular point since the variability of this path-averaged value will be less than the variability at a point. The path-averaged value of K which is exceeded for 99.9 per cent of the time is shown in Fig 2.20 as a function of path length for continental temperate climates [10]. Clearly, the longer the radio path, the greater is the minimum path-averaged value of K. In hot and wet climates the value of K will be higher, whereas in a dry climate it will be lower. For example, over a 50km path the average value of K commonly taken as the lower limit for path-checking calculations is K = 1 for wet climates, K = 0.8 for temperate climates and K = 0.6 for desert climates.

Line-of-sight path calculations

Whether or not a line-of-sight path exists between two sites can obviously be checked by plotting the path profile between them and determining if the signal ray would meet any obstruction. This tends to be a time-consuming operation, especially if a range of values for K needs to be considered. Fortunately, simple calculations can indicate quickly the likelihood of a line-of-sight path being present and therefore if a more detailed examination of the path profile is justified.

Consider the path shown in Fig 2.21. From simple geometrical considerations it can be shown that the distance from an elevated site to the horizon is given by:

$$d_1 = \sqrt{(2Ka(h_1 - h_3))}$$

where

a = the real radius of the earth

h1 = height of the site, asl

h3 = height of terrain at the horizon, asl

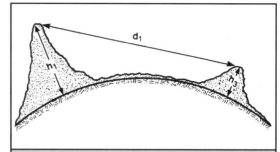

Fig 2.21 : Distance to the horizon for a line of sight path

K = the ratio of the earths effective to its real radius

Note that d, h and a are in the same units.

In the case where the horizon is at sea level, then h3 is zero. In other cases it is necessary to estimate the value of h3. This can usually be done by inspection of a map to find an approximate value for h3, using this value to calculate the position of the horizon more precisely, from which a better value for h3 can be determined and so on.

Assuming a mean value for the earths radius of 6,371km, and specifying h in metres, then the equation for the distance to the horizon becomes:

$$d(km) = 3.57\sqrt{(K(h_1 - h_3))}$$

The equivalent equation in imperial units is:

$$d(miles) = 1.22\sqrt{(K(h_1 - h_3))}$$

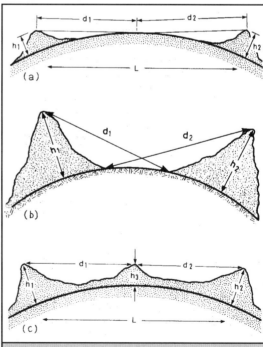

Fig 2.22 : Preliminary path checks for line of sight conditions where (a) signals from both sites share the same horizon; (b) signals from the two sites overlap; (c) signals from each site do not reach the other's horizon

Table 2.7 : Values of distance for various values of K and h

h(m)	Distance to horizon d (km)		
	K=1	K=1.17	K=1.33
100	36	39	41
200	51	55	58
300	62	67	71
400	71	77	82
500	80	87	92
1,000	113	122	130
3,000	196	212	226

Note: K=1 corresponds to no atmosphere; K=1.17 corresponds to the average optical value; K=1.33 corresponds to the average microwave value

Actual values of the distance d for various values of K and h are given in Table 2.7.

In checking whether or not a line-of-sight path exists between two sites, three main cases can be distinguished, as shown in Fig 2.22. These are:

- Signals from both sites share the same horizon. That is, signals can pass from one site to the other by just grazing the intermediate terrain. In this case, the overall path length is equal to the sum of the two distances to the horizon, and this represents the maximum value for d that can be achieved. Thus:

$$d\max = d_1 + d_2 = \sqrt{\left(2Ka\times(h_1 - h_3)\right)} + \sqrt{2Ka\times(h_2 - h_3)}$$

- Signals from the two sites overlap. In this case, d is less than the sum of d1 and d2 and signals can pass directly from one site to the other above the intermediate terrain.

- Signals from each site do not reach the others horizon. In this case, d is greater than the sum of h1 and h2 and the terrain forms a significant obstruction in the path.

The equations given above can be used to estimate the absolute maximum line-of-sight path potentially available in a given country. The greatest distance to the horizon is from the highest mountain to sea level, and the maximum path length cannot exceed twice this value. Practical line-of-sight paths may be considerably shorter; the longest available in the UK is in the region of 260km.

If the simple checks outlines above indicate that a given path may be marginal then it is sometimes necessary to draw a detailed terrain profile to check where potential obstructions may occur and to enable an estimate of additional path losses to be made. Techniques for constructing such profiles are given in Chapter 2 of [11].

Anomalous propagation

As discussed previously, the effective earth radius K is determined from:

$$K = \frac{1}{1 + \left(\dfrac{a}{n} \times \dfrac{d_n}{d_h}\right)}$$

where

a = real radius of the earth

dn/dh = gradient of n below 1km

Taking n and a to have values of 1.0003 and 6,371km, respectively, and substituting:

n = 1 + (10^{-6}) x N

Then K is given by:

$$K = \frac{157}{157 + \dfrac{dN}{d_h}}$$

where (dN/dh) is the gradient of the refractivity expressed in N units per kilometre. On average (dN/dh) has been found to have a value of -40N units per kilometre. This results in K = 1.33 (4/3) and hence normal refraction of the signal ray in passing from the transmitter to the receiver. Under normal propagation conditions, therefore, the presence of the troposphere results in the radio horizon being about 15 per cent further away than if the

earth were *in vacuo*. For light waves, however, the corresponding value is only about seven per cent, since at optical frequencies N is independent of the presence of water vapour in the troposphere. Hence a transmitter which is invisible to the eye may be visible over the horizon to a receiver.

In practice the gradient of N can vary over a wide range of values. It is of interest, therefore, to examine the variation of K with (dN/dh) since this will establish the tropospheric conditions necessary for long-distance microwave propagation to occur.

If (dN/dh) has a value of less than -40 N units per kilometre, the value of K is increased and above-average refraction will occur. Hence the radius of curvature of the microwave ray passing from the transmitter to the receiver will become more nearly equal to the radius of curvature of the earth, i.e. the distance to the radio horizon will be increased. When (dN/dh) has the critical value of -157 N units per kilometre then, from the equation given earlier, K becomes infinite, i.e. the radii of curvature of the microwave ray and the earth are equal. Under these conditions, if the earth were smooth, the radio horizon would be unlimited!

If the value of (dN/dh) becomes less than -157 N units per kilometre, K becomes negative. When this happens, ducting is said to occur and a microwave signal propagates as though confined vertically inside a type of waveguide. This confinement in the vertical plane results in a reduced transmission loss, which in the case of a perfect duct (i.e. complete vertical confinement), would be proportional to d, the path length, rather than d^2 as in the case of normal free-space propagation. It is this property of ducting which is so attractive to the microwave dx enthusiast.

Finally, if the value of (dN/dh) is greater than -40 N units per kilometre then below-average propagation conditions will occur. For example, if K has a value less than unity (i.e. (dN/dh) becomes positive) then the radius of curvature of the microwave ray will become smaller than that of the earth and the situation may arise that a distant transmitter can be seen with the eye but no signal can be

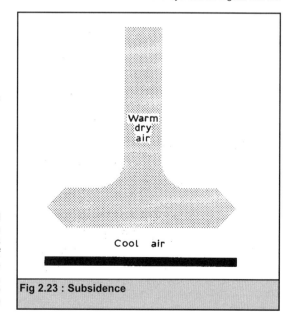

Warm dry air

Cool air

Fig 2.23 : Subsidence

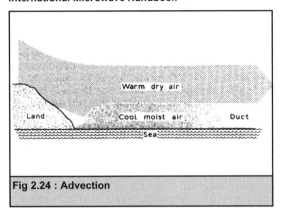

Fig 2.24 : Advection

received from it since the radio horizon is now less than the optical one.

Meteorological conditions associated with anomalous propagation

Having discussed the manner in which the gradient of the refractive index of the atmosphere close to the earths surface can influence microwave propagation conditions, it now remains to examine how such gradients can arise in practice, since an appreciation of the meteorological conditions which, for example, give rise to super-refraction and ducting can enable the amateur to make the best use of such phenomena.

The meteorological conditions which appear to be associated with the formation of super-refractive layers in the lower atmosphere are those of subsidence, advection, radiation, evaporation and the passage of weather fronts.

Subsidence

The air which descends slowly within an anti-cyclonic weather system (i.e. a region of high air pressure) becomes progressively heated by compression and, due to its high-altitude origin, it contains virtually no moisture. As shown in Fig 2.23, the subsidence is normally halted at some height when opposed by forces in the air underneath, and the warm air then spreads out in an elevated layer without significant mixing with the cooler air below. This results in a temperature inversion across the boundary region, and a corresponding sharp drop in the dew point.

If the subsidence inversion exists within about 2,000m of the surface, it imparts stability to the air below by suppressing convection air currents. This stability greatly increases the probability of ducts being formed through the advective and radiative effects described below.

Occasionally over the British Isles, and more frequently over the continental mainland, the subsidence inversion descents to within a few hundred metres of the surface and has a more direct influence on propagation. The very pronounced inversion layers which result at low levels are frequently super-refractive and form laterally extensive, elevated ducts. Land and sea areas are equally affected in this way and, due to the elevated position of the duct, it is reasonably insensitive to the roughness of the transmission path underneath.

Subsidence often continues for several days at a time and, when present at low levels, can enable long-distance contacts at microwave frequencies to be achieved.

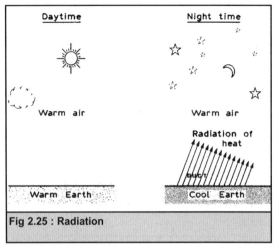

Fig 2.25 : Radiation

Advection

Advective conditions are characterised typically by warm, dry air passing over a colder, wetter surface. For example, the air flow around anti-cyclonic weather systems over central Europe produces a drift of warm, dry air from the mainland out over the cooler, very moist air of the sea, as shown in Fig 2.24. There is little convection turbulence over the sea and, when the winds are light, the overlay of warm air continues without mixing. The result is the formation of surface ducts which can extend for hundreds of kilometres. Over the southern North Sea protracted periods of advection are known to produce enhanced propagation at 11GHz for more than 10 per cent of the time, often persisting through both night and day. While advection is considered to be the most important over-sea propagation mechanism, it also affects low-lying coastal areas where the sea breeze draws cool moist air inland to meet air which is warmer and drier. The upper surface of the sea breeze where it meets the land air is thus a region where suitable humidity and temperature gradients may form, giving rise to a super-refracting layer.

A different advective situation exists when cold, dry air blows over a warmer, wet surface. Evaporation into the dry air produces a steep humidity lapse so that, despite the temperature lapse that is also set up, a duct is still formed. Examples include dry Arctic air blowing over the warmer sea, or cold, dry air behind a cold front that crosses land that has been wetted by heavy rain.

Radiation

Ducts are often formed due to nocturnal cooling of the earth's surface by radiation. As shown in Fig 2.25, conditions for duct formation are most favourable when the land has been heated by the sun to a high temperature during the day and there is little or no cloud cover during the subsequent night, together with little wind. A temperature inversion starts to form after sunset as the earth radiates heat rapidly into space, and it intensifies throughout the night to become most pronounced at about dawn. The cooling of the air near to the surface of the ground results in mist, fog or ground frost, all of which are indications of moist air within the inversion layer. As the sun rises, the ground surface temperature increases, there being little or no cloud, and gradually the inversion is destroyed and the moisture becomes dispersed throughout the atmosphere as the mist, fog, dew or ground frost disappear.

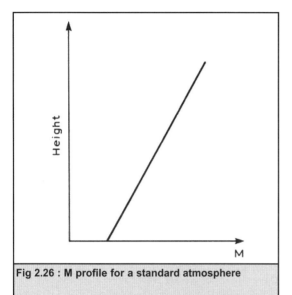

Fig 2.26 : M profile for a standard atmosphere

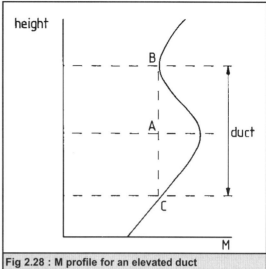

Fig 2.28 : M profile for an elevated duct

Ducts due to radiation are most prevalent over barren, dry land such as desert. However, even though other types of land surface may exhibit less extreme diurnal temperature variations, these may still be large enough to produce a temperature inversion in the lowest layer of the air due to contact with a cooler earth's surface. Nocturnal radiation is the most frequent cause of enhanced propagation conditions over land areas in the UK but, as the events tend to be of fairly short duration, it is likely that they will be less available to the average amateur than the periods of subsidence.

The influence of radiation cooling in coastal areas has been found to extend over sea paths between the Netherlands and the UK. However, over areas of deep water, for example oceans, ducts are not produced in this manner since the diurnal changes of surface temperature are too small.

Evaporation

A shallow duct exists frequently over most oceanic areas of the world in the form of the so-called evaporation duct. This arises from the fact that the air layer which is in immediate contact with the sea surface is saturated with water vapour due to evaporation, whereas the air some distance above the sea surface generally is not. This decrease of water vapour concentration with increasing height gives rise to a duct, the mean thickness of which ranges from about 6m in the North Sea up to about 15m in tropical regions. Such a duct will have a profound effect on nautical microwave propagation at frequencies up to about 40GHz. At higher frequencies, however, it is expected that the normally lower path losses due to duct propagation will be offset by scattering and absorption effects.

Weather fronts

Recent evidence from both amateur [12] and professional [13] experiments at microwave frequencies tends to suggest that propagation enhancement can also result from the movement of weather fronts into the area of a weakening anti-cyclonic weather system. The period of enhancement tends to be short and seems to result from super-refraction occurring somewhere within the frontal structure. However, the exact nature of the enhancement mechanism is still under investigation.

Propagation in ducts

A good insight into the way that microwaves are influenced by the presence of a duct can be gained by tracing out the paths taken by a microwave ray when launched at various angles with respect to the horizontal in the vicinity of the duct. In the discussion which follows the duct is assumed to extend uniformly in the horizontal plane, the only variation in refractivity being with height. It is convenient at this stage to introduce another variable in order to describe the refractive properties of the atmosphere. This is the modified refractive index M, defined as:

$$M = N + 10^6 h/a$$

where h is the height above the ground in kilometres, and N and a have already been defined. When dealing with

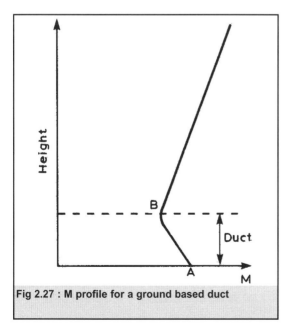

Fig 2.27 : M profile for a ground based duct

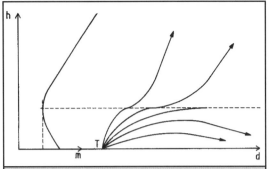

Fig 2.29 : Paths of rays launched from the ground at various angles into a ground based duct (from The Physics of Microwave Propogation, Donald C Livingston. © 1970. Reprinted with permission of Prentice Hall Inc, Englewood Cliffs, NJ, USA)

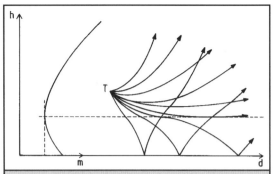

Fig 2.30 : Path of rays launched at various angles from above a ground based duct (from The Physics of Microwave Propogation, Donald C Livingston. © 1970. Reprinted with permission of Prentice Hall Inc, Englewood Cliffs, NJ, USA)

propagation in ducts it is more convenient to work in terms of M rather than N since the former is so defined that its vertical gradient (dM/dh) vanishes at any height h (h≈a) for which the path of a ray launched horizontally is a circular arc concentric with the surface of the earth, i.e. (dM/dh) = O when K = infinity. This method of defining M means that (dM/dh) = 0 at all heights in a homogeneous atmosphere over a flat earth. M itself would thus be constant, having at all heights the same value that it has at ground level. Hence a procedure based on the use of M rather than N is termed an earth-flattening procedure.

It was noted previously that under average propagation conditions (dN/dh), the refractivity gradient close to the earths surface, has a value of about -40N units per millimetre. Substituting this into the above equation and letting a = 6,371km gives for (dM/dh), the gradient of M, the value of 117N units per kilometre. The M profile for the so-called standard atmosphere is thus a straight line with a positive slope, as shown in Fig 2.26. In contrast, the idealised M profile for a typical ground-based duct is shown in Fig 2.27. Here the gradient of M vanishes at the point B and the approximate width of the duct is given by the length AB, since below point A the gradient of M is negative.

As in the case of a ground-based duct, the occurrence of an elevated duct is possible only when the gradient of M is negative over a range of heights. A typical idealised M

profile for this case is shown in Fig 2.28, where the edges of the duct are marked approximately by the points B and C.

Returning now to the case of a ground-based duct, Fig 2.29 shows the paths of a number of microwave rays launched at various elevation angles into the duct from a transmitter T on the ground. It can been seen that there is a critical angle at which the ray must be launched in order for it to reach the (dM/dh) = 0 level without passing through it and thus escaping from the duct. When the launching angle exceeds this critical angle, the ray escapes; when the angle is less than the critical angle, however, the ray reaches a maximum elevation which is closer to the (dM/dh) = 0 level the nearer the launching ground from its point of maximum elevation via a path which is a mirror image of that taken on its upward journey. If the ground is horizontal and specularly reflecting, a reflected ray will then traverse a path which is symmetrical with respect to the incident ray, reaching the same elevation as the incident ray and returning to the ground for a possible second reflection. Such a ray is said to be trapped inside the duct that can then be considered to extend from the surface up to the elevation at which (dM/dh) = 0.

Fig 2.30 shows the situation in which a transmitter is situation above a ground-based duct and a receiver is inside the duct. Again it is possible for signals to be

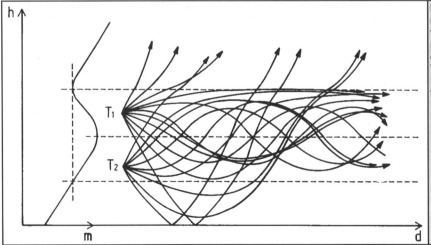

Fig 2.31 : Paths of rays launched at various angles from points witin an elevated duct (from The Physics of Microwave Propoga-tion, Donald C Livingston. © 1970. Reprinted with permission of Prentice Hall Inc, Englewood Cliffs, NJ, USA)

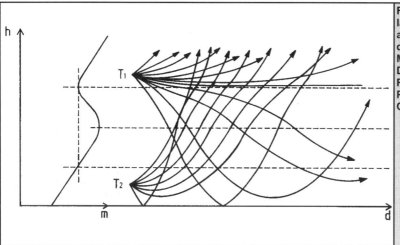

Fig 2.32 : Paths of rays launched at various angles above and below an elevated duct (from The Physics of Microwave Propagation, Donald C Livingston. © 1970. Reprinted with permission of Prentice Hall Inc, Englewood Cliffs, NJ, USA)

trapped inside the duct, providing that the launching angle is below some critical value.

Ray path diagrams for a transmitter located inside and outside an elevated duct are shown in Figs 2.31 and 2.32, respectively. When the transmitter is located inside the duct, rays which are launched over a wide range of elevation angles above and below the horizontal may be trapped. When the transmitter is outside the duct, however, only the ray which is launched at a critical angle below the horizontal from a point above the top of the duct can become trapped. It is for this reason that elevated ducts may appear to be of less interest to the amateur since signals launched from below the duct cannot be trapped, but there is also the possibility that signals might be reflected from the underside of the duct to give enhanced range or to clear obstacles, as shown in Fig 2.33.

Having established the existence of a critical angle for the trapping of a microwave signal inside a duct to occur, some estimate of the size of this angle will be of interest. In fact it depends on the total change in refractivity across the width of the duct, i.e.:

Critical angle +/- 0.0002√(dN) radians

For a typical maximum value of dN of 40, the critical angle is about +/-0.5, thus only those signals which are launched almost horizontally will be trapped by the duct.

Although the ray tracing procedure outlined above explains in qualitative terms how a duct can cause guiding of an electromagnetic wave, it also tends to suggest that any duct will guide waves of any frequency. In fact this is not so and a more exact analysis of wave propagation inside ducts shows that a wave of a given frequency can only be guided by a duct if the latter has a large enough lapse of N across it. For horizontally polarised waves, the minimum duct thickness is related to the wavelength of the signal and the gradient of N by:

$$d^{(3/2)} = \frac{\lambda}{2.5\sqrt{\dfrac{dN}{dh}}} \qquad \text{metres}$$

This expression has been evaluated for various values of dN/dh and frequency, the resulting curves are shown in Fig 2.34. Since shallow ducts are more prevalent than deep ones, it is to be expected from Fig 2.34 that duct

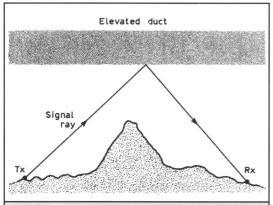

Fig 2.33 : Propagation over an obstructed path via reflection from an elevated duct

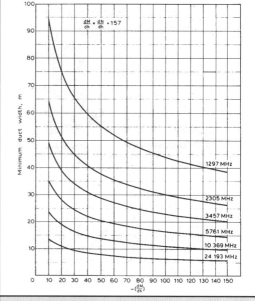

Fig 2.34 : Dependence of minimum duct width on the lapse rate DM/dh

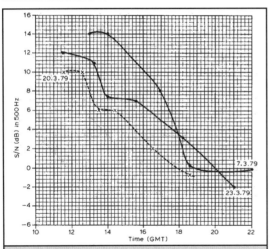

Fig 2.35 : Variation of troposcatter signal strength with time, showing diurnal variations

propagation should be much more frequent at microwave frequencies than at, say vhf. Thus microwave propagation conditions may be very good even at times when the vhf and uhf bands appear to be normal.

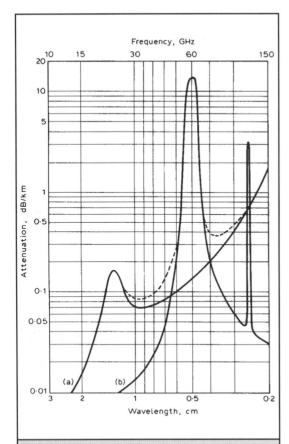

Fig 2.36 : Microwave attenuation due to water vapour and oxygen (from The Services Text Book of Radio - volume 5. Reproduced with the permission of The Controller of Her Majesty's Statinery Office)

Propagation via troposcatter

Paths that involve only line-of-sight propagation are not very common, and usually the signals will have been scattered off or diffracted around several obstacles on the way. As the length of the path increases, so does the number of obstructions or the angles through which the signals have to be diffracted. Under these conditions signal levels will decrease very rapidly with distance, and signals arriving by other propagation mechanisms may be stronger. Propagation beyond the horizon can occur by a variety of methods, such as ducting, as already discussed, or by the signals being scattered off an object which is high enough to be visible to both stations, such as an aircraft. However, these phenomena are only short-lived and a more permanent mechanism would be desirable. Satellites or eme are more predictable but, apart from the case of geostationary satellites, can only be used for some of the time.

Troposcatter uses the weak but reliable reflections that can be obtained from the dust particles, clouds, and refractive index variations that occur in the atmosphere in the region 300 to 15,000m asl, and this mechanism can be used for reliable dx working over distances of many hundreds of kilometres. As has already been discussed, the refractive index of the atmosphere depends on such properties as its temperature, pressure and humidity so variations in any of these can scatter the signals. The scattering process is more efficient at lower altitudes where the atmosphere is denser, and where turbulence associated with the weather can have marked effects on the signal levels and characteristics.

Characteristics of troposcatter signals

Several different types of fading are experienced on troposcatter signals [14]. The effects are more severe at high frequencies, and so are easier to observe and describe. At 10GHz the note of the carrier can appear quite rough, being modulated by the scattering process at frequencies up to about 50Hz.

There is also fading over a period of minutes and, in the longer term, signals tend to show a diurnal variation of about +/-5dB, often peaking in the afternoon when atmospheric turbulence caused by convection currents from the warm ground is at a maximum. Plots of signal level showing this effect are given in Fig 2.39. There is also an annual variation of similar amplitude, with signals peaking in the summer and being at a minimum in the winter. The daily and annual variations are probably the result of corresponding variations in the average value of N over the path.

The rapid fading is caused by the signal being scattered from various regions of air, each of which may be in turbulent motion, and moving relative to each other. This motion can cause both frequency and amplitude modulation of the signals. Frequency modulation results from the signals being scattered from air masses that are moving at different velocities, so there will be random Doppler shifts on the signals. At 10GHz a velocity of 50kmph will produce a shift of about 500Hz, and this effect can spread the energy of the carrier out over 1kHz or more, heavy rainstorms producing a sound rather similar to an auroral signal. These storms can also increase the signal levels by around 10 to 20dB, as the raindrops scatter the signal more effectively. Amplitude modulation results from variations in the scattering efficiency or interference effects

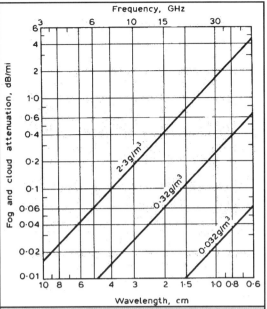

Fig 2.37 : Microwave attenuation due to fog and clouds (from The Physics of Microwave Propogation, Donald C Livingston. © 1970. Reprinted with permission of Prentice Hall Inc, Englewood Cliffs, NJ, USA)

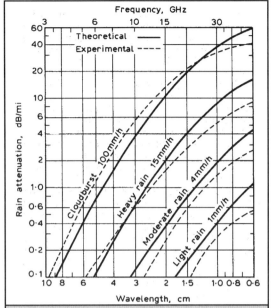

Fig 2.38 : Microwave attenuation due to rain (from The Physics of Microwave Propogation, Donald C Livingston. © 1970. Reprinted with permission of Prentice Hall Inc, Englewood Cliffs, NJ, USA)

between signals arriving by different paths.

Tropospheric propagation above 10GHz

So far it has been assumed that the troposphere is perfectly transparent to microwaves. In practice, however, some absorption or scattering of a signal will occur due to the presence of the constituent gases and hydrometers, i.e. rain, snow or hail. Those gases which have the greatest effect at microwave frequencies are water vapour and oxygen. Water has a permanent electric dipole moment and oxygen a permanent magnetic moment; hence both exhibit resonance absorption effects. Fig 2.36 as a function of frequency. It can be seen that absorption losses are negligible at frequencies below about 10GHz but becomes significant at higher frequencies due to oxygen resonances around 60GHz and 119GHz, and water resonances at about 22GHz and 183GHz.

The amateur band most affected by oxygen absorption is

thus likely to be that at 120GHz where the range of signals even in dry air is unlikely to exceed a few kilometres. Significant additional attenuation can also be caused by water vapour. It should be noted, however, that it is the absolute mass of water held in the atmosphere that matters, thus there may be more water in the atmosphere on a fine summer day than on a wet winter one. If the air temperature is below zero then the water content will be very low, long distance communication will be much easier, therefore, in arctic conditions or between mountains above the snow line.

At frequencies up to about 20GHz the attenuation of microwaves by hydrometers is proportional to the liquid water content. Thus rain and wet snow will cause the greatest attenuation, whereas hail and dry snow are of less significance. Below 10GHz the dominant loss becomes that due to scattering. Figs 2.37 and 2.38 show the attenuation due to fog, clouds and rain as a function of

Table 2.8 : Performance of simple equipment on the millimetric bands

Band (GHz)	plc	Margin (dB)						Max range (km) for 10dB margin		
		1km		10km		100km				
		dry	humid	dry	humid	dry	humid	space	dry	humid
24	177	57	56	37	34	15	-15	220	120	50
47	187	63	62	43	37	3	-32	350	70	48
76	197	67	65	47	34	-3		700	65	30
120	205	71	41					1,100	2	2
142	208	72	69	51	21	22		1,300	70	13
241	216	76	71	56	3	33		2,000	500	9

These figures were obtained for equipment having the following parameters. Transmitter: output power = 1 mW. Antennas: 20cm diameter dishes with low efficiency (25%) feed. Receiver: noise figure = 20 dB; i.f. bandwidth = 10MHz. In all cases the signal to noise ratios shown are pre detection.

frequency. It can be seen that only the attenuation due to rain is of significance except at the higher frequencies. Even this is not so high as might appear at first, since in general a given path will not be subjected simultaneously to the same rainfall intensity over the whole of its length due to the cellular pattern of rainfall.

Bearing in mind the above, the estimated potential for long-distance communication in the millimetric amateur bands is summarised in Table 2.8. From this the following conclusions may be drawn:

- It is expected that amateurs should be able to work paths at least tens of kilometres long on most of the millimetric bands even under conditions of high humidity.

- Ranges of hundreds of kilometres can be expected on most bands under conditions of low humidity, for instance over deserts of where the air temperature is below 0°C.

- Due to the high attenuation of signals by oxygen in the atmosphere at frequencies around 120GHz, propagation ranges on this band are likely to be restricted to a few kilometres.

- Although rain will cause high path losses on all the millimetric bands, these can often be avoided in amateur operation by judicious timing.

Propagation over earth-space paths

Many of the factors which affect microwave propagation over terrestrial line-of-sight paths will also be present over earth-space paths, especially if the elevation angle from the ground station to the satellite is small. At such angles, which will occur when a ground station, situated at a high latitude, is communicating with a geosynchronous satellite or is maintaining communication with a non-geosynchronous satellite for the maximum time that it is above the horizon, ground reflections, variations in refractive index and ducting may all be of significance. At much larger elevation angles, however, their effects will become small compared with other mechanisms such as absorption and attenuation due to hydrometers.

The extent to which a steady decrease in refractive index with increasing height causes refraction of a microwave ray has been discussed already. For a path that extends throughout the atmosphere, however, refraction causes the apparent elevation angle of a ray, as seen from the ground, to be greater than if the atmosphere were not present. The effects due to ray bending will be important only at very small elevation angles and especially when narrow-beamwidth antennas are being used.

As on a terrestrial path, the free-space transmission loss on an earth-space path will be increased by absorption due to oxygen and water vapour, and by scattering due to hydrometers. This additional loss will depend on the elevation angle since this defines the effective thickness of the atmosphere. To a first approximation, then, the effective distance for oxygen absorption can be taken as the distance the wave would have to travel if the atmosphere were replaced by one having constant density reaching upwards from the ground to a height of 4km, with a vacuum above. The corresponding height for water vapour is 2km. This gives a theoretical total one-way attenuation due to the combined effects of oxygen and water vapour of approximately 1.5dB at 1GHz, 2dB at 5GHz and 6dB at 15GHz, for a horizontal path (i.e. one

tangential to the earths surface). These values reduce as the elevation angle increases to become, in the limit, when the signal take-off direction is vertical, about 0.1dB at 15GHz, with only 1dB at the 22GHz, water absorption peak.

References

[1] The identification of trans-horizon interference propagation conditions from meteorological data, M T Hewitt and A R Adams, *Proc URSI Commission F,* Symposium on effects of the lower atmosphere on radio propagation at frequencies below 1GHz, Lennoxville, Quebec, Canada, May 1980, pp7.2.1-7.2.6.

[2] Tropospheric scatter propagation, J Gannaway, *Radio Communication,* Vol 57, 1981, pp710-714 and 717.

[3] Some experiments on the propagation over land of radiation of 9.2cm wavelength, especially on the effect of obstacles, J M McPetrie and L H For, *JIEE,* Vol93, IIIA, 1946, pp531-538.

[4] Multiple knife-edge diffraction of microwaves, J Deygout, *IEEE Trans on Ant and Prop,* AP14, 1966, pp-480-489.

[5] *VHR/UHF Manual,* ed G R Jessop, Chapter 2 (by R Flavell), RSGB.

[6] The radio refractive index over the British Isles, J A Lane, *J Atmos Terr Phys,* Vol 21, 1961, pp157-166.

[7] Forecasting television service fields, A H Lagrone, *Proc IRE*, Vol 48, 1960, pp1009-1015.

[8] *Radio Meteorology,* B R Bean and E J Dutton, NBS Monograph 92, 1966.

[9] *CCIR Green Book.*

[10] *Planning and Engineering of Radio Relay Links,* H Brodhage and W Hormuth, Siemens-Heyden, 1977.

[11] Microwave path checking, B Chambers, *Radio Communication,* Vol 54, 1978, pp122-126.

[12] *RSGB Microwave Newsletter.*

[13] Frontal disturbances in anti-cyclonic subsidence a cause of microwave interference propagation beyond the horizon, M T Hewitt, A R Adams and R G Flavell, *URSI XXth General Assembly,* Washington DC, USA, Aug 1981.

[14] Results of propagation tests at 505 and 4,090MHz on beyond the horizon paths, K Bullington, W J Inkster and A L Durkee, *Proc IRE,* Vol 43, 1955, pp1306-1316.

Chapter 3

Microwave antennas

In this chapter :

- Types of antenna
- Yagi type antennas
- Parabolic dishes
- Factors affecting practical dish design
- Practical dish construction
- The design of feeds for dishes
- Practical feed systems
- Practical omni-directional antennas
- Aligning and checking antennas

If you are interested in working dx, you are interested in high-performance antennas. This means antennas which give high gain with good radiation patterns, and what this means in practice should become apparent in this chapter.

A special characteristic of microwave antennas is the relative ease with which high gains can be achieved from manageable-sized hardware, compared with lower frequencies. Values between 20 and 45dB with respect to an isotropic source are quite common. Gain is normally quoted in dBi, referred to an isotropic source. Since the isotropic antenna is a theoretical concept and does not exist, another reference, used mainly by amateurs, is dBd, in which case the gain is referred to a dipole. A dipole in free space has a gain of 2.15dB relative to the point source of the imaginary isotropic radiator.

The temptation to take advantage of antenna gain is great, since it is the most socially aware method of increasing ones station effectiveness on both transmit and receive. By concentrating the radiated energy into a beam, several advantages result. The operator seeking to work at the extremes of propagation is always dealing with signals near the noise threshold. Here, even one or two extra decibels becomes significant, making the difference between a contact or no contact. Using a well designed beam allows concentration of energy, not only in the desired direction, but also at a low angle onto the horizon where it is more effective over terrestrial paths. For eme the target is, in angular terms, very small and radiated power not hitting the target is power wasted. The use of beams can lead to less interference to and from other stations using adjacent frequencies and can also minimise rf interference to other services, including domestic electrical equipment. However, the beam-widths become quite small and therefore extra skill is required in handling these antennas effectively. This skill is rapidly acquired and it is very soon realised that many of the remote locations are not quite where you thought they were.

Antenna gain is achieved by concentrating the radiated energy within angular confines to form a beam. In general, the smaller the angle, the higher the gain. The angular size of this beam reduces as the size of the array is increased, either the diameter of a dish or the boom length of a Yagi type antenna. Gain may also be increased by combining a number of individual antennas. By increasing the gain of the antenna, the effective power of a transmitter or the sensitivity of a receiver is increased, but the price to be paid is the need to align the antenna more precisely in the desired direction.

The ideal form of the polar diagram of a typical high-gain antenna is shown in Fig 3.1(a). This ideal state is never achieved and a poorly designed antenna might have a polar pattern like that of Fig 3.1(b). The general form of the polar diagram of a practical, well-designed antenna is shown in Fig 3.2, where the 1dB, 3dB and 10dB beamwidths represent the angles through which the antenna may be rotated before the power transmitted or received falls respectively by 26 per cent (-1dB), 50 per cent (-3dB) or by 90 per cent (-10dB) of the maximum value. The 10dB beamwidth should be considered the limit for normal pointing accuracy. The approximate relationship between gain and beamwidth is shown in Fig 3.3, normalised to permit performance on any frequency to be assessed.

The assumption has been made that the beamwidth in both planes is the same, as this is normal for long arrays or large-diameter dishes. This figure gives a guide to the accuracy with which the antenna needs to be pointed.

For example, the 3dB beamwidth of an antenna having a gain of 35dB is about 3°. To receive weak signals close to the threshold requires that the antenna be pointed within

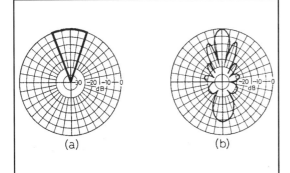

(a)   (b)

Fig 3.1 : (a) Idealised radiation pattern of high gain antenna, with flat nosed main lobe and no side lobes at all. (b) Unacceptable radiation pattern, with pointed main lobe and excessive side and back lobes.

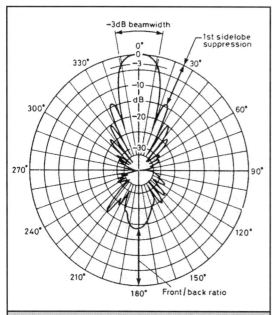

Fig 3.2 : E plane pattern of a good long Yagi. Note the generaol symmetry and the deep nulls at ±90°

half this angle, i.e. within the 1dB beamwidth or about 1.5°. For strong signals, the 10dB beamwidth may apply, but even this is still only about 5°. The fact that this is regularly achieved by amateurs, even under contest conditions, is a measure of the progress that has been made in acquiring the necessary skills. It should be stressed that if the direction of the station to be worked has been determined by the use of a lower frequency talkback, this can lead to errors that may mean no contact

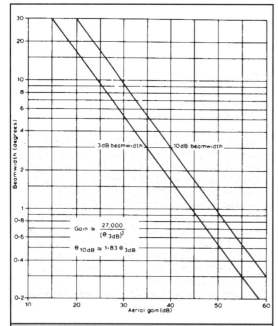

Fig 3.3 : Approximate relationship between antenna gain and beamwidth.

is made. The main problems are the comparative beamwidths and the likelihood of the deep nulls on either side of the main lobe being directed at the remote station. These nulls are located at approximately the main beams 3dB angular width (3dB beamwidth) either side of the true heading. The recommended method uses good maps and a calculator or computer programmed to give the results and accuracy required. The availability of such devices now means that complex calculations may be made with little or no effort, even when out on portable sites.

However, these factors do not remain constant and consideration of which antenna to use as the frequency increases is recommended. It becomes evident that when the highest bands are considered the obvious choice is a dish. Then the question is which f/D (focal length to diameter) should I use? The feed becomes the next choice to be made; several are discussed and are described for specific bands but the theory applies to any other band.

Types of antenna

The five main types

The five main types of antenna used on the microwave bands are:

- Yagi types, where only one element or a small group of elements is fed with power, the rest being passive.

- Phased arrays, including log periodics and waveguide omnis, where all elements are fed.

- Horns, which are developed from waveguide and may be flared in one or both planes.

- Parabolic reflectors, which are the best-explained type, due to their similarity to optical reflectors.

- Omni slots, now used on microwave frequencies as beacon and repeater radiators.

Antennas suited to particular applications

Antennas of the Yagi type are very popular on 1.3GHz and 2.3GHz, where their low windage and weight compare well with dishes having the same gain. Both are, of course, difficult to design and to adjust for optimum performance but both will repay the effort with good results, the choice being left to the users needs. However, on the higher bands dishes present an easier choice when the optimum result is required. It is not suggested that the design and adjustment of the Yagi types is impossible, as has often been demonstrated, but better results can be achieved with less work using a dish. Antennas such as helicals, corner reflectors and cylindrical paraboloids are now little used, mainly due to their low gain relative to their volume. They do, however, have specific applications to which they are well suited.

As already mentioned, various forms of Yagi antennas are now in use at 1.3 and 2.3GHz, where formerly dishes were used. Dishes less than about 10λ in diameter are rather inefficient and other simpler or less wind catching types are recommended for 1.3GHz as even these smaller dishes may be inconvenient in many installations. Dishes are almost mandatory for applications such as eme. This is because high gain is essential and several bands can be catered for by just exchanging the feed. Large arrays of Yagis can be used, but their use will be limited to a single band.

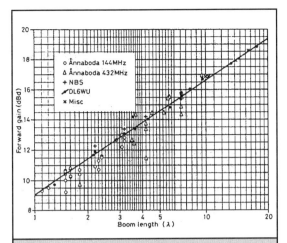

Fig 3.4 : Forward gain is proportional to boom length.

Helicals are used in applications such as satellite communication where circular polarity is often required. However, even this case may be better served by cross-polarised Yagi types that permit polarity to be selected. One exception is the eme worker who wishes to optimise a circularly polarised feed horn. To do this requires a pair of opposite sense eight turn helicals. By using these as measuring antennas, it is possible to obtain very good circularity while adjusting the dish feed. The same sense helix will give maximum output when used in this way.

Yagi type antennas

High-performance standard Yagis

This section presents a description of the general design principles of standard long Yagis, based on an article by G3SEK [1].

Fig 3.4 shows claimed and measured gain figures for 144MHz, 432MHz and 1.3GHz long Yagis, plotted against boom length in wavelengths. All gains are expressed in dBd, i.e. decibels above a half-wave dipole. With one or two exceptions, the gains claimed by antenna manufacturers lie far above the independent measurements of the same Yagis, and have not been included. Doubling the boom length of a long Yagi, and hence nearly doubling the gain, would reduce both the E-plane and H-plane beamwidths to about 70 per cent of their previous values. This follows from the so-called Kraus formula relating gain to E-plane and H-plane beamwidths:

$$B_e \times B_h = 25{,}000 / (\text{antilog}(G/10))$$

where the beamwidths are in degrees and G is in dBd. Since the two beamwidths are almost equal for a long Yagi, you can estimate them both with reasonable accuracy.

The Kraus formula is *not* accurate enough to estimate the gain from the beamwidths. The presence of side lobes, together with quite minor inaccuracies in beamwidth measurement, can totally blur the difference between a good Yagi and a poor one.

Several conclusions emerge from detailed analysis of the data summarised in Fig 3.4:

• Gain *is* proportional to boom length. The solid points and the straight line are from one consistent series of

measurements, and verify the proportionality to almost 20λ. Variations between measurements from different sources obscure the trend but do not negate it.

• Quads, loop Yagis, quagis, parrot-perches and motley-beams are all represented, but they do not stand out from the ordinary Yagis.

• For a given boom length, the difference in gain between a good yagi and a mediocre one is only a decibel or two.

• The same gain can be achieved with various numbers of elements on the same overall boom length. The number of elements does affect many other characteristics of the antenna, and more will be said about this later.

• The Yagis represented by the solid points will be worth a closer look.

When assessing a Yagi design, you can get far more information from a critical study of the directional radiation pattern than you can from simple tables of claimed gain and front/back ratio. Fig 3.2 shows a typical pattern, calibrated 360° around the compass, and inwards in decibels. Maximum power is radiated at the peak of the main lobe; this point is always plotted at the top of the diagram (0°, 0dB). Radiated powers in other directions can be plotted down to typically 40dB below the peak of the main lobe. Although usually called a radiation pattern, Fig 3.2 could equally well be called a reception pattern, for antennas behave the same in receive and transmit modes alike, i.e. they are reciprocal.

For dx working, the antenna should be horizontally polarised not from mere convention, but because horizontal works better for dx. Thus the most important radiation pattern, the one you observe when you turn the rotator, is the one in the same plane as the Yagi, which is called the E-plane. The pattern in the H-plane (the vertical plane for a horizontally polarised Yagi) is very similar to the E-plane pattern. The longer the Yagi, the more alike the E-plane and H-plane patterns become, the main differences being that the E plane pattern always has nulls at 90°, and the H-plane pattern always has a slightly broader main lobe and higher sidelobe levels.

Fig 3.2 also summarises the points to look for in a good radiation pattern. The main lobe is distinct, and the pattern is generally clean with good suppression of minor lobes. The pattern should be symmetrical and there should be deep nulls at 90 in the E-plane if not, there is something wrong with either the Yagi or the measurements.

Low levels of minor lobes in the polar pattern are almost as important as gain. The two features are in fact closely related, since any beam works by concentrating rf energy into the wanted direction at the expense of radiation in unwanted directions. The better the suppression of the minor lobes, the more energy can go into the main lobe. On receive, good suppression of the minor lobes means a stronger signal from the wanted direction, and less interference with stations in other directions. You cannot judge the minor lobe suppression by simple magic numbers like front/back ratio. The problem with front/back ratio is that it is too easy to optimise that one aspect of performance while letting the other sidelobe levels run wild. A better general indicator of minor lobe suppression is the level of the first side lobes on either side of the main lobe. These rabbit ears are present in all long Yagis, though in a good design they would be suppressed to about 15dB.

The higher the gain of a beam, the harder it becomes to aim it in the right direction. The reason is very simple; you cant get away from this, but you can considerably ease your aiming problems by choosing a Yagi design which gives a well-shaped main lobe.

The ideal would be the fan-shaped main lobe shown in Fig 3.1(a), with maximum gain across the whole beamwidth for ease of aiming, and no side lobes at all. Such perfection is not even attainable in theory; we have to accept that the edges of the main beam will always be rounded, and side lobes will always be present.

Even so, there is no reason to accept a main lobe like the one in Fig 3.1(b), which is sharply pointed and has less than 10dB of sidelobe suppression. If a Yagi has a pattern like that, there is something fundamentally wrong with it, no matter how high the maximum gain may be.

This is not just a matter of aesthetics; a beam with a pointed main lobe is really awkward to use. It is too difficult to peak a station right on the point of the main lobe, and too easy to peak on a sidelobe by mistake. The best achievable shape for the main lobe is shown in Fig 3.2; there is no doubt about finding the main lobe, yet its fairly flat nose makes peaking the signal quite uncritical.

To sum up all the desirable aspects of antenna performance, and optimised Yagi will have:

- high forward gain, for its boom length;

- a generally clean polar pattern (in particular, first side lobes at about 15dB;

- a well-shaped main lobe which causes no undue problems in aiming.

There are no fundamental conflicts between good performance in all of these areas, provided we do not go blindly chasing the maximum possible gain. In a modern optimised long Yagi, all the above objectives can be achieved together.

Modern long Yagi designs have evolved through a long process of trial and error mostly error. Designs for shorter Yagis have been adopted and extended, in the hope of producing workable longer Yagis with higher gain. The early work on long Yagis was done in the fifties and sixties, by investigators who seemed largely unaware of each others activities. Subsequently, the most successful designers are those who straightened out the historical tangles and built upon the work of their predecessors, rather than starting yet again from scratch. Thus there has evolved a definite mainstream of long Yagi design.

The mechanical dimensions required for optimum performance are usually found by experiment. Countless thousands of man-hours and tons of aluminium have gone into the optimisation of high-performance long Yagis. Computer analysis is increasingly used to assess potential improvements, and can save a great deal of experimental effort, but successful computer optimisation is a lot harder than it might seem. Yagi design still begins with ideas in someones head, and ends with measurements on real antennas.

Throughout the whole history of amateur band antennas, there have been frequent claims of miracle beams that break the gain/size barrier. Electrically small (e.g. two-element) beams can beat the odds to a small extent, but the reasons for this are clearly understood, and they do not apply to long Yagis. Unfortunately, since accurate gain measurements are so difficult, it is all too easy for people to obtain results that they proclaim with fervour and total sincerity, but which are in fact wrong.

Far more is known about Yagis using simple rod-type dipole elements than about other types of Yagis using loops or more elaborate element shapes. The latter may offer slightly higher theoretical gain, but even in theory this advantage dwindles away as the Yagi gets longer.

The existing body of knowledge about conventional elements can provide only hints about the behaviour of other element shapes, so the designer who chooses to use unconventional elements is starting again from scratch. In accurate comparisons, few of the more elaborate Yagis have come even close to competing with well-designed conventional Yagis. The G3JVL loop-quad Yagis are a rare example of a well-developed alternative approach, and the performance of the later versions is running neck-and-neck with the best of the plain Yagi designs.

Mainstream long Yagi design is based on some definite ideas about how a successful long Yagi should work. It is sometimes useful to think of a long Yagi launching a travelling wave along its structure and away into space. Thus you can divide the Yagi into two parts: a launcher consisting of the driven element, reflector and first few directors; and a travelling-wave structure consisting of all the rest of the directors. Although that is not necessarily the best all purpose description of how a Yagi works [2], it eliminates a lot of random cut and try by focusing on those combinations of element lengths and positions that show the most promise of launching and propagating a travelling wave.

The history of long-Yagi design starts with the work of Ehrenspeck and Poehler [3], who made a major investigation of uniform long Yagis whose directors were all the same length and equally spaced along the boom. Although short uniform Yagis can work very well, Ehrenspeck and Poehler found that they fail to achieve the expected increase in gain with boom length. In other words, uniform long Yagis are an evolutionary dead end. The next step forward came from investigations of the effects of tapering. In the language of long Yagi design, tapering has two alternative meanings. Tapering the element spacing means that successive directors are spaced further and further apart, going forwards along the boom (Fig 3.5(a)). Tapering the element lengths means that each director is shorter than the one before (Fig 3.5(b)). The two kinds of tapering should have very similar effects on a travelling wave. So, to keep the experimental work within manageable bounds, it seemed sensible to optimise either director spacing or director length, leaving the other one constant at some initially chosen value. This generation of long Yagis has since been called singly optimised.

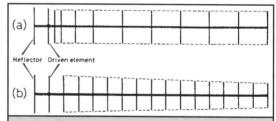

Fig 3.5 : Single optimisation schemes. (a) Optimised tapering of director spacing. (b) Optimised tapering of director lengths

Most experimenters have agreed that tapering should be quite pronounced in the launcher section, close to the driven element. Directors further along the boom need not be so strongly tapered. Tapering of spacing was investigated by W2NLY and W6QKI [4], who developed some quite successful Yagis in which all directors were the same length, and the director spacing initially increased and then became constant. The optimum value for this constant spacing seemed uncritical between 0.3λ and 0.4λ, though the wider spacing obviously involve fewer elements and less wind load.

Tapering of director lengths, with the spacing held constant, is the basis of the well-known NBS Yagis, derived from the work of Peter Viezbicke of the US National Bureau of Standards in the fifties [5]. This mammoth effort included full investigations of length tapering schemes and of the effects of different element diameters and mounting methods. It culminated in a set of designs including long Yagis with boom lengths of 3.2λ and 4.2λ. Joe Reisert, W1JR, and others have explained how to design NBS beams from Viezbickes charts [6, 7]. NBS long Yagis are good, but they cannot be extended; and much has happened in the time since they were developed.

The two single optimisation techniques produced the best long Yagis available at the time, and these designs have been enshrined in the amateur literature ever since. But single optimisation fails to produce Yagis that can be extended satisfactorily to meet todays requirements for longer Yagis with higher gain. So the next step was to combine the experience from tapering either the lengths or the spacing, and to taper them both. W2NLY and W6QKI found some improvement in bandwidth from length tapering their spacing optimised designs. Carl Greenblum [7] independently investigated double-tapering at about the same time as the single optimisation work was going on. Together, all these experiments formed the basis of some classic Yagi designs of the sixties and seventies.

The most promising of the modern-day long Yagi designs come from Gunter Hoch, DL6WU. He has thought carefully about ways of improving earlier designs, and his extensive experimental results have been obtained using professional test facilities. DL6WU considered first the director-spacing aspect of tapering. W2NLY and W6QKI showed that this is important in the launcher section, but can be stopped once the spacing has been gradually increased to about 0.4λ. Further increases in spacing in the travelling-wave part of the Yagi are counter productive.

This spacing scheme seemed close enough to optimum, so DL6WU turned his attention towards optimising the element length tapering as well. The result was to be a highly successful family of doubly optimised, long Yagis [8]. The simplest form of director length tapering would be to make each director shorter than its predecessor by a constant amount. This works if the Yagi is not extended too far, but eventually the forward gain bleeds away into increased levels of minor lobes. If you extend the Yagi far enough, subtracting a constant amount from each successive director, you eventually arrive at a director length of zero, a dead end if ever there was one!

DL6WUs answer to that problem is one of those brilliant ideas which seem so obvious in hindsight: taper each successive director length by a constant *fraction*. This logarithmic tapering makes the director lengths decrease quite sharply in the launcher section, but less markedly further along the array. No matter how far the Yagi is extended, the director lengths can never reach zero, and the logarithmic tapering seems to confer some degree of frequency independence. Various people have come close to this approach in the past, but only DL6WU has spelt it out as a design rule, and then followed it through to develop a whole family of successful long Yagis.

The shortest in the DL6WU family of long Yagis consists of only the launcher section with its gradually increasing director spacing. Longer Yagis can then be designed simply by adding more directors at a constant spacing of 0.4λ, with the appropriate logarithmic tapering of element lengths. The performance of the DL6WU Yagi is a tribute to Gunter Hochs clear thinking about the problem. Without any further optimisation, all the Yagis built from DL6WUs design charts have clean patterns and well-shaped main lobes, and therefore their forward gains are as high as their chosen boom lengths will allow. These are the long Yagis represented by the solid points in Fig 3.4, and the straight line shows every promise of extending beyond 20dBd that is, if you care to build a Yagi 24 wavelengths long! Perhaps the most remarkable feature of the whole family of DL6WU Yagis is the way that the patterns remain so consistent as the Yagis are extended. Naturally the main lobes become narrower as the Yagis become longer, but the patterns remain clean and the suppression of the first side lobes remains consistently good at about 15-20dB. In short, DL6WU long Yagis show every sign of working *properly*.

The success of the DL6WU family of long Yagis has demolished several cherished myths and moans.

- Long Yagis can't be extended. They can, if you do it right.

- Ordinary Yagis don't work on microwave. Oh yes, they do! The longest Yagis on Fig 3.4 were all developed on 1.3GHz. It is not suggested that Yagis are the best antennas for the higher microwave bands, but there is absolutely no reason why they should not work *if scaled properly* to the wavelength.

- Element dimensions for long Yagis are extremely critical. If some people were to be believed, the element lengths of the 1.3GHz Yagis would each need to be accurate to 0.1mm. In fact the logarithmic director taper appears to be very forgiving of constructional errors [8].

- Bandwidths of long Yagis are extremely narrow. With logarithmic tapering, the Yagi itself is quite wideband [1]; the frequency conscious part is usually the method of matching the feeder to the drive element.

- You can't feed a long Yagi with a simply dipole. Yes you can, and it does not need help from quad-loops, log periodic or ZL specials either.

Yagis based on the work of DL6WU are now appearing world-wide, both from individual amateurs and from professional manufacturers. They are highly recommended for use at 1.3GHz.

The design of 27- and 32-director singly optimised Yagis was discussed in the forth edition of the RSGB *VHF/UHF Manual* and these still represent simple, tried and tested designs for the beginner. However, the standard Yagi included here is a 26-element DL6WU design. The parameters were calculated using a computer program written by KY4Z, W6NBI and G4SEK based on articles

Design frequency = 1296·0 MHz   Number of elements = 26

Boom diameter = 12·700mm   Element diameters: driven = 2·500mm
                                                    parasitic = 2·500mm

Electrical boom length = 1975mm

Suggested stacking distances:
  horizontal = 754·7mm = 29·7 inches = 3·26 wavelengths
  vertical = 747·6mm = 29·4 inches = 3·23 wavelengths

Elements are SECURELY CONNECTED to the metal boom

CUMULATIVE SPACING			ELEMENT LENGTH	
mm	inches		mm	inches
Zero	Zero	REFL	120·32	4·737
50·50	1·988	D.E.	108·66	4·278
69·53	2·737	D 1	108·12	4·257
108·28	4·263	D 2	106·72	4·201
158·56	6·243	D 3	105·27	4·145
217·03	8·545	D 4	103·94	4·092
281·85	11·097	D 5	102·78	4·046
351·86	13·853	D 6	101·78	4·007
426·25	16·782	D 7	100·92	3·973
504·45	19·860	D 8	100·17	3·944
585·99	23·071	D 9	99·51	3·918
670·53	26·399	D 10	98·91	3·894
757·79	29·834	D 11	98·37	3·873
847·51	33·367	D 12	97·88	3·854
939·52	36·989	D 13	97·43	3·836
1033·64	40·694	D 14	97·01	3·819
1127·75	44·400	D 15	96·62	3·804
1221·86	48·105	D 16	96·25	3·789
1315·98	51·810	D 17	95·91	3·776
1410·09	55·516	D 18	95·58	3·763
1504·21	59·221	D 19	95·28	3·751
1598·32	62·926	D 20	94·99	3·740
1692·44	66·632	D 21	94·71	3·729
1786·55	70·337	D 22	94·45	3·718
1880·67	74·042	D 23	94·20	3·709
1974·78	77·748	D 24	93·96	3·699

Fig 3.6 : Construction details of a 26 element DL6WU long Yagi.

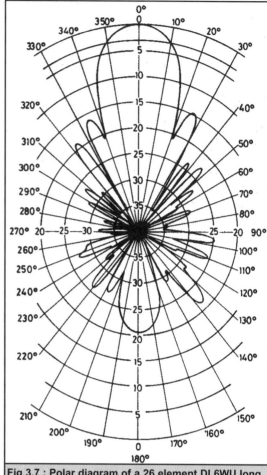

Fig 3.7 : Polar diagram of a 26 element DL6WU long Yagi.

and design data by Gunter Hoch, DL6WU. The boom length is 8.45λ or 1,955mm plus overhang. The estimated performance figures are:

Gain   = 18.5dBi

Horizontal beamwidth = 22.4°

Vertical beamwidth   = 22.8°

Suggested stacking distances are 528mm horizontally and 518mm vertically. Fig 3.6 gives the detail required to construct this antenna. Fig 3.7 shows its polar diagram. The equivalent length 27-element loop quad has 20dBi gain.

The G3JVL loop quad antenna

The design of this antenna is shown in Figs 3.8 and 3.9, and example dimensions for 1,296MHz and 2,320MHz in Table 3.1, which should be used in conjunction with Fig 3.9. The example values shown in the table will provide detail for any boom length, e.g. a 27-element version will include all elements up to D24. The construction of the loop quad is quite straightforward, but the dimensions must be closely adhered to, after all, at 1.3GHz 2.5mm (0.1in) error represents about one per cent of a wavelength, or a change in frequency of 13MHz.

When drilling the boom, for example, measurements of the position of the elements should be made from a single point by adding the appropriate lengths; if the individual spaces are marked from the preceding position, then errors may accumulate to an unacceptable degree. Elements other than the driven element may be made from flat aluminium strip. The two holes are drilled or punched before bending, with a spacing equal to the circumference specified in Fig 3.9

The radiator may be brass or copper strip. This form is by far the easiest to home construct but the mechanical strength is rather poor. A much superior form uses mild or stainless steel wire (welding rods) for all the elements. The parasitic element lengths should be cropped, folded

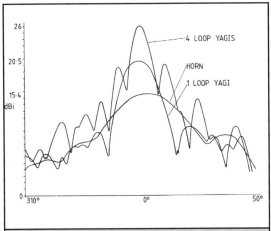

Fig 3.8 : Polar diagram of the loop quad antenna.

and brazed or silver soldered onto M4 countersunk screws. The entire assembled antenna may be zinc plated, at low cost, to protect and possibly improve conductivity.

The drive element differs in that it uses a countersunk M6 by 25mm long screw that has been drilled through at 3.6mm diameter. The element is brazed or silver soldered with its open end around the hole. The hole may need clearing or re-drilling carefully after brazing. To allow the connection of the semi-rigid coaxial cable, saw through the element at its top mid-point. The cut ends should be shaped to accept the coaxial cable as shown in Fig 3.9. To enable soldering it will be necessary to use an acid flux. Examples are Bakers Fluid or Multicores ARAX fluxed solder. Both of these require that any residue is washed off to prevent long term corrosion effects. Once the steel has been properly tinned, use normal non-corrosive flux to complete the joint to the cable. The

coaxial cable outer should be pre-trimmed to allow the connector to be fitted.

The whole assembly, including the coaxial cable, may be plated after completion. However, it is advisable to leave the ptfe covering the inner. This is because the zinc plating is very difficult to solder. Even after plating the radiator and all screws and soldered joints should be protected with polyurethane varnish, followed by a coat of paint on all surfaces. If inadequate attention is paid to this protection, then the gain of the antenna may decrease with time as a result of corrosion. However, there is only a very small chance of this occurring with the brazed wire elements, as no surface-to-surface contact is involved. The zinc plating is intended to retain a good appearance but its long term properties can be a little unpredictable. The additional protection of varnish or paint will help to maintain the finish and is strongly recommended.

Where the specified materials are not available, changes may be made to the diameter of the boom and the thickness and width of the elements. Compensation may be achieved by altering the length of all elements by applying the correction factors CF as detailed in Table 3.2. It may be found an advantage to use thicker, wider elements to increase the strength of the antenna, for example a 19mm (0.75in) diameter boom, and 6.35mm by 1.6mm (0.25in by 0.063in) loops. The element dimensions will therefore need correcting by (0.9 - 0.3 + 0.6) = 1.2 per cent. Thus all elements (the reflector, the driven element and the directors) should be made 1.2 per cent longer, but the element spacing is not altered. Provided that the antenna is constructed carefully, its feed impedance will be close to 50Ω.

If a vswr or impedance bridge is available, very small final adjustments may be made by altering the spacing of the reflector loop a little by bending with respect to the driven element. However, it is recommended that the method of adjustment is by altering the driven element length very carefully. As an alternative the G3JVL designed log

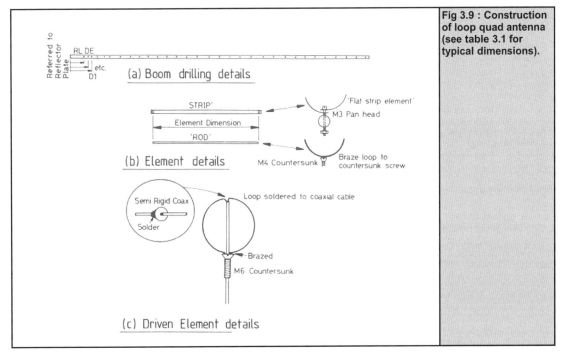

Fig 3.9 : Construction of loop quad antenna (see table 3.1 for typical dimensions).

Table 3.1 : Typical dimensions of loop quad antenna

1,296MHz				2,320MHz		
		Boom dia				Boom dia
		0.5in	0.75in			0.5in
R_p	R_w	R1	R1	R_{pw}	R_{ph}	R1
140	115	252	254	80	65	143
Elements	D_e	D_e		Elements	D_e	
2.5mm dia	230	232		1.6mm dia	130	
Directors				Directors		
D01-D12	213	215		D01-D12	121	
D13-D20	206	208		D13-D20	117	
D21-D30	198	200		D21-D30	113	
D31-D40	193	195		D31-D40	110	
D41-D55	190	192		D41-D55	108	
D56-D63	189	191		D56-D63	107	

Cumulative spacings				Cumulative spacings			
RP	0	RL	79	RP	0	RL	44
DE	103	D1	131	DE	58	D1	73
D2	152	D3	198	D2	85	D3	110
D4	243	D5	275	D4	136	D5	153
D6	333	D7	424	D6	186	D7	237
D8	514	D9	605	D8	287	D9	338
D10	695	D11	785	D10	388	D11	439
D12	876	D13	966	D12	489	D13	540
D14	1,057	D15	1,147	D14	590	D15	641
D16	1,238	D17	1,328	D16	691	D17	742
D18	1,418	D19	1,509	D18	792	D19	843
D20	1,599	D21	1,690	D20	893	D21	944
D22	1,780	D23	1,871	D22	994	D23	1,045
D24	1,961	D25	2,051	D24	1,095	D25	1,146
D26	2,142	D27	2,232	D26	1,196	D27	1,247
D28	2,323	D29	2,413	D28	1,297	D29	1,348
D30	2,503	D31	2,594	D30	1,399	D31	1,449
D32	2,684	D33	2,775	D32	1,500	D33	1,550
D34	2,865	D35	2,956	D34	1,601	D35	1,651
D36	3,046	D37	3,136	D36	1,702	D37	1,752
D38	3,227	D39	3,317	D38	1,803	D39	1,853
D40	3,408	D41	3,498	D40	1,904	D41	1,954
D42	3,589	D43	3,679	D42	2,005	D43	2,055
D44	3,769	D45	3,860	D44	2,106	D45	2,156
D46	3,950	D47	4,041	D46	2,207	D47	2,257
D48	4,131	D49	4,222	D48	2,308	D49	2,358
D50	4,312	D51	4,402	D50	2,409	D51	2,459
D52	4,493	D53	4,583	D52	2,510	D53	2,560
D54	4,674	D55	4,764	D54	2,611	D55	2,661
D56	4,855	D57	4,945	D56	2,712	D57	2,762
D58	5,035	D59	5,126	D58	2,813	D59	2,863
D60	5,216	D61	5,307	D60	2,914	D61	2,964
D62	5,397	D63	5,487	D62	3,015	D63	3,064

All dimensions are in millimetres except where stated

Table 3.2 : Correction values for loop quad antenna.

Thickness (mm)	(in)	CFx100 (%)	Width (mm)	(in)	CFx100 (%)	Boom dia (mm)	(in)	CFx100 (%)
0.71	0.028	0.0	2.54	0.10	0.4	0.00	0.00	-0.7
1.63	0.0625	0.6	4.76	0.1875	0.0	12.7	0.50	0.0
			6.35	0.25	-0.3	19.0	0.75	+0.9
			9.53	0.375	-0.95	25.4	1.00	+2.1

Other values of Thickness are determined by plotting a straight line through the values given. A graph can be constructed to extend these values.

necessary. The parameters concerned are the thickness, width and circumference of the elements and the boom diameter.

The completed array can be mounted using a standard clamp from tv antenna suppliers. However, it is essential that the antenna be mounted on a vertical mast. Horizontal supports in close proximity to the antenna can cause severe miss tuning. It has been determined that the spacing from any such objects should be at least equivalent to the largest element diameter much less than was previously suggested. Note that when the antenna is mounted with the loops to one side the polarisation is vertical and the spacing data just detailed also applies but, in this case, from the mast. It is recommended that the antenna be mounted to produce horizontal polarisation in order to benefit from the improvement in performance when working beyond the horizon paths. This shows up in the form of a less rapid rate of fading (QSB) and thus easier communication, especially over troposcatter paths.

Stacking Yagi antennas

More than one antenna may be stacked to achieve extra gain. However, several points should be borne in mind when this is done, or the results obtained may be rather poor. Perhaps the most critical factor is the impedance matching of the antennas. This is illustrated by Table 3.3 which shows how the stacking gain is reduced when two antennas of differing impedances are combined, assuming the worst case condition for the antenna impedances.

Note that no power is actually lost in either of the antennas. All that happens as the mismatch between the antennas becomes worse is that the gain and radiation pattern of the array just tends to that of one antenna.

It can be seen from Table 3.3 that even quite small vswrs can seriously degrade the stacking gain, and that for near optimum results the antennas should possess a vswr of better than 1.1:1. The actual reference for the vswr measurement is less critical it could be anywhere around 30-70Ω since the vswr of the array as a whole can be largely compensated for in the preamplifier and the transmitter tuning. The important criterion is the relative vswrs of the two antennas. Great care should therefore be taken in their construction and their impedance should preferably be measured and optimised.

It is also most important to ensure that the antennas are correctly phased. This means that, for example, the outer of the feeders should go to the same side of all the driven elements when the antennas are mounted in the array. It is worth checking that this is the case with the antennas installed, as it is easy to make mistakes. Needless to say,

periodic wideband feed, as shown in Fig 3.10, may be used. This should result in a reasonable vswr which is less critically dependent on frequency and which permits much wider coverage of the 1.3GHz allocation, for example.

Several versions of this antenna have been constructed for 432MHz, 1.3, 2.3, 3.4 and 10GHz. The dimensions of these have been derived by scaling linearly all the antenna dimensions and applying correction factors (derived from experimental work by G3JVL in 1974-5) as

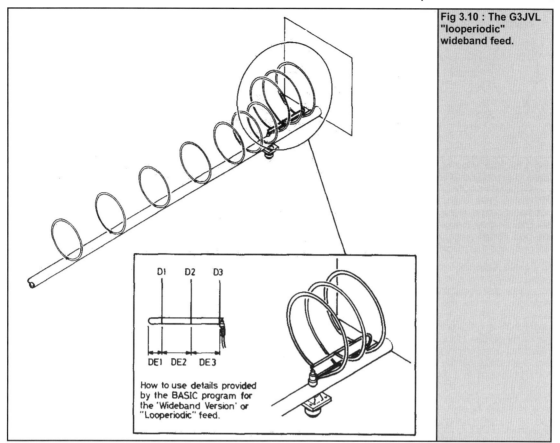

Fig 3.10 : The G3JVL "looperiodic" wideband feed.

How to use details provided by the BASIC program for the 'Wideband Version' or "Looperiodic" feed.

equal lengths of cable (multiples of half wavelengths) should be used to join each antenna to the power combiner or divider.

The stacking distance is the one parameter that may require some experiment. If the antennas are out of phase the resulting pattern will display a null in the desired direction of radiation. The result of unequal cable lengths will be an offset to the main beam. If the cables are as much as half a wavelength different, the combined pair will be out of phase and will produce a null on the boresight. Either side there will be two equal but narrower than expected lobes, both having gain. This effect can also be caused by combining the antennas out of phase due to a reversed driven element. It is easily detected when mounting side by side or baying (Fig 3.11(a)). However, when the antennas are vertically combined or stacked (Fig 3.11(b)) the resulting lobes will be above and below the horizon, giving rise to a very poor performance

indeed, which is difficult to understand. Too large a stacking distance will result in an increased side lobe level and an excessive narrowing of the main beam. Conversely, too small a stacking distance may result in lower than optimum gain for the array.

G3JVLs findings on stacking distance

When two antennas are stacked vertically or bayed horizontally, the distance between them is usually set by recommendation or by guesswork. This is due to a lack of definitive theoretical references dealing with this aspect. A number of references covering the combination of arrays

Table 3.3 : Stacking gain for antennas with different impedances

VSWR of each antenna	Max Z / min Z	Worst case stacking gain (dB)
1.0	50	3.0
1.1	55/45	2.6
1.2	60/42	2.2
1.5	75/33	1.6
2.0	100/25	1.0
3.0	150/17	0.5

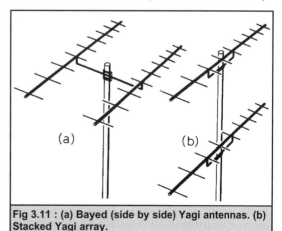

(a)   (b)

Fig 3.11 : (a) Bayed (side by side) Yagi antennas. (b) Stacked Yagi array.

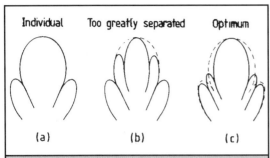

Fig 3.12 : The effect of baying distance on the polar pattern of loop quad antennas.

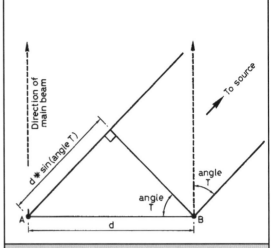

Fig 3.14 : Diagram of two spaced antennas for pattern analysis

of dipoles have been available for a long time. The optimising programs that have been developed require a starting design and may improve it where possible. There are now design programs as previously mentioned that will give good theoretical results.

In practice it has been found that in order to get the best compromise the closer spacing should be chosen. This is to ensure that the pattern obtained is not excessively narrowed and the first pair of side lobes enhanced. Associated with this, the main lobe width is less than it ought to be and the gain may also be reduced. The lost gain may be due to too much energy being radiated in unwanted directions in the side lobes. These side lobes are unavoidable but the amplitude of them should be as low as possible. In a good design the optimum compromise is the aim. The factors relating to gain and side lobe levels are in opposition. Care should be taken to ensure that the phase relationship between the individual antennas is correct. Deliberate reversal of the phasing will result in the cancellation of the main lobe and transference of the power almost completely to the first pair of side lobes, which both then have some gain and also a reduced beamwidth.

The desirable spacing gives both the increased gain and the expected beam narrowing, i.e. when one is vertically over the other the width in the azimuth plane (around the horizontal) will remain unaltered. But in the elevation plane (vertically) it should be reduced to 50 per cent of its previous value for two antennas. It is also likely that gain may be lost if too close a spacing is used. This value is very dependent upon the presence of nearby objects (other similar antennas in this case). If practical, it is preferable to lengthen the antenna to realise more gain. Practical limitations are very individual to the users needs, but boom lengths of 4 to 6m are quite practical. The extra components required when stacking antennas will detract from the advantage expected. Very careful design and

construction of a combiner or divider and cable preparation is essential. A suggested stacking or baying distance for the 27-element loop quad is 2.6 wavelengths and the 44/47-element version should be tried at 3.3 wavelengths. Fig 3.12 shows the effect of increasing separation and Fig 3.13 the stacking distances for arrays of antennas.

A more theoretical approach to stacking by G8DIC gives a better idea of what to expect, but still only provides a guide which may be less than optimum.

G8DIC offered the following observations concerning stacking antennas in order to improve overall system gain. Generally speaking, an array of two antennas will give a gain increase of 3dB. A further 3dB gain is obtained each time the number of antennas stacked is doubled. However, if antennas are stacked too far apart then large side lobes may be generated. If the spacing is too small there may be excessive mutual coupling or de-tuning between the individual antennas which can cause the potential 3dB increase to be reduced or even lost completely. The following analysis attempts to explain how to

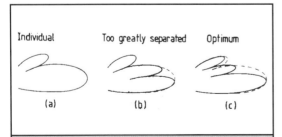

Fig 3.13 : The effect of stacking distance on the polar pattern of loop quad antennas.

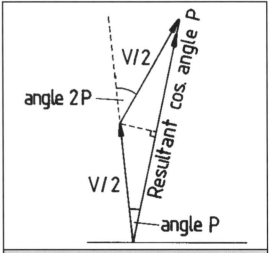

Fig 3.15 : Vector diagram of spaced antennas for pattern analysis

determine the optimum spacing. The spacing is, generally speaking, not very critical and it is usually best to use the second null of the array pattern to split the first side lobe of the individual antenna pattern.

Although the following is difficult to grasp at first, it is well worth reading and re-reading with reference to the diagrams. The array polar diagram is not the final result, but is the stacking effect polar diagram which is used to determine the final result when added to the actual polar diagram of one of the individual antennas that make up the stack.

Refer to Fig 3.14 in which A and B are two antennas spaced d apart and angle T is the bearing of the source relative to the main beam direction. The array's polar diagram can be deduced by adding the two received signals at A and B then observing how this sum varies with angle T. Since the source is at a great distance from the array, the two paths can be considered parallel and thus the two signals received are equal in amplitude. Departure from this is a reason why scattered signals frequently do not appear to be stronger when received on such an array.

With reference to the vector diagram, Fig 3.15, the signal from each antenna is half voltage and the phase difference between the two signals is twice angle P degrees. It is readily shown that the resultant varies as the cosine of angle P. We now need to know how angle P varies with angle T. In the diagram it is shown that the signal must travel the product of d and the sine of angle T further to get to antenna A than to antenna B, therefore the signal will be:

$$\frac{360\times(d\times\sin(angleT))}{\lambda}$$

in electrical degrees later at A than at B, and is the wavelength. Hence:

$$angleP = \frac{d\times\sin(angleT)\times180}{\lambda}$$

The array polar diagram is given by:

$$\frac{\cos(180\times d\times\sin(angleT))}{\lambda}$$

Fig 3.16 shows the magnitude of cos(angle P) plotted against angle P and also:

$$\frac{180\times d\times\sin(angleT)}{\lambda}$$

plotted against angle T for various values of d. To find the magnitude of:

$$\frac{\cos(180\times d\times\sin(angleT))}{\lambda}$$

for a given value of d and angle T, look up angle T on the vertical axis of the lower graph, then move horizontally until the curve for the value of d to be used is encountered, then vertically until the curve for cosine angle P is also encountered. The value of cosine angle P is then read from the vertical axis of the upper graph. This process shows that for large values of d a number of peaks and nulls occur in the array polar diagram. The overall stacking polar diagram, Fig 3.17, of an array of antennas is given by the product of the array polar

diagram and the individual antenna polar diagram. If you are working in decibels you simply add the two.

From this it is obvious that when either polar diagram drops into a null then the composite polar diagram will have a null, and also when the array polar diagram is at a maximum (at angle P 0, 180°, 360° etc) the composite polar diagram will have the same value as the individual antenna polar diagram. For a normal antenna with the first side lobes at around 10dB down it is usually best to position the second null of the array polar diagram (angle P = 270° at the bearing of the first side lobe of the antenna. However, some antenna types with inherent good first side lobe figures (dishes and horns) may benefit from consideration of the first side lobe level of the composite antenna (approximately for angle P = 180°). In either case:

$$d = \frac{angleP\times\lambda}{180\times\sin(angleT)}$$

gives the stacking distance required to get a particular feature of the array polar diagram at angle P to correspond to a bearing of angle T. For example an antenna with the first side lobe at 30° bearing will require stacking at:

$$d = \frac{270\times\lambda}{180\times\sin30} = 3\lambda$$

for the second array null to coincide with the first side lobe. In this case the first side lobe of the array polar diagram will be approximately the first side lobe of the composite antenna and correspond to angle P = 180°, or a bearing of:

$$angleT = asn\left(\frac{angleP\times\lambda}{180\times d}\right) = asn\left(\frac{180}{180\times3}\right) = 19.5°$$

The first array null will correspond to angle P = 90°, or a bearing of approximately 9.6°.

$$angleT = asn\left(\frac{90}{180\times3}\right) = 9.6°$$

Stacking more than two antennas

It can be shown that for n antennas spaced d apart the stacking polar diagram of the array is given by:

$$\frac{\sin(n\times angleP)}{n\times\sin(angleP)}$$

This function has a value of 1 at angle P = 0, 180°, 360°, 540° etc. It has nulls at n x angle P = 180°, 360°, 540°, 720° and 900° etc, except where the main lobes occur. (Note that n = 6 in this example).

$$\frac{1}{n\times\sin\left(\frac{90\times3}{n}\right)} \qquad \frac{1}{n\times\sin\left(\frac{90\times5}{n}\right)}$$

$$\frac{1}{n\times\sin\left(\frac{90\times7}{n}\right)} \qquad \frac{1}{n\times\sin\left(\frac{90\times9}{n}\right)}$$

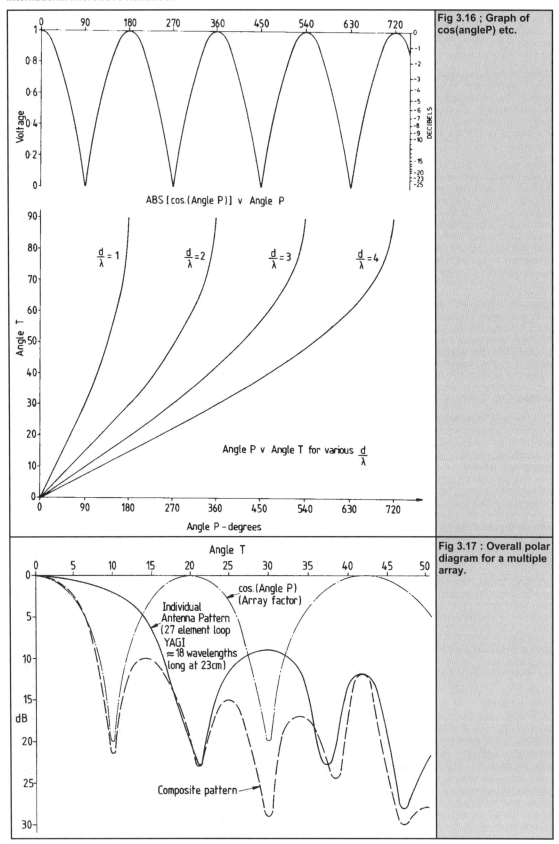

Fig 3.16 ; Graph of cos(angleP) etc.

ABS [cos.(Angle P)] v Angle P

Angle P v Angle T for various $\frac{d}{\lambda}$

Angle P – degrees

Fig 3.17 : Overall polar diagram for a multiple array.

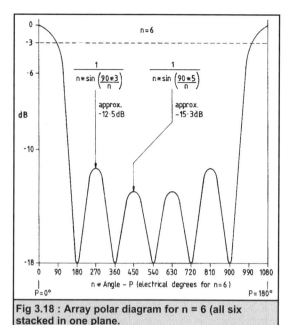

Fig 3.18 : Array polar diagram for n = 6 (all six stacked in one plane.

Fig 3.18 is a diagram showing the plot for n = 6 as an example (all antennas to be in one plane). It has side lobes at n x angle P = 270°, 450°, 630° and 810° etc with level:

Power splitters or combiners

The power splitters shown in Fig 3.19 enable either two, four, six or eight antennas of a given impedance to be fed from a single coaxial cable. The unit consists of a length of fabricated coaxial line that performs the appropriate impedance transformations. The inner is made exactly one half wavelength (electrically) long (or any odd multiples of half wavelengths) between the centres of the outer connectors. The outer is made approximately 38mm (1.5in) longer. The outer may be made from square section aluminium tubing. The open ends and the access hole, allowing the centre connector to be soldered, may be sealed by covering with aluminium plates secured with an adhesive. Alternatively copper or brass tubing may be used, and the plates soldered.

Any size of inner or outer may be used provided that the ratio of the inside dimension of the outer to the diameter of the inner conductor is unchanged.

The maximum size that may be used is governed by the requirement that *no waveguide modes are to be permitted in the structure*. Practically, this means that the outer must not approach λ/2 (i.e. below cut-off).

Other forms of coaxial line should work just as well. The most adaptable inner is one cut from sheet metal. The width of the required strip may be calculated with reference to Fig 3.20. Fig 3.21 gives the design details for a number of configurations.

Two-way (half-wave)

The impedance of the centre conductor to match two 50Ω antennas back to 50 may be determined from $Z_t = Z_0 \times Z_1$, where Z_t is the transforming section impedance, Zo is the coaxial cable impedance and Z_1 is the load to be matched. In this case the transforming-section impedance

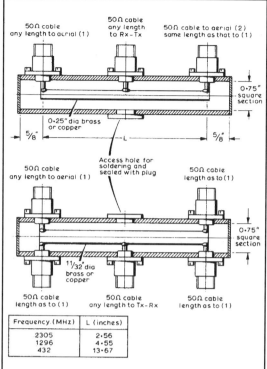

Frequency (MHz)	L (inches)
2305	2·56
1296	4·55
432	13·67

Fig 3.19 : 50Ω splitters for connecting two or four antennas to a common feeder.

should be 71Ω. When using a square section outer that measures 25.4mm and has walls of 18swg or 1.2mm, the internal dimension is 23mm. From Fig 3.20, the ratio required is seen to be 0.33. The centre conductor is therefore calculated from 23x0.33 = 7.59mm diameter.

Four-way (half-wave)

This version requires a transforming impedance of 50Ω which makes the centre conductor to outer ratio 0.47. The centre conductor is therefore determined from 23 x 0.47 = 10.81mm. The diameter of the centre conductor is quite critical, and it has been shown that a 10 per cent error will result in a poor match even when the loads (antennas) present a perfect 50Ω impedance. For this reason it is not good enough to use the nearest standard-size rod.

Table 3.4 : Examples of a flat strip conductor in a square outer

	Frequency (MHz)	Flat inner (mm)	Round inner (mm)	Outer length (half wave) (mm)
Two way	432	344	340	349
Four way	432	344	340	366
Two way	934	164	160	169
Four way	934	164	160	186
Two way	1,296	119	115	124
Four way	1,296	119	115	141
Two way	2,230	67	63	72
Four way	2,230	67	63	89

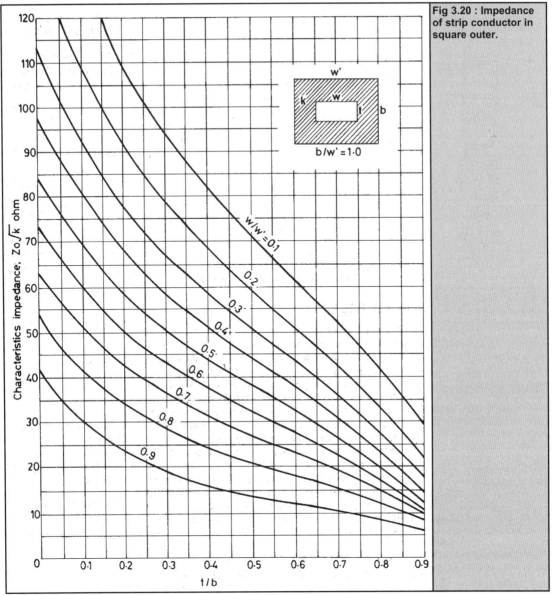

Fig 3.20 : Impedance of strip conductor in square outer.

$b/w' = 1.0$

Lengths of inner and outer

The dimensions shown in the examples shown include an extension of the outer of 14mm beyond the centre conductors theoretical length. This allows N type connectors to be mounted on the side wall.

Flat-strip conductor in a square outer

The flat strip version is indicated in Fig 3.20 and is discussed further here. It is an alternative design that is much less demanding in terms of machining as it uses a thin flat centre conductor. This version eases the problem imposed by standard size materials. The width required is established using Fig 3.20 and may be cut from sheet copper or brass. In this case the inner should be extended by 4mm to allow for soldering to the connector spills. Examples are given in Table 3.4.

A prototype four-way combiner/splitter, using an outer of 25.4mm square with 16swg or 1.6mm walls, required the centre conductor width to be 17.5mm and used copper strip for a 50Ω centre conductor impedance. The result obtained was good and a return loss of 40dB was measured. This performance is very close to optimum for this design.

When the strip width is 17.5mm, the gaps either side are just enough to accommodate the N type connectors but this is only satisfactory for power levels up to around 25W. If higher power operation is required then the line should be filed to leave a larger gap near to the connector body. Alternatively, trim the connector body so that only the ptfe protrudes in the vicinity of the inner line. The areas to be joined were pre tinned. The end connectors were joined using the soldering iron in the normal way, as access is easy. With the 1.6mm thick outer wall, the two-way (71Ω) inners width should be 13mm. The outer can be virtually the same without the extra connector mounting holes drilled.

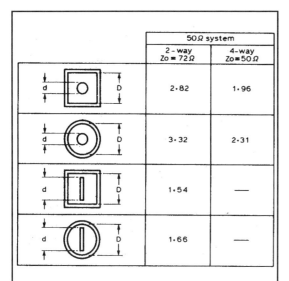

Fig 3.21 : Ratios o/d i/d for other useful coaxial configurations.

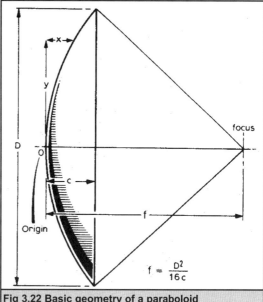

Fig 3.22 Basic geometry of a paraboloid

Parabolic dishes

Antennas based on paraboloidol reflectors are the most important type for the bands above 5.7GHz. Their main advantages are that, in principle, they can be made to have as large a gain as is required, they can operate at any frequency and they should require little setting up. Disadvantages are that they are not the easiest things to make accurately (which limits the frequency at which a given dish can be used), large dishes are difficult to mount and are likely to have a high windage.

The basic property of a perfect paraboloidal reflector is that it converts a spherical wave emanating from a point source placed at the focus into a plane wave, i.e. the image of the source is focused at an infinite distance from the dish. Conversely, all the energy received by the dish from a distant source is reflected to a single point at the focus of the dish.

The geometry of the paraboloid

A paraboloid is generated by rotating a parabola about a line joining its origin and focus. Two methods for constructing a parabola are given below and the geometry of a paraboloid is given in Fig 3.22.

Constructing a parabola by calculation

Convenient forms of the equations of a parabola re, using the notation of Fig 3.22:

$$y^2 = 4 \times f \times x \qquad (1)$$

$$y^2 = 4 \times D \times x \times (f/D) \qquad (2)$$

where
$$f = \frac{D^2}{16 \times c}$$

y has both negative and positive values

D = diameter of corresponding dish

f = focal length

c = depth of parabola at its centre

Suppose we wish to construct the profile of a dish 0.914m (36in) in diameter having a f/D ratio 0.6. The procedure is as follows:

Using metric units

• From equation (2)

$$x = \frac{y^2}{4 \times D \times (f/D)} = \frac{y^2}{4 \times 0.914 \times 0.6} = 0.000456 \times y^2$$

• Tabulate the calculations for as many points as accuracy requires in the form shown in Table 3.5.

• A plot of y from -457 to +457 against values of x will produce the required curve (both x and y in millimetres).

Using imperial units

• From equation (2)

$$x = \frac{y^2}{4 \times D \times (f/D)} = \frac{y^2}{4 \times 36 \times 0.6} = 0.01157 \times y^2$$

• Tabulate the calculations for as many points as accuracy requires in the form shown in Table 3.6.

• A plot of y from -18 to +18 against values of x will produce the required curve (both x and y in inches).

Graphical method

A simple graphical method for constructing a parabola is shown in Fig.3.23. The value of c is calculated from:

$$c = \frac{D^2}{16 \times f} \qquad \text{or} \qquad c = \frac{D}{16 \times (f/D)}$$

Both axes are divided into the same number of equal parts in the way shown and numbered. Points where corresponding lines intercept describe the required parabola.

Factors affecting practical dish design and use

Size

This is the most important factor since it determines the maximum gain that can be achieved at a given frequency and the beamwidth resulting. The actual gain obtained is given by:

$$G = \frac{4 \times \pi \times A}{\lambda^2} \times n \quad \text{or} \quad \frac{(\pi \times D)^2}{\lambda} \times n$$

where A is the projected area of the dish and n is the efficiency, which is determined mainly by the effectiveness of illumination of the dish by the feed, but also by other factors which are discussed.

Each time the diameter of a dish is doubled, its gain is quadrupled, i.e. increased by 6dB. If both stations double the size of their antennas, signal strengths can be increased by 12dB, a very substantial gain. A given dish used at twice the frequency also quadruples its gain if other factors do not intervene: accuracy and mesh size etc.

The relationship between the diameter of a dish and its gain at frequencies of amateur interest is shown in Fig 3.24. An efficiency of 50 per cent is assumed, which seems to be typical of better amateur practice. Antennas with a diameter less than 10 wavelengths will generally have a significantly lower efficiency than this value. The corresponding beamwidths are also given in Fig 3.24.

Two factors tend to limit the maximum gain that can be achieved. On the lower microwave bands (1.3, 2.3 and 3.4GHz) the physical size of the antenna is the limiting factor. Thus a 30dB antenna on 1,296MHz will have a diameter of just over 3.2m (10ft), which could well cause problems in fabrication and mounting. This gain can be achieved using four 50 element loop quad antennas with booms at 18λ long, properly constructed and combined. On the higher microwave bands (5.7GHz and above) the physical size of the antenna is less of a problem. Instead, the narrowness of the beamwidth tends to be the limiting factor. For example, a dish 0.9m (3ft) in diameter used at 24GHz will have a gain of up to 44dB and a beamwidth of 1°. Considerable skill is required in effectively handling such a directional antenna.

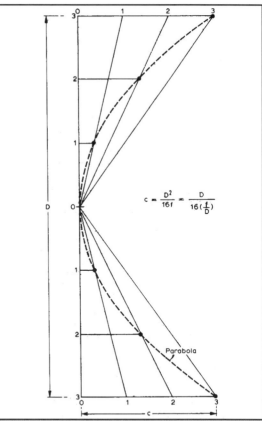

Fig 3.23 : Simple graphical construction for a parabola.

The ratio of the focal length of the dish to its diameter

This ratio f/D, using the notation of Fig 3.22, is the fundamental factor governing the design of the feed for a dish. The ratio is, of course, directly related to the angle subtended by the rim of the dish at its focus, and also therefore to the beamwidth of the feed necessary to illuminate the dish effectively. Two dishes of the same diameter but different focal lengths require different designs of feed if both are to be illuminated efficiently.

Practical values for the f/D ratio range from about 0.2 to 1.0. The value of 0.25 corresponds to the common focal

Table 3.5 : Tabulation of parabola calculations in metric units.

y(mm)	y^2	0.01157y^2=x(mm)
0	0	0
50	2.5E02	1.14
100	10.0E05	4.56
150	22.5E05	10.26
200	40.0E05	18.24
250	62.5E05	28.50
300	90.0E05	41.04
350	12.3E06	55.86
400	16.0E06	72.96
450	20.3E06	92.34
457	20.9E06	95.24

Table 3.6 : Tabulation of parabola calculations in imperial units.

y(in)	y^2	0.01157y^2=x(in)
0	0	0
2	4	0.05
4	16	0.18
6	36	0.42
8	64	0.74
10	100	1.16
12	144	1.67
14	196	2.27
16	256	2.96
18	324	3.75

plane dish in which the focus is in the same plane as the rim of the dish. However, values of the f/D ratio which produce a deeper dish are frequently used commercially where it may be important to minimise side lobe response. With such dishes, the unwanted interaction between antennas is reduced, albeit at the expense of antenna gain and extra difficulty in designing efficient feeds. Such considerations are unimportant in an amateur context. Indeed, the greater the side lobe response the better, provided overall gain does not suffer.

As will be seen, there are a number of factors that influence the choice of the f/D ratio. For most amateur applications the range 0.5 to 0.75 would appear to be optimum, although satisfactory feeds for dishes outside this range can be constructed. There are difficulties associated with shallower dishes, both mechanical and electrical blockage (shadowing). The shallow dish requires a higher gain and therefore a larger feed horn. Also, this horn must be mounted at a greater distance, resulting in a heavier and unbalanced mechanical structure.

Changing the geometry of a dish

As a given dish is reduced in size, so its f/D ratio is reduced proportionately. Thus, if a focal plane dish for which f/D = 0.25 is trimmed to half its original diameter (or a half-size moulding is made from an existing dish), then the smaller dish will have a f/D of 0.5. This approach has been used to convert a virtually unusable dish into a more easily illuminated and efficient dish. As would be expected, the reverse is also the case but does not find so much favour as the f/D is reduced, and extending is not as simple.

The effect of dish accuracy on performance

An understanding of the degree of accuracy required in a dish is important for two reasons. It enables the maximum frequency for efficient operation of a given dish to be determined. It also enables an estimate to be made of how construction tolerances influence the gain and hence the ease with which dishes may be constructed. The reduction in gain due to surface irregularities depends upon two factors:

- The amount by which the surface deviates from a true paraboloid. This will be expressed as the mean value of the deviation as a factor related to wavelength.

- The ratio of any such deviation to the wavelength of operation. In this way a distinction is made between short range and long range irregularities. Short range irregularities correspond to a bumpy surface on a good average shape. This limitation may be due to a mesh surface, for example, or a rough finish. Both will limit the upper frequency of operation.

The relationship between loss in gain, the peak deviation and the periodicity is shown in Fig 3.25. This assumes that the deviations occur uniformly over the surface of the dish. If they are restricted to a limited area, then the loss will be proportionately lower. It can be seen that when the periodicity of the irregularities is a small fraction of the wavelength, the irregularities can be fairly large before the gain falls off too much. Checking the long range accuracy requires a well made template. When checking remember that perhaps the parabola used is not quite that intended. This may show up as a steady deviation from the template. If the space is in the centre then the dish has a

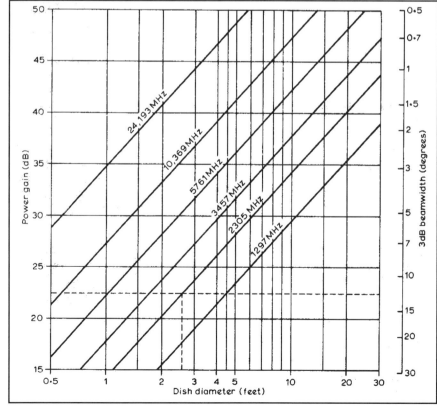

Fig 3.24 : Relationship between the size of a dish, its gain and beamwidth as a function of frequency. An overall efficiency of 50% is assumed. As an example, a dish 2.5ft in diameter at 2,305MHz will have a gain of 22dB and a beamwidth of about 13°.

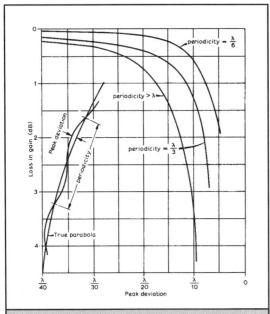

Fig 3.25 : The effect of dish accuracy on perform-ance.

greater f/D than the template. Similarly, if the gap is at or near the rim then the dish has a smaller f/D. It is therefore reasonable to try a slight variation in the template and retest until a satisfactory conclusion is reached.

Materials of construction

Dishes up to about a metre are usually made from solid material. Aluminium is frequently preferred for construction due to its weight advantage. It also has the added advantages of its durability and good electrical characteristics. Obviously, windage increases rapidly with dish size and soon becomes quite a sever problem. For example, at a wind speed of 50mph the force on a flat surface would be about 44kg/sq m (9lb/sq ft). The structure supporting a 1.2m (4ft) dish would be subjected to a wind loading force of 45kg (100lb). Good engineering practice requires that a safety factor of five should be used. The result of this is that a structure which can survive a loading of 227kg (500lb) is required.

Fortunately, amateurs can use dishes that have a reflecting surface that uses an open mesh. The resulting poorer front-to-back ratio is unavoidable but it is a small price to pay for safety or continued operation. The loss in gain as a function of mesh size is illustrated in Fig 3.26. For example, if the mesh-to-wire ratio is 10:1 and the square size (or wire spacing if only in one plane) is at 0.1 wavelength a gain loss of 0.2dB or 5 per cent power loss will occur. Please note that if polarisation in *one plane* is satisfactory then the spacing between any supporting members need only satisfy mechanical requirements for stability of shape. It is unfortunate that as the velocity of the wind increases the loading will approach the value of a solid surface. This should be borne in mind.

Copper, aluminium, brass, galvanised (zinc-plated) or tinned steel and iron are suitable mesh materials. Unplated materials (iron and steel) are likely to corrode and perhaps become more lossy. Practical dishes can be made from a number of shaped pieces (petals) of solid metal or mesh. It is not essential that there be good

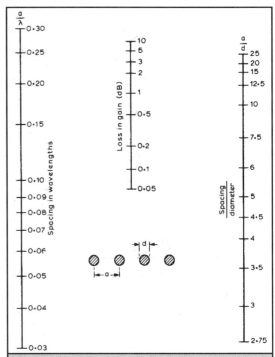

Fig 3.26 : Loss in gain for a mesh reflector as a function of mesh size.

electrical contact between sections but it is advisable to overlap or keep the gap small and insulated, as intermittent contact can introduce unwanted noises. Keep the joins parallel to the major plane of polarisation.

Positioning the feed off the axis of the dish

Fig 3.27 shows the loss of gain of dishes having several f/D ratios plotted against the amount of feed offset. Two important points are easily seen from this plot:

- The effect of construction errors on gain is not very great. For example, a focal plane dish 1.5m (5ft) diameter, having a focal length of 380mm (15in) and used at 2.3GHz, will have a 3dB beamwidth of 7°. If the feed is offset by 51mm (2in), which is a large constructional error, then the offset angle is about 7° or one 3dB beamwidth. The resulting gain loss will be about 1dB. However, note that if the f/D was 0.5 this loss would be very much lower. This offset will result in an angular error equal to, but in the opposite direction to, the resulting beam. The practical effect of an offset error on the horizontal plane would be compensated for by pointing the dish for maximum signal. Unless elevation control is available, an error in the other plane would not be correctable. When the gain is high and thus the 3dB beamwidth even less, then this loss may become very serious.

- For a dish with high f/D, the feed may be offset by a large number of beamwidths before a significant loss of gain occurs. This effect can be used to advantage, as several feeds may be mounted side by side, providing the offset is compensated for by the appropriate angle of rotation. It is also, perhaps, a useful fine search method for large dishes when used on the bands at 10GHz and above.

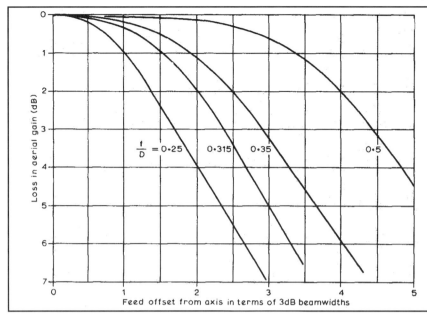

Fig 3.27 : Loss in gain due to positioning feed off the axis of the dish.

Graph axes: Loss in aerial gain (dB) vs Feed offset from axis in terms of 3dB beamwidths. Curves labelled $\frac{f}{D}$ = 0·25, 0·315, 0·35, 0·5

Dish obstruction

Fitting a feed in front of the dish inevitably obscures part of the dish and therefore causes some loss in gain. For example, when the diameter of the obstruction is one third of the diameter of the dish the gain loss is about 1dB or a 26 per cent power loss. Most practical feeds (except perhaps those based on the Cassegrain system) are usually much smaller and therefore will have a negligible effect.

The optimum ratio of focal length to diameter of a dish

The main parameters of interest, of course, are gain and efficiency. However, the effects of deficiencies in the design, adjustment or construction of both the dish and the feed are also very important. Side lobe level and front-to-back ratio are of less importance to most amateur operation. This is very dependent on the intended application of the system under construction. The main advantage that favours the use of a low f/D ratio is the compactness of the antenna. This is because the feed is placed effectively at the focus, and the smaller f/D ratio results in a shorter focal length and therefore a less bulky antenna.

An important factor favouring a high f/D ratio may be the ease of construction. It seems a fair assumption that the errors involved in making a curved surface are related to the degree of curvature. However, this only applies to some methods of construction as, for example, a moulding taken from another dish should not suffer in this way. The flatter the dish (i.e. the higher its f/D ratio), probably the more accurately it can be made when using a mesh covering. As has already been shown, the higher the f/D ratio the less critical also is the positioning of the feed. For dishes which are approximations to paraboloids, there is a clear relationship favouring a high f/D ratio.

A more important factor may be the ease of feeding. As is shown by Fig 3.28, the beamwidth of the feed required to illuminate a dish efficiently increases rapidly as the f/D ratio of the dish is decreased. However, producing this

beamwidth becomes more difficult. The predictability of a wide beamwidth feed is poorer as its beamwidth increases, which tends to compound the difficulties. As can be seen from Fig 3.28, most of the advantages in this respect are achieved when the f/D ratio is around 0.5 to 0.6.

The use of a long focal length for the reflector minimises the losses due to differences in the position of the phase centres of a feed in the horizontal and vertical planes. Similarly, the shift with frequency of the phase centre of multiband feeds such as log periodic arrays is less important. It also allows the siting of feeds alongside one another as an alternative to the latter form of multiband feed.

Finally, for a short focal length dish, there is a relatively large space loss. For a focal plane dish, for example, only the power contained within the 4dB beamwidth of the feed is reflected by the dish if an edge illumination of 10dB is specified. For a dish of 0.7, the power contained within a 9dB beamwidth is reflected by the dish, which is rather more efficient.

This statement suggests that the larger the f/D ratio the better, but the mechanical problems in supporting a feed for a long focal length dish becomes the main disadvantage. Practically, the obstruction of the aperture caused by the feed is only significant when small dishes are used on the lower microwave bands. For most amateur purposes the optimum f/D ratio, therefore, would appear to be in the range 0.5 to 0.75. One case where this mechanical problem is not so important is when the dish in question is being used to feed a flyswatter antenna.

Practical dish construction

Glassfibre reflectors

Perhaps the simplest dish construction method available to amateurs makes use of glassfibre faced with aluminium kitchen foil or fine wire mesh. This form of construction offers a practical method for making dishes with sufficient accuracy for use on the bands up to 24GHz. Some form of mould is required; this can be an existing solid dish,

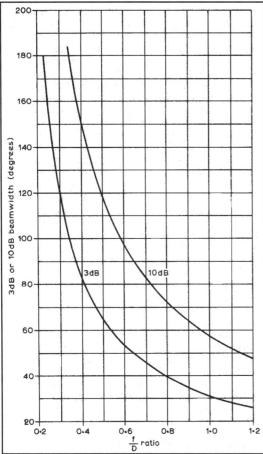

Fig 3.28 : Beamwidths of feeds for efficient illumination of dishes as a function of their f/D ratio. An edge illumination 10dB down compared with the centre is assumed.

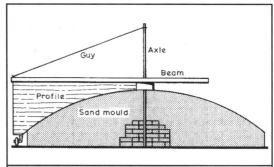

Fig 3.29 : Preparation of a mould for constructing a glassfibre dish.

A modified version of this technique is to use glassfibre as the main material, but to leave large holes which are later filled with a fine wire mesh. This method significantly reduces the windage compared with a solid dish but unless a suitable mesh is available the advantage may be lost. A suitable mesh is one with a thin wire and holes as big as possible determined by the wavelength in use. Yet another method is to make two thinner mouldings 2-3mm thick which are then joined together using stiffeners between them (as before, include mounting bushes inside). This form requires more preparation but results in a more rigid structure and perhaps a lighter dish. See [9] for further ideas on glassfibre reflectors.

For the adventurous

Make your own long focus dish from aluminium alloy sheet. Blank discs of suitable size are available from most non ferrous metal suppliers up to 0.6m (2ft) diameter. The softer alloys should be chosen for use. Due to the work hardening that occurs during hammering to shape, it will be necessary to soften the workpiece by heating regularly. This may be by using an oven set to just less than 600°C, or (with much more care) a flame or hotplate. Rub soap onto the cold metal surface as this will act as an indicator that the desired temperature has been reached by changing colour to a brown or dark shade. Decide on the f/D that is to be aimed for and calculate the central depth from:

$$C = \frac{D^2}{16 \times f}$$

where

       c is the depth

       D is the diameter

       f is the focal length

Mark and cut out a template for later use. It is useful to make or be prepared to make another template to one side of the chosen f/D. G3JVL made a dish with a 0.8 f/D and a diameter of 0.46m (18 in) and it is used to illuminate a flyswatter on 10GHz.

A sand filled, robust bag is essential to the constructor, as is a large helping of patience! Begin by hammering in the centre and while slowly rotating, increase the working diameter. Do not expect the result at this early stage to look very attractive. Continue in the described fashion, softening at regular intervals when progress slows. When the rough, distorted shape of the dish has a slightly greater depth at the centre than required, it is the time to

sand or a modern filler (substitute for plaster) cast taken from it. A method of making a mould is illustrated in Fig 3.29.

A template of the desired shape is rotated about a central pivot to shape the filler that is then left to set very hard so it will have a reasonable surface finish. The template can be used as a checking jig after setting. Smooth off with the fine sandpaper where necessary. To assist with the releasing of the glassfibre from the mould, prepare it with a wax or silicone polish. Allow this release agent to harden, then paint on a thin gel coat, which is a resin mixed with a coloured filler. Follow this by applying strips or shaped pieces of aluminium kitchen foil. Ensure the foil closely follows the contours of the mould. A further layer of gel or clear resin should then be applied. Layers of glassfibre impregnated with resin are laid on until the required thickness is built up. This is typically 6mm for a dish 1.2m in diameter. During the build up, plywood or aluminium stiffeners cut to shape may be incorporated into the structure as may any mounting fixtures required. The glassfibre should be allowed at least a week to harden before it is removed from the mould. After removal trim the edges and remove any excess resin. This time can be reduced if a large, *low-temperature* (80°C) oven is available.

work from the other side and hammer onto the outer surface of the dish. Place the roughly shaped dish face down on a paving stone or other suitable firm, flat area. Now, using the template previously make, compare the shape achieved. This is a much more rewarding time, as progress appears to be faster. By hammering radially from the centre and comparing frequently with the template, obtain a reasonable approximation, rotate a small amount and repeat. Remember to soften frequently to speed progress.

When the dish has a reasonable average shape, determined using the template, the final stage where the finer details are dealt with may begin. It is likely that some work will be required from both sides of the dish but only the minimum force should be used. When working on the inside surface, use a ball peined hammer and a flat one from the outside. Remember to keep the surface of the hammer pein free from scratches. Work only from the inside surface to remove bumps. Progress slowly round the dish with the template and hammer until you have a satisfactory match. The quality of finish required depends on the upper frequency of intended use.

Although this project is laborious it is rewarding if you can see it through. The skill acquired will also be useful and will also enable the repair of quite badly damaged dishes to be undertaken.

Construction of a mesh dish

Fig 3.30 shows a method for construction of dishes from shaped ribs covered with mesh. In the original version, which was 1.83m (6ft) in diameter, the dish was made in two halves to facilitate transportation. Eight ribs cut from waterproof (marine) plywood define the main shape of the dish, and inner circles of wire help in making the mesh conform to a near paraboloidal shape. The perimeter is made from paper rope impregnated with resin. Ordinary rope should suffice but the resin must be well worked in. Shaped sections of chicken wire are clipped to the spider with 22swg tinned copper wire and the final structure is protected by paint.

The design of feeds for dishes

In a perfect feed system for a parabolic dish, all the energy would appear to emerge from a point source placed at the focus, and would be contained within a cone which just intercepts the rim of the dish. All the energy would then be reflected by the dish in the form of a plane wave. This ideal picture is complicated by several factors:

- Practical feeds do not cut off very sharply. Generally the power density is at a maximum on the axis of the feed and then falls off on either side. Clearly there is no absolute value for the beamwidth of the feed. However, what can be specified is the beamwidth at which the power density is either a half or a tenth of the maximum value. There is therefore a judgement to be made on what the optimum beamwidth for the feed should be for a given dish. If too low, as is illustrated by Fig 3.31(a), most of the energy radiated by the feed is reflected by the dish but, since the energy is concentrated at the centre of the dish, the overall gain of the antenna suffers. The pattern produced is very clean with the minimum of side lobe amplitude. If the beamwidth of the feed is too high, as illustrated by Fig 3.31(b), much of the energy radiated by the feed is not reflected by the dish and is lost. This may result in large side lobes. There is clearly an optimum feed beamwidth that results

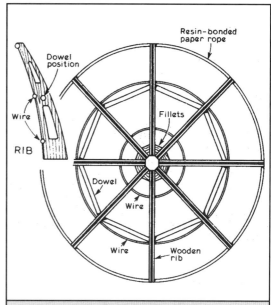

Fig 3.30 : Framework for a mesh dish.

in a fairly uniform illumination of the dish. This is achieved when the illumination at the edge of the dish is approximately 10dB down on that at the centre. This value will be assumed throughout this section.

- Because the rim of any parabolic dish is further from the feed than is the centre, there is already some loss built in the system. This loss is called the space loss, and it varies according to the f/D ratio of the dish. For a focal plane dish, it is 6dB. Therefore, to produce an Illumination at the edge of the dish 10dB down on that at the centre, the feed must have a 4dB beamwidth equal to the angle subtended by the rim of the dish, which is 180° in this example. For a dish of longer focal length, the space loss is smaller. For example, a dish having a f/D ratio equal to 0.6 has a space loss of 1.5dB at the rim. Therefore a suitable feed should have a 8.5dB (10 - 1.5dB) beamwidth at 88° to optimally illuminate this dish. This is obviously an easier task.

- To achieve high efficiency, the dish must be illuminated evenly over its surface, and it is therefore desirable to control the beamwidth of the feed in both the horizontal and vertical planes. Unfortunately, the commonly used dipole and reflector and tin can feeds are generally not very good in this respect. Pyramidal horns are much more satisfactory and their use will be described in a later section.

- The phase centre of a feed is defined as the point from which the energy appears to emanate. In practical feeds the phase centre is rarely a point since the size of the feed is always significant in wavelength terms. The situation is further complicated by the fact that the phase centre in the horizontal (E) plane may differ from that in the vertical (H) plane. Multiband feeds such as log periodic arrays suffer the additional disadvantage that the phase centres will move significantly as the frequency of operation is changed. The reduction in antenna gain due to variations in the position of the phase centres are likely to be significant, especially so for dishes of short focal length since the effect will be

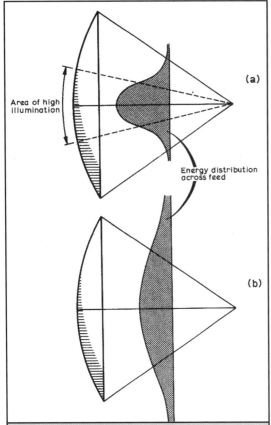

Fig 3.31 : Non-optimum dish illumination. (a) Under illuminated. (b) Over illuminated.

proportionally greater. This is yet another factor favouring the use of dishes with a relatively high f/D ratio.

Practical feed systems

Direct feed

In this method, which is illustrated by Fig 3.32, the phase centre of the feed is placed at the focus of the dish. Power radiated by the feed as a spherical wave is converted by reflection at the paraboloidal reflector into a plane wave.

The characteristics of the feed required to illuminate the dish correctly are determined by measuring the f/D ratio by the methods described earlier, and determining the beamwidth of the feed from Fig 3.28. Thus the feed for a focal plane dish for which f/D = 0.25 should have a 3dB beamwidth of 155°. For a dish having a f/D ratio of 0.7, the 3dB beamwidth should be 46°. Alternatively, the 10dB beamwidth may be specified, and the corresponding value is 83°. Fig 3.28 was derived assuming an edge illumination 10dB down on that at the centre of the dish. Due allowance has been made for space loss. It will be noted that the beamwidth required changes very rapidly as the f/D ratio is reduced, which makes the design of suitable feeds for short focal length dishes rather more difficult. In practice, under illumination is the result.

The main advantage of this method of mounting the feed is its simplicity in conception and construction. It can have a high overall efficiency, and it leaves the back of the dish clear for mounting on the mast or tripod. Its main

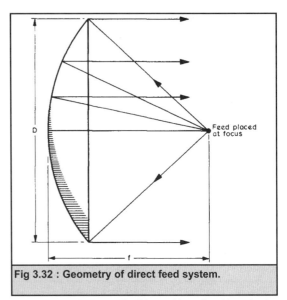

Fig 3.32 : Geometry of direct feed system.

disadvantage is that with dishes of long focal length, the feed support structure becomes quite bulky.

Indirect feed

Fig 3.33 illustrates the geometry of this method. Power radiated by the feed is reflected by a plane sub reflector onto the main reflector. A spherical wave generated by the feed is converted by the sub reflector into a spherical wave of the same radius of curvature but moving towards the dish. This, in turn, is converted by the dish into a plane wave.

The sub reflector should preferably be a minimum of a few wavelengths in diameter, but should not exceed a third of the diameter of the main dish in order not to incur more than 1dB loss of performance. Once the size of the sub reflector is chosen, then its position is fixed.

It must just intercept lines drawn from the real focus of the dish, F_r, to its rim. The position of the feed is also fixed. It is set at the virtual focus, F_v, which is as far in front of the sub reflector as the real focus is behind, meaning that the lengths m are equal. Because the sub reflector is planar, the virtual dish has the same focal length as the real dish, and therefore the design of the feed is the same as if the direct feed method were used.

The sub reflector is usually made from solid material since it is so small, but mesh or wires running parallel to the plane of polarisation and spaced by about a tenth of a wavelength offer a suitable alternative construction.

The main advantages of this method are that the feed can be supported from the centre of the dish (although this may complicate mounting the dish), and only a relatively light structure is needed to support the sub reflector. The disadvantages of the method are than an extra component, the sub reflector, needs to be aligned accurately, and that extra losses are involved compared with the direct feed method due to diffraction around the sub-reflector.

Cassegrain feed

The geometry of a Cassegrain system is shown in Fig 3.34. It is similar to the indirect feed method already described, the essential differences being that the planar

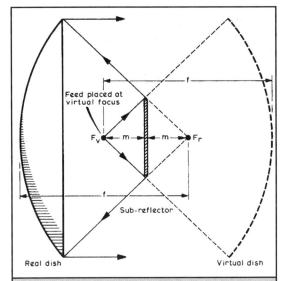

Fig 3.33 : Geometry of indirect feed system.

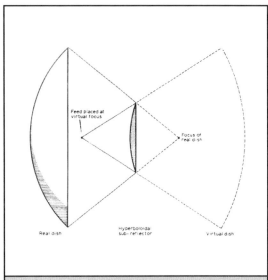

Fig 3.34 : Geometry of Cassegrain feed system.

sub reflector is replaced by a shaped reflector in the form of a hyperboloid. The main result of this change is that the virtual dish seen by the feed has a longer focal length than the real dish. Thus a dish of short focal length which can be difficult to illuminate efficiently can be converted into one of longer apparent focal length. A second feature is that there is more flexibility in the positioning of the feed. The feed may be mounted behind the dish firing through a suitable hole. This form has a great mechanical advantage, and is commonly used on optical telescopes.

Pyramidal horn feeds

Pyramidal horns have significant advantages over most other types of feeds that makes them especially suitable for use by amateurs. First, they offer a virtually perfect match over a wide range of frequencies and are therefore uncritical in their design and construction. Even quite large dimensional errors do not affect the quality of this match, but only the efficiency of illumination of the dish. A second advantage is that these horns are designed to produce optimum illumination of the dish in both planes. With other types of feed, there may be little or no control of the ratio of beamwidths in each plane.

The form of a horn is shown in Fig 3.35. It consists of a length of waveguide that is flared in one or both planes to produce the beamwidth required. At the higher microwave frequencies, the horn will normally be fed by waveguide. The body of the feed horn will therefore usually consist of a length of waveguide that matches the rest of the system. On the lower microwave bands (1.3, 2.3 and 3.4GHz) the horn is often fed by using coaxial cable. A waveguide to coaxial cable transition will be needed to serve as the method of launch. The horn design method is otherwise exactly the same. It is not necessary in this case to use a standard waveguide.

The dimensions of the pyramidal horn as a function of the f/D ratio of the dish to be illuminated are given in Fig 3.35. They are based on an edge illumination of 10dB down on that at the centre of the dish. Due allowance has been made for space loss. The dimensions are given in terms of wavelength, so enabling any frequency of operation to be catered for. The actual dimensions are, of course,

determined by multiplying the values of A/λ and B/λ by the wavelength in air at the design frequency.

It can be seen from Fig 3.35 that the aperture of the horn decreases as the f/D ratio of the dish is reduced. The limit of the design data is reached first with the A dimension at a value of A/λ equal to 0.8, which corresponds to a minimum f/D ratio of 0.48. To feed dishes of smaller f/D ratio, the Cassegrain system may be used.

Alternatively, the end of the waveguide may be suitably shaped to increase its beamwidth but, as this shaping has to be determined experimentally, much of the advantage of horn feeds is lost.

As an example of the design of a horn for a specific dish, consider a dish of diameter D or 0.914m (36in), which has a depth at its centre of c or 108mm (4.26in). It is to be fed at 10,400MHz, for which λ is 28.846mm (1.136in). The focal length of the dish is given by $D^2/16c = 0.483$m (19in), and the f/D ratio therefore is 0.53. From Fig 3.35, the values corresponding to this ratio are:

H plane aperture = A/λ = 0.96 (ratio)

A is broad wall = 28.846 x 0.96 = 27.69mm (1.09 in)

E plane aperture = B/λ = 0.78 (ratio)

B is narrow wall = 28.846 x 0.78 = 22.50mm (0.886 in)

L = A x A/λ = 27.69 x 27.69 / 28.846 = 26.58mm
(1.046 in)

At this frequency a convenient waveguide is WG16, so a practical horn would have an aperture of 27.69 by 22.50mm (1.091in by 0.886in), tapering to 22.86 by 10.16mm (0.9in by 0.4in). This design is shown in Fig 3.36.

The same design procedure is applicable to any dish with the same f/D ratio. For example, the feed horn at 1,296MHz would have an aperture of 222 by 180mm (8.74in by 7.10in), tapering to 165 by 82.5mm (6.5in by 3.25in), if WG6 were used. Similarly, one for use at 24GHz would have an aperture of 11.99 by 9.8mm (0.472in by 0.384in), tapering to 10.67 by 4.32mm (0.42in by 0.17in) for WG20.

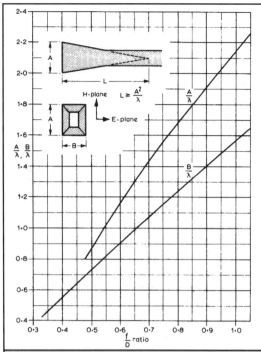

Fig 3.35 : Form of horn antenna to feed dish and dimensions as a function of dish f/D ratio.

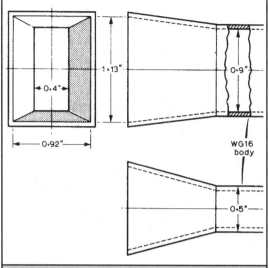

Fig 3.36 : A typical pyramidal horn feed designed for a dish having f/D ratio of 0.53, for use at 10GHz.

Sectoral horns

The sectoral horn is a form of pyramidal horn but flares in only one plane. The result of this is that a wider beamwidth is formed in the un-flared plane. Fig 3.37 shows the general form of this type of antenna.

A small horn with dimension B fixed at the width of the narrow wall of the waveguide would give a symmetrical polar pattern and low gain when it uses the ratios described above. However, if the broad wall is flared to form a long rectangular mouth, a very non-symmetrical but higher-gain pattern will result.

It should be possible to achieve a little wider coverage by tapering the narrow wall dimension to a narrower aperture than that of the waveguide. The length must still be close to that which would be required for a normal pyramidal horn with the broad wall dimension suggested.

If this form of horn is to produce a horizontally polarised wave, then the broad wall should be vertical. The result will be better coverage around the horizon and the beamwidth will be quite wide at perhaps several tens of degrees. The extra gain is achieved by compressing the vertical pattern. The gain achieved will appear to be rather moderate when compared to a dish but will give extra coverage in the favoured sector.

This is particularly useful where an omni-directional pattern would waste power in directions of no particular interest, e.g. when a beacon or repeater is located on a site which has some badly screened directions, or maybe when located at an extreme of activity geographically. But remember that, just because it has not been achieved before, this does not mean that under some circumstances there may not be a path to open. (There must be activity though!).

Biconical horns

This form of omni-directional antenna is rather more difficult to construct and to feed. However there is a form of launch called the Alford loop that should prove satisfactory at least up to 10GHz. It may be rather difficult to construct for the higher bands.

The bi-cone is self explanatory in form, but the points of the cones must be truncated by enough to allow the launch to be fitted in the gap. Some examples for 10GHz are shown in Fig 3.38. As with the Alford slot (see later), a coaxial cable will be needed to energise the launching loop. The form of this loop is shown in Fig 3.39.

Circular horn feeds

This type of feed is quite common on the lower microwave bands, its main advantage being that it can be made from readily available materials. Its efficiency is fairly high, but it does require some setting up. As is shown in Fig 3.40, the feed consists of a length of short circuited, circular waveguide, often made from tin cans. A simple form of coaxial cable to waveguide transition generally used consists of a probe just less than a quarter of a wavelength long, spaced approximately a quarter of a guide wavelength from the closed end. The exact distance and length should be established experimentally, and remember that the required length will alter with the diameter of the probe.

The design factors are as follows:

- The diameter of the feed D must exceed 0.58 wavelength in air at the design frequency. If it is smaller, the waveguide is then operating in its cut off region and proper operation will be difficult. A reasonable practical value for D is between 0.65 to 0.7 wavelength.

- The length of the horn L preferably should exceed a guide wavelength, λ_g, where

$$\lambda_g = L/\sqrt{(1-(\lambda/1.706 \times D) \times 2)}$$

For D = 0.65 x λ_g, L should be greater than 2.4 x λ_g

For D = 0.70 x λ_g, L should be greater than 1.8 x λ_g

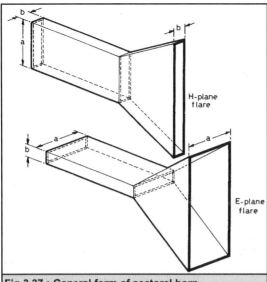

Fig 3.37 : General form of sectoral horn.

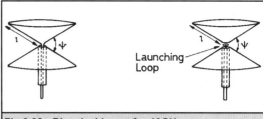

Fig 3.38 : Biconical horns for 10GHz.

When the value suggested for D is 0.65L the respective beamwidths are 46° and 78°. Reference to Fig 3.28 shows that these values would optimally illuminate a dish with a f/D of 0.69 and 0.43 respectively. The mean value of 0.56 represents, perhaps, the minimum f/D ratio of the dish that can be illuminated with reasonable efficiency. However, a real compromise will depend on the application. This is because one plane would be over-illuminated and the other under-illuminated.

There is very little flexibility of design. If the diameter of the horn is increased, then its length may be reduced, but the beamwidth of the horn will also decrease. By using an elliptical waveguide perhaps better symmetry can be obtained. The amount of distortion may be adjusted to optimally illuminate the dish. Typical dimensions of horns for the lower microwave bands based on D = 0.65L are given in Table 3.8. The equivalent phase centre of the feed will be just inside the mouth and this should be adjusted to coincide with the focus of the dish.

Dipole and reflector feeds

This type of feed is shown in Figs 3.41 and 3.42. The radiation pattern makes it suitable for feeding a dish with an f/D ratio of 0.25 to 0.35. The feeds are built around a length of fabricated rigid coaxial line; the ratio of the inside dimension of the outer to the diameter of the inner must be 2.3:1 to produce an impedance of 50Ω or 3.3:1 to

Note that as the diameter of the feed approaches $0.586\lambda_g$, the length of the horn required increases very rapidly. At cut off it would be infinitely long!

- The correct spacing between the probe and the short circuit should be determined experimentally. The spacing required is around a quarter of a waveguide wavelength, so the position of either the probe or the short circuit can be made adjustable.

- The E-plane (horizontal) and H-plane (vertical) 3dB beamwidth, θ degrees, is given by the values of 29.4 L/D and 50 L/D respectively. That these differ significantly is the main disadvantage of this type of feed since it results in an uneven illumination of the dish.

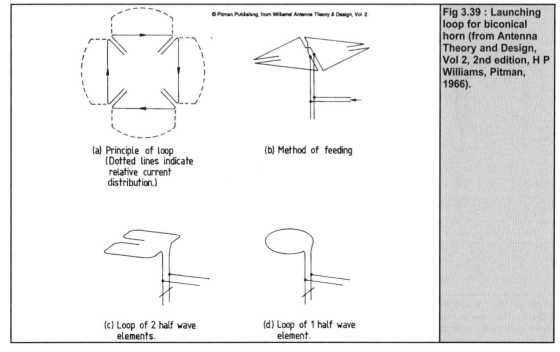

© Pitman Publishing, from Williams' Antenna Theory & Design, Vol. 2.

(a) Principle of loop (Dotted lines indicate relative current distribution.)

(b) Method of feeding

(c) Loop of 2 half wave elements.

(d) Loop of 1 half wave element.

Fig 3.39 : Launching loop for biconical horn (from Antenna Theory and Design, Vol 2, 2nd edition, H P Williams, Pitman, 1966).

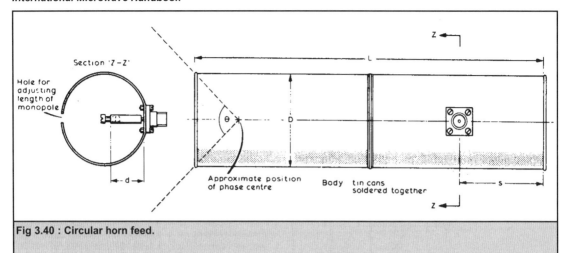

Fig 3.40 : Circular horn feed.

produce an impedance of 72Ω. The dipole to reflector spacing should be increased to return to a good match at 72Ω. It is also possible to obtain a good match to 72Ω by using a ptfe slug as a transforming section within the airline coaxial cable.

Balance to unbalance devices

The slot in the outer shown in Figs 3.41 and 3.42 is used to provide the balance to unbalance (balun) action required. Associated with this form of balun is an impedance transformation of 4:1. It behaves in exactly the same manner as the one used and described in the Alford slot section (see later) but this one is twice the length and provides a convenient method of mounting a dipole and reflector feed. It may be considered as two quarter wave sections end to end. This permits the inner to be physically joined to the outer at the dipole as shown. The impedance transformation in this case is 2:1 as there are two baluns in parallel.

This form of balun is analogous to the type that uses a

half wavelength of coaxial cable on the lower frequencies. The form used with this feed produces the balancing and transforming action at the outer rather than the inner, as does the flexible coaxial cable form. The outer is split and one side is energised by connection to the inner. The other half of the outer is thus forced to have an equal voltage, but of opposite phase, imposed on it. In this way double the voltage is present across the two halves and this explains the transformation.

The same explanation may be used when the standard coaxial cable type is considered, but read inner for outer. Of course in both cases this action occurs totally within the cable. Due to the skin depth being very small, none of this current flows on the outside. The radiation from the slots is extremely small providing they are narrow. A heat shrink sleeve should be used to strengthen this form of balun when made from semi rigid cable.

Waveguide feed for 2,320MHz

This feed for use with a dish of short focal length is shown

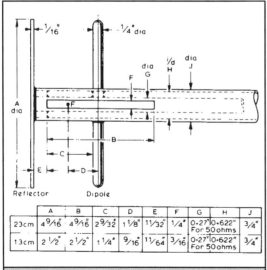

	A	B	C	D	E	F	G	H	J
23cm	4⁹⁄₁₆″	4⁹⁄₁₆″	2⁹⁄₃₂″	1⅛″	11⁄32″	¼″	0·27″ For 50ohms	0·622″	¾″
13cm	2½″	2½″	1¼″	9⁄16″	11⁄64″	3⁄16″	0·27″ For 50ohms	0·622″	¾″

Fig 3.41 : Dipole and reflector feeds for 1.3 and 2.3GHz. The feed is mounted so that point F coincides with the focus of the dish.

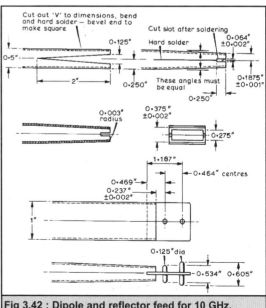

Fig 3.42 : Dipole and reflector feed for 10 GHz.

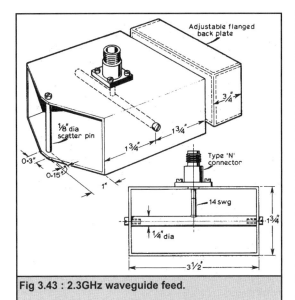

Fig 3.43 : 2.3GHz waveguide feed.

in Fig 3.43. By the use of a scatter pin, and by cutting the waveguide in the way shown, the angle of radiation has been increased significantly when compared with a plain aperture. The feed may be tuned by adjusting the position of the rf short which is then firmly bolted or soldered in place. The plane of polarisation is parallel to the scatter pin. Suitable materials for construction for the body are brass, copper or tinplate.

The penny feed a simple waveguide feed for short focal length dishes

Most dishes that amateurs inherit have a short focal length, which means that the ratio of the focal length to the diameter of the dish is typically within the range 0.15-0.3. The polar diagram of an example of this feed has been measured and had a double humped main lobe in the horizontal plane: see Fig 3.44.

It is suggested that the spacing between to two dipoles can be used to produce a beam shape to suit other f/Ds. This will require either flaring or narrowing the waveguide. In the vertical plane, the pattern is that of a single dipole. This feed could, perhaps, better illuminate a dish if this effect also occurred in the vertical plane. To achieve this a further pair of slots in the same plane are required.

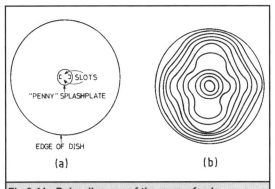

Fig 3.44 : Polar diagram of the penny feed.

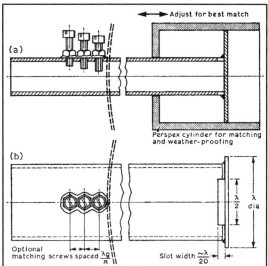

Fig 3.45 : Detail of the penny feed for dishes having an f/D ratio of 0.25-3. Matching can be done either by matching screws as shown or as in (a) by using Perspex matching (and weather proofing) sleeve. (b) Side view of feed.

The feed described is shown in Fig 3.45. It is constructed by cutting two slots at the end of a length of waveguide of suitable size, then soldering on a circular disc. The length and width of the slot formed alter the vswr so they are quite critical. Obviously, care should therefore be taken when cutting and filing the slots to size. It is easy to get a reasonable match over a fairly narrow bandwidth, and it is possible to cover the whole of the 10GHZ allocation for example. However, it is recommended that the important area between 10.35 and 10.45GHz is optimised to a very good match.

Values for λ_o and λ_g at the frequency of interest are given in Table 3.9. Horizontal polarisation is produced when the broad faces of the waveguide are mounted vertically.

A method by which both weatherproofing and matching can be achieved is shown in Fig 3.45. This uses a Perspex sleeve that is a sliding fit on both the disc and the waveguide. By adjusting its position, a proportion of the power may be reflected in the correct phase to cancel the mismatch. The sleeve to waveguide gap should then be sealed using a silicone rubber sealant.

Another method of weatherproofing uses small scrap pieces of Duroid or Copper Clad printed circuit board, with the copper removed, glued to the waveguide over the slots. In this case the slot length should be reduced by approximately 1 mm, depending on the thickness of pcb.

Table 3.8 : Typical dimensions for horns.

Frequency (MHz)	D (mm)	D (in)	L (mm)	L (in)	d (mm)	d (in)	s (mm)	s (in)
1,297	150	5.9	560	22.0	46	1.8	140	5.5
2,230	84	3.3	312	12.3	25	1.0	76	3.0
3,457	56	2.2	208	8.2	18	0.7	51	2.0
5,760	34	1.3	125	4.9	11	0.4	31	1.2
10,400	12.5	0.5	69	2.7	6	0.24	17	0.7

Table 3.9 Values of λ_0 and λ_g for penny feed

Centre frequency (MHz)	Suitable waveguide	λ_0 (mm)	λ_g (mm)
1,297	WG6	231	324
2,305	WG8	130	162
3,457	WG10	86.7	109
5,761	WG14	52.0	78.2
10,050	WG16	29.8	39.4
10,369	WG16	18.9	37.3
24,193	WG20	12.4	15.2

A good match will be obtained if the slot length and the board thickness are chosen correctly. Any plastic material can be used but common plastics generally require to be very thin due to their higher dielectric constant and losses.

Periscope or flyswatter antennas

Operating microwave equipment from domestic locations gives rise to new problems especially when mounting the antenna. Waterproofing, feeding to, and receiving power from, the antenna are problematical. Both waveguide and good coaxial cable are very expensive and even good quality coaxial is quite lossy (up to 3dB/m at 10GHz) so it should not be used in long lengths for serious operation. Even the better Andrews cable may result in many decibels loss and will be quite hard to handle due to its size and weight. One method for avoiding feeder problems is simply to mount the essential equipment at the top of the mast and feed the dc, a.f. and i.f. supplies instead. This approach can still present problems, of course. If the equipment is to be mounted semi permanently, then weatherproofing will need careful attention. On the other

hand, if it is fitted to the mast only when it is to be used, then some ingenious engineering will be required to ensure that this can be done speedily and reliably as a matter of routine. An alternative approach is the periscope or flyswatter antenna, as it has been dubbed.

An example is shown schematically in Fig 3.46. Although this antenna type is widely used professionally, it had received little attention from amateurs until G3JVL began using one in 1979. In this system, the feed is usually a parabolic dish (although not necessarily so), and this directs a signal upward at a reflector mounted so as to reflect it toward the horizon. The area of the reflector illuminated normally is an ellipse which will ideally have the ratio of the major to minor axes of 1.414:1 if the reflector is set at 45°. An elliptical reflector could obviously be used, but a rectangular one is to be preferred to ease of manufacture and actually produces a little extra gain. However, the shape is unimportant and the dimensions should be chosen to suit available materials. If a non ideal ratio is used that a non symmetrical beam shape will be the result. If a longer, narrower reflector is used the vertical beam will narrow and similarly vice versa, if in the other plane.

The reflector may be planar or parabolically shaped, the former case being that most likely to meet home station requirements. Detail of the expected performance is indicated by reference to Fig 3.47. Although interest in this antenna stems from 10GHZ operation, it works equally well on other frequencies if the guidelines are adhered to. Fig 3.48 shows the gain of a specified system as a function of frequency.

It should be considered also for use at some portable sites in order to avoid obstacles like trees or shrubs. A

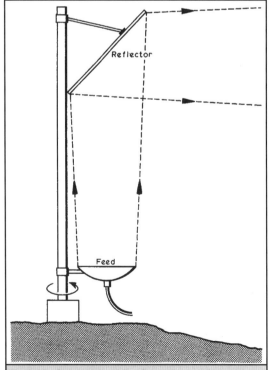

Fig 3.46 : Schematic of the flyswatter or periscope antenna.

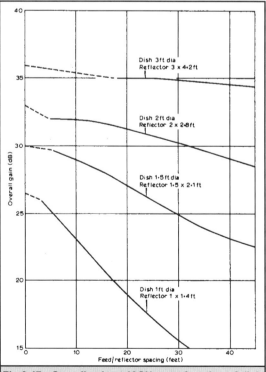

Fig 3.47 : Overall gain at 10GHz as a function of dish size, plane reflector same size as dish.

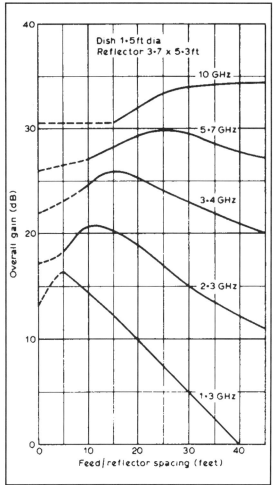

Fig 3.48 : Gain of flyswatter as a function of frequency.

system for portable use may be very much simpler than a home based version. A standard length of mast at 6m (20ft) with a reflector using flat sheet and stiffened by a bracing structure could be quite useful. The feed dish mounted at the base of the mast may be inside a tent or just above the car roof.

Design data

The generalised data from which the performance can be determined are given in Figs 3.49 and 3.50 for the flat or curved forms. The gain over and above the feeds gain when used with a parabolically shaped reflector is very hard won, and is not worth the trouble that will be needed to get it.

The overall system gain is the important factor and is dependent on the size of reflector used only, when considered at the simplest level. As can be seen from Fig 3.47, the choice to be made after deciding on the size of reflector to be used is what to feed it with.

The parabolic reflector will return a good gain with a small feed antenna but the distance between them is quite critical and rather close. This form of reflector is very much harder to make (Fig 3.51).

The flat reflector, however, can be fed with a dish that has a diameter of around 80 per cent of the reflectors width and will be much less critical as to spacing.

It is, however, possible to get a higher gain from a reflector by using a larger dish to illuminate it. Obviously, it may be impossible to use a larger dish as the feed for mechanical reasons.

Practical omni-directional antennas

Omni-directional horizontally polarised antennas

The slot antenna that has become known in the amateur world as the Alford slot actually derives from work by Alan D Blumicin of London and is detailed in his patent number 515684 dated 7 March 1938. The work by Andrew Alford

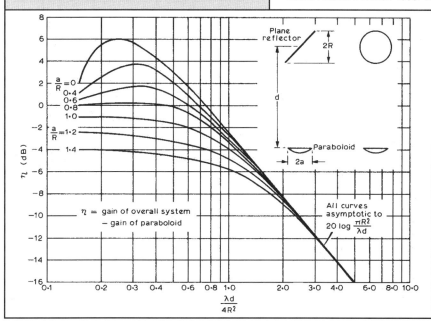

Fig 3.49 : Relative gain with planar reflector.

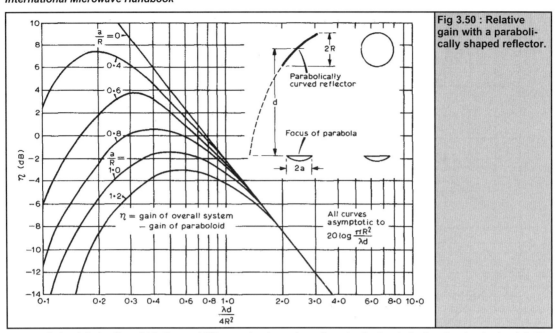

Fig 3.50 : Relative gain with a parabolically shaped reflector.

was carried out during the mid forties and fifties, and not applied to microwave bands but to vhf/uhf broadcasting transmitters in the USA.

G3JVLs development was carried out during 1978 when designs for the GB310W 1.3GHz beacon were being investigated. The initial experiment was carried out at 10GHz as the testing was found to be much easier, especially when conducted in a relatively confined space. A rolled copper foil cylinder produced results close to those suggested in the reference. The work published to date has been confined to the solid cylinder with a slot cut along its length. Models for 0.9, 1.3, 2.3, 3.4, 5.7 and 10GHz have been constructed and performed as expected.

Initially, it was thought that the skeleton version would be best used at the lower frequencies only. However, several models for use of 144MHz have been constructed and performed very well, but the design was at the time not regarded by G3JVL as being of interest. Further developments have resulted in working models being constructed for the 50MHz, 432MHz, 900MHz and 1.3GHz bands. However, some aspects were not easy to explain and valuable assistance was provided by G3YGF. This assistance provided more than just an explanation for the fact that early skeleton versions worked at a lower than designed frequency.

A working operational theory was developed as a direct result, along with a better understanding of the strange effects that were observed.

Propagation advantages influenced the RSGB Microwave Committees decision when insisting on the use of horizontal polarisation. There was a genuine desire to avoid the conflict, experienced at the lower frequencies, while still at the proposal stage of narrow band microwave beacons and repeaters. The Alford slot was offered as a solution to the objections voiced about the lack of suitable omni-directional antenna.

The use of these slots has been further fostered by the current world-wide and rapidly growing interest in atv

where the standard polarisation is also horizontal. For mobile use the skeleton version presents lower windage without losing performance, thus having a great appeal. This type does require a little more care during construction and final adjustment, as it has a narrower bandwidth.

Operation

It is well known that a vertical slot in an infinite plate produces a magnetic dipole and has the opposite polari-

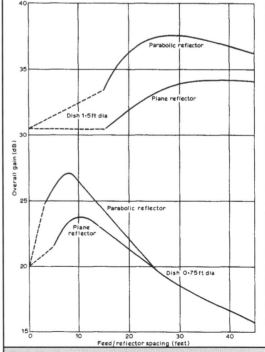

Fig 3.51 : Comparison of plane and parabolic reflectors.

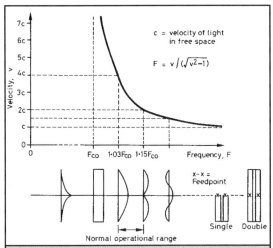

Fig 3.52 : Velocity of the wave along a slot antenna as a function of frequency.

sation to that expected from the physical appearance. It is also true for this type of slot but with the advantage that more gain can be obtained with only small physical differences. This novel feature is due to the ability to make the wave appear to travel along the slot faster than the speed of light.

This results in a field distribution over the slots length that has the appearance of a dipole but may be many times longer than the free space half wavelength value. In this way a net gain, equivalent to that obtained by feeding several dipoles in phase, is obtained but without the need for a complicated phasing harness. The gain obtained is directly proportional to the length of the slot in free space half wavelengths. The idea that waves are travelling faster than light would at first sight seem to be impossible, but in fact it is only a standing wave pattern that appears to travel at this speed; the actual wave travels at a lower velocity than that of light. The velocity of the wave along the slot varies with frequency as shown in Fig 3.52.

This is very similar to the effect that occurs in waveguide near its cut off frequency (F_{co}). The slot behaves like a transmission line shunted by inductive loops (the solid cylinder is equivalent to an infinite number of closely spaced loops). Cut off occurs when the shunt inductance resonates with the capacitance of the slot. Below the cut off frequency, waves cannot propagate at all. However, above the cut off frequency limit, the wavelength eventually returns to the free space value.

In principle, any velocity factor could be used by the higher the velocity factor, the more critical are the dimensions. Velocity factors of much greater than 10 are impractical for this reason and the normal operating range is around three per cent to 15 per cent above cut off, resulting in a velocity factor of between two and six. In the designs previously published the velocity factor has been up to around four. The gain achieved for these dimensions will be about 6dBi for the end fed version and 8dBi for the centre fed version. The dimensions are quite interdependent. The velocity factor will be increased by decreasing the tube diameter, or by increasing the slot width. However, the range of tube sizes is rather limited, especially when a high velocity factor is to be obtained. Reference to Fig 3.52 shows the usable range as 1.03 x F_{co} to 1.15 x F_{co}, or between the point where the

distribution is one or two dipoles across the length of the slot. However, note that the velocity factors indicated are true only when the physical length of the slot is two wavelengths for the double-length version.

It is equally practical to obtain more gain with a longer slot arranged to be within the equivalent limits of one or two dipoles (electrical), but this requires the normal range to move to the left where it is seen that the curve rises steeply, thus limiting the useful frequency range. As the velocity factor increases this range gets narrower. Designs are included which make use of these factors, thus permitting more gain to be obtained, but at the expense of bandwidth and ease of construction. Departure from the recommended dimensions means that to achieve the desired results careful adjustment is essential.

A further limit to the range of overall dimensions permissible is imposed when the deviation from the perfect omni pattern is required. The best pattern is obtained when the tube diameter is the smallest but the slot width is then becoming perhaps too small to control. Conversely it becomes much easier to produce the required conditions when the tube is larger and the slot wider. The disadvantage is that the pattern becomes distorted in a manner related to this increase. Fig 3.53 shows the form this distortion takes. It may be that for some applications the less than perfect omni pattern will actually be that desired!

When performance is being assessed it must be stressed that vswr and radiation characteristics are not closely related in quite the way that is expected. In practice, primary importance should be placed upon the distribution of the power along the slot and thus the radiation pattern. Then, finally, ensure that a good vswr is presented by some means of matching. For example, the solid cylinder version normally proves to have a good match to 200Ω (4:1 balun), but the skeleton versions have needed a short matching section. In practice this has been achieved by using two short parallel wires from the slot feed point to the coaxial balun (see Fig 3.54). The wire diameter, length and spacing are used as the final trimming method.

Blumlein pointed out long ago that as the length of an end fed vertically mounted slot is increased, the angle at which the wave is radiated will be increasingly above the horizon. This is due to the power distribution being unequal. Power is radiated as it progresses along the structure. Unless some means of restricting radiation initially is employed, the far end does not receive enough. This problem is avoided by centre feeding the structure. (Using this method, this effect is thereby compensated for). This is also very likely to be one of the limiting factors in waveguide slot arrays. Centre feeding the reduced height waveguide section may prove to be impossible. This may be explained as being due to the reduced power arriving at the more distant slots or slot position.

Slotted-waveguide antennas

This type of antenna [10] is made from rectangular waveguide and consists of a series of resonators (slots) cut into one or both broad faces (see Fig 3.55). If an omni-directional pattern is essential, reduced-height waveguide is a must: see Fig 3.56(b). The slots or dipoles are required to be parallel with, but alternately offset from, the centre line of the broad face of the guide. This ensures that the phase relationship between adjacent slots is correct. Horizontally polarised radiation is obtained when the slots are vertical. If standard waveguide is used then the pattern shown in Fig 3.56(a) is obtained,

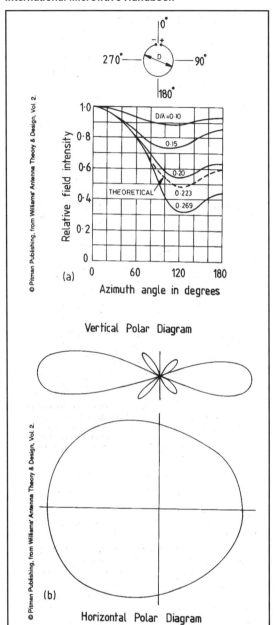

Vertical Polar Diagram

Horizontal Polar Diagram

Fig 3.53 : (a) Pattern distortion related to an Alford slot antenna diameter and slot width (from Antenna Theory and Design, Vol 2. 2nd edition, H P Williams, Pitman, 1966). (b) Polar diagram of an Alford slot antenna.

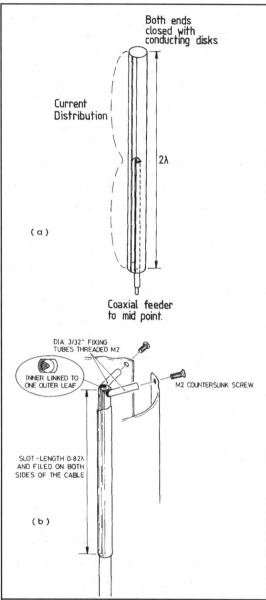

Fig 3.54 : (a) Construction of Alford slot antenna using a dual slotted cylinder. The impedance at the feedpoint is 200Ω. Dimensions for 1,296MHz are : slot length 510mm, slot width 4mm, tube diameter 31.75mm by 18swg (1.25in). Dimensions for 2,320MHz are : slot length 280mm, slot width 3mm, tube diameter 19mm by 18swg (0.75 in). (b) Construction of suitable balun.

having nulls adjacent to the narrow faces. In either case a narrow beam is formed in the vertical plane with the radiation concentrated on the horizon, making this antenna ideally suited to general-coverage beacons. On the lower bands the Alford slot is perhaps more manageable.

Departure from truly circular coverage is mainly dependent on the following factors:

• The internal height of the waveguide.

• The thickness of the broad wall material.

• Whether the slots are machined in one or both broad faces of the guide.

• The accuracy and the relative positioning of the resonators.

The resonator length is related to its width. A convenient ratio is when the width is approximately 1/20th of the waveguide wavelength. The length will then be about 0.85 x λ_0, the free-space wavelength.

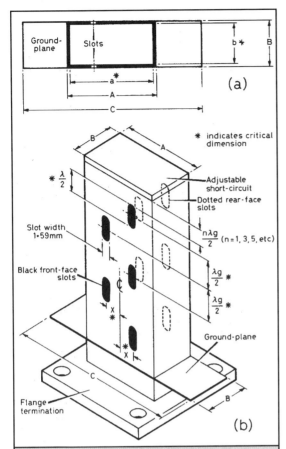

Gain is dependent on the *total* number of slots used on the antenna and also on how the power is distributed between them. An array using 16 slots (i.e. N = 16) is likely to produce a gain in the region of 10 to 12dB.

Reduced height waveguide

When both broad faces contain slots the pattern will approach optimum when the height is minimum.

The height that can be used is perhaps dictated by the ease of feeding the slots. A value of around 0.1 is recommended as minimum. Due to power being progressively radiated from each slot the ideal situation is departed from along the length of the array. It would be preferable if the power could be introduced to the slots centrally so as to obtain the best symmetrical pattern. See the discussion at the end of the next section which applies generally.

Standard waveguide

In this case, when both broad faces have slots the pattern is bidirectional, producing a four leaf clover shaped pattern but with the side nulls being very deep and the front less so.

In practice the perfect field pattern is never achieved even in an amateur installation. The tolerances for instance may be as high as 0.2mm (8/1,000in) and still give an acceptable pattern. In the design example given (for 10GHz), the width used is 1.6mm (1/16in). In practice, the use of a 1.6mm end mill or slotting drill has yielded acceptable results both at 10 and 24GHz, when combined with the somewhat wider tolerances mentioned.

The gain of an antenna using a total of 116 slots (as in the sample calculation) is 10 to 12dB. In order to benefit from a longer array some steps must be taken to ensure that the power arriving at the furthest elements is still sufficient to allow them to contribute in the correct manner. A possible explanation for this requirement is linked to the findings of Blumlein which show that a long structure, when end fed radiates power progressively (exponentially). This causes the angle of the radiated wave to shift away from the perpendicular axis of the slots. To avoid this disadvantage centre feeding the array is suggested.

Fig 3.55 : General configuration of a slotted waveguide antenna (not to scale). A ground plane is shown in the diagram, this is only needed if the antenna is made from full height waveguide. If dimension b is 0.1λ or less, the ground plane is omitted but a tapered section (at least 3λ long) from standard height guide to the reduced section will be required.

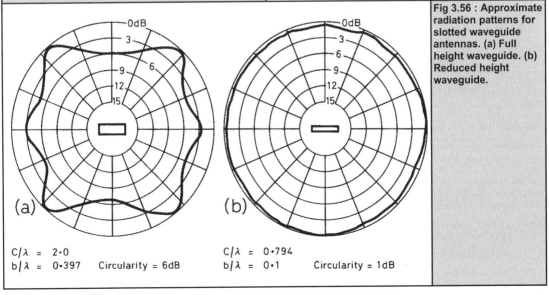

Fig 3.56 : Approximate radiation patterns for slotted waveguide antennas. (a) Full height waveguide. (b) Reduced height waveguide.

By combining two half length sections the problem is greatly reduced by introducing equal but opposite angle shifts on each half. One practical way of achieving this may be to construct two practical length arrays and combine them with a two way power splitting device. Then with the slot arrays mounted one up and one down. The resulting array will have both increased gain due to using more better fed slots and also due to the radiation being maintained on the horizon.

Matching from standard waveguide to reduced height waveguide is achieved by using a tapered section, the length of which is as long as possible, with three waveguide wavelengths being recommended. An alternative would be to include a matching iris or screws. This will produce the same result but over a much restricted bandwidth. The remote end of the antenna is closed by means of a sliding short circuit positioned $\lambda/4$ away from the end of the uppermost resonator. The final position may be adjusted on test for best vswr (minimum return loss). It may be advantageous to fit three or four matching screws, spaced approximately $\lambda_g/8$ apart, near the base of the antenna. A combination of adjustment to these screws and the sliding short should then result in a good match. However, please note a good match is *not* a sign of a good performance. A poorly made array is likely to be capable of a good match! The variation in vswr when an object is moved parallel to the polarisation plane will indicate approximately where the power is being radiated.

Design formulae

N = total number of slots

i.e. twice the number of slots in one face of the guide, and

$g = g_1 \times \sin^2(\pi \times x/a)$

where g = 1/N and x = displacement of slots from centre line, in millimetres.

L_0 = free space wavelength

L_g = waveguide wavelength

a = internal dimension of broad face of guide (mm)

b = internal dimension of narrow face of guide (mm)

Note that angular functions are in radians.

Then
$g_1 = 2.09 \times (L_g/L_0) \times (a/b) \times \cos^2(\pi \times L_0/2 \times L_g)$

Worked example for 10GHz

L_g (at 10,368MHz) = 37.322mm

L_0 (at 10,368MHz) = 28.911mm

N = 16

a = 22.86mm

b = 3.0mm

slot width = 1.59mm

$$g_1 = 2.09 \times \left(\frac{37.322}{28.911}\right) \times \left(\frac{22.86}{3}\right) \times \cos^2\left(\pi \times \frac{28.911}{2} \times 37.322\right) = 2.47$$

Then
$1/16 = g_1 \times \sin^2(\pi \times x/a) = 2.47 \times \sin^2(\pi \times x/22.86)$

$\sin^2(\pi \times x/22.86) = 0.025$

$\sin(\pi \times x/22.86) = 0.159$

giving   x = 1.16mm (slot displacement)

Slot length = $L_0/2$ = 28.911/2 = 14.46mm

Slot spacing = Lg/2 = 37.322/2 = 18.66mm

Spacing of top slot from end of antenna = 45mm (min)

The formulae and calculations given are directly applicable to other waveguide sizes, and antennas for other frequencies and bands can be designed along the same principles. Such an antenna can be mounted within a thin plastic or glassfibre tube for weather protection, and it has been found that such housings have little effect on the radiation pattern of the antenna but do cause vswr changes. This may be compensated for experimentally by either resetting the matching screws or perhaps by altering the design a little to return to optimum performance when the radome or housing is present. Thus the simplest 10GHz beacon could consist of a Gunn or dielectric tuned oscillator coupled directly to a stacked, slotted waveguide antenna, with both items of equipment housed within the one protective tube and the modulated supply voltage being fed to the oscillator via an ordinary coaxial cable of any convenient length.

PCB material version

This will behave in exactly the same manner but ought to be easily reproduced once the exact pattern is determined. A single artwork will be needed to allow the slots to be etched on both sides of a thin double sided board. The negative can be reversed to provide the other sides pattern. The edges may be connected using copper foil soldered along the edges in the same manner as is used to join ground planes together. The width of the broad face will be reduced by the square root of the dielectric constant as it forms a filled waveguide, and thus will have the same cut off with a narrower broad face dimension.

Aligning and checking antennas

An antenna may be aligned by using some form of power detector, field strength meter or receiver, and maximising its output from signals received from a relatively distant transmitter. There are several ways in which this apparently simple task is made difficult. Try to ensure that the antenna responds only to the direct signal from the transmitter, and not to any signals reflected from intermediate objects such as the earth's surface. The choice of the test site is of great importance when trying to avoid this problem. This risk of receiving reflections usually increases as the test antennas are spaced further apart, but also there is a minimum spacing that can be accepted: each antenna should be operating outside their near field or within their far field. This is determined by twice the diameter of the dish (D), squared, divided by λ, i.e. 2 x D^2/λ.

When the antenna is of the Yagi type, D should be determined by assuming a value for its gain and converting this to an equivalent dish size by reference to Fig 3.24.

A good test site is one where the test antenna is located on one side of a valley with the transmitter located at a similar height on the other side at less than a mile away, with the valley between broken up by trees, houses or rough ground. When testing high gain antennas, the transmitting antenna should have at least a comparable performance to minimise unwanted reflections. The effect of reflections is to make the performance appear to be inferior, due to the non uniform illumination of the test antenna. Ideally the signal received at the test antenna should not change by more that 1dB over more than its capture area. This is true in both planes and is the reason for the suggestion that the valley bottom should be cluttered to prevent or reduce ground reflections causing a vertical pattern of peaks and troughs. If these conditions cannot be realised, then the requirements are altered and the path length should be reduced to the smallest permitted, determined using the equation in [11].

In order to reduce the effect of ground reflections causing changes over the capture area, the signal source should be mounted close to the ground. This ensures that the reflected signal cancels well up from the ground at the test end. Also, if possible, alter the aim of the transmitting antenna, both elevation and azimuth and also the height, to ensure the signal is the greatest possible.

If large antennas and a moderately powerful signal source is used over a short range a diode detector can be used with a sensitive meter. This can be a conventional mixer or detector mount at the higher frequencies. The best indicator to use is a sensitive power meter, as the readings can be taken directly in decibels with good accuracy. However, the results with a diode detector used as described should include an attenuator between the antenna and the detector to minimise the effects of any mismatch. Over longer paths it may be necessary to use a receiver as the detector. Remember that the receiver needs to have a linear response, so if an fm receiver is to be used reduce its gain so that limiting does not occur. For a dish, the operations needed to return the optimum gain differ in detail from that required by a Yagi type of antenna.

The dish operations are as follows:

- The adjustment of the feed with respect to the dish. This is necessary because there is always some uncertainty as to the precise position of the phase centre of the feed, and possibly that of the focus of the dish. The feed should be adjusted by altering its distance from the dish surface to optimise the received signal and then clamped or soldered in place.

- It is not safe to assume that the antenna is free from squint, however accurately it has been made. It must be remembered that the vertical beamwidth of even a small dish may only be very small when used at the higher frequencies. Whereas squint in the horizontal plane (azimuth) is relatively unimportant, a squint in the vertical (elevation) plane will result in a permanent loss. An azimuth error just means the direction to point the dish differs from the mechanically obvious one. On the other hand, an elevation error not observed and corrected may considerably degrade the performance unless a means of controlling its elevation is provided.

Also, it is important to recognise that the axis of rotation should be truly vertical, otherwise the antenna will be tilted with respect to the horizon in some directions of rotation. If a tripod is being used on the higher frequencies the use of a spirit level is recommended to ensure that this problem does not occur.

- When checking the gain of an antenna the use of a calibrated attenuator or a power meter is recommended. It is possible to provide a direct measurement of antenna gain by substituting a second antenna of *known* gain and adjusting the attenuator to produce the same detected signal or reading on the power meter. For an antenna of this type, an efficiency of 50-60 per

cent of the theoretical value is the normal practical limit. This may be calculated for a dish by using $(\pi \times D/\lambda)^2$. The gain of a conventional pyramidal horn determined from Fig 3.38 can be predicted with sufficient accuracy to be used as a standard. A horn used as a reference antenna is not too large, even at 1.3GHz where a horn with a gain of 10-13dB may be quite easily constructed for this purpose. Horns normally exhibit gains well within a decibel or so of design and across a range of frequencies without the need for matching devices.

- An invaluable facility is an optical sighting method that may be aligned by sighting the signal source after optimising the received signal. A suitable device can be made using a small-diameter tube that should have cross wires fitted. With a Yagi array the sighting may be by use of the boom direction. If the signal source cannot be seen directly, use a landmark determined from a map. When this has been done and the sighting device is fixed firmly, then on arriving at a new site the same method may be used and checked by the use of a sighting compass. It is also of very great value to use a fairly local beacon to give a reference direction. If the site is not clear in the direction of the beacon do not be fooled by false headings or even by the broadening effect of a range of hills. It is likely the best signal will be coming from a reflection or just be broadly spread by knife edge refraction over the top. For this reason it is essential that as many directions as possible should be used to check the heading reference. Especially if several beacons are available, the headings to them from a portable location can be determined prior to being on site. But do not ignore the fact that the best signal or even the only signal may be on a wrong heading due to the direct path being blocked.

When this method is employed and an apparently broad heading is observed, the first thoughts are that the high gain antenna is not better, or it is not working. This is common and it becomes more obvious as the frequency rises due, in part, to the narrower beamwidths obtained with the same physically sized antenna. Also, the objects reflecting signals are bigger in terms of wavelengths and are therefore likely to be more effective. When it is possible to separate these individual sources of reflection, it is perhaps easier to comprehend why gain measurements attempted on this type of signal source produce doubtful results. A high gain antenna has a narrower beamwidth and therefore sees only part of the spread out signal that a wider beamwidth antenna would receive. The result is that when the two antennas received signal strengths are compared there may be very little difference. This concept is rather difficult to accept at times.

The way to avoid this is to choose a test site carefully and use it with these difficulties in mind. A short distance of 1 to 2km (around a mile) across a fairly steep sided valley would be satisfactory.

References

[1] High performance long Yagis, Ian F White, G3SEK, *Radio Communication* April 1987, pp248-252.

[2] Yagi antennas, Gunter Hoch, DL6WU, *VHF communications* 9 (March 1977), pp157-166, UKW Berichte.

[3] A new method of obtaining maximum gain from Yagi antennas, H W Ehrenspeck and H Poehler, *IRE Trans on Antennas and Propagation* October 1959, p379.

[4] Long Yagis, Kmosko, W2NLY, and Johnson, W6KQI, *QST* January 1956.

[5] Yagi antenna design, P P Viezebicke, *NBS Technical Note 688,* December 1976.

[6] How to design Yagi antennas, J H Reisert, W1JR, *Ham Radio* August 1977, pp22-30.

[7] Notes on the development of Yagi arrays, C Greenblum, *QST,* and other ARRL publications, edited by Ed Tilton, W1HDQ.

[8] Extremely long Yagi antennas, G Hoch, DL6 WU, *VHF Communications* (14 March 1982), pp131-138. Also More gain from Yagi antennas, *VHF Communications 9* (April 1977), pp 204-211.

[9] Designing paraboloids, T C Jones, G3OAD, *Radio Communication* April 1971.

[10] X-band omnidirectional double-slot array antenna, T Takeshima, *Electronic Engineering* October 1967, pp617-621.

[11] Antenna performance measurements, R Turrin, W2IMU, *QST* November 1974.

Chapter 4

Transmission lines and components

In this chapter :

- Basic theory of transmission lines
- Coaxial Lines
- Stripline
- Microstrip
- Waveguide
- Measurements and matching
- Simple passive components

- Splitters and combiners
- Directional couplers
- Tuners
- Resonators
- Non reciprocal componenents
- Construction techniques

P robably the most fundamental distinguishing characteristic of microwave circuitry is that the dimensions of the components are of the same order as the wavelength of radiation. This means that transmission lines at microwave frequencies are used not just to carry signals to and from the antenna as at lower frequencies the majority of components are actually made from, or built inside, sections of transmission line. The whole subject of transmission lines is therefore of great importance in understanding microwave components, devices and techniques.

In contrast to lower frequencies, where coaxial cable (and occasionally twin-wire transmission line) are the only common types of line used, there is a much wider variety of transmission-line media used at microwave frequencies; certainly much more than the rectangular waveguide conventionally associated with microwave plumbing! These include certain types of coaxial line (usable to 18GHz and beyond), circular and elliptical waveguide, stripline and microstrip, and various millimetre wave transmission lines including finline. Some common transmission lines are illustrated in Fig 4.1.

This chapter sets out to introduce the various types of lines and passive components used at microwave frequencies, and to give an explanation of how they work. Many of these are peculiar to microwaves and much of the fascination and challenge to the amateur is related to their exploitation. Indeed some microwave techniques are closer to optics than to rf.

The treatment is at a fairly basic level with an emphasis on the practical aspects and a minimum of mathematics, though references are given where appropriate for those seeking a more complete treatment or further background reading. In addition, a considerable amount of practical advice on the use of different types of transmission lines and connectors is also given.

Basic theory of transmission

A transmission line is any structure that guides electromagnetic energy along itself, and in reality an enormous variety exists. It can be either conductive, e.g. coaxial, open-wire feeder and metallic waveguide, or dielectric, e.g. fibre-optic cables, dielectric rod and atmospheric ducts caused by temperature inversions.

In all cases, energy is carried by varying electric and magnetic fields that have instantaneous values, which are functions of both position and time. The most fundamental properties of a transmission line, common to all types, are the characteristic impedance, attenuation and velocity of propagation.

The characteristic impedance is a function of the geometry of the line and the dielectrics used within it. In the case of coaxial lines, it represents the ratio of the voltage

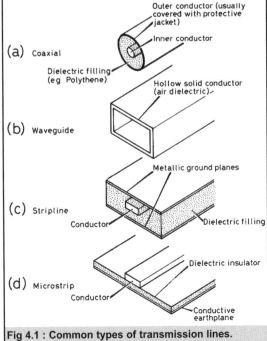

(a) Coaxial — Outer conductor (usually covered with protective jacket), Inner conductor, Dielectric filling (eg Polythene)

(b) Waveguide — Hollow solid conductor (air dielectric)

(c) Stripline — Metallic ground planes, Conductor, Dielectric filling

(d) Microstrip — Dielectric insulator, Conductor, Conductive earthplane

Fig 4.1 : Common types of transmission lines.

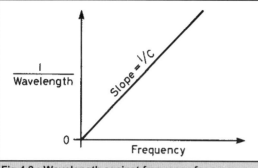

Fig 4.2 : Wavelength against frequency for a non dispersive line.

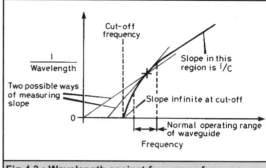

Fig 4.3 : Wavelength against frequency for a dispersive line.

across the line to the current in each conductor. In the case of a waveguide, it represents the ratio of E field to H field strengths. Its absolute value is often not important, but variations along a line are usually undesirable.

Losses in a line can occur either in the series resistance of the conductors or in the dielectric between them. In some types of line where the fields are not contained inside a metallic screen, e.g. open wire feeder, losses can also occur due to the field radiating energy into space.

Fields in transmission lines, like those in free space, cannot travel faster than the velocity of light c, which is 3×10^8m/s. In a simple coaxial line with air dielectric, the fields travel at this speed, but in practice most cables are filled with dielectric and consequently the fields travel more slowly. The velocity is about 0.7c for ptfe or polythene dielectric.

While it is simple to measure voltage and current on lines such as coaxial or twin wire, this is not really possible with either metallic or dielectric waveguides. For this reason, power is the simplest quantity to measure.

Modes

There are a number of ways that energy can propagate down any transmission line these are referred to as modes. A mode describes the pattern in which the field strength varies across the transmission line. Normally only one mode exists in a well designed transmission line, this is known as the dominant mode. However, if the transverse dimensions of the line are large enough, usually greater than about a half wavelength, a variety of patterns can exist. Common modes are illustrated for each type of line later in this chapter. A more comprehensive set is included in appendix 1.

Higher order modes can be thought of as resulting from giving the wave room to propagate across as well as along the line. In general, if it is theoretically possible for more than one mode to exist, then the performance of the line may be unpredictable. This situation is best avoided as any discontinuities in the line may generate other modes.

Modes can be divided into two types tem and non tem. TEM stands for transverse electric and magnetic wave, as both the electric and magnetic field components are always at right angles (transverse) to the direction of propagation. The tem mode requires at least two separate conductors and therefore cannot exist in waveguides. In general, tem lines are quite straightforward and simple to analyse. The properties of non tem lines may be frequency dependent and so are much more complex.

Examples of tem lines are coaxial and stripline; examples of non tem lines are waveguide and microstrip.

In most microwave lines, there is usually only a possibility of one or two modes existing. However, in optical fibres whose dimensions can be many hundreds of wavelengths across, the mode pattern across the fibre might consist of a random speckle pattern of thousands of dots that change randomly with time or movement of the cable.

Dispersion

If the propagation velocity varies with frequency, then the line is said to be dispersive. This means that if pulses of different frequency were simultaneously sent off down the line, after a while they would each have travelled different distances, or dispersed.

Most coaxial and twin wire lines are examples of non dispersive lines. Small amounts of dispersion can arise from changes in dielectric properties with frequency; much larger ones can be caused by resonant or periodic structures in the line. Waveguides are a good example of dispersive lines.

One way of illustrating these properties is to plot the wavelength against frequency, as shown in Fig 4.2. In this graph the vertical scale is the reciprocal of λ, being the number of cycles per metre, also known as the wave number. In non dispersive media, such as free space propagation and most coaxial or twin conductor lines, this graph is a straight line whose slope is related to the velocity of propagation, i.e.:

$$c = F \times \lambda$$

An example of this plot for a dispersive line such as waveguide is shown in Fig 4.3. This only propagates frequencies above a certain frequency, the cut off frequency. At frequencies well above the cut off, the graph is very similar to the previous diagram, i.e. the properties of the guide are similar to free space. However, as you approach the cut off the wavelength is longer than before, and at the cut off frequency the wavelength becomes infinite. The guide cannot propagate waves in the conventional sense at frequencies below this; the fields are then what is known as evanescent. This is a non propagating field which decays exponentially with distance, an effect used in piston attenuators and some special types of microwave filter.

The problem that now arises with a dispersive line is that there are two possible slopes on the graph at any point one is the line from the origin to the point, and the other the tangent to the curve at the point. These two had the same value for the non dispersive line, but are now different. What does this mean in practice? Well, it is possible to measure the velocity of propagation in two ways.

- Measure the distance between adjacent peaks or troughs in the field at any instant. This could be said to be the distance that the wave has travelled during the period of one cycle. The time taken to travel this distance is one period of the frequency of the wave. Thus velocity = distance/time. This gives what is known as the phase velocity, since you are effectively observing a point at a particular phase on the sine wave, and seeing how far it travels in a known time.

- Another approach is to produce a pulse of rf with a very fast rise time, and measure how long it takes to travel down a known length of cable. Again, velocity = distance/time. This is known as the group velocity, as you are taking a specific group of cycles of the wave, and timing them over a known distance.

The reason for these two velocities can be illustrated by considering how waves propagate in waveguide. Fig 4.4 shows three different frequencies being sent down a rectangular waveguide. In Fig 4.4(a) the frequency is much higher than cut off, and the waves, being much shorter than the dimensions of the guide, travel in almost straight lines down its centre. The phase and group velocities are nearly equal to the free space values.

If the frequency is, say, a factor of two above the cut off frequency, the waves can be thought of as bouncing off the sides of the guide in a zig-zag course, as in Fig 4.4(b). The wave now has to travel further, along an oblique course, and so takes longer to make headway along the guide. The delay before it appears at the other end will therefore be longer than in the first case, so the group velocity is lower.

When the frequency is only a few per cent above the cut off frequency (Fig 4.4), the waves are launched across the guide, almost at right angles to it, and so travel to and fro across the guide, making very little progress along it. Here, they take a very long time to come out of the far end, corresponding to a very small group velocity.

Each time the wave crossed the centreline of the guide, it is almost exactly in the same phase as the last time it crossed it. (The path from centre to edge is a quarter wave long, and there is a 180° phase change at the reflection.) Thus the phase of the field in the centre of the guide changes only very slowly with distance along the guide, and so the phase velocity appears to be very large. The phase might change by a few degrees for every free space wavelength travelled along the guide.

In the limit when the guide is exactly a half wave across, the phase of the field does not change with distance along the guide, and the phase velocity appears to be infinite. The group velocity is then zero, as the wave is just bouncing to and fro across the guide at all points along the guide. The amplitude will still decay as the wave travels along the guide, but at a much greater rate than when operating well above the cut off frequency.

This non propagating wave, which consists of a field whose polarity will oscillate at the rf frequency, but whose oscillation is in phase over a large number of wavelengths, is called an evanescent wave. It is essentially a short range effect, involving distances of a few wavelengths, and is often employed in filters and antennas.

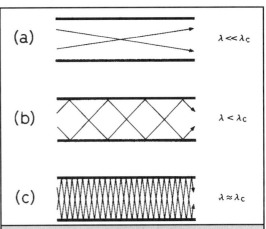

Fig 4.4 : Illustration of dispersion in waveguide.

The piston attenuator works on this principle to give a linear increase in the number of decibels attenuation as two coupling loops are moved apart in a metal waveguide that is below cut off.

The delay in the signal appearing at the end of the line also implies that energy is stored in the line, the amount corresponding to the power fed into it for the time the wave takes to travel through the line. There are many parallels between transmission lines and filters; in particular, energy storage - the signal has to be present for some time before an output is obtained. A filter operates by comparing the phase of the signal at several instants in time, and to do this is has to store energy.

If the velocity in the original straight line was the free space velocity c, then we have just shown that the phase velocity V_p has gone faster than c, and the group velocity V_g slower than c. In fact this is a general principle, that the product of phase and group velocity in a line must always equal the square of the speed of light in that dielectric:

$$V_g \times V_p = \frac{c^2}{\xi_r}$$

The phase velocity is thus generally used to predict the position of maxima and minima of the rf field on a line. The group velocity is used to predict how long changes in signal level will take to travel through a line or circuit.

When using highly dispersive lines, such as waveguides near their cut off frequency, you must be careful to use the correct velocity for the type of calculation. Thus, to calculate the length of guide needed to produce a 90° phase shift when making a balanced mixer, you must use the phase velocity. However, to calculate the propagation delay down a guide, e.g. to calculate the time a pulse took to arrive at the end, as in a radar system, the group velocity would be needed.

All this talk of infinities and velocities greater than light may appear contradictory to the Theory of Relativity, which says that nothing can travel faster than the speed of light in a vacuum. However, the phase velocity is not the speed at which anything tangible travels it is rather an abstract but useful concept to describe the wave pattern that has been set up. The fields and energy that form that pattern, and this includes any modulation waveforms, always travel slower than light, i.e. at the group velocity.

Model of a transmission line

The theoretical treatment of transmission lines is usually heavily mathematical. The exponential notation is commonly used because it is a very concise way of describing the waves behaviour. This section attempts to explain the practical meaning of some of the more commonly used terms, without delving too deeply into the maths. The beginner should concentrate on the qualitative explanations rather than the maths in the first instance and perhaps skip this section on a first reading.

The basic properties of a line are:

• The field at any point on the line varies with time at the frequency w, where w = 2π x F, F being the frequency in hertz and w the frequency in radians per second.

$$V = V_0 \times \sin(w \times t)$$

• The wave is attenuated as it travels along the line. This is represented by the constant α, which defines the fraction by which the voltage decreases over a distance of one wavelength along the line. This is related to the familiar losses quoted in decibels per 10m.

$$V = V_0 \times e^{\alpha \times l}$$

If α is zero, then V does not change with length the line is lossless. If it is not zero, the V decreases exponentially with distance another way of saying a number of decibels per metre. As will be seen later, the loss of a cable can be used to predict the Q that will be obtained if it is used as a resonator, e.g. a quarter wave line.

• The phase of the wave changes as it travels along the line. This is represented by the constant β, which determines the rate of change of phase with distance along the line, in radians per metre. The wavelength is the distance between two points where the phase is the same.

$$V = V_0 \times e^{j \times \beta \times l}$$

These two are combined into one complex constant γ, which describes how the amplitude and phase of the voltage vary with distance; it is known as the propagation constant.

$$\gamma = \alpha + j\beta$$

When put together, the overall expression for the voltage on the line is:

$$V = e^{\alpha \times l} \times e^{j \times \beta \times l} \times e^{j \times w \times t}$$

The simplest line to consider is balanced twin wire line, the equivalent circuit of which is shown in Fig 4.5. Each wire has a certain amount of self inductance per unit length, which is represented by the series L in each wire.

There is also distributed capacitance between the two wires, represented by C. There is a series resistance due to the resistance of the conductors, represented by R, and a shunt resistance due to dielectric losses, represented by G. Provided that a sufficiently large number of these elements are used, e.g. many per wavelength, this lumped constant circuit is an accurate model of a transmission line.

A quick check can be made on this model by considering it at dc at very low frequencies all the reactances can be ignored, so it simplifies to two wires. However, the model only applies up to the frequency where the line acts as a low pass filter. It therefore only applies to the lowest mode on the line.

In terms of the constants shown in Fig 4.5,

$$\gamma = \sqrt{((R + jwL) \times (G + jwC))}$$

and

$$Z_0 = \sqrt{\left(\frac{R + jwL}{G + jwC}\right)}$$

So far, this may look quite heavy going. However, for a lossless line, G and R are zero, so these simplify considerably:

$$Z_0 = \sqrt{\left(\frac{L}{C}\right)}$$

$$\gamma = j \times w \times \sqrt{LC}$$

$$\alpha = 0$$

$$\beta = w \times \sqrt{LC}$$

So the fundamental properties of a line its characteristic impedance, velocity and wavelength are all determined by the inductance of, and the capacitance between, the conductors.

In non dispersive lines, the L and C are independent of frequency. Dispersion is caused by L or C varying with frequency.

The way in which a cut off occurs in a line can now be seen if the shunt capacitance can be made zero, e.g. by placing inductors across the shunt capacitors so that they form parallel resonant circuits, the wavelength and phase velocity become infinite. From the previous section, this means that the group velocity must be zero. So the line does not propagate. This is described in more detail later when waveguides are discussed, these can in fact be thought of as a twin wire line with shunt inductors across it.

Waves in free space can be though of as being in a transmission line whose dielectric properties are determined by the permittivity, ε_0 and its magnetic properties by the permeability, μ_0, which correspond to the values of C and L in the twin line.

$$\xi_0 = 8.85 \times 10^{-12}$$

$$\mu_0 = 4\pi \times 10^{-7}$$

$$Z_0 = \sqrt{\frac{\mu_0}{\xi_0}} = 377\Omega$$

$$V = \sqrt{\mu_0 \times \xi_0} = 3 \times 10^8 \, m/s$$

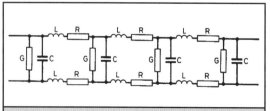

Fig 4.5 : Equivalent circuit of a transmission line.

For waves propagating in space filled with a dielectric, these can more usefully be expressed in terms of relative permeability and permittivity, where unity represents the free space values:

$$Z_0 = 377 \times \sqrt{\frac{\mu_r}{\xi_r}}$$

$$V = c \times \sqrt{\mu_r \times \xi_r}$$

Where c is the free-space velocity, 3×10^8 m/s.

The ratio of V to the speed in free space, c, is known as the velocity factor. Thus a block of dielectric can both slow down a wave and also reduce the impedance, just as it would if it were a dielectric in a coaxial line.

Transmission lines as impedance transformers

The input impedance of a length (1) of line whose characteristic impedance is Z_0, as shown in Fig 4.6, is given by:

$$Z_s = Z_0 \times \frac{Z_l + Z_0 \times \tanh(\gamma \times l)}{Z_0 + Z_l \times \tanh(\gamma \times l)}$$

Where Z_1 is the terminating impedance and is the propagation constant defined earlier. For simplicity $\gamma \times 1$ can be thought of as electrical length in degrees, i.e. 90° is equal to $\lambda/4$, 45° to $\lambda/8$ and so on.

This equation can be used to explain the operation of many components. For example, open circuit and short circuit stubs are simply lines where the terminating impedance is either infinity or zero.

Coaxial lines

Theory of coaxial lines

Coaxial lines are the most familiar type of transmission line. They consist of concentric inner and outer conductors on a common axis, separated by a dielectric. Various types of cable are available flexible cables, semi rigid and Heliax, for example, and these are described later.

Coaxial lines are a simple form of transmission line where the electric and magnetic fields are always at right angles to the director of propagation; they are thus known as transverse electric and magnetic lines or tem lines. They have no cut off frequency and so theoretically operate from dc to infinity. The electric and magnetic fields in a circular coaxial line are shown in Fig 4.7 for the lowest tem mode.

However, higher order, non tem modes having more complicated field patterns can exist at sufficiently high

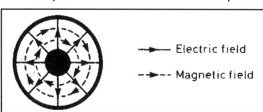

Fig 4.7 : Fields in a circular coaxial line.

frequencies. The first of these can appear when the mean circumference is greater than one wavelength. Thus, at high frequencies, there is a limit to the diameter of coaxial lines that can be used if overmoding is to be avoided. This is only likely to be a problem on the higher bands; for example at 10GHz the maximum diameter before over-moding occurs is about 9mm, so the larger diameters of Heliax should be used with caution. However, 0.5in Heliax, FHJ4, has been used successfully at 10GHz with a measured loss of 4dB per 20ft.

Neglecting losses, the expression for the characteristic impedance Z_0 simplifies to:

$$Z_0 = \sqrt{\frac{L}{C}}$$

where L and C are the inductance and capacitance per unit length of the line. Similarly, the velocity of propogation is :

$$V = \sqrt{(LC)}$$

The inductance and capacitance depend upon the geometry of the line. The capacitance also depends upon the dielectric filling the cable; this is typically about 100pf/m. For a circular line, Z_0 is :

$$Z_0 = 138 \times \sqrt{\xi_r} \times \log\left(\frac{b}{a}\right)$$

and the velocity V is :

$$V = \frac{C}{\sqrt{\xi_r}}$$

Thus the wavelength in coaxial lines is simply that which would exist in space filled with a medium of the same dielectric constant as that filling the line :

$$\lambda_g = \frac{\lambda_{fs}}{\sqrt{\xi_r}}$$

The characteristic impedance for a round line is plotted in Fig 4.8. See appendix 1 for expressions and graphs of Z_0 for other line cross sections.

50Ω has evolved as standard impedance for coaxial lines, although several other values are still used. It is interesting to see why this is so. For coaxial transmission lines there are three main important properties.

• Attenuation

• Power handling capability

• Breakdown voltage

Of these, the attenuation and power capacity are the most important. The power capacity will be determined by the breakdown voltage at low frequencies (where the attenua-

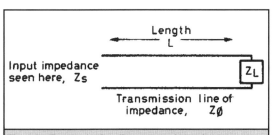

Fig 4.6 : Transmission lines as an impedance transformer.

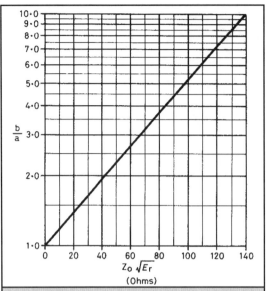

Fig 4.8 : Characteristic impedance of a round coaxial line.

tion is low) and by the heating effect due to the losses at high frequencies (where the attenuation is higher).

Coaxial lines of the same material have higher conductor losses than waveguides and also have dielectric losses as well. The latter stem from the fact that, apart from short lengths of line for special components, the centre conductor needs supporting concentrically. For short lengths, spaced insulating beads can be used, but usually this support is provided by completely filling the line by a solid or foamed dielectric.

Loss occurs in the conductors due to their resistance (so called ohmic losses). The current flows mainly within a thin layer on the surface of the conductor, known as the skin depth, and defined as the depth where the field has decreased to 1/e of the surface value, i.e. -8.7dB.

$$\text{Skin depth} = \sqrt{\left(\frac{1}{\pi \times F \times \sigma \times \mu}\right)}$$

where σ is the conductivity, μ the permeability, and F the frequency.

This depth decreases as the square root of frequency, and so the loss increases as the square root. The losses in the dielectric are much smaller than the conductor losses at low frequencies, but usually increase directly with frequency. As a result the conductor loss dominates

Table 4.1 : Flexible cables

Type	Impedance (ohms)	Diameter (mm)
RG178B/U	50	1.8
RG174A/U	50	2.8
UR43	50	5.0
RG58C/U	50	5.8
UR70	75	5.8
UR67	50	10.3
RG213/U	50	10.3

at low frequencies and the dielectric losses dominate at high frequencies. This means that at microwave frequencies, dielectric losses are the more significant and are prohibitively large for long lengths of line.

The following comparisons of properties apply to air cored cylindrical coaxial cables. The dielectrics used in cables will probably stand quite high voltages (typically 50kV/mm for polythene, 40 to 80kV/mm for ptfe). However, in practice the breakdown in dry air (1.2kV/mm) will usually be the limiting factor as there are almost inevitably gaps in the dielectric at connectors. The following relations give the breakdown voltage V and the power-carrying capacity P:

$$V = \frac{E_{max} \times b \times \ln\left(\frac{b}{a}\right)}{1.414 \times \left(\frac{b}{a}\right)}$$

$$P = \frac{E_{max}^2 \times b^2 \times \ln\left(\frac{b}{a}\right)}{1{,}920 \times \left(\frac{b}{a}\right)^2}$$

where a is the diameter of the inner, b the inside diameter of the outer conductor and E_{max} the break down voltage of the dielectric (assumed to be air).

For a given cable outside diameter, properties vary as shown in Fig 4.9. From these:

Minimum attenuation occurs at $Z_0 = 77\Omega$

Maximum breakdown voltage occurs at $Z_0 = 60\Omega$

Maximum power handling occurs at $Z_0 = 30\Omega$

As a reasonable compromise, 50Ω was chosen as a standard. The attenuation is 10 per cent higher than the minimum at 77Ω, the breakdown voltage 5 per cent lower than the maximum at 60Ω, and the power capacity is 20 per cent lower than the maximum at 30Ω. 75Ω cables are

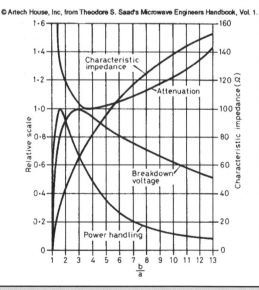

© Artech House, Inc, from Theodore S. Saad's Microwave Engineers Handbook, Vol. 1.

Fig 4.9 : Attenuation, power handling and breakdown voltage of a coaxial line.

sometimes used where low attenuation is important, but the improvement obtained will be small, and will only reduce to 0.9dB. The 75Ω cable also happens to be closest to the input impedance of a half wave dipole, which can thus be fed directly, although a balun should be used.

A number of cables are quoted as 52Ω impedance; the reason for this is that the impedance of a coaxial line changes slightly with frequency. As the frequency is increased from audio to hf/vhf, the current changes from flowing in all of the inner conductor to only the outer surface, due to the skin effect. This change alters the self inductance of the inner and causes a small impedance change. There is about a 4 per cent decrease as the frequency is increased through the transition region. Theoretical expressions for the impedances of lines usually refer to the low frequency value; this must be allowed for when using them at higher frequencies. The transition depends on the diameter of the inner, but usually occurs between 1MHz and 20MHz.

Flexible cables

Flexible cables are available in a variety of sizes, materials and characteristic impedances. The most common impedances are 50Ω and 75Ω, but others, eg 93Ω, are available.

The cheaper types use a polythene dielectric and a pvc outer sleeve; higher quality and more expensive types use ptfe for both the die and sleeving, which is heat resistant and has a slightly lower loss. A few common types are listed in Table 4.1; for a more complete list plus graphs of loss versus frequency see appendix 1.

The very small sizes are suitable for wiring up pcbs and relays inside equipment. The larger cables will be needed for connecting equipment, together or for feeders, although at microwave frequencies it will often be necessary to use semi rigid coaxial cable and Heliax for these applications. PTFE types are recommended when the cable has to be soldered directly to circuitry so that the insulation does not melt. Other dielectrics such as polythene will melt but can be used with care.

Semi rigid cable

This consists of a solid inner conductor (usually copper or

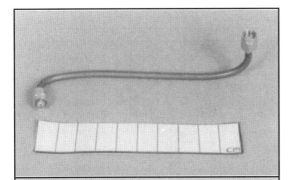

Fig 4.10 : 0.141in semi rigid coaxial cable fitted with SMA plugs. Note that the outer of the cable is solid copper. The inner is usually copper plated steel and is used to form the connecting pin when used with this form of connector. The dielectric is solid ptfe.

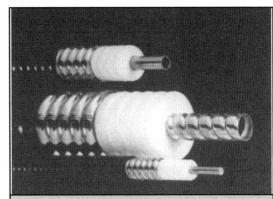

Fig 4.11 : Heliax cables (courtesy of Andrew Antennas).

silver/copper plated steel), a solid ptfe dielectric and a solid copper outer tube. Its characteristic impedance is usually 50Ω, although 10Ω, 25Ω and 75Ω are also available. It is made in a wide variety of outer diameters, from 0.020 to 0.25in, 0.141in being the commonest (see Fig 4.10). It is widely used professionally, since it is sealed and its rigid construction means that it cannot flex and so alter its electrical length (it is phase stable). However, its loss is still relatively high because of its small diameter and it is therefore generally used for connections between modules inside equipment rather than as a feeder to an antenna.

Heliax

Heliax cables are basically larger diameter flexible versions of semi rigid cable, but have a corrugated solid outer conductor to enable them to flex more easily. They are available in diameters of up to 5in; see Fig 4.11. These cables are used for frequencies of up to several gigahertz, where their loss is not significantly inferior to waveguide, and they are rather cheaper than long and heavy waveguide runs.

We have remarked that air is the least lossy dielectric (unless you can evacuate the cable and have a vacuum). For this reason, air spaced cable is the best way of achieving low loss. The problem remains of how to keep the centre conductor accurately in position since any sideways movement of it will alter the characteristic impedance of the line and cause a mismatch. The two common solutions to this used in Heliax are:

• Use a helical supporting membrane

• Use a foam dielectric

While the foam dielectric is slightly more lossy, the losses in the air spaced cable can be greatly increased if water vapour leaks in. Professionally, the air dielectric cables are pressurised with dry air so that any leakage lets dry air out rather than moist air in. Leaks are detected by monitoring the cable pressure. Such cables have valves on their connectors, which allow pumps or pressure gauges to be connected. When pumping them up, a silica gel filter is used to dry the air entering.

Although special connectors are required for Heliax cable, it is possible to modify standard N types to fit some of the smaller diameter cables.

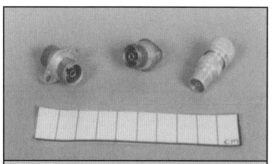

Fig 4.12 : Various types of Belling Lee connectors. Note that these are normally 75Ω and are intended to work up to a few hundered megahertz.

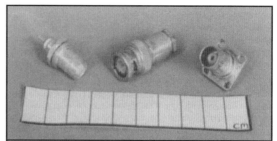

Fig 4.14 : Various types of BNC connectors. These are obtainable as 50Ω or 75Ω types, offer constant impedance and are usable well into the gigahertz region. However, beware of overmoding and impedance changes caused by flexible cables!

Coaxial connectors

The function of a connector is to provide a low loss cable to equipment or cable to cable joint, whose performance does not deteriorate with time. For a lead which is in everyday use this means withstanding repeated connection and disconnection and flexure of the cable close to the connector. For a joint at the top of the mast, it is more important that the joint is waterproof. The choice of connector type depends on several factors, but is governed mainly by the size of cable to be used.

The mechanical aspect of the joint is extremely important. It cannot be stressed too highly that intermittent connections can cause all sorts of problems, and it is well worth spending time and trouble to make connectors up properly. Crimped connectors, where the cable braid is held to the connector body by a crimped sleeve, are recommended but need a special tool to make them up. Suhner make a particularly good crimped BNC connector. This has a tapered plastic sleeve which protects the cable/plug joint, and also prevents the cable flexing near the plug and fracturing the joint. A similar protection can be provided by taping the cable/plug joint and the first inch or two of the cable with several layers of self amalgamating tape.

There are two separate aspects of outdoor connections that should be sealed against moisture. First, there is the joint between the cable and the connector. If this leaks, and moisture enters the cable, the cable loss rises rapidly. It is impossible to reclaim cable once this has happened, as the conductors will have become corroded. Some connectors (e.g. N type, BNC) incorporate rubber sealing rings for this purpose. Second, there is the plug/socket joint. Screw type connectors provide better protection than bayonet types and are to be preferred, but the whole

joint should be protected with tape to be certain. For this purpose, self amalgamating tape is highly recommended (available from RS Components in the UK, stock no 512-042). This chemically amalgamates to itself when applied, and produces a strong, waterproof joint with excellent insulation properties.

The use of a plug not designed for a particular cable will result in poor clamping and early mechanical failure of the connection. A frequent failure mode is that the braid or clamp loosens, which allows the whole cable to rotate in the plug body; the inner connection fails soon afterwards. To prevent this the cable clamp must be done up very tightly with spanners. Taping the plug/cable joint for a few inches also helps to avoid this problem, since it supports the cable close to the plug and makes the plug/cable joint waterproof as well.

The following list includes most types of connector likely to be encountered by amateurs.

Belling Lee: a very simple, cheap, push-fit connector, much used on tv antenna cables which have an impedance of 75Ω. The plug is mechanically unreliable unless made up properly with the correct diameter cable; the inner must be soldered to the pin. These connectors are not waterproof and do not grip the cable well. Not really recommended. See Fig 4.12.

PL259/SO239 (uhf): a fairly common screw locking connector. While their electrical performance is poor above 100MHz as they are not of constant impedance,

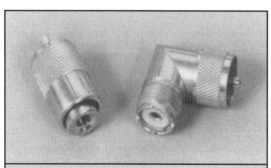

Fig 4.13 : PL259 / S039 connectors.

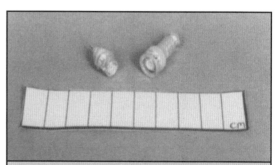

Fig 4.15 : Miniature BNC connectors. These have similar characteristics to the standard BNC connectors but are designed to be used with miniature flexible cables.

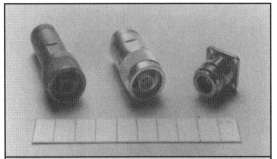

Fig 4.16 : N type connectors. These are available in 50Ω or 75Ω impedance. Adaptors for different sizes of cable can be fitted. Special or modified N type connectors are required for Heliax semi rigid cables. Usable to 10GHz, they are probably the connector of choice for general microwave purposes.

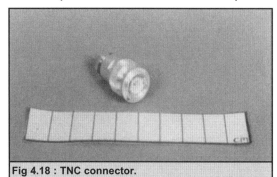

Fig 4.18 : TNC connector.

they are quite useful at hf because they can accommodate UR67 sized cable. They are quite reliable mechanically as long as the plugs are correctly made up, but are not waterproof, and only really grip UR67 sized cable satisfactorily. They were originally used for video connections in radar (only a few megahertz). Not really recommended at uhf. See Fig 4.13.

BNC (bayonet Navy connector): a very common bayonet locking connector, some makes being better than others. They are used primarily with UR43 sized cable and their cable grip is quite adequate. The plug/cable joint is waterproof, but the plug/socket joint is not. Works well up to about 2GHz, above that it can show intermittent tendencies and may radiate energy. Highly recommended for all but the most demanding applications. See Fig 4.14

Miniature BNC: a bayonet locking connector similar to the BNC, but about two-thirds of the size; intended for use with miniature cables (e.g. RG174/U) where the small size is important. Not very common. The electrical performance is similar to the BNC. See Fig 4.15.

N type (Navy connector): these screw locking connectors were originally developed during the second world war. Their specification varies from manufacturer to manufacturer, but more recent precision types are usable up to 18GHz. They are very common and are available for almost all types of cable. They provide an effective cable clamp and both the plug/cable and plug/socket joints are waterproof. Highly recommended. See Fig 4.16.

C-type: a bayonet locking connector that looks rather like an N sized BNC. Not particularly common. Their perform-

ance is roughly equivalent to BNC, but they can cope with larger diameter cables. See Fig 4.17.

TNC (threaded Navy connector): this is a screw on version of the BNC connector and, though they are not particularly common, they are sound, both mechanically and electrically. They provide an effective cable clamp, and both the cable/plug and plug/socket joins are waterproof. They are usable up to 16GHz as the threaded clamp makes a more reliable joint than the bayonet connector and reduces stray radiation. See Fig 4.18.

SMA (sub-miniature A connector): the professional standard microwave connector, originally developed for use with 0.141in semi rigid cable, although available for miniature flexible cables. Usually gold plated, occasionally stainless steel. Small size and high performance. Also known as OSM (Omni Spectra miniature) and SRM by various manufacturers. Highly recommended. See Fig 4.19.

SMB (sub-miniature B connector): a miniature snap on connector for use with miniature coaxial cable (RG188/U etc). They are gold plated and provide an effective cable clamp but are not waterproof. Not intended for frequent connection and disconnection. See Fig 4.20.

SMC (sub miniature C connector): the screw on version of the SMB connector. These are slightly more rugged with higher performance. In general the same comments as for SMB apply. See Fig 21.

GR (General Radio connector): a rather old fashioned type of hermaphrodite connector (i.e. plug and socket are identical), made by the General Radio Company and used mainly on their test equipment. Not very common. See Fig 4.22.

Fig 4.17 : C type connectors, they resemble large BNC connectors.

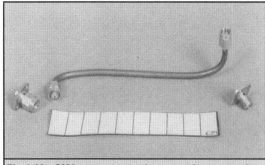

Fig 4.19 : SMA connectors, these are the amateur's choice for use at any microwave frequency up to and including 24GHz. Designed for use with semi rigid cable.

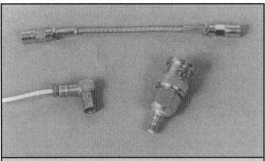

Fig 4.20 : SMB connectors.

Apc-7 (Amphenol precision connector, 7mm): a very high precision hermaphrodite connector used primarily on microwave test instruments. Very expensive and unlikely to be encountered by the amateur. See Fig 4.23.

BNC-HT: a connector made by Radiall which looks very similar to the BNC, but has a longer plug and a recessed pin. Both the plug and socket have extended ptfe insulation around the inner. It is intended for eht applications and is rated at 10kV test voltage, from dc to 2GHz. The plug takes UR43 sized cable. These connectors are recommended for taking eht supplies into high power amplifiers, as the recessed pin cannot be touched accidentally. They are not compatible with ordinary BNC connectors. See Fig 4.24.

PET 100: a screw on connector which is intended for the same sort of applications as the BNC-HT connector (mainly in instruments such as photomultiplier and Geiger counter types) but is rather more common. Rated to 3kV working voltage, 9kV proof, 2.5 A rms, the plug takes UR43 sized cable. See Fig 4.25.

Some connectors are available in either 50Ω or 75Ω impedance, in particular N types, BNC and C types. Be careful not to mix them, as this can cause damage to the connector as well as increasing the vswr. The impedance is usually stamped on the connector body, but for type N and C connectors, the difference in the diameter of the centre pin on the plug is also very obvious. For 50Ω plugs it is about 1.6mm, which for 75Ω ones it is about 1.0mm. The two different types will apparently fit together, but a 50Ω plug with a 75Ω socket splays out the inner contact of the socket, and a 75Ω plug with a 50Ω socket makes poor contact. With BNC, the difference is less obvious, and the 50Ω and 75Ω versions do not damage each other when

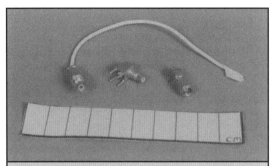

Fig 4.21 : SMC connectors, somewhat similar to SMA but the frequency range is slightly more restricted. Ideal for interconnecting modules within equipment when using small diameter cable.

Fig 4.22 : General Radio to BNC adaptor.

Fig 4.23 : APC-7 connector.

mixed. The difference is mainly in the socket; 50Ω sockets have a dielectric tube around the centre pin, while those of 75 ones have air dielectric. There is very little visible difference in the plugs, but the impedance is usually marked. The diameter that the cable clamping parts of the plug are designed to accommodate also differs in the two types.

Some N type connectors allow a certain amount of choice in how far the centre pin protrudes. If one or other protrudes too far, the shoulder on the plug centre pin can damage the socket fingering when the plug is done up tightly (see Fig 4.26). A guide is that the pin in a plug should be positioned with the tip very slightly behind the outer fingering, but the critical distance is from the

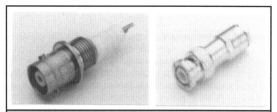

Fig 4.24 : BNC-HT connectors.

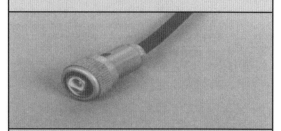

Fig 4.25 : PET-100 connector. This is suitable for high voltage and rf connections up to about 1.5GHz. Usually used in nucleonic equipment such as photomultipliers and Geiger detectors. Occasionally found on surplus microwave equipment, these connectors are very expensive.

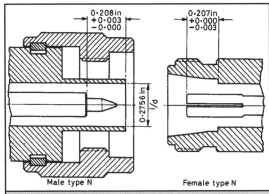

Fig 4.26 : Dimensions of mating parts of N type connectors.

shoulder on the pin to the outer fingering which should be about 5.5mm. In a socket, the centre pin should be 4.5mm behind the open end of the outer body. Note that some precision N type plugs do not have a rubber sealing ring and the fingering around the centre pin is replaced by a solid tube. If you hear or feel a grating noise as the connector is done up the last turn, stop and check the centre pins are in the right place.

More detailed specifications for some of those connectors listed are given in Table 4.2.

Table 4.2 : Connector specifications

Connector	VSWR	Insertion loss (dB)
BNC	<1.12 + 0.007 x F, dc to 2GHz	<0.1 x F
N	<1.06 + 0.007 x F, dc to 10GHz	<0.05 x F
TNC	<1.07 + 0.007 x F, dc to 15GHz	<0.05 x F
SMA	<1.02 + 0.005 x F, dc to 18GHz	<0.03 x F
SMB	<1.08 + 0.017 x F, dc to 4GHz	<0.12 x F
SMC	<1.08 + 0.017 x F, dc to 10GHz	<0.06 x F
APC-7	<1.003 + 0.002 x F, dc to 18GHz	
PET 100	<1.3, dc to 1.5GHz	

These figures are taken from a connector catalogue by Omni Spectra, and refer to top specification connectors. Second hand ones may well not meet these specs, but the figures should serve as a good guide. It would seem

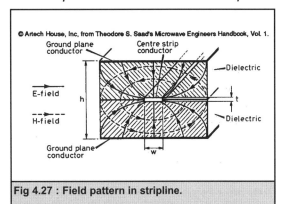

Fig 4.27 : Field pattern in stripline.

to be worth standardising on BNC and N connectors for all but the most demanding applications, though above about 2GHz the more specialised microwave connectors (e.g. SMA, TNC) should really be used.

Stripline

Stripline is also known as triplate and can be thought of as a rectangular section of coaxial line with the side walls removed. Provided the open edges are not too close to the centre strip, its properties are similar to coaxial line. The dielectric can be air, but is more commonly a solid material and it can be made by sandwiching together a strip conductor between two pieces of single sided pcb.

Stripline is a transmission line where the wave propagates in a transverse electromagnetic (tem) manner. The electric and magnetic fields are shown in Fig 4.27. The most common dielectrics used are glassfibre, polythene or ptfe, or combinations of these. More complicated stripline circuits can be produced by etching the required conductor pattern on one side of double sided pcb, the other side being left as a ground plane. A sheet of single sided, copper clad pcb is then placed on top of this and the two pieces firmly clamped together.

The main advantage of stripline is that components and circuitry can be built in the same thickness of line. The circuit is essentially two dimensional and most of the components are realised by altering the width of the line. This results in a considerable saving in volume over coaxial lines and waveguide, especially at the lower end

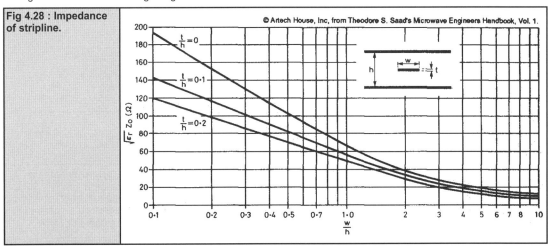

Fig 4.28 : Impedance of stripline.

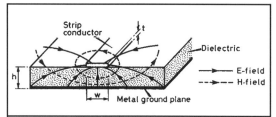

Fig 4.29 : Field in microstrip.

of the microwave spectrum. However, the dielectric losses can be significant.

There are no simple expressions for the characteristic impedance. The thickness of the inner line is a parameter, and when this is particularly thick the line is often called slabline. Very accurate expressions are complicated (see [4]), but for most amateur purposes the following approximation will suffice:

$$Z_0 = \frac{94.2}{\sqrt{e}} \times \ln \left(\frac{1 + \frac{W}{b}}{\frac{W}{b} + \frac{t}{b}} \right) \quad \text{ohms}$$

where W is the conductor width, t is the conductor thickness and b is the spacing between the ground planes.

The accuracy of the formula is approximately one to two per cent for W/b>1, and around five per cent for W/b = 0.75, provided that t/b is less than 0.2.

Results from more accurate equations are plotted in Fig 4.28 for W/b ranging from 0.1 to 10, for several values of t/b.

As with coaxial lines, if the frequency is sufficiently high, higher order modes can propagate. This can occur when the distance between the ground planes is greater than a half wavelength in the dielectric.

Microstrip

The construction of microstrip is essentially like stripline but with only one ground plane (see Fig 4.29). It can easily be made by etching one side of double sided pcb, and is more widely used than stripline.

Microstrip is a non tem type of transmission line as the field is not entirely contained within the dielectric. The electric and magnetic fields are shown in Fig 4.29. Because not all of the electromagnetic field is confined within the dielectric, part of the wave travels in air and part in the dielectric, and consequently the effective dielectric constant is lower than that of the substrate. This constant is a function of both the substrate dielectric constant and the ratio of the conductor width to dielectric thickness. Thus velocity of propagation and hence the wavelength are functions of this ratio W/h. See Fig 4.30 for graphs of

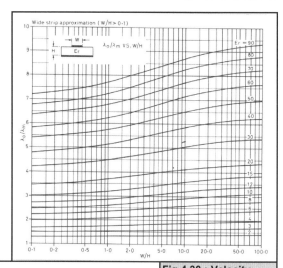

Fig 4.30 : Velocity factor of microstrip.

© Artech House, Inc. from Theodore S. Saad's Microwave Engineers Handbook, vol 1.

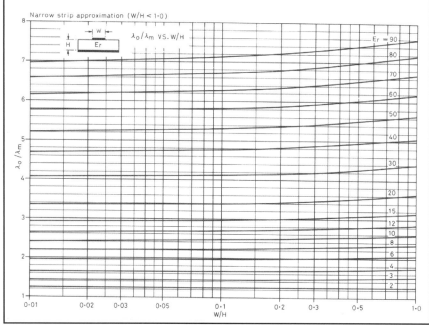

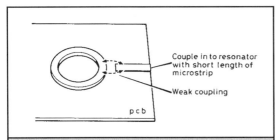

Fig 4.31 : Measurment of velocity factor using a ring resonator.

the wavelength against W/h for various dielectrics.

The line is also dispersive. It is therefore important to have accurate data on the substrate properties for any critical designs.

These effects are relatively unimportant with most microwave dielectrics, but are more significant with glass epoxy board, where a typical permittivity of 4.55 at 1MHz may fall to 4.25 at 500MHz. Manufacturers permittivity data tends to be unreliable as it is usually based on simple measurements on the capacitance of an area of board and it is difficult to take account of edge effects. The standard technique is to measure the resonant frequency of a ring resonator as shown in Fig 4.31, and to derive the effective value of the permittivity from that.

In order to confine most of the energy to the dielectric, and so to minimise the amount of stray radiation, a high dielectric constant is desirable. Professionally, such materials as alumina with a dielectric constant of 9.8 and a loss tangent of 0.0002 at 10GHz are used. However, these high dielectric constants also mean that the wavelength is much shorter and the line widths also become very narrow. Very high accuracy is therefore necessary in defining the line widths and a very smooth surface necessary on the substrate. These materials are not very suitable for amateur use and so lower dielectric material is more suitable. Thus, be aware that there will be significant radiation from the lines.

As with stripline, accurate expressions for the impedance are very complicated. However, some approximations exist, and the most useful for amateur applications is one which is valid for dielectric constants between two and six:

$$Z_0 = \frac{60}{\sqrt{\left(0.475 + \dfrac{0.67}{e}\right)}} \times \ln\left(\frac{h}{0.134W + 0.168t}\right)$$

where h is the dielectric height, e the dielectric constant, W is the conductor width and t the conductor thickness.

For W/h less than 1.25, a dielectric constant between 2.5 and 6.0, and t/W between 0.1 and 0.8, this formula will give results within five per cent.

The effective dielectric constant k for calculating the velocity factor for the transmission line is given approximately by:

k = 0.475e + 0.67

Again, this equation applied for e between 2.0 and 6.0.

See Fig 4.32 for graphs of the characteristic impedance against w/h for various dielectrics.

In addition to microstrip, there are other configurations

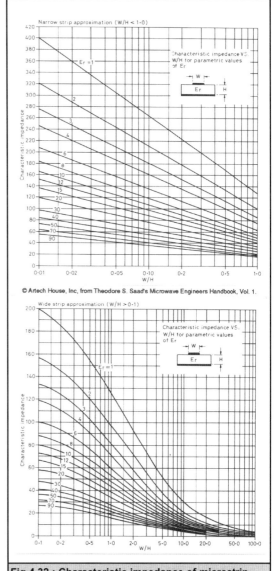

© Artech House, Inc, from Theodore S. Saad's Microwave Engineers Handbook, Vol. 1.

Fig 4.32 : Characteristic impedance of microstrip.

which look similar and are useful in some applications. Some examples are the coplanar line, slot line and suspended substrate line (see Fig 4.33).

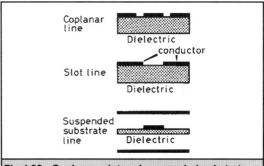

Fig 4.33 : Coplanar, slot and suspended substrate lines.

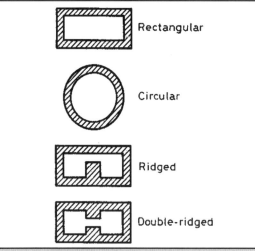

Fig 4.34 : Common types of waveguide.

Waveguide

Although all transmission lines are wageguides, since they guide waves, the term is generally used to describe hollow metal tubes with the field propagating down the middle. Four of the commonest types are illustrated in Fig 4.34.

When any electromagnetic field is confined in this way, its propagation characteristics are different from those in free space. The conducting boundaries only permit the field to exist if it conforms to specific patterns and there is no electric field along the surface of the metal. The properties of the waveguide are therefore related to its shape and size. Any discontinuities within the waveguide will alter its properties as a transmission line and this effect is used to produce inductive or capacitive reactances. This is the basis of many microwave components.

All waveguides of this type behave as high pass filters, and will only carry frequencies above a certain limit known as the cut off frequency. This behaviour can best be illustrated by taking extreme examples.

Any attempt to connect the two leads of a dc supply to the metal waveguide will result in a dead short across the supply, so it is obviously incapable of propagating very low frequencies. Conversely, since the waveguide is a hollow pipe which you can look down, very high frequencies such as light must be able to propagate down it.

The lowest frequency which such a guide will propagate depends on its shape, but is roughly when the diameter is

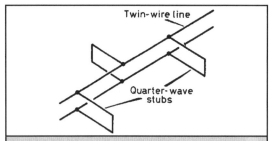

Fig 4.35 : Twin wire line with shunt quarter wave stubs.

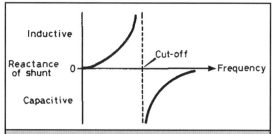

Fig 4.36 : Equivalent shunt impedance across a twin wire line for waveguide.

equal to half a wavelength. The reason waveguides are restricted to microwaves is now obvious: that of size. For example, at 50MHz the necessary size for a rectangular guide could be used as a garage for a small car.

The parameters of voltage and current have no practical significance in waveguides as there is no unique value for either them. Instead, energy is carried by sinusoidally varying electric and magnetic fields that have instantaneous values, these being functions of both position and time. The quantities measured in practice are always power and impedance.

To explain the operation of waveguide, consider it to be made up of a twin wire transmission line, with the wires running along the centre of each of the broad faces. This would naturally give an E field in the correct sense, i.e. between top and bottom, between the wires. Propagation along the twin line would not be affected if shorted quarter wave stubs were connected across it from each side (Fig 4.35). An infinite number of these would join up to form a closed pipe.

The reactance of these quarter wave lines will change with frequency. The net result is that the shunt impedance across the twin line varies from a short at dc, and inductance up to the cut off frequency and a capacitance above it (see Fig 4.36). At dc the shunt inductors short out the signal, so it cannot propagate low frequencies. Above the frequency where the shunt C and L are parallel resonant, the shunt reactance returns to the normal capacitive values and the line will propagate again.

Enclosing an electromagnetic field within conducting boundaries alters the wavelength of propagation from that of free space; it is made longer, and is a function of the guide dimensions. The general expression for any shape of guide is:

$$\lambda_g = \frac{\lambda}{\sqrt{\left(\xi_r - \left(\frac{\lambda}{\lambda_c} \right)^2 \right)}}$$

Where λ_c is the cut off wavelength of the waveguide, λ is the free space wavelength and ξ_r is the dielectric constant of the medium filling the guide (usually air, for which ξ_r is approximately one).

So far we have just considered using the waveguide near the bottom limit of its frequency range. The normal operating range is from about 1.25 to 1.9 times the cut off frequency. Higher order modes exist whose cut off is just above this range and, when more than one mode can propagate, it is difficult to control in which way the energy is distributed between them. The most serious problem of using a guide which allows many modes to propagate is

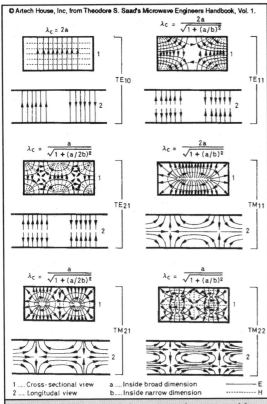

© Artech House, Inc, from Theodore S. Saad's Microwave Engineers Handbook, Vol. 1.

$\lambda_c = 2a$

$\lambda_c = \dfrac{2a}{\sqrt{1 + (a/b)^2}}$

TE$_{10}$ TE$_{11}$

$\lambda_c = \dfrac{a}{\sqrt{1 + (a/2b)^2}}$

$\lambda_c = \dfrac{2a}{\sqrt{1 + (a/b)^2}}$

TE$_{21}$ TM$_{11}$

$\lambda_c = \dfrac{a}{\sqrt{1 + (a/2b)^2}}$

$\lambda_c = \dfrac{a}{\sqrt{1 + (a/b)^2}}$

TM$_{21}$ TM$_{22}$

1 ..... Cross-sectional view   a .....Inside broad dimension   ——— E
2 ..... Longitudal view   b.....Inside narrow dimension   ---------- H

Fig 4.37 : Mode patterns in rectangular waveguide.

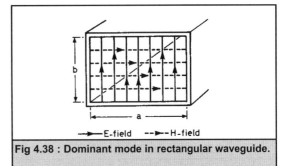

—→ E-field --►-- H-field

Fig 4.38 : Dominant mode in rectangular waveguide.

long run of guide is used, then unless its mechanical dimensions are perfectly accurate it is likely that the mode pattern may change as energy moves from one mode to another. This is known as mode conversion.

Waveguide losses are comparatively low, usually one or two orders of magnitude lower than coaxial cables when used at the same frequency. Also, at higher frequencies the physical size is such that many components can be built inside it or immediately around it, and a complete transceiver may consist of a number of waveguide components bolted together.

Rectangular waveguide

The most common form of waveguide consists of a hollow rectangular tube. The dominant mode is the TE-10 mode; the electric field is transverse to the guide, extending between the two broad walls (see Fig 4.38). The intensity of the electric field is a maximum at the centre of the guide, and drops off sinusoidally to zero intensity at the edges. The magnetic field is in the form of loops which lie in planes parallel to the broad faces of the guide. The plane of polarisation is parallel to the electric field: for horizontal polarisation the broad faces of the waveguide should be vertical.

For an air filled guide (ξ_r approximately 1), the waveguide wavelength is given by:

$$\lambda_g = \dfrac{\lambda}{\sqrt{\left(1 - \left(\dfrac{\lambda}{(2a)^2}\right)\right)}}$$

where λ is the free-space wavelength and a is the larger inside dimension.

Rectangular waveguides are available in a range of standard sizes to cover all microwave frequencies in bands, with some overlap between each band. Waveguides commonly used to cover the amateur bands are listed in Table 4.3 with their dimensions and cut off frequencies. For a more complete listing see appendix 1.

that they can all have different velocities and so, if energy were launched into several modes at the same time, it would not all arrive at the far end at the same time. This can result in interference effects between the various modes that will produce peaks and nulls in the line attenuation at certain frequencies, as well as causing dispersion.

The field patterns for these modes are usually different and some mode patterns are shown in Fig 4.37 for rectangular waveguide. There is a standard notation for describing the different modes, shown with each one. They are named firstly according to whether the electric or magnetic field is zero along the direction of propagation, and there are three possibilities: TE, TM and TEM for the electric, the magnetic and both being zero. In waveguide only TE and TM modes exist. Suffixes are then added to describe the number of peaks and troughs in the pattern. They represent the number of times the field goes from zero through a maximum and back to zero again along an axis of the guide. For rectangular waveguide these directions are parallel to the broad face and parallel to the narrow face respectively. In circular waveguide, the directions are around the circumference, and radially.

Thus the dominant mode in rectangular waveguide is known as TE-10 because there is one variation along the broad face and none along the narrow face.

Components which are designed to work with one mode will be unlikely to work with another. Overmoding also results in unpredictable performance because a probe which couples into one mode may not couple into another as a result of the different field patterns in the guide. If a

Table 4.3 : Waveguide specifications.

Band (GHz)	WG no	External dimensions (mm)		Cut off frequency (GHz)
1.3	WG6	169.16	86.61	0.908
2.3	WG8	113.28	58.67	1.372
3.4	WG10	90.42	47.24	2.078
5.7	WG14	38.10	19.05	4.301
10	WG16	25.40	12.70	6.557
24	WG20	12.70	6.350	14.05
47	WG24	6.807	4.420	31.39

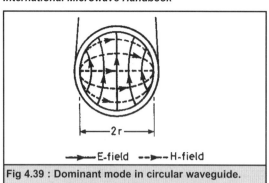

Fig 4.39 : Dominant mode in circular waveguide.

Circular waveguide

Circular waveguide can also be used to transmit energy. However, it possesses no characteristic that positively prevents the plane of polarisation rotating about the axis of the guide as the wave travels along it. A signal with one plane of polarisation injected into a circular waveguide may not necessarily have the same polarisation at the other end. For short lengths of guide this is not a severe problem, and circular guide is often used for dish feeds.

Like rectangular waveguide, a given size of circular waveguide has a cut off frequency below which signals will not propagate, and a frequency above which higher modes may also propagate. The dominant mode for circular waveguide is the TE-11 mode, shown in Fig 4.39. The cut off higher modes may also propagate. The dominant mode for circular waveguide is the TE-11 mode, shown in Fig 4.39. The cut off wavelength for this is given by:

$$\lambda_c = 3.412a$$

where a is the internal radius of the waveguide. Signals with a larger wavelength than this will not propagate.

The wavelength in the guide λ_g, is given by:

$$\lambda_g = \frac{\lambda}{\sqrt{\left(1-\left(\frac{\lambda}{\lambda_c}\right)^2\right)}} = \frac{\lambda}{\sqrt{\left(1-\left(\frac{\lambda}{3.412a}\right)^2\right)}}$$

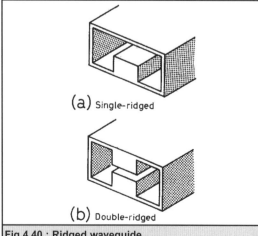

(a) Single-ridged

(b) Double-ridged

Fig 4.40 : Ridged waveguide.

Fig 4.41 : Flexible waveguide.

The next higher mode is the TM-01 mode for which $\lambda_c = 2.613a$. This represents a frequency that is only about 30 per cent higher than that of the dominant mode. Thus circular waveguide can be used only over a much narrower frequency range than rectangular waveguide if the risk of overmoding is to be avoided. Any cylindrical tubing with a reasonably uniform bore can be used as circular waveguide, e.g. copper pipe or empty tin cans.

Ridged waveguide

In wideband systems, it may be difficult to use ordinary rectangular waveguide, which has a typical bandwidth of 0.4. Coaxial cables and components may be used but an alternative is ridged waveguide where the bandwidth is increased by adding a centred ridge to one or both of the broad walls (see Fig 4.40).

The double ridged type has the widest bandwidth, and will typically cover two ordinary waveguide bands, e.g. 8 to 18GHz. Standard sizes are available and components such as bends, switches etc are available in these sizes.

Flexible waveguide

Short lengths of flexible waveguide are needed in a system, e.g. when it is difficult to get exact mechanical alignment between two components or when movement is required.

There are two types of flexible rectangular guide that can be used flexible and flexi twistable.

Flexible guide can be bent in both E and H planes, but not twisted. It is made in a concertina like manner from corrugated tube (see Fig 4.41). It is almost always silver plated and encased in rubber for strength.

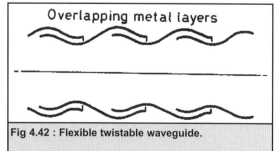

Overlapping metal layers

Fig 4.42 : Flexible twistable waveguide.

Fig 4.43 : Flexible elliptical waveguide (courtesy Andrew Antennas).

Flexi twistable waveguide is made by winding silver plated tape with interlocking rings round a rectangular former (see Fig 4.42). It is encased in rubber to hold it together.

Both types are more lossy than plain waveguide, the twistable type being the most lossy. Also, they are particularly susceptible to water once damp they are difficult to dry out and can be very lossy.

In addition, there is a type of waveguide available called elliptical waveguide, designed for long feeder runs, e.g. to dishes mounted on a tower. It is made from a corrugated copper tube, and covered with a plastic protective jacket, see Fig 4.43. It is rather like a very stiff flexiguide. Standard sizes are available, together with flanges which mate with rectangular waveguides. Some, available from Andrews in the UK, that could be used on 10HGz are listed in Table 4.4.

Waveguide bends and twists

Bends are sometimes required in the assembly of equipment or connection to an antenna. The simplest form is one in which a piece of waveguide is simply bent with a radius of curvature of several wavelengths. The change in impedance of the guide in the bent section (compared to a straight section) tends to be quite small, and the wideband nature of the guide is scarcely affected. This approach is an example of the general rule that is applicable to waveguide, you can get away with many things provided you do it gradually over a distance of a number of wavelengths.

Commercially, bends are made by filling a straight piece of guide with sand or a soft, low melting point alloy to prevent the sides buckling, and bending around a former. Alternatively, they may be made by electro depositing metal onto a shaped mandrel (electroforming). For the amateur, gently bends can be produced by clamping one

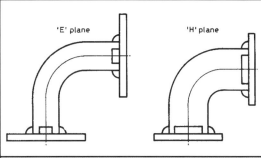

Fig 4.44 : E and H plane bends in rectangular waveguide.

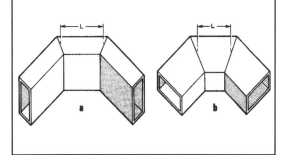

Fig 4.45 : Short bends.

end of a length of guide, applying a light load to the other end, and heating with a brush flame until the correct local deformation has occurred, before moving to the next section. E and H plane bends are shown in Fig 4.44; the plane refers to the plane in which the angle of the bend is measured.

The bends made in this way are necessarily large. Bends having a small radius of curvature can be made by fitting straight lengths of guide into specially shaped corner pieces that are available commercially. They may also be constructed as shown in Fig 4.45. The mid point length L should be an odd number of quarter guide wavelengths so that reflections at the two discontinuities tend to cancel. Mitred corner bends are also possible as shown in Fig 4.46. Twists are pieces of waveguide which, as their name implies, are twisted through an angle so that equipment using one plane of polarisation may be connected directly to another of differing polarisation. Commonly, the angle of rotation is made 90 (see Fig 4.47). Twists may be made using the same techniques as for

Table 4.4 : Elliptical waveguide sizes.

Type	Frequency range (GHz)	Outer dimensions (mm)
EW85	7.7 - 10.0	33.5 by 22.9
EW90	8.3 - 11.7	33.5 by 20.3
EW122	10.0 - 13.25	27.2 by 18.3

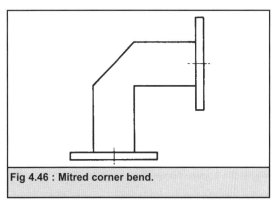

Fig 4.46 : Mitred corner bend.

Fig 4.47 : Waveguide twist.

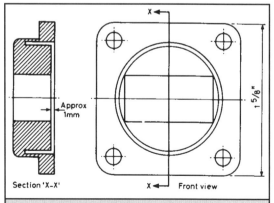

Fig 4.49 : Round flange for waveguide 16.

large bends, and their length should again be several wavelengths.

Waveguide flanges

Flanges are used to connect lengths of waveguide and waveguide components together. For each size of waveguide these are standardised, but often there are two or three types. For example, there are two common types used to join WG16, the square type with four holes for bolts, and the round type which are joined together with a pair of coupling/locating rings (see Fig 4.48 and 4.49).

Within each type there are also a number of variations. Often provision is made for an O ring for sealing, with a circular groove cut in the face of one flange. This type should be used for outdoor joints with a suitable O ring. Also, the flange may be of the choke type, whereby it is arranged that the waveguides do not quite meet at the join but are held slightly apart, see Fig 4.50. One flange of this type is used together with a plain flange (do not use two choke flanges together) and it is intended to provide rf electrical continuity without physical continuity. This is to prevent arcing in high power equipment across what could otherwise be an intermittent joint, and it also reduces leakage; the powers used in amateur equipment are too low for this to be a problem and plain flanges are satisfactory.

Plain flanges may be fitted to waveguide so that a small amount of the guide protrudes, and then soldered or brazed in place. The excess waveguide may be trimmed off by turning in a lathe or alternatively by filing them and rubbing the end on carbide paper. Choke flanges must either be fitted exactly in the first place or the end trimmed using a lathe.

Dielectric waveguide

Most people are familiar with the idea of using metallic conductors to confine and route electric currents and fields. However, insulating dielectrics can also be used to guide waves. Familiar examples of this are atmospheric ducting at vhf and microwaves, and optical waveguides using glass fibres.

The guiding action can be explained in a number of ways, but in general terms waves can be trapped inside areas of high dielectric constant. In the same way that waves can be reflected by sudden changes in conductivity, e.g. at the surface of a metal plate, they can also be reflected by changes in the dielectric constant. At optical frequencies an example is the reflection seen in a plate glass window, which is particularly strong when the light just glances off the surface at a small angle.

However, these reflections are not as strong as those when the light passes from a more dense medium, e.g. glass, into a less dense medium, e.g. air. Under the right conditions 100 per cent reflection can be obtained, in which case it is known as total internal reflection. The

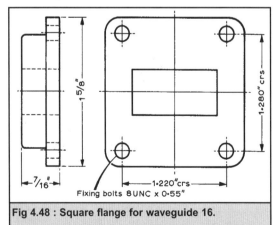

Fig 4.48 : Square flange for waveguide 16.

Fig 4.50 : Choke flange for waveguide 16.

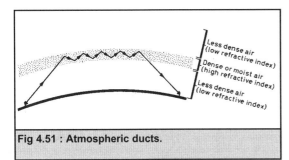

Fig 4.51 : Atmospheric ducts.

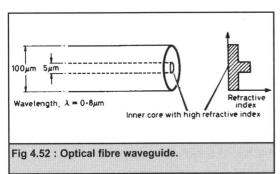

Fig 4.53 : Dielectric waveguide antenna.

same effect can still occur if the boundary between the different dielectric constants is not sudden, but a gradual change.

In summary, waves travelling in a medium whose dielectric constant varies will tend to be bent towards the more dense medium. The paths of waves in the atmosphere under ducting conditions demonstrate this, as shown in Fig 4.51.

Just as with metal waveguides, there is a limit to how small the dielectric waveguides can be and still guide the waves effectively. In waveguides that are only a few wavelengths across, there are only a limited number of ways (modes) in which the field can propagate down it.

When it is sufficiently small, only one mode can propagate down it, and the waveguide becomes known as single mode. The dimensions of a typical single mode optical fibre are given in Fig 4.52.

Unlike metal waveguides, the fields are less well confined, and a considerable amount of energy travels in the cladding as well as in the centre. It is possible to use a single piece of dielectric surrounded by air as a waveguide, but there will be a large amount of energy travelling in the air just outside the dielectric. Provided that no other structure is put within a few wavelengths of the guide to disrupt the field, it will work satisfactorily.

An antenna can be made by introducing irregularities in this dielectric, e.g. corrugations or a taper. This will cause the wave to leave the guide, and radiate into space. An example of such an antenna is shown in Fig 4.53.

Measurements and matching

VSWR

One of the main reasons for wanting unity vswr is for interchangeability of components. A matched component can be placed in a matched system without disturbing the operation of either. A few general points are:

• Few amateur measurements of vswr below about 1.3

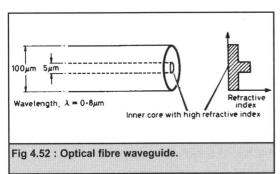

Fig 4.52 : Optical fibre waveguide.

are accurate above vhf.

• Power reflected from a load which presents a low vswr (eg <2) is neither lost nor dissipated in the transmitter. Any transmitter that is reasonably efficient will appear as a bad match to the line, and so will return reflected power to the load again. The line becomes part of the matching network between the p.a. and the load, so the transmitter just sees a slightly different load impedance. By re-tuning the transmitter output and received input circuitry, the original performance into a 50Ω load can be restored. This is easier with valve p.a.s than semiconductor types.

• A vswr on a feeder does not make it radiate. Radiation results from connecting a balanced antenna to an unbalanced feeder (e.g. connecting a dipole to coaxial cable, so that currents flow on the outer of the cable). A line can have any vswr on it and still not radiate, or have unity vswr and still radiate. The fields inside and outside the line are completely separate.

• Using a length of line that is the wrong characteristic impedance, but a multiple of λ/2, does not prevent a vswr from existing on that piece of line. It means that its input impedance is the same as the load at its far end, so the whole line appears to be matched at that frequency.

• The vswr in a system of several components, each introducing their own small vswr, will increase rapidly with the number of components, as the overall worst case vswr is the product of all the individual vswrs.

• The value of a vswr does not tell you what impedance the mismatched line will present. It specifies the degree of uncertainty in the mismatch there are a large number of combinations of R, L and C that will give a particular vswr. You will go through them all as you increase and decrease the length of line by λ/2. A poor vswr can mean an impedance greater or less than 50Ω.

Situations where vswr is important

When many antennas are used in a stacked array, the splitting of power between the antennas has to be done accurately, and poor vswr (e.g. each antenna presenting a mismatch, or a badly matched phasing harness) causes uncertainty in the impedances. This can upset the power distribution between the antennas and prevent the full stacking gain from being realised, or produce an undesirable radiation pattern.

On lines carrying high power (near the maximum rating of the cable), a large vswr (e.g. greater than 3) will cause peaks of I and V every half wavelength which can cause breakdown (sparking) in the dielectric, or heating of the inner conductor which may melt the dielectric.

In systems with a large bandwidth and long feeder runs, e.g. in amateur television, a vswr will cause the impedance at the input of the feeder to vary greatly across the bandwidth. The line length represents a large number of wavelengths, and wavelength changes with frequency, so that even over a small frequency band, the effective line length can vary by more that λ/2, presenting the corresponding range of impedances. In very long feeder runs, the multiple trips up and down the line can cause ghosting on the picture if the vswr is poor.

Any filters, attenuators or power meters will be designed to work in a matched line, and different results may be obtained if there is a vswr on the line because the load and source impedances may be different to the design values. Bear in mind, too, that an antenna usually terminates the line properly only at is design frequency, it may present any impedance at harmonics or spurious frequencies. The same is true of filters that usually work by presenting a mismatch outside their passband.

In very high performance systems the slight increase in loss due to vswr may be important. But if it is a high performance system, the losses or load vswr should not be high enough to increase the line loss in the first place!

Very low noise preamplifiers are designed to work from a specific source impedance for best noise figure, and are usually adjusted for a 50Ω source.

The preamplifier may need to mismatch the feeder and this will create a vswr, but this is intentional and will not degrade the performance in any way, provided that both the antenna and feeder are 50Ω. A vswr due to the feeder or connectors not being 50Ω will degrade the noise figure, unless the preamp matching is re-optimised on that system with that particular piece of feeder.

VSWR due to incorrect line characteristic impedance

If the characteristic impedance and length of each piece of line in a system are known, the vswr can be calculated fairly easily.

First, consider a short mismatch as shown in Fig 4.54. For a piece of line with a vswr of S relative to the rest of the line, if its length L is less than 0.1x(wavelength in the cable), the vswr that it introduces to the system will be:

$$VSWR = 1 + 2\pi \times \left(S - \frac{1}{S}\right) \times \frac{L}{\lambda_c}$$

Where $S = Z_1/Z_0$. This can be used to estimate the effect of using connectors where parts of them have a different impedance. The impedance Z_1 can be calculated from the dimensions of the conductors and the dielectrics used. The results show that at frequencies up to perhaps 100MHz the effect is not serious, and therefore the mechanical properties of the connector are more important. Above this, however, a constant impedance is

desirable.

Second, if the length of mismatched cable is much longer, the vswr can vary between unity and S^2, being unity when the electrical length is an even number of quarter wavelengths, and S^2 when it is an odd number of quarter wavelengths. If the actual length of the mis-matched section is not known, the worst case values should be assumed.

When there are several components introducing mismatches, the actual vswr can be calculated If the lengths of line are accurately known. There is, however, a simple expression for the worst case vswr:

VSWR = S1 x S2 x S3 x ....

where S1, S2 , S3 etc are the worst case vswrs of each component. It is evident that a large vswr can build up quite rapidly. If there are four components, each with S = 1.3, the total worst case vswr would be 3. Depending on the relative spacing of the mismatches, the overall vswr can vary from unity to the worst case value, and this effect will be frequency dependent.

If there are just two mismatches in a piece of cable, the worse case vswr will be S1 x S2 and the best case will be S1/S2, where S1 is the larger and S2 the smaller vswr.

Measurement of vswr and impedance

In most cases it is not the existence of an vswr on a line that is important, but the effect that the vswr has on the impedance looking into the input of the line. VSWR can either be measured directly by observing the fields along the line, or be inferred by measuring the input impedance of the line if the load impedance is known. Techniques are:

- Use a slotted line to measure the peaks and troughs of the field along the line.

- Sample V and I on the line with a coupler.

- Measure the input impedance of the line with an impedance bridge.

Slotted lines

This is probably the most direct and basic method of measuring the vswr on a line. At microwave frequencies the wavelength is only a few tens of centimetres, so it is quite practical to observe the actual variations of voltage along the line with a probe.

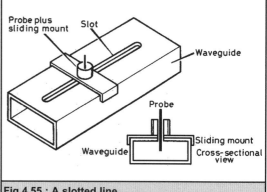

Fig 4.55 : A slotted line.

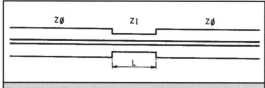

Fig 4.54 : Effect of a short length of mismatched line.

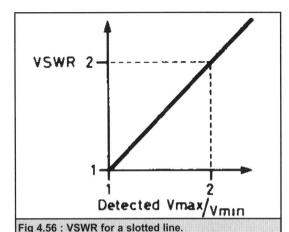

Fig 4.56 : VSWR for a slotted line.

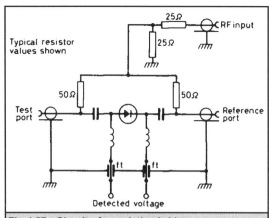

Fig 4.57 : Circuit of a resistive bridge.

A slotted line consists of a section of rigid transmission line, either waveguide or coaxial cable, with a thin slot cut along its length so that a probe can be inserted to sample the field (Fig 4.55). It is important that the geometry of the probe in the line should remain constant as it moves along the line, so that any variations seen are due to a vswr and not to mechanical defects of the line.

The relation between maximum and minimum values of the field and the vswr is given below, and plotted in Fig 4.56.

$$SWR = \frac{1+r}{1-r} = \frac{V_{max}}{V_{min}}$$

where r is the scalar reflection coefficient, the fraction of the voltage wave reflected from the load.

The term return loss is another way of quoting a reflection coefficient. It is the loss in decibels between the incident and reflected waves:

$$return\ loss = 20\ \log(r)$$

This method is capable of great accuracy provided that the section of line used for sampling Z_o is accurately known. The accuracy comes mainly from precise mechanical construction, which can be verified with fairly simple instruments such as vernier calipers or a micrometer.

Impedance bridges

There are two types of impedance bridge. The full vector bridge measures both the resistive and reactive parts of the load. The scalar bridge effectively measures the magnitude of the reflection coefficient of the load, which is closely related to the impedance expressed as a vswr.

Often, all that is required is to adjust the impedance of a piece of equipment, e.g. an antenna, to 50Ω. In this case the scalar bridge is perfectly adequate, and in fact far easier to use.

At hf/vhf it is quite easy to build an accurate scalar bridge which does not require calibration. By using chip resistors and careful construction, a return loss of about 25-30dB can be measured up to uhf. One example of this type of bridge is the noise bridge, which is often used to measure antenna impedance.

The principles of operation are as follows. The bridge is

driven with a fixed amplitude signal of a few milliwatts at the frequency at which the impedance is to be measured. A detector must measure the signal level across the bridge without disturbing the balance.

While ferrite baluns are often used to couple out the signal to be detected, it is difficult to make them with the high degree of balance required. It is much easier to use a simple diode detector across the bridge, and bring out the dc voltage through chokes or high value resistors which will not upset the balance. A design for such a bridge is shown in Fig 4.57. A low source impedance for the rf is needed so that the excitation voltage across the bridge does not change with load.

In practice the bridge is set up in the following way. The drive power is increased to give fsd on the detector with an open or short circuit on the test point, and a known load on the other. Then the unknown load is applied and the new detected voltage noted. The return loss of the load is the ratio of the two detected powers. At low levels the detected voltage is proportional to power, and so the ratio of voltages is often used. It is always a good idea to cross check measurements on any test gear that is available. It is not unknown for even the most expensive equipment to go wrong sometimes. The relation between vswr and detected power is shown in Fig 4.58.

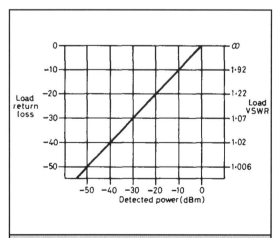

Fig 4.58 : Return loss and vswr for detected power.

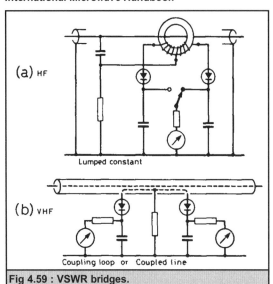

Fig 4.59 : VSWR bridges.

Couplers

Even though standing waves are produced by interference effects between a forward and a reflected wave, at any instant there is only one value of V and I (or E and H) at any point on the line. Knowing the line Z_0, the swr can be calculated from this one simple measurement.

All conventional swr meters operate by sampling the line current and voltage. At hf it is easy to use lumped components, e.g. a current transformer in the centre conductor, and a potential divider to get the voltage (Fig 4.59(a)). At vhf/uhf a current transformer would disturb the main line too much, and a short coupled line is used (Fig 4.59(b)). This also samples V and I, the inductive coupling with the inner induces a voltage proportional to the line current. The capacitance between the line and the loop, and the loop and the outer, then provides the potential divider to sample this voltage. Forward and reflected power are then obtained by adding or subtracting the rf voltages derived from the V and I samples, as shown in Fig 4.60, and detecting the result. If you consider all possible combinations of phase and amplitude of the two signals, you will see that there is only one combination where the two signals are equal in amplitude and 180 out of phase, so they cancel to give zero output. This is arranged to correspond to zero reflected power by choice of the component values as below.

The voltages induced in the coupling loop are given by:

$$V_i = j \times w \times M \times I$$

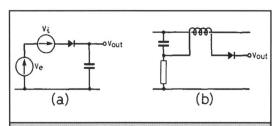

Fig 4.60 : Equivalent circuits of vswr bridge.

$$V_e = j \times w \times C \times R \times E$$

This assumes that the reactance of C is much greater than that of R. The condition for the two voltages to cancel each other is that:

$$M \times I = C \times R \times E$$

The ratio E / I is the characteristic impedance of the Line Z_0, giving as the condition for balance:

$$R = \frac{M}{Z_0 \times C}$$

The same exercise for the case where a separate coupled line is used shows that a null is obtained when the resistive load on the coupled line is made equal to

The Z_0 of the coupled line, so that it is matched. This also applies to designs which use coupled lines that are short compared with a wavelength, and these are the main types used at microwaves.

The coupling is relatively independent of frequency when the coupled part of the lines is a quarter wavelength. When the coupled lines are much shorter than a quarter wavelength, the coupling increases with frequency. Once they are longer than a quarter wavelength, the coupling becomes more narrowband again. A graph of coupling coefficient against length of the coupled section is shown in Fig 4.61.

Thus any accurate directional coupler or swr bridge either needs to be built to a high degree of precision, or needs setting up with a known load on the main line. These instruments are easier to build and calibrate if the coupling between any probes or lines and the main line is fairly weak, e.g. less than 20dB.

The instruments used professionally to measure rf impedances and vswr are the time domain reflectometer (tdr) and the network analyser. Both essentially make the same measurement on the device to be tested, but do so in different ways. They both send energy out of their test port at all frequencies in the band of interest, and measure and display what is reflected back. Although these instruments are very expensive, it is often possible to make use of these techniques at home albeit with lower accuracies or reduced frequency coverages.

The time-domain reflectometer

The tdr is basically a pulse generator with a 50Ω source impedance and a high input impedance oscilloscope (see Fig 4.62). The pulse generator produces a series of

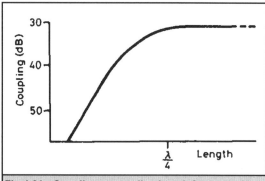

Fig 4.61 : Coupling versus line length for a coupler.

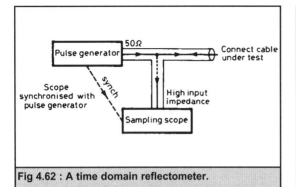

Fig 4.62 : A time domain reflectometer.

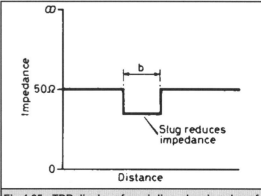

Fig 4.64 : TDR display of an incorrectly assembled rf connector.

rectangular pulses at a low repetition rate, say, hundreds of kilohertz, with very fast rise times, from 1 to 0.01ns. The scope has a rise time similar to that of the pulses, and would in this case be a sampling scope. The pulse generator has a source impedance of 50Ω and sends pulses down the line to be examined. Any discontinuity in this line reflects some of the pulses energy, and both the outgoing pulses and reflected signals are seen on the scope. The pulse generator matches the line, so that signals returned to it are not sent down the line again, which would confuse the display.

The scope sweep is synchronised to the pulse repetition frequency, so a display such as Fig 4.63 would be seen. Thus the pulse is seen on the scope as it is generated, the voltage corresponding to a 50Ω surge impedance of the cable. The signal reflected from the end of the cable will return after the time taken to travel twice the length of the cable. An open circuit reflects the voltage in phase with the incident signal, a short circuit in anti phase.

The main application of this instrument is in examining connectors and short lengths of cable to check their characteristic impedance, as the scope display is essentially a plot of Z_0 against distance along the cable. It will show up incorrectly assembled plugs, e.g. where an air gap has been left if the dielectric has been cut too short, resulting in a higher Z_0. An example is shown at a in Fig 4.64. An air line with a slug of dielectric in it would reduce the Z_0 at the slug. This would be seen at b in Fig 4.65. The magnitude of the reflection coefficient can be measured by:

height of reflected wave / V_{out}

To get the resolution needed to examine the small

distances involved in connectors, very fast rise times are necessary, of the order of 10 to 20ps, which implies frequencies up to 10 or 20GHz are present. This can only be used with short lengths of low loss line, typically up to a few feet long, otherwise the high frequency components will be attenuated by the line and lost, which effectively increases the rise time of the reflected pulse and impairs the resolution.

It is also used with much slower rise time pulses to look at much longer lines, and locate faults in cables. The resolution obtained along the length is approximately one tenth of the wavelength of the frequency corresponding to the rise time, e.g.:

20ps	10GHz	0.1in	(approximately 2.6mm)
10ns	200MHz	5in	(approximately 127mm)
100ns	20MHz	4ft	(approximately 1,219mm)

A simple tdr for use at lower frequencies could be built using ttl or ecl gates to generate pulses, and an ordinary scope for display. In many ways, a tdr is like a radar system, looking down a cable instead of out into space.

The pulses which the tdr sends down the line contain energy at almost all frequencies up to a limit defined by the pulse rise time, so it is probing the line at all frequencies simultaneously, and displays the result as a plot of impedance against line length. The only slight disadvantage with it is that it only works on lines that have a response down to dc, so it will not work in waveguides (which are dispersive anyway).

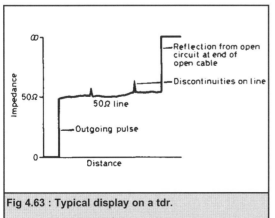

Fig 4.63 : Typical display on a tdr.

Fig 4.65 : TDR display of an air line, showing slug of dielectric in the line.

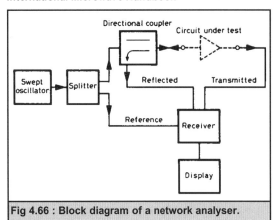

Fig 4.66 : Block diagram of a network analyser.

The network analyser

The other instrument, the network analyser, has an oscillator which it sweeps over the band of interest, and a directional coupler which samples the reflected signals, so their phase and amplitude relative to the forward signal can be measured (see Fig 4.66).

This information is displayed on a screen as the frequency is swept and is usually presented in one of two display modes: return loss or polar (Smith chart) display. Return loss plots the magnitude of the reflected signal vertically on a logarithmic display, and frequency horizontally. From this, the reflection coefficient and hence the vswr can be calculated. The phase information is discarded as in an ordinary vswr bridge.

For the polar display, the phase information is utilised, and a Smith chart overlay of co-ordinates is often provided. The magnitude of the reflected signal deflects the spot radially, and the phase deflects it around the circle so that it plots the impedance at its test port directly on the Smith chart. This is the basic instrument used to measure S parameters of devices.

So far only its application to a single port network has been mentioned, but it can also be used to measure the gain or attenuation (i.e. transmission characteristics) of two port devices. If a signal from the output of the network under test, rather than the signal reflected from its input, is compared with the forward (reference) signal, the amplitude and phase response from the input to the output of the network can be measured. Thus the amplitude and phase response of amplifiers, filters and attenuators can easily be plotted against frequency.

This can also be used with either type of display, depending on whether the phase information is required. So the instrument can be used either to measure impedances of single port or two port networks (e.g. loads or the input and output impedances of an amplifier), or the gain and phase characteristics of any active or passive two port networks (e.g. amplifiers, attenuators etc).

The network analyser can be used over any band of frequencies, and does not require the line or device to respond down to dc, as is the case with the tdr. This instrument also probes the device to be tested at all frequencies, but does so sequentially rather than simultaneously, and presents the information in a completely different way to the tdr - the display is a plot of impedance against frequency.

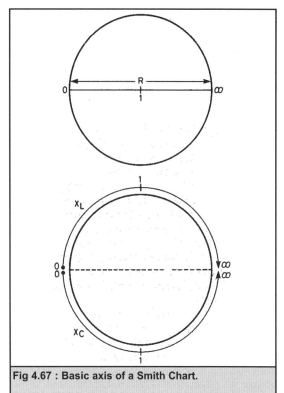

Fig 4.67 : Basic axis of a Smith Chart.

In principle it would be possible to convert the information from either a tdr or network analyser display to the other. The same information is there in both displays, but the process is quite complicated and would require the services of the infamous microprocessor.

Smith charts

These are often though of as being rather esoteric and theoretical, and they can certainly be used in that way. However, they can also serve as a useful aide memoire for the whole subject of transmission lines, matching and vswr, as well as being useful for calculations and as a means of displaying impedances. They will be explained quite simply from this viewpoint, starting from a few basic principles about vswrs.

By using a seemingly rather strange set of non linear scales, the Smith chart manages to represent all of the infinite number of possible combinations of reactance and resistance (i.e. vswrs) within a single circle. Those interested in a mathematical treatment should consult the standard reference text [10].

The main axes of the chart are shown in Fig 4.67. Any purely reactive impedance can be represented by a point somewhere on the outer circle. Capacitive reactances ranging from zero to infinity are on the bottom semicircle, inductive reactances from zero to infinity are on the top semicircle. A pure resistance between zero and infinity will be a point along the horizontal diameter line. These are the basic axes of the chart. All combinations of resistance and reactance will be points somewhere within this circle. All the other lines correspond to specific values of resistance or reactance, and circles, or parts of circles, which all pass through a point on the right hand edge of the chart. The values are given in units normalised to the

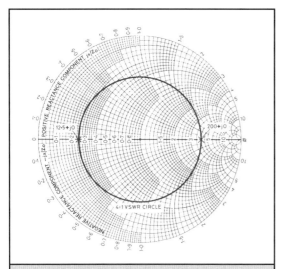

Fig 4.68 : Smith chart showing a 4:1 swr.

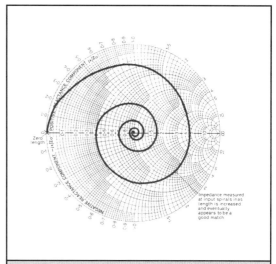

Fig 4.70 : Smith chart of a long length of lossy line.

line impedance used. Usually this is 50Ω, so an impedance of 100 + j50 would be given by the coordinates 2 + j1. The chart is therefore useful as a means of presenting a complex impedance as a function of frequency, many transistors now have their parameters presented as a line on a Smith chart in the data sheets. The line has markings to indicate frequencies at which the impedances were measured.

The next step is to apply the chart to transmission lines. Since the centre of the chart represents the lines characteristic impedance, the input impedance of a length of line when correctly terminated will be at the centre of the chart. If the line has a mismatched load, say, a 200Ω resistor, this corresponds to a load of 4 + j0, and a vswr of 4:1. However, as the length of cable between the load and the impedance meter increases, the impedance will vary through all the range of impedances which correspond to a vswr of 4:1, one particular value being 12.5 + j0 at a quarter-wave from the load. In fact, the range of impedances seen as the line length varies is a circle on the

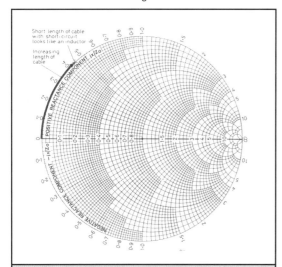

Fig 4.69 : Smith chart plot of a shorted stub.

Smith chart centred on the centre of the chart (Fig 4.68). The radius becomes larger the larger the vswr. A vswr of 1:1 corresponds to a radius of zero, i.e. a dot in the centre, always 50Ω; an infinite vswr is a radius of the outside of the chart, i.e. purely reactive impedances. The vswr can be easily seen by the value of resistance where the circle crosses the resistive diameter a resistance of 12.5 or 200Ω.

The radius of the vswr circle is another axis of the chart, and can be used to specify a particular range of impedances.

It is well known that voltage and current values repeat themselves every half wavelength on a mis-matched line. For this to occur, one trip round the circle must correspond to travelling one half wavelength along a cable.

A quarter wavelength then corresponds to 180 around the chart. This illustrates the quarter wave transformer effect, where a short circuited quarter wave line looks like an open circuit, and vice versa.

Take a short length of cable with a short circuit across it. You would expect it to look like a small inductance, the reactance is given by rotating the corresponding amount clockwise from the short circuit point on the chart (Fig 4.69).

Likewise, a short length of cable with an open circuit looks like a capacitance, rotate a small amount clockwise from an open circuit, and you see a small capacitive reactance.

So putting a length of cable between the load and the test port is equivalent to travelling around a circle of constant vswr (the load). In real life cables are lossy, and this loss improves the apparent vswr of the load. In fact a very long length of lossy cable looks like a perfect match (Fig 4.70). Losses in the cable improve the apparent vswr as the cable length increases and thus correspond to spiralling in to the centre from the outside (a perfect mismatch).

This has hopefully provided an introduction to the Smith chart, and made it possible for readers to at least get a feel for the impedances presented on it. If you want to use the chart for some of the theoretical calculations then you should refer to other textbooks where these applications are explained.

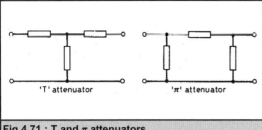

Fig 4.71 : T and π attenuators.

Simple passive components

Coaxial loads and attenuators

There are two types of coaxial attenuators. The first and most common, uses discrete resistors in series and shunt with the inner conductor. This type works down to dc, but is limited at high frequencies by inductances in the resistor leads. At microwave frequencies, it is not possible to use normal-leaded components, and chip or coaxial resistors are arranged in either a T or π form within the coaxial line (see Fig 4.71), so that the correct characteristic impedance is seen from either end.

The second type of attenuator consists of either a lossy inner conductor or dielectric (Fig 4.72). Provided the losses are low, in terms of decibels per wavelength, the match will be very good; however, this type does not work down to dc and the attenuation varies with frequency.

Loads may similarly be built from either discrete resistors or long lengths of lossy line. This second type is particularly convenient for dissipating high powers, and a long length of coaxial cable (with 10 or 20dB of attenuation) is often used as a dummy load for a transmitter (see appendix 1 for cable losses). For a high power load, the main problem is getting heat away from the lossy element. This conflicts with the requirement that the load should be small to minimise stray reactances. Often special cooling techniques are used, e.g. oil or beryllia.

Variable attenuators are often made by switching fixed attenuators, e.g. in steps of 1dB or 10dB. At microwave frequencies this becomes difficult to do without the switch degrading the vswr and a more practical approach is the piston attenuator, as shown in Fig 4.73. This relies upon using a length of metal tube as a waveguide below cut off, in which signals decay exponentially with distance. The signals are coupled in and out of the tube using small loops. These are capable of providing smoothly adjustable attenuation over a range of 100dB, with fairly linear calibration.

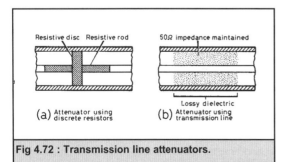

Fig 4.72 : Transmission line attenuators.

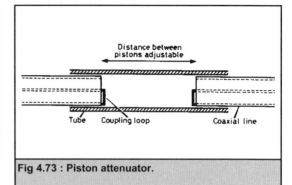

Fig 4.73 : Piston attenuator.

Waveguide attenuators and loads

The function of an attenuator is to absorb power from a transmission line without reflecting it, i.e. by converting it to heat. The most common form in waveguide is a sheet of lossy material (e.g. carbon coated card) mounted parallel to the E field, as shown in Fig 4.74.

For maximum attenuation, the material should be mounted in the region of maximum E field, between the middle of the two broad faces. The attenuation can be increased by moving the lossy card into the centre of the guide from either one side or the top, or by rotating the card from being perpendicular to being parallel to the E field. Some methods of construction are shown in Fig 4.75.

A piece of lossy material in the guide represents a discontinuity that would reflect energy. This is avoided by tapering each end over a wavelength or so. This provides a gradual, matched transition between normal waveguide and the lossy section of line. Suitable materials are wedges of wood or conductive foam, or card with graphite powder bonded to it. Be warned that the resistivity of foam can vary considerably between types and can be very high. Lower resistivity types are usually needed for attenuators and loads. Home made resistive card can be

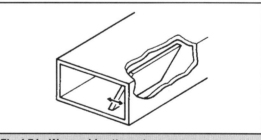

Fig 4.74 : Waveguide attenuator.

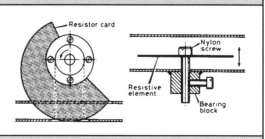

Fig 4.75 : Variable wavegide attenuator.

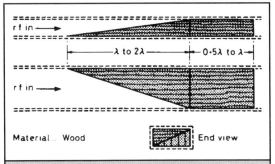

Fig 4.76 : Matched waveguide load.

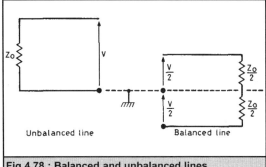

Fig 4.78 : Balanced and unbalanced lines.

prepared with a soft pencil on stiff card, and then sprayed with lacquer. The supports for the absorbing material should, of course, be insulators.

A matched load is an attenuator where all (or virtually all) the power is absorbed in the lossy material before it reaches the end, where there is generally a short circuit. The most common form is a length of tapered wood or lossy ferrite loaded material mounted in the guide with a shorting plate at the far end, as shown in Fig 4.76. The material usually tapers in thickness across both the broad and narrow faces of the guide until it occupies the whole cross section of the guide. This taper is normally done over a distance of several wavelengths, the longer the taper, the better the match. The loss should be greater than half the return loss required of the load, i.e. the loss should be 15 to 20dB.

Sliding shorts

Sliding shorts can be used to form an adjustable cavity and can be a useful item of test equipment. If a microm-eter is fitted to the moving short, it can be used as a simple wavemeter.

The simplest form of short consists of a block which is clamped in place after adjustment. If the block is a good fit in the guide then its length is uncritical. However, it should be about $\lambda_g/4$ in length if not. Fingering, or braiding fitted in a groove in the block, can be used to ensure good contact with the waveguide.

A better form of short includes a choke as shown in Fig

4.77. Here two blocks are used, each $\lambda_g/4$ and spaced $\lambda_g/4$ apart, which are insulated from the waveguide, forming quarter wave chokes. Using this arrangement there will be no intermittent contact between the short and guide, thus making it more reliable.

Baluns

These are used to convert signals between lines that are balanced and unbalanced with respect to earth. Their most common use is to match between a balanced antenna, e.g. a dipole, and an unbalanced feeder, e.g. coaxial cable. They are also used in some components, e.g. a balanced mixer, and may provide an impedance step up or step down.

Examples of unbalanced components are coaxial cable, microstrip, stripline, quarter wave whips on a ground plane, and any transmission line with only one conductor

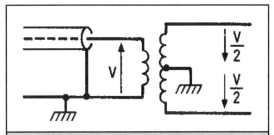

Fig 4.79 : Schematic diagram of a 1:1 balun.

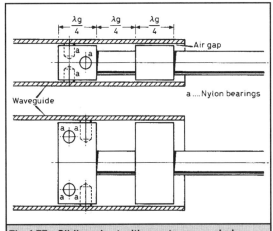

Fig 4.77 : Sliding short with quarter wave chokes.

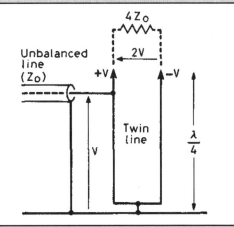

Fig 4.80 : Schematic diagram of a quater wave 4:1 balun.

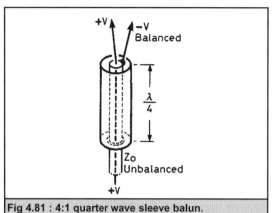

Fig 4.81 : 4:1 quarter wave sleeve balun.

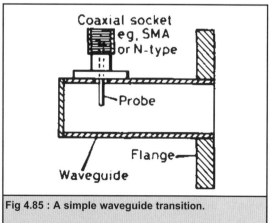

Fig 4.83 : A half wave balun.

Fig 4.84 : Operation of a half wave balun.

that works against a ground place. Balanced components include twin wire transmission line and dipoles.

Fig 4.78 shows a balanced and an unbalanced line which have to be connected together. Each has the same Z_0 so, for the same power to exist on each, the voltages have to match as shown. A 1:1 balun therefore has to convert a single voltage V into two voltages each V/2, an 180° out of phase with each other. These are represented as phasors in the figure. A 1:1 balun is therefore a transformer with a centre tapped output, as shown in Fig 4.79. At hf and vhf quite broadband baluns are possible using transmission line type transformers on ferrite cores, and bandwidths of over a decade are common for small signal devices. At uhf and microwave frequencies the stray reactances make it difficult to use transformers, and resonant baluns are used.

The quarter wave balun consists of a twin wire line with one end short circuited. This end is connected to earth, as shown in Fig 4.80. Due to the symmetry, any voltage which appears across the two ends must be symmetrical about earth. It can be likened to a parallel tuned circuit with an earthed centre tap. The unbalanced input signal is connected between earth and one of the ends of the twin line, and the balanced output is taken from the two ends of the twin line. This balun produces a voltage step up of two, and so an impedance ratio of 4:1.

The Z_0 of the twin line is not critical, but should be of the same order of magnitude as the input line. The ratio between the two affects the bandwidth of the balun.

Practical examples of this type are the sleeve balun, and one made from semi rigid cable.

The sleeve balun is shown in Fig 4.81. In this case the quarter wave stub is made from a coaxial line, not a twin line. The outer metal sleeve is connected to the coaxial cable outer conductor at a quarter wavelength from the end of the cable. This quarter wavelength has to be in the dielectric between the outer metal sleeve and the cables outer conductor. It may be air or the plastic jacket of the cable, but is not related to the velocity factor of the coaxial cable itself.

The other type makes the twin line by cutting slots in the cable outer for the last quarter wavelength, also shown in Fig 4.82. In this case the velocity factor will be somewhere between that of the cable and that of air.

The half wave balun is also very similar. This uses a half wavelength of line to provide the 180° phase shift to generate the other balanced output, as shown in Fig 4.83. This is also a 4:1 impedance ratio. The relevant voltages are shown in Fig 4.84. The load on the far end of the half wave line is half the balanced output load, so the line must have this Z_0, which will be twice that of the input, ie 100Ω in a 50Ω system.

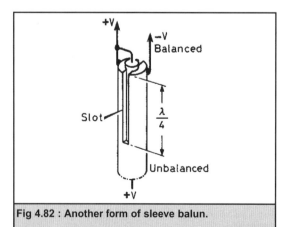

Fig 4.82 : Another form of sleeve balun.

Fig 4.85 : A simple waveguide transition.

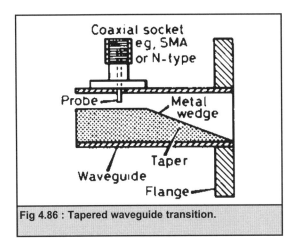

Fig 4.86 : Tapered waveguide transition.

Waveguide transitions

Very often it is necessary to change between waveguide and coaxial line. This can be done quite simply using a transition, which basically consists of a probe from a coaxial socket protruding into a shorted length of waveguide. A simple design is shown in Fig 4.85.

The position of the probe relative to the short, the length of the probe and its diameter determine the operating frequency and the impedance seen looking into the coaxial socket. For a given impedance, which will normally be 50Ω, only certain combinations of these three parameters are possible. Some commercial transitions also have a dielectric surrounding the probe. The bandwidths can be increased by the dielectric sleeve, or using a larger diameter probe.

The simple probe type of transition is usually narrow band but broader band transitions may be made by tapering the guide as shown in Fig 4.86. This reduces the impedance of the guide from its normal value of a few hundreds of ohms to a similar value to that of the coaxial line. A third type uses a T-bar across the guide, as shown in Fig 4.87. These also operate over a wider band than the simple probe.

Rotary joints

These are useful at microwave frequencies because the cables and waveguides used are much less flexible than at hf or vhf. Their main application is to connect fixed equipment to a rotatable antenna.

Coaxial joints may have fingering to maintain a connection at dc, but a moving metallic contact is undesirable as

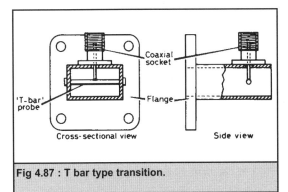

Fig 4.87 : T bar type transition.

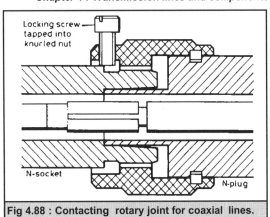

Fig 4.88 : Contacting rotary joint for coaxial lines.

it is likely to generate noise and become unreliable. The best coupling relies on capacitance between the two conductors.

The contact type is illustrated in Fig 4.88. This can be made using an N type plug and socket, and not tightening the outer knurled nut fully; the nut can be locked against the socket body with a grub screw. Although not waterproof, and of limited life, such a joint has carried several hundred watts at uhf while slowly rotating.

At uhf and microwave frequencies it is relatively easy to make a capacitance whose reactance is small compared with 50Ω, either by a simple pin and sleeve, or a more complex set of concentric cylinders with a thin dielectric, rather like a beehive trimmer (see Fig 4.89). If the overlap is a quarter wave long, then the open circuit at the inside end will appear as a low impedance between the lines at the outside end. An alternative for the outer would be two large diameter discs. These quarter wave chokes can provide a very effective connection with low rf leakage.

In circularly polarised waveguide the relative angular position of the two guides does not matter provided the rotational speed is small compared with the frequency (this is not usually a problem!). It is then only necessary to provide enough capacitance between the ends of the

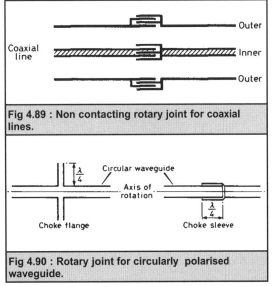

Fig 4.89 : Non contacting rotary joint for coaxial lines.

Fig 4.90 : Rotary joint for circularly polarised waveguide.

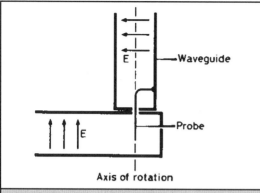

Fig 4.91 : Rotary joints for linearly polarised waveguide.

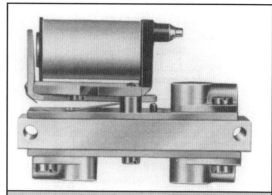

Fig 4.93 : Conventional coaxial changeover switch.

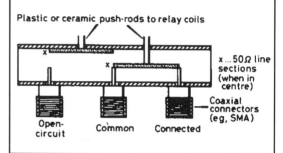

Fig 4.94 : Coaxial relay with improved isolation and match.

guides to reduce the leakage. This can be done with a choke flange, which could be implemented as a sleeve overlap, as shown in Fig 4.90.

In linearly polarised waveguide the fields have to be transferred between the two guides by a means which is independent of angle; one such method commonly used in rectangular waveguide is shown in Fig 4.91. It uses a loop to couple the signal into a coaxial line which then passes through the rotating joint and couples into the new guide as an E field probe, its coupling being independent of angular position. Another method is to make a transition to circular guide, passing through the rotary joint in circular guide, and returning to linear on the other side. A waveguide rotary joint of the latter type is shown in Fig 4.92.

Coaxial relays

Coaxial relays are required primarily for changing over from transmit to receive. On the lower bands this probably means switching the antenna from the power amplifier output to the preamp input, and possibly switching over a transceiver or transverter.

Coaxial relays vary from simple types capable of handling a few watts with acceptable losses up to perhaps 500MHz or 1GHz, to precision high power types that can be used up to 10GHz. The performance is determined by the internal construction and, in addition to the loss and vswr, it is also important what happens during switching.

The simple types do not maintain a true 50Ω characteris-

tic impedance through the relay. Often the relay will work by having a line between the common port and the two other ports, sprung so that it rests against one or the other port when un-energised, and a solenoid is used to move it to the other port when activated. A relay of this type is shown in Fig 4.93. The principal disadvantage of this type is that, when switching, the moveable contact will momentarily lie between the two ports and there will be very poor isolation between them. If one port has a high power amplifier connected to it and the other a low noise preamp, there could be a serious problem!

A particularly good type is shown in Fig 4.94. This type uses two moveable line sections, such that the common port is connected to one or other port with one of these line sections in place. When the relay is energised this

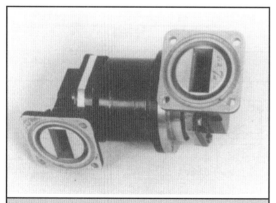

Fig 4.92 : Waveguide rotary joint.

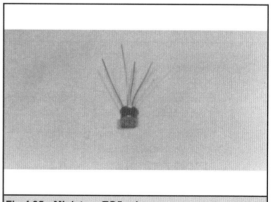

Fig 4.95 : Miniature TO5 relay.

Fig 4.96 : Professional waveguide switch.

line section is moved away and the other section to the other port moved into position. Using this method it is possible to keep the characteristic impedance inside the relay close to 50Ω, and to provide a much higher isolation between ports compared to the simple type. These relays, when properly constructed, are capable of carrying hundreds of watts at microwave frequencies, and have isolations of around 50dB.

Another useful type of relay is the miniature TO5 can type, for use at i.f (see Fig 4.95). These can be used up to about 500MHz and several watts, for example to switch a vhf transceiver on the input to a transverter. Unfortunately they are rather expensive in amateur terms, and their use is mainly restricted to professional equipment.

Waveguide switches

Waveguide switches are also useful for transmit/receive changeover. Commercial switches are usually four port devices as shown in Fig 4.96. The switch consists of an outer block, usually square, with one port in each face. Inside this is a round inner block with two bends cast so that two pairs of adjacent ports are joined together. This inner block is free to rotate so that the ports connected together can be swapped over. Stops are provided to limit the rotation to exactly 90° and ensure exact alignment of the bends with the ports.

With this simple arrangement it is possible for a small amount of power to leak round the gap between the rotating block and outer block.

To overcome this, high performance switches have grooves machined in the circumference of the rotating block. These are a quarter wave deep and act as chokes, and may be filled with a lossy material to improve the isolation. Usually there will be two of these, spaced a quarter wave apart for maximum attenuation, which can be around 70dB.

If waveguide relays are required, the basic switch is fitted with a solenoid operated stepper motor (known as a Ledex). When operating, these relays take large pulses of current and provide sufficient torque to turn the switch round quickly. They can produce high voltage, high energy transients which can damage power supplies or other equipment, and should be suppressed with diodes or voltage dependent resistors (vdrs).

This type of switch is not easy for the amateur to build, and an alternative is shown in Fig 4.97. Here a flapper is used to switch one common waveguide port between two other ports. This type will not have a very good isolation, but this is not important for many applications.

Phase shifters

These are used to add a variable phase shift to a signal. They operate by introducing dielectric into the guide, which reduces the velocity and makes the waveguide appear longer. The delay is usually variable over the range 0 to 360°.

The construction is similar to that of waveguide attenuators, as shown in Fig 4.98. A thin strip of dielectric is mounted on rods so that it can be moved from the edge into the centre of the guide. Its effect is negligible when against the conductive wall, and maximum when in the middle of the guide in the maximum part of the E field. The ends are usually tapered to minimise reflections from the discontinuity.

Splitters and combiners

These devices are used to divide power from one source amongst a number of other loads. A two way splitter will be used for the purposes of illustration, but it could apply to any number of ports.

A number of types of splitter are used, differing in the insertion loss and isolation they provide between each of the ports.

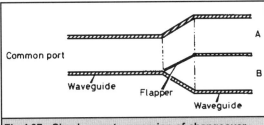

Fig 4.97 : Simpler amateur version of changeover switch.

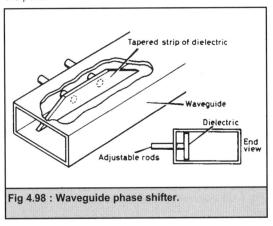

Fig 4.98 : Waveguide phase shifter.

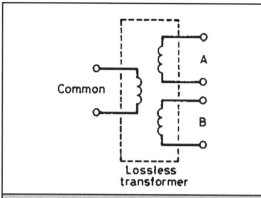

Fig 4.99 : 3dB splitter using a transformer.

An ideal, lossless, two way splitter would split the signal into two parts, with a loss of 3dB into each port. No power would be lost, as it would all end up at one port or the other. Such a splitter could be built from a transformer, as shown in Fig 4.99. However, there would be little isolation between the ports and, if one of the loads had a poor vwsr, it would alter the amount of power reaching the other load. Hybrid splitters have been developed to give a minimum loss and good isolation between the ports.

The losses of matched resistive and hybrid splitters are compared in Table 4.5.

Resistive splitters

Several types of resistive splitters are used, shown in Fig 4.100. Resistor values for the two resistor splitter are given by:

$$R = (N-1) \times Z_0$$
$$Loss(dB) = 20\log(N)$$

where N is the number of output ports.

While the input presents a good match with both loads connected, the impedance looking into each output port is not a good match. This type is only meant for use as an output to a detector in a levelling loop, where the loop will affect the output impedances anyway by varying the generator output level.

Resistor values for the three resistor combiner can be calculated from the expression:

$$R = Z_0 \times \frac{(N-1)}{(N+1)}$$

Table 4.5 : Losses of matched resistive and hybrid splitters.

No of ways	Loss resistive (dB)	Loss hybrid (dB)	Excess loss for resistive combiner (dB)
2	6.02	3.01	3.01
3	9.54	4.77	4.77
4	12.04	6.02	6.02
6	15.56	7.78	7.78
8	18.06	9.03	9.03

Insertion loss is given by:

$$Loss(dB) = 20\log(N)$$

This type gives a good vswr at all ports when the other ports are matched. It can also be built as a delta instead of a star network. This type of splitter must be used if a number of pieces of equipment are to be connected together, so that they all see each other equally.

Hybrid splitters

If you want to ensure that each port always presents a matched load when the load on the other port varies you must use a hybrid splitter. This has a 3dB insertion loss and theoretically infinite isolation between the output ports. If one load is mismatched, this does not alter either the impedance seen by the source, or the power reaching the other load. In practice, the isolation is around 30dB.

The basic device is shown in Fig 4.101. It has two inputs and two outputs. The outputs are the sum and difference of the two input ports. In normal operation the difference port is connected to a 50Ω load.

As a splitter, it divides the input power equally between the two output ports, with no other losses. If one port is mismatched, this does not affect the power appearing at the other port. The imbalance is concealed by the fact that some power is now being dissipated in the internal termination. When the device is used as a combiner, the signal from each input is split equally between the internal load and the output. Thus there is a loss of 3dB from each input to the output.

Hybrid splitters can be built in either coaxial or waveguide circuitry. In coaxial circuitry the splitter is made using transformers to generate the in phase and out of phase signals.

In waveguide the hybrid splitter is known as a magic tee or hybrid tee, which probably gives some idea that its operation was not well understood. It has four ports (Fig 4.102), all of which are accessible. In normal operation a

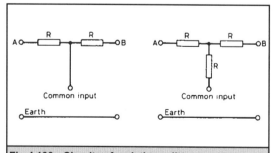

Fig 4.100 : Circuits of resistive splitters.

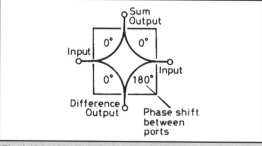

Fig 4.101 : Circuit of a hybrid splitter.

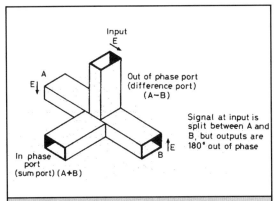

Fig 4.102 : The magic tee.

load is placed on the sum or difference port, depending whether an in phase or 180° splitter is required. The operation is shown in Fig 4.103, where the splitting of power is represented by E field vectors.

Directional couplers

A directional coupler consists of two transmission lines which can interact electrically so that a proportion of the power flowing in one line is fed to the second in one direction only, see Fig 4.104. If power is supplied to port A, then a fraction appears at port D and the remainder at port B. In practice some power will also appear at port C. The coupling factor, which is the loss between the power in the main line and than in the coupled line, is defined as:

$$\text{Coupling factor (dB)} = 10 \times \log\left(\frac{P_a}{P_d}\right)$$

Typical values range from 3 to 60dB. With a 3dB coupling

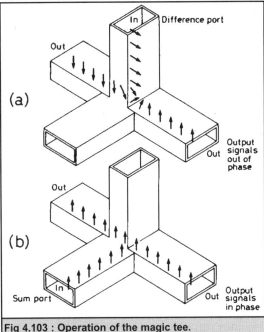

Fig 4.103 : Operation of the magic tee.

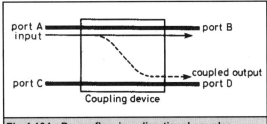

Fig 4.104 : Power flow in a directional coupler.

factor there will also be a 3dB loss from port A to port B. As the coupling is reduced, this loss will also decrease. A graph of insertion loss versus coupling is shown in Fig 4.105.

The directivity is defined as the ratio of the powers at each end of the coupled line, when the main line is correctly terminated:

$$Directivity(dB) = 10 \times \log\left(\frac{P_d}{P_c}\right)$$

In a well designed coupler the directivity will be typically 30dB. Directional couplers can be constructed from a variety of transmission lines and some common types are described below.

Coaxial couplers

Coaxial couplers can be made by placing two coaxial lines side by side with a part of the outer of each removed. An example is shown in Fig 4.106, built from two semi rigid lines soldered together.

The slots, which are less than a quarter wave long, allow the field to leak out into the coupled line and reproduce the forward and reflected waves in the coupled line. A high directivity can be obtained, provided the coupling is not too great, and lines of equal characteristic impedance are used. This means in practice a maximum coupling of around 30dB, otherwise the slots become too wide and begin to alter the impedance of the lines.

Stripline and microstrip couplers

Stripline and microstrip couplers may be made in a similar manner to the simple coaxial line coupler described above. An example is shown in Fig 4.107(a), where it is

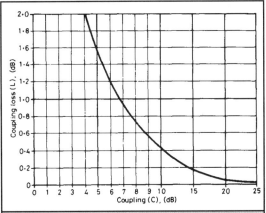

Fig 4.105 : Insertion loss versus coupling coefficient.

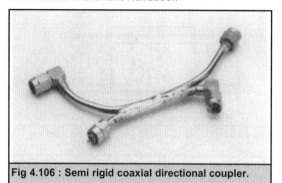

Fig 4.106 : Semi rigid coaxial directional coupler.

arranged for two sections of microstrip to run close to each other to allow coupling to take place. Coupling factors up to 10dB can be readily obtained. With stripline couplers, higher coupling factors of up to 3dB can be readily achieved.

An alternative method of coupler construction is shown in Fig 4.107(b). Here two quarter wave lengths of line join the main transmission lines together. The operation is as follows:

Consider a signal entering port A. A portion of it will pass to port D via the first interconnecting branch, and a similar portion via the second branch. The two signals will thus arrive at port D in phase and reinforce each other. Next consider the signals arriving at port C. A portion from port A will pass to port C via the first branch. A similar portion will arrive via the second branch, but it will have travelled an extra half wavelength and be of opposite phase to the first portion. The signals tend to cancel, hardly any signal leaves port C, and thus the coupler is directional.

The directivity is limited by the fact that the amplitudes of the signals travelling by the two different routes will not be quite equal. This problem is more serious for closer values of coupling. To overcome this, more complicated structures are used, with more inter connecting branches, and errors in construction then become more important.

Waveguide couplers

The transfer of energy between two waveguides may take place at openings made into the common wall of the two guides. These may be run parallel to each other for a sufficient distance, either on top of each other (broad wall coupling) or side by side (side wall coupling), as shown in Fig 4.108. Alternatively they may simply cross, in which case the coupler is known as a cross coupler, as shown in Fig 4.109. For low coupling coefficients, round holes may be used. However, for closer coupling (10dB or closer), the round holes would have to be so big that they would merge into each other and so cross-shaped holes are used.

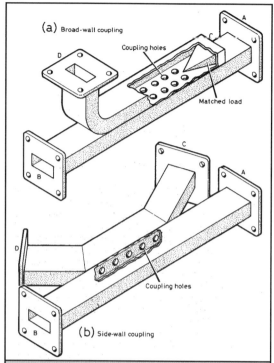

Fig 4.108 : Waveguide directional coupler.

To understand how these couplers operate, consider the coupler shown on Fig 4.110. Two holes are spaced $\lambda_g/4$ apart connecting the two guides. The size of these two holes determines the coupling. Waves moving from port A pass through both holes and arrive at port D in phase, regardless of the distance between the two holes, since they both travel the same distance. The two sets of waves thus reinforce each other. However, the two sets of waves arriving at port C travel distances which differ by twice the distance between the coupling holes. If the spacing is made $\lambda_g/4$, the path difference will be $\lambda_g/2$ and consequently the waves will cancel.

Since this distance depends directly on λ_g, the cancellation in a given design (i.e. the directivity) will change somewhat if the coupler is used at another frequency. In most applications this is not an important factor, port C will often be terminated in a matched load. The degree of

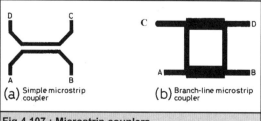

(a) Simple microstrip coupler

(b) Branch-line microstrip coupler

Fig 4.107 : Microstrip couplers.

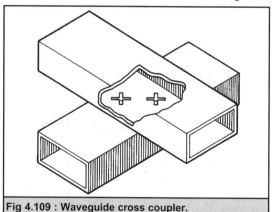

Fig 4.109 : Waveguide cross coupler.

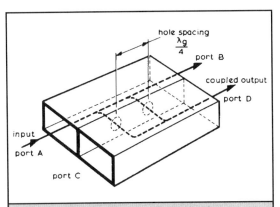

Fig 4.110 : Operation of a two hole coupler.

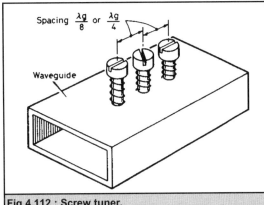

Fig 4.112 : Screw tuner.

coupling will also change, which may be a problem when using a coupler over a wide frequency range.

The coupling is determined by the size and positioning of the common openings and may be anything from 3dB (i.e. half power) down to 60dB. A very large directivity (up to 50 or 60dB) may be obtained by using a large number of holes, although the structure then becomes very long and very accurate manufacture is needed.

Tuners

Several types of tuner are used to provide adjustable matching between components. They are particularly useful as items of test gear as they allow minor adjustment to matching between pieces of equipment to optimise its performance, without disturbing adjustments within the equipment.

Two examples of this use are in aligning low noise preamplifiers and high power amplifiers. The noise figure of low noise preamps is very dependent on the input matching. The tuner allows you to alter this matching without disturbing the matching networks inside the preamp. Many high power valve power amplifiers have a limited range of adjustment on the p.a. loading, due to the difficulty of making adjustable, very low loss components in circuitry handling very high voltages. A tuner allows you to experiment with the output loading so that you can determine what changes need to be made to the p.a. matching network.

There are three types of tuner: the screw, stub and slug tuners. The screw and stub tuners use shunt reactances across the transmission line; the screw tuner is used in waveguide and the stub in coaxial line. The slug tuner uses lumps of dielectric to form moveable quarter wave transformers.

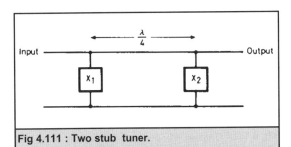

Fig 4.111 : Two stub tuner.

In theory it should be possible to match between any two impedances by using two shunt reactances spaced $\lambda/4$ apart on a transmission line. This can be likened to the π-network matching circuit. An example of such matching is shown in Fig 4.111. In practice the components producing the reactances are not lossless, and so have a resistive component as well as a reactive one. It may also be difficult to obtain the complete range of reactances. To ease the stringent requirements on their performance, three reactances are usually used instead of two, the extra one being put midway between the outer two, the net result being three, each spaced by $\lambda/8$.

A slightly different approach is to insert two sections of line with a different characteristic impedance in series with the main line. These operate as quarter wave transformers. This technique is normally used in coaxial lines, but could also be applied to waveguide.

Screw tuners

These are used in waveguide, since a convenient way of producing a variable reactance is to use a metal rod protruding from one of the broad walls of the guide part of the way across to the other (Fig 4.112). This is easily achieved by drilling and tapping a hole in the broad face of the waveguide wall and inserting a screw into the guide. During experimental work, the screws may be spring loaded to eliminate play, but after final adjustment the screws should be fixed using a lock nut.

The reactance as a function of penetration into the guide and the screw diameter is shown in Fig 4.113. For small penetrations the reactance is capacitative, since it is effectively bringing the two walls closer together.

At a certain length, depending on the screw diameter, the screw becomes a series tuned circuit across the guide, i.e. a reactance of zero ohms. The screw can then be thought of as a quarter wave whip or stub. As it protrudes further, it becomes inductive.

In the case of the three screw tuner, usually only one or two screws will have a significant effect. The screws should be inserted the minimum amount necessary to achieve a good match. If screws have to be inserted well into the waveguide, then the component is probably a bad match; if all three screws have little effect, then the component is probably well matched.

The screw tuner in waveguide is an excellent example of why waveguide construction appeals to the amateur constructor. By using cheap, simple components, e.g.

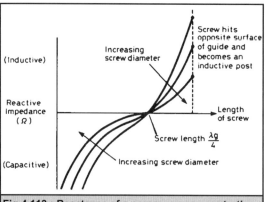

Fig 4.113 : Reactance of a screw versus penetration into guide.

brass screws and nuts, you can build a cheap, versatile, low loss matching network which is very easy to adjust. While microstrip is easier to mass produce, it is less easy to adjust. Adjustments can be made, however, by sliding small metal tabs or discs along the line to provide extra capacitance, and then soldering or silver epoxying them in place.

Stub tuners

In coaxial lines, the most convenient form of variable reactance is the sliding short. The variable reactance consists of a piece of rigid transmission line (stub) with a sliding short circuit which can be moved along the line so that its length can be varied from zero to a quarter wave. The impedance at the open end of this line varies from zero to infinity ohms of inductive reactance. If the line can be made longer than a quarter wave, then the reactance becomes capacitive (Fig 4.114).

Slug tuners

The operation of the slug tuner can best be described in terms of the familiar quarter wave transformer which uses a quarter wavelength section of line to match between two resistive impedances. A load impedance Z_l can be matched to a line whose impedance is Z_o by using a quarter wavelength of line whose impedance is $\sqrt{(Z_o \times Z_l)}$, i.e. the geometric mean of the two.

This is illustrated in Fig 4.115(a). The impedance that would be seen looking towards the load at any point along the cable is shown in Fig 4.115(b). The dotted lines show the characteristic impedance of the quarter wave line segments. A logarithmic resistance scale is used as it illustrates the symmetry of the impedance transformations

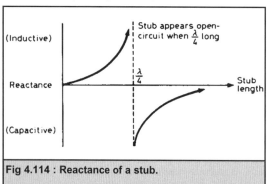

Fig 4.114 : Reactance of a stub.

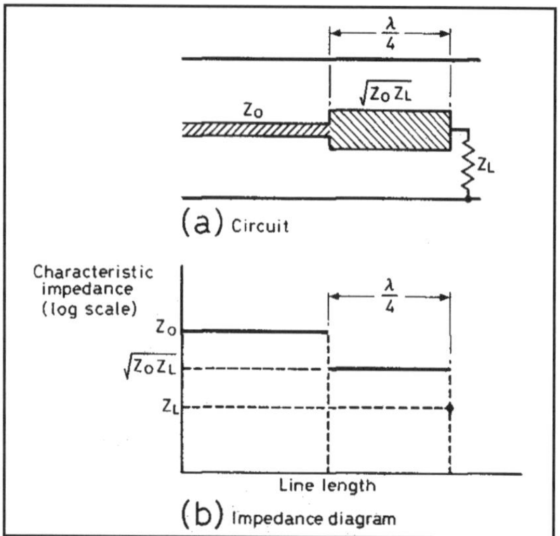

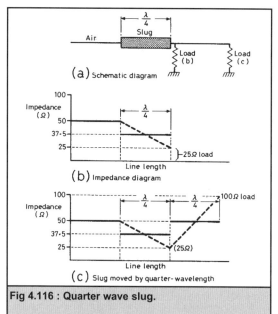

Fig 4.115 : Quarter wave transformer.

about the line characteristic impedance, a given vertical distance represents a fixed ratio of resistances.

The impedance of a line is determined by the geometry of its conductors and the dielectric constant of the material between them. In a 50Ω air line, a section (slug) with a ptfe dielectric constant of approximately 2 will have a characteristic impedance of $50/\sqrt{2}$, ie 37.5Ω. Referring to Fig 4.116(a), a slug that is quarter wave long will transform a 25Ω resistive load up to a 50Ω resistive impedance. This is shown in Fig 4.116(b) for a 50Ω line. If a quarter wave length of the original 50Ω air line is inserted between the load and the slug, this will transform between 25Ω and 100Ω, as shown in Fig 4.116(c). The slug can thus generate the two extremes of a 2:1 vswr, namely 25Ω and 100Ω. By varying the length of air line between the load and the slug from zero to one quarter wavelength, a 2:1 load vswr of any phase can be transformed to 50Ω resistive.

Fig 4.116 : Quarter wave slug.

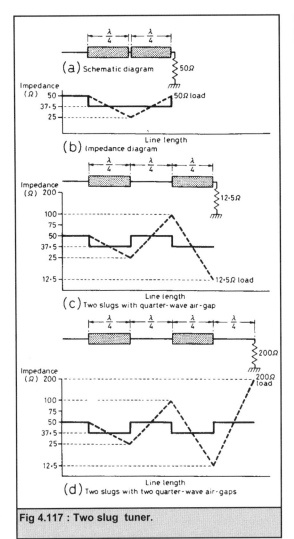

Fig 4.117 : Two slug tuner.

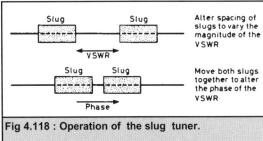

Fig 4.118 : Operation of the slug tuner.

residual mismatches, i.e. between pieces of equipment or between the transmitter and the antenna feeder.

The range of dielectric materials available limits the matching range than can be covered by a single tuner. However, it is possible to make slugs with a larger effective dielectric constant by building them from concentric cylinders of metal and dielectric.

The slug tuner can be used to provide independent adjustments of phase and magnitude of an swr as shown in Fig 4.118. By terminating one end of the tuner with a load equal to the characteristic impedance of the coaxial line used to make the tuner, and looking in the other, the magnitude of the swr can be varied from 1:1 to 4:1 by moving the two slugs apart symmetrically about their midpoint. By moving both of them equally in the same direction, the phase of the swr can be varied.

One advantage of a dielectric tuner is that there are no sliding metallic contacts which can arc or produce unpredictable results if they are intermittent.

Resonators

Resonators are important microwave components used in filters and in other applications such as narrowband amplifiers. Whereas resonant circuits are usually made from coils and capacitors at hf and vhf, at microwave frequencies these are formed from sections of transmission line or resonant cavities.

Transmission line resonators

The simplest type of transmission line resonator is a quarter wave line shorted at one end. At the other end, this short will appear as a high impedance at the resonant frequency. More commonly used is a length of line shorter than a quarter wave, tuned to resonance by an end capacitor (see Fig 4.119).

The impedance of such a line at the open end is given by:

$$Z_{in} = Z_0 \times \tanh(\gamma \times 1)$$

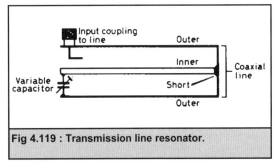

Fig 4.119 : Transmission line resonator.

The snag with the single slug tuner is that the magnitude of the vswr cannot be varied, a single ptfe slug can only match loads whose vswr is located actually on the 2:1 circle on a Smith chart. This can be overcome by using two slugs.

Consider the various configurations that can be used. If two quarter wave slugs are placed adjacent to each other in the line, there is no net impedance transformation (Figs 4.117(a) and 4.117(b)). This is equivalent to inserting a half wavelength of line of different impedance in a system, which has no effect.

Next consider what happens when the slugs are moved so that they are one quarter wavelength apart. There are now three sections of line, each one terminated in a mismatch (Fig 4.117(c)). This represents the greatest possible transformation, of 50Ω to 12.5Ω. This is twice that which can be obtained from one slug, i.e. 4:1. Again if a further quarter wave of air line is inserted between the load and the slug, the 12.5Ω is transformed to 200Ω (Fig 4.117(d). Thus the two slug tuner can transform from 50Ω to any impedance whose vswr is between 1:1 and 4:1, at any phase.

This should be more than adequate for mopping up small

where Z_0 is the characteristic impedance of the line, γ is the propagation constant, and l the line length.

For a lossless line, γ is simple.

$$\gamma = \frac{j \times 2\pi}{\lambda}$$

Thus, the impedance simplifies to :

$$Z_{in} = Z_0 \times \tanh\left(j \times 2\pi \times \frac{l}{\lambda} \right)$$

or :

$$Z_{in} = j \times Z_0 \times \tan\left(\frac{2\pi \times l}{\lambda} \right)$$

For a short line (less that $\lambda/4$), the input impedance of the line is inductive, signified by the j. When I is exactly a quarter wave long, the input impedance is very high, the above expression gives a Z_{in} of infinity, but losses in any any practical transmission line reduce the impedance to finite value. Above $\lambda/4$, the input impedance appears as a capacitance, signified by a -j. The impedance is plotted in Fig 4.120.

Note that when the line is filled with a material of dielectric constant ξ_r, the wavelength will be reduced:

$$\lambda = \frac{\lambda_{fs}}{\sqrt{\xi_r}}$$

An inductive line can be made to resonate at a frequency F by placing a capacitance C in parallel with it :

where :

$$C = \frac{1}{(2\pi \times F)^2 \times L}$$

or :

$$C = \frac{1}{2\pi \times F \times Z_{in}}$$

Transmission line resonators may be constructed from coaxial lines, striplines and microstrip. For example, consider an air spaced 50Ω coaxial line, 5cm long, at 1.3GHz, which is a little under a quarter wave long :

$$Z_{in} = j \times 50 \times \tan\left(\frac{2 \times 3.142 \times 5}{23} \right) = j \times 50 \times \tan(1.366)$$

i.e. Zin = j x 240.7 ohms. To resonate at 1.3GHz, the required C is :

$$C = \frac{1}{2 \times 3.142 \times (1.3 \times 10^{-9}) \times 240.7} = 0.51 pf$$

If the above line were now filled with ptfe dielectric, where ξ_r is 2.1, the input impedance will be :

$$Z_{in} = j \times 50 \times \tan(1.980)$$

i.e. :

$$Z_{in} = -j \times 115.3\Omega$$

Thus, the input impedance is now a capacitance, whereas before it was inductive.

With microstrip the velocity will be reduced below that of free space, due to the dielectric, but the amount it is reduced, and hence the wavelength, is not so easily calculated. In this case, the graphs in Fig 4.30 should be used to determine the effective velocity, and hence the effective value of λ.

As already stated, the quarter wave line shorted at one end is the simplest form of transmission line resonator, and it is normally made slightly shorter and tuned onto frequency with a small capacitance. This type of resonator will have higher resonances, the first being when it is three quarter wavelengths long, and these may be used by design.

It is also possible to use an open circuited length of line as a resonator. A short length will look like a capacitance but, as the length is increased beyond a quarter wavelength, it will also look like an inductor. The input impedance of such a line is given by:

$$Z_{in} = -j \times Z_0 \times \cot\left(\frac{2\pi \times L}{\lambda} \right)$$

Calculation of Z_{in}, and hence the required capacitance for the line to resonate at a particular frequency, is similar to the calculation for the shorted line. There is a simple and very useful relation between the loss of a transmission line and the Q that it will give when used as a resonator, which can be used to estimate the Q that can be obtained from the lines available. Alternatively, the loss of a cable can be predicted by making a resonator out of a short length and measuring the Q.

$$Q = \frac{0.091 Fx}{kL}$$

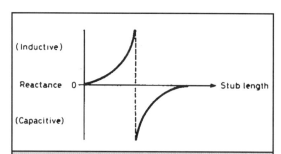

Fig 4.120 : Impedance of a shorted stub.

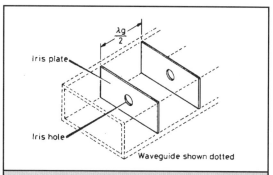

Fig 4.121 : Waveguide cavity resonator.

where F is the frequency in megahertz, k the velocity factor of the line, and L the loss in decibels of a cable of length x metres.

As a rule of thumb, the resistance across the open circuit end of a shorted quarter wave line is Q x Z_0, and across the open circuit end of a shorted half wave line is Z_0/Q.

Cavity resonators

Cavity resonators may be formed from a variety of hollow structures, including rectangular boxes, cylinders and spheres. The most common are rectangular cavities and, in particular, waveguide cavities.

A cavity can be formed in waveguide by partitioning off a portion of the guide with iris plates. One of both of these plates will have a hold (or iris) in it to provide coupling to the cavity (see Fig 4.121). However, it is not necessary to actually partition the waveguide in this manner and alternative methods of defining the cavity may be used, e.g. posts across the waveguide.

The lowest frequency of resonance of a cavity formed in this manner is when the wavelength (in waveguide) is twice its length. It is usual to make the length slightly small and tune the cavity onto frequency using a capacitive tuning screw protruding into its centre. There will be other resonance on high frequencies.

For example, a half wave cavity resonator is required at 10GHz, built in WG16 waveguide. For WG16, λ_c= 45.75mm. At 10GHz, λ= 30mm.

$$\lambda_g = \frac{30}{\sqrt{\left(1-\left(\frac{30}{45.75}\right)^2\right)}}$$

i.e. λ_g = 39.7mm. Thus a half wave cavity would be 19.8mm long, and a suggested length to allow for tuning would be 18mm long.

When a filter is constructed from such cavities coupled together the loading of the cavity due to coupling in and

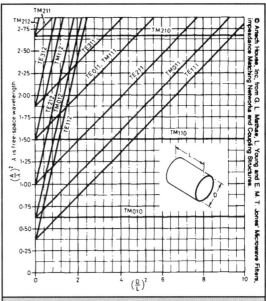

Fig 4.123 : Mode chart for right circular cavities.

out will affect the resonant frequency slightly, and this is compensated for by shortening the length slightly.

The cavity type just described uses the dominant mode of rectangular waveguide. Many resonators make use of higher order modes, usually to get a higher Q. The most common type is either a cylindrical or rectangular box. Figs 4.122 and 4.123 show the frequency ranges of various modes in such cavities. The mode is described in a similar manner to modes in waveguide, except that a third suffix is added to specify the number of field variations along the length of the cavity. For example, the dominant mode in a rectangular waveguide is TE-10, giving a TE-101 mode in a cavity.

Higher order modes are often used for wavemeter cavities where a high Q is important. The snag with these modes is that undesired modes may have resonances in the frequency range of interest. The mode chart is used to select a suitable mode to cover the frequency range so that this does not occur.

Dielectric resonators

Dielectric resonators consist of a ceramic resonator with a high dielectric constant (e.g. 30), usually cylindrical in shape, and resonating at the frequency of interest. They are essentially cavity resonators. Coupling into and out of the resonator is achieved by running a microstrip line next to it. At 10GHz, a typical size is 6mm diameter by 2.5mm high.

They are mainly used for stabilising GaAsfet oscillators. The cylinder is simply placed next to the fet oscillator, which is usually built on a ptfe or ceramic based sub-strate. Fine tuning of the resonant frequency is possible by moving a tuning screw near to one face of the cylinder.

YIG resonators

The yig resonator is constructed from an yttrium iron garnet (yig) sphere which is tuned by a magnetic field. Again, it is a form of cavity resonator.

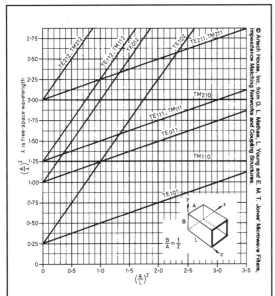

Fig 4.122 : Mode chart for rectangular cavities.

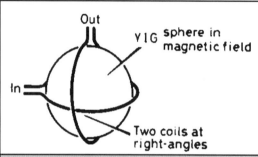

Fig 4.124 : YIG resonator.

The sphere is cut from yig crystals and is typically 0.3 to 1mm in diameter in order to resonate at the frequency of interest. It is placed in an electromagnetic field that is altered to tune the resonant frequency (see Fig 4.125). RF is coupled in and out of the sphere by two coupling loops at right angles. There is normally no coupling between the two loops except when the sphere provides it at resonance. The tuning is very linear, e.g. it is possible to tune the whole of X band (from 8 to 12GHz) with a worst error of only 10MHz.

The yig resonator may be used as a filter or more frequently as part of a GaAsfet oscillator. In this latter case the fet is placed near the yig sphere so that power is coupled into it before being coupled into the output circuitry. The yig is used to set the frequency of oscillation and a broadband oscillator can be produced with highly linear tuning. Frequency modulation can be applied to the signal, usually by using a second, smaller tuning coil which permits faster modulation rates than the main tuning coil.

Non reciprocal components

Several forms of non reciprocal devices exist which rely on the special properties of ferrites for their operation. Ferrites are magnetic oxides of iron usually doped with other metals such as manganese, nickel and rare earth elements. Depending on their exact composition, they can present high permeability and very high resistivity, and may thus be almost transparent to microwave. Conversely, other compositions can be made almost totally absorbent to microwaves.

The magnetic properties of ferrites are related to the

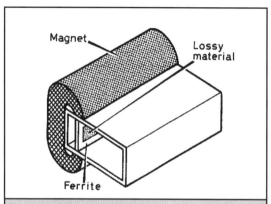

Fig 4.125 : Waveguide isolator.

existence of unpaired spinning electrons in the structure and, if the ferrite is placed in an externally generated magnetic field, the atoms within the structure of the ferrite will align themselves along the induced internal field, giving a net(volume effect) magnetisation. Magnetisation of the ferrite increases with the applied field until saturation occurs, when this point is reached, no further increase in induced magnetisation occurs with the increasing external field.

The saturation magnetisation determines the low frequency limit of the ferrite, and this expressed in megahertz approximates to the saturation magnetisation expressed in gauss. Such a saturated ferrite exhibits different permeabilities in opposite directions of propagation when magnetised in the transverse plane. If the ferrite is mounted within waveguide and suitably magnetically biased by means of an external permanent magnet, then the external magnet effectively controls propagation within the waveguide and propagation becomes non reciprocal. This property is exploited in both the isolator and the circulator.

Isolators

An isolator is a two port device in which the transmission loss is very low in one direction and high in the reverse direction. As its name implies, it is used to isolate one circuit from another. It is frequently used to protect oscillators, so that power generated by the oscillator can pass unimpeded to the external circuitry, but any power reflected by the circuit is heavily absorbed and therefore does not influence the oscillators frequency. The operation of an isolator depends on the special magnetic properties of ferrites described above.

Three types of isolator are available and are known respectively as the Faraday rotation isolator, the resonance isolator and the field displacement isolator. Each has its particular advantages and disadvantages from a professional point of view, although it matters little to the amateur which type is used.

The Faraday rotation isolator exhibits low insertion loss but operates over a comparatively narrow bandwidth of 5 to 10 per cent. The resonance isolator has low insertion loss and large bandwidth but requires a very high magnetic bias field strength. This can make it susceptible to malfunction if the biassing magnet is damaged by loss of strength caused by dropping, or exposure to other strong magnetic fields. The field displacement type offers low insertion loss and good isolation and is very compact. However, it cannot be used at high power levels, although this is unlikely to be a factor in most current amateur equipment. This type is probably the most useful on account of its small size and light weight.

A common form of waveguide isolator consists of a length of waveguide inside which is a ferrite bar (see Fig 4.125). This is biased by the field from a permanent magnet outside the guide. The rf fields corresponding to forward and reverse waves passing through the waveguide are distorted by the ferrite in different ways. A lossy material is placed within the waveguide at a point where the field due to the forward wave is zero, and where that due to the reverse wave is at a maximum. The forward wave therefore passes through with little attenuation, usually with a loss of 0.5 to 1dB, while the reverse wave is heavily attenuated, generally by 20dB or 30dB. Fig 4.126 shows a commercial waveguide isolator.

Fig 4.126 : Waveguide isolator.

Fig 4.128 : Waveguide circulator.

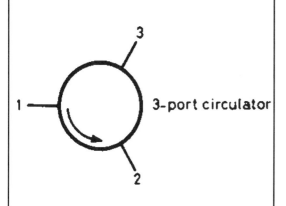

Fig 4.127 : Schematic of an isolator.

Coaxial isolators are also available, but these are really circulators with one port terminated with a matched load.

Although isolators can greatly simplify the construction and operation of equipment, there are a number of precautions which should be taken in their use. Clearly they must be connected the right way round. Their performance can be permanently degraded by a change magnetic properties of the biasing magnet caused by large mechanical shocks, or by allowing the isolator near steel components or strong magnetic fields. They provide isolation over a restricted range of frequencies, in some cases as low as 5 per cent bandwidth, and harmonics may be only poorly attenuated.

Circulators

A circulator is a non reciprocal device with three or more ports. It contains a core of ferrite material in which energy introduced into one port is transferred to an adjacent port, the other ports being isolated. This is illustrated in Fig 4.127 for a three port circulator.

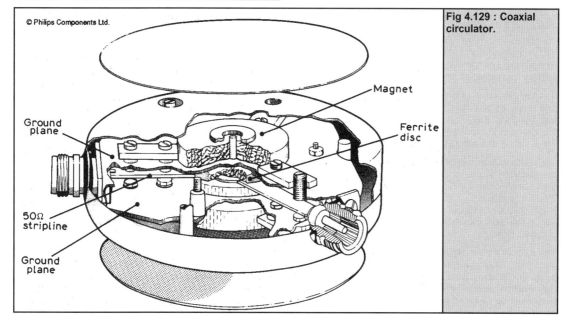

Fig 4.129 : Coaxial circulator.

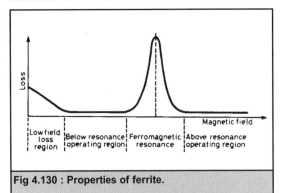

Fig 4.130 : Properties of ferrite.

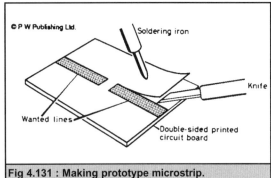

Fig 4.131 : Making prototype microstrip.

Energy entering into port 1 emerges from port 2 and energy entering into port 2 emerges from port 3 etc. In this direction of circulation the loss is typically 0.5 or 1db. In the reverse direction, the isolation is 20 or 30dB. Circulators may be constructed in either waveguide or coaxial line.

In the waveguide type, three (for a three-port device) waveguides usually intersect each other at 120°, as shown in Fig 4.128. Exactly in the centre of the intersection is a piece of ferrite, located between two magnets. Posts or tuning screws may also be fitted in the In the waveguide type, three (for the three port device) waveguides usually intersect each other at 120° as shown in Fig 4.128. Exactly at the intersection is a piece of ferrite, located between the two magnets. Posts or tuning screws may also be fitted in the waveguide branches to achieve a good match.

A coaxial circulator is illustrated in Fig 4.129. In this type three copper strips intersect each other at 120° in the centre of the circulator. These are mounted between two earth plates, thus forming striplines. In the centre of the circulator two ferrite discs are mounted, one above and one below the copper strips. These are then sandwiched between two magnets. Tuning screws may be provided for fine adjustment of the centre frequency.

When the signal enters the ferrite disk, it splits into two parts, one propagating in a clockwise and the other in an anticlockwise direction around the disc. The velocities in the two directions re different. The circumference of the disc is chosen so that when the signals meet at one of the other ports on the circumference they either add or subtract, giving either the coupled or isolated ports.

It is the presence of the external biasing magnetic field that produces a difference between the velocities in the two directions. Only certain values of field can be used if the losses in the ferrite are to be kept small, as shown in Fig 4.130. These are just below or above resonance.

The resonance occurs because the magnetic field causes the electrons in the ferrite to precess. Operation can be above resonance for frequencies from about 50MHz to 2.5GHz, and below resonance from 500MHz to around 24GHz. Above resonance operation gives a bandwidth of about half an octave, and below resonance operation about an octave or more.

The main use for a circulator is in duplexing, i.e. connecting a transmitter and receiver to the same antenna. Also, if one port is terminated with a matched load, the device may then be used as an isolator. Again, as with isolators, care must be taken to keep circulators away from other magnetic fields. They can be optimised to work over a narrow bandwidth or be made to cover a wider frequency range with a reduced performance.

Construction techniques

Microstrip

Professionally this is produced by photographic methods, as for normal pcbs.

However, this technique is rather expensive for amateur development work, and it is possible to obtain adequate results with rather more barbaric techniques, particularly if the circuit will need adjustment anyway.

Lines can be produced by much simpler techniques such as using black tape or ink directly on the board to cover the resist. Alternatively the track pattern can be cut out with a sharp knife, and the unwanted copper removed with a knife and hot soldering iron (Fig 4.131).

Fine tuning can be performed by sticking self adhesive copper foil on the track to form a stub; it can be bent up to adjust the capacitance to earth. Also small copper discs can be slid along the line to add capacitance at various places and soldered in position.

Do not be put off by the very high precision used in professional artwork. Most amateur circuitry will not require such critical tolerances and will often need optimising for individual devices anyway.

PCB materials

At lower frequencies glass loaded epoxy pcb is used, which has a dielectric constant of approximately 4.

At microwave frequencies this material is quite lossy and alternatives such as ptfe and ceramic-loaded types can be used. The ptfe types are available under several trade names such as RT-Duroid and CuClad, and have dielectric constants ranging from about 2.1 to 2.5. The disadvantage of these types is the cost, typically 10 times that of glass loaded epoxy types.

The various types are available in a range of thicknesses, from a fraction of a millimetre to several millimetres, with copper coatings on one or both sides. Common thicknesses are 0.01in (0.25mm), 1/32in (0.79mm) and 1/16th (1.59mm). The thickness of the copper coating(s) is normally specified in ounces per square foot, and 1 or 2oz/ft² is the most common. In addition, some types can be obtained with a uv-sensitive coating applied to the copper, ready for exposure, developing and etching.

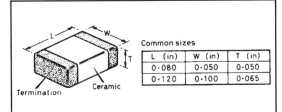

Fig 4.132 : Chip resistors and capacitors.

Screening

This is particularly important in the case of cavity resonators in high power amplifiers. Lids to such cavities should be earthed to the box every λ/10 or so by means of screws, which should be done up tightly. It is very difficult to maintain a good electrical contact along the length of a lid by any other means. Leads running inside such a cavity, e.g. power supplies to a valve, should be screened and the screens earthed at frequent intervals. Similar comments apply to the lids of filters and low noise amplifiers using high Q resonators. Otherwise, unwanted coupling can occur between resonators, altering the response. De-tuning may also occur as an intermittent contact between the lid and box changes.

Stray impedances

At microwave frequencies, the dimensions of many components are comparable with the wavelength and their size becomes significant. The inductance of leads can appear as a high impedance in series with the component and a quite small stray capacitance may appear as a low shunt impedance.

The internal construction of components is also important. For example, wirewound resistors are unsuitable even at vhf. At uhf, carbon composition or metal oxide types can be used, provided the lead lengths are kept to a minimum. At even high frequencies, e.g. 10GHz, it will be necessary to use unpackaged chip resistors.

Unpackaged chip resistors and chip capacitors are readily available for use at microwaves (see Fig 4.132). These include high Q types for better performance, although these are more expensive. Unpackaged semiconductor devises (diodes, transistors and fets) are used commercially but are difficult for the amateur to use and not easily obtained.

Radiation coupling and losses

When using open transmission lines, e.g. microstrip, or any components on the surface of a pcb, there is a risk of stray coupling due to radiation from the lines. The amount of radiation depends on the line length in wavelengths, and also on the degree to which the field is confined within the dielectric. Low dielectric constant boards are undesirable at higher microwave frequencies for this reason. However, higher dielectric constant board also means that the track widths will be narrower for a given impedence.

This radiation can cause instability, and also degrade the noise figure if it allows an amplifier to pick up noise. Such circuitry should be kept in a screened box, and screens also used between stages above the board if necessary.

References

[1] *Reference Date for Radio Engineers*, 6th edn., ITT, Howard W Sams, 1979. A general-purpose reference book.

[2] *Microwave Engineers Handbook,* ed T Saad, Artech House, 1971 (two volumes). A good collection of useful data on practical microwaves.

[3] The realm of microwave s (2) Microwave transmission lines, Wireless World, March 1973.

[4] *Microwave Transmission Line Impedance Date*, M A R Gunston, Van Nostrand Reinhold, 1972. A lot of data on impedances of lines, especially stripline and microstrip.

[5] Simple design of λ/4 stripline circuits, W Lerche, DC3CL, *VHF Communications*, 1/1980, pp25-28.

[6] Dimensioning of microstripline circuits, W Schumacher, *VHF Communications*, 3/1972, pp130 - 143

[7] Reflectometers and directional power meters, M M Bibby, G3NJY, *Radio Communication,* June 1968, pp362, 363, 372.

[8] The rotating loop relectometer, W H Elkin, *Marconi Instrumentation,* Vol 5, No 8, pp221-227.

[9] Using Smith diagrams, E Stadler, *VHF Communications*, 1/1984, pp23-28.

[10] *Electronic applications of the Smith chart, in waveguide, circuit and component analysis,* P H Smith, published by McGraw-Hill, 1969. A comprehensive treatment of Smith charts.

[11] Understanding coaxial circulators and isolators, B Sekhon et al, *microwave Systems News,* June 1979, pp84-103.

[12] Shielding barriers block electromagnetic wages, P Grang, *Microwaves,* June 1982, pp97-102 (Part 1); Microwaves July 1982, pp79-86 (Part 2).

[13] *Waveguide Handbook,* ed N Marcuvitz, McGraw-Hill 1951. A comprehensive study, though very mathematical.

[14] Taking the magic out of the Magic-Tee, A J Burwasser, *RF Design,* May/June 1983, pp44-60.

Catalogues

The following manufacturers have catalogues which contain a lot of useful information, including application notes.

Suhner, for precision connectors, including SMA, MB, SMC, APC-7 etc.

Sealectro, for a good range of connectors and also sem-rigid cable.

Greenpar, for a general range of coaxial connectors.

Andrew Corporation, for heliax cable and its connectors.

Anzac, for splitters and combiners (plus mixers and amplifiers, see next chapter).

Minicircuits, for splitters and combiners (also mixers).

Hewlett Packard have a wide range of application notes on a wide range of subjects, including rf components, and rf test equipment and its use (eg network analysers).

Gabriel, for rigid and flexible waveguide, flanges and waveguide components.

Chapter 5

Microwave semiconductors and valves

In this chapter :

- Diodes
- Transistors and FETs
- Detectors
- Mixers
- Multipliers

- Gunn diodes and oscillators
- Amplifiers
- Monolithic microwave integrated circuits
- Switches and variable attenuators
- Microwave tubes

Microwaves were first seriously used in radar during the second world war. These used pulse magnetrons and germanium point contact detectors. Since then, considerable progress has been made, with the move away from free running oscillators to crystal controlled sources. Devices operating in excess of 100GHz are now available, and the frequencies continue to increase into the realms of optics.

A great variety of special purpose valve and semiconductor devices have been developed for microwave use because conventional devices are not able to function effectively at these frequencies, mainly because of transit time effects. Their applications are outlined below.

Detectors: These are used to convert rf to dc, usually to look at the envelope of an rf signal. They are often referred to as video detectors because they were originally used in that application, but the term has been used to cover any application where the bandwidth of the detected signal ranges from dc to a few megahertz. A variety of types of diode are used as detectors, including germanium point contact diodes, silicon Schottky diodes and backward diodes.

Mixers: The same range of diodes that are used as detectors are also used for mixer applications, with the exception of the backward diode. For best noise figure performance, GaAs Schottky diodes are often used. Diodes used as mixers rather than detectors may be optimised for slightly different parameters, though.

Oscillators: These can be divided into two and three terminal devices. At lower microwave frequencies, conventional three terminal devices such as valves and transistors can be used; GaAsfets operate up to tens of gigahertz. At higher frequencies there are a variety of special two terminal devices such as tunnel diodes, Gunn diodes, impatt and trappatt diodes. However, these solid state devices are generally limited to mean powers of a few watts. Any amplifying device can also be used as an oscillator, so at the lower frequencies microwave valves such as the 2C39 can be used to produce a few tens of watts. A number of special valves use magnetic fields to either focus or control the path of the electron beam, examples being the magnetron, gyratron and backward wave oscillator. Some of these are capable of very high mean or peak powers. The bwo is essentially a broadband, travelling wave tube type of amplifier with internal

feedback to produce oscillation.

Multipliers: Multipliers are used to generate power at microwave frequencies when fundamental oscillators are not suitable. They operate by converting the fundamental energy into harmonics. For a low order of multiplication, e.g. two, three or four times, varactor diodes are commonly used to give output powers of tens of watts. For a high order of multiplication, snap or step-recovery varactors should be used instead.

Amplifiers: Amplifiers at lower frequencies can be built using transistors, fets and valves. At higher frequencies GaAsfets can still be used, and there are a number of devices peculiar to microwaves such as travelling wave tubes and klystrons. Other types of amplifiers include low noise parametric amplifiers and reflection amplifiers.

Attenuators: Variable attenuators and switches can be made with pin diodes. These are effectively rf resistors whose resistance is determined by the dc current flowing through them. Limiters are a special type of attenuator which is controlled by the rf level and limits the power passing through it to a specific level. They are used to protect receivers, and were originally developed to protect radar receivers against the transmitted pulses. This chapter introduces the more common devices and their applications.

Diodes

The first microwave diodes were germanium point contact detectors. These are still used, particularly at frequencies above 100GHz, where other types are difficult to manufacture. There are now a variety of other detector and mixer diodes, silicon Schottky barrier types being the most popular. Gallium arsenide is used for ultimate low noise and high frequency performance.

There are now a number of oscillator diodes available based on negative resistance effects; these include impatt and trapatt devices. The most popular is the Gunn diode, which is fully described in a separate section.

Point contact diodes

The point contact diode is formed by bringing a pointed metal wire into contact with a piece of semiconductor (Fig 5.1(a)). A pulse of current is sometimes passed through the junction to fuse the wire to the semiconductor to make

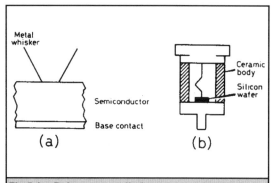

Fig 5.1 : Point contact diode.

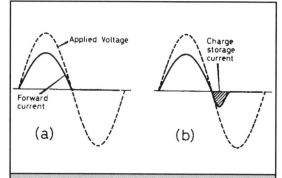

Fig 5.3 : (a) Ideal junction with zero recombination time. (b) Typical junction showing effects of finite recombination time.

it more robust. Since there is a metal to semiconductor junction it is in effect a type of Schottky barrier junction.

The junction capacitance is determined by the area of contact, and so diodes for use at microwave frequencies are quite delicate devices, both electrically and mechanically. They tend to be of lower capacitance than the diffused junction types. A typical construction for a 10GHz diode is shown in Fig 5.1(b).

The equivalent circuit of a junction diode is shown in Fig 5.2. R_s is the series resistance of the bulk semiconductor material. R_j and C_j are the junction resistance and capacitance.

The junction diameter is usually small compared with the thickness of the semiconductor. For microwave diodes, the junction radius should be as small as possible. The highest cut off frequency will be obtained with gallium arsenide, followed by germanium and then silicon, since this is related to the carrier mobility in the material.

When the junction is forward biased, majority carriers cross the depletion region and become minority carriers on the opposite side. They diffuse away from the junction where they recombine in under 10^{-8} seconds.

If the bias reverses before all the carriers have recombined, then the uncombined ones will flow back across the junction and form a pulse of current in addition to the normal reverse leakage current.

This is shown in Fig 5.3. This delay in recombination has the effect of added capacitance, and is called diffusion or storage capacitance.

Schottky barrier diodes

Schottky barrier diodes (also known as hot carrier diodes)

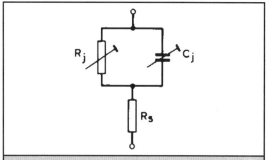

Fig 5.2 : Equivalent circuit of junction diode.

are more efficient rectifiers at high frequencies. This is because they use only majority carriers which do not suffer from the recombination times of minority carriers.

Modern diodes are made using a planar process, where a thin layer of insulating silicon dioxide is grown over the semiconductor, and the junction is formed by etching a small hole in the insulator and depositing a metal film onto the semiconductor through the window formed. This disc of metal might be 5μm diameter (20 times the wavelength of green light). A wire is then bonded to this film to take the connection out to the package. The construction is illustrated in Fig 5.4.

This type of construction has the following advantages over the point contact diode.

• A lower series resistance.

• Better burn out rating and lower low frequency noise due to the larger junction area and increased junction capacitance.

• More rugged and reproducible since the mechanical construction is more robust.

The net result of the lower series resistance and higher capacitance is an upper frequency limit similar to that of point contact diodes, since the limit is determined by the produce $R_s \times C_j$. The junction radius has no effect on the cut off frequency in practice. The theoretical expression for the V-I characteristic is:

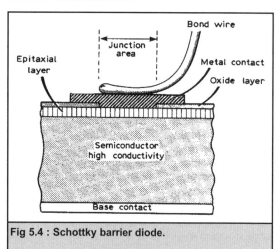

Fig 5.4 : Schottky barrier diode.

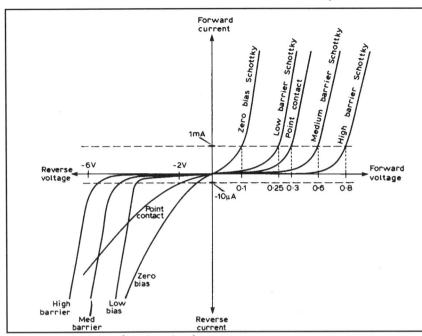

Fig 5.5 : V-I characteristics of the Schottky barrier diode and point contact diode.

$$I = I_s \times \left(\exp\left(\frac{eV}{nkT} \right) - 1 \right)$$

where :

I_s is the saturation current

e the electronic charge, 1.6 x 10⁻¹⁹ coulombs

T is the absolute temperature, in Kelvin

k is Boltzmann's constant, 1.38 x 10⁻²³ joule/K

V is the voltage across the diode junction in volts

n is a factor allowing for the diode being non ideal, n=1 for an ideal diode

e/kT for a perfect diode is therefore 1/25mV

Fig 5.5 illustrates the V-I characteristic for the Schottky barrier diode, together with the characteristic for point contact diodes for comparison. Table 5.1 lists typical values for reverse leakage currents and forward voltage drops.

Tunnel diodes

This diode is of similar construction to the conventional pn junction diode, but has a much higher doping level (level of wanted impurities) and a very abrupt junction with a narrow depletion layer. This heavy doping reduces the width of the depletion layer to around 10-100 atom diameters. Electrons can therefore pass from one side to the other without going through the usual diffusion and recombination process, which gives it a very fast response, but the junction area must be kept much smaller than usual because the narrower junction increases the capacitance per unit area.

Tunnelling is a term that describes how the electrons can pass from one side of a potential barrier to the other, when they would not normally have sufficient energy to get over the barrier. The doping levels are such that the increase in tunnelling current is followed by a decrease as the voltage is increased. The net result is a region of negative resistance in the diode characteristic. Fig 5.6 shows the V-I characteristic.

Backward diodes

A backward diode is similar to a tunnel diode, except that the tunnelling current is reduced to a very low value so that the negative resistance region does not exist. The V-I characteristic is shown in Fig 5.7. It is mainly used as a low level, broadband detector or mixer. It has a very low value of slope resistance at zero bias. The name backward diode arises because the detected voltage is of the

Table 5.1 : Reverse breakdown currents and forward voltage drops for Schottky barrier diodes.

Diode type	Typical reverse leakage current	Typical forward voltage drop	Typical video impedance
Zero bias Schottky	100µA at 1V	0.12V at 1mA	1kΩ
Low barrier Schottky	1nA at 1V	0.25V at 1mA	10kΩ
Point contact	10nA at 1V	0.3V at 1mA	100kΩ
Medium barrier Schottky	1nA at 1V	0.45V at 1mA	above 1MΩ
High barrier Schottky	100pA at 1V	0.65V at 1mA	above 1MΩ

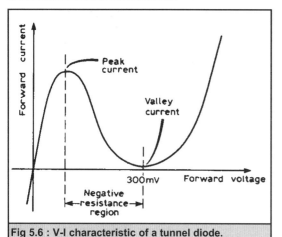

Fig 5.6 : V-I characteristic of a tunnel diode.

opposite polarity to normal.

Unlike the Schottky diode, it has very low flicker noise, so it can be used for receivers with i.fs in the audio frequency range-doppler receivers. However, it does only have a very limited dynamic range.

PIN diodes

PIN diodes are used as current controlled resistors at rf. They have an incremental resistance that is controlled by the dc current flowing through them, in a similar manner to the ordinary junction diode. However, their resistance is much more linear and does not produce the distortion that a normal exponential diode characteristic would. They are therefore useful in rf switching and current controlled attenuators.

The pin diode consists of p-type and n-type regions that are separated by an intrinsic (i) region, i.e. a region in which there are negligible impurities or charge carriers (Fig 5.8). Charge carriers from the doped p and n regions have to cross this intrinsic region, and at low frequencies the diode acts as a rectifier. At high frequencies the

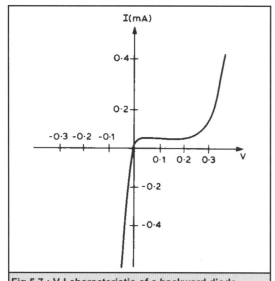

Fig 5.7 : V-I characteristic of a backward diode.

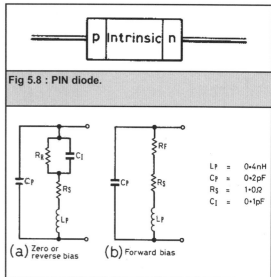

Fig 5.8 : PIN diode.

L_P = 0•4nH
C_P = 0•2pF
R_S = 1•0Ω
C_I = 0•1pF

(a) Zero or reverse bias   (b) Forward bias

Fig 5.9 : Equivalent circuit of package pin diode of (a) zero or reverse bias and (b) forward bias.

storage time in the intrinsic region is so great (i.e. the transit time is so long compared to the period of the signals) that the diode ceases to act as a rectifier, and instead acts as a linear resistance and conducts in both directions. Its effective resistance is inversely proportional to the amount of charge in the intrinsic region, and this can be controlled by the dc bias current. In this way the conductivity can be varied by several orders of magnitude.

The complete equivalent circuit of a packaged diode is shown in Fig 5.9, and at microwave frequencies the package reactances are very significant. At microwave frequencies the junction capitance C_j is constant and purely a function of the junction geometry. At zero or reverse bias the intrinsic region is depleted of charge and has a relatively high resistance of several thousand ohms. Under forward bias electron and hole charge carriers are injected into the intrinsic layer and the resistance of the layer drops to below an ohm. The variation of resistance with current is shown in Fig 5.10, and the minimum value is limited by R_s.

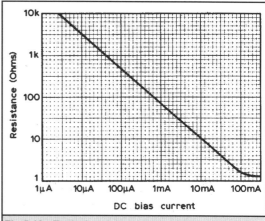

Fig 5.10 : Typical resistance versus bias current for a pin diode.

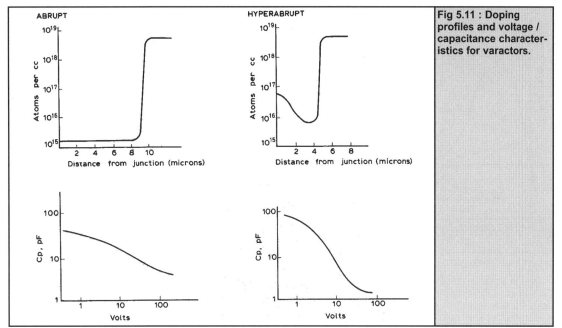

Fig 5.11 : Doping profiles and voltage / capacitance characteristics for varactors.

These properties only apply at high frequencies, where the period of the rf is smaller than the recombination time (t) of the carriers this is typically 10-300ns. At frequencies below $f = 1/(2\pi \times t)$ the injection and removal of charge can follow the rf waveform, producing inefficient rectification of the signal. Above it, the charge level is determined mainly by the dc bias current, and is only modulated very slightly by the rf signal. Conversely, the rf may be modulated by applying a low frequency signal superimposed on the dc bias.

A short carrier lifetime allows very fast switching of the diode, but limits use of the diode to higher frequencies. A compromise must be chosen between switching speed and lowest rf operating frequency. Compared with other types of diode the pin type has a low junction capacitance and high reverse breakdown voltage, and so can handle quite high powers at high frequencies.

Tuning diodes

Tuning varactor diodes are optimised to use the variation of junction capacitance with reverse voltage, instead of the variation of junction resistance with forward voltage. They are basically similar to pin diodes, but do not have the intrinsic region between the p and n layers.

The capacitance change is not a result of stored charge as in the pin diode, and so can be made to change much more rapidly. However the diode reactance is now a function of the rf signal voltage, and so much more intermodulation will occur. They are lower power devices because the junction is much thinner. The continuously variable capacitance can be used to produce continuously variable phase shifts.

As tuning diodes, they can tune oscillators or tuned circuits over several octaves in low power applications at lower frequencies. The junction capacitance is related to the applied voltage:

$$C_j = (V + V_o)^{-n}$$

where V_o is the built in junction voltage and n a function of the junction doping. For an abrupt junction, n = 0.45 to 0.48, for a hyperabrupt junction n is greater than 0.5. Examples of doping profiles and voltage / capacitance characteristics are shown in Fig 5.11. The tuning range is given by :

$$C_{max} / C_{min} = \frac{(V_{max} + V_o)^n}{(V_{min} + V_o)^n}$$

Typical values of V_{max} and V_{min} are 20V and 2V

A lower doping level will give a higher breakdown voltage and a larger tuning range, but at the expense of a lower Q. Heavier doping will give a higher Q but a smaller tuning range.

The tuning range that a varactor will give in a real circuit depends on the proportion of the total energy in the tuned circuit that is stored in the varactor. In a lumped component circuit, this is simply a matter of the fraction of the total capacitance across the coil that is provided by the varactor. In a cavity resonator this is more difficult to

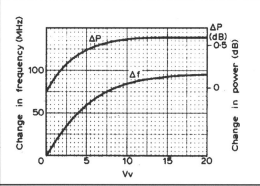

Fig 5.12 : Variation of power and frequency with tuning voltage in a half wavelength WG18 cavity oscillator at 13GHz.

119

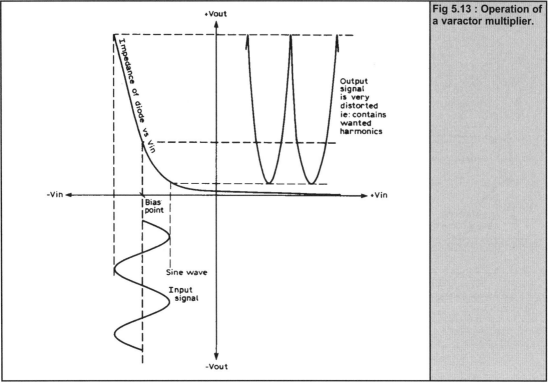

Fig 5.13 : Operation of a varactor multiplier.

identify, but the varactor should be very highly coupled to the cavity for maximum tuning range.

A typical tuning curve for a varactor in a waveguide cavity is shown in Fig 5.12. Note that the losses increase rapidly at the higher capacitances.

Multiplier varactor diodes

Another application of the variable capacitance diode is as a frequency multiplier (or mixer). This uses the variation of junction capacitance rather than resistance with applied voltage to provide the desired modulation, as illustrated in Fig 5.13. This process is more efficient as the variable impedance is almost lossless.

As multipliers, they can handle powers of tens of watts at around 1GHz, and hundreds of milliwatts at 10GHz. The main difference between these and tuning varactors is that the multipliers are built to dissipate much higher powers.

Step recovery diodes

Step recovery varactors (snap diodes) are a variant of the normal multiplier varactor diode. The diode has a low impedance while forward biased which, when the voltage reverses, remains low until the carriers are swept out of the depletion region.

At this point it rapidly changes to a high impedance (in typically 50ps), shunted by the reverse capacitance of the diode. This rapid switch off is used to generate high order harmonics. The characteristic of the srd is shown in Fig 5.14.

Consider the circuit in Fig 5.15. With suitable choice of the circuit values, it can be arranged that there is maximum current from the low frequency drive source in the inductor at the instant the diode changes to a high impedance. In practice, the bias voltage is usually supplied by the diode rectifying the rf across the resistor. The voltage pulse across the diode travels down the line and,

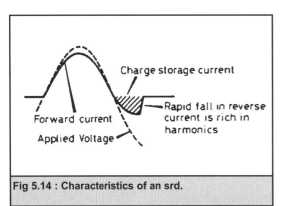

Fig 5.14 : Characteristics of an srd.

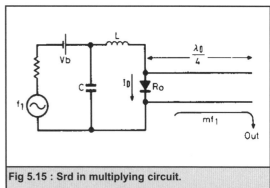

Fig 5.15 : Srd in multiplying circuit.

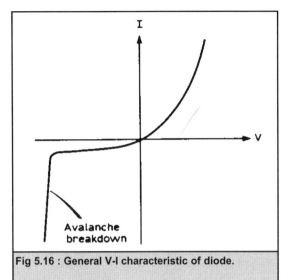

Fig 5.16 : General V-I characteristic of diode.

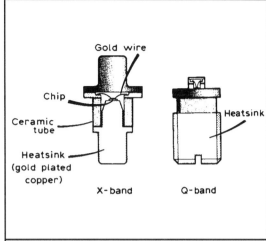

Fig 5.17 : Typical impatt diode package.

if the line is tuned to a harmonic of the input frequency, these pulses will reinforce.

IMPATT diodes

The V-I characteristic of a semiconductor diode has the general form shown in Fig 5.16. Under sufficient reverse bias, avalanche breakdown occurs, and this effect is utilised in the familiar zener voltage reference diode. In the impatt diode, the combination of the build up time of the avalanche current and the time taken for the carriers to drift across the depletion region causes a phase delay between the current and voltage, which can result in a negative resistance, and hence the possibility of oscillation. This term impatt stands for impact avalanche transit time.

The efficiency of the impatt is about 10 per cent for silicon diodes, and about 15 per cent for GaAs devices. This compares with about two per cent for Gunn diodes. At frequencies above 70GHz, the impatt is the only solid state microwave source available, and is useful as a pump source in parametric amplifiers. The principal disadvantages of the impatt compared with the Gunn diode are its noise performance and the need for a high bias voltage. The impatt (and trapatt) are fundamentally noisier than the Gunn diode, the noise originating in the avalanche multiplication process. A typical X band impatt (single drift) will produce a cw output power of 1.5W at an efficiency of 10 per cent.

This corresponds to a dc bias supply of 80V at about 200mA, and a dissipation of bout 15W. Thus the diode must be constructed in such a way that the heat generated can be efficiently removed, and it is usually built with an integral heatsink (see Fig 5.17 for a typical impatt package).

TRAPATT diodes

In 1967 a second mode of avalanche oscillation was discovered. This new mode remained unexplained for some time, and was initially referred to as the anomalous mode.

If the voltage is applied sufficiently rapidly, a different type of breakdown occurs because it is possible to over drive the field and create a dense hole electron plasma. This state, in which the diode is filled with plasma and the voltage has dropped to a low value, is referred to as a trapped plasma state. The diode then recovers gradually as the holes and electrons drift slowly out of the active region, thereby restoring the field and voltage to their initial level. This has been called the trapatt (trapped plasma avalanche trigger transit) mode.

The efficiency of the trapatt is much higher than that of the impatt being up to 60 per cent at low frequencies (around 1GHz) and dropping to about 25 per cent at 10GHz. The trapatt is best suited to high power pulsed applications.

Burn out rating

This is determined by the amount of energy that a diode can absorb without damage. High frequency diodes will obviously be more delicate since their junctions occupy a smaller area, so the energy will be more concentrated. They are also more susceptible to damage from static charges since their junction capacitance is lower, and thus a higher voltage will exist across the capacitance for a given amount of applied charge. In general terms, point contact diodes have better resistance to short transients, and Schottky diodes have a better sw rating. Silicon diodes are more robust than germanium, while the germanium backward diode is the most delicate.

The circuit in which the diode is used can also be a major factor in burn out, since it can determine how much of the incident power is absorbed by the diode. As the power in the diode increases, its impedance decreases. The diode will protect itself to some extent if it is already matched to the line impedance at low signal levels, since it will reflect most of the incident power as its impedance decreases. The diode circuitry should thus be arranged so that the match between the diode and the power source always degrades as the power level increases. In practice these effects are probably more important than the type of diode chosen.

In the case of very high-power devices such as varactors, heatsinking is very important, and it is necessary to maintain a very low thermal resistance path between the package and the outside world, as well as keeping stray reactances down to a minimum.

Transistors and FETS

Transistors

Bipolar transistors are useful up to the low microwave frequencies. Fairly low noise figures can be obtained, but the gains are not as high as can be achieved with GaAsfets. Virtually all the types that are of use at microwaves will be in packages designed for mounting on microstrip. Popular type numbers for low noise devices are BFR90, BFR91, BFR96 and NE021.

Power devices exist, but are somewhat expensive. Cellular construction can be used to combine any low power chips to give a high power device. At 1GHz, up to about 10W can be achieved fairly cheaply. These packages again are usually designed for mounting on microstrip, and often have heatsink studs. Many power devices at this frequency on the surplus market are designed for common base circuits and are difficult to use in a linear fashion, though useful for Class C operation.

The characteristics of microwave transistors (and fets) are usually specified by S parameters. A typical set of values for a low noise transistor and for a high power one are shown in Tables 5.2 and 5.3. The input impedance of the low noise device is close to 50Ω. High power devices normally have an input impedance of under 10Ω, though the example given has some matching built into the transistor which makes it easier to use. External matching

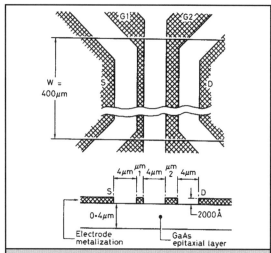

Fig 5.18 : Schematic view of 3SK97 GaAsfet.

is essential if internal matching is not present. Similar comments apply to the output.

GaAsfets

Gallium arsenide fets (GaAsfets) have become so popular because they not only have quite high gains up to 10 or 20GHz, but they also have extremely low noise figures. They have provided a great step forward in the performance of solid state microwave devices. One of the reasons for their improved performance is that gallium arsenide has a much higher carrier mobility than silicon; this reduces the transit times in the material and hence gives better performance at high frequencies.

The construction of a dual gate GaAsfet is shown in Fig 5.18. A layer of gallium arsenide is deposited on a substrate by vapour phase deposition passing GaAs over the substrate as a hot gas, which condenses and grows as a crystalline layer on the cooler substrate. The gate is isolated by a Schottky diode which is reverse biased when the gate is negative with respect to the channel. The dopants and metallising are then diffused in the usual way.

In order to achieve a good microwave performance, the structure must be extremely small. This accounts for its delicate nature. The channel length is about 1 micron, and the gate metallising 0.5 micron wide and of similar thickness. The cross sectional area of the gate conductor is thus 0.25^{-6} sq mm. A current of 1mA in this gate

Table 5.2 : S parameters for low noise transistor.

Frequency (MHz)	200	500	1,000	1,500
S11	0.72	0.51	0.47	0.51
Angle (deg)	-65	-125	155	135
S22	0.83	0.62	0.51	0.49
Angle (deg)	-25	-35	-40	-50
S21	5.25	3.06	1.70	1.20
Angle (deg)	130	95	65	50
S12	0.08	0.11	0.14	0.17
Angle (deg)	55	45	55	65

Device : BFR91 V_{ce} = 5.0v, I_c = 2mA, nf = 2.5dB typical at 1GHz

Table 5.3 : S parameters for high power transistor.

Frequency (GHz)	1	2	3	4
S11	0.74	0.64	0.32	0.32
Angle (deg)	-178	153	129	-145
S22	0.37	0.51	0.73	0.80
Angle (deg)	-92	-119	-148	-177
S21	3.91	2.32	1.86	1.38
Angle (deg)	59	10	-49	-113
S12	0.06	0.07	0.09	0.08
Angle (deg)	7	-8	-42	-98

Device : HP HXTR-5102 V_{ce} = 1.8v, I_c = 110mA, pwr o/p = 0.8W at 2GHz (11dB gain), 0.5W at 4GHz (7dB gain)

Table 5.4 : Absolute maximum ratings for the MGF1402 (T_a = 25°C).

Symbol	Parameter	Limits
V_{GDO}	Gate to drain voltage	-6V
V_{GSO}	Gate to source voltage	-6V
V_{DSX}	Drain to source voltage	8V
I_D	Drain current	100mA
PT	Total power dissipation	300mW
T_{ch}	Channel temperature	150°C
T_{stg}	Storage temperature	-55 to + 150°C

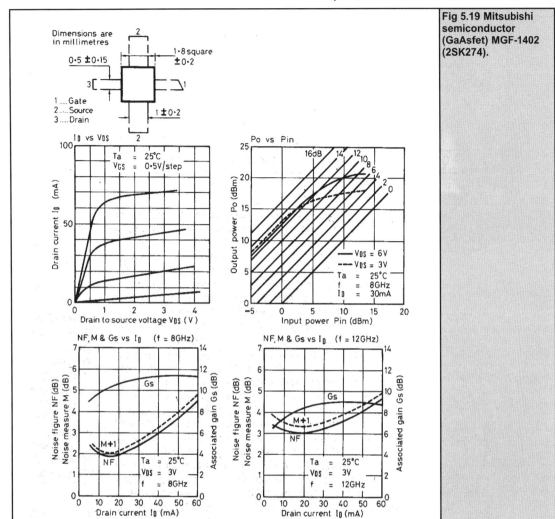

Fig 5.19 Mitsubishi semiconductor (GaAsfet) MGF-1402 (2SK274).

metallising corresponds to a current density of 4,000A/sq mm! Similarly, a voltage of 5V between the gate and drain, a distance of 0.5 micron, is a voltage gradient of 100kV/cm, which is well above the break down field for air. The dielectric strength of gallium arsenide is many times that of air, however, but it is evident that even when running these devices at 5V and currents of milliamps, they are operating as close to catastrophic breakdown as some high power devices, so the manufacturers limits must be taken seriously. In particular, the gate must not

Table 5.5 : S parameters for the MGF1402 (T$_a$ = 25°C, VDS = 3V)

ID (mA)	f (GHz)	S11 Magn	S11 Angle	S12 Magn	S12 Angle	S21 Magn	S21 Angle	S22 Magn	S22 Angle
10	2	0.935	-47.4	0.040	54.1	2.848	133.4	0.758	-33.1
10	4	0.835	-88.4	0.065	21.3	2.469	92.3	0.710	-62.6
10	6	0.800	-119.1	0.054	13.6	2.333	65.8	0.713	-75.8
10	8	0.709	-164.5	0.048	-9.5	2.286	27.7	0.643	-97.6
10	10	0.658	155.7	0.052	-59.3	1.805	-13.9	0.599	-138.7
10	12	0.713	130.0	0.044	-37.4	1.488	-37.1	0.480	177.3
30	2	0.916	-53.1	0.032	55.2	3.864	130.8	0.676	-32.5
30	4	0.787	-198.0	0.048	25.1	3.203	88.3	0.632	-60.9
30	6	0.742	-129.2	0.034	39.4	2.928	61.8	0.657	-71.3
30	8	0.661	-176.7	0.035	27.4	2.751	23.3	0.604	-91.9
30	10	0.640	144.1	0.037	-29.3	2.160	-15.7	0.580	-134.0
30	12	0.720	121.2	0.046	9.7	1.770	-37.6	0.433	-178.0

All angles are in degrees

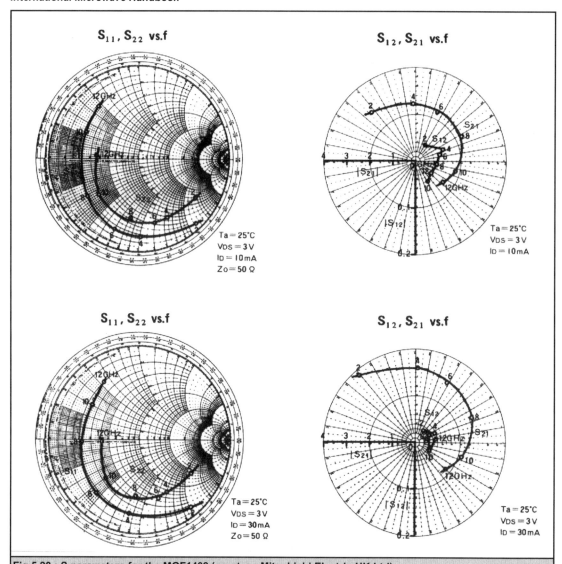

Fig 5.20 : S parameters for the MGF1402 (courtesy Mitsubishi Electric UK Ltd).

be taken positive because the diode becomes forward biased and then currents of many milliamps can flow. The results of exceeding the ratings can be a catastrophic failure the resulting crater looks like there has been a small explosion; in a way, there has. Similar damage can occur due to discharge of static into the device.

Being fets, the input and output impedances of the devices are very high at hf/vhf//uhf. They reduce with frequency, and are in the region of 50Ω at 10GHz, but are generally higher than bipolar transistors at the same frequency. Fig 5.19 and Table 5.4 give some data for a typical GaAsfet, the Mitsubishi MGF1402. Fig 5.20 and Table 5.5 give S parameter data for the same device.

The packaged fet is intended for use in microstrip applications, but it can be mounted in waveguide with a little ingenuity. Professionally, naked chips are often mounted directly on alumina and connections from the chip to the lines made direct by bondwires.

Detectors

A wide variety of types of diode are used as detectors, but they can be grouped into two types those with an exponential V-I characteristic and those with other special characteristics.

The first group includes conventional junction diodes, point contact diodes and zero, low, medium and high bias Schottky diodes. The second includes backward diodes.

Germanium point contact diodes were the first type made and are still used. They have a low barrier height. Silicon Schottky diodes are more commonly used nowadays, and are available in zero bias, low, medium and high barrier versions. For best performance at low signal levels backward diodes are used.

Operation of the diode detector

Junction diodes are used to detect low level rf signals by

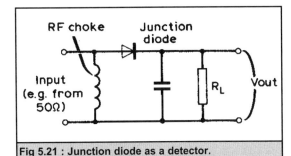

Fig 5.21 : Junction diode as a detector.

operating as a normal half wave rectifier. They are essentially very simple broadband power detectors, and are often mounted in coaxial line or waveguide to monitor the power in a transmission line, or across a simple antenna to indicate radiated power. The basic circuit is shown in Fig 5.21.

As can be seen from the dc V-I characteristic (e.g. Fig 5.5), the diode has a low resistance when forward biased and a high resistance when reverse biased. This circuit will detect rf power over a wide range of levels, but its characteristics vary over the range. At rf levels below about a few hundred millivolts, the detected voltage or current is directly proportional to the rf input voltage, i.e. V_{out} is proportional to the square root of the power.

The operation at high power levels is fairly easily understood, in that the diode operates as a switch and the output follows the maximum positive value of the rf waveform. This is therefore a peak voltage detector. However, at lower levels the operation has to be studied more closely. The characteristic of a Schottky diode will be used to illustrate the operation. Its exponential V-I characteristic is shown in Fig 5.22 together with the incremental resistance of the diode (i.e. the slope of the V-I curve).

Consider how current flows through the diode during one complete cycle of sinusoidal rf input.

As the input voltage goes through its positive half cycle, current flows forwards through the diode, increasing from zero to maximum and back to zero. The total amount of

charge is the current multiplied by the time, ie the area shown under the V-I curve. This charge would be stored in the capacitor. However, in the negative half cycle the same process occurs, but this time the current is in the opposite direction and a net charge is taken out of the capacitor. After this single cycle, the capacitor holds the difference between the amount of charge supplied and removed. This net amount of charge passed through the diode represents the rectified current, and the voltage on the capacitor the detected voltage.

This operation is very obvious when large input voltages are used, because the ratio of the forward and reverse resistances is very high, and in practice charge only travels in the forward direction almost none is removed in the negative half cycle. However, at low levels the changes in resistance between forward and reverse bias are very small, and consequently almost all the charge given to the capacitor is taken back in the second half cycle. The net difference between the two is a very small fraction of the total charge moved. The diode is therefore behaving as a non linear resistor, its mean value being the slope of the diode curve at zero applied voltage.

The reason why the diode produces an output proportional to power at low levels is that, although the diode characteristic is an exponential, the difference between the currents for an equal positive and negative voltage is proportional to the square of the voltage. It follows that the difference between the forward and reverse currents averaged over a cycle is a square law, i.e. the mean current is proportional to the square of the input voltage.

This can be verified by putting a value for V into the expression for the diode characteristic and calculating the current for +V and V.

The difference between these two will be found to be proportional to V2 provided V is small, i.e. under 100mV, but the exact values will differ between types of diodes.

Video impedance and tangential signal sensitivity

The difference between various types of detector diode should now be clear. Almost all diodes have basically the same exponential characteristic; the main difference between the various zero bias, low, medium and high bias diodes is the position of this curve relative to the origin. This manifests itself in two ways. The reverse leakage current Is is higher in the lower bias diodes, and the slope resistance at zero bias voltage, which is known as the video impedance, is lower in the low bias diodes. Diodes are often biased with a small dc current to lower the video impedance by shifting the operating point into the forward bias region.

Diodes used in this way are much less sensitive than a narrowband receiver. The limit is about 50 to 60dBm and is known as the tangential signal sensitivity, tss.

The sensitivity of a variety of detector diodes is compared below. Fig 5.23 compares the sensitivity of diodes with an infinitely large load resistance with and without dc bias; Fig 5.24 shows the effects of lower load resistances. In general, a low load resistance is needed if a wide bandwidth is required, e.g. many megahertz, but the sensitivity then suffers. The low barrier diodes are best suited to operating into low resistance loads. The load resistance relative to the type of barrier height also has a marked effect on the linearity of the detector.

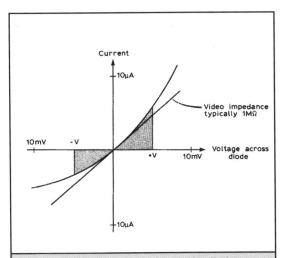

Fig 5.22 : V-I characteristic of Schottky diode.

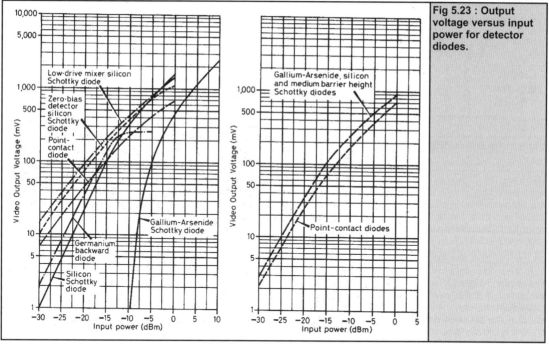

Fig 5.23 : Output voltage versus input power for detector diodes.

Figure of merit

The quality of a detector diode can be compared in terms of the short circuit current sensitivity and video impedance, by using an arbitrary figure of merit M:

$$M = B \times \sqrt{Z}$$

where B is the short circuit current sensitivity and Z is the video impedance. The video impedance is usually given on a manufacturers data sheet.

This only refers to the diodes operation at zero bias, and does not take into accounts its noise performance.

Selecting a diode

For high efficiency, the diode capacitance must be minimised. This is particularly important as the sensitivity depends on the square of the capacitance.

For maximum detector sensitivity the curvature of the characteristic at the origin should be as large as possible, and thus low or zero bias diodes are best. The back diode has one of the highest slopes at the origin, and is thus a

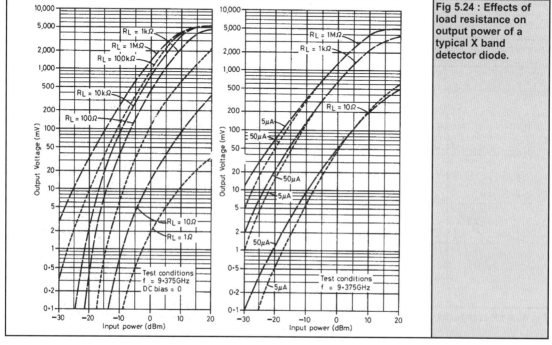

Fig 5.24 : Effects of load resistance on output power of a typical X band detector diode.

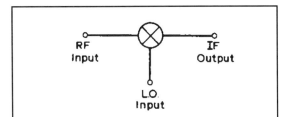

Fig 5.25 : Basic mixer.

very sensitive detector. However, there is a price to be paid for these high sensitivities. The greater the slope, the greater effect temperature has on the sensitivity.

The values of the components in the detector circuit have to be related to the diode impedance to obtain best sensitivity, particularly the rf source impedance and the dc load resistance. Low barrier devices operate with rf and dc resistances of a few hundred ohms, which high barrier devices need tens of kilohms or megohms.

For optimum efficiency and linearity from a low level detector, the rf source resistance should be equal to or lower than the diode video impedance. The load resistance should be much greater than the video impedance for a voltage detector, and equal to or less than it for a current detector.

Mixers

Mixers are mostly used to translate signals up and down in frequency without changing them in any other way. However, the same basic function also appears in modulators, phase detectors, multipliers, gain controlled stages and switches anywhere where the circuit requires that one signal controls the amplitude of another. In all these cases, the amplitude of one signal is being multiplied by another.

Most practical mixers at rf use the properties of semicon-

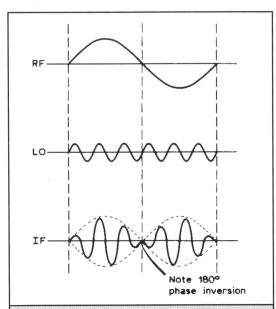

Fig 5.26 : Operation of ideal linear mixer.

ductor diodes to perform this function, and consequently have many shortcomings. Transistors, valves or fets are also used as mixers, but they still operate on the same basic principles that will be described for the diode mixer, in that they all have a non linear transfer function. It is perhaps surprising that such good results can be obtained from devices based on the humble diode, despite its non linear nature.

The main shortcomings of real mixers are that they generate outputs at other than the wanted sum and difference frequencies, and they produce non linear distortion on the signals passing through them. Better performance can be obtained by combining a number of diodes to form single or double balanced mixers.

The basic types of mixer in common use are listed below in increasing order of complexity and quality of performance:

• Single diode

• Balanced

• Double balanced

• Product return

• Termination insensitive

• Parametric

• Transistor/fet

However, in many applications no one type of mixer is obviously the best; there are often many parameters which need to be traded off against each other, e.g. conversion loss, noise figure, 1dB compression point, degree of balance, local oscillator power and noise sidebands, and bandwidth, to name a few! Their characteristics will be described in more detail after explaining the basic operation of a mixer.

Theory of an ideal mixer

Shortcomings in mixer performance are often blamed on non linearity. This may seem to be rather contradictory since it is also said that a mixer must have non linear properties to produce the mixing effect. First we will explain what properties an ideal mixer should have.

Fig 5.25 shows the representation of a mixer with two inputs, rf and local oscillator (lo), and one output, i.f.

An ideal mixer has the property that there is a perfectly linear path from each input to the output, but the fraction transmitted through one path is controlled linearly by the voltage at the other input. Mixers can operate in either a switching or linear mode. In a linear mixer the attenuation of the rf to i.f. path is smoothly varied from zero to infinity by a sine wave lo signal. In a switching mixer, the rf to i.f. path is either turned fully on or fully off by a square wave lo. The mixer is therefore rather like a voltage controlled attenuator, but with the important difference that if the controlling input goes negative the polarity of the output signal is inverted. This is illustrated in Fig 5.26.

This ideal mixer produces outputs only at the sum and difference frequencies of its two inputs. Two input signals W_a and W_b would produce outputs at $W_a + W_b$ and $W_a - W_b$ (e.g. as in Fig 5.27). The amplitude of the outputs would be proportional to the amplitude of the inputs.

This operation can be described mathematically by using two sine wave input signals at frequencies W_a and W_b,

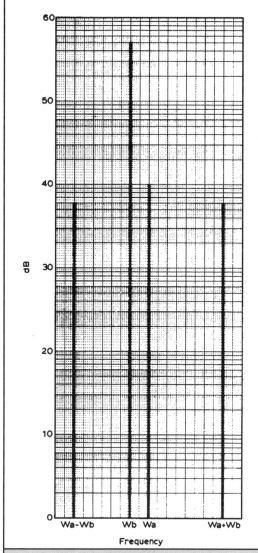

Fig 5.27 : Output spectrum from ideal mixer.

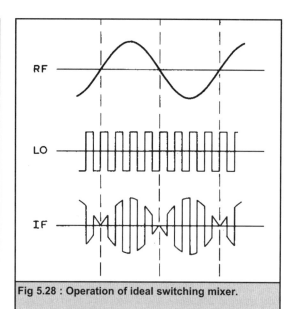

Fig 5.28 : Operation of ideal switching mixer.

sideband suppressed carrier signal in an ssb transmitter.

In most rf applications mixers are used in the switching mode, and the lo input is usually a square wave so that the diodes are either fully on or off. The lo signal effectively switches alternate pairs of diodes into conduction, reversing the polarity of the rf input signal at the lo frequency. The operation is then simply that of a reversing switch, as shown in Fig 5.28. The switching mode usually gives better linearity of the rf to i.f. path, and a lower conversion loss. Linear mixers are more common at audio frequencies than at rf.

The single-diode mixer

The simple single diode mixer is still in common use, as well as being the basic building block of the more advanced types. Its operation is very easy to explain, and this will illustrate some of the problems, and reasons for developing the more complex circuits.

The operation of a simple diode mixer can be described from the V-I characteristic of an ideal diode. The relationship is fairly accurately exponential for Id between a few nonoamps and a few millamps. The diode current is give by:

$$I_d = I_o \times (\exp(eV/kT)-1)$$

where : e is the charge on an electron, 1.6×10^{-19}; k is Boltzmann's constant, 1.23×10^{-23}; V is the voltage across the diode; T is the absolute temperature; I_o is the reverse leakage current. At room temperature (T = 290°K), e/kT is about 40 per millivolt. For an ideal diode, I_o is constant, and a function of barrier height; in practice, of course, it increases due to breakdown at high reverse voltages.

The resistance of the diode is given by the ratio V/I at any point on the curve, and thus varies with diode current. The incremental (small signal) resistance is the tangent to the curve, and is given by $25/I_d$ ohms for a silicon diode, where the current is in milliamps. So, by varying the diode current, the small signal resistance can be varied from a very high to a very low value. By using the diode as part of a potential divider circuit, it can act as a variable attenuator (Fig 5.29). The on resistance will be limited by the

and amplitudes A and B. The voltage of each of these signals varies with time t in the following way:

$$A \times \cos(W_a t)$$

and

$$B \times \cos(W_b t)$$

The mixing process consists of multiplying these two input voltages, giving :

$$A \times B \times (\cos(W_a t) \times \cos(W_b t))$$

This product of the two cosine functions can also be written as :

$$(A \times B/2) \times (\cos((W_a+W_b)t) + \cos((W_a-W_b)t))$$

This represents two products on frequencies which are the sum and difference of the two input frequencies, and corresponds to the output from a double.balanced mixer. The output waveform is similar to that of a double

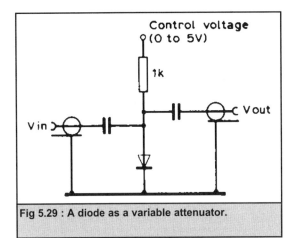

Fig 5.29 : A diode as a variable attenuator.

series resistance of the diode, which is typically 10Ω.

This circuit will function as a mixer provided that the local oscillator power is sufficient to change the diode current and hence impedance significantly, and that the signal is much weaker than local oscillator. This deviates from our ideal mixer mentioned earlier in a number of ways.

- The signal path is not very linear it has the same characteristic that we are using with the local oscillator to produce the variable attenuator effect! So intermodulation products and harmonics will also be generated by the input signals.

- The three ports will all be very poorly isolated from each other. High levels of lo signal will be present at the rf and i.f ports. The input signal will appear at the i.f port. RF band pass filtering will be necessary to prevent the energy from each of the input signals from being wasted in the sources connected to the other ports. Fig 5.30 shows a typical arrangement.

- The input impedance to the large local oscillator signal will vary with I_d, and hence will vary during the lo cycle. Thus the lo sine wave will become very distorted as harmonics of it are generated within the mixer diode.

The single diode mixer, while not optimal in many respects, is still popular in waveguide circuitry because of its simplicity. A diode is either mounted directly across the waveguide, or in a coaxial mount adjacent to it.

A variety of types of diode with different barrier height is available. These are chosen according to the dynamic range required and the local oscillator power available. A low barrier height will give poor strong signal handling.

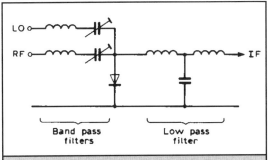

Fig 5.30 : Typical arrangement of single diode mixer.

Conversion loss of a diode mixer

The overall noise figure of a mixer/i.f combination is given by:

$$F = L_c \times (F_{if} + Nr - 1) \quad \text{(as a ratio)}$$

Where L_c is the conversion loss, F_{if} is the noise figure of the i.f amplifier and Nr is the noise temperature ratio of the diode. All values are as ratios, not in decibels.

In this simple circuit the diode will be switched off for approximately half a cycle and, depending on the drive level, the conduction angle may vary from 120° to 170°. The effect that those other diode parameters not present in an ideal model have on the conversion loss is summarised below. The conversion loss (in decibels) is given by:

$$L_c = 3.9 + 17 \times F/F_{co} + 9 \times R_b/Z_0$$

F_{co} is the cut off frequency which is derived from the time constant of the junction series resistance and capacitance. R_b is the forward resistance of the diode when the lo current is at a maximum (typically a few ohms). Z_o is the impedance of the line in which the diode is mounted, and also the impedance of the signal and lo sources as seen by the mixer (typically 50-100Ω) Note that the ratios are not converted to decibels. If the package reactances are significant, they may transform Z_o to a different value at at the mixer chip. The factor of 3.9dB depends slightly on what is done with the other mixing products, as mentioned earlier.

The minimum possible conversation loss depends on the slope of the log-log plot of the diode V-I characteristic. At a frequency w (w = $2\pi \times F$), the loss is given by:

$$L_c = \frac{R_s + \dfrac{R_b}{1 + \left(w \times C_b \times R_b\right)^2}}{\dfrac{R_b}{1 + \left(w \times C_b \times R_b\right)^2}} \times L_O$$

where L_o is the If conversion loss determined from the V-I curve; R_b is the barrier resistance; C_b is the barrier capacitance; and R_s is the series resistance.

Thus, at any frequency where the barrier capacitance is significant, the product $C_b \times R_s$ must be minimised for lowest loss.

Noise included in the noise temperature ratio includes thermal noise in the series resistance, shot noise in the barrier (due to the charge being quantised), and flicker noise. Shot noise dominates at rf, but flicker noise becomes more important at audio frequencies.

Switching mixers

The simple diode mixer can be improved by a number of measures. The simplest one is to use a high lo drive level. Taken to the limit, this is equivalent to driving it with a square wave lo. This means that the diode spends most of its time either reverse biased and open circuit, or heavily forward biased, the on resistance being limited to a few ohms by the series resistance. Thus the diode always appears as a fairly linear resistor, and the non linearity in the signal path is reduced. This also minimises the conversion loss for a single diode mixer, as the input signal is then being 100 per cent modulated.

Further increases in the lo power will only increase the noise figure, either due to the oscillator side band noise,

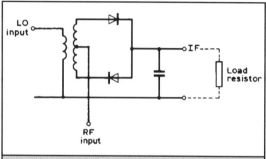

Fig 5.31 Single balanced mixer.

or broadband shot noise associated with the rectified current flowing in the diode (2e x I x B). While this may increase the maximum signal power that the mixer can handle, it will reduce the dynamic range.

At very low frequencies the ideal element for a mixer is a mechanical relay. This is lossless and extremely linear. However, in practice some solid state device or valve has to be used, and these are far from linear. In fact all the devices used for mixers are inherently non linear, and have either a square law or exponential characteristic, so they have to rely on the rf signal only making small changes to the device operating point.

Balanced mixers

The first improvement over the simple diode mixer is to balance it to one port, as shown in Fig 5.31. This improves the linearity and suppresses one of the inputs at the i.f output, e.g. the lo and its a.m. noise side bands. This technique is often used in ssb transmitters to generate dsb suppressed carrier, the balance suppressing the carrier.

The lo is fed 180° out of phase to two mixing elements (in this case diodes). Such a mixer has a slightly more linear signal path as there are two diodes in series or parallel in the signal path, with the rf current flowing in opposite directions in them. However, it needs more lo power than a single diode. Any changes in diode current due to the signal are of opposite sense in each diode, so, as the resistance of one increases, the other decreases. Over a small range these cancel each other out, reducing the even order distortion products. The diodes have to be matched for junction capacitance, series resistance and forward voltage drop at the particular current. Ideally they

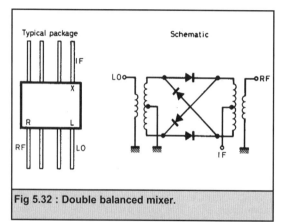

Fig 5.32 : Double balanced mixer.

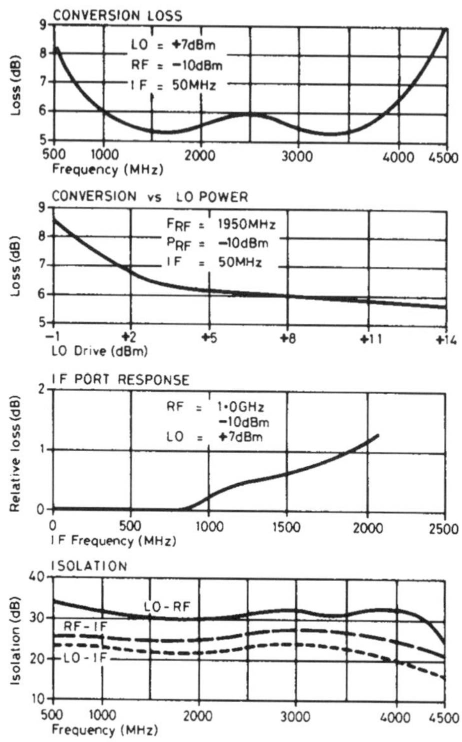

Fig 5.33 : Characteristics of a 800 to 4,000MHz double balanced mixer.

would be on the same chip.

Double balanced mixers

In this type (see Fig 5.32) both the lo and signal inputs are cancelled at the i.f port. The diodes act as a single pole changeover switch which reverses the connections at the lo frequency. This is a very close approximation to an ideal mixer, and characteristics of a typical device are shown in Fig 5.33. It is a good design, since all the ports are well isolated from each other over a broad frequency range.

The double balanced mixer is particularly popular at vhf and uhf, and usually purchased in an eight pin rectangular metal can. They are available from a number of manufacturers, e.g. Minicircuits, Anzac and Avantek. Alternatively, you can make your own by mounting diodes in a microstrip circuit; this method is more common at microwave frequencies.

Product return mixer

This is also known as the image rejection mixer. Any product at the i.f port which is derived from the rf input signal must have taken energy from the signal input to generate it. This energy detracts from that available for the wanted output. Thus all unwanted outputs represent wasted power and increased conversion loss. An ideal mixer would be able to convert all the signal power to the i.f. In practice, the losses vary from 4 to 10dB.

The product return mixer reduces the conversion loss

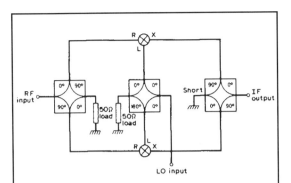

Fig 5.34 : Product return mixer.

between the rf input and the i.f output by returning any unwanted mixer products to the mixer, so the energy in them is reflected back for another attempt at conversion to the wanted output frequency. The equivalent circuit is shown in Fig 5.34.

If F_{lo} and F_{rf} are mixed, most of the energy goes into F_{lo} + F_{rf} and F_{lo} - F_{rf}. However, because of the high level of the lo, its second and third harmonics are also present in the mixer at quite high level. These will also mix with the input signal to give outputs at 2 x F_{lo} + F_{rf}, 2 x F_{lo} - F_{rf}, 3 x F_{lo} + F_{rf} and 3 x F_{lo} - F_{rf}, as in Fig 5.35. The correct impedance must be presented to all the unwanted products to minimise the conversion loss. Since many of them are well above the normal operating frequencies of the mixer, it is evident that for high efficiency the mixer circuitry must be designed to function well at many times the normal working frequencies. One of these is the frequency that

would correspond to the image response if the mixer were used in a receiver; hence the name image rejection mixer is also used.

When correctly set up, this type of mixer will only produce either a sum or a difference output frequency the unwanted one will be suppressed by about 20dB. It is therefore somewhat similar to an ssb generator.

There are several ways of recovering the energy from these unwanted products. They can all be thought of either as reflecting them back into the mixer in the appropriate phase, or terminating the mixer ports in the appropriate impedance at all the unwanted product frequencies. Some of the techniques used are:

• Separating the unwanted products from the wanted output using filters, and returning them to the i.f port by reflecting them in the appropriate phase with a mis matched load.

• Designing the circuitry in the immediate vicinity of the diode so that the unwanted products are terminated in open or short circuits.

• Using broadband phase shifter and combiner techniques to return the products.

The familiar G3JVL transverter used on 10GHz applies these methods to achieve a very low conversion loss from a single diode mixer. The diode is mounted across the waveguide between two narrowband filters one for the local oscillator and one for the signal frequency. The image frequency is reflected by both of these filters back to the diode, and hence the mixer is often called an image recovery mixer.

Termination insensitive mixers

This class of mixers is similar to the product return mixer, in that the impedance presented to unwanted mixer products at the i.f output does not affect the intermodulation performance of the mixer. One way of achieving this is to prevent the unwanted products from emerging at the i.f port.

Fig 5.36 shows such a mixer. Isolation between each hybrids opposite ports allows the lo to control independently the switching action of alternately conducting diode sets.

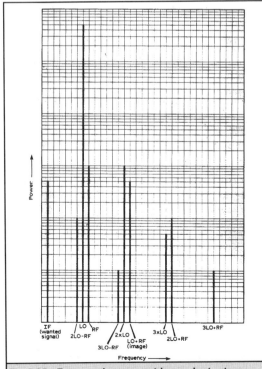

Fig 5.35 : Frequencies present in product return mixer.

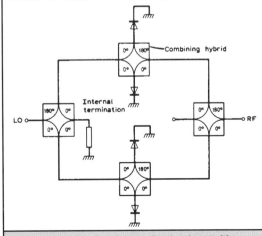

Fig 5.36 : Schematic of a termination insensitive mixer.

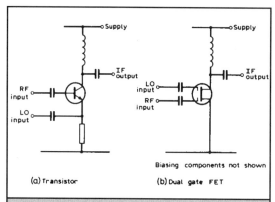

Fig 5.37 : Using transistors and fets as mixers.

Owing to its complexity, this type is generally only used at vhf and uhf in high-performance equipment.

Parametric mixers

All the mixers described so far have been passive, in that the mixer has always been a lossy, variable resistance device. If a reactive device such as a varactor diode were used, then losses could be further reduced as no power could be dissipated in the diode. (In fact, under the right conditions to lo can act as a pump and some gain obtained, as in a parametric amplifier). These are particularly popular for high level mixers in the microwave area, as multiplication and mixing can be performed in the same diode. Otherwise they are similar to a single diode mixer.

Transistor and fet mixers

Transistors and fets are most often used as mixers in the same way as the simple diode mixer (see Fig 5.37). The main difference is the slightly lower lo power requirements and conversion gain provided.

Transistor mixers have much in common with the simple diode mixer, as the mixing action occurs in the base emitter diode of the transistor. They are essentially the same as a diode mixer followed by an amplifier. FET mixers are somewhat different, as their gate source characteristic is closer to a square law, in contrast to the exponential diode characteristic.

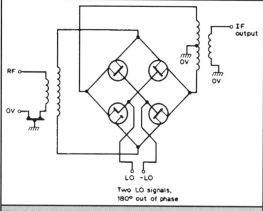

Fig 5.38 : FET ring mixer.

A real benefit comes from using a dual gate fet, and applying the signal to one gate and the lo to the other. This gives a considerable amount of isolation between the two ports.

Transistors are often used as transmit mixers when high level mixing is required. Such a mixer can give an output of several watts from an input of several hundred milli-watts.

Another configuration is to use the fet as a voltage controlled switch, as in Fig. 5.38. The drain and source are treated as a switch and connected in the same circuit, as for example a balanced diode ring mixer. The lo is then applied to the gates and used to control the resistances of each of the channels. This technique can give good lo isolation and very wide dynamic range mixers, provided a suitable type of fet is used, generally a medium power device. Such circuits have been described by Ulrich Rohde in [2].

Dual-gated GaAsfets are available quite cheaply for use at around 1GHz, mainly intended for tv tuners and they can be used at the lower microwave frequencies. At higher microwave frequencies single gate GaAsfets are becoming popular as mixers, mainly because of the low lo power required and the low noise gain they provide.

Multipliers

Multipliers are used to convert power at low frequencies, where it can be easily generated, to higher frequencies where it cannot! Conventional varactors are used as doublers or triplers, while for higher harmonics step recovery varactors are used. Typical applications are to produce several watts at a few gigahertz or tens of milliwatts at 10-20GHz, starting from uhf (typically around 400MHz).

Transistor amplifiers are often used as multiplier stages. One reason for their high efficiency is that the collector base junction is acting as a varactor diode, and the harmonics are generated more by this effect than by signal clipping in the amplifier stage.

A band pass filter is always needed on the output of a multiplier to select the required harmonic of the input signal. A filter is also needed on the input to stop the energy in the harmonics from getting back into the source. In a low order varactor multiplier, other tuned circuits (idlers) are sometimes used to put a specific impedance across the diode at the intermediate harmonics to improve the efficiency.

The multiplier is thus built combined with a set of resonant circuits. A simple doubler is illustrated in Fig 5.39. For high order multiplication the input will normally be in coaxial line and the output in waveguide. A typical design is shown in Fig 5.40.

Selecting a step recovery diode

In selecting a diode for an application there are five important parameters to consider. These are the diode capacitance C_t, the snap time T_s, the lifetime T_l, the breakdown voltage V_b, and thermal resistance θ_j.

The capacitance at the normal dc operating bias should have a reactance of between 30 and 60Ω: see Fig 5.41. For example, at a 10GHz output frequency, a good value for C_t is 0.5pF.

The snap time is the time taken for the diode to switch

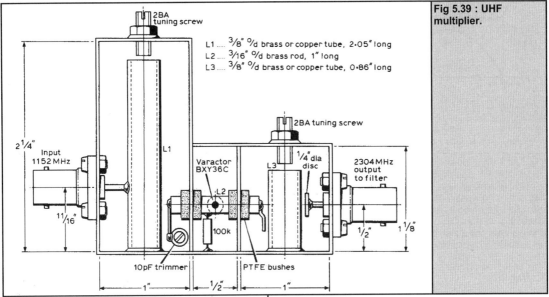

Fig 5.39 : UHF multiplier.

L1 ..... ³/₈" °/d brass or copper tube, 2·05" long
L2 ..... ³/₁₆" °/d brass rod, 1" long
L3 ..... ³/₈" °/d brass or copper tube, 0·86" long

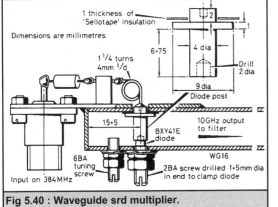

Fig 5.40 : Waveguide srd multiplier.

from a conducting to a non conducting state. It should be less than the reciprocal of the output frequency (see Fig 5.42).

The lifetime is a measure of the time required for the stored charge to be recovered, and must be long enough for the rf current to reach a negative peak before it snaps. The lifetime should be at least 10, and preferably 20 to 30, times the period of the input frequency. See Fig 5.43.

The breakdown voltage, along with thermal resistance, determines the power handling. Note that a low thermal resistance increases the diode capacitance and thus lowers its maximum frequency.

Bias resistors

All diodes require a bias resistor to provide a path for the dc current resulting from the diode rectifying the rf. This current flowing in the resistor must produce the correct bias voltage across the diode. This value can be estimated for srds using the following expression:

$$R_{bias} = \frac{5 \times T_1}{N^2 \times C_t}$$

Where T1 is the lifetime, N is the order of multiplication and Ct the diode capacitance. The resistor should be non inductive.

Efficiency

Efficiences of around 30-50 per cent can be obtained for doublers and triplers. To obtain these efficiencies the impedances around the diode need to be defined at all the frequencies being generated for a tripler a circuit is

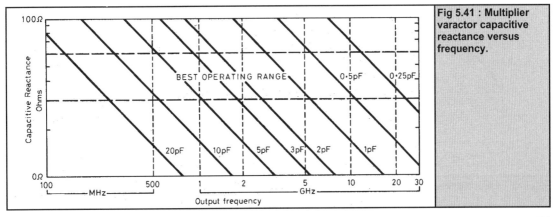

Fig 5.41 : Multiplier varactor capacitive reactance versus frequency.

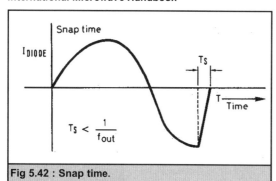

Fig 5.42 : Snap time.

particularly necessary at the second harmonic, known as the idler. For higher order harmonics the efficiency is lower, e.g. at the 20th harmonic the efficiency is around one per cent using a single diode. A higher efficiency can be obtained by using several low order stages in cascade. This is at the expense of complexity and possibly stability.

Reliability

In order to guarantee reliable operation of a multiplier, a number of tune circuits are needed to ensure that energy at the various frequencies is confined to the correct places. In general the input requires a band pass filter on the input frequency, the output a band pass filter on the output frequency.

This prevents changes in load impedance at other frequencies from altering the circuit operation.

For good reliability the multiplier must be isolated from changes in the impedance seen at the input and output. Isolators (or circulators) can provide this with minimal loss, but may not be easily obtainable by the amateur. Note that filtering is still required, as isolators or circulators will not work at harmonics of their nominal frequency. Alternatively, a resistive attenuator on the input can be used, if sufficient drive power is available, to isolate the input. At least a 3dB attenuator is desirable; a 6dB one is preferred. Such isolation is particularly important when several multipliers are cascaded, otherwise they interact, become very difficult to set up and may become unstable as thermal and other changes occur with prolonged use.

Heatsinking is important for medium and high power diodes. Note also that a bad mismatch on the output when running at high efficiency might double the dissipation in the device, causing failure if it is operating near its maximum ratings.

There is not much difference between a parametric amplifier and a varactor multiplier in theory, and the varactor multiplier can thus have regions of negative resistance which may produce outputs on a variety of spurious frequencies. This effect is known as the output breaking up, and usually results in the device producing a very broad output spectrum. The total output power can be similar to that produced in normal operation, and so adequate test equipment is needed to check operation of the multiplier before it is connected to an antenna. It should be noted that an improperly matched load could lead to such instability.

Gunn diodes and oscillators

The Gunn effect

The Gunn effect was discovered in 1960 by the British physicist J B Gunn:

> *"When I pushed the electric field up to the neighbourhood of 1,000 to 2,000V/cm something entirely unexpected happened. Instead of simple variation of current with voltage, all hell broke loose the current started to jump up and down in a completely irregular way that very much resembled electrical noise mechanisms I knew. The current variations were the order of amperes rather than the nano amperes you normally see..."*

The Gunn effect proved to be the first of several solid state microwave oscillator effects, which have since become well established in many radar and communications applications.

Gunn diodes are made from gallium arsenide (GaAs). The electrons in GaAs can be in one of two conduction bands. In one band, the electrons happen to have a much higher mobility than the other. The electrons are initially in the higher mobility band but, as the electric field is increased, more and more are scattered to the lower mobility band. The average electron velocity lies somewhere between the individual band velocities. As more electrons are scattered to the lower band, the average velocity drops. The electric field at which the velocity begins to drop is called the threshold field, and in GaAs is 320V/mm. Since current is proportional to electron velocity, and voltage is proportional to electric field, the device characteristic will have a region of negative resistance. This is illustrated in Fig 5.44.

Negative resistance is a somewhat strange concept, so a brief explanation will be given. Resistance is defined as the slope of the line from the origin to the point on the V-I

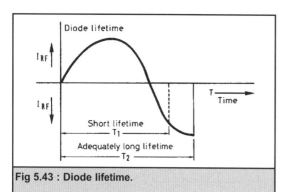

Fig 5.43 : Diode lifetime.

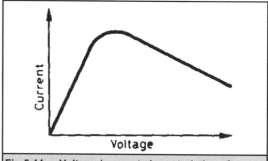

Fig 5.44 : Voltage / current characteristics of a Gunn diode.

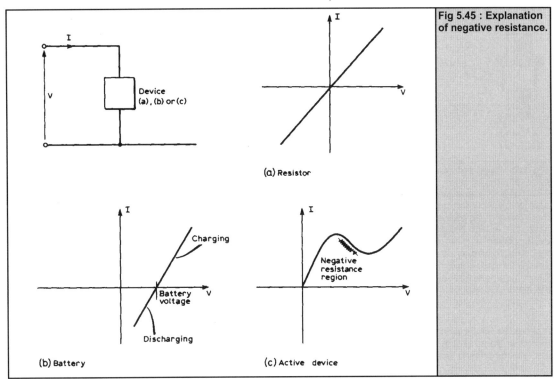

Fig 5.45 : Explanation of negative resistance.

(a) Resistor

(b) Battery

(c) Active device

characteristic, R = V/I. For a resistor this slope is constant and positive, as in Fig 5.45(a), so the resistor always absorbs power. A device that generates power, e.g. a battery, has negative resistance and this is shown in Fig 5.45(b). Active devices normally only exhibit negative resistance over a small part of the V-I characteristic (i.e. negative differential resistance), and so considered over their whole characteristic do actually absorb power. This is shown in Fig 5.45(c).

The current through the device takes the form of a steady dc current upon which are superimposed pulses of current (Fig 5.46). These pulses are due to the transit of so called domains of electrons across the device, and propagate at the electron velocity (10^5m/s). The repetition frequency of the pulses is governed mainly by the thickness of the layer of GaAs, so the diode must be made such that its thickness is close to a transit time length at the frequency of interest. This ranges from 20µm for a C-band (4GHz)

device to 1.5µm for a Q-band (40GHz) device (the upper practical limit). A transit time length is defined as the distance an electron travels during one rf cycle. It is expressed as:

$$L = v \times T$$

or $$L = v/f$$

where v is the electron drift velocity and f is the frequency. This shows that the frequency of oscillation of a device may be varied by changing v with the bias voltage.

The active part of the device consists of a thin layer of n-type GaAs grown epitaxially on a substrate of low resistivity GaAs about 60µm in diameter. This substrate acts as a connection to one side of the epitaxial layer (anode), and in turn is bonded to the metallic base that forms a heatsink. A lead bonded to a metallic film evaporated onto the other face of the epitaxial layer, or to

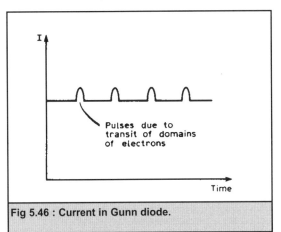

Fig 5.46 : Current in Gunn diode.

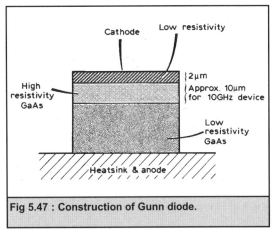

Fig 5.47 : Construction of Gunn diode.

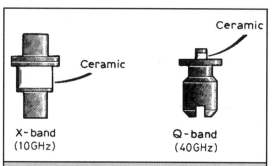

Fig 5.48 : Typical Gunn diode packages.

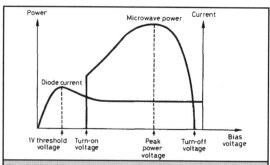

Fig 5.50 : Characteristic shape of a Gunn oscillator's bias power curve.

a second layer of low resistivity material (which produces more reliable devices), forms the second connection (cathode). In high power devices, the heatsink is formed by depositing a thick (50μm) layer of metal with high thermal conductivity, such as gold, silver or copper, directly on the surface of the semiconductor slice. This is known as the integral heatsink process. Fig 5.47 shows the active part of the device, and Fig 5.48 shows typical packaging for X-band (10GHz) and Q-band (40GHz) diodes.

It should be appreciated that the device is not a diode in the conventional sense, since it does not consist of a pn junction, but rather a GaAs resistor biased into the region of negative resistance.

Gunn oscillators

A Gunn oscillator is made by mating a diode with a microwave circuit that provides a positive resistance equal to the negative resistance of the diode. The frequency of operation is determined by the resonance of this circuit. A simplified rf equivalent circuit of such an oscillator, excluding the dc paths, is shown in Fig 5.49. It consists of a lossless resonant circuit (L and C) in parallel with three resistors

- R_1 represents the losses in the tuned circuit.

- R_0 represents the load to which the oscillator is coupled. The power dissipated in this resistor is the output power.

- R_d represents the negative differential resistance of the Gunn diode. It will usually be negative.

The L-C circuit would oscillate continuously if it were lossless (all resistors of infinite value). In reality losses will be present, so oscillations, once started, would die away. The circuit is made to appear lossless by placing a negative resistance (the Gunn diode) in parallel with it, so

that the negative resistance cancels out the positive ones corresponding to the circuit losses and load. The condition for stable oscillation is then that $R_1//R_0 = -R_d$ (//denotes in parallel with). When no power is taken out R_0 is infinity, so $R_1//R_d = 0$, i.e. $R_1 = -R_d$. This is the condition for oscillations to just start, and corresponds o the turn-on voltage in Fig 5.50. It depends on the Q of the tuned circuit, i.e. the lower the Q, the greater the difference between $V_{threshold}$ and $V_{turn-on}$. This is all just another way of saying that the Gunn diode replaces power dissipated in the circuit losses and the load.

The value of R_d varies with the bias on the diode. If R_d can be made smaller than R_1, then R_0 can be decreased to the point where $R_0//R_1 = -R_d$ again. So the smaller the negative resistance, the more power can be taken from the circuit.

A particularly important characteristic of a Gunn oscillator is its power bias swept response. Fig 5.50 shows a typical response. Power turn on occurs at some voltage above the V-I threshold voltage, depending on the Q. For voltages above turn on the power increases until the power peak voltage is reached, when the negative resistance is at a maximum. Going still higher in voltage causes the power to drop, and eventually oscillations cease at the turn off voltage. Obviously it is better to operate close to peak power, except that this may make it difficult to apply frequency modulation superimposed on the dc bias.

Gunn oscillator cavities

There are three main types of Gunn oscillator cavity, based on different ways in which the waveguide cavity is formed: (a) the post coupled cavity; (b) the iris coupled cavity; and (c) the reduced height waveguide cavity. These are illustrated in Fig 5.51.

Of these, probably the most widely used and most successful is the iris coupled waveguide cavity. It offers many advantages, such as high stability, low fm noise, wide mechanical tunability and low susceptibility to load variations pulling the frequency (load pulling). It is a relatively simple circuit to scale in frequency and power level, and also simple to build with basic metalworking facilities.

The fundamental mode of the cavity is excited when the distance from the iris to the effective rear wall of the cavity is half a guide wavelength. Lower power diodes (less than 25mW) work best when mounted centrally, while higher power ones (greater than 200mW) work best when offset from the centre of the guide. However, the post position is not usually a critical parameter.

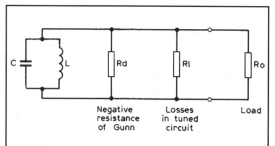

Fig 5.49 : Simplified equivalent circuit of Gunn oscillator.

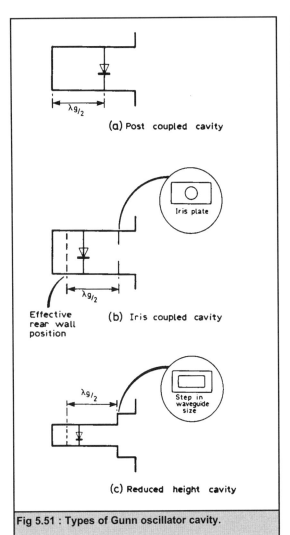

(a) Post coupled cavity

(b) Iris coupled cavity

Effective rear wall position

(c) Reduced height cavity

Fig 5.51 : Types of Gunn oscillator cavity.

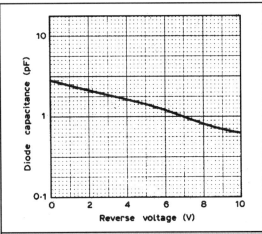

Fig 5.52 : Typical diode capacitance versus reverse voltage for microwave tuning varactor.

Output coupling is proportional to the iris size. Although circular irises are very popular, elliptic and square ones work equally well. Over enlargement of the iris will over couple the cavity, causing the power to drop, and eventually oscillations cease. Enlarging the iris also lowers the cavity frequency, and this change in frequency can be 5 to 10 per cent between very loose coupling and critical coupling (where the power output is a maximum). A good rule of thumb is to design the cavity to operate initially 10 per cent above the highest frequency required; it can then be tuned down to frequency using the methods described below.

Oscillator tuning

There are three accepted methods of tuning Gunn osciallators:

- metallic tuning screw
- dielectric tuning screw
- varactor tuning.

Metallic tuning screw

In this method of tuning, a metallic screw is introduced

through the top wall parallel to the E-field plane, which has the effect of lengthening the cavity and hence reducing its resonant frequency. The screw may be introduced into the side wall of the cavity, in which case it has the effect of shortening the cavity and hence raising its resonant frequency. Unfortunately, in this second case the electric field in the region of the side wall is relatively small so that the tuning range is also small, and hence broad wall tuning is usually used.

The tuning rate of a metal screw depends on the degree of penetration, and ranges from a few megahertz per turn as it just enters the guide to tens of megahertz per turn when several millimetres are inserted. Consequently it is desirable that the tuning screw should be operated with the minimum penetration for the most reliable operation. This implies that the highest frequency required should be set by other means (e.g. by a separate bandset screw).

The chief problem with metallic screws is obtaining a reliable, precise point of contact between the screw and the cavity wall, as any mechanical shortcomings can cause erratic tuning and poor stability.

Dielectric rod tuning

In this method the cavity may be loaded with a low loss dielectric material, the behaviour of which may be considered to be similar to a dielectric capacitor acting in parallel with the cavity capacitance. Dielectric tuning screws have the great advantage over metal screws in that they require no contact with the waveguide and cause less of a disturbance to the field in the cavity, so that they afford a very smooth tuning characteristic.

A dielectric rod inserted midway between the iris and the post serves as a smooth, convenient method of tuning. The widest tuning range (up to 30 per cent) is obtained with materials of high dielectric constant (e.g. ceramic), though ptfe (dielectric constant 2.1) allows up to 10 per cent range that is usually satisfactory for amateur use. Nylon can also be used. Tuning range may be increased by using a larger diameter tuning rod, but excessively large rods can cause loss of power by acting as circular waveguides, radiating the signal to the outside world and degrading the stability. At X-band the rod diameter should not exceed about 5mm. The tuning rate of the dielectric screw is linear (decreasing frequency with increasing

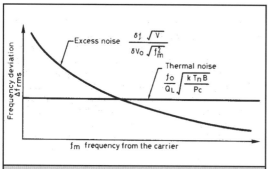

Fig 5.53 : Excess and thermal components of a Gunn oscillator's fm noise spectrum.

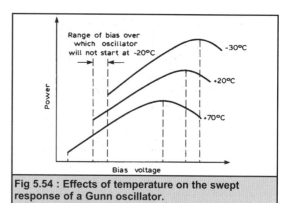

Fig 5.54 : Effects of temperature on the swept response of a Gunn oscillator.

depth of penetration) except at the extremities of the cavity, i.e. the screw first entering the cavity or approaching the opposite wall.

Varactor tuning

The third method of tuning is by varactor. Varactor diodes have a capacitance characteristic that is continuously variable with bias voltage, from a small value at large reverse bias to a large value at forward bias (Fig 5.52). The Q of the diode is maximum at high reverse bias, decreasing steadily towards zero as the bias is reduced and reversed. A varactor may be coupled to the cavity either by a post or by a loop, though posts are most common as they make use of the same feedthrough hardware as the Gunn diode post.

The amount of tuning that a varactor can provide is controlled in two ways:

• the capacitance of the varactor

• the coupling between the varactor and cavity

Raising the capacitance reduces the amount of tuning. This is because the varactor chip, which acts as a variable capacitor, is in series with the package ribbon inductance. If the chip capacitance is very large, its reactance is quite small compared to the reactance of the ribbon. This means the series combination looks almost like the ribbon inductance alone. Since tuning is accomplished by changing the reactance of the series combination, a large chip capacitance provides very little tuning. Only when the reactance of these elements is nearly equal can considerable tuning range be achieved. When they are equal, a resonance occurs which results in the varactor being very lossy. This is thought to be due to the high rf voltage, which occurs at resonance across each element, shifting the varactor into forward conduction where, as has already been noted, its Q is low. This effect is observed as a suck-out of the rf power at some point in the oscillators tuning curve. The cure is to use a higher capacitance varactor in order to lower the resonant frequency below the operating frequency. At 10GHz, varactors should have zero bias capacitances exceeding 1pF. At 6GHz, the capacitance should exceed 2pF.

A post coupled varactor will provide maximum tuning when the post is near the cavity centre. As the post is moved towards the side wall, coupling into the varactor can cause considerable power variation across its tuning range. This can only be cured by decoupling the varactor, or by using a lower loss varactor.

On balance, the dielectric rod tuning is probably the most

attractive tuning method for the amateur, being simple and easy to build.

Starting problems

Turn on problems can plague a cavity oscillator and are caused by oscillations occurring at spurious cavity resonances. The problem is to identify the causes of these resonances, which are most likely to be due to:

• higher order cavity modes

• coaxial modes associated with the diodes mounting post

• radial modes which are set up in the bias feedthrough structure

Higher order cavity modes are controlled by moving the mode cut off frequency as high as possible and a simple way to accomplish this is to use a reduced height waveguide. The guide height has no effect on the fundamental mode cut off frequency, but higher order modes are cut off at progressively higher frequencies by reducing the height. However, a price is paid in terms of a lower cavity Q. A good rule of thumb is to use the standard waveguide height for the next waveguide band above the desired operating frequency.

Coaxial modes have a TEM field pattern in the immediate vicinity of the diode and its post. The resonant frequency of this mode is determined by equating the combined post and diode height to half a wavelength. Coaxial modes may be interrupted by placing the diode in the middle of the post as this will double the resonant frequency, but the price to be paid is lower output power.

Another source of starting problems is the bias feedthrough structure. Ideally the feedthrough should appear as a short circuit at all frequencies above dc. Unfortunately many choke structures can support resonant radial modes inside their sections. There is no easy solution to this problem; each situation must be treated individually. Another type of feedthrough that has been used successfully is the simple rf capacitor, either the disc or tubular type. The trick seems to be to get the capacitor as close as possible to the microwave circuit, which with a waveguide cavity oscillator means getting the capacitor very close inside the guide cavity.

Gunn oscillator noise

A limitation of Gunn oscillators is their high level of fm noise. Measurements have shown that this originates from two sources. The first is a thermal source that contributes a white deviation spectrum, and the second is

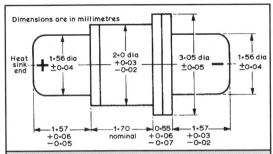

Fig 5.55 : The outline of a typical low power Gunn diode encapsulation.

a low frequency flicker source. Gunn oscillators are bias tunable and so bias fluctuations will randomly push the oscillators frequency.

Analysis of thermal fm noise leads to an rms deviation which is inversely proportional to the cavitys loaded Q multiplied by the square root of the carrier power. Also, measurements have shown that the fm noise close to the carrier is directly proportional to the oscillators voltage pushing. Thus, to minimise the noise here, the bias voltage should be adjusted to a region where the bias frequency curve is most nearly flat. This region often occurs near the maximum safe bias voltage. A typical plot of noise against distance from the carrier for an X-band oscillator is shown in Fig 5.53.

Temperature effects

A sad story of erratic starting goes something like this. Someone builds an X-band oscillator which works well at room temperature. However, when taken out to a cold, windy hilltop, all oscillations in X-band suddenly cease. A check with a spectrum analyser would show that the only output was a little power up in Ku-band. Let us consider what happens to the swept bias response as the temperature is varied. Fig 5.54 gives the results of sweeping a diode through its bias range at -30, +20 and +70°C. No circuit changes have been made. Notice that both the turn on voltage and peak power voltage creep up as the temperature is lowered. The creep rate is about 2V/100°C for X-band diodes. For reliable operation, you must ensure that you are operating sufficiently high above room temperature turn on to guarantee turning on at the lowest operating temperature. Generally, a 20 per cent margin at room temperature is sufficient to assure starting down to minus 40°C. A related point is that the operating voltage should not exceed the high temperature power peak voltage. This is because the power will drop very quickly above its peak, and in many cases the a.m noise will also then become excessive.

A Gunn oscillator will also drift with temperature. There are two causes of this drift:

• metal expansion

• diode capacitance change

Metal expansion causes drift because the cavity grows physically longer as it is heated and this lowers the resonant frequency. The ratio of frequency change to centre frequency is the same as the ratio of length increase to original length. Copper expands at a rate of 15ppm/°C

Which means a frequency drift of 150kHz/°C. This is very close to that observed in an uncompensated cavity, but

the agreement is only close if the cavity Q is high.

A Gunn diodes capacitance changes at a rate of approximately 10^{-4}pF/°C. This will tune the cavity by an amount that is inversely proportional to Q. For instance, a typical diode will cause a temperature frequency coefficient of -1,000kHz/°C in an X-band cavity with a Q of 10. However, for a Q of 100 the drift is reduced to -100kHz/°C. If the cavity's Q is maintained in the 100 to 1,000 region, the diodes contribution to the oscillator drift will be insignificant.

In many applications, a high degree of frequency stability with temperature is necessary. A technique called temperature compensation allows Gunn sources to achieve a stability approaching that of a quartz crystal.

Temperature compensation turns metal expansion to advantage for cancelling out the basic causes of drift. This is done by mounting the dielectric tuning rod inside a metal shaft that expands with temperature. If the tuner and the shaft are only connected at the one place (top of shaft), the expanding shaft will withdraw the tuner as the temperature rises. This withdrawal tunes the frequency upward while cavity expansion is tuning the frequency downward. If everything is right, one effect just cancels the other and there is not net drift. This never happens in practice, but drift under 50kHz/°C is not difficult to achieve.

Heatsinking

Referring again to Fig 5.48, it can be seen that Heatsinking from the chip is specifically from one end. Fig 5.55 shows the outline dimensions of a typical low power diode. With such devices, heatsinking presents no major problems. Diodes may therefore be used either way up to suit the polarity of the rest of the equipment, although the thermal stability will be improved if the heatsink end is connected to the body of the cavity rather than the mounting post. To improve heatsinking, the heatsink end of the diode should be a firm fit in its mount and the contact area should be maximised by removing burrs. A tiny drop of heatsink compound may be applied to this joint. Although the encapsulation is quite strong in pure compression, shear forces should be kept very small. The flanged end of the diode may therefore be a loose mechanical fit in its post to allow for any misalignment.

With high power devices, heatsinking is crucial since several tens of watts are to be dissipated. The heat extraction end of the diode is usually mounted in a collet arrangement, and heatsink compound used to minimise the thermal resistance of the joint. Unfortunately, this joint cannot be soldered, since this would damage the diode (though soldering can be used to solve the same problem with high power varactor diodes). Since a mounting post would be incapable of conducting the heat away fast enough, the heatsink end of the diode and its mounting collet are up against the broad face of the waveguide cavity, with suitable cooling fins on the waveguide outer surface. The usual rule with power semiconductors applies: if it is too hot to hold, the heatsinking is inadequate. Overheating degrades the stability of oscillator and reduces the life of the diode.

Amplifiers

Amplifiers can be generally classed as either small signal or power amplifiers. Low noise types invariably use semiconductors nowadays, the most common being the

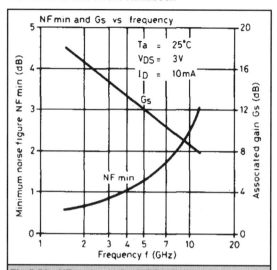

Fig 5.56 : NF and gain for MGF1402 fet.

GaAsfet. Valves are still the only means of generating really high powers, though at the lower frequencies semiconductors can now produce medium powers.

Low noise amplifiers

In most terrestrial microwave systems, thermal noise from the earth or lossy components in circuitry limit the sensitivity of receivers to a noise figure of around 3dB. For applications where antennas point out into space, benefits can be had from using receivers with still lower noise figures. The term low noise generally refers to amplifiers with a noise figure comparable with these limits; i.e. under about 10dB, and usually around 3dB or better.

In order to obtain the optimum noise performance from a low noise device, its operating conditions must be set up fairly accurately. Often, the results do not come up to expectation because one or more of the following basic requirements have not been met.

• The device must have a low inherent noise figure.

• The device must be biased to the correct operating point.

• The input matching must transform the source to the correct impedance at the device terminals.

• The input matching circuitry must have sufficiently low losses.

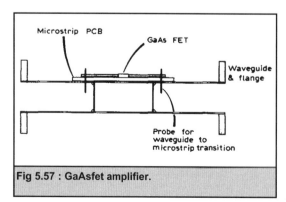

Fig 5.57 : GaAsfet amplifier.

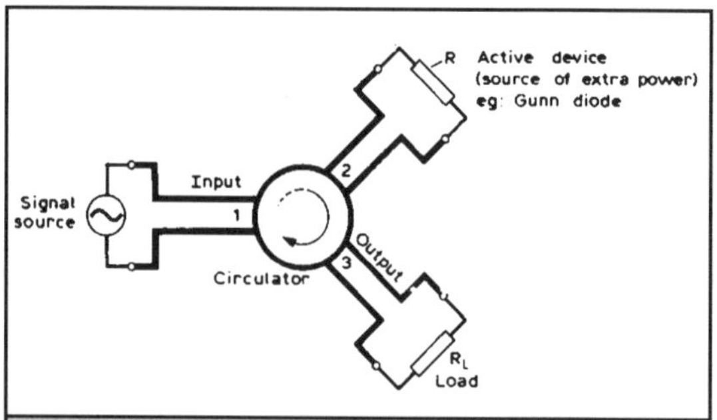

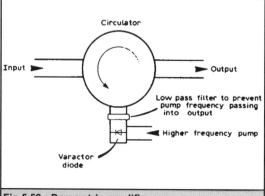

Fig 5.58 : Operation of reflection amplifier.

• The amplifier must be well screened to prevent variations in its surroundings either detuning it or introducing noise.

• The amplifier is stable.

In general, the matching conditions for optimum noise figure are not the same as those for maximum gain. Both gain and noise matching conditions are usually quoted in the manufacturers data sheets for the higher performance devices.

Vacuum devices such as travelling wave tubes can have noise figures down to about 8dB, but they cannot compete with the solid state devices.

Transistors can provide quite low noise figures of around 1dB up to a few gigahertz, while GaAsfets can provide noise figures of small fractions of a decibel up to a few gigahertz, and 1-2dB at 10GHz. Associated gains for both devices are about 10-15dB in their appropriate frequency ranges. Fig 5.56 shows, the variation of noise figure and gain for a typical fet, the MGF1402. FETs are becoming available for 10-20GHz and the progress upward in frequency is still continuing. All these devices obtain their good high frequency response by extremely small electrode geometries and spacing because of this they are fairly delicate devices.

The fet can be mounted on mirostrip on either glassfibre or ptfe-loaded pcb, but these materials are quite lossy and line lengths must be minimised. The level of loss depends on the application. For ultimate low noise figures, e.g. -30 50K around 1GHz, the device has to be mounted in the lowest loss circuitry possible. All input matching circuitry should be air line. At 10GHz it is possible to mount the

Fig 5.59 : Parametric amplifier.

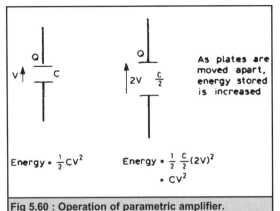

Fig 5.60 : Operation of parametric amplifier.

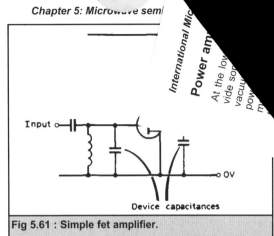

Fig 5.61 : Simple fet amplifier.

whole device and its circuitry in the waveguide, and this could give very low losses. However, it is a little more likely to be unstable as the waveguide cannot provide any defined impedances outside its normal operating range. One technique is to build an amplifier with very short input and output matching lines on a pcb mounted on the outside of the broad face of the guide, and use probes to couple through into the guide, as in Fig 5.57.

Reflection amplifiers

These rely on devices that exhibit negative resistance to provide gain. When signals are sent down a line which is terminated by a device which shows negative resistance, the power reflected from the device is larger than the incident power. This can be seen if negative values are used in the expressions for reflection coefficient. The only problem is to separate the incident and reflected waves. This is usually done using a circulator, as shown in Fig 5.58. The input signals enter port 1 of the circulator, and leave at port 2 to go to the negative resistance device, which may be a Gunn or avalanche diode. The reflected power then enters port 2 and leaves via port 3. The maximum gain is limited by the isolation of the circulator and the source and load matches, as power reflected from the load can get back to the input from port 3 to port 1.

Parametric amplifiers

A parametric amplifier produces gain by altering the parameters of a circuit at a frequency of many times that of the one to be amplified. This type of amplifier is capable of extremely low noise figures, e.g. 3K, if cooled to a low temperature e.g. with liquid helium. They are used where the ultimate in low noise figure is required, such as in radio telescopes. A typical arrangement is shown in Fig. 5.59. The low noise figure is mainly due to the diode being a pure reactance; it has few losses that would contribute thermal noise to the signal.

They operate by transferring energy from the high frequency pump signal to the lower frequency to be amplified. In practice this is usually done using voltage variable capacitors across a tuned circuit, in the form of variable capacitance diodes. The operation of the circuit can be more easily explained if the variable capacitance were made by mechanically varying the spacing of the capacitor plates.

The tuned circuit is fed with a constant amplitude sine wave, and the value of the capacitance can be varied either mechanically, by moving the plates apart, or electrically. When V is a maximum the capacitance is reduced e.g. by separating the plates. The charge on the plates does not change so, since Q = C x V, the voltage will increase. The capacitance is increased to its original value when the voltage is at zero. The plates will have been pulled apart against an attractive force, requiring work to be done: they are moved together against no force as there is not voltage across the plates. Thus, there has been a net input of energy to the circuit, as illustrated in Fig. 5.60.

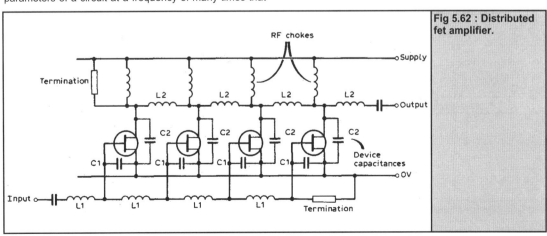

Fig 5.62 : Distributed fet amplifier.

...plifiers

...er microwave frequencies transistors can pro-
...e tens of watts at fairly low gains per stage, but
...m devices are still the best choice for higher
...ers, whether for amplifiers or oscillators. Powers of
...any hundreds of watts are possible, together with many
tens of decibels of gain.

At higher frequencies, e.g. 10GHz, GaAsfets give output powers of several watts, with gains of 5 to 10dB per stage. For more power, valves must be used; more information is given later in the section on valves.

Distributed amplifiers

This technique is a means of combining many devices in parallel to give increased output power, or to produce very broadband amplifiers.

Capacitance from input and output terminals to earth in the device sets an upper limit to the high frequency response by reducing the input and output impedances. The effects of this capacitance are usually removed by tuning them out with a parallel inductance, giving the familiar tuned amplifier configuration, Fig 5.61. The disadvantage of this method is that the Q required from the tuned circuit becomes very large at high frequencies, and the bandwidth is consequently very narrow. A point is finally reached where the losses in the tuned circuits become unacceptable.

A distributed amplifier can provide a constant gain from dc to the same upper frequency limit, but this time with a completely broadband, flat response. The circuit is shown in Fig 5.62. It consists of a number of similar devices effectively connected in parallel, but used in such a way as to produce a lumped constant transmission line using input and output capacitances.

The load impedance presented to each device is the characteristic impedance of the transmission line:

$$Z = \sqrt{\frac{L}{C}}$$

The upper frequency limit is determined by the resonant frequency of the L and C combination, which forms a multi element low pass filter.

The effect is to produce a wave travelling along past each device in turn, which at the same time produces an amplified version in the output line. The phase of the

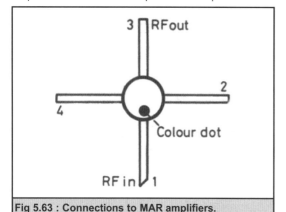

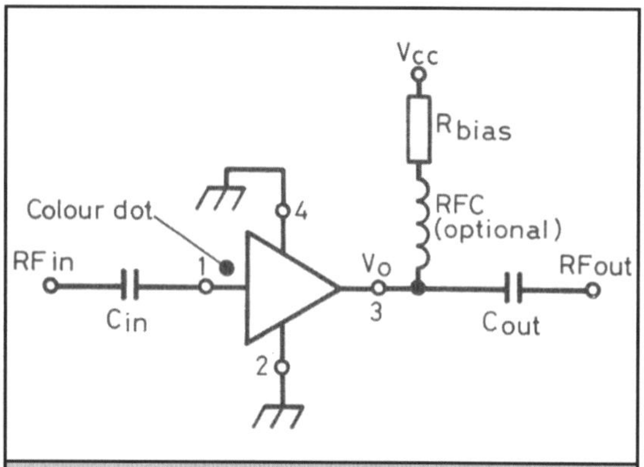

Fig 5.64 : Connections to the MMIC could hardly be simpler.

output signal is such that they combine to produce a wave travelling in the same direction in the output line.

One advantage of operating devices in parallel is that it reduces the output resistance, so that less matching has to be done in tuned circuits.

Monolithic microwave integrated circuits

Monolithic Microwave Integrated Circuits (MMICs) have appeared in a number of transmitter and receiver designs over recent years. They are easily available and offer a number of useful features for constructors. With just two capacitors and one resistor , it is possible to construct an amplifier for stable operation over a wide frequency range. (Fig 5.63 And 5.64).

The amplifiers listed in Table 5.6 and 5.7 Except MAR-8 have 50Ω input and output impedances. The MAR-8 has a complex impedance which is frequency dependant and readers are advised to consult the manufacturers data sheet for this one. Note that the load resistor is effectively in parallel with the output impedance and at low supply voltages could cause a mis-match. This can be corrected by adding an RF choke as shown.

The total impedance of resistor and choke should typically be 500 - 1000Ω at the lower end of the operating frequency range. Similarly, the coupling capacitors should have a low impedance. Suitable values might be 100 nF for up to 30MHz and 10nF to 300MHz and smaller value chip capacitors above this frequency. Some experimentation may be useful here.

Fig 5.63 : Connections to MAR amplifiers.

Table 5.6 : MMIC details.

Model Number	Colour Dot	Freq MHz	Gain dB (typ) at (MHz)			
			100	500	1000	2000
MAR-1	Brown	DC-1000	18.5	17.5	15.5	
MAR-2	Red	DC-2000	13.0	12.8	12.5	11.0
MAR-3	Orange	DC-2000	13.0	12.8	12.5	10.5
MAR-4	Yellow	DC-1000	8.2	8.2	8.0	
MAR-6	White	DC-2000	20.0	19.0	16.0	11.0
MAR-7	Violet	DC-2000	13.5	13.1	12.5	10.5
MAR-8	Blue	DC-1000	33.0	28.0	23.0	

Table 5.7 MMIC parameters.

Model Number	Maximum power, dBm at 1dB compression	Noise factor dB (typical)	3rd order Intercept dBm	DC power in at pin 3 Current (mA)	at Voltage
MAR-1	0	5.0	15	17	5.0
MAR-2	+3	6.5	18	25	5.0
MAR-3	+8*	6.0	23	35	5.0
MAR-4	+11	7.0	27	50	5.0
MAR-6	0	2.8	15	16	3.5
MAR-7	+4	5.0	22	20	4.0
MAR-8	+10	3.5	36	27	7.5

* +4dBm from 1 - 2GHz

A practical application of MMICs

This practical application is from an article produce by William A. Parmley, KR8L [12].

It has been said that "The future is up, in direction and frequency". Those of us who are active in the amateur satellite program certainly agree with the "direction" part of this statement. Given the steady progress of commercial communication services into the gigahertz bands, it would be difficult for anyone to argue with the frequency part.

The following paragraphs describe a preamplifier built using the MAR-6 monolithic microwave integrated circuit (MMIC) manufactured by Mini Circuits Labs. The MAR-6 is a four terminal, surface mount device (SMD) with an operating frequency range of dc to 2000 MHz (2 GHz), a noise figure of 3 dB, a gain of up to 20 dB, and input and output impedances of 50Ω. The most amazing feature of MMIC devices like the MAR-6 is that of the four terminals, two are for ground, and only three or four external components are required to make them work!

Circuit Description

The first reaction to a four terminal device with two grounds may be something like this: Lets see, ground, ground, signal in, signal out... Hey! Where does Vcc go? Take a look at Figure 5.65, which is the schematic for the preamplifier. C1 and C2 are simply dc blocking capacitors. Vcc connects to the device through the output lead via a resistor and RF choke (R1 and L1). The only other components are the bypass capacitors on the Vcc lead.

The component values are chosen as follows.

- C1 and C2 should present a low impedance at the lowest signal frequency of interest. For example, the amplifier was originally constructed as a preamplifier for use at 435 MHz, which is a downlink frequency for many amateur satellites, 220 pF disc ceramic capacitors were used (X_C = 1.6Ω at 435 MHz). Subsequently this was changed to use the preamplifier at 29 MHz for downlink signals from Russian RS series satellites, C1 and C2 were replaced with 0.001μF disc ceramic capacitors (X_C = 5.5Ω at 29 MHz).

- The power supply voltage determines R1's value. The

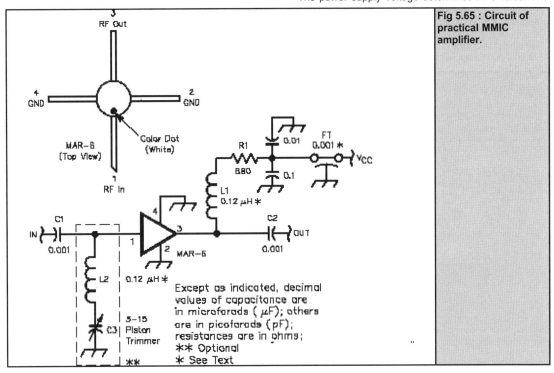

Fig 5.65 : Circuit of practical MMIC amplifier.

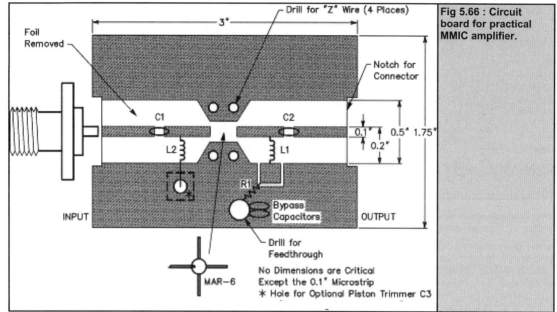

Fig 5.66 : Circuit board for practical MMIC amplifier.

MAR-6 draws about 16 mA, and needs a closeup Vcc of about 3.5 V. Use Ohms Law to calculate the necessary voltage drop from your power supply voltage to 3.5 V. The power supply used for the example provided about 14.6 V, so R1 was a 680Ω, 1/2 W resistor. Its tempting to use a 3.5 V power supply and omit R1. Dont! R1 isolates the power supply from the MMIC output. Since the MMIC output impedance is 50Ω, R1's resistance should be considerably greater than 50Ω. L1 is listed as optional on the data sheet; it helps isolate the MMIC and power supply. The recommended combined impedance of L1 and R1 is at least 500Ω. A value of 0.12 mH was selected for L1 (X_L = 328Ω at 435 MHz). The inductor is 8 turns of #30 enameled wire around the shank of a 3/16 inch drill bit, spaced for a total length of 0.3 inches. (Remove the drill bit, of course; its only a winding mandrel!) This value of L1 was adequate even when the preamplifier was used at lower frequencies.

• The remaining three essential parts are bypass capacitors. Because capacitors have self resonant frequencies (resulting from unavoidable inductances in the devices and their leads), it is a common practice to use capacitors of several different values. This design uses a 0.001μF feedthrough capacitor that passes through the circuit board ground plane. The 0.01μF and 0.1μF capacitors are disc ceramics. L2 and C3 are optional components preventing desensitisation.

Circuit Construction

It is certainly possible to build this circuit using conventional construction techniques. Such a circuit might even perform satisfactorily up through the 70 cm band. However, since one of the reasons for undertaking this project was to study and practice microwave construction techniques, that is the approach described here

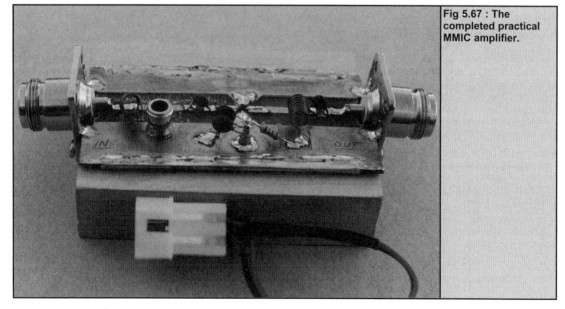

Fig 5.67 : The completed practical MMIC amplifier.

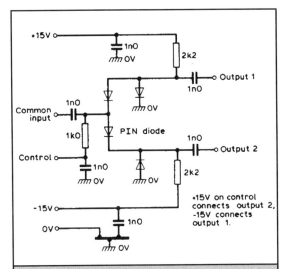

Fig 5.68 : PIN diode two way switch.

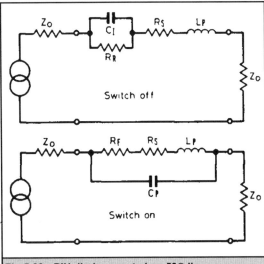

Fig 5.69 : PIN diode mounted on 50Ω line.

Figure 5.66 shows the circuit board layout. The material is a double sided, glass epoxy board with a thickness of 0.0625 inches, known as FR04 or G-10. This is the least expensive board material suited for microwave use. (The board used was a product of GC Electronics in Rockford, Illinois.) Notice that most of the top of the board, and all the bottom of the board, serve as circuit ground.

The signal conducting part of the circuit is a microstrip, the line width, board thickness and board dielectric constant determine the microstrips characteristic impedance. A 0.1 inch wide line and the ground plane on 0.0625 inch thick G-10 form a 50Ω transmission line, which matches the MMICs input and output impedance.

The MMIC is tiny, it should be connected it to the tracks with the shortest possible distances between the tracks and the body. Also, the device leads are very delicate; if possible, do not bend them at all. To fit the MMIC leads flat on the PC board tracks without bending, a small depression can be ground in the board dielectric for the MMIC body. Alternatively drill a hole completely through the board, but the first method provides better mechanical support for the device. The coloured dot (white on the MAR-6) on the body marks pin 1, which is the input lead. The other leads are numbered counterclockwise; pin 3 is the output lead.

Mount the blocking capacitors as close to the board as possible. To do this, cut the capacitor leads to about 1/16 inch long. Then tin both the capacitor leads and circuit tracks and solder the capacitors in place. This method of mounting minimizes lead inductance.

Because the preamplifier was intended for use at UHF, N type connectors were used. To achieve a zero lead length, the ends of the board are notched to fit the profile of the connectors and the connectors soldered directly to the board.

It is important that all portions of the ground foil be at equal potential, particularly near the MMIC and the board edges. To achieve this, wrap the long edges of the board with pieces of 0.003 inch thick brass shim stock and solder them on both top and bottom. Thinner or thicker material is suitable (up to about 0.005 inch), as is copper

flashing. Drill two small holes on either side of each MMIC ground lead, and fit a small Z wire in each hole and solder the Z wires to each side of the board. (A Z wire is a short, small gauge, solid copper wire bent 90°, inserted through the hole and then bent 90° again.) The inductor is mounted for minimum lead length. One lead connects to the microstrip, and the other to the square pad. The resistor connects from the pad to the feedthrough capacitor, and the other two bypass capacitors connect from the feedthrough to the ground foil.

Connection and Operation

For the basic preamplifier design there is nothing to align or adjust. Simply connect the preamplifier between your antenna and receiver and apply power. If you connect the preamp to a transceiver, take precautions to prevent transmitting through the preamp!

Switches and variable attenuators

Switches and attenuators referred to here are electrically controlled switches and attenuators, they usually consist of pin diodes. They use the ability of the diode to change resistance under the control of the dc bias. Devices which are operated from the rf level passing through them are known as limiters or t-r cells and are made from either pin diodes or gas discharge tubes.

PIN diode switches

Switches can be designed simply to reflect the incident power back to the source when in the off state, very little power is then dissipated in the diodes. This is the simplest design, but often signal sources do not like the reflected power, and so more frequently switches, attenuators and modulators are designed to present constant impedance to the source and loads. This may be done using a π or T network of diodes, or using a circulator or isolator. In these cases, the unwanted power is either absorbed in the diodes or routed to some other load by the circulator or isolator. PIN diodes are capable of controlling microwave power from microwatts to kilowatts, in either pulsed or cw operation.

Examples of the diode used as a series and shunt

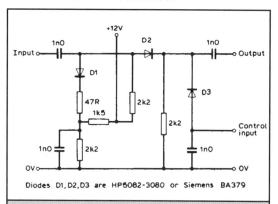

Diodes D1,D2,D3 are HP5082-3080 or Siemens BA379

Fig 5.70 : Typical pin diode attenuator at uhf.

element in a switch are shown in fig 5.68. For the switch to be on, the series diode must be of low resistance, and the shunt diode of high impedance. In the off state, these two change over. To obtain the best compromise of low insertion loss and high isolation, the values of the on and off resistances must be chosen to be suitably distributed either side of the impedance of the transmission line in which they are used.

Consider the diode in the circuit shown in fig 5.69, mounted in a 50Ω line used at 1GHz. The transmission loss is given by :

$$\text{Loss} = 10 \times \log(R^2 + X^2/(4 \times Z_0^2) + R/Z_0 + 1)$$

When forward biased at 50mA, R_f is 1Ω, so $R=R_f+R_s=2\Omega$, and $X_1=2\pi \times f \times L = 2.5\Omega$. In this case C_p can be ignored compared to X1. This gives the insertion loss of 0.2dB. When reverse biased, $R_f = 10k\Omega$ and, ignoring C_p, the attenuation is about 24dB. The junction capacitance degrades the isolation by shunting Rf, without it the isolation would be about 40dB and independent of frequency. In reality both the insertion loss and isolation degrade as the frequency is increased. The performance can be improved for narrow band operation by adding external reactances to cancel out the diode reactances. C_p and L_p should be parallel resonant at forward bias, and C_i and L_p should be in series resonance at reverse bias. The tuned switch might typically have a bandwidth of 5%. Higher isolation can be obtained by cascading these circuits, at the expense of higher insertion loss.

PIN diode attenuators

The same circuits can be used as variable attenuators by

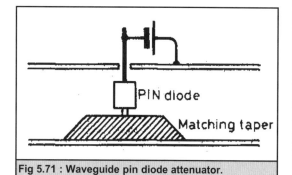

Fig 5.71 : Waveguide pin diode attenuator.

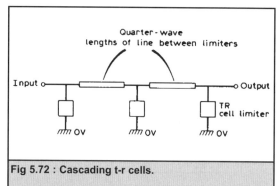

Fig 5.72 : Cascading t-r cells.

varying the dc bias smoothly between the extremes used for switching. They are particularly useful as modulators and agc stages. Another application is to control rf levels remotely, so that the rf does not have to leave the circuit board to travel to the panel controls. If designed correctly, they are unlikely to introduce any significant intermodulation distortion - amplifier and mixer stages will usually be the limiting factor.

A pi attenuator constructed with pin diodes is shown in Fig 5.70. A simple waveguide attenuator is shown in Fig 5.71. The waveguide impedance is matched to the diode impedance by the taper. The diode is placed in a position of maximum electric field and absorbs energy from the passing wave in the waveguide. How much energy it absorbs and dissipates as heat is determined by the resistance of the diode, which is in turn determined by the bias.

Limiters and t-r cells

PIN diodes can be used to limit the level of rf power that they pass to protect receiver front ends, for example. They can do this automatically without bias. When a signal of sufficiently high power appears, it will saturate the intrinsic region with charge when it forward biases it, but will not be able to remove it when the voltage reverse biases it, The average current increases and the diode impedance drops, shorting the line and reflecting most of the incident power. Effectively the diode rectifies the incident power which then biases itself on. The response time is several times the carrier lifetime, so a very short

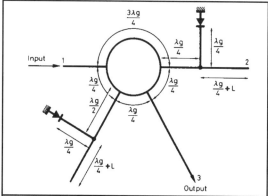

Fig 5.73 : Hybrid ring phase shifter makes use of directional properties of the coupler and uses pin diodes to switch reactive lengths of line in and out of circuit.

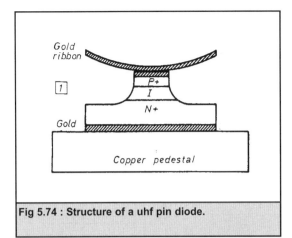

Fig 5.74 : Structure of a uhf pin diode.

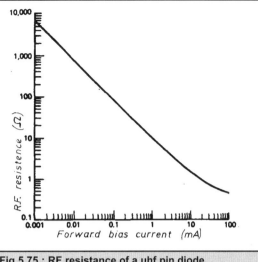

Fig 5.75 : RF resistance of a uhf pin diode.

pulse of the incident power will still get through before limiting takes place, typically this time is 100nS. The diode can however cope with quite high powers - up to several kilowatts.

T-R cells perform a very similar function, and are used to protect a receiver from the transmitter output during transmit. They were originally developed for radars which require very rapid receive transmit changeover switching, typically under a microsecond. It is not possible to perform such changeovers mechanically, and so a purely electronic means of changeover and protection was developed. They were basically gas discharge tubes, rather like neon bulbs, mounted across the waveguide or coaxial line. When the transmit power appears in the line, the tube strikes, and becomes a very low impedance which reflects most of the power, protecting the receiver. Special mixtures of gases and electrode geometries are used to minimise the time the tube takes to strike. Radioactive sources are sometimes included to lower the threshold.

An arrangement of quarter wave lengths of line are used to maximise the isolation on transmit, as shown in Fig 5.72.

Phase shifters

PIN diodes can be used to produce phase shifts by using them to switch extra lengths of transmission line. This has many applications in phased array antennas and modulators. A circuit to do this using a "rat race" (hybrid combiner) is shown in Fig 5.73, omitting the biasing details for the diodes. Without the diodes and stub lines, power entering the combiner at arm 1 splits equally between arms 2 and 4, and none appears at arm 3, as can be verified by summing the path lengths around the coupler.

When both diodes are low impedance, this is transformed up to a high impedance at the end of the quarter wave lines where they join the ring. When the diodes are biased off, and of high impedance, the lengths of each stub is increased by the length L. Switching the diodes changes the phase of the output by $180 + 2 \times L \times 360/\lambda_g$ degrees. By suitable choice of L, any phase shift between 0 and 360° can be produced.

Microwave T/R switches using low cost PIN diodes [14]

The use of PIN diodes is interesting because low cost diodes, such as those manufactured for consumer appli-

cations, can be used achieving both low insertion loss (a fraction of a dB) and a power handling capability up to 100 watts or more.

Such switches are, therefore, a valid solution for terrestrial communications, as the small increase of the noise figure does not change significantly the overall receiver sensitivity. Also these switches have the advantages of unlimited lifetime and no degradation when operated in extreme environmental conditions (i.e. when they are located near the antenna). There are many possible configurations for PIN diode switches, the classical Single Pole, Double Throw switch using low cost PIN diodes originally designed for UHF TV tuners is described here.

Properties of PIN diodes

The PIN diode is so called because the semiconductor material has a sandwich structure with three layers and manufactured using Silicon crystal (Fig 5.74).

- One layer P doped at one end.

- One Intrinsic layer in the middle, therefore a pure non doped layer characterised by very high resistivity.

- One N doped layer at the other end.

When the diode is forward biased both negative charges (electrons) and positive charges (holes) are injected in the intrinsic layer and these charges combine very slowly. Due to this fact conduction also continues for a short time if the polarity is reversed. This means that the rf period is shorter than the relaxation time of the injected charges, the diode loses its rectifying properties and its behaviour is just like a linear resistance.

Fig 5.75 Shows the equivalent series resistance of a typical PIN diode v.s. the biasing current. The rf resistance drops to a few ohms with the biasing current of a few mA. Obviously there is a minimum operating frequency below which the PIN diode behaves like a rectifier, this frequency is the inverse function of the carrier lifetime in the intrinsic layer. Modern PIN diodes have a low minimum operating frequency since they are used in the front ends of top class hf communications receivers, withstanding large signal levels as switches of pre selection filters or electronic attenuators.

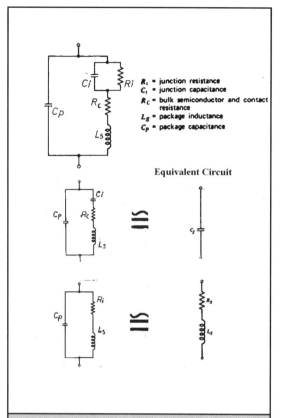

Fig 5.76 : Equivalent circuit of pin diodes.

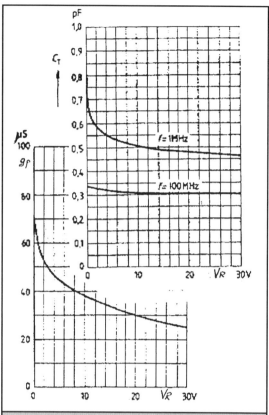

Fig 5.77 : Conductance and capacitance.

The upper frequency limit of a PIN diode is determined bt two major parameters :

- The diode capacitance when the diode is in the open state (i.e. without the biasing current).

- The parasitic series inductance when the diode is in the closed state (i.e. with biasing current).

The capacitance is mainly the intrinsic one associated to the PIN junctions, with minor contributions due to the package of the device. On the other hand, the parasitic inductance is due only to the package itself and to the interconnecting wires to the semiconductor chip.

Both the capacitance and the parasitic inductance of the PIN diode limit the degree of isolation between RX and TX ports and the SWR achievable when the device is used as an rf switch. The series resistance and the thermal properties of the package determine the insertion loss characteristics and the maximum power level that can be handled.

Fig 5.76 Shows the equivalent circuit of a PIN diode and its approximation for on and off states. The series resonance F_c, or cut off frequency, is the absolute upper limit. It is possible to design T/R switches with good performance up to about 1/3 of the cut off frequency by

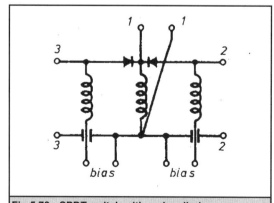

Fig 5.78 : SPDT switch with series diodes.

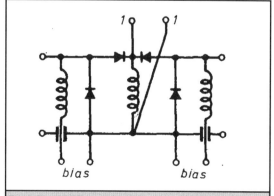

Fig 5.79 : SPDT switch with series / parallel diodes.

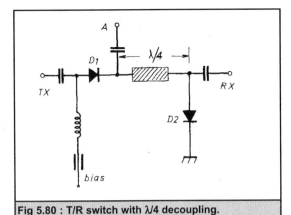

Fig 5.80 : T/R switch with λ/4 decoupling.

implementing compensation techniques.

PIN diodes are widely used in professional applications up to and beyond 100GHz and only in low level applications are the alternative of MESFETS used. The main advantages of these devices are that they can be integrated with other devices on a single monolithic chip.

Among various PIN diodes easily available, some of the consumer types are very cheap and can be used to build solid state T/R switches up to about 3GHz. These are BA379 and BA479 or equivalents in other packages for surface mount. The main characteristics of the BA379 are:

- Capacitance (at Vd = 0 volts), Cj - 0.3pF

- Series resistance (Id = 10mA), Rs - 3 - 4Ω

- Parasitic inductance, Ls = 2nH

The upper useful frequency limit of this device is basically limited by the plastic package that sets the cut off frequency to 6.5GHz. The same semiconductor chip assembled in a coaxial package would give a cut off frequency of 20GHz and therefore would be useful for applications in the 6cm band.

Another interesting feature of the BA379 is its very low capacitance at zero volts. This makes the device usable witout reverse bias in the off state. However a reverse bias of a few volts, reduces the parallel conductance to negligible levels, minimising the insertion loss (Fig 5.77).

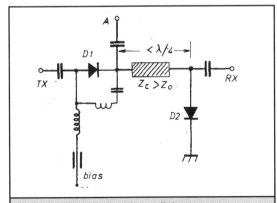

Fig 5.81 : T/R switch with parallel resonance.

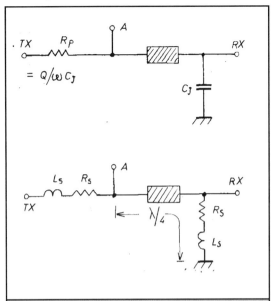

Fig 5.82 : Compensated SPDT switch.

SPDT switch and general guidelines

PIN diodes for rf applications can be classified into two categories :

- Broad band switches.

- Narrow band switches.

The broad band types have a large frequency coverage but have lower performance compared to those optimised for narrow frequency ranges. Therefore the second category is more suited to amateur use.

A simple broadband SPDT switch is shown in Fig 5.78

- The isolation of the open port (cross talk) is limited by the diode capacitance Cj. At 1.3GHz, using a BA379, it is just 12dB which is too small to avoid the destruction of the receiver front end even with low power levels.

- For the diode to be in the off state it requires a reverse bias voltage, to avoid conduction the maximum transmitter power is limited by the breakdown voltage (V_{break} - 30v for BA379).

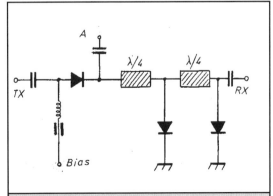

Fig 5.83 : Improved isolation by two shunt sections.

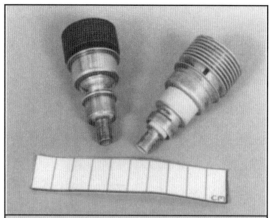

Fig 5.84 : The 2C39 triode, showing glass version (left) and ceramic version (right). The anode cooling fins on the ceramic version are removable, clamped in place with allen bolts.

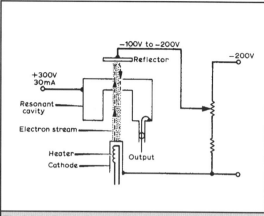

Fig 5.85 : Configuration of a single form of klystron.

The isolation is improved using a series / parallel connection as shown in Fig 5.79, but the limitation due to diode breakdown remains unchanged.

With narrow band switches the power limitation can be avoided (Fig 5.80). All diodes are forward biased when the switch connects the antenna to the transmitter and impedance inverters are used to isolate the receiver input. D1 connects the transmitter to the antenna and D2 shorts the receiver input to ground whilst the quarter wave line translates this short circuit to an open circuit at the D1 junction.

In receive mode, both diodes are open circuit therefore the antenna is connected directly to the receiver. The useful bandwidth of such a switch is limited by the impedance inverter, with a simple quarter wave line this is approximately 10 - 15%

There are some undesirable effects to note :

• In transmit mode, the parasitic inductance of D1 is series connected with the transmitter output and therefore affects the tuning of the output stage. The parasitic inductance of D2 translates to a capacitive load at the D1 junction.

• In receive mode, the capacitance of D1 couples a fraction of the input signal to the transmitter. The capacitance of D2 to ground, mis-matches the receiver.

To remove these constraints the compensation techniques shown in Fig 5.81 can be used.

• D1 capacitance resonates with a parallel inductance in the open state, thus the isolation of the receiver port is increased by one order of magnitude or more. In fact more diodes could be connected in parallel to handle high power without deterioration of receiver sensitivity.

• The length of the line is lower than a quarter wavelength and the characteristic impedance is higher. In this way the receiver isolation is unchanged and the capacitive susceptance at the junction is cancelled plus reflection losses at the receiver port are minimised.

Fig 5.82 Shows the equivalent circuit of a compensated SPDT switch.

It is worth noting that introducing this kind of compensa-

tion the losses measured at the receiver port are within 0.1 - 0.2dB at 1.3GHz and less than 0.3 - 0.4dB at 2.3GHz. For higher power, the leakage to the receiver is reduced by using two sections as shown in Fig 5.83.

Microwave tubes

The conventional valve still has a useful role to play where power is required at the lower microwave frequencies. The ubiquitous 2C39 and its many variants are capable of outputs of tens of watts at these frequencies.

At higher frequencies, travelling wave tubes can provide high broad band gain with high output power. Klystrons and magnetrons can provide higher powerr but are narrow band. These devices are most common at frequencies from a few gigahertz to 10GHz or so, although they are available at frequencies from uhf to several tens of gigahertz.

Valves

The most common microwave valve is the 2C39 triode (see Fig 5.84). It has very small electrode spacings to minimise the transit time delays and the structure is designed so that direct, low inductance connections can be made to all the electrodes. In fact, at these frequencies it is essential to make the valve structure part of the resonant cavity, so that the valve actually sits with its cathode in one cavity, the grid forming a screen common to both, and the anode forms the inner line of the output cavity. This makes grounded grid the natural mode of operation.

One problem with the 2C39 is that of cooling, the anode fins that are supplied with it are very difficult to force air through and it needs some structure around it to force the air to go through the fins. The alternative is to make a more suitable structure. This is often quite easy since the anode either unscrews or unclamps on some variants of the tube. The anode cavities operate at a high Q, and tuning drift as the valve warms up is often a problem, this is because the anode to grid capacitance is the only capacitance across the cavity. The only solution to this is to cool the valve adequately so the anode temperature does not change significantly, possibly using water instead of air.

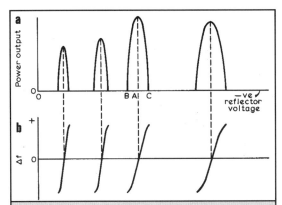

Fig 5.86 : The relationship between reflector voltage of a klystron and (a) power output, (b) frequency of oscillation.

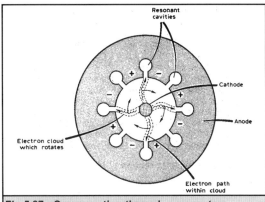

Fig 5.87 : Cross section through a magnetron.

Klystrons

A klystron is a valve like device which usually has the necessary frequency determining components built in, so that merely applying the appropriate working voltages is all that is required to produce rf power. The output may be via a coaxial connector, a coaxial line terminating in a waveguide probe, or via waveguide. The overall operating efficiency of a klystron is typically a few per cent.

The construction of a simple form of reflex klystron is shown schematically in Fig 5.85. A fairly conventional heater/cathode assembly is used to produce a focused electron beam that is directed at a second electrode called a "resonator". This consists of a cavity that is resonant at the design frequency of the klystron and is operated at normal valve anode voltages and currents. Built into many klystrons is a mechanism by which the dimensions of the cavity can be altered so its resonant frequency can be changed. The resonator has two central holes through which part of the electron stream can pass. Beyond the resonator is a third electrode that is biased a few hundred volts negative with respect to the cathode. Electrons approaching this electrode are repulsed and return either to the resonator body or back through the holes in the resonator. For obvious reasons, this third electrode is called a "reflector" or "repeller".

The mode of operation of a klystron is as follows. Electrons emitted by the cathode have a wide range of velocities. If the klystron is assumed to be oscillating, then the rf field existing in the holes in the resonator will affect the velocity of the electrons passing through, slower moving electrons being retarded and faster electrons being accelerated. In other words, the electron stream is velocity modulated at the frequency of oscillation of the klystron. Because of their higher velocity, the faster electrons travel further into the resonator/reflector space and therefore cover a relatively long path before returning into the vicinity of the resonator. However, slower electrons travel a shorter path, albeit at low velocity. The overall effect is that the time of flight of the electrons tends to be the same irrespective of initial velocity, so that electrons which enter the resonator/reflector space in a random manner return in bunches.

If these bunches pass back through the holes in the resonator at a point in the oscillation cycle such that the electrons are slowed by the rf field, then they will deliver power to the resonator and the klystron will oscillate. This

oscillation will be strongest when the time of flight of the electrons in the resonator/repeller space corresponds to (n+3/4) cycles of the resonator frequency, where n is an integer. If the bunches pass through the resonator when the field is trying to accelerate them, then energy will be removed from the resonator and oscillation will be suppressed.

The time of flight of the electrons is dependent on the reflector voltage, and therefore oscillation will occur only when the voltage is set to a number of particular voltages. This is illustrated in Fig 5.86. The reflector voltage that produces the highest peak is normally selected. Note that if the resonator cavity is tuned to a different frequency, then the reflector voltage will also have to be altered to maintain output.

Changes of the order of a few volts to the reflector voltage when the klystron is oscillating have two effects: the output power varies from zero to a maximum, and the frequency of oscillation varies in a fairly linear manner as shown by Fig 5.86. The frequency range of this electronic tuning is of the order of 10MHz before the output power falls to half its maximum value. These characteristics may be used in modulating the device by the appropriate choice of reflector voltage and modulating voltage as the following examples show.

- If the reflector voltage is set for maximum power output, i.e. operating point A in Fig 5.86, then a small modulating voltage will generate fm with little a.m. This is the operating condition usually used in amateur practice.

- If the reflector voltage is set midway between A and B, or A and C, then an additional modulating voltage will produce mixed a.m/fm output.

- If the reflector voltage is set at or just below B, and pulse modulation is applied with a peak voltage sufficient to reach point A, then a pulse/fm output will be obtained. Alternatively the reflector may be set at or just above point C.

Adjustment of the reflector voltage can also be used as a fine tuning control of limited range. This may be done either manually or by an afc voltage. The latter technique may be used to lock the oscillator on to an incoming signal or on to a local frequency standard. Because of the dependence of frequency on reflector voltage, and to a lesser extent on reflector voltage, the dc supplies should be stable and hum free.

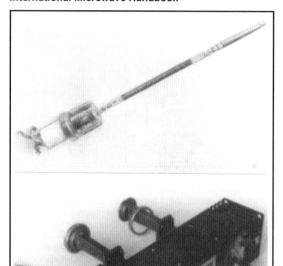

Fig 5.88 : A 10W output X band twt by STC (upper) and the same tube in its mount (lower). Note the controls for adjusting the focusing and the waveguide input and output.

Magnetrons

Magnetrons are high power oscillator tubes developed during the second world war, and widely used in radars and (more recently) in microwave ovens.

Physically, a typical magnetron consists of a central cylindrical cathode and a concentric cylindrical anode, the latter comprising a solid copper block with a number of cavities resonant at the frequency of operation. Fig 5.87 shows a section through the magnetron.

The device is operated with a strong axial magnetic field provided by a permanent magnet, and a potential of 10 to 50kV on the anode. An electron on its way from the cathode to the anode experiences a magnetic force perpendicular to its direction of motion, and an electric force radially towards the anode. Thus under normal conditions of operation, but in the absence of oscillation, an electron would travel approximately in a circle such that its farthest point from the cathode is about half way across the cathode anode space. However, if at this point the electron is retarded by the rf magnetic field associated with one of the cavities, it gives up some of its energy and moves in a path which brings it closer to the anode (where the rf field is stronger) than it would be in the absence of the rf field. If it arrives opposite another cavity at the moment when it is again retarded, it again gives up energy to the rf field and moves still closer to the anode.

Hence, if the right phase relationship can be maintained, some of the electrons will give up energy to several cavities in succession. Equally, of course, some electrons will be accelerated by the rf field and will return to the cathode, but on balance the latter take much less energy from the rf field than the former give to it, and the net

transfer of energy maintains oscillation. The power is extracted by means of a coupling loop or through a waveguide slit in one of the cavities.

Radar magnetrons are operated in pulsed mode, in pulses of the order of 1 second duration at a pulse repetition frequency of the order of 1ms. High efficiency (up to 70 per cent) is obtained in this mode of operation. Since the frequency of operation is determined by the resonant frequency of the anode cavities, magnetrons are fixed frequency devices.

While radar magnetrons are unlikely to be of use to amateurs, microwave oven magnetrons are cw devices, working at frequencies around 2.4GHz (the top end of the 13cm band) and delivering powers of the order of lkW. In principle it should be possible to phase or injection lock such a tube, which would give a high power narrow band transmitter of considerable potency.

Travelling wave tube amplifiers (twts)

The history of the twt goes back to the second world war, when research into radar devices and techniques was at a particularly intense level. The twt was invented in the Nuffield Laboratory Physics Department, Birmingham University, by Rudolf Kompfner. He was seeking an alternative to the klystron with better noise performance, and in a 1946 paper he explained his reasoning:

"One of the main reasons for the lack of sensitivity of the klystron as an amplifier is the inevitable energy exchange between the electron beam and the electric field in the rhumbatrons (resonators). It was therefore a very inviting thought to use the signal in the form of a travelling electric field (instead of a stationary one) and utilise the energy exchange between the travelling field and electrons which travel at about the same velocity."

In December 1943 the first tube gave a gain of about 8dB at a 9.lcm wavelength, with a 13dB noise figure. The work was later transferred to the Clarendon Laboratory, Oxford. Much of the mathematical theory of twt operation was developed by John R Pierce, of Bell Labs, and in 1947 Kompfner joined Pierce to continue twt research.

Nowadays, twts are by far the most widely used microwave tubes, and are employed extensively in communication and radar systems. They are especially suited to airborne applications, where their small size and light weight are valuable. Satellite communication systems are another extremely important application for the same reasons.

Practical travelling wave tube amplifiers (twtas) have applications in both receiver and transmitter systems, and come in all shapes and sizes, but they all consist of three basic parts - the tube, the tube mount (which includes the beam focusing magnets) and the power supply.

When used as receiver rf amplifiers, they are characterised by high gain, low noise figure and wide bandwidth, and are known as twt "lnas" (low noise amplifiers). These usually come with tube, mount and power supply in one integral unit, with no external adjustments to make - just input socket, output socket and mains supply connections. A typical twt lna would have an octave bandwidth (e.g. 2 to 4GHz), 30dB gain, 8dB noise, and a saturated power output of 10mW, within a volume of 2 by 2 by 10in (see Fig 5.88).

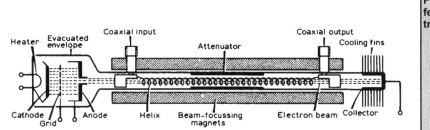

Fig 5.89 : Essential features of a typical travelling wave tube.

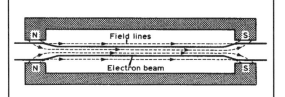

Fig 5.90 : The focusing of the electron beam by a magnetic field.

Transmitter twtas are naturally somewhat bulkier, and often have the power supplies as a separate unit. Medium power tubes have power outputs of up to about 10W, while high power tubes deliver several hundred watts. The major manufacturers of twts are EMI-Varian, Ferranti, EEV, Hughes, STC, Litton, Raytheon, Siemens, Watkins-Johnson and ThomsonCSF.

Construction

The features of a typical twt are shown in Fig 5.89. The electron beam is provided by an electron gun which is very similar to those used in crts, though the beam current is much larger. Electrons from a heated cathode are accelerated towards the anode, which is held at a high potential with respect to the cathode, and a proportion pass through a hole in the anode to produce the beam. Some tubes have a grid between the cathode, at a few tens of volts (adjustable) and negative with respect to the cathode, to control the beam current. The electron beam travels down the tube, inside the helix, to the collector, which is maintained at a high voltage with respect to the cathode. The helix is also held at a high potential but the helix current is low because of the beam focusing.

As is shown in Fig 5.90, this focusing is achieved by a magnet (either a solenoid electromagnet or permanent magnets) round the outside of the tube. An electron with a component of velocity perpendicular to the magnetic field lines experiences a restoring force tending to bring its direction parallel to the field lines.

A very large magnetic field is required to achieve good focusing by this method, which can mean a bulky, heavy magnet. However, the arrangement usually employed is called "periodic permanent magnet" (ppm) focusing, in which a number of toroidal permanent magnets of alternating polarity is arranged along the tube, as is shown in Fig 5.91, this figure also shows the contour of the beam. This arrangement greatly reduces the weight of the magnet (under ideal conditions by a factor l/N^2, where N is the number of magnets used). The alternative method, solenoid focusing, is generally only used in high power, earth station twts where size and weight are unimportant.

The input to, and output from, the helix are via coaxial connectors or occasionally via waveguide. In practice, it is impossible to provide a perfect match at these transitions, especially over a wide bandwidth, so an attenuator is used to prevent the energy reflected back down the helix causing instability. This usually takes the form of a resistive coating on the outside of the tube, though a physical discontinuity in the helix is also used in some cases. The attenuator reduces the rf input signal, as well as the reflected signal, to nearly zero but the electron bunches set up by the signal are unaffected.

The helix itself is a delicate structure and must be provided with adequate thermal dissipation to prevent damage. In medium power tubes the helix is often supported in between three beryllia or alumina rods, but for high power twts alternative slow wave structures are employed (e.g. coupled cavities), though usually at the expense of bandwidth. In this form, the twt resembles a klystron amplifier.

Theory

The essential principle of operation of a twt lies in the interaction between an electron beam and an rf signal. The velocity v of an electron beam is given by:

$$v = \sqrt{(2e \times V_a / m)}$$

where V_a is the accelerating anode voltage; e is the electron charge (1.6×10^{-9} coulomb); and m is the electron mass (9.1×10^{-31} kg).

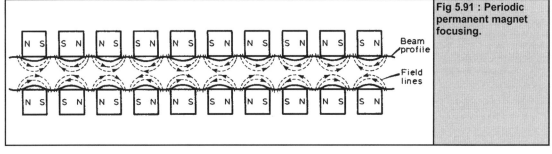

Fig 5.91 : Periodic permanent magnet focusing.

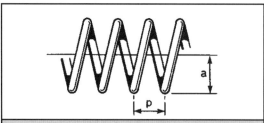

Fig 5.92 : The geometry of the helix.

An anode voltage of 5kV gives an electron velocity of 4×10^7 m/s. The signal would normally travel at c, the velocity of light (3×10^8 m/s), which is much faster than any "reasonable" electron beam (relativistic effects mean that the electron mass actually increases as its velocity approaches c, so that achieving electron velocities approaching c is a complicated business). If, however, the signal can be slowed down to the same velocity as the electron beam, it is possible to obtain amplification of the signal by virtue of its interaction with the beam. This is usually achieved using the helix electrode, which is simply a spiral of wire around the electron beam.

Without the helix, the signal would travel at a velocity c. With the helix, the axial signal velocity is approximately:

$$c \times \frac{p}{2\pi \times a}$$

where c is the velocity of light, and a and p are as shown in Fig 5.92, so the signal is slowed by the factor $p/(2\pi \times a)$. Note that this is independent of the signal frequency. The signal travelling along the helix is known as a "slow wave", and the helix is referred to as a "slow wave structure". The condition for equal slow wave and signal velocities is therefore approximately:

$$\frac{c \times p}{2\pi \times a} = \sqrt{\frac{2 \times e \times V_a}{m}}$$

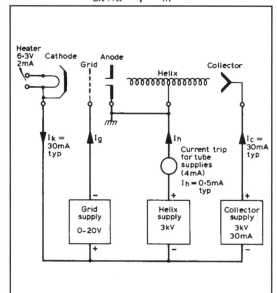

Fig 5.93 : Power supply arrangements for a typical travelling wave tube amplifier.

The interaction between the beam and the slow wave takes the form of "velocity modulation" of the beam (i.e. some electrons are accelerated and some retarded) forming electron bunches within the beam, as with the klystron. The beam current therefore becomes modulated by the rf signal, and the bunches react with the rf field associated with the slow wave travelling down the helix, resulting in a net transfer from the beam to the signal and consequent amplification. Since there are no resonant structures involved in this interaction, amplification is obtained over a wide bandwidth. In fact the principal factors which limit bandwidth are the input and output coupling arrangements.

Operation

A schematic circuit for a power supply for a typical twt is shown in Fig 5.93. The voltages and currents given are for a 10W output tube, but the alignment details apply to most tubes. The manufacturers data regarding electrode voltages should be referred to before running up a particular tube. It is important that a matched load be connected to the output of the amplifier, as the power reflected from any mismatch is dissipated in the helix and can burn it out.

The beam current is controlled by the grid cathode voltage. In modern twts, the magnetic beam focusing is preset and no adjustment is necessary, but if the focusing is adjustable the tube should be run initially at a low beam (collector) current, and the beam focusing magnets adjusted for minimum helix current. The helix voltage should also be set for minimum helix current.

With the tube running at its specified collector current, rf drive can be applied. The collector current will hardly change, but the helix voltage should be set for maximum output consistent with not exceeding the tube voltage or helix current ratings. If the focusing is adjustable, this should be re adjusted for minimum helix current since the rf drive will defocus the beam slightly.

As the helix is fragile and will not dissipate more than a certain power without damage, the helix current should be metered and a current trip incorporated to cut the power supplies to the tube if it becomes excessive. The eht supplies to the tube, especially the helix, should be well smoothed since the ripple will phase modulate the output and give a rough note.

If the collector dissipates more than about 100W it may be necessary to use a blower to cool the collector end of the

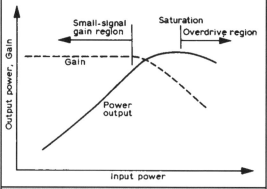

Fig 5.94 : The transfer characteristic of a travelling wave tube amplifier.

tube. Typical efficiency of the twta is about 10 per cent, though some modern tubes may reach 40 per cent. The transfer characteristic is essentially linear (see Fig 5.94), which permits the tube to be used to amplify ssb - one of its great advantages in an amateur context. However, as the input is increased the amplifier saturates. There is no harm to the tube in operating at saturated output power, though amplification is no longer linear, but if appreciable harmonic power is generated this may be reflected at the output transition and damage the helix through over dissipation.

Backward wave oscillators (bwos)

Analysis of wave propagation along a slow wave structure shows that, as well as the forward wave used in the twt, there is the possibility of propagation of a so called backward wave, whose energy travels in the reverse direction. In mathematical terms, the phase and group velocities of the forward wave are both in the positive direction while, for the backward wave, the phase velocity is in the positive direction and the group velocity is in the negative direction. (Phase velocity = w/k, group velocity = dw/dk, where w = $2\pi f$, k = $2\pi/\lambda$.) This type of device is called a "backward wave oscillator" (bwo).

As long as the phase velocity of the wave and the electron beam velocity are synchronised, energy transfer from the beam to the wave can occur. Although for a given beam velocity there will be a number of backward wave modes capable of interaction with the beam, in practice only one is generated. The oscillator is tuned by varying the electron beam velocity, which is achieved simply by varying the beam accelerating voltage.

In common with twts, the primary advantage of bwos is their wide bandwidth, since a precisely tunable output can be obtained over a full waveguide band (for example 8 to 12GHz) or more. BWOs have been extensively used in swept frequency sources (sweepers), which is where the amateur is most likely to encounter them, but they are nowadays increasingly being replaced in this application by solid state sources.

References and further reading

[1] Introduction to Microwave Electronics, T C Edwards, Edward Arnold, 1984. A good introduction to microwave semiconductors and valves.

[2] "Optimum design for high frequency communications receivers", U L Rohde, DJ2LR, *Ham Radio* October 1976, pp I0-25. Contains some fet mixer circuits.

[3] "Performance capability of active mixers", U L Rohde, DJ2LR, *Ham Radio* March 1982, pp30-35 and April 1982, pp38-43.

[4] "Design ehf mixers with minimal guesswork", S A Maas, *Microwaves* August 1979, pp66-70.

[5] "Diode applications in frequency multipliers for the microwave range", H Fleckner, DC8UG, *VHF communications* 3/1978, pp I45-153.

[61 "Harmonic generation using step recovery diodes and srd modules", Hewlett Packard Application Note 920.

[7] "High dynamic range transistor amplifiers lossless feedback", D E Norton, *Microwave Journal* May 1976, pp53-57.

[8] "Low-noise cooled GaAsfet amplifiers", S Weinreb, *IEEE Transactions on Microwave Theory and Techniques,* Vol MIT-28, No 10, October 1980, pp I04I-I054.

[9] "L-band cryogenically cooled GaAsfet amplifier", D R Williams and S Weinreb, *Microwave Journal* October 1980, pp73-76. Explains use of source inductance for gain and noise matching.

[10] *GaAs FET Principles and Technology,* ed J V DiLorenzo and D D Khandelwal, Artech House, 1982.

[11] *MicrowaveDiode ControlDevices,* R V Garver, Artech House, 1976.

[12] From an article produced by William A. Parmley, KR8L and published in QST Nov 1997.

[13] Mini Circuits Laboratory, PO Box 350166, Brooklyn, NY 11235_0003. Data sheets for the MAR-6 and countless other MCL products are available at http://www .mini-circuits.com.

[14] From an article by G Sabbadini, I2SG, published in Dubus 1/1998

Catalogues

The following manufacturers publish catalogues containing useful applications notes: Anzac, Avantek, Minicircuits, Watkins Johnson, Alpha Industries, Microwave Associates (MA), Hewlett Packard, Mullard.

Construction Techniques

Microwave equipment construction is often regarded as an esoteric mixture of plumbing and precision engineering requiring extreme accuracy and an arsenal of sophisticated machine tools for its execution.

While this might in part be true it is still possible for the less well equipped amateur to construct worthwhile items of equipment with patience, some manual skill and a small range of hand tools - the sort to be found normally in an amateur's home workshop. Accordingly, microwave construction should be regarded as the art of the possible rather than the techniques of the impossible

Even with minimum resources, good constructional practices and sensible design, together with proper operation techniques, can bring the performance of even modest equipment up to an acceptable high standard.

The purpose of this chapter is to introduce the materials, tools and techniques which in the author's experience are most likely to lead to such results and which are probably somewhat different from the more usual chassis and cabinet work with which the reader may be more familiar. It is not intended to be a comprehensive guide to workshop techniques but it will suggest practices which are of most use to the microwave constructor. Much of what follows concerning tools and techniques outlines what some may regard as perfectionist. This information should not be read as absolute or definitive but rather as guidance towards a target to be aimed at in the fullness of time and with increasing skill and experience. It is possible however to construct fully functional microwave equipment almost literally on the kitchen table, using cruder construction techniques. The increased accuracy and finer finish resulting from better engineering techniques and tools, exercised in the garage workshop rather than on the kitchen table, will usually result in improved performance and it is for this reason that such methods are discussed here. The reader should not be discouraged from having a go using existing tools and skills for it is only by attempting to make equipment that he or she will acquire new skills.

For more detailed information on workshop techniques see the references listed at the end of the chapter. Manuals [1] to [4] have been written especially for the amateur engineer and model maker and cover all levels from beginner to advanced.

It is often possible to acquire skills or gain access to machine tools by attending short course hobby night classes run by local education authorities at various schools or colleges throughout the country

Construction materials - conductors

The most commonly used metallic materials for construction are as follows

• Copper and its alloys (particularly brass)

• Aluminium and its alloys

• Silver and its alloys (hard solders and plating)

• Zinc and its alloys

• Lead and its alloy (soft solders)

• Tin plated steel sheet.

Each metal or alloy has its own particular characteristic and therefore requires to be worked in its own particular way. The constructor will meet most materials in rod, bar, tube or sheet and should learn to recognise their characteristics and uses at an early stage. Table 6.1 gives some of the more important characteristics of these metals. Relative resistance is important in considering the choice of material for the construction of particularly high Q components such as filters and resonant cavities, since quite high rf currents can flow on and near to the surfaces of such structures (the rf skin effect). Similarly, the linear coefficient of expansion can be important when considering the effects of temperature on the tuning of resonant structures. In some applications (for instance the heat sinking of active or passive devices such as transistors or varactors) high thermal conductivity is another important factor to take into consideration at the materials selection stage of a design. Inspection of the table will allow the constructor to rank the materials to be used in terms of relative resistance the ranking is silver, copper, aluminium, zinc and brass. It can be seen that electrical losses associated with lead, tin and lead / tin alloys (soft solders) can become too high to be acceptable, hence the emphasis laid on the use of the minimum of soft solder in the construction of microwave components and particularly within waveguide. Similarly steel and tin (tinplate) are rather too lossy to be considered for general construction. Apart from electrical or physical considerations, the cost and workability of the metals must be taken into account. The cost of silver prohibits its use in construction (other than as a very thin plating applied to the final, assembled component) and the amateur is thus limited to copper,

Table 6.1 Properties of some metals.

Metal (1)	Relative resistance (2)	Coeff. of expansion (3)	Thermal conductivity (4)	Remarks
Aluminium	1.64	25.5	48	Soft white metal oxidises rapidly
Brass	3.90	18.9	26	Yellow metal works well
Copper	1.00	16.7	92	Soft red metal works less easily
Lead	12.80	29.1	8	Very soft grey metal oxidises rapidly
Silver	0.95	19.5	100	Soft white metal tarnish conductive
Tin	6.70	21.4	16	Soft white metal oxidises slowly
Zinc	3.40	26.3	27	Brittle grey metal
Iron (5)	6.10	60.0	18	Hard grey metal rusts rapidly

(1) All figures should be taken as representative only, absolute values will vary with purity of metal or composition of alloy.

(2) Copper arbitrarily given unity value.

(3) Parts per million per degree centigrade.

(4) Values at 20°C, referred to silver = 100%

(5) Values for cast iron. Figures may vary by a factor of ten. For example stainless steel - relative resistance 53, coefficient of expansion 10, thermal conductivity 7%.

aluminium, zinc and brass. Copper, second only to silver in its electrical and thermal properties, is commonly available and relatively inexpensive. It is a soft, malleable (deforms under pressure without fracture) and ductile (stretchable) metal and, as such, must be handled with a certain amount of care - it is easy to distort by over zealous punching or by the use of taps and dies applied to holes or stock of the wrong size. It takes both soft and hard solders freely when thoroughly cleaned and exhibits quite reasonable resistance to corrosion.

For the highest quality of performance, copper can be easily and economically silver plated; because of the skin effect, even a few microns thickness of silver plate can improve component performance to be comparable to that of solid silver. At signal and low power levels the cost of this added performance is often unwarranted, but at high power levels, for instance in linear amplifiers (and perhaps in high Q filters even at low power levels), the gain in performance can be quite significant. Pure copper is not particularly easy to turn or mill because of its softness and for such operations the constructor may choose to substitute brass despite its inferior characteristics.

Pure aluminium is inexpensive but, on account of its even greater softness, is perhaps more difficult to work than copper. It also suffers from several other serious disadvantages in common with all its alloys. It is subject to very rapid atmospheric oxidation and the oxide so formed is electrically non conducting. Both the oxide (alumina) and the metal are relatively good conductors of heat. Because of this rapid, natural formation of the oxide, aluminium and its alloys are extremely difficult for the amateur to solder and impossible to plate.

Zinc alone is too brittle to be used directly in construction. When alloyed with small amounts of aluminium, copper and other metals, it is commonly used for die casting. It is in the form of die cast boxes for equipment housing and shielding that the constructor is most likely to come across zinc alloys, although some 10GHz (and higher) Doppler modules may be die cast also, such is the accuracy to which these alloys can be manipulated. Such alloys cut, drill, mill, tap and turn quite easily and cleanly

but usually solder with great difficulty (if at all) and are still somewhat brittle. Die cast items may also be fabricated from aluminium alloys, which are brighter and whiter in colour than the zinc alloys, which tend to be a duller, greyer colour.

Brass, which is an alloy of copper (58% to 90%, commonly 70%) and zinc, is readily available and works well but has electrical and thermal properties markedly inferior to copper. Where optimal electrical performance is needed, brass should, by preference, be silver plated. Like copper, it will take soft or hard solders very freely when clean.

To illustrate the effect that the choice of metal can have on equipment performance the reader should consider G3JVL's calculations of the Q of cavity filters constructed in waveguide, for instance the multi section iris coupled filters typical of some current amateur narrow band equipment. The unloaded Q-values are calculated as being:

97% pure silver 23,339

Pure copper 22,759

Pure aluminium 18,181

Brass 11,804

To summarise briefly, copper is the most electrically desirable of the economically priced metals and may be the choice for all microwave applications regardless of frequency or power. Brass being somewhat harder and easier to work is commonly used by the amateur for many purposes, but for the best performance and power handling should be silver plated after assembly. Lead and tin alloys (solders) are very lossy and in general their use should be minimised, certainly within waveguides and cavities. Silver solder and silver loaded soft solder are to be preferred for the majority of applications.

Little has been said about tin plate. Tin is a relatively lossy metal usually applied electrolytically as a thin overlay on steel, which is itself equally lossy. The use of tin plate is thus usually restricted to the construction of horn anten-

Table 6.2 Properties of some insulators (dielectrics)

Matrial (1)	Dielectric constant	Power factor (2)	Dielectric strength (3)	Resistance ohms / cm	Softening temperature °C
Cellulose acetate	6 - 8		250 - 1000	4.5×10^{10}	70
Glass (Pyrex)	4.5	0.2 - 0.7	335	10^{14}	600
Mica	7.0 - 7.3	0.02 - 0.03	600 - 1500	5×10^{17}	1200
Nylon	3.6	2.0	305	10^{13}	71
PTFE	2.0	0.001	500	10^{19}	250
Phenol-formaldehyde	5.0	1 - 2.8	400 - 475	1.5×10^{12}	
Polythene	2.25	0.03	1000	10^{17}	104
Polypropylene	2.25	0.02	1000	10^{17}	150
Polystyrene	2.5	0.02	500 - 760	10^{17}	80
PVC (4)	2.9 - 3.2	1.6	400	10^{14}	
Urea-formaldehyde	5 - 7	2 - 5	300 - 550	10^{12}	200

(1) Figures may vary considerably from sample to sample, according to the degree of polymerisation, cross linkage, "purity" of polymer and the nature of "fillers" (see text). Figures should be taken as representative only.

(2) Power factor usually measured at up to 50MHz, values may increase with frequency, i.e. losses increase. For many of the modern plastics (polyolefins and polyflourolefins) the increase may be insignificant.

(3) Values in volts / thou inch (0.025mm).

(4) Values very variable, according to grade of material.

nas, dish feeds of the beer can variety or to screening enclosures, although in the case of the latter, the use of double sided copper clad pcb material can make an even better enclosure when properly soldered together. When making enclosures from double sided pcb material it is most important that *all inside* seams are soldered together and the constructor should be especially careful to ensure that connector bodies, coaxial lead screens and other earthing connections are made to the *internal* cladding of the enclosure rather than to the external cladding. Whilst on the subject of screening, the constructor is advised to consider the use of die cast boxes for housing microwave circuitry. Although these may seem expensive they provide very good mechanical rigidity and thermal stability for the circuits housed within them, as well as providing good electrical shielding when the leads into and out of the enclosure are properly routed and decoupled.

One further material, so far not mentioned, is phosphor bronze. This is an alloy of copper and tin containing a small percentage of phosphorous and which is electrically fairly similar to brass. In strip or sheet form (when tempered) it is springy and is used in the construction of fingering to contact the electrode rings of disc seal valves commonly used at the lower and middle microwave frequencies. It is fairly common for the amateur to make such fingering and it is possible to do so using phosphor bronze strips which are easily and cheaply available in the form of household draught excluders. Although these are usually not very highly tempered, this is no disadvantage since it allows the strip to be cut and bent without the real risk of fracture which exists with highly tempered strip; even with these soft tempered strips, repeated bending through sharp angles will rapidly cause fracture and the stock should be bent with care. The fingering is made by sawing [5] or, more elegantly, by constructing a simple slotting press [6]. Fingering should not be hard soldered since the high temperatures involved will cause annealing (de tempering) which leads to loss of springiness, thus defeating the object of using phosphor bronze in this application. Re-tempering is difficult and messy, involving

heating to a high temperature and quenching in oil it is easier and more convenient to avoid the need for re -tempering by soft soldering the fingering into position, taking care not to cause local overheating during soldering. Similar remarks apply to beryllium copper strip, which is another alternative to phosphor bronze for this type of construction.

Construction materials - insulators (dielectrics)

The constructor is faced with a large choice of insulating materials, some of natural origin but many more man made. Those which the constructor is most likely to handle are listed here:

- Polyamides (Nylons)
- Polystyrenes (Perspex*)
- Polyesters (Terylene*, Mylar*, and catalytic casting resins)
- Polyolefins (Polyethylene-Polythene* and Polypropplene-Propathene*)
- Polyfluorolefins (polytetrafluorethylene, ptfe-Teflon* and fluoroethylene-propylene copolymer, fep)
- Polyvinyl chloride (pvc).
- Rubbers or elastomers (Nitrile, Butyl, silicone and natural)
- Glass
- Mica (including composites such as Mycalex*)
- Ceramics (including ferrites)

Note: a * indicates a proprietary name. Such names are given here because they are names commonly associated with the particular material described. The use of proprietary names does not imply preference for those products.

The use of other plastics, such as phenol-formaldehyde, urea-formaldehyde, cellulose acetate and the elastomers, can be generally discounted either because they possess inferior qualities or have limited uses. For instance the older formaldehyde based thermosetting plastics have poor electrical characteristics; cellulose based polymers tend to be moisture sensitive and the elastomers, by definition, are elastic and dimensionally unstable and, although reasonable insulators, are usually restricted to sealing duties where weatherproofing of equipment is needed.

Of the list, the last three do not lend themselves to manipulation by the amateur. They are extremely hard and brittle materials, which simply cannot be worked without dedicated, specialised equipment. Mica in sheet form can be cut and split into thin leaves and possibly drilled and punched to provide insulating plates with high electrical, mechanical and thermal stability but cannot, for instance, be bent or otherwise formed. In the selection of dielectric or insulating materials for a particular application the constructor is usually concerned with six parameters:

• Dielectric strength, the breakdown voltage under dc and/or rf stress.

• Resistivity, how good an insulator.

• Dielectric constant, defines the change in capacitance or capacitive reactance when the dielectric replaces air in a capacitor, transmission line or resonant structure.

• Power factor, a measure of lossiness in an ac (rf) field.

• Mechanical strength.

• Workability.

Table 6.2 gives some electrical and thermal information for most of the commonly used and more easily obtainable materials. Where significant heating effects occur, then a seventh parameter needs to be considered, the softening or melting point of the insulator. It should be noted that many of the figures given in Table 6.2 may be radically altered (usually detrimentally as far as rf performance is concerned) by the presence of fillers, for instance carbon black, titanium dioxide and other substance added to the plastic to stabilise it against the effects of weather or sunlight or to make it decorative. This is particularly true for domestic grades of polyethylene and polyvinyl chloride and in the foamed (expanded) versions of plastics. Carbon filled expanded plastics used for the protection of ics from static can make useful low powered waveguide loads, whereas the unfilled version can be almost perfectly rf transparent. For further reading, [7] will provide a wealth of data.

Examination of the table indicates the use of glass, mica, ptfe, polystyrene and polythene/poly-propylene. Glass, as already indicated, is unworkable only to a limited extent. The choice, therefore, is effectively limited to the latter three. Polyethylene is soft, polystyrene is brittle and both have low softening points, leaving ptfe (or fep which is broadly similar in properties) as the dielectric of choice for many purposes. Polyethylene and polypropylene in sheet form can be used where mechanical strength is relatively unimportant and the higher softening point of polypropylene can be advantageous. PTFE has better strength, a much higher softening point and excellent electrical performance in all respects.

The constructor may sometimes see the use of self adhesive cellulose acetate tape (for instance Sellotape)

recommended for low voltage insulation purposes. The writer has not found this really satisfactory since the tape appears susceptible to ageing and moisture absorption that can ultimately lead to local corrosion problems especially on brass and copper surfaces. A much better substitute is plumbers pipe-wrap (ptfe) tape available in a number of widths and thicknesses. Polyester film or tape are also satisfactory substitutes where good mechanical strength is required.

The application to which insulators are put can be divided into five main categories:

• Insulation/dielectric in high dc and rf stress situations where considerable heat generation maybe involved (e.g. linear amplifiers)

• As dielectric tuning elements in microwave oscillators such as Gunn oscillators

• As rf transparent weather shields for antennas (radomes) or other equipment

• As tuning slugs in slug tuners

• As supports for the inner conductor of fabricated coaxial lines such as those described as an integral part of antenna feeds for 1.3, 2.3GHz and 3.4GHz in the relevant chapter.

The qualities of mica or ptfe suit the first category well, with mica possessing exceptional mechanical strength and dimensional stability when used in the fabrication of bypass capacitors. However ptfe sheet is easier to obtain and work and can be substituted with the assurance that the electrical performance will be little different, although rather more care will have to be taken to ensure the softer plastic is not punctured by a sharp metal component during assembly, and that the dielectric constant is taken into account when calculating the effective capacitance required if this is critical.

PTFE screws are particularly useful in both the assembly of such fabricated bypass capacitors and, in the second application, as the tuning elements in Gunn oscillators. In this use, metal screws suffer the disadvantage of contact problems, which cause erratic tuning or frequency jumping. This does not occur with a dielectric screw.

The choice of material for the third type of application is difficult. Temporary weather protection can be afforded by the judicious use of polythene household containers, polythene bags or even cling film. More durable housings are made from stock pipe (waveguide antennas) or thermally or adhesively bonded heavy sheet. The constructor should try to use stabilised but unfilled plastics for the reasons given in an earlier paragraph, fillers in domestic grades of plastics can be an unknown quality and it could be that the constructor may be well advised to consider making a weatherproof enclosure from glass fibre reinforced catalytic castings resins (grp). These materials are used under the most rigorous environmental conditions to construct boats, canoes and the like. Instruction on the use of such materials is usually available from the supplier of the resin or by reference to any amateur boat builders manual. The advantage of using grp is that the constructor knows the nature of the filler (glass) and can control the thickness of the enclosure to optimise rf transparency where this is important. A further use for grp is in the construction of dish antennas which can be made both light and rigid, the reflective surface being provided by facing the inside of the dish with carefully applied

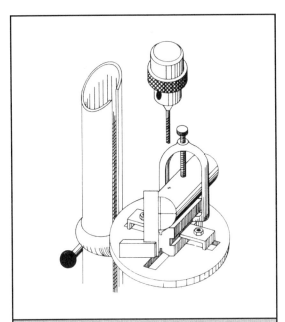

Fig 6.1 : Using a V-block with work marks to ensure the accurate drilling of round stock.

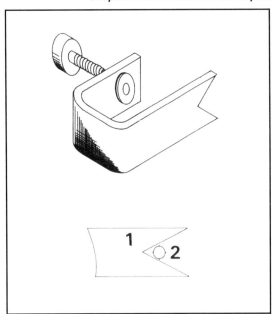

Fig 6.2 : Special G-clamp for use with sheet metal. 1 is the v-arm of the clamp, 2 the hole being drilled.

aluminium cooking foil.

For the last two applications the best insulator/dielectric is, without doubt, ptfe.

Tools - selection and care

Many of the tools required for microwave construction may already be in the amateurs toolbox or workshop. The only advice that needs to be given is that cheap tools are a waste of money; they are either inaccurate (in the case of measuring tools) or manufactured from inferior materials, which blunt quickly (in the case of cutting tools). *Always* purchase good quality, branded tools of reputable manufacture, for such tools are an investment. The writer is still using a few, mainly measuring tools, known to have been made around 1920! Names such as Dormer (drills, reamers, taps and dies). Eclipse (saws and other cutting and forming tools), Moore and Wright, Rabone-Chesterman and Mitutoyo (measuring instruments) are typical of the better class of tools made to close tolerances from good materials by well established manufacturers of high reputation and should always be selected in preference to cheaper tools of indeterminate origin.

The choice of imperial or metric standards is left to the constructor, for it is pointless to go metric if good imperial sized tools are available or vice versa. Often the substitution of the nearest comparable BA and metric threads is acceptable, as might be standard metal stock sizes of bar or rod.

Four items of equipment which should ideally *not* be omitted from any workshop include:

• A good, general purpose vice with detachable soft jaws, mounted on a sound, solid bench.

• A pillar or stand drill which, if using a domestic electric drill, should be capable of several speeds or preferable continuous electronic speed control. It should be fitted with a chuck of at least 10mm (3/8 inch) and preferably 13mm (1/2 inch) capacity.

• A machine vice, preferably of the type with one or more vertical V grooves machined into the jaws, will be suitable for accurate location of the work piece whilst drilling. It should be capable of being clamped to the drill table.

• Two flat-surfaced plates.

Using a good pillar or stand drill and machine vice it is possible to carry out a number of operations usually associated with a lathe. For instance, facing off and centre drilling round bar stock to produce filter elements, as well as the more normal drilling and reaming.

Other tools are optional depending on the types of work to be undertaken V blocks for drilling diametrically through round bar or tube (Fig 6.1), G-clamps with a V jaw for clamping sheet metal whilst drilling (Fig 6.2) and so forth.

The final, essential (and innate) tools for successful microwave construction are thought and pre-planning. By this is meant a logical and ordered sequence of operations leading to good fit, good finish and a piece of equipment that not only works well but also looks professional.

Having selected and purchased good quality tools, the constructor is urged to look after them properly. Bench management is the first essential keep the workbench clean, tidy and uncluttered. Regularly sweep away swarf, filings and abrasive particles that cannot only damage the work piece but also the constructors hands and measuring instruments. In engineering terms, cleanliness is next to Godliness that is to say that even in the simplest of workshops, tidiness and method are complementary and conductive to good engineering techniques.

When not in use, tools benefit from either being kept indoors in a dry atmosphere or, if this is not possible, by being very lightly smeared with a film of light machine oil after use. This will help prevent rust and corrosion. The use of vpi (vapour phase inhibitor) paper is also to be encouraged where tools are kept under less than ideal conditions. Again, measuring tools should not be dumped into a drawer along with heavy tools such as hammers and saws, for this is the surest way to ruin good measuring tools. Everything has its place and everything is in its place is another old engineering adage, which is worth noting. It may seem obvious that the reference surfaces of measuring tools should be kept clean and free from blemishes never use the tip of a steel rule as a screwdriver or the butt of a try square as a hammer. These are but two examples of how tools *do* get misused! Damage can also be caused by laying measuring tools down on a dirty bench surface coated, perhaps, with oil and swarf. Cutting tools should be kept sharp. Frequent cleaning of files, replacement of saw blades and the sharpening of drill bits is essential if the very best quality of work is to be carried out. Accurately sharpened drill bits can, these days, almost be regarded as throw away if the amateur has neither the means nor the skill to regrind them. It is advisable *not* to use the metal working drills on glass fibre pcb. Materials like this will blunt (particularly) the smaller drill bits extremely quickly. If much work is to be carried out with glass fibre, it is advisable to obtain special tungsten carbide tipped drills available for this purpose.

Mention was made of the desirability of having two flat faced plates available. A cheap and, these days, very accurate surface plate (for classical marking out) is widely available in the form of a sheet of plate glass of, say, 300mm by 230mm by 13mm thick. Such a piece of glass is best mounted in a shallow wooden frame or box, lined with thin felt for resilient support. One plate should be used solely for marking out, the techniques for which are described in most introductory workshop practice manuals, and the other for honing operations. Alternatively, for this second plate, it is sometimes possible to salvage a cast steel plate from an old electric cooker of the black heat (solid plate) type. This can also double up as an anvil for centre punching or other light forming operations. Never use glass plate for these functions!

Bypassing and decoupling

The need for thorough bypassing and decoupling particularly of power leads, cannot be stressed enough. Small, amounts of stray rf or noise carried to the oscillator stage of a drive source which is subsequently high order multiplied can be the cause of unwanted sideband noise or instability. Techniques used should be apparent from the designs in other chapters in this manual and the copious use of solder in or screw in feed through capacitors, ferrite or quarter wave chokes and "filtercons" is to be recommended. Screw in types can be conveniently fitted into holes tapped directly into die cast boxes where supplies need to enter leave the enclosure. On-board decoupling may use either feed through or leadless discs. Component lead lengths must always be minimal and pick up on power leads can be minimised by using miniature flexible coaxial cable for such supplies, at least in lower voltage applications. The outer sheath of the cable can removed to enable frequent gounding of the braid along the length of the cable as well.

Earthing in hf and microwave circuits, a case for puff [9]

In any circuit with active components, we are confronted with the problem that certain lines or parts of circuits do not belong to the purely AC voltage circuit. Using appro-

priately adjusted earthing points, we can ensure that these components fulfil only DC tasks. And these earthing points are the subject of this article. In practice, we see every possible structure on printed circuit boards, and ask ourselves what advantages this structure is intended to convey. Which is precisely where this article is intended to be of use.

Definition of term correct earthing

We should first agree on the definition below for the term correct earthing.

A circuit point or a connection line can be said to be correctly earthed if, within the frequency range under consideration, the impedance (Z) between this point and earth is lower than 5 Ohms.

Let us now put the miscellaneous standard measures under the microscope, taking this point of view, originating from developer's practice.

The earthing capacitor

This measure is certainly the oldest in electronics. In theory this method is first class, since it is indeed well-known that the AC resistance of a capacitor decreases as the frequency increases, and thus the connection to earth can only become even better.

Aluminium electrolytic capacitors

In practice, though, things look rather different. At very low frequencies, polarised aluminium electrolytic capacitors are used, but these display a series resistance that rises with the frequency and a reactive component of the impedance which also rises, due to inherent inductance. Moreover, it is not officially mentioned that the quality of electrolytic capacitors has deteriorated. The mechanical dimensions have been continuously reduced, but any radio or television technician will be happy to confirm that, for example, modern electrolytic capacitors are the cause of the problem in 50% of repairs of TV sets. They no longer break down, but usually slowly dry out, because the seals become porous. The unfortunate result of this is that, in addition to the capacity drop, which usually would not be so awful, there is very rapidly a sharp rise in the series resistance and that's bad!

Tantalum electrolytic capacitors

Tantalum electrolytic capacitors behave considerably better here. Sintered from powder, they basically represent semi-conductors (i.e. diodes) which are operated in the filter attenuation band. Their advantages lie in their small dimensions, their low series resistance (approximately 1-2 Ohms) and their low inherent inductance of a few nH. This gives excellent coupling capacitors and/or wide-band earthings. In particular, in parallel circuits with several capacitors, they cover a range from low frequency to far above 100 MHz.

Let us not conceal their negative characteristics. Like any semi-conductor, they react extremely sensitively to overvoltage and/or pole reversal. In the most favourable case, disruptive short-circuits then occur. In the most unfavourable case, however, there can be scarcely predictable reductions, great or small, in the insulation resistance, which are to some extent dependent on temperature.

One factor here is often misunderstood. It will certainly say somewhere in the data sheet that, in the interests of

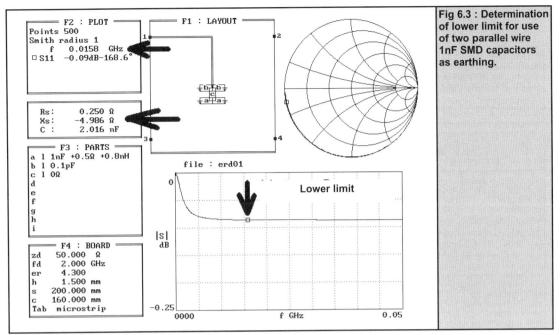

Fig 6.3 : Determination of lower limit for use of two parallel wire 1nF SMD capacitors as earthing.

service life, standard tantalum electrolytic capacitors should not be subjected to a switching current pulse exceeding approximately 0.3 A.

The reason for this is very simple. If this value is exceeded, it can quickly become so hot at individual points within the sintered material (in which, indeed small grains are in contact with each other) that a melt-on occurs, leading to a short circuit.

It makes no difference whether the current limitation required is brought about through the power supply itself or using a pre-resistance.

Ceramic capacitors

From 100 MHz upwards, the only capacitors still used are essentially ceramic models. Their quality is adequate for all requirements up to 2 GHz, even with the cheapest standard formats.

However, one should, as far as possible, attempt to use only SMD solutions, for only their low inherent inductance (< 1 nH) eliminates the tiresome inherent resonances of the wired-up copies. Here it is particularly advantageously to have parallel circuits of at least two capacitors, which means that the Ohmic resistance and the inherent induct-ance are smoothly halved.

We can use Puff to show how such a layout behaves at various frequencies. To do this, four components are parallel-wired in a simulation:

- Two SMD capacitors, each with 1nF + 0.5 Ohms series resistance + 0.8 nH inherent inductance, and

- Two capacitors, each with 0.1 pF, each simulating the size 0805 SMD pad.

It is recommended that analyses should be distributed into an upper and a lower range for such very broad-band layouts and for the Puff representation, restricted to 500 dots. Fig. 6.3 therefore shows the behaviour from 0 to 50 MHz, in the form of the reflection factor S11, for this parallel circuit, which will supply the values for the impedance curve of the layout.

For this purpose, use Page up or Page Down to go to a specific frequency, place the cursor in field F2 on S 11 and key in the equals sign, =.

The active and reactive components of the input resist-ance immediately appear in the dialogue window, and the associated Substitute dummy component is displayed as an inductance or capacity value. Now we look for the frequency at which the capacitive reactive component of this layout undershoots 5 Ohms. In Fig. 6.3, this happens at 15.8 MHz.

In a second pass, we take the range from 10 MHz to 5 GHz amd there we find, at precisely 2 GHz, 5 Ohms for the (now inductive) reactive component (Fig. 6.4). It specifies the upper limit of the usable frequency band.

Anyone who wants to know why the actual value zero was not taken as the lower frequency limit for Puff can repeat the simulation using this value. The blemish arising in the representation in the Smith chart, due to the fact that the step widths at very low frequencies are now far too great, is avoided using the setting proposed.

The parallel wired layout of ceramic capacitors is thus effective and correspondingly popular. To set the bottom frequency limit still lower, we simply replace the ceramic SMD capacitors with a parallel circuit of several tantalum electrolytic capacitors, in which the loss resistances and inherent inductances are approximately twice or three times as great.

Notes on sources of errors

Unfortunately, we can find some circuit technology earth-ing errors in power supply systems not only in DIY projects, but even in industrially produced circuits.

Fig. 6.5 shows a wrong and a right format for a multi stage broad band amplifier as an example:

- There are no protective resistors for the tantalum elec-trolytic capacitors.

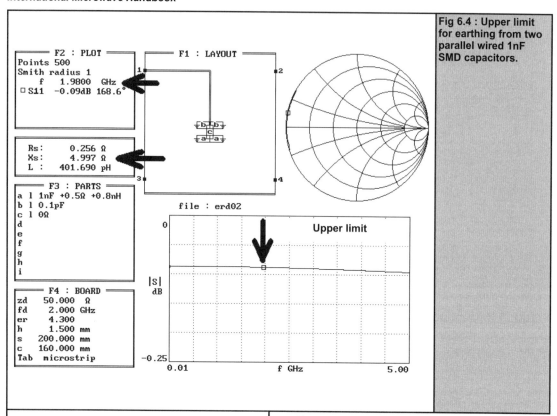

Fig 6.4 : Upper limit for earthing from two parallel wired 1nF SMD capacitors.

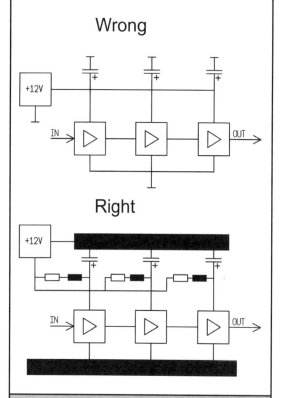

Wrong

Right

Fig 6.5 : Wrong and right layout for a multi stage broad band amplifier.

• The power supply lines of the individual stages must have star connections to the power supply and must be de-coupled from one another. With the help of a choke coil and / or an Ohmic resistor, we can thus then create an LC or RC low pass in each feeder with the tantalum electrolytic capacitor. Only in this way can we reliably prevent part of the high level of the last stage from going back to the input section and causing the layout to oscillate.

The HF choke used, however, must not be of too high a quality (to avoid resonance effects). Here it is thoroughly normal to wire Ohmic resistors into the circuit. This not only attenuates but improves the low-pass effect, especially at lower frequencies, where the inductive resistance is known to be still low. This simultaneously provides the protective resistors for the tantalum electrolytic capacitors.

• The points connected directly to earth (= without capacitor) within an amplifier stage must not just be looped on to the next stage, or an unnecessary oscillation risk will be created.

A version mounted on printed circuit boards in accordance with the laws of microwave engineering is ideal, and almost oscillation-proof, even at low frequencies. Coated on both sides, this version thus has an underside which is an integrated earthing and earth level. In the illustration it is shown as a blocked-in rectangle. The direct earthings are brought about through suitable throughplatings, which can be implemented in the form of full tubular rivets (silvered, diameter 0.8 mm.).

The screening action of such choke/capacitor layouts at various frequencies can also be determined by means of Puff. But you can find out from [10] how expensive

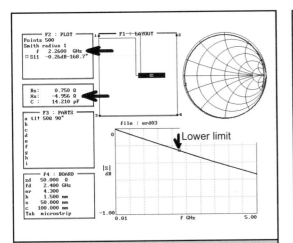

Fig 6.6 : Determination of lower limiting frequency using a 50 ohm microstrip line on no load as earthing.

Fig 6.8 : Using a square pad with an edge I = $\lambda/4$ as earthing.

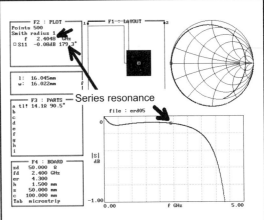

this can be, for example, for a low-noise oscillator.

Earthing through microstrip lines

At higher frequencies, if discrete components slowly fail, we resort to sections of line working at no-load as earth connections. It is well known that such a line represents a short circuit at I = $\lambda/4$.

Thus, for example, we can take such a lossy $\lambda/4$ line with 50 Ohms for 2.4 GHz and simulate its behaviour for a printed circuit board made from epoxy material FR4, using the data:

- Thickness 1.5 mm, r = 4.3 and

- Loss factor It = 0.02

We are less interested in behaviour under resonance here, for it is well known that only the small loss resistance remains behind there (approximately 0.8 Ohms). Use Page Up or Page Down to move the cursor until keying in the equals sign at S11 gives you an inductive or a capacitive reactive component of 5 Ohms (Figs. 6.6 and 6.7). Frequencies associated with this are 2,260 MHz and 2,560 MHz earthing would thus be usable

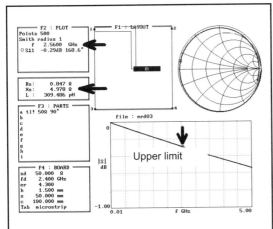

Fig 6.7 : Upper limit using 50 ohm microstrip line on no load as earthing.

only in a narrow range 300 MHz wide. This method is therefore also deliberately used only for narrow-band applications or oscillators.

Things work considerably better with a broader band, if the section of line working at no load is altered to such an extent that the length and width are the same and form a square. If we also make the area of this square sufficiently big, the line, because of the low impedance level, acts like a big capacitor with a correspondingly low reactive impedance, even at really low frequencies (i.e. at I < $\lambda/4$).

If the frequency is then increased, the $\lambda/4$ resonance follows, with the short circuit at the input. Not until this resonance is exceeded does the reactive component increase again, due to the inductive behaviour, but with a low-Ohm line the inductance is indeed also very low. This leads us to expect a decidedly wide usable frequency range, in which the impedance is sufficiently low.

This can be tested for 2,400 MHz using Puff. With the same printed circuit board data as before, the length and width at this frequency are to be given values which correspond to a quarter of the wavelengths. These include lengths and widths of approximately 16 mm. and an impedance level of 14.1 Ohms (Fig. 6.8).

If we now simulate the behaviour of this line and look again for the points with a reactive component of 5 Ohms, we find the frequency values 2,867 MHz and 1,973 MHz are suitable. This layout would also fulfil the requirements laid down in a range of approximately 900 MHz. For a direct comparison, the impedance curves (Z) = f(frequency) for the two sections of line have been plotted in a joint diagram (Fig. 6.9).

The advantage of the square section of line in broad-band applications is once again easy to see here. Unfortunately, it is very cumbersome for practical application at low frequencies.

Earthing through radial line stubs

Now let us look more closely at the mysterious circuit segments which are found in nearly all microwave circuits from approximately 2GHz upwards. Fig 6.10 shows such a layout, from application note 1091. How do such layouts work, and what advantages do they offer?

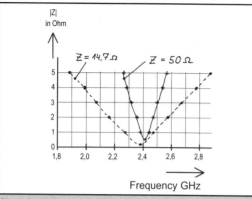

Fig 6.9 : Impedance curve for the two line sections investigated.

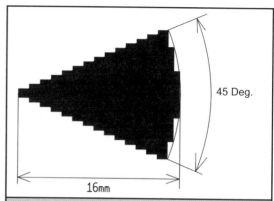

Fig 6.11 : Radial line stub assembled from 16 elements.

Interestingly, they are used by almost all microwave circuit developers, although there is scarcely any literature about it. Only an application note from HP [11] taken from the Internet gives them their correct name (radial line stubs) and names some literature sources [12], [13].

It does, though, become apparent very quickly that the calculation formulae given there are very lavish and are tailored more for use on mainframe computers. The HP note also describes the application in a very precise and enlightening manner:

"...problems of location and parasitics of low impedance shunt stubs were solved by using fan-shaped open stubs with the narrow end connected to the main transmission line..."

Consequently, we can introduce a low-Ohmic resistance at a three-dimensionally limited point in the circuit and that is exactly what we want for earthing. I am reminded here of the concept of tapering, which in the past was an important method for broad-band transformations and matchings. We are referring here to lines or cavity conductors which continuously alter their dimensions and thus their wave resistance, with the usable frequency range being markedly increased through this measure and the line length required for this reduced.

First a few basic observations:

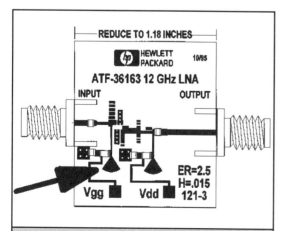

Fig 6.10 : Application example for radial line stubs (12GHz LNA).

• Because the line end is under no-load here too, the associated series resonance must consequently be introduced somewhere (recognisable at the 180° phase angle at S11) with the short circuit at the input. The knife-sharp tip of the structure is very advantageous here, so that the earthing can be positioned precisely at the desired spot on the printed circuit board.

• If we make the central angle very small, the construction becomes more and more like a normal but narrow and also high-Ohmic microstrip line, with its narrow-band resonance. So increasing the central angle must produce the opposite effect, namely broad-band low impedance.

• And, assuming the function of tapering has been described correctly, the layout would even have to turn out markedly shorter than a standard λ/4 line!

We can now use a little trick in our investigation with Puff:

One of these radial line stubs is chopped up into many short line sections, but all with the same length, which are wired in series. The line width of each partial section is then selected to be small enough to obtain the best approximation to the original structure.

For a frequency f = 2,400 MHz, we select the radius of the associated full circle to be precisely the same as the λ/4 line length in the previous example, i.e. approximately 16 mm.. Let a piece be cut out of the full circle with a central angle of 45 degrees.

It is helpful if the entire structure can be drawn out with a scale of 10:1 on millimetre paper or squared paper, and the dimensions determined for the strip line conductor widths required or they can simply be worked out on a pocket calculator.

We are working here with 16 elements in all (N.B.: you can list a maximum of only 18 components in field F3 of Puff!). Each element has a length of 1 mm. and the widths are graduated in such a way that the best approximation to the ideal stub is obtained (Fig. 6.11). The following individual data for the simulation can be measured out from the drawing of the line sections or calculated with the help of the geometrical formulae. The numbering for the line sections starts on the left, i.e. at the tip of the circle sector (Table 6.3).

Now the hard work begins. You can certainly key in the lengths of the microstrip line sections in the Puff F3 field

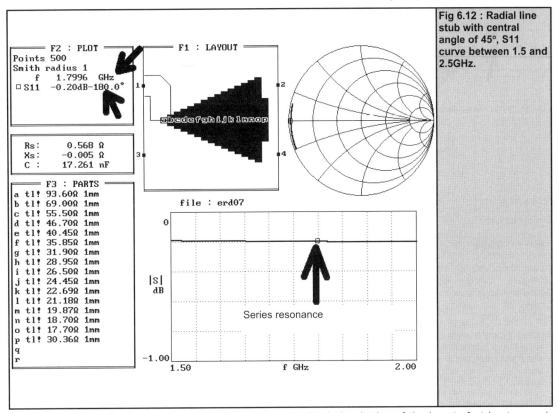

Fig 6.12 : Radial line stub with central angle of 45°, S11 curve between 1.5 and 2.5GHz.

directly in mm. but not their mechanical widths. To do that, you first have to assign any impedance to each of the 16 sections and vary it until the desired conductor width is set as a reaction when the equals sign is keyed in. Then the stub structure is assembled in field F1, and the simulations are begun for the range 1.5 2.5 GHz.

Not only does Fig. 6.12 show the complete list of stripline sections, together with their associated impedances, in field F3, but also the resonance frequency of the earthing (= reactive component precisely 0 Ohms, and thus phase angle of 180° at S11) can be read off at approximately 1.8 GHz. Thus, for a central stub angle of 45°, we have a

mechanical reduction of the layout of at least approximately 25%, as against the standard λ/4 line.

Just for interest, the calculations and simulations were repeated again for the central angles 30°, 60° and 90°, in order to obtain a feel for optimal dimensioning. Figs. 6.13, 6.14, 6.15 now show both the required impedances of the line sections for the three central angles investigated and also the associated simulations with the resonance frequencies, which now differ only insignificantly.

To be able to carry out a direct comparison between the characteristics of the 4 stub versions investigated, a summary diagram was drawn up for the impedance curve

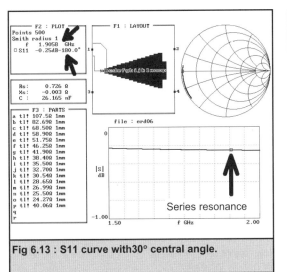

Fig 6.13 : S11 curve with 30° central angle.

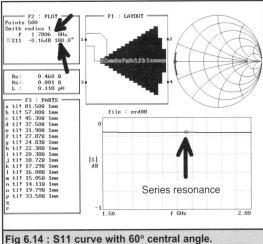

Fig 6.14 : S11 curve with 60° central angle.

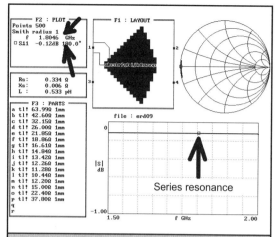

Fig 6.15 : Radial line stub with 60° central angle.

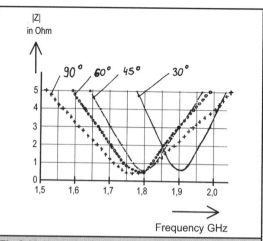

Fig 6.16 : Impedance plot for radial line stubs with 30°, 40°, 60° and 90° central angle.

of the four different stub versions, using Puff, in the frequency range from 1.5 to 2 GHz (Fig 6.16). If we look closely at this picture, we can obtain the following knowledge from it:

- With a central angle of 30°, the taper effect is already weakening markedly, and the behaviour corresponds more and more to the normal λ/4 line. This can be seen, above all, from the fact that the reduction is no longer so effective and the band width is diminishing.

- Between 45° and 90°, the reductions are almost identical in the order of 25% - but the broad-bandness naturally increases with the central angle, as was conjectured.

Practical tips:

As a developer, you can lay in a stock of such radial stubs for the main frequency ranges in which you work.

The procedure required for this can be demonstrated using the example of the design frequency of 2,400 MHz already used previously.

Step 1, Decide on a central angle between 45° and 90° - for example, 60°.

Step 2, The design frequency can now be selected to be higher by a factor of 1.25. Thus we can enter 3.0 GHz in field F4.

Step 3, Decide on the mechanical length required for a standard λ/4 line for the printed circuit board material used and for the frequency of 3 GHz. This gives the radius of the full circle from which the stub will be cut out.

Step 4, Now sketch the stub, replace it by a series of max.

18 series-wired microstrip line sections see above and enter these sections in field F3. Then comes the wearisome task of determining the matching impedance for each line section.

Step 5, Now assemble the stub in field F1 from the individual line sections and then start the simulation for S 11.

N.B, Dont be afraid of carrying out two separate simulations, one for a narrow band width of 2 to 3 GHz and another for a broad-band range of 0 to 10 GHz. Thus for a subsequent practical application you can have the option of investigating its characteristics even outside the range for which it is intended to be used and, for example, checking for oscillation tendencies. If you have used only the narrow-band version of stub here, Puff immediately reacts with a corresponding error message if the pre-set frequency range is exceeded.

Step 6, Now switch to a text processing program and print out the *.puf file just produced, together with any device file existing for an FET or transistor from the Puff directory. If we lay the two print-outs directly next to one another, we can recognise which parts of the stub file must be altered or deleted to make a device file from it.

The device file created in this way can then be stored under a suitable name (ending in *.dev) in the Puff directory. It will then be available as a component for future campaigns.

Step 7, If, though, you want to turn your attention to another frequency range, you have to repeat the entire procedure using the corresponding data.

Step 8, Anyone who, like the author, tends to make more

Table 6.3 Details of line sections.

Line section															
1	2	3	4	5	6	7	8	9	10	11	12	13	14	15	16
Width in mm															
0.816	1.63	2.45	3.265	4.08	4.90	5.71	6.53	7.35	8.16	8.98	9.80	10.61	11.43	12.24	6.12
Length in mm															
1.0	1.0	1.0	1.0	1.0	1.0	1.0	1.0	1.0	1.0	1.0	1.0	1.0	1.0	1.0	1.0

and more use of the ARRL Radio Designer (in parallel with Puff) to determine the noise figure, the stability factor, k, or an impedance curve, etc., will now have rather more work to do. You can certainly integrate the S-parameter file for the stub generated by Puff into the Radio Designer, but it does need rather more extensive preliminary work:

Not only do all superfluous parts of the Puff file have to be deleted but the unit GHz has to be added to every line of the S-parameter listed following the frequency value. The altered file is then copied into the ARD file currently being worked on to do this, see [14].

Miscellaneous constructional techniques

Under this heading come the "odds and ends" of practical technique not covered elsewhere. Some are of personal preference and others are suggestions that the constructor might like to explore.

Silver plating

Professional electroplating involves the use of carefully prepared plating solutions usually containing cyanide (which is highly toxic) and precisely controlled plating currents as well as scrupulous cleaning of the item to be plated. Such techniques are best left to the professional plater and reference to the "Yellow Pages" or a local telephone directory should reveal such services. Alternatively a proprietary silver plating solution may be used following the maker's instructions carefully.

Simple plating of flat or rounded surfaces can be accomplished without the use of cyanide solutions and electrolysis and the following method due to G4PMK and G3SEK [15] can be used to produce a coherent, if thin, plate finish which will help to preserve the good appearance of the equipment. Tarnish will form more slowly than on copper and the tarnish is, in any case, almost as conductive as the metal.

Mix together two parts by weight of finely ground sodium chloride (common salt), two of potassium hydrogen tartrate (cream of tartar) and one of silver chloride. Store the mixture away from moisture and strong sunlight (preferably in a dark brown glass, screw topped bottle). To silver plate an article, dampen a little of the powder with water and apply the resulting paste with a cloth using a vigorous rubbing action. Wear rubber gloves and avoid eye and skin contact. When finished. wash the article thoroughly and dry it.

Lacquer protection

Some degree of metal protection can be afforded by the judicious use of lacquers both on pcb surfaces and on the external surfaces of waveguide and other components. Care must be taken that the lacquer does not penetrate areas where metal to metal contact is needed. The lacquer can be applied by brush or by spraying. Clear car aerosol lacquer, may be used and, if applied thinly is "solder through" on surfaces such as pcbs

Paint protection

Polyurethane or epoxide based paints can be successfully used, for instance to protect horn or other antennas from the effects of weather. Again precautions should be taken to prevent ingress of paint into areas where metal to metal contact is essential.

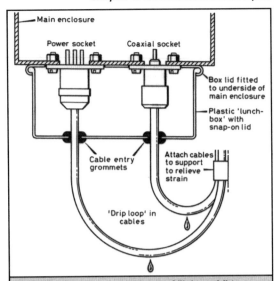

Fig 6.17 : Forming a splash proof "labrynth" to protect cable entry sockets on mast head equipment.

Weatherproofing

The success of weatherproofing is largely dependent on the careful choice of the weatherproof "container" and the care taken in attention to detail. Heavy plastics containers and the use of glass reinforced plastics (grp) have already been mentioned, as have die cast boxes. The latter are available with a gasket seal which renders them virtually submersible and, although expensive, are recommended for housing masthead equipment. All holes drilled into such boxes must be sealed against the weather by the application of auxiliary gaskets or silicone rubber sealant.

Silicone rubber "bath caulk" such as Dow-Corning is very suitable for this but contact with the equipment housed within the box should be avoided since the uncured rubber contains acetic acid and acetic anhydride which are corrosive to most metals and irritant to the skin, eyes and throat. When cured by air drying, the acid disappears and the risk of corrosion with it. The entry of power or signal cables into such an enclosure also presents the constructor with problems which are probably best tackled by the wise choice of connectors. N type coaxial connectors are gasketted to be water resistant and there are several types of multi pole connectors available which can be made waterproof. Reference should be made to manufacturers' or suppliers' literature before choosing connectors for this type of application. Whether waterproof or merely water resistant, such connectors should be placed in a position where protection against direct water entry is afforded, for instance recessed inside an auxiliary "splash guard" (a plastic box or skirt) mounted on the underside of the main box as shown in Fig 6.17. Cable routing, as shown, should also be designed to prevent water running into the connectors or connector box. This is one area of microwave construction where there is considerable scope for ingenuity and thought. Having made the housing as water tight as possible, it is still advisable to drill a small hole (1.5 mm or 1/16 inch) at the lowest point to allow the enclosure to "breathe" with changes of temperature and let water drain out! Too large a hole will allow the entry to insects which can also do untold damage to circuitry.

Printed circuit boards

The production of printed circuit boards (and in particular microstrip lines) at least for the lower microwave frequencies, is feasible at home. For the higher bands, say above 3.4GHz, stripline production is not really feasible unless the constructor has access to photographic equipment capable of precise reduction since the dimensions tend to become highly critical with increasing frequency, whereas at the lower frequencies rather more tolerance is permissible and slight inaccuracies can often be "tuned" or matched out.

The important factor in producing successful pcbs is the selection of correct materials of known and reproducible characteristics and, in general, these characteristics will be defined by the design, i.e. a particular type of board material will be specified in the design description. Ordinary (G-1 0) polyester glass fibre board is usable at 1.3GHz but for better performance at this, and particularly higher frequencies, ptfe glass fibre or ptfe board is to be preferred and is available in a range of thicknesses, both of the substrate and the copper cladding. Such board has better defined manufacturing tolerances and superior dielectric properties and, as a consequence, is considerably more expensive than the more conventional board which most constructors will have used. Several makeshift methods of fabrication can be used whatever the material. Conventional masking and etching is feasible and, with careful drafting on paper, it is possible to make transparencies from these drafts on many modern photocopying machines of the type to be found in most offices. Make sure, however, that the photocopying machine copies at exactly 1:1 ratio - some of them certainly do not! Alternatively, self adhesive masking strips (for example Circuitape) of the correct width can be used with transparent sheet to produce suitable transparencies. Photosensitive resist lacquer is available in aerosol spray form and the pcb material will be found to be quite easy to coat with such sprays. Alternatively, ready coated board suitable for use up to 1.3GHz can be purchased and used. Exposure to sunlight or use of a simply constructed ultra violet light box will allow the constructor to produce quite accurate and well defined pcbs from the transparencies. Correct exposure and subsequent development is a matter for experiment. [16] is a comprehensive account of amateur pcb techniques and should give the microwave constructor many useful ideas to work on. (Caution : exposure of the eyes to ultra violet radiation is extremely dangerous and can lead to permanent, irreversible corneal damage).

An alternative method is to mark off, for instance a stripline, and using a straight edge and sharp scalpel, to cut along the edges of the line taking care to cut through the cladding but not through the dielectric. Areas of unwanted copper can then be physically stripped off the board, a process which is aided by the careful application of a soldering iron to the unwanted copper until it is possible to lift the edge of the cladding. It is then quite easy to "peel" the cladding off the board. More complicated shapes than stripline may not be achievable by such simple techniques.

Static precautions

Some types of microwave components, for example Schottky diodes (mixers and detectors), microwave bipolar transistors and GaAsfets can be damaged or destroyed by static charges induced by handling and thus certain precautions should be taken to minimise the risk of damage.

Such sensitive devices are delivered in foil lined, sealed envelopes in conductive (carbon filled) foam plastic or wrapped in metal foil. The first precaution to be taken is to leave the device in its wrapping until actually used. The second precaution is to ensure that the device is always the last component to be soldered in place in the circuit. Once in circuit the risk is minimised since other components associated with the device will usually provide a "leakage" path of low impedance to earth that will give protection against static build up.

Before handling such devices the constructor should be aware of the usual sources of static. Walking across nylon or polyester carpets and the wearing of clothes made from the same materials are potent sources of static, especially under cold dry conditions. Earth leakage potentials from an improperly earthed mains voltage soldering iron are also a source of trouble. Static to a potential of several thousand volts may be carried by the body although much lower leakage potentials existing on improperly earthed soldering irons are still sufficient to cause damage. Some precautions are listed below:

- Avoid walking across synthetic fibre carpets immediately before handling sensitive devices.

- Avoid wearing clothes of similar materials.

- Ensure that the soldering iron is properly earthed whilst it is connected to its power supply. This is a common sense precaution in any case. Preferably use a low voltage soldering iron.

- Use a pair of crocodile clips and a flexible jumper wire to connect the body of the soldering iron to the earth plane of the equipment into which the device is being soldered.

- If the component lead configuration allows (and the usual flat pack will), place a small metal washer over the device before removing it from its packing in such a position that all leads are shorted together before and during handling. Alternatively, it might be possible to use a small piece of aluminium foil to perform the same function, removing the foil once the device has been soldered in place.

- A useful precaution which will minimise the risk of heat damage, rather than static damage, is to ensure that the surfaces to be soldered are very clean and preferably pre tinned.

- Immediately before handling the device, touch the earth plane of the equipment and the protective foil to ensure that both are essentially at the same potential.

- Place the device in position, handling as little as possible.

- Disconnect the soldering iron from its power supply and quickly solder the device in place. It may be necessary to repeat some of the operations if the soldering iron has little heat capacity.

- Finally, when assembling items of equipment to form a complete operating system, for instance when installing a masthead pre amplifier and associated transmit/receive switching, it is important to keep leads carrying supply voltages to the sensitive devices well away from other leads carrying appreciable rf levels or those leads which might carry voltage transients arising from inductive (relay) switching. Such supply lines should be well screened and decoupled in any case, but physical separation can minimise pick up, thus making the task of decoupling easier.

Conclusion

To be successful in microwave construction the amateur does not necessarily need to be a highly skilled engineer. Logical techniques, planning and the execution of such a plan of construction - given a little care and patience - will certainly lead to functional equipment which cannot otherwise be obtained. A number of simpler techniques have been outlined which, when applied to the designs contained within this manual, will ensure success. The proof of the constructor's success will lie in the results obtained from sensible, good operating practices such as those outlined in chapter 1.

References

[1] The Beginner's Workshop, by Ian Bradley. Model and Allied Publications, Argus Books Ltd. ISBN 0 85242 428 0.

[2] The Amateur's Workshop, IBID. ISBN 0 85344 049 2.

[3] The Amateur's Lathe, by Lawrence Sparey. IBID. ISBN 085242 288 1.

[4] Lathework Questions and Answers, by J A Oates, Newnes Technical Books. Butterworth Press. ISBN 0 408 00065 1.

[5] VHF/UHF manual, 4th Edition, pp 9.12 and 9.13, RSGB Publications.

[6] "Home made finger stock", by J Nilson, SM6FHI. VHF Communications Vol 9, no 2(1977).

[7] Modern Plastics Encyclopaedia, 1983-84, McGraw-Hill Publication,,,.

[8] Schneider, M.V., Microstrip Lines for Microwave Integrated Circuits, Bell System Technical Journal, Vol. 48, no. 5, May-June, 1969

[9] From article by Gunthard Kraus, DG8GB published in VHF Communications 3/2000 pp 167 - 178

[10] Kraus, Gunthard, DG8GB,Design and realisation of microwave circuits, Part 9, VHF Reports, no. 2 / 1998, pp. 119 ff.

[11] Broadband Microstrip Mixer Design, Application Note 976 from Hewlett Packard, On Internet under:www./w.hp.com/HP-COMP/rf

[12] Winding, J.R.,Radial Line Stubs as Elements in Strip Line Circuits Nerem Record, pp. 108-109, 1967

[13] Schneider, M.V. Microstrip Lines for Microwave Integrated Circuits, Bell System Technical Journal, Vol. 48, no. 5, May-June, 1969

[14] Kraus, Gunthard, DG8GB and Zimmermann, Andreas, DG3SAZ,Low-noise pre-amplifier for 137 MHz NOAA weather satellite range and / or 145 MHz 2-m. amateur radio band / Part 2, VHF Reports, no. 1 / 1999, pp. 37 ff.

[15] "More gain from 1.3GHz power amplifiers". by R Blackwell, G4PMK and I White, G3SEK. Radio Communications June 1983, pp 500-503.

[16] "Printed circuit techniques for the amateur". Cliff Sharp, G2HIF. Radio Communications December 1979, pp 1128-1136.

Chapter 7

Common Equipment

In this chapter :

* Local oscillator sources
* Frequency multipliers

* Amplifiers

T here may be many items of common equipment in a typical microwave enthusiast's station. An excellent example is an 1152MHz signal source. Such a source may be used either as the local oscillator in a 1296 MHz transverter or multiplied to 3456, 5760 or 10,368MHz as the driver for a cw or fm transmitter; see Fig 7.1. The same basic design of oscillator may also be used to generate local oscillator signals for converters and transverters for these same higher bands.

Many converters and transverters use similar intermediate frequency (i.f) circuitry such as low noise i.f preamplifiers. A number of tried and proven designs will be presented in this chapter.

Some items that might be considered common equipment, for example couplers, attenuators and loads, are covered in chapter 8, Test Equipment.

Local oscillator sources

For stability reasons, most local oscillator sources for the microwave band are crystal controlled. Generation directly at the final frequency is usually considered impractical due to the absence of suitable high frequency crystals;

therefore the crystal oscillator/multiplier approach is the accepted method of reaching the final frequency.

The crystal frequency chosen to start the chain will obviously depend on the required final frequency. If too low a starting frequency is chosen then a large order of multiplication will be needed, which may cause problems with stability. It may also prove difficult to provide sufficient inter stage filtering to reduce the level of unwanted products at the output. Too high a crystal frequency and the crystals become very expensive and fragile. The optimum starting frequency for many amateur microwave applications is in the range 90 to 120 MHz.

Crystal oscillators operating in the region of 100MHz use overtone crystals, usually operating on their fifth or seventh overtone. There are a number of suitable oscillator circuits, which can be used, but for reasons of reproducibility, stability and noise performance, the choice is reduced to just a few. Perhaps the best know of these is the series resonance crystal Colpitts oscillator using a power fet as the active element [1].

The circuit of Fig 7.2 has justifiably become the standard oscillator used in almost all current European amateur microwave designs. It does, however, suffer from one drawback. To achieve good phase noise performance it is necessary to use a power fet such as the Texas Instruments P8000. This type of device is notoriously difficult to obtain. As a result there has been a tendency to use lower power fets such as the Siliconix J310. Although these work well in this circuit, the noise performance is not especially good and this may be of considerable concern in sources for 10GHz and higher.

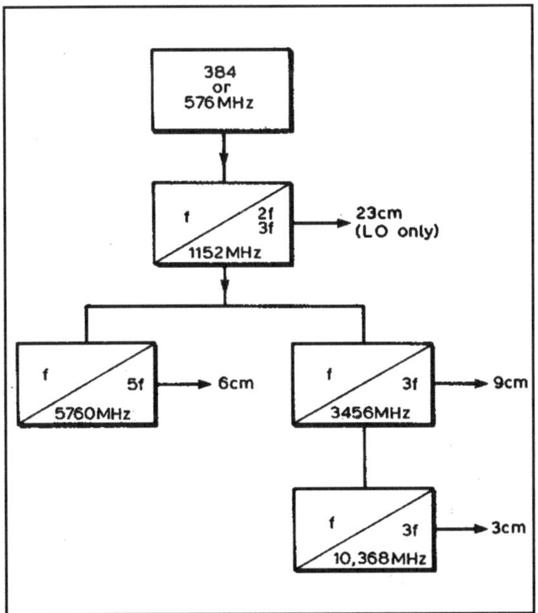

Fig 7.1 : 1152MHz multiplier chains.

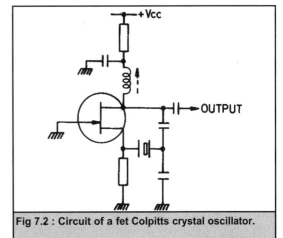

Fig 7.2 : Circuit of a fet Colpitts crystal oscillator.

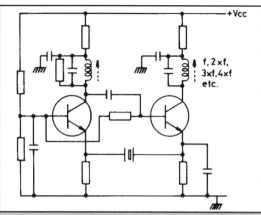

Fig 7.3 Circuit of a Butler oscillator using bipolar junction transistors (bjt).

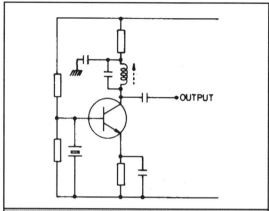

Fig 7.4 : Circuit of a bjt Colpitts oscillator.

Setting the crystal oscillator on exactly the required frequency can occasionally prove difficult. This may be tuned out by adding a small amount of inductance across the crystal. The inductance needs to resonate with the crystal holder capacitance at the crystal series resonant frequency.

An alternative oscillator circuit that is capable of phase noise performance at least as good as the power fet Colpitts circuit is the Butler oscillator of Fig 7.3. In this circuit the function of amplitude limiting is performed by a second stage, leaving the function of the first stage as an amplifier. A further advantage of the Butler oscillator is the inherently high harmonic output. The limiter stage output circuit is usually tuned to the required harmonic of the overtone oscillator. For example a 100MHz crystal oscillator stage may have an output tuned to 300MHz, at which frequency several mW will be available, hence eliminating the need for one additional stage of multiplication after the oscillator stage.

The Butler oscillator is also particularly easy to modulate as shown in the following section. However the modulation produced by this simple circuit is a mixture of fm and phase modulation. For many applications this does not matter.

The bipolar transistor Colpitts overtone oscillator circuit, so often used in early amateur designs, has now almost entirely fallen from favour as its relatively poor noise performance has now been widely recognised; see Fig 7.4.

An oscillator source for 1.0 to 1.4GHz

The source described in the following section was originally designed [2] using discrete components, to provide two 10mW outputs at 1152MHz for use in a 1296MHz transverter with a 144MHz intermediate frequency (i.f) - see circuit diagram Fig 7.5. Later work showed it was possible to use the same board, with a suitable crystal, anywhere in the range 1000 to 1400MHz with only minor changes in component values and output spectral purity. In 1993 G4JNT worked on an updated SMD version of the G4DDK-001 1.1 /1.3GHz oscillator source. Sam, G4DDK, the original designer of this popular unit, pointed out that he had already explored this avenue not long after the original design was published. Although it worked, it was not as good as the original design, several reasons were found for this, including more critical coupling between the SM trimmer capacitors, poorer through board grounding and (at that time) the wrong choice of SM transistors. It was also found, after discussion with many other amateur microwave enthusiasts, that not everyone was in favour of using fiddly SMDs for all components. Some constructors choose to use existing components for their own version of the design. A compromise was reached where the size of the design was reduced to fit into one of the popular tin plate boxes (rather than diecast box) and some modern SM devices used to substitute for older, more critical standard devices: notably decoupling capacitors and transistors. The difficult trapezoidal capacitors and right angled coaxial output connectors were eliminated and the results obtained were entirely consistent with those of the

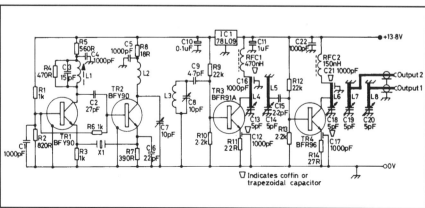

Fig 7.5 : Circuit of the original G4DDK-001 1.3GHz source.

∇ Indicates coffin or trapezoidal capacitor

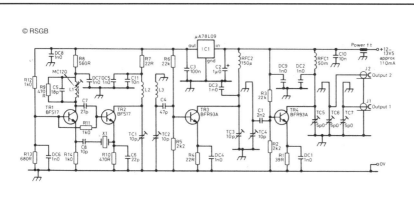

Fig 7.6 : Circuit of the revised 1.3GHz source, G4DDK-001B, designed to use smd components.

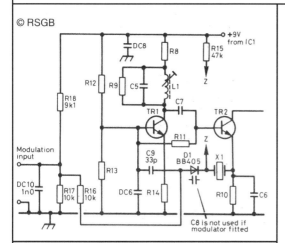

Fig 7.7 : Modulator circuit for G4DDK-001B source.

original design, if anything, the spectral purity, already excellent, was slightly better!

The circuit

The circuit of the revised design, known as the G4DDK-00 1B, (Fig 7.6 and Table 7.1 for components) retains, the original, well established Butler oscillator but with the addition of an improved on board modulator which is now suitable for 9.6kb/s FSK modulation. The varicap circuit (Fig 7.7) pulls the crystal directly and is biased to optimise linearity. If modulation is not needed simply omit R15 - R18, C9, DC10 and DI, but note that C8 is still required. C8 was chosen to suit most types of fifth overtone crystal, so that it can be pulled accurately onto frequency. Even so, some crystals may refuse to pull onto frequency. In this case, the value of C8 may be increased to 1000pF. If this does not work, change the crystal! The Toko S18 coil

has been changed to one from the Toko MC120 range with a screening can. An aluminium core *must still* be used and, since these are not normally supplied with this range, must be purchased separately. The BFY90 transistors are replaced with more predictable SM transistors, type BFS17 and the BFR91 multiplier transistors by SM BFR93A. SKY (black) 10pF trimmers are preferred for positions TC1 - TC4 and SKY (green) 5pF for TC5, 6 and 7. Murata TZ03 20pF (red) could be used for TC1 and 2 and 10pF (white) for TC3 and 4. It should be possible to use Murata 2.3pF (black) forTC5, 6 and 7, although this has not yet been tried. Note that the Murata black trimmers will *not* work for TC1 and TC2!

Sam tried using a two pole helical filter in place of the three pole microstrip filter and found the output spectrum to be noticeably less clean. Because of the use of a tin plate box, it is now possible to use ordinary panel mount SMA, SMB or SMC connectors. The two output sockets are too close together to use larger types, if a single output is required (cut the output track 2 where it leaves the line, as in the original design), it is possible to fit a BNC or TNC socket.

You could produce your own PCB from the artwork given in Fig 7.8, but art work should be available from Sam via his web site [3]. Although a ground plane mask was included in Sam's original design, it has not been reproduced here because it is easier for home etching to avoid the use of double masking and to make clearance holes, where needed, after etching and drilling.

Full details of the construction methods are not given here since these should be familiar to most constructors. The PCB should be soldered into the tinplate box (see the Components List) with the ground plane 17mm below one rim of the box after marking its position and drilling holes for the output socket(s) and feed through capacitor(s). Take your time with these operations and get them right! Fit the output socket(s) by soldering the spill(s) to the track(s) first, then the body(s) to the box.

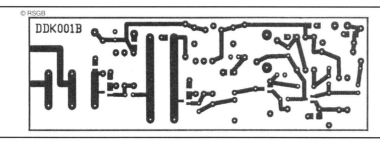

Fig 7.8 : Artwork for the G4DDK-001B source. (not to scale)

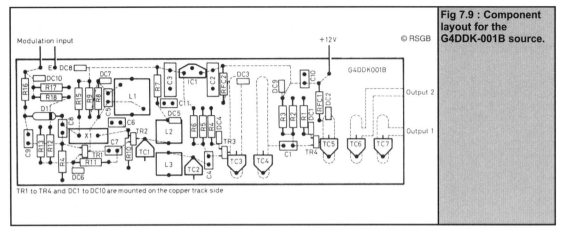

Fig 7.9 : Component layout for the G4DDK-001B source.

TR1 to TR4 and DC1 to DC10 are mounted on the copper track side

Next, solder all resistors and ceramic plate capacitors, followed by tantalum capacitors, chokes, trimmer capacitors, [CI and DI (if fitted)]. Positions are shown in Fig 7.9. Make ground plane earth connections where needed.

Where the chip decouplers are to be fitted, through board grounding links are needed. These can be short lengths of discarded component leads (such as those cut off the resistors) or you might use small PCB pins such as those used in all the G3WDG designs. Whichever is used, be sure to solder both sides! Fit both the InF decoupling capacitors and the transistors on the track side of the board where indicated. The transistors have a right way up as shown in Fig 7.10. Fig 7.11 shows the suggested layout of the connectors and feed through capacitors. Actual placing is by measurement from the soldering line of the board in the box and the position of tracks relative to the end walls of the box.

Alignment

Alignment is similar to the original design, except that

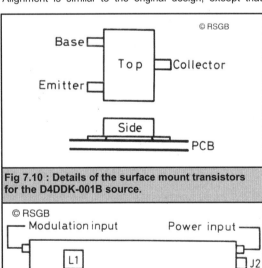

Fig 7.10 : Details of the surface mount transistors for the D4DDK-001B source.

Fig 7.11 : Top view of the approximate positions of the feedthrough capacitors and connectors for the G4DDK-001B source.

since LI is in a screening can, you can no longer easily couple an absorption wavemeter to the coil to check the crystal oscillator is working!

Original alignment instructions :- In the following description it is assumed the unit is to be tuned to 1152MHz.

Connect dc power and check the current drawn is no more than about 150mA. If significantly more, then check for short circuits or wrongly placed components. When all is well, proceed with alignment. Place an absorption wavemeter pick up coil close to L1 and tune to 96MHz. A strong reading should be indicated on the meter. Peak the indication by turning the core of L1. Turn the oscillator on and off to check that it re starts satisfactorily. If it doesn't, then turn the core of L1 about a quarter turn and try again. Don't worry about getting the frequency to exactly 96.0000MHz yet; a few KHz errors doesn't matter at this stage.

Set a moving-coil multimeter to the 2.5 volt range (or nearest equivalent) and measure the voltage across R11. This should be no more than a few hundred mV. Peak the reading by tuning C7 and C8. Confirm the frequency selected is 288MHz by placing the coil of the wavemeter close to L3.

Transfer the meter leads across R14 and peak the reading by tuning C13 and C14. Again check the correct harmonic (576MHz) has been selected by using the wavemeter.

Finally connect a low power wattmeter (+10 to +20dBm full scale) to the output and tune C18, 19 and 20 for a maximum reading. Confirm that the correct harmonic (1152MHz) has been selected by using the wavemeter. It may now be necessary to go back and *slightly* re peak the trimmers for an absolute maximum reading at the 1152MHz output.

Final setting of the frequency of the crystal oscillator can now be done by either using a known high accuracy frequency counter or by connecting the source as the local oscillator of your 1296MHz converter and listening for a beacon whose frequency is known. L1 can then be adjusted to bring the signal onto the correct receiver dial calibration.

If difficulty is experienced in pulling the frequency to that marked on the crystal, then it is very likely you have a non standard crystal. Pulling the frequency too far can result in the oscillator failing to restart after switching off and then on. The cure is to put a small value ceramic plate

Table 7.1 G4DDK-001B source component list.

Resistorsall 1/4W carbon or metal film.

R1	39R
R2	2k2
R3	22k
R4	22R
R5	2k2
R6	22k
R7	22R
R8	560R
R9,10	470R
R11,12	1k0
R13	680R
R14	1k0
R15	47k
R16,17	10k
R18	9k1

Ceramic plate capacitors: Philips 682 Series, 2.5mm lead pitch.

C1	2p2F
C4	4p7F
C5	18pF
C6	22pF
C7	27pF
C8	10pF
C9	33pF

Philips 629 series, 2.5mm lead pitch.

C10,C11	10nF

Tantalum bead capacitors:

C2	1µF 35V wkg, AVX, Kyocera etc.
C3	0.1µF 35V wkg, AVX, Kyocera etc.

SMD decoupling capacitors: all 0805 size D

C1-10	1000pF, AVX, Philips etc.

Trimmer capacitors:

TC1-4	10pF SKY (black), see text
TC5-7	5pF SKY (green), see text

Semiconductors

TR1 and TR2	BFS17 (Philips) SOT23 outline
TR3 and TR4	BFR93A (Philips) SOT23 outline
1C1	78L09 (Philips, Motorola etc)
D1	BB405 (Philips)

Miscellaneous

X1	96MHz fifth overtone, HC18/U case, overtone freq. = output freq. /12

L1 MC120 Toko part number E526ANA100075 (green) with screening can and aluminium core. It may be necessary to order with a ferrite core and buy an aluminium core separately. Alternatively, use S18 type, part number 301AN0506

L2/L3: 3 turns of l mm diameter silver plated wire. Internal diameter of coils, 3mm, turns spaced one wire diameter. Coil mounted 2mm above ground plane.

RFC1/2: 150uH axial lead choke (Cambion, Toko)

Tinplate box: type 7754 (37 x 111 x 30mm) from Piper Communications

Coaxial sockets: 3, type SMA/B/C (see text)

Lead through capacitor for power 1000 to 5000pF

Lead through for modulation (low capacitance), Oxley

capacitor, say 10 to 33pF in series with the crystal by cutting the pcb track near to the crystal. If you have to do this modification, then please keep the leads of the new capacitor short and use a zero temperature coefficient capacitor or frequency may drift unacceptably as the crystal oscillator warms up.

However, there is greater coupling into TR3 than before and the emitter voltage of TR3 increases from about 100mV to about 300mV when the oscillator is working correctly. When remaining stages are correctly aligned, TR3 emitter should be at 500 to 700mV and the emitter of TR4 at 1.0 to 2.0V. The only critical tuning operation is TC2.

Sam acknowledges the help of Dave, G8KKB, in developing the new modulator circuit and the local packet fraternity for proving the modulator in 9.6kb/s packet links in the Suffolk Data Group area.

An oscillator source for 2.0 to 2.6GHz

This oscillator source is a direct development of the previous discrete component design, but incorporates an additional multiplier stage to provide output in the 2 to 2.6GHz frequency range. This range encompasses such uses as the local oscillator in receive and transmit converters operating in the 2.3GHz amateur band.

Several modern design 10GHz narrow band transverters require local oscillator drive at 2556MHz [4]. This source could also be used for this purpose. This was the original reason for developing the unit. Versions of this unit have been built with outputs in the range 2176MHz to 2592MHz. Based on the available capacitance swing left in the tuned circuits, the unit should be usable with outputs between 2.0 and 2.6GHz. It may even be possible to reach 2.8GHz, opening up the possibility of using the source to drive a doubler to 5.6 or 5.7GHz.

Circuit description

The circuit of the source is shown in Fig 7.12. Component values are given in Table 7.2. The crystal oscillator is designed around the Butler circuit used in the previous design. This circuit provides good phase noise performance and inherent multiplication. Only first grade BFY90 transistors should be used in the oscillator. Surplus devices have been known to cause problems. Several people have experience problems getting the oscillator to reliably re start after switching off if the crystal frequency has to be pulled significantly. The cure for this problem is simple and requires only that a 10 to 33pF capacitor be inserted in series with the crystal. Provision has been made for the capacitor on the board. If it is not required, then a short wire link should bridge the Cx pads or a ceramic 1nF capacitor used.

SKY (green) trimmers were used for C13 and 14 because they were available. Any small (5mm diameter) 5pF trimmer could be used but, since you will need green SKY trimmers in the output filter, it makes sense to use them here as well. Similarly, SKY trimmers should be used for C18, 19 and 20, although the black Murata type could be used in this stage depending on the required frequency.

The final multiplier uses a BFG91A and operates well at these frequencies, easily achieving the specified output. This type of transistor has two emitter connections and both must be thoroughly decoupled to ground to ensure stable operation; however, only one emitter has dc connection to ground through R19.

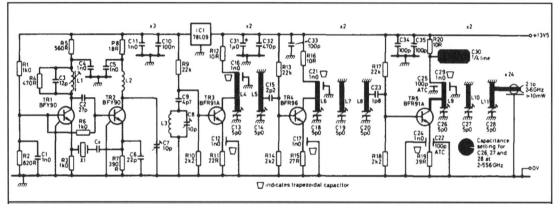

Fig 7.12 : Schematic diagram of the 2.5GHz source G4DDK-004.

Additional decoupling has been provided in the collector supply circuit of this stage using a ¼ wave open circuit low impedance transmission line. The open circuit at the end of the line is transformed to a short circuit ¼ wave away at the junction of R20 and the ¼ wave choke line formed by the track from C29 to R20. Although this arrangement can only be optimum at one frequency, in practice the bandwidth of the line is such that it still remains effective over the entire range 2.0 to 2.6GHz.

The opportunity has been taken in this design to eliminate the collector supply chokes used in the 1152MHz unit. These chokes have caused problems in the past due to difficulty in obtaining the required axial lead types specified. The 10R resistors are not as effective as chokes but the decoupling capacitor values have been carefully chosen to be as effective as possible in the frequency ranges encountered in their respective stages.

Construction

The unit has been designed to fit into the type 45 tinplate box (also known as type 7768). This box is 55.5 x 148 x 30mm and retails at low cost compared with diecast boxes of similar size.

A pcb layout has been provided for the oscillator unit together with a component overlay. These are shown in Fig 7.13 and 7.14 respectively.

The pcb material used for the oscillator is ordinary 1.6mm thick epoxy glass, double clad with 1oz copper. Slight differences in ε_r between board materials may affect

performance slightly at the high frequency end of the range. A high quality pcb (G4DDK 004) is available, ready drilled, slotted and tinned from the RSGB Components Service.

Start construction by drilling the etched board. 0.8mm holes are needed for all components except the transistor mounting holes (5mm) for TR3, 4 and 5. 1.2mm holes are needed for the trimmer capacitors. Except for component leads that should be earthed, remove the copper on the earth plane side of these holes, using a larger drill. Slots 1mm wide need to be cut for the grounding strips at the ends of L5, 7, 8, 10 and 11. 1.2mm wide slots are needed for trapezoidal capacitors C12. 16, 17, 21, 24, 25, 29.

File a small area of board to clear the two overlapping corners of the tin plate box. Next place the drilled board inside the L shaped half of the specified tin plate box. The box should not be soldered together at this stage. Position the board such that the top of LI is 5mm below the rim of the box. Mark all round the inside of the box where the board will be soldered into place. Transfer the line to the inside of the other half of the box. Mark where the output socket is to be mounted, allowing the spill of the connector to lie flush with the output track on the pcb. It is better to use a single hole mounting socket and solder it flush to the outside wall of the box with its spill protruding into the box. The socket should be an SMA, SMB, or SMC (CONHEX) type. N types are too large and BNC connectors can prove unreliable at these frequencies. Drill the socket mounting hole, also drill a hole in the same end of the box to take the feed through capacitor that will be used to bring dc power into the box. If a crystal heater is

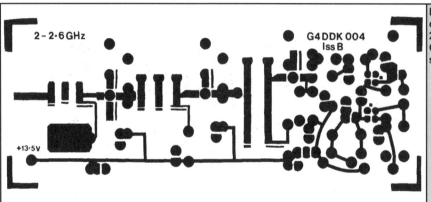

Fig 7.13 : Printed circuit board for the 2.5GHz source, G4DDK-004. (not to scale).

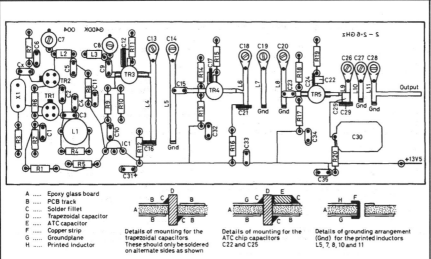

Fig 7.14 Layout diagram of the components for the 2.5GHz source, G4DDK-004.

A ..... Epoxy glass board
B ..... PCB track
C ..... Solder fillet
D ..... Trapezoidal capacitor
E ..... ATC capacitor
F ..... Copper strip
G ..... Groundplane
H ..... Printed Inductor

Details of mounting for the trapezoidal capacitors
These should only be soldered on alternate sides as shown

Details of mounting for the ATC chip capacitors
C22 and C25

Details of grounding arrangement (Gnd) for the printed inductors
L5, 7, 8, 10 and 11

to be used then also drill holes for the power feed for this in the other end wall of the box.

Carefully solder the board into one half of the tin box. Next solder the other half into place, ensuring a good fit. It may be necessary to carefully file a small amount of the pcb away at the ends or sides to get a comfortable fit. The soldering should be along the whole of the sides and ends

of the box to give a good rf tight connection.

Cut five short lengths of thin copper strip and solder them at the ends of L5, 7, 8, 10 and 11, as shown in the component overlay diagram Fig 7.14. Solder all the resistors and capacitors into place where shown, taking care to solder grounded leads both top and bottom of the board.

Table 7.2 Component list for the G4DDK-004 source.

RI,3,6	1 k
R2	820R
R4	470R
R5	560R
R7	390R
R8	18R
R9,13,17	22k
R10,14,18	2k2
R11	22R
R12,16,20	10R
R15	27R
R19	39R

All resistors 0.25W miniature carbon film or metal film

Capacitors

C1,4,5,11	1000pF high K ceramic plate e.g. Philips 629 series
C2	27pF low K ceramic plate e.g. Philips 632 series
C3	12pF
C6	22pF
C9	4p7
C15	2p2
C23	1p8
C31	1µF tantalum bead 16 volt working
C32	470pF medium K ceramic plate e.g. Philips 630 series
C33,34,35	100pF low K ceramic plate e.g. Philips 632 series
C10	0.1µF tantalum bead 16 volt working
C12,16,17,	1 nF trapezoidal capacitor from RSGB or Cirkit
C21,24,29,C25,29	22pF trapezoidal capacitor from Cirkit

C7,8	10pF miniature trimmer (5mm diameter) e.g. Cirkit 06-10008
C13,14,18,19 20,26,27,28	SKY trimmer (green) from Piper Communications
C30	PCB track
Cx	10 to 33pF type as C3 (see text)

Inductors

L1 - TOKO S18 5.5 turn (green) with aluminium core

L2,3 - 2 turns of 1 mm diameter tinned copper wire. Inside diameter 4mm. Turns spaced to fit hole spacing. Exceptionally 3 turns at the low frequency end of the range

L4 to 11 Printed on the PCB.

Semiconductors

TR1,2	BFY90
TR3	BFR91A
TR4	BFR96
TR5	BFG91A source
IC1	78L09

Miscellaneous

X1   fifth overtone crystal in HC18/U case.
       Frequency of crystal = Fout/24

Tinplate box type 45 from Piper. Also known as 7768. Size 55.5mm wide, 148mm long and 30mm high. A box could also be made from off cuts of double clad PCB material. In this case make the box bottom slightly longer to allow for mounting holes.

Output socket single hole mounting type SMA, SMB or SMC.

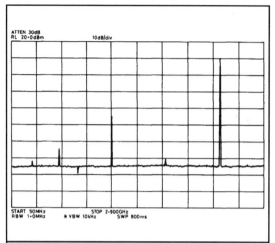

Fig 7.15 : A spectrum analyser plot of the output spectrum of a 2.556GHz source, G4DDK-004.

Solder the coils into place, remembering to solder the grounded end of L3 both top and bottom. Solder TR1 and 2 into place, making sure they are seated well down onto the board but leaving just enough room to solder the case lead of both transistors to the ground plane of the board. Solder IC1 into place remembering to ground the centre lead. Solder the trapezoidal capacitors into place as shown in the component overlay diagram.

Carefully bend the flat connection lead of the SKY trimmers out at 90 degrees to the capacitor body. Place the trimmer with the round lead in the hole in the tuned line. Solder to the tuned line and then solder the flat lead to the ground plane. C7 and 8 can be treated in the same way but, unless black SKY trimmers are being used, the mounting arrangement may be different. Use the shortest possible leads on the trimmers, whichever type you choose to use.

Solder TR3, 4, and 5 into place taking care to get the leads the right way round. Solder the crystal into place last, seating it well down onto the board. It may be advisable to earth the case of the crystal, especially if a heater is to be used.

Alignment

In this section it is assumed that the unit is to be aligned to 2556MHz. It is possible to align the oscillator / multiplier with nothing more than a multimeter. However, this will give little information as to what frequency the unit is tuned to, or what output level has been achieved. As far as aligning the unit is concerned, the most essential items of test equipment are a moving coil multimeter and absorption wavemeter(s) to cover the range 106 to 2600MHz. A digital multimeter is not advised for tuning up as it can give *very* misleading results. The wavemeter is also preferable to a frequency counter since it not only indicates the required output but also the presence of any unwanted signals.

Connect a 13.5V supply and check that the unit takes no more that about 150mA. If the consumption is much higher than this switch off and check for faults. When all is well, check the regulated voltage is 9V.

Place the wavemeter pick up coil close to L1 and tune to 106.5MHz. Adjust the core of L1 for a strong reading.

Switch off then on and check that the oscillator restarts. If not, adjust L1 core a quarter turn and try again. The exact frequency does not matter at this stage.

Set the multimeter to its 2.5V range (or nearest equivalent) and connect across R11 and adjust C7 then C8 for a maximum reading. When you adjust C7 initially you may only see the slightest movement of the meter. This is normal and what you are looking for and a digital meter will probably miss it due to the up dating method used. Peak the reading with C8. Confirm with the wavemeter that you have set L2 and 3 to the required third harmonic of the crystal frequency at 319.5MHz

Transfer the meter to R15 and adjust C13 and 14 for a maximum voltage reading across R15. Confirm that you have selected the correct harmonic at 639MHz. Transfer the meter to R19 and adjust C18, 19 and 20 for a maximum reading. Because there are three tuned circuits this time, the initial meter movement may be very small. The middle tuned circuit tunes especially sharply. Confirm with the wavemeter that you have tuned these circuits to 1278MHz.

Connect an absorption wavemeter, preferably the type with built in diode detector, to the output connector. Tune the wavemeter to 2556MHz. Adjust C26, 27 and 28 to obtain a maximum reading. This should be close to 10mW if your wavemeter is also calibrated for output power. As mentioned earlier use an insulated trimming tool to adjust the output trimmers. Any metal bladed trimming tool will affect the tuning point and may well prevent the circuits resonating as high as 2556MHz.

It is now worth going back over the previous adjustments and re peaking to obtain maximum output. If any adjustment requires more than the slightest tweak, be suspicious of the alignment of that stage. In these circumstances it is often worth starting alignment from the beginning again.

If the output frequency is a little low and pulling it up onto frequency results in reduced output or failure to restart after switching off and then back on, it may be necessary to add the capacitor Cx as previously mentioned. This should cure the problem. If not change the crystal!

Due to component tolerances, there can sometimes be instability in the final multiplier stage, making the unit difficult to align. The remedy for this is to replace the BFG91A with a BFR91A and to add, on the upper surface of the board, additional decoupling in the form of two l00pF ATC porcelain chip capacitors soldered across C17 and C24. It is possible, with a little care, to fit these extra capacitors between the ground plane and the silvering of the existing trapezoidal capacitor, making modification to the board unnecessary. It has sometimes been found advantageous to shorten the centre line by 1 to 1.5mm, especially if operation towards the high end of the specified range is contemplated. This can be achieved either by soldering a pcb pin through a drilled hole in the ground plane and line or by cutting another narrow slot at the appropriate point and using copper foil in the usual manner. These modifications have often resulted in an output of between +11 and +l3dBm. To realise highest power output it has also been found essential to use type BFR91A transistors, not the older BFR91.

One unexpected use for the board has emerged since production of the prototypes. The tuning range of the final multiplier filter is such that it has been found possible to resonate it as low as 1.2GHz, allowing the final multiplier

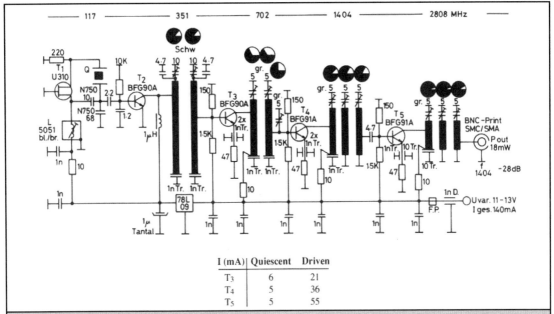

I (mA)	Quiescent	Driven
T_3	6	21
T_4	5	36
T_5	5	55

Fig 7.16 : Schematic diagram of the DC0DA 2.8GHz source, courtesy of DARC and J Dahms, DC0DA.

to operate as an amplifier producing 50 to 70mW output. In this form, the board may be used as a low power transmitter in the 1.3GHz band. Fig 7.15 is a spectrum analyser trace of the output spectrum of this design when aligned for output at 2.56GHz.

An oscillator source for 2.8GHz

The source described in this section is typical of the designs by several German amateurs, notably J Dahms, DC0DA. The source was designed to produce about 20mW output at 2808MHz for use with a harmonic mixer operating at 5760MHz [5]. A fet Colpitts overtone crystal oscillator, of the type described at the beginning of this chapter, operates at 117MHz using a U3I0 junction fet.

The oscillator is followed by a tripler to 351MHz and then by three frequency doubler stages to 2808MHz. A feature of the design is the use of all printed inductors in the multiplier stages, in contrast to the mixture of inductor types used in the two previous designs presented in this chapter.

The circuit of the 2.8GHz source is shown in Fig 7.16

while Fig 7.17 is a photograph of the 2.8GHz source.

Phase locked oscillators

Another type of oscillator source used in amateur microwave designs is the PLL. For many years, professional microwave system engineers have used a versatile phase locked loop technique for generating signals anywhere in the range from 1 to over 20GHz. The technique uses a voltage controlled oscillator (vco) which is phase locked to a crystal oscillator. As an example of this technique, the following section show the details of a PLL used in frequency synthesiser designed by Bernd Kaa, DG4RBF [6].

The PLL

The core of the circuit is the modern PMB2306T PLL circuit from Siemens, which was developed for mobile radio applications. The circuit comprises three freely programmable counters: a reference counter (16-bit R counter), an N counter (14-bit N counter), and an A counter (7-bit A counter), the phase detector with charge pump, and a lock detector, and can be programmed by

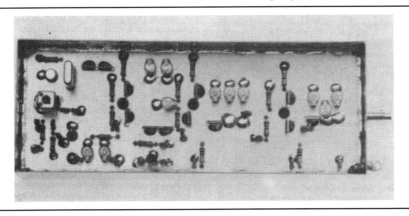

Fig 7.17 : Photograph of the 2.8GHz source, courtesy of DARC and J Dahms, DC0DA.

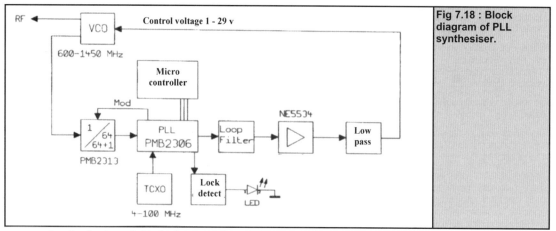

Fig 7.18 : Block diagram of PLL synthesiser.

means of a 3 wire bus. A block diagram of the PLL assembly is shown in Fig.7.18.

The PMB 2306 can handle frequencies from 0.1 to 220 MHz directly. The reference oscillator frequency can be selected between 4 and 100 MHz. The frequency data is sent serially to the IC in the form of data blocks, thus a specific program is required for the micro controller.

Since the PLL IC can only handle frequencies up to 220 MHz, a pre scaler (PMB2314 or PMB2313) must be added for our applications.

Pre scaler

The two pre scalers, PMB2314 and PMB2313, are dual mode pre scalers, with the dividing ratios 1:n and 1:n+1, the mode for which is switched to using pin 6 (MOD). Pin 3 (SW) is used to set a divider ratio of 1:64/65.

Actually, the PMB2314, which is specified for up to 2,100 MHz, should be used here. Unfortunately this IC was not available at the time when the synthesizer was being developed, thus the PMB2313 counter was used.

The PMB2313 pre scaler is specified for 1,100 MHz, and is given a maximum frequency of 1,400 MHz in the data sheet. Trials carried out by the author have indicated that

the PMB 2313 is still operating reliably at 1,450 MHz, provided the input level remains in the range between - 20dBm and - 1 dBm.

The signal from the VCO is fed into the PMB2313 pre scaler at pin 1 (I1), with a maximum level of - 10 dBm. Pin 3 (SW) is connected to + 5 V, which sets a scale factor of 64/65. The divided signal is available at the output pin 4 (Q), and arrives at pin 8 (FI) of the PLL IC through the coupling capacitor C4.

PLL IC circuit

The signal from the reference oscillator is connected at pin 1 (RI) of the PMB 306 PLL IC through a capacitor (Fig 7.19). A small standard TCXO makes the most suitable reference oscillator.

Since the reference oscillator frequency can be selected to be almost anything (depending on the software), there are certainly no procurement problems. The reference oscillator must, of course, meet the following conditions:

• The frequency must lie between 4 MHz and 100 MHz;

• The frequency must be divisible by 5 kHz;

• The oscillator should operate at a stable frequency

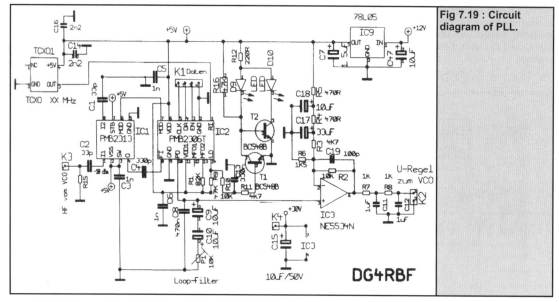

Fig 7.19 : Circuit diagram of PLL.

(hence the TCXO!), since as the reference it is responsible for the precision of the synthesizer.

All the PLL programming is carried out through a 3 wire bus. This consists of the lines "Enable" (pin 3), "Data" (pin 4) and "Clock" (pin 5), which are linked to the data lines of the micro controller.

The MOD output (pin 7) controls the dual mode pre scaler.

The PD current is set at pin 13 (MFO2) through an 82 kOhm resistor. The VCO signal, divided by 64/65, is at the FI input (pin 8).

The loop filter is connected to the PD output (pin 10 - phase detector). It consists of two bi polar inter connected 10µF tantalum capacitors (C9, C10) or of a 4.7µF foil capacitor, in series with a 10kΩ trimmer. Parallel to this is a 470nF capacitor (C8). High quality capacitors must be used here, with a low loss factor, since the loop filter is largely responsible for the performance of the PLL.

High quality foil capacitors would be best suited for this. But as these are difficult to obtain in an acceptable size with a value of 4.7µF, the option was to use two good tantalum capacitors.

A low noise DC amplifier with NE5534N (IC3) is used after the passive loop filter, and takes the control voltage for the VCO to the 28 - 30 Volts required. The amplification factor is set through the resistors, R2, R3 and R6.

The + 5V operating voltage is fed through a low pass filter to reduce the noise. There is another low pass filter for noise reduction at the operating amplifier output. This low pass filter also influences the control behaviour of the PLL, so high quality capacitors must be used here as well.

The LD signal from pin 14 is used to indicate whether the PLL is locked or not.

Direct Digital Synthesiser (DDS) [7]

Since direct digital synthesis, (DDS) was invented [8] in 1970, it has become more and more prevalent in the communications world including Amateur Radio. Some interesting articles have been published in amateur radio magazines explaining how DDS works. [9] [10]

DDS chips such as the AD9850 convert a reference oscillator input into a sine wave output at a frequency selected by the user. A block diagram of a DDS VFO is shown in Fig 7.20 with a complete schematic is shown in Fig 7.21. The rotary shaft encoder shown is for dialing in a

frequency. A micro controller monitors the outputs from the rotary encoder. The micro controller then translates the rotary encoder signals into frequency control data, which is loaded into the DDS chip. Likewise, the micro controller translates the selected frequency into data for display on the liquid crystal display. The reference oscillator provides a digital clock to the DDS chip. The purpose of the low pass filter is to smooth the digitised sine wave output of the DDS chip.

Micro controller

The control program stored on the micro controller contains most of the complexity involved with this VFO. A complete listing of the source code for this project is available at ftp:// ftp.arrl.org/publqex.

The DDS Chip

The AD9850 is a complete DDS chip. It contains a 32 bit phase accumulator, a 14 bit look up table and a 10 bit digital to analogue converter (DAC). It can be clocked at 125MHz to produce a 41MHz sine wave output. The spurious free dynamic range is greater than 50 dB at 40MHz with a 125MHz reference clock. A complete data sheet for the AD9850 can be downloaded from the Analog Devices web site at http://www.analog.com/ pdflad9850. pdf.

The frequency control word for the AD9850 can be loaded byte wide or serially. The serial mode is slower but is used in this VFO to minimise the number of output pins required on the micro controller. Serial mode is selected as the default by wiring pin 2 of the AD9850 to ground while pins 3 and 4 are wired to the supply voltage. Pin 25 is the serial data input, and pin 7 is the data write clock. After shifting in 40 data bits, pin 8 is used to transfer the data from the chip's input register to the DDS core. The 40 bits are a 32 bit frequency control word, 3 control bits and 5 phase modulation bits.

The AD9850 data sheet gives :

$$F_{out} = (\Delta Phase \times ClkIn) / 2^{32}$$

Where: $\Delta Phase$ = value of the 32 bit tuning word, ClkIn = reference clock frequency in MHz, and F_{out} = frequency of the output signal in MHz.

for calculating the required control word. In order to minimise program size, the actual calculation uses the algorithm :

$$\Delta Phase = \Sigma_n \text{ LCD-Digit} \times \text{Digit-Weight}$$

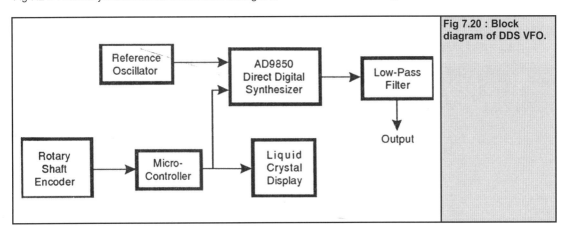

Fig 7.20 : Block diagram of DDS VFO.

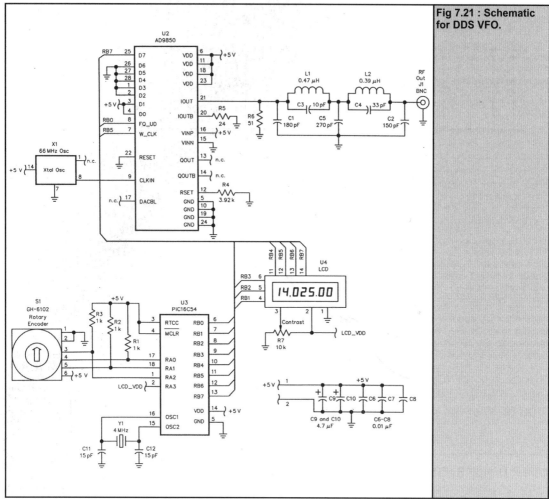

Fig 7.21 : Schematic for DDS VFO.

Where : ΔPhase is the control word sent to the AD9850, n is the range of 1 to 7, LCD-Digit is one of the seven digits being display on the LCD, and Digit-Weight is a precalculated value given in Table 7.3.

This algorithm has some round off error, but the error is less than 1 Hz, which is small enough to ignore in this application.

The AD9850 DAC output is a differential current on pins 20 and 21. A resistor placed from pin 12 to ground determines the full scale output current for the DAC as given by :

$$I_{out} = 32(1.248V / R)$$

Table 7.3 Digit weights for a 66.66MHz reference clock.

LCD-Digit	Digit-Weight (in hex)
10's	284
100's	192A
1k	FBA9
10k	9D49B
100k	624E13
1M	3D70CC1
10M	26667F90

The current equation is valid provided the voltage across the DAC output pins is less than 1.5 V. Setting the resistor to 3.92k yields a DAC current of about 10.2 mA. With the parallel load of the filter terminator and an external 50 load, this current results in a voltage swing of about 250 mV peak to peak.

Reference Oscillator

The reference oscillator is a standard clock oscillator module. The accuracy of the reference oscillator directly determines the VFO output accuracy. If the reference oscillator has a 100 ppm tolerance, so does the output. Clock oscillator modules up to 66.666 MHz are readily available. For many vendors, higher frequencies are special order items. Choice of a reference frequency will depend on the application. Changing the reference frequency requires updating the control program values in Table 7.3. One factor in the choice of reference frequency might be the locations of spurs. All DDS systems will have low level spurious outputs [11]. The frequency of these spurs is very predictable. They are related to the reference clock frequency and harmonics of the output frequency.

Low Pass Filter

The output of the DDS chip is a digitised or sampled sine

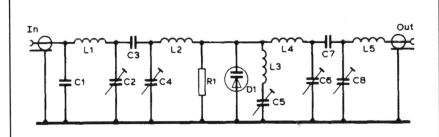

Fig 7.22 : Schematic diagram of the 1152MHz tripler by Microwave Modules and G8AGN.

wave. Such a wave shape has strong frequency components at the reference clock frequency plus or minus the output frequency. Filtering out these components produces a clean sine wave. For this project, with a reference clock near 66 MHz and a maximum output frequency of 20 MHz, the low pass filter must cut off frequencies above 46 MHz while passing frequencies below 20 MHz. The fifth order elliptic low pass filter [12] shown in the schematic has 55 dB or greater attenuation at frequencies above 46 MHz. The filter requires a 50Ω termination.

The calculated pass band ripple of the filter is about 1 dB. However, a more significant amplitude variation can occur as the output frequency increases. As the output frequency goes higher, the digitised sine wave is constructed from fewer samples per cycle. At the DAC output pins, the wave shape begins to look less and less like a sine wave. As this happens the spurious frequency components constitute a larger portion of the total output power; meanwhile the desired output component is less.

DDS for use with microwave transceivers

The heart of a DDS design is the control program used. There are a number of designs available to the radio amateur, one of these comes from the Australian supplier, MiniKits [13],criteria for the design were :

- DDS frequency output 0 - 40MHz

- TX/RX switching using the RA4 pin to do to TX mode

- Frequency readout offset by 1Hz to 11GHz

- No requirement for band switching, but able to program upper and lower limits

- Memorise the last frequency on power up

- Use low cost mechanical rotary encoder instead of the optical type

- Variable rate tuning not required

- By grounding the RA2 pin and turning the rotary shaft encoder sets the step size e.g. 100Hz, 1kz, 10kHz etc.

- Use 16 x 2 LCD with or without R/W pin. (use delays rather than busy checks)

- Software fine tuning of the crystal oscillator frequency

The software for this DDS VFO can be downloaded from the Mini Kits web site, kits for this project are also available.

Frequency multipliers

The sources described in the previous section all produce relatively low output power. There are some applications where higher power is required and the use of amplifiers is not justified or even possible. In these situations it is often desirable to use diode multipliers driven by high power, lower frequency sources.

A high power 1152MHz tripler

High power varactor multipliers have been used for many years to produce large amounts of output power on harmonically related bands. Many of the published designs left a lot to be desired in terms of stability and spectral purity.

The following varactor tripler design is one of the most successful ever produced. Originally manufactured, by Microwave Modules in the UK, for use as a multiplier from 432 to 1296MHz, it was later made available, re tuned, for 384MHz input and 1152MHz output. Commercial produc-

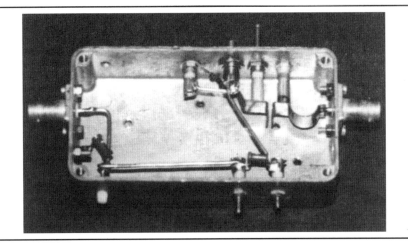

Fig 7.23 : Photograph of the 1152MHz tripler. Note the input socket and the connection to C1 / L1 are different to those shown in Fig 7.24, this is not significant.

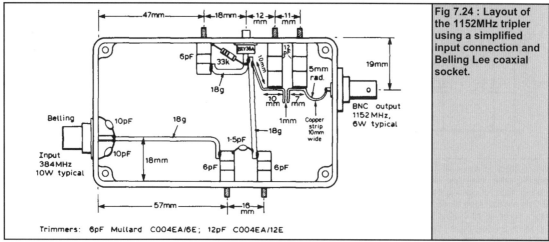

Fig 7.24 : Layout of the 1152MHz tripler using a simplified input connection and Belling Lee coaxial socket.

Trimmers: 6pF Mullard COO4EA/6E; 12pF COO4EA/12E

tion of the multiplier has ceased, but Microwave Modules have kindly given permission for details of the tripler to be published in this manual. Fig 7.22 shows the circuit diagram of the high power tripler and the component values are given in Table 7.4. A photograph of the unit is shown in Fig 7.23.

Many of the components used in the commercial version have proven difficult to obtain. G8AGN has made a number of small modifications to the design in order to allow the use of more readily available parts. Typically 6W output can be obtained with a drive power of l0W.

Circuit description

Input matching to the varactor diode is accomplished in the two stage filter consisting of L1, 2 and capacitors C1 to C4. R1 is the bias resistor. L3 and C5 form the idler circuit, which is resonant at the second harmonic:

$$2 \times 384\text{MHz} = 768\text{MHz}$$

This circuit is crucial to the efficiency of the tripler. Second harmonic currents must be developed in the diode so that mixing with the fundamental can take place to produce usable output power at the third harmonic. L4 and 5, together with C6, 7 and 8, filter the 1152MHz output and

provides matching from the diode to the 50Ω load.

Construction

The tripler is built into an Eddystone 7134P diecast box. Component layout is shown in Fig 7.24. Holes are drilled in the box where indicated. The sockets, trimmers and diode are mounted in the holes by their fixing nuts. Interconnection of these components is made by the inductors. Carefully shape L5 by wrapping it round a 10mm diameter mandrel for half a turn.

Alignment

Varactor diode multipliers should, ideally, be aligned using swept frequency and swept power sources. However, when the multiplier is to be used only at a single drive level, such as would be the case in a local oscillator source, swept power at least can be dispensed with. With care, this tripler can be aligned without the swept frequency source as well. Initially alignment should be attempted at low power only because the diecast box is a very poor heat conductor. Until alignment is complete most of the power going into the diode will be converted into heat. The diode is therefore in danger of over dissipating and self destructing.

Connect the tripler into a test set up like that shown in Fig 7.25. The coupler at the output ensures that the tripler only sees the low vswr of the 50Ω load. If the wavemeter were connected directly at the output there is a possibility that instability could be induced when the wavemeter is tuned. Adjust C2 and C4 for a low SWR at the tripler input. Tune the wavemeter to 1152MHz and adjust C6 and C8 for a peak reading on the meter. If an output indicator is used it should be indicating some output signal. Retune the wavemeter to 768MHz and adjust C5 for a minimum

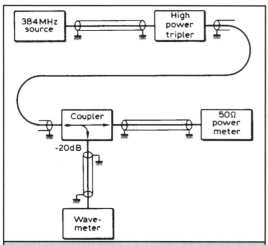

Fig 7.25 : Recommended alignment set up for the 1152MHz tripler.

Table 7.4 : Component list for the high power tripler.

D1	BXY35A
C1	20pF ceramic disc
C2,4,5	6pF tubular ceramic trimmer
C3	1.5pF ceramic
C6,8	12pF tubular ceramic trimmer
C7	Formed by ends of L4 and L5
R1	22k 0.25W carbon
L1,2,3	18swg tinned copper wire bent to fit
L4,5	20swg copper strip 10mm wide

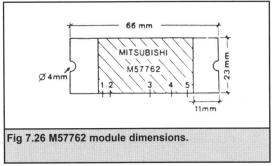

Fig 7.26 M57762 module dimensions.

reading. This will coincide with a peak in the reading on the output meter. Go back and re adjust all trimmers for a maximum reading on the output meter. Increase the drive level and repeat all adjustments.

It should be possible to drive the multiplier with up to l0W, but only if at least 5W is present at the output, otherwise the diode may fail.

Amplifiers

Amplifiers are used extensively in microwave equipment for raising the level of various system signals, such as the input to receive converters or the output from oscillator or transmit stages. Some of these amplifiers may use similar designs even if they are to be used at different frequencies.

They therefore qualify as common equipment. UHF amplification can be provided at relatively low cost compared with the same amount of amplification in the microwave region. Generally the higher the frequency, the more expensive the power amplifying devices become. At the lower microwave frequencies, amplifier modules such as those from Mitsubishi are commonly used.

A power amplifier using the Mitsubishi M57762 module [14]

These modules are characterised by simplicity and simple drive requirements. The construction requires no special PCB materials, such as PTFE, and also no special components. It can, in fact, be constructed as soon as the M57762 module has been obtained, from materials available from most radio amateur workshops.

The dimensions of the basic module are shown in Fig 7.26, with the circuit and wiring diagram in Fig 7.27 and Fig 7.28. The power output vs frequency is shown in Fig 7.29.

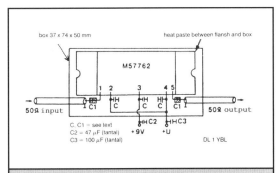

Fig 7.27 : Wiring diagram of M57762 amplifier.

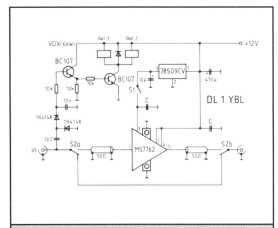

Fig 7.28 Single M57762 module amplifier circuit.

Construction

A proprietary tin plate box, dimensions 37 x 74 x 50mm is used to house the module. No cut outs should be made in the cover of the box, since poor earthing could encourage oscillation, thus reducing the output power. The lower half of the tin plate box should be soldered all around its perimeter to the wall of the box to ensure good electrical contact and mechanical strength. The ground tabs of the module are secured with two M4 screws to the enclosure cover and a suitable heatsink.

The supply voltage should be connected via feed through capacitors. The capacitors marked C1, on the wiring diagram, are constructed from small (5 x 5mm) pieces of double sided PCB. They are soldered directly to the bottom cover of the enclosure and form an rf connection point and decoupling capacitor. The capacitors marked C, on the wiring diagram, are a 1nF plate capacitor plus a 4.7nF and 10mfd tantalum capacitor, soldered directly to the bottom cover of the enclosure under connections 2,3 and 4 of the module. For AM TV and SSB working, C2 and C3 (tantalum capacitors) are required. The input and

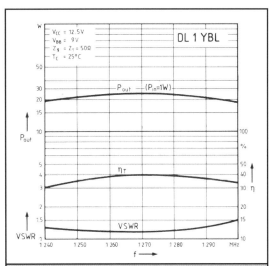

Fig 7.29 : Output power, efficiency and vswr as a function of frequency for the M57762 amplifier.

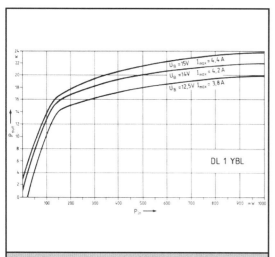

Fig 7.30 : Characteristics of M57762 amplifier at 1260MHz.

output matching are so good that the usual series preset capacitors used for matching purposes are not required.

Power characteristics

Fig 7.30 shows the power output data for a power amplifier constructed as described above. It can be seen that the maximum peak input driving power for SSB working is 200mW. The maximum amplification in the linear region and at 14 volts supply, is approximately 20dB. If the amplifier is required for FM ATV it can be run into the saturation region with a power output up to 22W without any fear of the ratings being exceeded.

Monolithic microwave integrated circuit (mmic) amplifiers

Mmics are now widely used in amateur radio designs and are available from several manufacturers including Mini-Circuits and Agilent (Hewlett Packard), at a price that makes them very attractive for many applications.

Keeping track of the devices that are available can be a

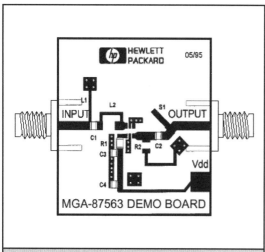

Fig 7.31 : Agilent pcb for 2400 MHz LNA.

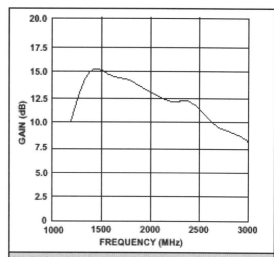

Fig 7.32 : 2400MHz LNA associated gain at maximum noise figure. with V$_d$ = 5v

problem, two useful sources of information are the Agilent web site [15] and the Mini-Kits web site [13]. Some useful information from these suppliers is reproduced here

2400 MHz LNA Design from Agilent

The 2400 MHz LNA was designed to provide an optimum noise match from 2400 through 2500MHz, making it useful for applications that operate in the 2400 to 2483 MHz ISM band. The component labels appearing in the following paragraphs refer to positions shown in Fig 7.31. The input match consists of a shunt inductor at L1 and a series inductor at L2. Both of these inductors use the traces as originally etched on the circuit board without modification. The output is conjugately matched with a simple shunt open circuited stub (S1) on the output 50Ω microstripline. 22 pF capacitors were used for both the input (C1) and output (C2) blocking capacitors.

A 16Ω chip resistor placed at R1 and decoupled by a 100 pF capacitor at C3 provides a proper termination for the device power terminal. An additional bypass capacitor (100 to 1000 pF) placed further down the power supply

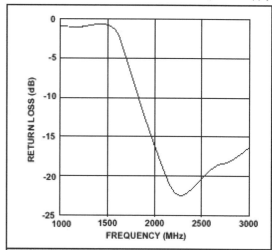

Fig 7.33 : 2400MHz LNA output return loss with V$_d$ = 5v.

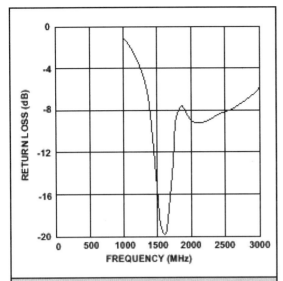

Fig 7.34 : 2400MHz LNA input return loss with V_d = 5v

Table 7.5 : 2400MHz Noise Figure and gain with V_d = 3v

Frequency (MHz)	Gain (dB)	Noise Figure (dB)
1700	10.4	2.60
1800	13.8	2.57
1900	11.3	2.45
2000	11.9	2.38
2100	13.3	2.06
2200	12.8	2.02
2300	12.9	2.12
2400	11.5	2.05
2500	11.5	2.14
2600	10.5	2.25
2700	10.9	2.29
2800	10.3	2.33
2900	9.8	2.35
3000	9.6	3.42

line at location C4 may be required to further decouple the supply terminal especially if this stage is to be cascaded with an additional stage. Proper decoupling of device VCC terminals of cascaded amplifier stages is required if stable operation is to be obtained.

If desired, a 50Ω resistor placed at R2 will provide low frequency loading of the device. This termination reduces low frequency gain and enhances low frequency stability.

The MGA-87563 has 3 ground leads, all of which need to be well grounded for proper RF performance. This can be especially critical at 2.4 GHz where common lead inductance can significantly decrease gain.

The performance of the LNA as measured on the HP 8970 Noise Figure Meter is shown in Table 7.5. At 2.4 GHz, the loss of the FR-4/G-10 epoxy glass material can add several tenths of a dB to noise figure and lower gain by double the amount. The swept plots, Figs 7.32 - 7.34 were taken on a scalar analyser and show the performance of the amplifier.

References

[1] Overtone crystal oscillators in series and parallel resonance, H J Brandt, DJ1ZB, VHF Communications 1/77 pp38-43.

[2] A high quality UHF source for microwave applications, RSGB Microwave Committee, Radio Communication, October 1981. pp906-910.

[3] www. btinternet.com /~jewel l/G4DDK001.htm

[4] 10GHz Transverter module portable, J Dahms, DC0DA, Belgian VHF/UHF Convention, November 1988.

[5] Ein 6cm Transvertersystem moderner konception, R Wesolowski, DJ6EP and J Dahms, DC0DA, CQ-DL 12/87 pp755-759.

[6] From article by Bernd Kaa, DG4RBF, VHF Communications 2/98 pp 103 - 121.

[7] From article by Curtis W. Preuss, WB2V, QEX July 1997 pp 3 - 7.

[8] IEEE Transactions on Audio and Electoacoustics, Vol. Au-19, No. 1, March 1971 pp 48 - 56.

[9] Bergeron, Direct Digital Synthesis, an Introduction, Communications Quarterly summer 1993 p 13.

[10] Cahn, Direct Digital Synthesis - An Intuitive Introduction, QST August 1994 p 30.

[11] Bar-Giora Goldberg, Digital Techniques in Frequency Synthesis, pp 107 - 118.

[12] Stefan Niewiadomski, Filter Handbook - A Practical Design Guide, Chapter 2.

[13] Mini Kits - http://www.minikits.com.au

[14] From article by Joachim Berns, DL1YBL, published in VHF Communications issue 4/1989 pp 211 - 213.

Table 7.6 : Data for Agilent MMICs.

Product Family	Min Freq (GHz)	Max Freq. (GHz)	Pout (dBm)	Gain (dB)	Noise Figure (dB)	Isolation (dB)	Bias Current (mA)	@ Vdd (v)
MGA-641	1.2	10	12	12	7.5	35	50	10
MGA-725	0.1	6	17	14	1.4	23	20	3
MGA-815	0.5	6	14	12	2.8	24	42	3
MGA-825	0.5	6	17	13	2.2	22	84	3
MGA-835	0.5	6	22	22	6	32	142	3
MGA-855	0.5	8	9	18	1.6	41	15	3
MGA-865	0.8	8	6	20	2	46	16	5

Table 7.7 : MMIC data, an extract from data on the Mini Kits web site.

Model Mini-circuits	Equivalent MAR/MAV	Equivalent Avantek	Alphanumeric	Dot Colour
MAR-1	MAV-1	MSA0185	A01	Brown
MAR-2	MAV-2	MSA0285	A02	Red
MAR-3	MAV-3	MSA0385	A03	Orange
MAR-4	MAV-4	MSA0485	A04	Yellow
MAR-6		MSA0685	A06	White
MAR-7			A07	Violet
		MSA0735		
MAR-8		MSA0885	A08	Blue
		MSA0835		
MAV-1	MAR-1	MSA0104	1	
MAV-2	MAR-2	MSA0204	2	
MAV-3	MAR-3	MSA0304	3	
MAV-4	MAR-4	MSA0404	4	
		MSA0504	5	-
		MSA0604	6	-
		MSA0704	7	-
		MSA0804	8	-
MAV-1 1		MSAO 1 104	A	-
ERA-1			EI	
ERA-2			E2	
ERA-3			E3	
ERA-4			E4	
ERA-5			E5	
ERA-6			E6	

Model	Gain Typical dB at Frequency (GHz)							Max power out 1dB comp, 1GHz	Noise Figure	IP3 dBm
	0.1	0.5	1.0	2.0	3.0	4.0	6.0			
MAR-1	18.5	17.5	15.5	-	-	-	-	+1.5dBm	5.5	+14.0
MAR-2	12.5	12.3	12.0	11.0	-	-	-	+4.5dBm	6.5	+17.0
MAR-3	12.5	12.2	12.0	11.5	-	-	-	+10.0dBm	6.0	+23.0
MAR-4	8.3	8.2	8.0	-	-	-	-	+12.5dBm	6.5	+25.5
MAR-6	20.0	18.5	16.0	11.0	-	-	-	+2.0dBm	3.0	+14.5
MAR-7	13.5	13.1	12.5	11.0	-	-	-	+5.SdBm	5.0	+19.0
MAR-8	32.5	28.0	22.5	-	-	-	-	+12.SdBm	3.3	+27.0
MAV-11	12.7	12.0	10.5	-	-	-	-	+17.SdBm	3.6	+30.0
ERA-1	-	-	-	11.6	11.2	-	10.5	+13dBm (2GHz)	7.0	+26.0
ERA-2	16.0	-	-	14.9	13.9	-	11.8	+14dBm (2GHz)	6.0	+27.0
ERA-3	22.2	-	-	20.2	18.2	-	-	+11dBm (2GHz)	4.5	+23.0
ERA-4	13.8	-	14.0	13.9	13.9	13.4	-	+19.IdBm	5.2	+36.0
ERA-5	20.4	-	20.0	19.0	17.6	15.8	-	+19.6dBm	4.0	+36.0
ERA-6	11.1	-	11.1	11.8	11.5	11.3	-	+18.5dBm	8.4	+36.5

Application	Model
High Freq Gain	ERA1 Usable to
10GHzLow Noise Amp	MAR6/ MAR8/ MAV11
Medium Noise	ERA3/ERAS
High Dynamic range	MAV11
Stable High Gain	MAR1/ERA3
Medium Output	MAV11/ MAR3/ MAR4
High Output	MAV11/ERA4/5
Multiplier	ERA3 Clean Harmonics

Test Equipment

In this chapter :

- Amplitude Measurements
- Design and construction of simple attenuators and loads
- Directional couplers

- Waveguide directional couplers
- Frequency Measurement
- Signal sources and alignment aids

Radio Amateurs have demonstrated that it is possible to construct and align quite complex pieces of equipment without recourse to sophisticated test equipment. Access to a spectrum or network analyser can undoubtedly be a major advantage when it comes to identifying problems in the development of new equipment, but they are by no means indispensable. Simple test equipment, used intelligently, can often be used to achieve the same results. Analysers, however, are usually quicker and more accurate.

Basically, test equipment can be divided into two types: that used to measure the amplitude of signals, and that used to measure signal frequency. Examples of amplitude measuring equipment are power meters, noise figure meters and swr meters. Frequency measuring equipment includes frequency counters and wavemeters. In addition, other items of test equipment are often needed such as signal sources and test amplifiers.

The following chapter describes easy to build items of test equipment in each of these categories.

Amplitude measurements

A Simple VHF/UHF power meter

The simple power meter described [1] is capable of remarkably accurate power measurement in the range of 100mW to 2W, up to at least 500MHz. The circuit is shown in Fig 8.1 and component values in Table 8.1.

The meter measures the peak rf voltage across the load, minus the forward voltage drop across the diode. Divided by root 2, it gives the rms value, and the power is then calculated using the familiar equation:

$$P = \frac{V^2}{2 \times R}$$

Where

  V = peak voltage = rms x 1.414

  P = Power in watts

  R = Load resistance

For Germanium diodes such as the OA47, the relationship will be:

$$P = \frac{(V + 0.25)^2}{100}$$

Table 8.1 : Component values for the simple VHF UHF power meter.

R1,2	100Ω 1/8 watt miniature carbon
C1	1000pF ceramic disc
D1	1N914, 1N916, 1N4148, OA47
V	Voltmeter

while for silicon diodes such as the 1N914, 1N916 or 1N4148, the relationship will be:

$$P = \frac{(V + 0.7)^2}{100}$$

Although the circuit shown in Fig 8.1 (a,b) is easily built, it is not suitable for use much above 500MHz. The terminating load, formed by the two 100Ω resistors tends to be the main limit to frequency response. Its voltage/power calibration curve is given in Fig 8.2. By using a commercial 50Ω BNC or N type termination, a diode mount and a T piece as shown, it is possible to extend the range to several GHz (Fig 8.3). Whilst it is fairly easy to obtain the termination and T piece, it can be much more difficult to

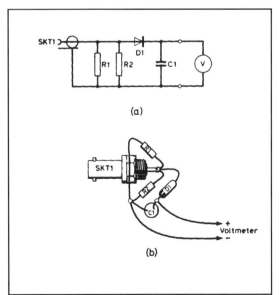

(a)

(b)

Fig 8.1 : (a) Circuit, (b) Construction of the simple VHF/UHF power meter.

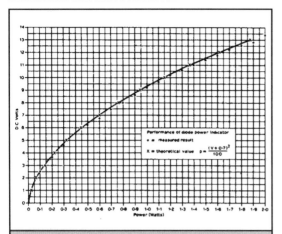

Fig 8.2 : Voltage/power calibration curve of the simple power meter.

find a suitable diode mount. These are occasionally found at rallies, sold as video detector mounts. It will often be necessary to use a silicon point contact diode in a cartridge type package when using this type of mount.

If a commercial diode mount cannot be found, then it is possible to construct a suitable alternative using a BNC plug as the diode housing. A leaded diode is soldered to the centre pin of the plug, and a decoupling capacitor soldered between the other end of the diode and the plug outer body. Fig 8.4 shows the form of construction.

The meter circuit can be either a dedicated voltmeter or the shack multimeter, but it must have a reasonably linear scale if power measurements are to be accurate. Some cheap imported meter movements can be very non linear, leading to large errors unless the meter scale is carefully calibrated before use.

A wideband, linear reading power meter

A circuit described here has a linear law over a very wide range, at power levels from 10nW, but it is not particularly accurate in absolute terms [2]. Possible applications for the power meter include:

• The calibration of attenuators

• The measurement of received signal to noise levels, ground or sun noise.

• As a power indicator, when connected to a directional

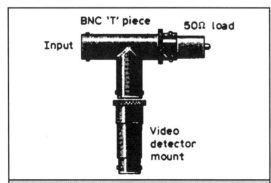

Fig 8.3 : Alternative form of construction to extend the frequency range of the power meter.

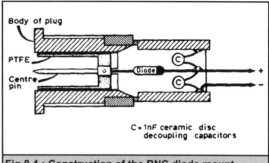

Fig 8.4 : Construction of the BNC diode mount.

coupler or when connected to an antenna, it can be used to measure rf levels over short ranges.

• The power meter is also capable of giving an output voltage proportional to input power in dB. This facility can be very useful for plotting filter characteristics when combined with a tuneable / swept oscillator, or when there is a requirement for a large dynamic range indication on the meter or oscilloscope.

Theory

Under certain conditions, the dc current flowing in a diode detector operating into a low resistance load varies linearly with the applied rf power over a wide range. This effect is used to advantage in the power meter described.

The dynamic range of this type of power meter depends on diode linearity at the power limit, and dc amplifier drift at the low end.

The diode used in the power meter is a zero bias Schottky, Alpha type DDC4561 (1N23 type package) or DDC4562 (wire ended type). These diodes have a forward voltage drop of 80mV at 1mA. When using these diodes the deviation of linearity is a maximum of 1dB at the upper power limit (1mW) and is considerably better at lower power levels.

Diodes with a slightly higher forward voltage drop such as the HP HSCH3486 (160mV at 1mA) can be used, but will not be so linear at high power levels (100µW to 1mW). This effect can be partly compensated by inserting a resistor in series between the dc output of the diode and the input to the operational amplifier. For the HSCH3486 the resistor should be 180Ω. Point contact diodes which have a forward voltage drop of 200-300mV can also be used, but they have a much lower limit of linearity at about 1µW, and so will only give a linear dynamic range of 20-30dB. Other diodes with higher forward voltage drops are not suitable for use in this circuit.

Several types of microwave diodes have been tested, such as the CS10B and the zero bias Schottkys mentioned, and they appear to give a flat frequency response (±1dB) up to about 2GHz. It has not been possible to check the sensitivity at higher frequencies, but it should remain fairly constant; the linearity certainly does. For the diodes specified above, the rf source impedance should be 50Ω and the dc load impedance zero ohms.

The diode mount

The details of the diode mount are shown in Fig 8.5. The upper frequency limit of the meter will be set by the method of construction and so all components should

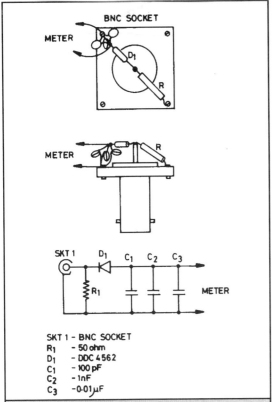

Fig 8.5 : Construction of the diode mount.

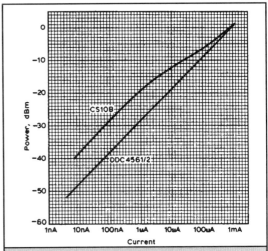

Fig 8.7 : Power/current characteristics of the DDC4561/2 diode with CS10B for comparison.

have the shortest possible leads. The lower frequency limit is set by the value of the smoothing capacitor, but if it is physically too large, its inductance will degrade the high frequency performance. The components should be soldered directly to the body of the socket, or to the copper board next to the socket, if it is mounted through a piece of double sided pcb.

For best results a chip resistor should be used for the 50Ω load, but below 1GHz there is little difference if a

miniature film resistor with very short leads is used. Please remember that the centre pin of the socket can move, so do not solder delicate components directly to the pin, but use a short link of wire.

The vswr of this detector using a chip resistor is shown in Fig 8.6, and is less than 1.2:1 at all power levels. The vswr will become slightly worse as the power is increased due to the diode impedance decreasing and shunting the 50Ω rf load resistor. The return loss is only a few dB worse at 1GHz if an ordinary film resistor is used. A diode in a standard waveguide mount should also be satisfactory, provided the resistance in the dc return is less than 50Ω. The power/current characteristics are shown in Fig 8.7. A dc return in the signal source is not necessary for the circuit to function, but for best linearity the source impedance should be 50Ω both at dc and rf.

The meter circuit

The meter circuit in Fig 8.8 and Table 8.2 presents a very low impedance load to the diode and produces a voltage output, which is either proportional to power, or the log of power. In linear mode the meter indicates 1V fsd for each of the selected decade ranges, which cover 10nA to 10mA. In the log mode, the feedback network of IC1 uses the exponential characteristic of the base/emitter junction of a silicon transistor to produce a voltage, which is proportional to the log of power. A change of 60mV indicates a change in power of approximately 10dB. The exact value will depend on the transistors used.

The output of the first amplifier can be zeroed using the offset null control connected between pins 1 and 5 of IC1. This control should be mounted on the front panel, as adjustment will be required when working at low input levels. One position of the range switch is used to indicate the battery voltage and has an fsd of 12V. IC2 is a unity gain inverter, which enables a positive output to be obtained for the meter when using the external, switch selected input, with a diode mounted with either earthed anode or cathode.

At low input levels with the log mode selected, the meter reading may fluctuate wildly, or go negative; this is due to the input signal level going very near to, or through zero.

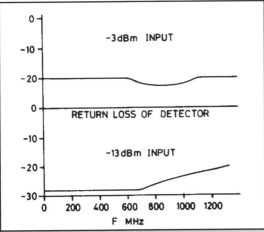

Fig 8.6 : Return loss characteristic of the detector.

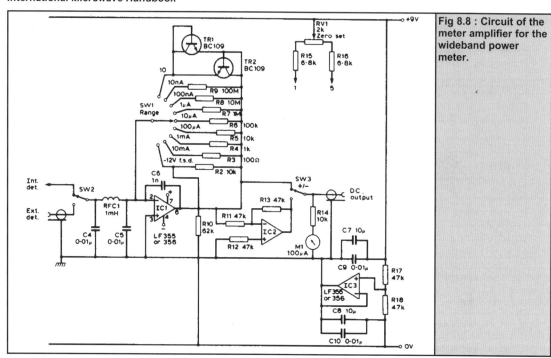

Fig 8.8 : Circuit of the meter amplifier for the wideband power meter.

This behaviour is normal, and can be compensated by careful adjustment of RV1, the set zero control.

The values of the smoothing capacitors across the diode output may need to be reduced if a rapid response from the meter is required. This may be necessary when sweeping a signal source. The values of the capacitors may also be increased when detecting power levels at very low frequencies, e.g. <100kHz. There is no reason why the power meter should not be used at audio frequencies. The low pass filter at the input to IC1 has a cut off frequency of about 50kHz.

Table 8.2 : Component values for the power meter circuit for the wideband power meter.

R1	50Ω miniature carbon film or two parallel connected 100Ω
R2,5,14	10k
R3	100Ω
R4	1K
R6	100K
R7	1M
R8	10M
R9	100M
R10	62k
R11,12,13,17,18	47k
R15,16	6.8k
RV1	2k ten turn miniature pot
All resistors 0.25W	miniature carbon or metal film
C1	100pF ceramic disc
C2,6	1000pF ceramic disc
C3,4,5,9,10	0.01µF ceramic disc
C7,8	10µF tantilum bead 16v
IC1,2,3	LF335, LF356
D1	DDC4561/2
M1	100µA meter
SW1	1 pole, 9 way
SW2,3	SPDT

A fet operational amplifier is used for IC1 because of its low offset current, but IC2 type is not critical. It would be better to use fet operational amplifiers in all ic positions as their output can go very close to the supply rails, and the circuit will be less affected by low supply voltages.

The detector and current meter are best mounted in the same screened box, and a battery used to supply the op. amps. This helps to avoid earthing problems, which can occur, as the input to IC1 is very sensitive. A socket can be provided to allow IC1 input to be switched to an external detector via a switch. Any external detector should be connected to the meter via coax. cable to limit the stray pickup of rf.

Waveguide power meters

Conventional point contact mixer diodes can be used to provide an indication of rf power with an accuracy sufficient to prevent gross mistakes in the construction of microwave equipment. The relationship between rf input and dc output for a typical diode is shown in Fig 8.9, and was derived from data given in the 1972 edition of the Microwave Engineers Handbook. Although this applies specifically to X band, the results obtained at other microwave frequencies can be expected to be similar. Each curve corresponds to a particular total series resistance. The resistance of meters, typically 100Ω for 1mA fsd, 1000Ω for 100µA fsd, is significant and must be allowed for. As an example of the use of this data, suppose a mixer is fed, via a 20dB (x100) directional coupler, from a Gunn oscillator and a current of 100µA is observed on a meter having a resistance of 100Ω. The power indicated by the diode corresponds to 200µW of rf, which implies the Gunn oscillator is delivering in the region of 20mW of rf. Comparing this value with the manufacturers data will give a guide to the efficiency of the system.

The corresponding relationship of rf input to dc output for Schottky diodes is given in Fig 8.10.

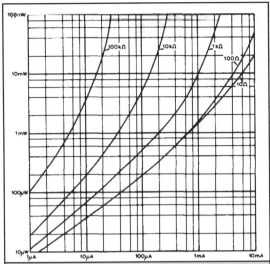

Fig 8.9 : The approximate relationship between rf power input and dc output for a conventional point contact mixer diode.

The power meters described are only able to measure relatively low power. In order to increase the range of power that can be measured, it is necessary to use either attenuators or directional couplers. Broadband amplifiers, with accurately known gain, can be used to extend power meter sensitivity.

Design and construction of simple attenuators and loads

Good attenuators capable of working at frequencies of up to several hundred MHz or higher are relatively easy to construct using standard resistors and can be put to a variety of uses. In addition to extending the range of power meters, they can also be used as pads between interacting stages such as varactor multipliers or to follow noise or signal generators to bring their output impedance close to 50Ω.

At i.f they can be used for calibrating microwave attenua-

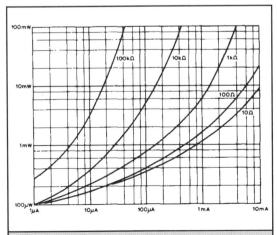

Fig 8.10 : The approximate relationship between rf power input and dc output for a typical Schottky barrier mixer diode.

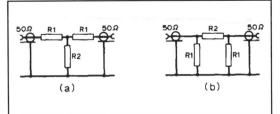

Fig 8.11 : (a) T type attenuator, (b) π type attenuator.

tors, since their attenuation is fairly predictable at lower frequencies. They may also be used for calibrating S meters and as a reference for noise measurements of sun or ground noise.

Two simple configurations of symmetrical attenuators are the T and the π type depicted in Fig 8.11 (a) and (b).

The design formulae for these two types are :

$$T - attenuation(dB) = 20\log\left(\frac{50 + R1}{50 - R1}\right)$$

where
$$R = \frac{2500 - R1}{2 \times R1}$$

$$\pi - attenuation(dB) = 20\log\left(\frac{R1 + 50}{R1 - 50}\right)$$

$$R = \frac{5000 \times R1}{R1 - 2500}$$

The resistor values for some useful attenuation levels are given in Table 8.3. It will be seen that in most cases the values do not coincide with preferred resistor values. This restricts the levels of attenuation that can be obtained, and the best procedure seems to be to choose the nearest preferred values to R1 and R2. This usually results in an acceptable and predictable performance, but where the approximations are too great it may be better to try the other configuration, or to use series or parallel combinations of resistors for R1 and R2. A microcomputer programme utilising standard range resistors is given in *Amateur Radio Software* by GM4ANB, published by the RSGB [3].

Two methods of construction are illustrated in Fig 8.12. A π type is shown in (a) and a T type is shown in (b) but, of course, either version could be useful in either design. The resistors should be of low inductance with the common form of carbon film resistor being particularly suitable. Lead lengths should be as short as possible. For higher power attenuators at lower frequencies, parallel combinations of 0.5 watt carbon resistors can be used to increase dissipation.

Provided care is taken, these attenuators can be used up to 1-2GHz. The biggest error is likely to arise in the higher value attenuators, where stray coupling may reduce the attenuation below the expected value. For this reason it is better to use several low value stages in cascade when a high value of attenuation is required.

Low value attenuators are able to handle more power that higher ones due to the ratio of transmitted power to dissipated power. 10 watts applied to a 3dB attenuator will result in 5 watts dissipated and 5 watts transmitted through, whilst 10 watts applied to a 10dB attenuator will result in 9 watts dissipated and 1 watt transmitted through. The 10dB attenuator has therefore to dissipate

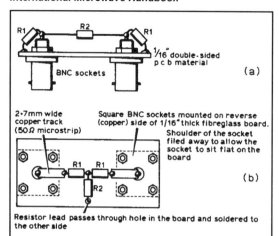

Fig 8.12 : Method of constructing attenuators. (a) method of construction below 500MHz. (b) method of construction above 500MHz.

80 percent more power in the form of heat.

Because of the way in which power is dissipated within the attenuator, higher attenuators can be advantageously constructed using a combination of high and low rated resistors.

Low powered 50Ω coaxial (BNC) loads and attenuators may feature in components suppliers catalogues and be suitable for amateur use in the 1.3 and 2.3GHz bands. Such inexpensive devices may be only rated to 1GHz (or less), but experience and measurement have shown their frequency response to extend (with acceptable vswr) well beyond their makers specification.

A high power l0dB attenuator

From Table 8.3, the resistor values for a π type attenuator are 96Ω for the parallel resistors and 71Ω for the series resistor. Fig 8.13 shows the attenuator schematic.

Assuming the maximum input power, when properly terminated, is to be 10W, then 22.4V rms will appear across the 50Ω input. One watt will appear at the

Table 8.3 Design data for 50Ω T and π type attenuators.

dB	T R1Ω	T R2Ω	π R1Ω	π R2Ω
1	2.9	433	870	5.8
2	5.7	215	436	11.6
3	8.6	142	292	17.6
4	11.3	105	221	23.9
5	14	82	178	30.4
6	16.6	67	150	37.4
7	19.1	56	131	44.8
8	21.5	47.3	116	53
9	23.8	40.6	105	62
10	26	35.1	96	71
12	30	26.8	84	93
14	33.4	20.8	75	120
16	36.3	16.3	69	154
18	38.8	12.8	64	196
20	40.9	10	61	248
25	44.7	5.6	56	443
30	46.9	3.2	53	790

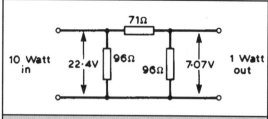

Fig 8.13 : Schematic diagram of the high power 10dB attenuator.

attenuator output, giving 7.07V rms across the load. The power dissipated in each of the three resistors can now be calculated from:

$$Power = \frac{V^2}{R}$$

The input resistor dissipates 5.19W, the 71Ω resistor dissipates 3.29W, whilst the third resistor dissipates only 0.52W. An attenuator, to the above design, could conveniently be constructed using a parallel combination of two 300Ω and one 270Ω, 2W resistors for the input; a 150Ω and 130Ω parallel combination of 2W resistors for the series resistor; and a 91, 1 W resistor for the output. A l0dB attenuator built in this way to handle l0W, measured 9.5dB attenuation at 432MHz, with an acceptably low swr. Fig 8.14 shows the resistor combinations for the high power attenuator.

A low power 50Ω load

Low power 50Ω loads of quite reasonable vswr can be made from unwanted crimp type BNC plugs. When a plug on the end of a piece of cable goes intermittent, it is usually not possible to re-use it, as spare crimp sleeves will not fit.

Choose a 50Ω film or composition resistor that just fits inside the sleeve to which the outer is normally crimped. Cut the resistor lead short at one end so it just fits into the pin and solder it into place. Insert the pin and resistor from the rear of the plug, then solder the other wire end of the resistor to the sleeve, filling up the hole with solder. Depending on the type of resistor used, the vswr can be better than 1.05:1 up to 1GHz.

Waveguide attenuators

Attenuators for use on the higher bands can be constructed in waveguide. Similar construction can also be

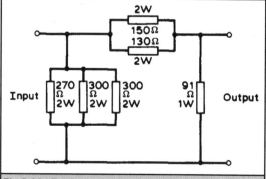

Fig 8.14 : Resistor combinations for the high power attenuator.

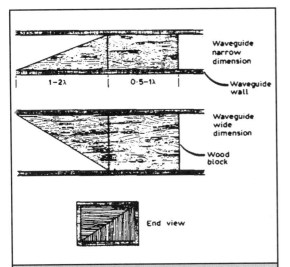

Fig 8.15 Waveguide dummy load, long taper absorbing element.

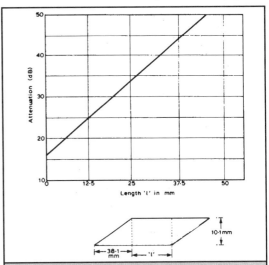

Fig 8.17 Approximate attenuation of 200Ω per square resistive card.

used to make matched waveguide loads and attenuators for I0GHz. Wood is used as the absorbing element in the following designs of waveguide attenuator and matched load. A good match is ensured in the design shown in Fig 8.15, by the long taper at the 'hot end'. Its efficiency as a load may be checked by connecting it in a system, followed by a sliding short. If the short is now moved over half a wavelength, there should be no effect on the system performance. This shows that the load is absorbing all the power.

A compact low power termination is shown in Fig 8.16. A piece of resistor card with a typical resistivity of 200Ω, is passed through slots cut in the narrow wall of the waveguide and is glued in position. Any rf passing through the resistor is reflected from the short and attenuated a second time. For 10,050MHz, $\lambda/4 = 9.84$mm (0.388in); for 10,369MHz, $\lambda/4 = 9.33$mm (0.367in).

Attenuators are also made by fitting lossy material into the waveguide. In this case both ends are tapered to provide a good match in both directions. A convenient form of resistive material consists of graphite powder bonded on

to paxolin sheet about 0.5mm (0.02in) thick. This is commercially available and is described in terms of ohm per square.

A convenient resistivity is 200Ω per square, and the attenuation measured at 10GHz for this material in WGI6 is shown by Fig 8.17. Note that as the resistivity is increased, so the attenuation decreases. Homemade resistor card can be prepared by heavily marking non glossy card with a very soft pencil, or by dipping the card in colloidal graphite. A coating of cellulose lacquer can be used as protection.

For maximum attenuation, the resistive sheet should be fitted half way across the broad dimension of the guide and fit from wall to wall in the other direction. A convenient method of supporting it is to prepare a length of expanded polystyrene (which is transparent to microwaves) to fit inside the guide, slice this into two and then sandwich the attenuator between the two halves. Dummy loads can obviously be made in the same way. Even quite low value attenuators make good loads: a 10dB attenuator should have a swr of less than 1.3:1 and a 15dB attenuator less than 1.05:1, due to power passing back through the attenuator.

Waveguide variable attenuators

Variable attenuators are particularly valuable pieces of test equipment. They can often be found at rallies, although usually without calibration charts.

In an fm receiver, the audio signal to noise ratio varies strongly with the input signal strengths only over a narrow range of inputs. In practice one tends to hear signals with a high signal to noise ratio, or nothing at all. It is only by chance that signal strengths are in the narrow range of a few dB, when noisy signals are produced such that the audio quality may be used as a guide for adjusting the equipment for optimum performance.

The effective signal strength can of course be changed by altering the antenna and/or pointing it away from the optimum direction. A more satisfactory method is to insert a variable attenuator into the antenna connection. This is then adjusted so that alignment can be done under the most sensitive conditions and with a more direct measure

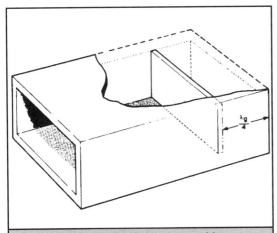

Fig 8.16 : A compact low power waveguide termination.

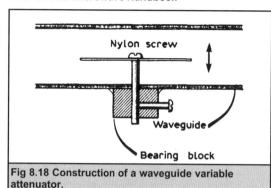

Fig 8.18 Construction of a waveguide variable attenuator.

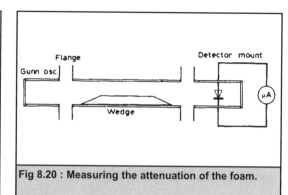

Fig 8.20 : Measuring the attenuation of the foam.

of improvements in performance.

A widely used method of construction is to mount a sheet of resistive element as shown in Fig 8.18, so that it can be moved across the guide. The attenuation is at a maximum when the element is half way across the guide and reduces as it approaches the side wall. If the element can be made a close fit against the side wall, then the attenuation becomes insignificant and the attenuator can be permanently installed. For details of a suitable resistance card and dimensions of the element, see Fig 8.17.

The push rod may be metallic if its diameter is small compared with the height of the guide (say 3.2mm (0.125in) diameter compared with 10.1mm (0.4in) for WG 16) and if fitted at the centre of the element so that any reflections from it are also attenuated.

Alternative designs are shown in Fig 8.19(a) and (b). In both of these, the resistive element is inserted progressively through a slot in the broad face of the guide, in Fig 8.19(a), the element is clamped to an adjustable hinged arm. In Fig 8.19(b), the element is mounted eccentrically on a rotatable shaft. These types of attenuator are in some ways easier to make, but there is a slight risk of the slots resonating and producing spurious effects.

The maximum attenuation available can be increased if a second slot is cut in the lower face so that the resistive element can pass completely through the guide. A suitable tool for cutting the slot is a small hacksaw blade with

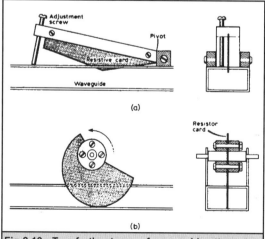

Fig 8.19 : Two further types of waveguide attenuator.

the sides of the teeth ground away to a width of about 0.8 mm (0.032in).

Some sort of scale for indexing the position of the attenuator must, of course, be fitted. For the type shown in Fig 8.18, a small micrometer is frequently used. A knob and a scale is all that is required for that shown in Fig 8.19(b). Calibration is straightforward if one has access to suitable calibrated attenuators, but difficult if not. One possible method would be to make use of directional couplers with known values of coupling.

Attenuators and dummy Roads for 24GHz

Basic designs of waveguide attenuators and dummy loads for 24GHz follow similar practice to those for 10GHz. The choice of a suitable lossy material can pose some problems. Wood generally gives acceptable results, but the conductive foam that is used for packing cmos integrated circuits has also been used successfully and is much easier to work. This material comes in a wide range of resistivities and this can sometimes give rise to unexpected results [4].

When measured by inserting the probes of a test meter into the foam a few mm apart, the usual cmos foam can have a resistance varying from a few megohms up to tens of megohms. Some foams have a much lower resistance, perhaps only tens of kohms, while the foam used to line anechoic chambers is in the region of a few kohms. The lower resistance foam will probably not give a very good match in waveguide and will reflect a lot of power. The medium and high resistance foams are suitable for loads and the high resistance ones for attenuators.

Loads are built either with the end of the guide open or more usually with a short circuit across the end. The load should be designed so that in the first type the signal is attenuated to a sufficiently low level before it is radiated into space, or in the second case so that the signal reflected back up the guide is sufficiently small after passing through the foam twice that it does not represent a significant swr. Measuring information is shown in Fig 8.20.

For a load to present a good match, the attenuation through the foam need only be about 15 to 20dB. This would give a return loss of 30 to 40dB, corresponding to a swr of 1.1 to 1.02. This should be more than adequate. If higher attenuation is used in relatively short pieces of foam it is likely that the discontinuity caused by the tapered foam wedge will reflect a significant amount of power and degrade the swr. The best way of measuring the performance of particular foam is to mount a tapered wedge of it in the guide between a source of a few mW

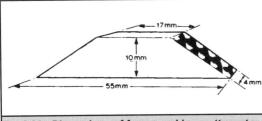

Fig 8.21 : Dimensions of foam used in an attenuator for 24GHz.

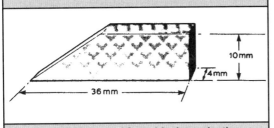

Fig 8.22 : WG20 tapered foam block termination.

and a diode detector and to measure the amount of power that is transmitted through it.

If the diode detector is connected to a low resistance meter then at power levels below a few hundred µW the meter current will be linear with power.

Start with a fairly small block of foam in the guide and measure the power transmitted through. Gradually increase the width and length of the centre section until the attenuation is about 10 to 20dB.

The foam should be tapered at an angle of about 30 degrees at both ends for an attenuator, but for a load the taper need not be present at the closed end. The angle of 30 degrees ensures that the taper occurs over a distance of at least a wavelength. Figs 8.21 and 8.22 give starting values for the dimensions, but may need to be changed

according to the type of foam available. Although attenuators are a good way of extending the range of power meters, when the dissipation exceeds a few tens of watts, it becomes very difficult to devise a form of construction that allows the heat to be lost, while retaining good matching.

A general purpose high power dummy load

An article in Ham Radio Magazine [5] described a 150W dummy load based on a CTC TA-150-50 terminating resistor. However, this resistor is usable only to 1.5GHz and the design required the use of a milling machine for its construction. The present design has a vswr of better than 1.25:1 across most of the range dc to approximately 5GHz, a maximum continuous rating (cw or pep) of 40W and uses a CBT40 power terminating resistor. Its construction does not need a milling machine and it can easily be made using only handtools. Its performance is shown in Fig 8.23 and Fig 8.24 is a photograph of the finished load. Using this simple method of construction there are some unwanted resonances. These would almost certainly disappear if the microstrip line is made to *exactly* the right dimensions and perhaps if the spill of the socket were tapered and extra grounding screws fitted near the ends of the board, i.e. near the socket and the load.

It is not recommended that the connecting spill of the CBT40 be bent up or down, to enable it to be soldered to the stripline. The original Ham Radio design used a milled hole in the heatsink to accomplish this alignment.

The present design uses a piece of 1/16in (1.6mm) aluminium as a spacer plate to pack the pcb to the right height for soldering without bending. Provided that several screws, tapped into the heatsink, are used to hold the board and spacer in place, there are few problems with producing an adequate ground.

Heatsinking *must* be provided and the recommended component is the Marston Palmer (Farnell 148-126 or identical RS Components 403-128) heatsink. It is a heavy,

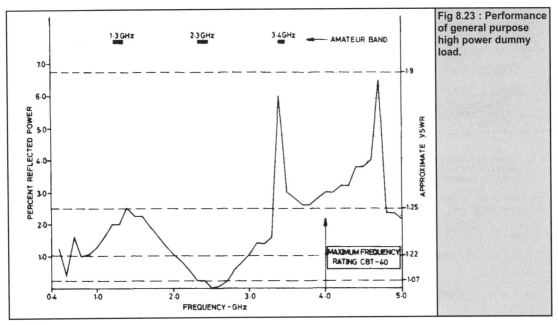

Fig 8.23 : Performance of general purpose high power dummy load.

Fig 8.24 : General purpose high power dummy load.

extruded, black anodised aluminium sink, rated at 0.75°C/ W. This is adequate rating for this type of load resistor. The finish of the mounting surface is quite smooth but will benefit by being polished with 30 micron abrasive paper where the resistor package is to be mounted.

The microstrip transmission line is theoretically 2.28mm wide on either CU Clad 233 or RT Duroid 5780 ptfe/glass double clad board. Both have a dielectric thickness of 0.79mm, an ε_r of 2.33 and are 1oz clad. The length of the line is uncritical in this application. However, if other materials are used, the microstrip line *width* will need to be recalculated according to the characteristics of the material used.

The major engineering operations are drilling and tapping holes in the heatsink. Take care to tap threads properly,

without raising a lip or bell mouth around the edge of the tapped hole. This is especially important in soft metals, such as aluminium. The formation of lips will, in effect, destroy the plane surfaces and make good overall contact almost impossible. It is worth taking a little extra trouble to avoid the problem in the first place!

Find the centre of the heatsink mounting surface and mark the mounting hole centres as shown in Fig 8.25. Drill and tap them to take 3mm bolts, to a depth of approximately 8mm. Cut the pcb to the size shown in Fig 8.25 (not critical) and etch the stripline on one side. The reverse side is an un etched ground plane. The prototype used a piece of 2.5mm self adhesive pcb drafting tape as an etch resist. In practice some undercutting occurs during the etching process and the resultant line is sufficiently close to 50Ω to give acceptable results for all but the most exacting amateur use.

Alternatively, accurately mark the stripline width and then cut carefully through the copper along the lines with a sharp scalpel blade or modelling knife. Unwanted copper can be removed by heating it with a soldering iron and stripping it off whilst the bonding is still soft. With care, either method will yield a line sufficiently accurate to ensure a reasonable vswr in the finished load. Drill 3mm clearance holes in the board, as shown in Fig 8.25 from the ground plane side to ensure the copper is neither split nor is pushed up into a lip by the drilling operation. It would, if drilled from the stripline side! The positions of the holes are not critical. Any small burrs on the ground plane side can be removed by using, with your fingers, a sharp 4 or 5mm drill to produce a slight chamfer on the ground plane side. Ensure the board is still flat.

Cut the spacer from clean, unblemished aluminium sheet to the same size as the board. Cut it by sawing or guillotining rather than by using tin snips, which cause lipping and distortion. Smooth the edges and chamfer them slightly with a very fine file. Using the pcb as a template, mark and drill the 3mm clearance holes in the packing piece, again taking care to avoid lips or burrs. Polish the surfaces to remove blemishes by gentle

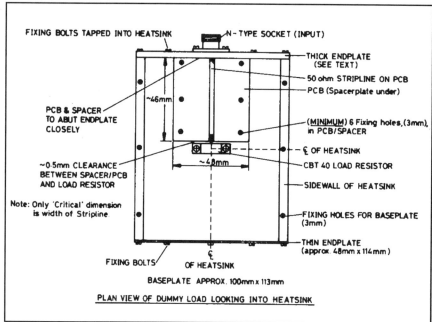

Fig 8.25 : Leading dimensions and layout of general purpose high power dummy load.

rubbing on fine emery paper laid on a flat surface such as a piece of plate glass. Lightly scribe the centre line as shown in the diagram. Temporarily mount the CBT40 on the heatsink, tightening the bolts just enough to hold the device in place.

Slide the spacer under the connecting spill and align the centre mark under the spill, by eye. Make sure that the input end of the packing piece is flush with the edge of the heatsink. Holding it in place, use the holes to mark the centres for the fixing holes in the heatsink. Remove the spacer and the load. Now drill and tap the six holes to take 3mm bolts, as for the load mounting holes.

Take a standard N type chassis mounting socket and measure the diameter of the spill and the height of the boss on the rear of the socket. Cut an endplate from aluminium plate, the thickness of which is chosen (approximately 6 to 8mm) to match the measured depth of the boss, i.e. so that the rear of the socket boss comes flush with the back of the plate. The dimensions are shown in Fig 8.25. Cut another to the same dimensions, this time from 1/16in (1.6mm) thick aluminium sheet. Drill all the 3mm clearance holes shown and de burr the plates.

Using the thicker endplate as a template, mark the fixing hole centres on that end of the heatsink, which will face the CBT40 connecting spill. Note that both endplates should be positioned so that 1/16in (1.6mm) stands proud above the bottom of the heatsink side plates. This will allow the base plate to sit recessed between the two endplates when it is fitted. Drill and tap the holes for 3mm screws.

Temporarily fix the pcb, spacer and thick endplate in place on the heatsink. Ensure that the edge of the pcb, packing piece and endplate abut closely. Mark the endplate with the position of the centre of the stripline (vertical mark) and the level of the surface of the stripline (horizontal mark). Remove the endplate, add half the diameter of the spill to the mark (get the direction right!), centre punch the new mark and drill it out to just clear the boss on the back of the socket.

Remove burrs and refit the endplate. Present the socket to the hole and rotate it until the longer edge of the spill touches the stripline. Mark the hole centres for the socket flange. Remove the socket and endplate. Drill and tap the four holes to take suitable bolts, right through the plate. The thinner endplate is fixed to the opposite end of the heatsink by similar means, but needs neither the socket hole nor the socket mounting holes.

Cut a piece of 1/16in (1.6mm) aluminium plate to fit as a base plate and arrange to fix this in place with small bolts tapped into the bottom of the heatsink sides. Rubber or plastic feet can be fitted to the plate. This completes the fabrication of all the parts for the load and assembly comes next.

Make sure that the heatsink contact surface is clean and free from grease and score-marks. Put a thin smear of zinc oxide based heatsink compound (e.g. RS Components 554-311) onto the underside of the resistor mounting flange. Place it in position and bolt it firmly into place but do not over tighten the bolts, as this could strip the threads in the heatsink. Slide the spacer and pcb into place under the resistor spill and bolt them into place.

Bolt the input endplate into place and then bolt on the socket with the spill touching the stripline. Solder both ends of the stripline to the respective connections. Fit the other endplate and base plate. This completes the assembly.

With an ohmmeter, check that 50Ω ± 2.5Ω (dc test) appears between the inner and the outer of the socket. If it is more, suspect bad soldered joints or bad contact between the pcb and the heatsink. If it is significantly less, or a short circuit, look for the cause and correct it!

Borrow an rf source and good vswr indicator, both functional at as high a frequency as possible, e.g. 430MHz or 1.3GHz. The rf source should be of adjustable power. Connect the source to the vswr indicator and the indicator to the load with short lengths of good quality 50Ω cable, using N type connectors. Beware of units using UHF connectors as these cause significant mismatches at the higher frequencies. They are just not suitable.

Starting with low power (a watt or two), measure the vswr of the load. A high reading could mean poor soldered joints, an rf short circuit or some other assembly problem. Search it out and correct it! If everything seems in order, gradually increase the power towards the maximum of 40W. Under no circumstances should 40W (cw or pep) be exceeded, or this may cause destruction of the load resistor.

Directional couplers

Directional couplers can allow greater power handling than attenuators. Most of the power is delivered direct to a load (which then has to dissipate the power!) whilst a small sample of the through power appears at the power meter. As long as the ratio of through power to that delivered to the power meter is accurately known, it is possible to indicate the power delivered to the load.

Directional couplers are occasionally found on the surplus market and are much sought after. Fortunately, it is possible to make directional couplers without too many problems, although calibrating them can be more difficult.

Principle of operation of the directional coupler

A directional coupler is a device that is able to couple power to or from a transmission line, depending upon direction of power flow in that line. In addition to extending the range of power meters, as already described, the directional coupler can also be used to:

- Couple together antennas or signal sources

- Combine outputs from power amplifiers

- Sample power in a transmission line in order to determine frequency or amplitude of signals

- Determine amplitude of reflected power (and hence vswr) of a load such as an antenna.

Although there are several different ways of constructing directional couplers, the principle of operation remains the same. Fig 8.26 shows the basic form of the coupler. It consists of a main transmission line, close to which is placed a secondary transmission line. The secondary line lies within both the electric and magnetic fields of the main line. Assuming all ports of the coupler are terminated in their characteristic impedance, power flowing in the main line will give rise to currents in the secondary line as follows.

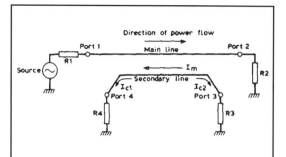

Fig 8.26 : Principle of the directional coupler.

The magnetic field induces an instantaneous current, I_m, in the secondary line. This flows in the direction shown. The electric field also couples a capacitive current, I_c, into the secondary line. This current has two components, I_{c1} and I_{c2}, which flow in opposite directions. Assuming symmetrical construction, I_{c2} will be equal and opposite to I_m, thus cancelling at port 3. Currents I_{c1} and I_m will add at port 4, resulting in power being dissipated in the load at port 4. The magnitude of the currents I_c and I_m depends upon the electrical length of the secondary line and upon its spacing from the primary line. Because of this, exact cancellation of the currents at port 3 can only occur over a limited band of frequencies, hence restricting the coupler bandwidth to, typically, one to two octaves.

The performance of a directional coupler is specified by several parameters, and it is useful to be familiar with these if you intend to build or use couplers.

• Insertion loss

The change in output port power due to the insertion of the coupler into a system. In the case of a 3dB coupler the insertion loss would be 3dB, although some excess loss due to connectors and the inherent loss of the transmission line may increase insertion loss by a fraction of a dB.

$$Insertion-loss(dB)=10\log\left(\frac{power-at-port-2}{power-at-port-1}\right)$$

• Coupling attenuation

The ratio of power at port 4 to that at port 1. It depends on length and spacing of the lines, as well as operating frequency. Typical values range from 3 to over 30dB. Also known as coupling coefficient.

$$Coupling-coefficient(dB)=10\log\left(\frac{power-at-port-4}{power-at-port-1}\right)$$

• Isolation

A measure of the isolation between the input to the coupler and the output at port 3. Ideally the isolation would be infinite, assuming correct termination, but in practice small imperfections in construction will limit isolation to values between 20 and 40dB.

$$Isolation(dB)=10\log\left(\frac{power-at-port-3}{power-at-port-1}\right)$$

• Directivity

Indicates the dynamic range of the coupler. Expressed as the difference between the isolation and the coupling coefficient. This parameter is a good indication of the quality of the coupler.

Directivity, D in dB is given by:

$$D=10\log\left(\frac{power-at-port-3}{power-at-port-1}\right)-10\log\left(\frac{power-at-port-4}{power-at-port-1}\right)$$

The above parameters are measured with all ports terminated in the correct impedance. The fraction of power appearing at port 4, when port 2 is incorrectly terminated, increases in proportion to the degree of mismatch but only within limits determined by the coupler directivity. It is this feature that is used in the measurement of mismatch. Coupler directivity limits accurate measurement of mismatch. G3WDG has calculated error limits for a given directivity of coupler when using it to measure the return loss (or vswr) of an unknown load [6]. These are summarised in Table 8.4.

For example, if a coupler with 20dB of directivity is used to measure the return loss of an antenna and the measured value is 26dB, then the true return loss can be anywhere between 16.5 and 26.0dB. Attempting to set up the antenna for zero reflected power would simply result in an antenna whose return loss is 20dB (or 1.2:1 vswr).

Although this is what is usually done, inserting different lengths of cable between the coupler and the antenna will quickly show that the match is not perfect, since the readings will change.

An alternative approach is to adjust the return loss to that of the directivity of the coupler such that inserting lengths of cable between the antenna and the coupler does not change the indicated reflected power. Under these conditions the antenna would be perfectly matched. However this method is much more laborious and assumes that the directivity of the coupler is accurately known.

Because of the difficulties in producing a coupler with good directivity over a wide range of frequency, it is desirable that the degree of capacitive or magnetic coupling is adjustable. This can be achieved either by incorporating some form of variable capacitor between the secondary line and the body of the coupler or by altering, slightly, the position of the secondary line with respect to the main line. By this means it is possible to optimise the performance over a limited range of frequencies, covering at least one amateur band.

Table 8.4 : Error limits for a given directivity of coupling.

Measured return loss (dB)	Range of true return loss (dB)	
	20dB directivity	25dB directivity
10	7.6 to 13.3	8.6 to 11.7
12	9.1 to 16.4	10.2 to 14.2
14	10.5 to 20.0	11.8 to 16.9
16	11.8 to 24.7	13.4 to 19.8
18	12.9 to 31.7	15.0 to 23.1
20	14.0 to INF	16.1 to 27.2
22	15.0 to 33.7	17.3 to 32.7
24	15.8 to 28.7	18.5 to 43.3
26	16.5 to 26.0	19.5 to 44.3
28	17.1 to 24.4	20.4 to 35.7
30	17.6 to 23.3	21.1 to 32.2
INF	20	25

A directional coupler using semi rigid coaxial cable

This design is an attempt to produce a coaxial directional coupler for use from hf to the low microwave bands [7]. Its main feature is a fairly high directivity without relying too much on precision construction. The directivity achieved depends on the accuracy of the characteristic impedance of both the main and coupled lines, as well as the load terminating the coupled line. The directivity is obtained by sampling a small fraction of the E and H fields in the main line using a small slot, which is less than a quarter wavelength long. These fields leak out into the coupled line and reproduce the forward and reflected waves in the coupled line. The accuracy of the sampling is improved in this design by using equal diameter lines with the same dielectric. Consider Fig 8.27, where the coupler has a load with a 1.2:1 swr (20dB return loss) on it. The forward power will couple -30dB into the termination, and the reflected power -50dB into the detector. If the coupled line termination has an swr of 1.2:1, then this will also reflect -20dB of the incident signal (-30dB), i.e. -50dB back to the detector. This is then producing a signal at the same level as the one to be measured. One of the main problems in obtaining a high directivity is that of making the lines, either coax or stripline, sufficiently accurately and also making the transitions from them to the loads and connectors. For this reason, any couplers made using home built lines are unlikely to be very directive unless they are checked or adjusted on known test gear.

Construction

Semi rigid coax of any diameter can be used, 0.141in, 0.250in and 0.325in being common values. They can be used with BNC SMA or N type connectors according to the size, but ensure that the connector is intended for use with that size of cable, or there may be a significant vswr introduced. The larger sizes are stronger mechanically and easier to work with and can be made to fit into N type connectors very easily. The construction is shown in Fig 8.28.

Hold the cable in a vice and file away the outer to form a flat on the cable of the desired length. It should be a quarter wavelength long in the cable dielectric (ptfe) at the highest frequency required. Dielectric constant of ptfe is 2.5 for the type used in this cable.

Be careful not to tear the copper outer away from the ptfe

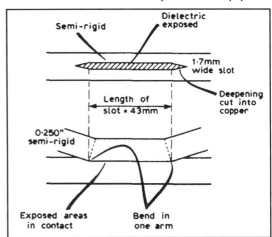

Fig 8.28 : Construction of the coupler.

by over enthusiastic filing. At each end of the slot taper the thickness of the copper so that the gap between the two cables is as small as possible. Bend the second line to the shape shown before filing the slot in it. Clamp the two together and solder them all the way round the join to seal the two-coax lines together. The slot width should be about 1mm on 0.141in cable and 2mm on the larger sizes.

The high directivity is only obtained for fairly weak coupling and the slot widths specified give about 35dB maximum at the high frequency end of their range. 30dB should be regarded as the maximum value to be used as the slots then become too wide and begin to alter the characteristic impedance of the lines.

The coupling coefficient and directivity vary with frequency as shown in Fig 8.29 for a typical example. A directivity of 20-25dB is typical, whilst the coupling varies at the rate of 20dB per decade of frequency. Connectors

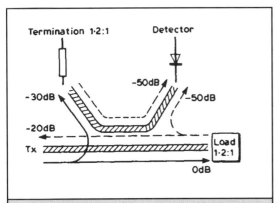

Fig 8.27 : Schematic of the coupler.

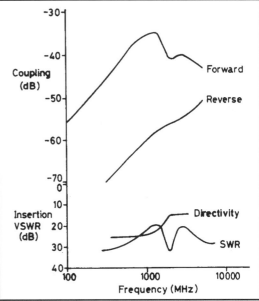

Fig 8.29 : Coupling coefficient and directivity of the coupler.

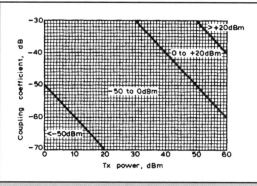

Fig 8.30 : Detector ranges for the coupler.

will be required at the ends of the main line, but on the coupled line there are two options. Either use connectors for flexibility, since its performance can then be measured on professional test equipment; alternatively the load and detector can be soldered directly onto the cable. A 50Ω chip resistor soldered across from inner to outer will provide a good load up to the low microwave region. The diode detector can then be soldered across another chip 50Ω resistor at the other end of the coupled line.

The type of detector must be chosen according to the power level and coupling coefficient used, as shown in Fig 8.30. A receiver or amplifier is needed at very low power levels. The wideband, linear reading power meter described earlier, is useful in the range -40 to 0dBm and the simple vhf/uhf power meter type, above 0 to +10dBm. Above about +20dBm the dissipation in the chip resistor becomes excessive and external attenuators become necessary. If connectors are fitted to the coupled line, the positions of the load and detector can be reversed, making it possible to indicate forward and reverse power without reversing the main line.

High power directional coupler

As an alternative to the all semi rigid coupler described above, this coupler uses a fabricated section of 50Ω main line, but retains the semi rigid coaxial sampling line [8]. Originally designed for use on 430MHz, it is equally suited to use on 1296MHz. The coaxial main line is formed by a concentric assembly of two lengths of copper or brass tubing. The ratio of the two diameters to give the required impedance is given by the formula:

$$Z = 138 \times \log (D/d)$$

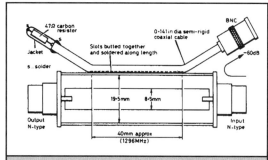

Fig 8.31 : Construction of the high power directional coupler.

where   D = inner diameter of the outer

and   d = outer diameter of the inner

For a characteristic impedance of 50Ω, the ratio is 2.3. The original used 19.5mm tube for the outer, and 8.5mm tube for the inner.

Construction

Cut the outer tube to the required length. From the centre, mark a slot 40mm long (20mm either side of centre) and about 2.5mm wide. Drill a series of holes along this slot outline. Join the holes together using a small Swiss file and then carefully file to form the slot.

Bend the length of semi rigid coaxial cable to the shape shown in Fig 8.31. With the semi rigid held in a vice, carefully file the outer screening away for a length of 40mm, to expose the ptfe dielectric. Do not file the ptfe.

Assemble the coaxial main line with N type sockets at either end. Soldering the inner tubing to the pin of the N type at one end poses no problem since this can be done with the inner tube clear of the outer, but the other end is not so easy. One solution is to cut a clearance slot in one end of the outer such that it is possible to gain access to the socket pin with a small soldering iron. When the connection has been completed, a small section of copper or brass sheet can then be soldered over the slot to restore the screening integrity of the coaxial line.

A suitable low cost load can be made by soldering a 0.5W carbon resistor, with very short leads, to one end of the semi rigid coaxial sampling line. A short length of outer tubing from the same cable is then slipped over the resistor. The end of this is then flattened and soldered to the resistor as shown.

For convenience a BNC socket can be mounted to the other end of the sampling line. To this socket a suitable power meter or detector can be connected in order to indicate the coupled power. The coupling is about -60dB, at 1296MHz, for the size of slot specified. Making the slot larger will increase the coupling, as will a smaller diameter main line. The directivity of the prototype was measured at 18dB on 1.3GHz. An alternative would be to build the coupler as a dual unit so that both forward and reflected power can be monitored.

On 430MHz, the length of slot should be increased by a factor of three. The exact length is, however, not critical.

Directional coupler for 2320MHz

Coaxial line can be constructed with various types of cross section from round to rectangular. In this simple design of coupler by G4LRT, the main section of line is constructed with a round inner and a rectangular outer [9].

Conveniently, a 50Ω characteristic impedance is obtained with a 15mm diameter round tube within a 25mm (1in) by 32mm (1.25in) trough. The round tube is a short length of 15mm diameter copper central heating tubing whilst the trough is an RS 509-923 or Eddystone 7969P diecast box. The coupling lines are 12mm (approximately) wide strips of 22SWG copper sheet. The width may need to be altered during setting up so it is best to start with a slightly wider line. Two strips are required to allow simultaneous measurement of forward and reverse power.

Construction

Drill both ends of the box centrally, to accept a flange

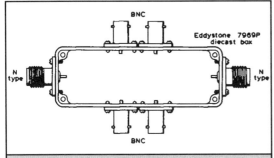

Fig 8.32 Position of the sockets in the box.

mounting N type socket. A small flange type such as the TRL95029 or its equivalent is easier to use. Cut the centre pins of the N sockets back until about 3mm (0.125in) is left protruding beyond the ptfe insulation. Drill the sides of the box to take two flange mounting BNC sockets either side, as shown in Fig 8.32. The flanges of the two sockets should be just touching. Cut the centre pins of these sockets back as above, but only 3mm should protrude beyond the inside surface of the box.

Cut the 15mm tube just long enough to fit between the remainder of the N socket centre pins. The ends of the tube should be cut at 45 degrees as shown in Fig 8.33. Bend the top of the tube down to form a beak like hook. The tip of the hook must be in line with the centre line of the tube.

The 22SWG copper strip should be cut as shown in Fig 8.34. Solder the coupling strips into place between the pairs of BNC sockets, so that they are spaced about 3mm from the box.

Position the 15mm tube centrally in the box with the hooks resting on the N type socket pins. Solder into place on the pins.

Low cost commercial 50Ω loads are used to terminate the sampling lines. RS 456-251 loads are only specified up to 1GHz but in practice they have proved to be satisfactory to at least 2.3GHz. The detectors are assembled using BNC plugs as shown in Fig 8.35. The meter movement can be located remotely from the coupler.

Adjustment

Connect a good 50Ω load to the antenna port of the coupler and the 2320MHz transmitter to the other port. Connect the sampling line load and detector to one pair of the sampling line sockets. The detector should be connected to the socket nearest the transmitter. Transmit through the directional coupler into a well matched load and check that a reading is obtained on the meter. Switch the transmitter off.

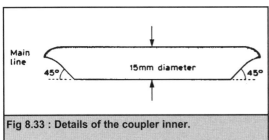

Fig 8.33 : Details of the coupler inner.

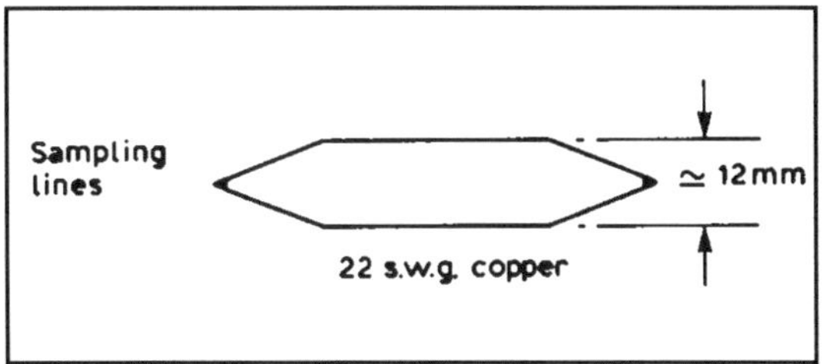

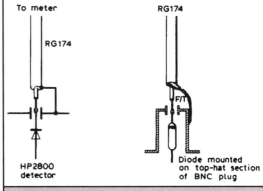

Fig 8.34 : Cutting details of the sampling lines.

Swap the load and detector over and switch the transmitter back on. The meter reading should now be much lower, if not, switch the transmitter off and adjust the position of the sampling line slightly. Switch the transmitter back on and check the reading is now lower. It may be necessary to alter the width of the sampling line to get the lowest possible reverse direction reading.

Repeat the above procedure for the other sampling line. It should be possible to obtain a directivity of better than 20dB with this design.

Other bands

The directional coupler may be used on bands other than 2320MHz. The dimensions of the main line section remains the same, of course, but the length of the sampling lines will need to be changed if optimum coupling is to be obtained. It is not recommended that this design be used on any higher bands than 2320MHz, but 430 and 1296MHz are practical. The length of the sampling lines is set by the separation of the BNC sockets. The centre to centre spacing should be 25mm for 1296MHz and as far apart in the box as possible for 430MHz.

Printed circuit directional coupler

A simple method of constructing a directional coupler is by use of a printed circuit board, illustrated in Fig 8.36. The line impedance can be made in accordance with the design information given in the chapter on transmission lines. The coupling lines may be of any convenient length to suit the meter in use, but they should be short compared with λ/4.

The diodes should be small signal types, such as point contact Germanium, and suitable for the frequency concerned. The terminating resistors should be, as far as

Fig 8.35 : Construction of the diode detector.

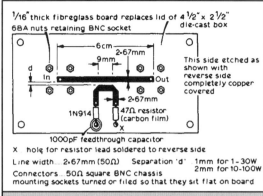

Fig 8.36 : Printed circuit directional coupler.

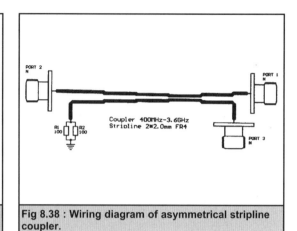

Fig 8.38 : Wiring diagram of asymmetrical stripline coupler.

possible, of a non inductive type such as carbon film. Bypass capacitors may be disc, plate or feed through, with the latter having the additional advantage of providing a suitable terminating connection for the meter. It is important to use a type of capacitor suitable for the frequencies to be used.

Stripline Directional Coupler for 400 MHz to 3.6 GHz [10]

A comprehensive description of directional couplers can be found in the book "Stripline Circuit Design" by Harlan Howe [11]. Why Stripline (Triplate)? It very quickly became apparent that Stripline has marked advantages, as against Microstrip, in relation to the directivity. A directional coupler which consists only of 2 λ/4 long coupled lines has a band width of one octave. If this coupler is extended by additional λ/4 couplers with an expanded interaction gap (Fig.8.37), the usable frequency range is expanded. At n = 3, we can already obtain a band width of 4:1 to 8:1, depending on the permitted ripple. If we want a coupler with a band width exceeding 9:1, we have to use a structure with n ≥ 4. The consequence is that the coupler becomes very long and this also means the losses in the substrate increase.

Another option is to use an asymmetrical structure. This saves almost half the λ/4 segments required, with scarcely any deterioration in the characteristics (Fig.8.38).

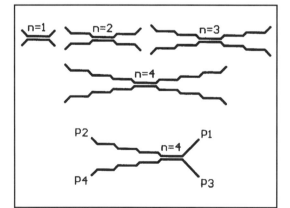

Fig 8. 37 : Symmetrical and asymmetrical couplers.

For a symmetrical coupler, the phase displacement to the de-coupled port is always 90°.

For an asymmetrical coupler, the phase displacement is frequency dependent, but this is irrelevant for amateur radio applications.

Are there differences between the 2/4 coupling and the 1/3? Simulations show that the coupling for the 2/4 port is more strongly frequency dependent than that for the 1/3 port. The reason for this is the transmission loss through the segments right up to the λ/4 segment, limiting for the high frequencies, with the smallest coupling gap.

Construction

Once these basic principles were clear, the original problem could also be solved very quickly. The result was a symmetrical coupler with n = 3 on Rogers RO3003 material.

Since this solution is very expensive, because of the base material used, the idea of a version that would give better value was pursued. So how well suited, or how badly suited, is FR4 epoxy material to these high frequencies?

Through simulation, using Super-Compact, it was found that, with limitations, a "normal" assembly could be used right up to 3.5 GHz. As a result of the simulation, the original structure was expanded (n = 4) by one λ/4 segment. The useful frequency range goes from 400 MHz to 3.5 GHz at 20± dB coupling.

The biggest disadvantage of FR4 is the attenuation, which here reaches approximately 1.5 dB at 3.5 GHz, and is thus close to maximum. A certain amount of uncertainty comes from the er, which depends on the manufacturer, between er = 4.0 and 4.6. In simulations the value of er = 4.2 for FR4 was used. The effects are displayed mainly by the fact that the operating range of the coupler is displaced (Fig.8.39).

2.0 mm. was selected as the thickness of each substrate. The Triplate was reinforced by a 2 mm. thick aluminium plate to give sufficient mechanical stability.

2 x 100-Ohm SMD 0805 resistors were integrated into the Triplate between two 2.0 mm. thick epoxy printed circuit boards. In order to obtain enough room for the resistors, some epoxy had to be "carved out" of the second board. The important thing here was that the earth contact for the resistors was soldered to all 3 layers with 2 pieces of wire (0.8 mm. long). You should also make sure that there is a

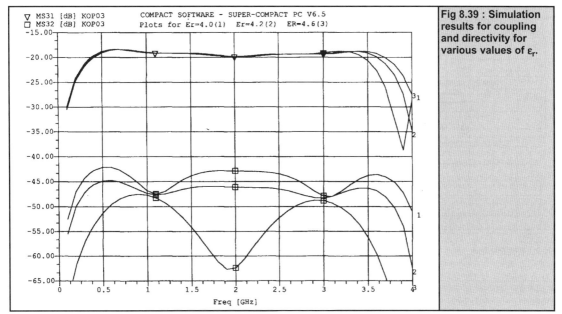

Fig 8.39 : Simulation results for coupling and directivity for various values of ε_r.

good contact between the screws and the two earth surfaces (Fig.8.40).

In a reflection measurement, the two chip resistors, which added up to 0.25 W, limited the preliminary power to 25 Watts.

Unexpected obstacles

Everything can be calculated perfectly for the coupler, use a high quality substrate and have a good seal, and still not obtain good directivity. The reason for this is usually the sockets used. These components, or the way they are matched, have or has a direct influence on the directivity. The best solution is to use special sockets for striplines with "pennants" as centre contacts (e.g. from Rosenberger).

As no such N sockets were available during assembly, "normal" N sockets were tried, which possess a thinner dielectric (d = 5.8 mm.) on the flange side.

After completing the coupler the return loss with this socket was still only RL = - 22 dB at 2.4 GHz. The reason for this is that this socket has a section about 2 mm. longer with only 45 Ω. Another type of socket was even

worse. It had an area 5 mm. long with an impedance of only 40 Ω.

Measurement results

No matter how attractive the simulation results may have been, the measurement results (Fig 8.41 and 8.42) for the prototypes looked completely different. However, this was mainly in relation to the return loss, which at times was only 14 dB (Fig.8.43). The reason for this lay in the sockets, or in the way they were mounted.

Making your own

For those who want to start making their own version straight away, here are the dimensions for a coupler, with n = 5, mounted on 2 x 1.5 mm. FR4 epoxy. All lines lie on a 5 mil (0.005°) basic grid.

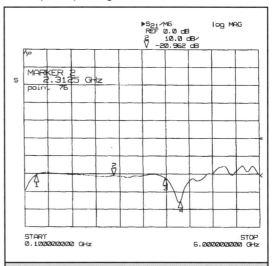

Fig 8.41 : Measurement of coupling, port1 to port 3, 400MHz to 3Ghz = 20dB ± 1dB. Marker 1 at 400MHz, marker 2 at 2.31GHz, marker 3 at 3.6GHz and marker 4 at 4.0GHz.

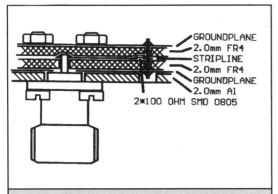

Fig 8.40 : Structure of triplate with aluminium ground plane, mounted N socket and soldered on SMD resistor.

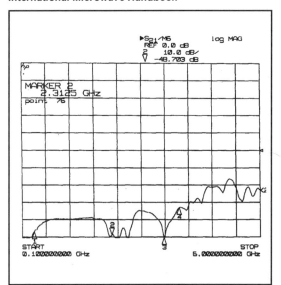

Fig 8.42 : Measurment of directivity, port 2 to port 3. 400MHz - 2.5GHz ≥ 18dB. 400MHz - 3.5GHz ≥ 14dB.

The printed circuit boards are always 56 mil wide, and the length of a coupler segment is 690 mil.

The length of the entire coupler is approximately 88 mm. and thus fits crosswise onto a Euro-card.

The intervals of the coupler lines (centre of track in each case) are S1 = 175 mil, S2 = 140 mil, S3 = 120 mil, S4 = 100 mil and S5 = 85 mil > coupling gap: 119 mil, 84 mil, 64 mil, 44 mil, 29 mil.

The layout was printed onto film on a 1:1 scale using a laser printer and engraved using normal engraving equipment. The assembly coupling is 21 ± 1 dB (350 MHz to 3.5 GHz). The directivity is better than 20 dB up to 2 GHz and better than 18 dB up to 3.5 GHz.

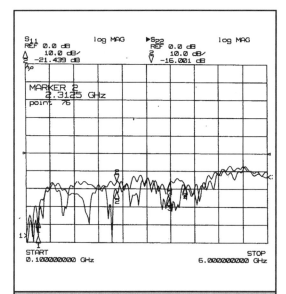

Fig 8.43 : Matching of ports 1,2 and 3. 400MHz - 2.5GHz R_L ≥ 16dB. 400MHz - 3.5GHz R_L ≥ 13dB.

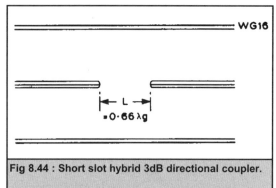

Fig 8.44 : Short slot hybrid 3dB directional coupler.

Waveguide directional couplers

Directional couplers for the higher microwave bands can be made relatively easily using lengths of waveguide.

A 3dB coupler for 10GHz

The construction of a short slot hybrid coupler is shown in Fig 8.44. Windows 17mm long and 10.16mm high are cut in a narrow face of two pieces of waveguide, which then are mounted so the windows coincide. For waveguide 16, the length L is 26.0mm (l.023in) at 10,050MHz, and 24.6mm (0.970in) at 10,369MHz.

A 9-30dB cross hole coupler for 10GHz

A feature of this design is that up to 9dB coupling can be obtained, which is sufficiently large for use in simple cross coupler receiver designs. The coupling can be made as small as required and data is given in Table 8.5 for down to 30dB. However for a coupling of less than 20dB or so other types, which are simpler to make, are usually preferred.

The form of the coupler is shown in Figs 8.45 and 8.46. The coupling slots may be machined in the wall of one waveguide, but a separate plate as shown is recommended for kitchen table constructors. Note that, because the thickness of the plate recommended (0.91mm, 0.036in) is slightly smaller than the wall thickness of the waveguide (1.27mm, 0.050in), then the assembly becomes self jigging. The coupling slots can be cut by clamping an oversize piece of brass sheet in a vice and using the edge of the vice as a reference for drilling and filing. The 25mm (1in) square outline is then marked as a best fit to the slots actually cut. This procedure is much more effective than trying to cut slots accurately in a 25mm square plate.

The walls of the guide can be removed by drilling and filing. A piece of wood fitted inside the guide prevents damage to the opposite face. By continually monitoring

Table 8.5 Dimensions of cross hole couplers.

Coupling (dB)	Dimension A (mm)	Slot width (mm)
9	12.9	1.83
12	12.8	1.65
15	11.73	1.65
20	10.4	1.65
25	9.32	1.65
30	8.33	1.65

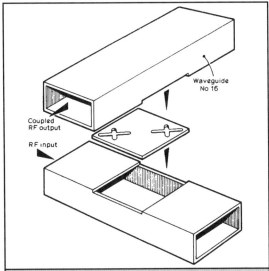

Fig 8.45 : Exploded view of a 9 to 30dB cross coupler.

with a micrometer or vernier gauge, an accuracy of 0.025mm (0.00lin) can easily be achieved with simple hand tools (and patience). For soldering this type of work where the assembly can be thoroughly washed, solder paint spread thinly over the joints to be soldered is recommended as this avoids the problem in feeding solder to the less accessible parts. A gentle gas flame or a hot plate is used for heating. Alternatively, merely clamping the components mechanically together is sometimes satisfactory.

The round hole coupler

When a directional coupler with a coupling coefficient greater than 20dB is required, a convenient design is the round hole coupler illustrated in Fig 8.47. Coupling is via three circular holes drilled at the corners of a square of side equal to $\lambda/4$. The degree of coupling varies with the ratio of the diameter of the larger holes to that of the broad internal width of the waveguide. The smaller hole has a diameter two thirds that of the larger holes.

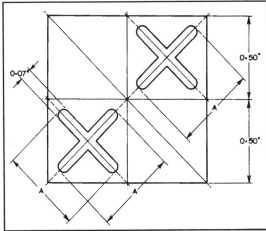

Fig 8.46 : Dimensions of septum plate. Material 0.036 in. (20swg) brass.

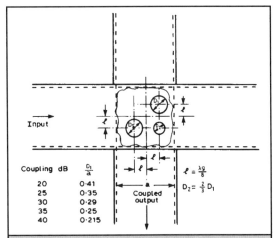

Fig 8.47 : Dimensions of a round hole coupler.

For waveguide 16, a = 22.86mm (0.9in) and λ_g is 39.37mm (1.551n) at 10,050MHz, or 37.34mm (l.47in) at 10,369MHz. A 25dB coupler for the lower frequency, for example, would require two holes 7.9mm (0.311in) in diameter and one 5.3mm (0.208in) diameter, spaced 4.93mm (0.194in) from the centre lines of the waveguide.

The holes may be drilled in the wall of one of the waveguides, preferably the input arm, the corresponding broad face wall of the second piece being removed. Alternatively, the holes can be drilled in a separate plate as described for the previous coupler.

Frequency measurement

Wavemeters

Many radio amateurs seem to regard a wavemeter as an item to be kept in the shack only to comply with the licence requirements! However, if any serious radio construction is contemplated, a wavemeter covering the frequency range of interest is absolutely indispensable (unless you have a spectrum analyser!).

Even a frequency counter cannot readily take the place of a wavemeter. The main reason for this is the wave meters ability to indicate the relative levels of signals, as well as their approximate frequency. Aligning a frequency multiplier, for example, is most easily accomplished by tuning the wavemeter to the required output frequency and placing it close to the multiplier tuned output. When the multiplier output is adjusted to the correct frequency, the wavemeter will indicate some output. Further tuning of the multiplier should then result in an increased reading. Tuning the wavemeter to other harmonics of the source frequency will result in an indication of the relative levels of these products.

Attempting to use a frequency counter in the same way will result in either no reading at all, or a reading that is constant over a wide range of multiplier tuning. This also assumes the counter is capable of working at the required frequency.

A wide range cavity wavemeter

Measurement of frequencies above about 500MHz becomes difficult using lumped circuit wavemeters, so that it is an advantage to use a cavity design [12]. These are

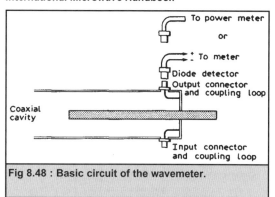

Fig 8.48 : Basic circuit of the wavemeter.

often constructed to cover relatively narrow bands, but for amateur purposes it is desirable to cover several bands, if possible, thus providing continuous coverage and allowing its use for second harmonics. The wavemeter described has been designed to cover from around 400MHz to 2.5GHz using a single cavity with an adjustable (inner) λ/4 element.

Basically, the direct measurement of a λ/4 element represents the wavelength at which it is resonant. An equivalent circuit of the wavemeter is shown in Fig 8.48. It comprises an input coupling, resonant circuit, and an indicator probe or circuit.

When a λ/4 circuit is energised, a current maximum will occur at the shorted end and a voltage maximum at the open end as shown in Fig 8.49.

A current resonance indicator must, therefore, be coupled to the low impedance end of the circuit; that is, as near as possible to the short circuit. Also, as the input will normally be connected to the low impedance output of an oscillator or transmitter, this will also dictate a tap close to the short circuit.

Both these couplings should be relatively loose so as not to foreshorten the length of the inner conductor by capacitive loading and so that the mechanical length of the inner conductor is substantially the electrical length of λ/4. There will, however, be some apparent shortening of the inner conductor compared with the free space length, due to the stray field from its end to the continuing outer. This will be most noticeable at the highest frequencies, and may be as much as 4mm at 2.5GHz.

The characteristic impedance of the cavity is of no significance in the case of a frequency meter and may be of a value convenient to the materials available. It may be either circular or square in section. The sensitivity will naturally depend to a large extent on the meter used. With a 50µA meter satisfactory indication at levels down to 5mW may be observed.

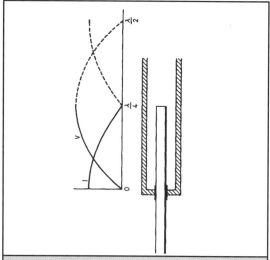

Fig 8.49 : Voltage and current distribution in the wavemeter cavity.

In the design illustrated in Fig 8.50, the outer consists of 25mm inside diameter tube with an adjustable length, inner conductor of 6.35mm (0.25in) diameter.

The outer, in this case, has a narrow slot running most of the outer tube length so that the position of the inner conductor may be observed directly and the calibration scale can be fixed along the slot (similar to a slotted line). An alternative method of fitting a calibration scale to the extension of the inner conductor, outside the cavity, is indicated in the diagram.

Construction

As mentioned, the precise dimensions of the cavity are not critical, though if materials permit, it can be made for 50 or 75Ω and may either be round or square in cross section.

For convenience the inner conductor should be 6.35mm (0.25in) in diameter, and the outer, 25mm (1 in) inside. The material may be copper or brass (the latter is more rigid and therefore preferable for the inner conductor). The outer may for preference be copper, as this is more likely to be obtained in the form of water pipe. It is important to provide a reliable sliding contact for the inner conductor. If 6.35mm material is used then two conventional shaft locks should be connected back to back, preferably with an extension tube between them, to provide a long bearing. The input coupling is arranged at the shorted end and consists of a strip drilled and soldered directly to the connectors.

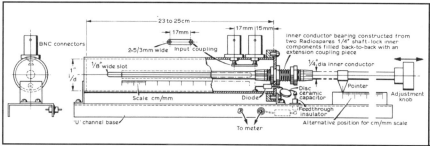

Fig 8.50 : Layout of the wide range cavity wavemeter.

Using the wavemeter

For frequency measurement the indicator should be connected to one port of the wavemeter and the other port connected to the output of the source whose frequency is to be measured. In the case of a source that is likely to be sensitive to load impedance it may be preferable to connect the wavemeter via an attenuator or directional coupler, since the wavemeter when tuned to frequencies other than that of the source output will appear as a variable reactance.

Slide the inner of the wavemeter slowly out whilst watching the indicator for a peak reading. In the case of a wavemeter that is poorly constructed there may be a residual reading whilst still far off the resonant frequency. This is usually due to over coupling between the input and output circuits. The resonance peak will, in this case, be preceded by a sharp dip in this residual reading.

It is possible with this type of wavemeter to obtain more than one indication due to the wavemeter resonating not only at $\lambda/4$ wavelengths but also at all odd multiples of $\lambda/4$ wavelength. For example, when measuring the output of a 1296MHz source a reading will also be obtained when the wavemeter is tuned to 432MHz.

This characteristic can be used when calibrating the wavemeter, since it is possible to use a single (but high) frequency to obtain several calibration points. These can be plotted on a graph and, by interpolation; the frequencies between can be calibrated. Obviously the more frequencies available when calibrating, the more accurate the overall calibration will be.

A self calibrating wavemeter for 10GHz

The wavemeter shown in Fig 8.51 enables frequencies to be measured between at least 9.5GHz and 11GHz, with an accuracy of ±10MHz.

An important feature is that it is self calibrating, which avoids the need to refer to a precision frequency standard.

Construction

It consists of a rod of adjustable length set coaxially in a cavity, which is loosely coupled to waveguide. Absorption of power occurs when the rod resonates, that is, when the rod is electrically (not necessarily physically) either $\lambda/4$ or $3\lambda/4$ long.

Because the wavelength is short at these frequencies, the tuning rates of this type of wavemeter tend to be high, in the region of 1300MHz/mm for the $\lambda/4$ mode, and 440MHz for the $3\lambda/4$ mode. The constructional problems that could be associated with these rates have been avoided by the use of a standard micrometer head (Moore and Wright type 952M or its variants), the spindle of which forms the resonating element.

The wavemeter body is fabricated from a block of brass through which a 12.7mm (0.5in) diameter hole is drilled. This single hole both locates the micrometer stem and forms the cavity, thus ensuring their alignment. The micrometer spindle passes through a $\lambda/4$ choke which electrically defines the position of the cold end of the resonating element more reliably than mechanical contacts such as fingering. To maintain a reasonably high Q, the gap between the spindle and the choke should be kept as small as possible, preferably less than 0.25mm

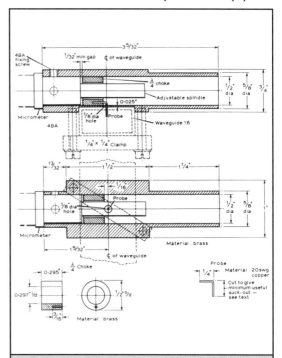

Fig 8.51 : Construction of a simple self calibrating wavemeter.

(0.01in) without their actually touching at any point. A short probe from this choke passes through a hole in the thinned wall of the cavity, and through a corresponding hole in the wall of the waveguide, to couple rf.

The choke and probe are best soldered in a single operation. The choke is fitted in the micrometer end of the body with the correct orientation and the probe, formed at the end of a 30.5cm (12in) length of wire, is inserted through the wall. The body is then pushed over the plain end of a 12.7mm (0.5in) drill held vertically in a vice until the choke is in its correct position with the probe located in the hole in the choke. The body is clamped using the micrometer fixing screws, and the extended probe wire supported externally. The choke and the probe are then soldered using the minimum amount of solder necessary. The probe is cut to a length of about 9.52mm (0.375in), and the cavity carefully cleaned.

The body of the wavemeter and the micrometer spindle may be plated with gold, copper or silver, although this is not really necessary. The plating on the spindle should not exceed 5μm (0.0002 in) or difficulty may be found in reassembling the micrometer.

Calibration

To calibrate the wavemeter, a source of rf and a means for detecting relative power levels are required. These will normally be part of the receiver or transmitter with which the wavemeter is to be used. The probe should first be trimmed to reduce the suck out to the minimum convenient, for example, 10 percent reduction in mixer current. For each of a number of unknown frequencies, the micrometer readings RI and R2, corresponding to the $\lambda/4$ and $3\lambda/4$ suck outs, should be noted. The difference between these readings is accurately $\lambda/2$ at the frequency measured.

Hence:

$$f(MHz) = \frac{C}{2 \times (R1 - R2)}$$

Where RI and R2 are in millimeters and C = 299,600 for air at 25°C and 30% humidity. A conventional calibration curve can be prepared using this technique, for both modes of operation. Normally, advantage should be taken of the slower tuning rate of the 3λ/4 mode.

An alternative self calibrating wavemeter

A second form of self calibrating wavemeter is shown in Fig 8.52. A piece of waveguide 16 coupled to the waveguide system, forms the body of a resonant cavity, the length of which can be altered by an adjustable rf short. As this short is withdrawn, then a suck out will be observed when the length of the cavity is approximately $\lambda_g/4$. A second suck out will occur if the short is further withdrawn, the distance between the two being precisely λ_g2.

Note that the coupling hole is in the cavity wall and that the blocks forming the rf shorts should be a sliding fit within the waveguide. The 2BA drive shaft could be replaced with one having a OBA thread, which has a 1mm pitch, or by a micrometer head.

A high Q wavemeter for 8 to 12GHz

The wavemeter shown in Fig 8.53 has a Q in the region of 5000 and therefore a resolution of about 2MHz. It can be calibrated to measure frequencies from 8 to 12GHz.

Construction

The wavemeter consists of a resonant cavity 17.8mm (0.7in) diameter and 12.7mm (0.5in) long, end coupled to waveguide. It is tuned by the spindle of a standard micrometer head, the electrical position of the cold end of

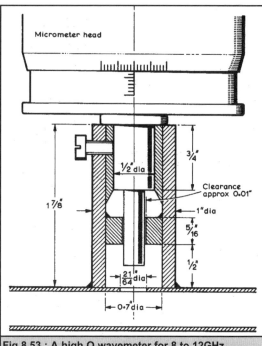

Fig 8.53 : A high Q wavemeter for 8 to 12GHz.

which is defined by a λ/4 choke. Fortunately the calibration is linear with frequency in the range 10 to 10.5GHz within about one per cent, the tuning rate measured being almost exactly 10MHz/0.01mm. If one accurate calibration point can be determined within this range, then a calibration for the remainder can be calculated without undue error.

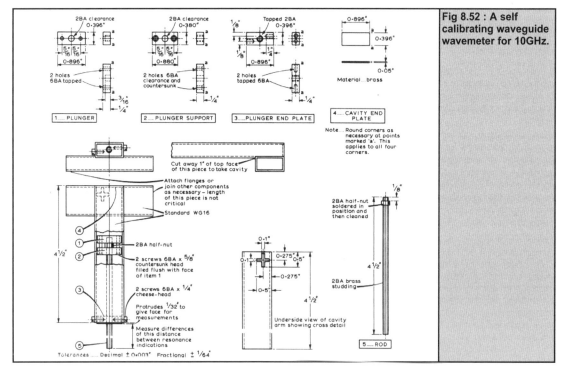

Fig 8.52 : A self calibrating waveguide wavemeter for 10GHz.

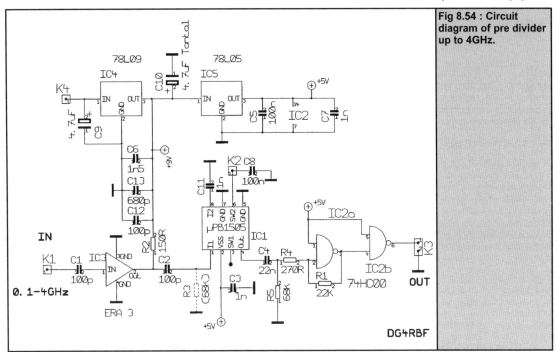

Fig 8.54 : Circuit diagram of pre divider up to 4GHz.

To take full advantage of the resolution of this wavemeter, a drum type micrometer (such as Moore and Wright type 480M) can be used. Alternatively, a smaller micrometer such as type 952M may be employed, perhaps with a reduction drive and a directly calibrated scale.

The original wavemeters were made from 25mm diameter brass bar, which was drilled and reamed to produce a tube. The choke was made as a separate part. To assemble, a 17.8mm (0.7in) drill was clamped vertically in a vice and the wavemeter body was positioned on this to locate the choke correctly and held in place by the micrometer clamping screws.

The choke was soldered into place, the joint checked visually and then the waveguide was soldered. An accurately turned sleeve served to locate the micrometer head. Other methods of fabrication may be used, for example machining the body, choke and sleeve from a solid bar, provided that the internal dimensions quoted are adhered to. Alternatively, a flange may be fabricated on the body of the wavemeter, so that it can be bolted, rather than soldered, to the waveguide. In the original versions, only the micrometer spindle was plated (5μm copper or gold), although even this is not necessary.

Counters and Prescalers

The frequency counter has become a standard piece of test equipment in the experimenter's shack. Various designs have been published in Radcom, QST and VHF Communications, the intending constructor does not have to look far to find a circuit to suit his/her needs and pocket. There are also many suitable instruments available on the surplus market. Most of these counters will have inputs suitable for measurements up to a few 10s of MHz or even a few hundred MHz.

For the microwave experimenter it may not be so easy to find a suitable instrument to measure into the GHz region. This is where the prescaler is required.

Frequency divider up to 4Ghz [13]

Suitable frequency dividers are often required for home-made circuits in the UHF-SHF range. Recently, though, there have been problems procuring frequency dividers which still function at frequencies exceeding 2 GHz. Either the integrated circuits can no longer be obtained or the prices are disproportionately high.

For example, the μPB581 and 582 pre dividers from NEC, well known to radio amateurs (divide by 2 and divide by 4) can be used up to 3.7 GHz. But there is now almost nowhere where these advantageously priced integrated circuits can still be obtained.

One very interesting alternative is the NEC divider, type μPB 1505. The μPB 1505 frequency divider is specified for up to 3 GHz, but functions up to 4 GHz and beyond, and is housed in an 8 pin SMD housing.

The sensitivity and dynamics are considerably better than in the older versions, μPB 581/582. The programmable divider factors of 64, 128 or 256 are also very interesting.

Here are a few of the frequency dividers data, taken from the manufacturers specifications:

• High actual frequency range, 0.5GHz to 3.0GHz

• Low current consumption, Typically 14 mA at 5 V

• High divider ratio, divide by 256, 128 and 64

• High input sensitivity, -14 to + 10 dBm (at 1.0 to 2.7GHz)

• High output level, 1.5 Vss (CL = 8 pF load)

The required divider factor can be set using the connections, SW1 and SW2, as follows :

• Divider factor, 256, SW1 = Low and SW2 = Low

• Divider factor, 128, SW1 = Low and SW2 = High

• Divider factor, 64, SW1 = High and SW2 = High

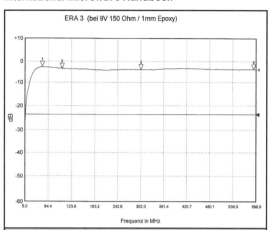

Fig 8.55 : Plot of ERA3 amplification from 5.0MHz to 589.9MHz. Reference level -16dBm. Markers : 1: 48.9MHz/21dB; 2: 100MHz/20db; 3: 301.0MHz/20dB; 4: 591.6MHz/20dB.

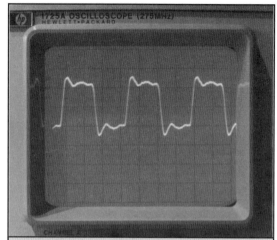

Fig 8.57 : Output signal of divider at frequency of 3.6GHz divided by 256.

The following typical application shows how little it costs to expand a relatively old frequency counter to 4 GHz.

Most frequency counters which function up to approximately 1,000 MHz have a pre divider with one of these three divider factors : 64, 128 or 256. These pre dividers are usually mounted directly at the input jack on a small printed circuit board or in a separate tinplate housing. This pre divider module is simply replaced by the new pre divider, which has a greater operational range.

The great advantage of a replacement, as against additional series connected dividers, is that the frequency to be counted is directly displayed. This dispenses with the burdensome conversion to the actual frequency necessitated by the odd divider ratio, together with additional modification measures to generate the gate time.

Pre-divider circuit description

The circuit essentially consists of three units: pre-amplifier, frequency divider and level converter (Fig. 8.54).

The preamplifier fulfils two requirements. Firstly, it raises

the level of the input signal, which increases the sensitivity, and secondly it protects the divider against destruction due to excessive input power.

The ERA 3 from MiniCircuits was used as a pre amplifier here. This MMIC gives a very uniform amplification of approximately 20 dB up to 2GHz, falling somewhat as this level increases, though it is still approximately 13dB at 4GHz.

In this connection, see the measurement curves in Figs 8.55 and 8.56 They were measured before the divider IC was inserted, and show an outstandingly uniform amplification of 19 to 20dB over the range in question.

The high amplification rather balances out the decrease in sensitivity of the pre divider below 200 MHz, so that the latter can be used with good sensitivity even at 100 MHz (- 26 dBm).

This is followed by the divider IC, which needs practically no external circuitry.

Finally, a 74HC00 provides the correct TTL level that is required for further processing in the frequency counter.

Fig. 8.57 shows the output signal for an input signal of 3.6 GHz with a division of 256. The divider factor 64 can not be used below 4GHz, since the resultant output frequency is too high for the level conversion and also for the subsequent TTL stages (4,000 MHz: 64 = 62.5 MHz).

Construction

The small circuit is constructed on a 1.0 mm. thick, double sided printed circuit board made from epoxy material (Fig. 8.58), which naturally already represents a certain compromise for the high frequencies. Table 8.6 shows a list of components.

The feed throughs are manufactured first. The best thing to use for this is small compression rivets with a diameter of approximately 1 mm, which are well soldered to the top and bottom sides of the printed circuit board. But make sure the compression rivets do not fill up with solder! Special attention should be paid to the low inductance feed throughs on the divider IC. The feed through directly below the divider IC must be kept flat, so that it will not touch the IC when it is inserted.

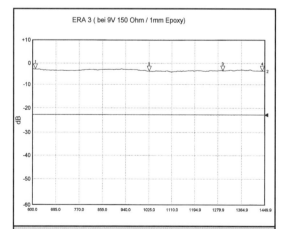

Fig 8.56 : Plot of ERA3 amplification from 600.0MHz to 1449.9MHz. Reference level -16dBm. Markers : 1: 609.2MHz/20dB; 2: 1025MHz/19db; 3: 1296.9MHz/19 dB; 4: 1439.5MHz/19dB.

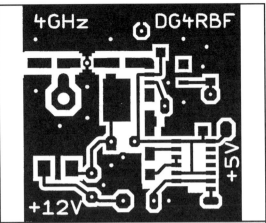

Fig 8.58 : Layout of pre divider printed circuit board.

The hole for the pre amplifier IC measures 2.2 mm. The MMIC (ERA 3) is inserted from the earth side, so that its earth connections can be soldered directly on the under-side of the printed circuit board. To this end, the input and output connection of the IC are bent up at 90 degrees and are fed through the hole for the MMIC and then soldered to the top side of the printed circuit board. If necessary, the hole must be slightly widened to do this.

The dot on the ERA 3 indicates the input. And make sure that the positive pole of the SMD electrolytic capacitor is also identified by a line or bar.

Two connection options have been provided for the high frequency input, so that the circuit can be directly sol-dered to the input jack of the frequency counter if required. Alternatively, it is possible to build the circuit into a little tinplate housing.

The components are soldered on the layout side of the printed circuit board, with the usual precautions against static charging.

The circuit has been laid out in such a way that there are

Table 8.6 : Component list for 4GHz pre divider6

SMD components:

R1	22 kΩ
R2	150Ω
R4	270Ω
R5	68 kΩ
C1, C2, C12	100 pF
C3, C7, C11	1 nF
C4	22 nF
C5, C8	100 nF
C6	1.5 nF
C9, C10	4.7 µF tantalum
C13	680 pF
IC1	µPB 1505
IC2	74 HC00
IC3	ERA 3

Wired components:

IC4	78 L09
IC5	78 L05
R3	68 kΩ

selectable options for the supply voltage.

The following types of supply are possible for the circuit.

Firstly, the circuit could be powered from the subsequent frequency counter with a stabilised voltage of + 5 V. In this case, the two fixed voltage regulators are not fitted, but are replaced by bridges. The protective resistor for the ERA 3 is calculated for a voltage of 5 Volts (Rbias for 5 V is about 39-43 Ohms).

The other option is to use an input voltage of 10 to 12 V. Here the two fixed voltage regulators (+ 9 V and + 5 V) are required. Now ther is the advantage that the resistor (Rbias) for the MMIC can be made larger so we can obtain approximately 2 dB more amplification (Rbias for 9V is approximately 150 Ohms).

In exactly the same way, we can also mount only the +9V voltage regulator which supplies the MMIC, the 5V section being supplied from the subsequent counter.

The ERA 3 is operated with a current of 35 to 38 mA. The 5V section of the circuit requires approximately 25 mA.

The circuit is calibration free and should function at first go.

Conclusion

Like many other pre dividers, this one also oscillates at the frequency of its maximum sensitivity if no input signal is present (approximately 2.5GHz). But since this oscilla-tion is unstable, a clear distinction should be made between it and a genuine input signal.

There is a course of action which can be helpful, but which can not be recommended unreservedly.

We can try to suppress the oscillation with a resistance of approximately 68 kOhms at the input of the divider (R3). However, a certain loss of sensitivity is also involved. The resistance must be individually matched, and should not be any lower than is absolutely necessary.

Since, in the upper frequency range in particular, this course of action leads to a detectable loss of sensitivity (from 3 GHz), it should be possible to switch off the resistance using a switch or a small relay.

The prototype is very sensitive, even in the upper frequency range.

Sensitivity of frequency divider with ERA 3 :

50MHz	- 19dB	1,800MHz	- 43dB
80MHz	- 23dB	2,000MHz	- 43dB
100MHz	- 26dB	2,200MHz	- 44dB
145MHz	- 30dB	2,400MHz	< - 50dB
200MHz	- 33dB	2,600MHz	< - 50dB
400MHz	- 37dB	2,800MHz	- 44dB
600MHz	- 41dB	3,000MHz	- 38dB
800MHz	- 42dB	3,200MHz	- 36dB
1,000MHz	- 43dB	3,400MHz	- 34dB
1,200MHz	- 43dB	3,600MHz	- 30dB
1,400MHz	- 44dB	4,000MHz	- 26

Signal sources and alignment aids

A crystal controlled frequency marker for 10GHz

A characteristic of operating equipment on 10GHz is that by the time one multiplies the uncertainties in frequencies

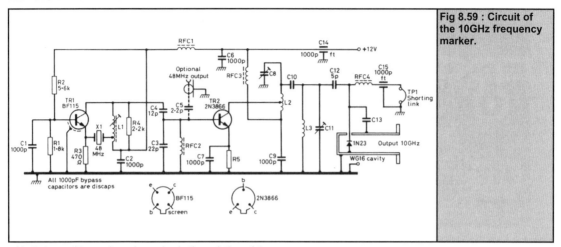

Fig 8.59 : Circuit of the 10GHz frequency marker.

of both transmitter and receiver, plus in the pointing of the highly directional antennas normally employed, the number of permutations that need to be covered before contact is established can be very considerable. The more skilled operators tend to be more successful mainly because they can set their frequencies more precisely and can direct their antennas more accurately and this can reduce very greatly the number of permutations that need to be covered.

When accurately set, the simple unit to be described generates a large number of signals at precisely known frequencies which can be used to calibrate receivers and transmitters accurately, thus effectively eliminating uncertainties from this source.

Although the output power of individual harmonics is very low, signals are detectable with efficient receivers even with 40 to 50dB of attenuation between the unit and the receiver. This means the unit can provide a rough check on the sensitivity of a receiver; signals should be detectable with even the most insensitive receiver. The output power is more than sufficient for use as a frequency reference for an afc system, but is insufficient for the unit to be employed as a signal source for tuning antennas. The range of the unit is only about 10 meters when antennas of 15dB gain are used.

Constructional details

The circuit diagram is shown in Fig 8.59, and the component layout in Fig 8.60. TR1 is a crystal oscillator on 48MHz, followed by a doubler to 96MHz and then a mixer diode used as the final multiplier. TR2 should be fitted with a small heatsink and its emitter lead kept as short as possible, typically about 3.3mm (0.125in). Its output is via a bandpass filter to reduce 48MHz feed through. This vhf circuitry is mounted on a piece of single sided printed circuit board which is bolted to the inside of the lid of a 111mm (4.375in) by 60.3mm (2.375in) diecast box.

The final multiplier consists of a length of WGI6 at least 60.3mm long which is closed at one end, with a 1N23 mixer diode mounted centrally 7.5mm (0.29in) from the closed end. The waveguide is clamped to the outside of the lid, connection to the diode being made using the inner of a Belling Lee socket which passes through holes drilled in the printed circuit board and the lid as shown in Fig 8.61. The component values for the 10GHz crystal controlled marker are given in Table 8.7.

In setting up the unit, L1 is adjusted to produce the maximum voltage across R5. C1 and C3 are adjusted and readjusted to maximise the diode current measured at TP1. The value of R is then changed as necessary to set

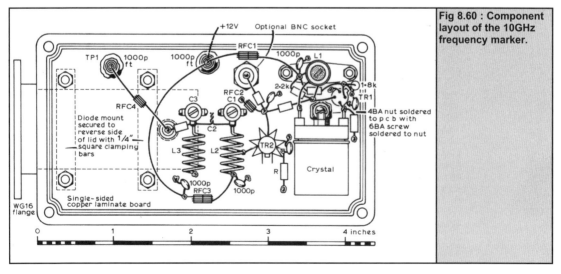

Fig 8.60 : Component layout of the 10GHz frequency marker.

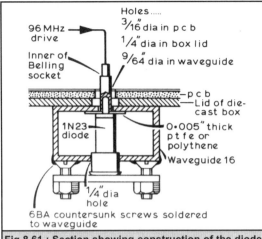

Fig 8.61 : Section showing construction of the diode multiplier signal source.

Labels in figure:
- 96 MHz drive
- Holes.....
- 3/16″ dia in p c b
- 1/4″ dia in box lid
- 9/64″ dia in waveguide
- Inner of Belling socket
- p c b
- Lid of die-cast box
- 1N23 diode
- 0·005″ thick p t f e or polythene
- Waveguide 16
- 1/4″ dia hole
- 6BA countersunk screws soldered to waveguide

this current to 20-25mA. Finally, the frequency of the crystal oscillator can be checked on a counter via the optional output connector or using, for example, a 144MHz receiver. Maximum output may not coincide exactly with maximum diode current.

Table 8.7 Component values for the 10GHz crystal controlled frequency marker.

R1	1.8kΩ
R2	5.6kΩ
R3	470Ω
R4	2.2kΩ
R5	68 - 100Ω, adjusted during alignment to set diode current to 20 - 25mA
C1,2,6,7,9	1000pF ceramic disc
C3	22pF ceramic disc
C4	12pF ceramic disc
C5	2.2pF ceramic disc
C8,11	35pF foil trimmer
C10	Two lengths of thin single strand insulated wire twisted together for 12mm
C12	4.7pF ceramic disc
C13	Formed by a 0.005in ptfe or polythene sheet between the end of the 1N23 diode and the waveguide wall
C14,15	1000pF feedthrough
RFC1,3	2.5 turns of 28swg ecw on two FX1115 ferrite beads
RFC2	As above, but one bead only
RFC4	10 turns 28swg ecw on one FX1115 ferrite bead
L1	10 turns of 28swg ecw on 8mm former tapped at one turn from C2 end
L2,3	5 turns of 18swg tinned copper wire 6mm inside diameter, 12mm long, L2 centre tapped
TR1	BF115
TR2	2N3866
X1	48MHz third overtone crystal
D1	1N23
WG16	50mm length of waveguide 16

Table 8.8 : Performance of the 10GHz crystal controlled frequency marker.

Voltage supply	12V
Current supply	25 - 40mA
Frequency pushing	8kHz/V
RF level of 96MHz harmonics in range 10.0 - 10.2GHz	-75dBm
RF level of 48MHz harmonics in range 10.0 - 10.2GHz	-95dBm

Instability may occur at maximum diode current, so it is worth checking with a 144MHz and 10GHz receiver that the unit is working correctly. Some useful performance parameters are given in Table 8.8. Using the frequency marker as described, the unit produces relatively strong signals at 96MHz spacings, with signals about 20dB weaker spaced every 48MHz. The choice of 48 and 96MHz as stage frequencies represents a compromise between generating a reasonable number of signals within the tuning range of most receivers while minimising the risk of confusion due to difficulty in identifying each of the harmonics. Other crystals may, of course, be used but there is an obvious advantage in using round number frequencies and especially those at such frequencies as 36, 54, and 72MHz which also produce harmonics at other amateur frequencies and are therefore more easily measured.

Note that no provision is made for modulating the output. This is unnecessary if the local oscillator of the receiver can be frequency modulated with a tone, which greatly assists finding signals. In calibrating a receiver it is essential that its local oscillator is already calibrated to within about 20MHz using a wavemeter. The self calibrating wavemeter described above is more than adequate for this purpose. The marker is preferably connected directly to the receiver via a variable attenuator, but alternatively it can be spaced from the receiver input by a few metres.

With 20-30dB attenuation, the receiver should detect weak signals which are harmonics of 96MHz only. If the attenuation is reduced to around 10dB, the 48MHz intermediate harmonics should be heard as weak signals, together with the now strong 96MHz harmonics.

It is possible more signals than expected will be heard as a result of the receiver having poor or even no image rejection. Thus a receiver with a 30MHz i.f. will respond to relatively strong harmonics at 9.984GHz (104 x 96MHz) when its local oscillator is tuned to either 9.954 or 10.014GHz, and to signals at 10.080GHz (105 x 96MHz) when tuned to either 10.050 or 10.110GHz. For the same reason, the receiver may also respond to the weaker harmonics at 48MHz spacing. The receiver local oscillator may therefore be calibrated precisely at several points, from which the (two) corresponding signal channels may be determined provided the i.f. is known accurately. If a larger number of calibration points is required, then either the i.f. may temporarily be changed, or a different crystal used in the marker.

It is possible to use the unit "in the field", since it is small enough to be waved about in front of the receiver prior to establishing a contact. Alternatively the unit could be built into the system, coupled to the receiver through a 10dB

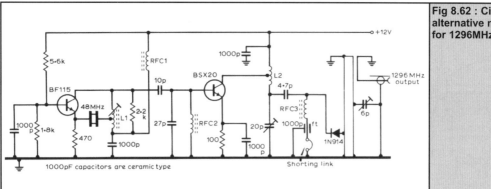

Fig 8.62 : Circuit of an alternative multiplier for 1296MHz.

directional coupler, to provide an instant check on its calibration and, in turn, that of the transmitter.

It is worth noting that the l0GHz multiplier stage can be replaced with one for any of the lower bands, hence producing a unit capable of providing calibration signals on any of the intermediate bands.

Crystal markers for other bands

It is possible to use the basic crystal oscillator and first multiplier stage to drive multipliers for other bands. Fig 8.62 shows an oscillator/multiplier and Fig 8.63 a diode multiplier suitable for 1296MHz (48 x 27). Although the output signal level is only of the order of a few micro watts, signals have been received, using high gain antennas, from a similar unit over a distance of 21km with a report of 569. A half wavelength line is tuned to the required harmonic with the 6pF trimmer. The multiplier diode is tapped onto the line 6mm from the ground end, where the impedance is low. Table 8.9 lists the components for the 1296MHz and also a 2320MHz multiplier.

Alignment is similar to that for the l0GHz multiplier. The shorting ink is removed and a milli ammeter inserted in its place. The 6pF trimmer is then adjusted for maximum current. If possible, monitor the output signal on a 1296MHz receiver to ensure the signal is clean. There

may be a tendency for the output spectrum to break up at maximum diode current, in which case it is best to reduce drive slightly by de tuning C11 by a small amount.

It is also possible to use the marker on 2320MHz by shortening the tuned line to the size shown in Fig 8.64. The strength of the output signal will be much less than that at 1296MHz and it may be necessary to change the multiplier diode to a Schottky barrier device such as the Hewlett Packard HP2800.

Table 8.9 : List of components for the 1296MHz and 2320MHz multipliers.

C15	1000pF feedthrough
C16	6pF ceramic tubular trimmer
D2	1N914, iN4148 for 2320MHz HP2800 may be better
RFC5	4 turns 22swg ecw, 3mm id closewound
L4,5	See illustration

Unless the 48MHz crystal is changed, the only useful harmonics will be at 2352, 2400 and 2448MHz. None of these are in the recognised narrow communications section of 2320 to 2322MHz. Changing the basic crystal frequency to 48.333MHz will give an output at 2320MHz that is likely to be much more useful.

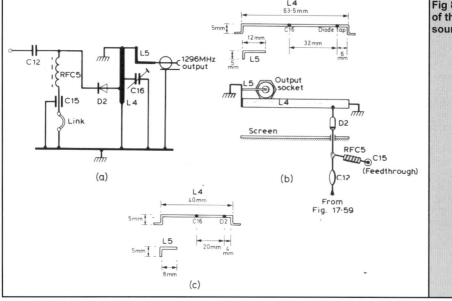

Fig 8.63 Construction of the 1296MHz signal source.

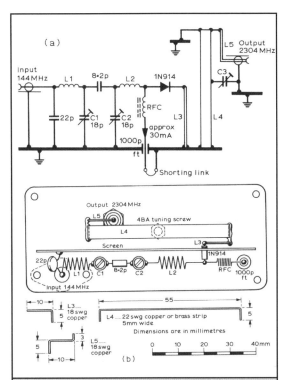

Fig 8.64 : (a) Layout of the 2320MHz signal source. (b) Dimensions of the tuned line and the coupling loop for use at 2320MHz

Receiver alignment aid

The limit to sensitivity in receivers for the higher bands is mainly due to the level of noise generated in the early stages of the receiver. This noise contribution can be minimised by the use of low noise components in well-designed circuits. Even the most professional design can produce indifferent results if it is not correctly aligned. By far the most common method of aligning amateur low noise receivers relies upon listening to a weak signal from a distant transmitter, such as a beacon, and adjusting the matching components of the receiver input stage for maximum signal to noise ratio.

Signals from distant sources are notoriously unreliable, varying rapidly in strength over a range of many dB. This makes it necessary to repeatedly check the strength of the beacon to ensure that an improvement in signal to noise ratio has been achieved.

A locally generated signal which can be adjusted in level down to barely detectable would appear to be ideal, since it would not suffer from the vagaries of propagation. In practice it can be very difficult to attenuate the test signal to the required level because of the amount of screening needed.

A second, and not often considered problem with this approach is matching between the source and the receiver. A well attenuated signal generator output will provide a good 50Ω match, whereas the antenna may not provide the same degree of matching. The result can be less than optimum.

A better approach to aligning low noise receivers is to use a noise generator in place of the signal generator [14].

With this technique, broadband noise is injected into the receiver input. The noise source is turned on and off and the receiver matching adjusted until the ratio of noise on to noise off at the receiver output is at a maximum. It can be very difficult to judge aurally when the ratio is maximum, so that some form of visual indicator becomes desirable.

The instrument described is known as an automatic noise figure meter when it indicates directly the true noise figure of the item under test. It is, however, necessary to use a source with an accurately known noise output in order to make an accurate measurement.

If the noise output of the source is not accurately known the instrument can still be used to adjust the receiver for best signal to noise performance, although actual noise figure will not be known.

The instrument described here can be used to adjust receivers operating at any frequency for optimum sensitivity. If the noise source is radiated into the receiver via the antenna, then the previously mentioned problem of matching is also solved.

The instrument provides a continuous readout of the difference between the audio output of a receiver with no rf input and the output when a wideband noise generator is connected to the receivers antenna socket. The meter indicates the ratio between the outputs under these two conditions.

By design, the meter reading is not affected by changes in audio level over a wide range of volume settings.

The meter has a logarithmic response so that the meter scale can be linearly calibrated in signal to noise ratio. Unless the absolute level of noise output from the noise source is known, the scale cannot be marked in noise figure.

Circuit description

The circuit is shown in Fig 8.65. Audio input from the loudspeaker socket of the receiver under test is connected to a small speaker at the instrument input. This speaker provides a means of monitoring the receiver output, which would otherwise be inaudible due to the muting action of most speaker circuits. A precision rectifier circuit using IC1 provides a dc output from ac inputs down to levels much less than that of conventional diode rectifier circuits. The gain of the circuit is given by:

$$G = \frac{R2}{R1 + RV1}$$

D2 and R3 prevent the operational amplifier saturating on negative half cycles of the input. R4 and C1 filter the input to IC2.

TR1 acts as a logarithmic feedback element around IC2. The voltage across the base/emitter junction of a transistor with its base connected to its collector is proportional to the logarithm of the current through the transistor.

The logarithmic amplifier is fed alternately with two voltages corresponding to the receiver noise output and the receiver noise plus signal output. The difference (in mV) between its output voltages under these two conditions is a function of the ratio between the two input voltages and this ratio is independent of the average input level.

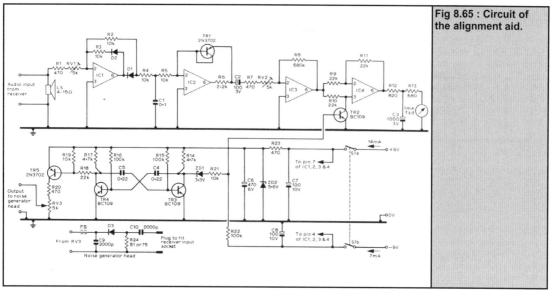

Fig 8.65 : Circuit of the alignment aid.

Provided the various stages of the receiver and the circuit around IC1 are working within their linear range, the ac output from the circuit around IC2 at the pulse frequency used will be dependant only on the overall signal to noise ratio. Since the ac output of this circuit is only a small fraction of a volt peak to peak, it is amplified by IC3. The gain of IC3 is given by:

$$G = \frac{R8}{R1 + RV2}$$

The output of IC3 is fed to a unity gain phase sensitive detector (PSD) circuit that uses the operational amplifier, IC4. The reference signal is provided by TR2 from the pulse generator TR3 and TR4. A psd is ideally suited to applications such as this, where an indication is required of the magnitude of an ac signal that has a known frequency and phase but a high accompanying noise level. In this application the psd gives a usable output when the signal is accompanied by so much noise that it is undetectable by ear.

Full scale deflection of the meter in the prototype was set at approximately I0dB signal to noise with the scale reading linearly in dB. IC4 has a low output impedance suitable for driving the 1mA meter. RI2, R 13 and C3 are chosen to give adequate damping for the meter that otherwise would have a very erratic response due to the nature of the noise inputs.

A conventional astable multivibrator operating at about 30Hz drives the generator on off switch, TR5, as well as providing the reference signal for the psd. The negative voltage supply is required to ensure that TR2 switches cleanly.

The noise generator uses a reverse biased diode mounted in a separate enclosure with matching and decoupling components. The switching voltage is fed to the noise head through a coaxial lead.

Construction

Construction of the alignment aid is not critical and audio

Fig 8.66 : Artwork for the alignment aid pcb.

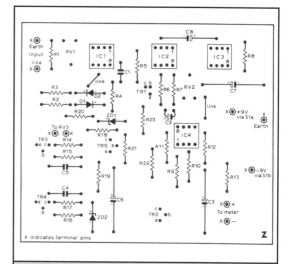

Fig 8.67 : Component layout for the alignment aid.

Table 8.10 : Component values for the alignment aid.

R1,7,20,23	470Ω
R2,3,4,5,19,21	10kΩ
R6	2.2kΩ
R8	680kΩ
R9,10,11,18	22kΩ
R12	820Ω
R13	680Ω
R14,17	4.7kΩ
R15,16,22	100kΩ
R24	51Ω or 75Ω
RV1,2	5kΩ skeleton preset, 0.1w Hor.
RV3	5kΩ carbon liner
C1	0.1µF polyester
C2	100µF tantalum bead 3v
C3	1000µF electrolytic 3v
C4,5	0.22µF polyester
C6	470µF electrolytic 6v
C7,8	100µF electrolytic 10v
C9,10	200pF ceramic disc
FB	FX1115 ferrite bead
ZD1	3.3v zener diode 400mW
ZD2	5.6v zener diode 400mW
D1,2	OA47,OA79,OA90
D3	see text
TR1,5	2N3705,2N3703,2N4126
TR2,3,4	BC109,2N2926
IC1,2,3,4	uA741 8pin DIL
LS1	4-15Ω miniature loudspeaker
S1	DPDT

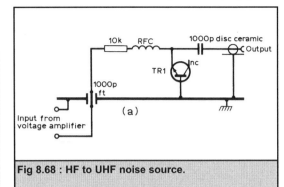

Fig 8.68 : HF to UHF noise source.

techniques can be used with the exception of the noise head, which must be built using vhf techniques if it is to operate reliably at the highest frequencies.

Artwork for a suitable printed circuit board for the main unit is shown in Fig 8.66. Veroboard construction would be equally acceptable if pcb production facilities are not available. The layout is given in Fig 8.67. See Table 8.10 for component values for the alignment aid.

Alignment

The unit requires little alignment and no test equipment is needed. Plug the noise head into a receiver and gradually increase the noise diode current until an audible purring sound is heard from the receiver's loudspeaker. Connect the audio output of the receiver to the input of the unit. The meter should now give a fairly steady reading, which can be varied by adjusting the noise diode current. Set RV1 so that the meter reading is constant over a wide range of receiver volume settings. RV2 is set to give full scale deflection of the meter at maximum diode current on the highest frequency band of interest.

Operation

Connect up the unit to the receiver under test and adjust RV3 for about half scale deflection on the meter. Any adjustment to the receiver that results in an improved signal gain with no change in the noise figure, or a reduced noise figure with no change in signal gain, or both simultaneously, will result in an increased meter reading. By noting the reading of the meter before and after any circuit adjustments, improvements in performance can readily be seen.

Although the unit is not especially sensitive to small

temperature changes, it is best to switch the unit on at least 10 minutes before use and ensure a constant ambient temperature.

Use of the alignment aid assumes reasonable linearity of the receiver, therefore care must be taken when aligning fm receivers to ensure that the receiver does not limit with the noise source on. With most receivers this will mean the level of noise injected must be as small as possible, consistent with still exceeding the fm threshold. In an ssb receiver the noise blanker and agc must be disabled if meaningful results are to be obtained.

Care must be taken if the aid is to be used for initial alignment of a converter or receiver. Noise output from the unit is constant over a wide range and it is therefore possible to inadvertently align on a spurious or the image frequency, especially if the receiver has a low intermediate frequency. A signal generator or off air signal should be used for initial alignment to avoid this problem.

Some receivers have been encountered that have a small dc voltage appearing at the loudspeaker socket. When connected to the alignment aid this voltage can bias IC1 input beyond its linear range, resulting in false readings on the meter. Connecting an electrolytic capacitor of about 47µF in series with the input cures this problem. The negative terminal of this capacitor should be connected to the junction of R1 and the monitor loudspeaker.

Considerable development work has been done to the alignment aid since it was first published and this has resulted in several very useful improvements [15] and [16].

The original noise head was designed primarily for vhf operation. An alternative design that can be used throughout the hf range and up to at least 1.3GHz is shown in Fig 8.68. Useful output may still be available at 2.3GHz when a suitable transistor is used in position TR1. It is best to select a transistor with a high ft for TR1. It may be necessary to try several before one with enough output is found. Table 8.11 gives component values.

A noise source suitable for 10GHz is shown in Fig 8.69. A 1N23 diode is mounted in a length of waveguide 16, with

Table 8.11 : Component values for the HF to UHF noise head.

R1	10kΩ
C1	1000pf ceramic disc
C2	1000pf feedthrough
RFC1	3 turns of R1 lead, 2mm inside diameter
TR1	2N2369,BFY90 etc. see text

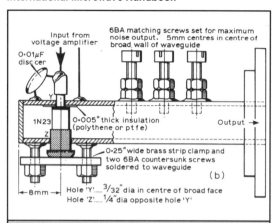

Fig 8.69 : 10GHz noise head.

matching screws to optimise output across the band. It is possible to scale this noise source for other. waveguide bands such as 5.7GHz and maybe even 24GHz, although the diode type will need to be chosen carefully for this latter band.

Both these noise sources require a higher bias voltage

Table 8.12 : Component values for the additional circuitry to provide a higher output voltage for the noise source.

R1	100Ω
R2,3,4,10	10kΩ
R5	100kΩ
R6	8Ω
R7	470Ω
R8,9	3MΩ
R11	4.7MΩ

All resistors 0.25W miniature carbon or metal film

RV1	10kΩ 10 turn preset potentiometer
C1	220µF electrolytic 10v
C2,4	0.1µF ceramic disc
C3	1000µF electrolytic 10v
IC1	LM324
M1	3mA meter
SK1	Loudspeaker break jack
SK2	Input socket to suit

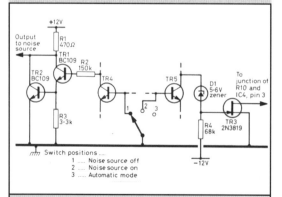

Fig 8.70 : Additional circuitry to provide higher output voltage for the noise source.

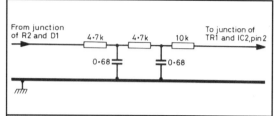

Fig 8.71 : An improved interstage coupling network between IC1 and IC2.

than the original aid is able to provide. An alternative output stage that is able to provide this higher voltage is shown in Fig 8.70. The component values are shown in Table 8.12. A switch has been provided to allow the noise source to be turned on or off manually.

Better phase detector performance is achieved at low levels with a fet in place of the bipolar transistor TR2 in the original circuit. Sometimes difficulties have been encountered with the meter reading not being independent of audio drive level. This can be just a matter of incorrect use, or it can arise when the alignment aid is used in conjunction with receivers possessing an odd audio frequency response. This can be cured by changing the interstage coupling components R4, R5 and C1 to the values shown in Fig 8.71.

A slight improvement to the performance of the logarithmic amplifier may be obtained by replacing R6 with a short circuit and increasing the value of C2 to 25µF. Fluctuating meter readings can also be a problem at times. Changing the meter to one of 50µA sensitivity and altering the time constant of the meter circuit can noticeably improve matters. Fig 8.72 incorporates these modifications as well as including a switch to give different full scale readings of signal to noise ratio.

Logarithmic amplifier up to 500MHz with AD8307 [17]

As a supplement to a good stock of measuring equipment, a logarithmic amplifier is described below with a useful dynamic range of 90 dB. Suitable outputs are provided so that the reading can be displayed, for example, on an oscilloscope

The AD 8307 logarithmic amplifier from Analog Devices [18] represents an almost ideal component, as a detector with dB linear output. A glance at the specimen circuit shows how few components are required (Fig. 8.73).

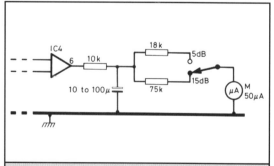

Fig 8.72 : Modified meter circuit.

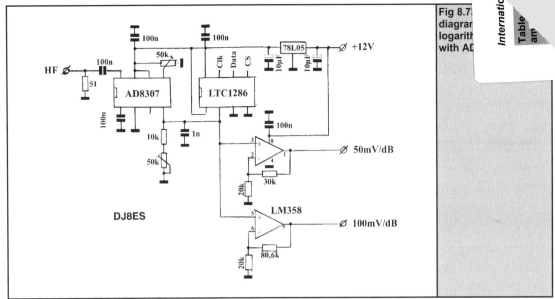

Fig 8.7
diagram
logarith
with AD

Circuit description

The core of the circuit is the AD 8307 logarithmic amplifier from Analog Devices. In the data sheet, the IC is specified as up to 500 MHz, with a dynamic range of approximately 90 dB. This is valid for input levels of +17 dBm to -75 dBm. The AD8307 can even be used (with restricted accuracy) up to 900 MHz and beyond.

As the frequency increases, the display voltage decreases slightly. The associated characteristics are shown as the frequency response of the AD8307 in Fig. 8.74.

The dB linear output (pin 4) has a gradient of 20 mV/dB in the given circuit and is high impedance. A type LM358 double op amp. is downstream as a buffer stage. Thanks to the two different feedback resistances, this makes useable outputs available with gradients of 50 mV/dB or 100 mV/dB.

The linearity and the offset (in relation to the input power) are set with two 50kΩ spindle trimmers. Table 8.13 shows the output voltage, plotted against the input power following balancing.

As an extension, space was provided on the printed circuit board (Fig 8.75 and Fig 8.76) for a 12 bit LTC1286 A/D converter manufactured by Linear Technologies [19].

This makes the signal to be measured available for further processing using a micro controller or PC.

Putting into operation

Feed in the + 12V DC power supply and an input signal of 0dBm (1 mW). You should already be able to obtain a reading from a measuring instrument connected to the two outputs.

Use the 50kΩ trimmer (through 10kOhms to pin 4 of AD8307) to set the linearity of the entire system.

The voltage differential between any test signal and the signal reduced by 10 dB must be exactly 1V. The simplest way of balancing this is with a switchable attenuator. During the balancing procedure, the absolute value of the measurement voltage displayed is of no importance. The only thing relevant is the differential of 1V to be established between the two measurement levels.

Finally, adjust the input signal offset. To do this, feed the test signal into the input at a level of 0 dBm and use the 50kOhm trimmer at pin 5 of AD8307 to set 100 mV/dB output to precisely 9 V.

That completes balancing. The logarithmic detector now measures outputs between + 10 dBm and - 80 dBm in accordance with Table 8.13. A reading of + 10 dBm (10

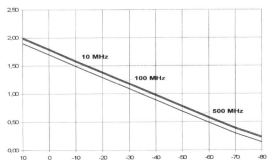

Fig 8.74 : Frequency response for AD8307. Display voltage v Input dB.

Table 8.13 Output voltage plotted against input power.

Input (dbm)	20mV/dB	50mV/dB	100mv/dB
10	2.00	5.01	10.01
0	1.80	4.50	9.00
-10	1.59	3.99	7.97
-20	1.39	3.48	6.96
-30	1.19	2.99	5.97
-40	0.99	2.48	4.96
-50	0.79	1.98	3.95
-60	0.59	1.49	2.97
-70	0.40	1.00	2.00
-80	0.24	0.60	1.20

8.14 Component list for AD8307 logarithmic amplifier.

1 x DJ8ES 047 printed circuit board

1 x tinplate housing 55.5 mm. x 74 mm. x 30 mm.

1 x AD8307 logarithmic amplifier, DIL 8-pole

1 x LM 358 operational amplifier, DIL 8-pole

1 x LTC 1286 A/D converter, DIL (optional)

1 x 78L05 fixed voltage regulator

2 x 10 µF/35 V, tantalum electrolytic capacitor

2 x 50-kOhm vertical spindle trimmer

1 x 1-nF Df capacitor, solderable

1 x 4 mm. soldering lug, turned down

5 x Teflon bushings

5 x 100 nF, RM 2.5 mm., ceramic

1 x 1 nF, RM 2.5 mm., ceramic

¼ W resistances, RM 10 mm.:

1 x 51 Ohms

1 x 10 kOhms

2 x 20 kOhms

1 x 30 kOhms

1 x 80.6 kOhms

mW) thus corresponds to a voltage of 10 V at the 100 mV/dB output or, to put it another way: 1 V per 10 dB.

References

[1] A high quality UHF source for microwave applications, RSGB Microwave Committee, Radio Communication, October 1981.

[2] A broadband linear rf power meter, J Gannaway G3YGF, Microwave Newsletter, 3/8/81.

[3] Amateur Radio Software, J Morris GM4AN, RSGB. Publications.

[4] Attenuators and dummy loads for 24GHz, Microwave Newsletter, 7/2/84.

[5] Lance Wilson WB6XQF, Ham Radio magazine, 9/76.

[6] Notes on using directional couplers, Microwave Newsletter, 3/5/82.

[7] A directional coupler, Microwave Newsletter, 4/4/82.

[8] High power directional coupler for 1296MHz, Microwave Newsletter, 3/2/82.

[9] Directional coupler for 2320MHz, Microwave Newsletter, 6/2/85.

[10] From article by Gregor Storz, ZL1GSG, DL2GSG, VHF Communications 1/98 pp 2 - 9

[11] Harlan Howe, Jr. Stripline Circuit Design (1974), Artech House

[12] Wide range UHF cavity wavemeter, Test Equipment, VHF/UHF Manual.

[13] From article by Bernd Kaa, DG4RBF, VHF Communications 1/2000 pp 2 - 9

[14] An alignment aid for VHF receivers, I R Compton G4COM, Radio Communication, January 1976.

[15] Microwaves, C Suckling G3WDG, Radio Communication, October 1979.

[16] Microwaves, C Suckling G3WDG, Radio Communication, March 1980.

[17] From article by Wolfgang Schneider, DJ8ES, VHF Communications 2/2000 pp 119 - 124

[18] Data sheet AD8307, Analog Devices, Inc., 1997

[19] Data sheet LTC 1286, Linear Technology Corporation, 1994

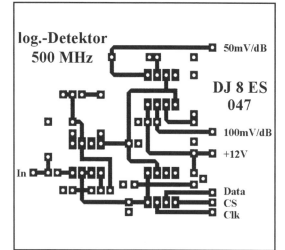

Fig 8.75 : PCB layout for AD8307 logarithmic amplifier.

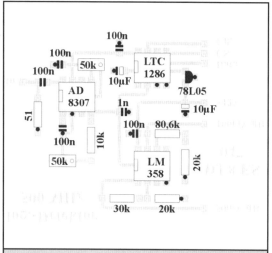

Fig 8.76 : Component layout for AD8307 logarithmic amplifier.

Chapter 9

The 1.3GHz (23cm) band

In this chapter :

- Preamplifier
- Transceiver
- Transverter

- Amplifiers
- Antennas
- Filters

T he 1.3GHz band is of particular interest because it represents a transition between VHF/UHF and microwaves, both in terms of equipment design and operating practices. At this frequency, any of the critical components are becoming a significant fraction of a wavelength in size, which means that the techniques involved tend to be those more typical of microwave practice. On the other hand most stages of the equipment, other than those at the final frequency, are at VHF or lower frequencies. In one sense, therefore, the techniques employed are only one step in advance of what many would regard as conventional radio; this forms part of the appeal of experimentation in this region of the radio spectrum.

The relatively short wavelength also has a powerful influence on antenna design, in that it is easy to make antennas that have high gain but which are physically much smaller than lower frequency antennas. It is also the lowest frequency band where dishes are used in ordinary operation.

Amateurs first made regular use of the 1.3GHz band in the late 1950s. However, the equipment in use at that time was not particularly powerful or effective. A typical system might have employed low gain antennas such as corner reflectors or small dishes. The receiver could well have had a noise figure as high as 15-20dB and the transmitter output power was generally limited to a few watts. Unless operated from particularly favourable sites, the overall performance of this kind of equipment was such that it limited the range under normal conditions to tens of kilometres only, which gave the false impression that the band was suitable only for local contact.

In the 1970s, a number of technical developments transformed attitudes to operating at this frequency. In particular, receivers improved as better converter designs became available, followed by the widespread adoption of preamplifiers. Transmitter powers increased and some good Yagi designs were developed, making high gain antennas practicable for all stations.

An important feature of these developments was that reasonable results could be obtained even from ordinary locations. As a consequence, operation on this band is now very well established. The ranges under normal propagation conditions are typically up to several hundreds of kilometres; during enhanced conditions, contacts over 1,000km are now quite common on this band.

More recently, there have been a number of interesting developments that will no doubt result in yet more activity on this band. These include:

Television

The 1.3GHz band is now the lowest frequency where there is sufficient spectrum space allocated to amateurs to permit full definition fast scan tv operation, without in band interference problems. In particular, the wider bandwidth of fm tv can easily be accommodated. As a consequence, this technique is becoming more popular as it simplifies the construction of equipment to some extent.

Repeaters

These are currently being established for single channel speech. TV repeaters are also in use in some countries, including the UK. Although the design of the RF sections of microwave repeaters is different to that of their lower frequency counterparts, the general design principles are not greatly different.

High speed packet radio links

A more recent use of specified frequencies within the band is high speed packet radio links between user access nodes operating in other (often lower) frequency bands, effectively forming cross band digital repeaters.

Moonbounce

With the improved techniques now available, it is possible to communicate world wide on 1.3GHz via moonbounce (eme), using antennas which are small enough to be fitted into most back gardens. The equipment required is within the capabilities of many stations.

Satellites

With the advent of the AMSAT satellites, a new type of transponder has become available, known as Mode L. The uplink and downlink frequencies for these transponders are 1,268MHz and 436MHz respectively and the transponder bandwidth is 800kHz. The wider bandwidth provided by this type of transponder (approximately five times that of lower frequency transponders) can support much higher activity levels.

Allocation to the amateur services

Most countries, world wide, have an allocation in the range of 1,240-1,300MHz. The primary use is for radio location and radio navigation satellite, with the amateur service designated as a secondary user. In the UK, the frequencies 1,300-1,325MHz are also available for use by amateurs on a secondary basis. Despite the use of relatively powerful equipment by both primary and sec-

1.3GHz (23cm)	Licence	Amateur Service: Secondary.

Notes: Amateur Satellite Service: 1260 - 1270MHz Secondary *Earth to space only*.
Power limit: 26dBW PEP.
Permitted modes: Morse, telephony, RTTY, data, fax, SSTV, FSTV
Unattended operation: Not permitted in Northern Ireland.
In the sub-based 1298 - 1300MHz, unattended operation is not allowed in Nothern Ireland or within 50km of SS206127 (Bude) or SE302577 (Harrogate)

IARU	Novice	U/A Rem Ctrl	U/A Digital	U/A Beacon	UK Usage	
1.240.000						
All modes					1240.150	*Packet radio (150kHz b/w)*
					1240.300	*Packet radio (150kHz b/w)*
					1240.450	*Packet radio (150kHz b/w)*
					1240.600	*Packet radio (150kHz b/w)*
					1240.750	*Packet radio (150kHz b/w)*
1.243.250						
ATV					1248.000	*RT1-3 FM TV repeater input*
					1249.000	*RT1-2 FM TV repeater input*
1.260.000						
Satellites						
1.270.000						
All modes						
1.272.000						
ATV					1276.500	*RT1-1 AM TV input*
1.291.000						
Repeater inputs					1291.000	*RM0*
						(UK) 25kHz spacing
					1291.375	*RM15*
1.291.500						
All modes						
1.296.000						
CW only					1296.000 - 1296.025	*Moonbounce*
1.296.150	✓					
SSB and CW					1296.200	*Narrow band centre of activity*
					1296.400 - 1296.600	*Linear transponder input*
					1296.500	*SSTV*
					1296.600	*RTTY*
					1296.700	*Fax*
					1296.600 - 1296.800	*Linear transponder output*
1.296.800						
Beacons exclusive					1296.800 - 1296.990	*Beacons*
1.297.000						
Repeater outputs - note 1					1297.000	*RM0*
						(UK)25kHz spacing
					1297.375	*RM15*
1.297.500						
FM simplex - note 1					1297.500	*SM20*
					1297.750	*SM30*
1.298.000						
All modes	✓	✓				*Remote control*
						Digital communications
1.299.000						
All modes			✓		1299.000	*Packet radio (25kHz b/w)*
					1299.425	*Packet radio (150kHz b/w)*
					1299.575	*Packet radio (150kHz b/w)*
					1299.725	*Packet radio (150kHz b/w)*
1.300.000						
TV repeater outputs (UK only)					1308.000	*FM TV repeater output*
					1310.000	*FM TV repeater output*
					1311.500	*FM TV repeater output*
					1312.000	*FM TV repeater output*
					1316.000	*FM TV repeater output*
1.325.000						

Notes:
1. Local traffic using narrow-band modes should operate between 1296.000 - 1296.800MHz during contests and band openings.
2. Stations in countries which do not have access to 1298 - 1300MHz (eg Italy) may also use the FM simplex segment for digital communications.

Fig 9.1 : UK bandplan for 23cm.

Frequency	USA Usage
1240-1246	ATV #1
1246-1248	Narrow-bandwidth FM point-to-point links and digital, duplex with 1258-1260.
1248-1258	Digital Communications
1252-1258	ATV #2
1258-1260	Narrow-bandwidth FM point-to-point links digital, duplexed with 1246-1252
1260-1270	Satellite uplinks, reference WARC '79
1260-1270	Wide-bandwidth experimental, simplex ATV
1270-1276	Repeater inputs, FM and linear, paired with 1282-1288, 239 pairs every 25 kHz, e.g. 1270.025, .050, etc.
1271-1283	Non-coordinated test pair
1276-1282	ATV #3
1282-1288	Repeater outputs, paired with
1288-1294	Wide-bandwidth experimental, simple ATV
1294-1295	Narrow-bandwidth FM simplex services, 25-kHz channels
1294.5	National FM simplex calling frequency
1295-1297	Narrow bandwidth weak-signal communications (no FM)
1295.0-1295.8	SSTV, FAX, ACSSB, experimental
1295.8-1296.0	Reserved for EME, CW expansion
1296.00-1296.05	EME-exclusive
1296.07-1296.08	CW beacons
1296.1	CW, SSB calling frequency
1296.4-1296.6	Crossband linear translator input
1296.6-1296.8	Crossband linear translator output
1296.8-1297.0	Experimental beacons (exclusive)
1297-1300	Digital Communications

Fig 9.2 : USA bandplan for 23cm.

ondary users, in general there has, until recently, been little, if any, mutual interference in most peoples experience, although the introduction of high power radars is starting to cause some problems to the amateur service at the present time in the UK and near continent.

Band planning

The current UK band plan is shown in Fig 9.1 and the USA band plan is shown in Fig 9.2.

Propagation and equipment performance

Most of the propagation modes encountered on the VHF/UHF bands are also present at 1.3GHz, with the exception of those which require ionised media, such as aurora, meteor scatter and sporadic E. The most commonly encountered propagation modes at 1.3GHz are line of sight, diffraction, tropospheric scatter, ducting, aircraft scatter and moonbounce. Futher information on some of these modes can be found in chapters 1 and 2.

Using the techniques described in these chapters, it is possible to predict whether communication is possible over a given path, when the performance parameters of the equipment are known. Fig 9.3 shows the path loss capability of five different sets of equipment, varying size from a relatively modest station, such as a commercial transverter feeding a single Yagi antenna, to a station with

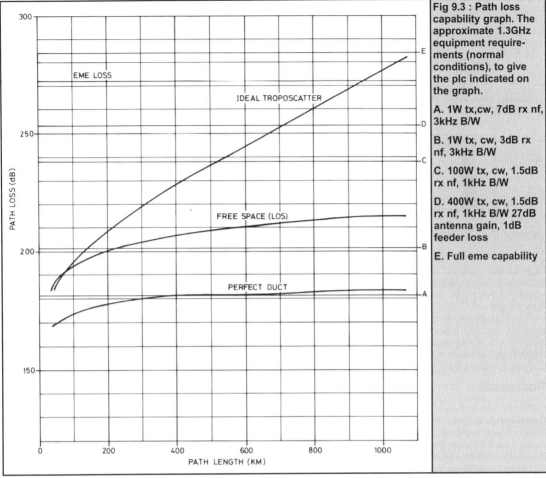

Fig 9.3 : Path loss capability graph. The approximate 1.3GHz equipment require-ments (normal conditions), to give the plc indicated on the graph.

A. 1W tx,cw, 7dB rx nf, 3kHz B/W

B. 1W tx, cw, 3dB rx nf, 3kHz B/W

C. 100W tx, cw, 1.5dB rx nf, 1kHz B/W

D. 400W tx, cw, 1.5dB rx nf, 1kHz B/W 27dB antenna gain, 1dB feeder loss

E. Full eme capability

full moonbounce capability. Also shown are the path losses of three types of propagation, line of sight (in general the lowest loss mode of propagation), tropo-spheric scatter (a reliable terrestrial mode of propagation) and moonbounce. It can be seen that even the modest equipment is capable of working very long line of sight paths easily (the signal to noise ratio can be estimated by subtracting the path loss from the path loss capability of the equipment). In practice, paths are not line of sight and propagation losses are higher. The tropospheric scatter loss is a reasonable upper limit for a terrestrial path and can be used to estimate the maximum range of the equipment under normal propagation conditions. It can be seen from the graphs that the modest equipment might be expected to work paths up to about 150km in length.

More powerful equipment will cover longer paths and equipment C, is typical of that used by well equipped stations, could be expected to work paths up to 500km on ssb, or 700km on cw. The limit for terrestrial communica-tion under normal conditions is about 1,000km, beyond which it becomes easier to use moonbounce. Of course, under "lift" conditions path losses are considerably re-duced, often to values near to line of sight and even modest equipment is then capable of working very long paths.

Preamplifiers

Even with modern receivers and transceivers there is still

a requirement for a low noise preamplifier which may be mounted as close to the antenna as possible.

A high performance GaAs fet preamplifier for 1.3GHz

Modern GaAs fets are capable of producing very low noise figures at 1.3GHz and preamplifiers using these transistors can give excellent performance. Indeed, the pick up of noise from the earth can almost be said to be the major factor limiting receiver performance when GaAs fet preamplifiers are used. However, in order to realise the full potential of these devices it is essential that the input circuit of a preamplifier has a very low loss, or the added noise may degrade the performance unacceptably. Also, any practical design must allow for the fact that these devices are potentially unstable at 1.3GHz and steps should be taken to ensure stability. The preamplifier described below was designed to take these features into account while being relatively easy to construct.

The circuit diagram of the preamplifier is shown in Fig 9.4 and component list in Table 9.1 The low loss input circuit consists of L1, C1 and C2. Source bias is used so that the preamplifier can be run off a single positive rail supply. The two source decoupling capacitors are leadless types, to ensure that the source of the GaAs fet is well grounded at RF. The output circuit is untuned and consists of R2 and C5. This configuration ensures that the amplifier is

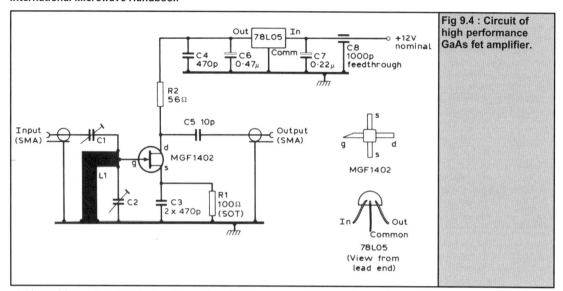

Fig 9.4 : Circuit of high performance GaAs fet amplifier.

stable and has a low output vswr. The value of C5 was chosen so that the capacitor is series resonant, i.e. it has a very low series impedance at 1.3GHz. A three terminal voltage regulator is included within the preamplifier housing. This not only provides the 5V supply required, but also affords some degree of protection from voltage "spikes" on the power supply line, which could otherwise damage the GaAs fet.

Construction

Constructional details of the preamplifier are shown in Figs 9.5(a) to 9.5(g). Firstly cut out a sheet of copper for the main box, mark out the positions for the holes, make scribe lines for the corners and to indicate the position of the centre screen. Cut out the corner pieces (preferably by sawing) and drill all holes with the exception of holes E, and tap hole D. Clean the sheet using a printed circuit rubber or 'Brasso' taking care to remove any residue of the cleaning agent at the end. Bend up all four walls and adjust the corners using pliers to ensure that the walls touch. The centre screen should be fabricated next. It is best to make this piece slightly oversize initially and then to file it to be a tight fit in the main box. The screen should be cleaned after drilling and deburring the hole. The lead less capacitors are then soldered to the screen, in the positions shown in Fig 9.5(c). The soldering is best done by clamping the sheet in a horizontal plane and applying a small flame from underneath. Tin each capacitor on one side using a soldering iron and small amount of solder and tin the screen in three places, again using only a

small amount of solder. Place the capacitors on the screen, tinned sides down, and reheat the screen under each capacitor in turn.

As soon as the solder melts, press down on the capacitor until it is flat on the screen. Using an ohm meter, check that the capacitors are not shorted to the screen. Remove the excess solder by filing if shorts are found. Fabricate the retaining plate as shown in Fig 9.5(d) and place it on to the screen so that the bumps align with the capacitors. Fix it in position using an M2 screw and nut.

Locate the screen in the box using the scribe marks as a guide to correct alignment. Measure the distance between the screen and the inside of the end wall of the larger compartment and check that this agrees with the dimension given in Fig 9.5(a). Move the screen if necessary. Clamp the screen in position using a toolmaker's clamp applied to the sidewalls of the main box around the screen. Jig a clean brass nut in position at hole D using a stainless steel or rusty screw to hold it in place. Tighten the nut slightly against the wall of the box. Mount the box in a vice so that the junctions of the sidewalls are horizontal and solder along the junctions using a small flame from underneath each corner in turn, re positioning the box each time. It is easiest to preheat the corner using the flame and to make the joints with a soldering iron. This reduces the chance of re melting the joints already made. The screen should be soldered next, along all three sides using the same technique, with the box mounted so that the open side is uppermost. Finally, invert the box and solder the nut in position.

When the assembly is cool, remove the jigging screw and run a tap through the nut and sidewall. Remove the tap and check that the tuning screw runs freely in and out. If necessary, file away any excess solder around the nut so that the output connector fits into position correctly. Drill the fixing holes for the input and output connectors. The lid is made next. Cut out the material for the lid and place the box symmetrically on the lid. Scribe around the outside of the box on the lid, cut out the corners and bend the lid over the box to form a tight fit. Using a 1.6mm bit, drill through the sidewalls of the lid and box (holes E). Tap the holes in the box and open out the holes in the lid to 2.1mm. The next stage in the construction is to make the input line, details of which are given in Fig 9.5(e). The line

Table 9.1 Component list for high performance GaAs fet amplifier.

L1	Stripline inductor, see fig 9.5e
C1	Input capacitor, see fig 9.5f
C2	Tuning capacitor (0BA or M6 screw)
C3	2 x 470pF trapezoidal leadless disc capacitor
C4	470pF trapezoidal leadless capacitor
C5	10pF miniature ceramic plate capacitor
C6	0.47µF tantalum, 16v
C7	0.22µF tantalum 16v
C8	1000pF bolt in feedthrough capacitor
R1	100Ω (select on test)
R2	56Ω

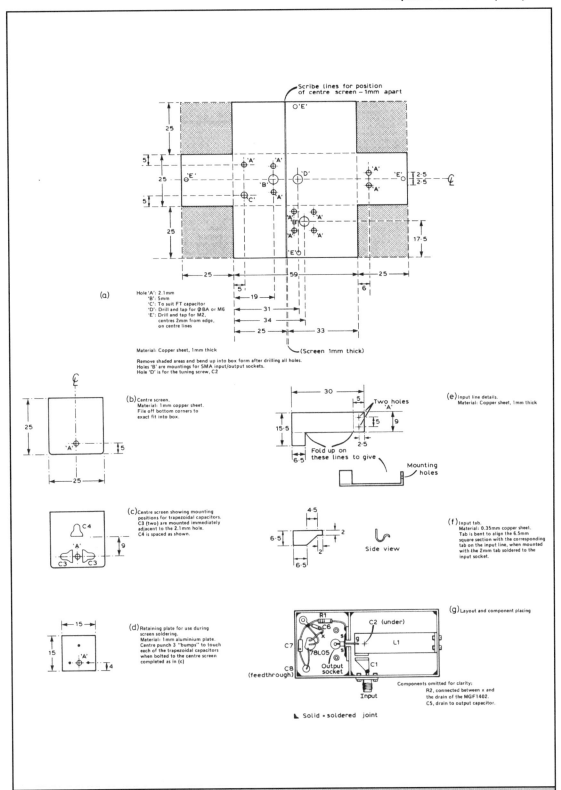

Fig 9.5 (a) Details of the main box for the GaAs fet amplifier. (b) Centre screen. (c) Centre screen showing mounting positions for trapezoidal capacitors. (d) Retaining plate for use during screen soldering. (e) Input line detail. (f) Input tab, material 0.35mm copper sheet, (g) Layout and component placing.

Table 9.2 : GaAs fet preamplifier performance.

Frequency (MHz)	Gain (dB)	Noise Figure (dB)
1,235	15.6	1.04
1,245	16.4	0.90
1,255	16.6	0.76
1,265	16.4	0.65
1,275	16.1	0.58
1,285	15.6	0.56
1,295	14.8	0.55
1,305	13.9	0.56
1,315	12.9	0.58
1,325	12.2	0.62

should be cleaned before bending up the tabs. The overall length of the line is quite critical and since errors can occur during bending, it is best to make the line slightly longer initially and then file off excess material at the "input" end of the line. Re check all dimensions of the line using vernier calipers before proceeding. The line can then be fitted into the box, using M2 fixings.

Measure the distance between the top surface of the line and the bottom of the box, at the fixing end. This should be 4mm (±0. 1 mm). If necessary, remove the line and file the holes in the box so that this dimension can be achieved. Similarly. measure the height of the input end and bend the line up or down as necessary. The input tab is made next and should be bent so that when located on the input connector, the sides and top of the tab align with the tab on the input line. The tab can then be soldered to the input connector, leaving a 0.5 - 1mm gap between the tabs. The assembly is completed by fitting the components. The layout is shown in Fig 9.5(g). After mounting the grounding tag, solder R1 into position. When making the joint to the source decoupling capacitor, do not allow solder to run along the capacitor or it will be difficult to mount the GaAs fet correctly. Cut the leads of C5 to 2mm length and solder one end to the output connector. Bend the leads of C5 so that the free lead is in line with the hole in the screen.

Mount the feed through capacitor and connect up the voltage regulator circuitry. Apply power and check that +5V is available at the output of the regulator (connect a lk resistor between the output of the regulator and ground during this measurement). Cut the source leads of the GaAs fet to 2.5mm length. Using tweezers twist the drain lead through 90° and, holding the end of the drain lead, mount the GaAs fet into position, with its source leads touching the bypass capacitors. Unplug the soldering iron and solder the drain lead to the free end of C5. Check that the gate lead passes centrally through the hole in the screen and adjust the position of the device by careful bending if necessary. The source leads of the GaAs fet are soldered next. Again, unplug the iron just before making the joints. Soldering the gate lead requires some care.

Firstly, slightly loosen the fixings of the input line, but not enough to allow the line to move. Using a minimum of a 45W iron, heat the end of the line at its edge until the solder flows. The body of the iron should be connected via a flying lead to the box during this operation. Move the iron to the centre of the line next to the gate lead. Then solder the gate lead to the line. Re tighten the fixings of the input line. The final operation is to fit R2. Cut one lead

to 2mm length and tin. With the iron unpluged, solder the other end of R2 to C4 such that the free end is touching the drain lead of the GaAs fet about 1 mm from the package. Then solder R2 to the drain of the GaAs fet.

Adjustment

With the preamplifier connected in circuit, apply power and check the current drawn. This should lie somewhere between 10 and 15mA. If not, the value of RI will need to be altered. Next, fit the tuning screw and adjust for maximum noise. Tune in a weak signal and adjust the position of the input tab and the tuning screw for optimum signal to noise ratio. Alternatively, an automatic noise figure alignment aid can be used (see chapter 8, "Test equipment"). The performance of the preamplifier is shown in Table 9.2.

For applications where noise figure is at a premium, the MGFI402 device specified can be replaced by the MGFI412-09. This device will yield a lower noise figure, by approximately 0.2dB.

Transceivers

Microwave SSB transceiver design [1]

When discussing SSB transceivers, the first question to be answered is probably the following - does it make sense to develop and build new SSB radios? Today SSB transceivers are mass produced items for frequencies below 30MHz. There is much less choice on the market for 144MHz or 432MHz SSB transceivers and there are just a few products available for 1296MHz or even higher frequencies.

Most radio amateurs are therefore using a base SSB transceiver (usually a commercial product) operating on a lower frequency and suitable receive and transmit converters or transverters to operate on 1296MHz or higher frequencies. The most popular base transceiver is certainly the good old IC202. All narrow band (SSB/CW) microwave activity is therefore concentrated in the first 200kHz of amateur microwave segments like 1296.000-1 296.200, 2304.000-2304.200 etc due to the limited frequency coverage of the IC202.

Transverters should always be considered a poor technical solution for many reasons. Receive converters usually degrade the dynamic range of the receiver while transmit converters dissipate most of the RF power generated in the base SSB transceiver. Both receive and transmit converters generate a number of spurious mixing products that are very difficult to filter out due to the harmonic relationships among the amateur frequency bands 144/43 2/1296...

However, the worst problem of most transverters is the breakthrough of strong signals in or out of the base transceiver intermediate frequency band. This problem seems to be worst when using a 144MHz first IF. Strong 144MHz stations with big antenna arrays may break in the first IF even at distances of 50 or 100km. Since the problem is reciprocal, a careless microwave operator may even establish two-way contacts on 144MHz although using a transverter and antenna for 1296MHz or higher frequencies.

Some microwave operators solved the above problem by installing a different crystal in the transverter, so that for example 1296.000MHz is converted to a less used

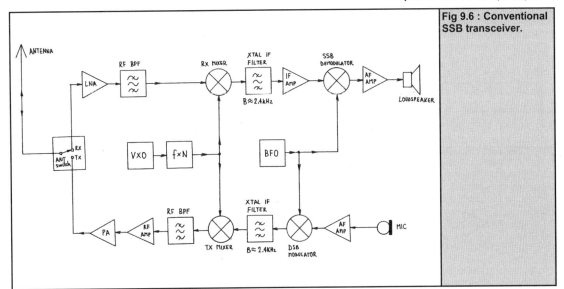

Fig 9.6 : Conventional SSB transceiver.

segment around 144.700MHz. Serious microwave contesters use transverters with a first IF of 28MHz, 50MHz or even 70MHz to avoid this problem. Neither solution is cheap. The biggest problem is carrying a large 144MHz or HF all mode transceiver together with a suitable power supply on a mountaintop.

Even the good old IC202 has its own problems. This radio has not been manufactured for more than a decade. New radios can not be purchased while the maintenance of the old ones is becoming difficult. Second hand radios are usually found in very poor conditions due to the many modifications and improvements made by their previous owners.

As a conclusion, today it still makes sense to develop and build SSB radios for 1296MHz and higher frequencies. Since the problems of the transverters are well known and are not really new, many technical solutions have already been considered by different designers. Most solutions were discarded simply because they are, too complex, too expensive and too difficult to build, even when compared to the already complex combination of a base transceiver and transverter.

Most commercial SSB transceivers include a modulator and a demodulator operating on a high IF, as shown on Fig 9.6. The resulting SSB signal is converted to the RF operating frequency in the transmitter and back to the IF in the receiver. Both the transmitter and the receiver use expensive components like crystal filters. Besides crystal filters, additional filtering is required in the RF section to attenuate image responses and spurious products of both receiving and transmitting mixers.

The design of conventional (high IF) SSB transceivers dates back to the valve age, when active components (valves) were expensive and unreliable. Passive components like filters were not so critical. Complicated tuning procedures only represented a small fraction of the overall cost of a valve SSB transceiver.

SSB crystal filters usually operate in the frequency range around 10MHz. A double or even triple up conversion is required to reach microwave frequencies in the transmitter. On the other hand, a double or triple down conversion is required in the receiver to get back to the crystal filter

frequency. Commercial VHF/UHF SSB transceivers therefore save some expensive components by sharing some stages between the transmitter and the receiver.

A conventional microwave SSB transceiver is therefore complicated and expensive. Building such a transceiver in amateur conditions is difficult at best. Lots of work as well as some microwave test equipment is required. The final result is certainly not cheaper and may not perform better than the well known transverter + base transceiver combination.

Fortunately, expensive crystal filters and complicated conversions are not essential components of an SSB transceiver. There are other SSB transceiver designs that are both cheaper and easier to build in amateur conditions. The most popular seems to be the direct conversion SSB transceiver design shown on Fig 9.7. A direct conversion SSB receiver achieves most of its gain in a simple audio frequency amplifier, while the selectivity is achieved by simple RC low pass filters.

The most important feature of a direct conversion SSB transceiver is that there are no complicated conversions nor image frequencies to be filtered out. The RF section of a direct conversion SSB transceiver only requires simple LC filters to attenuate distant spurious responses like harmonics and sub harmonics. In a well designed direct conversion SSB transceiver, the RF section may not require any tuning at all.

The most important drawback of a direct conversion SSB transceiver is a rather poor unwanted sideband rejection. The transmitter includes two identical mixers operating at 90 degrees phase shift (quadrature mixer) to obtain only one sideband. The receiver also includes two identical mixers operating at 90 degrees phase shift to receive just one sideband and suppress the other sideband. A direct conversion SSB transceiver operates correctly only if the gain of both mixers is the same and the phase shift is exactly 90 degrees.

A direct conversion SSB transceiver therefore includes some critical components like precision (1%) resistors, precision (2%) capacitors, selected or paired semiconductors in the mixers and complicated phase shifting networks. The most complicated part is usually the audio

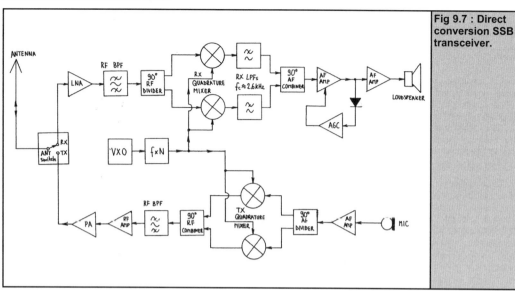

Fig 9.7 : Direct conversion SSB transceiver.

frequency 90 degree divider or combiner including several operational amplifiers, precision resistors and capacitors. Although using precision components, the unwanted sideband rejection will seldom be better than -40dB. This is certainly not enough for serious work on HF.

In spite of these difficulties, direct conversion designs are quite popular among the builders of QRP HF transceivers. At frequencies above 30MHz it is increasingly more difficult to obtain accurate phase shifts. Due to the low natural (antenna) noise above 30MHz, a low noise RF amplifier is usually used to improve the mixer noise figure. An LNA may cause direct AM detection in the mixers. An LNA may also corrupt the amplitude balance and phase offset of the two mixers, if the local oscillator signal is picked up by the antenna. A VHF direct conversion SSB transceiver is therefore not as simple as its HF counterpart.

On the other hand, a direct conversion SSB design has important advantages over conventional SSB transceivers with crystal filters, since there are no image frequencies and less spurious responses. Professional (military)

SSB transceivers therefore use direct conversion, but the AF phase shifts are obtained by digital signal processing. The DSP uses an adaptive algorithm to measure and compensate any errors like amplitude unbalance or phase offset of the two mixers, to obtain a perfect unwanted sideband rejection.

Additional AF signal processing also allows a different SSB transceiver design, for example a SSB transceiver with a zero IF as shown on Fig 9.8. The latter is very similar to a direct conversion transceiver except that the local oscillator is operating in the centre of the SSB signal spectrum, in other words at an offset of about 1.4kHz with respect to the SSB suppressed carrier frequency.

In a zero IF SSB transceiver, the audio frequency band from 200Hz to 2600Hz is converted in two bands from 0 to 1200Hz. Low pass filters therefore have a cutoff frequency of 1200Hz, thus allowing a high rejection of the unwanted sideband. A zero IF SSB transceiver therefore retains all of the advantages of a direct conversion design and solves the problem of the unwanted sideband rejection.

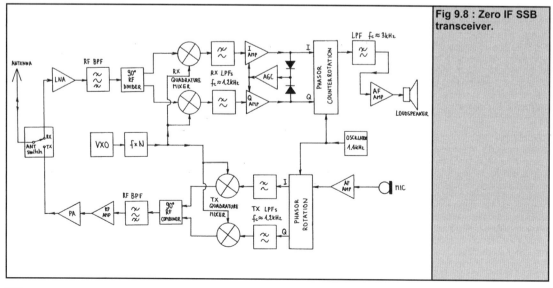

Fig 9.8 : Zero IF SSB transceiver.

The quadrature IF amplifier of a zero IF SSB transceiver includes two conventional AF amplifiers. Since the latter are usually AC coupled, the missing DC component will be converted in the demodulator as a hole in the AF response around 1.4kHz. Fortunately this hole is not harmful at all for voice communications, since it coincides with a hole in the spectrum of the human voice. In fact, some voice communication equipment includes notch filters to create an artificial hole around 1.4kHz to improve the signal to noise ratio and/or to add a low baud rate telemetry channel to the voice channel. Therefore a potential drawback of a zero IF design is actually an advantage for voice communications.

Like a direct conversion transceiver, a zero IF SSB transceiver also requires quadrature transmit and receive mixers. However, amplitude unbalance or phase errors are much less critical, since they only cause distortion of the recovered audio signal. Conventional components, like 5% resistors, 10% capacitors and unselected semiconductors may be used anywhere in a zero IF SSB transceiver.

Finally, a zero IF SSB transceiver does not require complicated phase shifting networks. Both the quadrature modulator in the transmitter and the quadrature demodulator in the receiver (phasor rotation and counter rotation with 1.4kHz) are made by simple rotating switches and fixed resistor/op-amp networks. CMOS analogue switches like the 4051 are ideal for this purpose, rotated by digital signals coming from a 1.4kHz oscillator.

Although the block diagram of a zero IF SSB transceiver looks complicated, such a transceiver is relatively easy to build. In particular, very little (if any) tuning is required, since there are no critical components used anywhere in the transceiver. In particular, the RF section only includes relatively wideband (10%) bandpass filters that require no tuning. The IF/AF section also accepts wide component tolerances and thus requires no tuning. The only remaining circuit is the RF local oscillator. The latter may need some tuning to bring the radio to the desired operating frequency.

Microwave SSB transceiver implementation

The described zero IF concept should allow the design of simple and efficient SSB transceivers for an arbitrary frequency band. Four successful designs of zero IF SSB transceivers covering the amateur microwave bands of 1.3, 2.3, 5.7 and 10GHz are described in this handbook, the basic design is given in this chapter with the RF sections for other bands included in the relevant band chapters. Similar technical solutions were first tested in PSK packet radio transceivers operating at 1.2Mbit/s in the 23cm and 13cm amateur frequency bands.

Of course several requirements and technology issues need to be considered before a theoretical concept can materialise in a real world transceiver. Fortunately the requirements are not severe for the lower amateur microwave bands. In this frequency range no very strong signals are expected, so there are no special requirements on the dynamic range of the receiver. Only a relatively limited frequency range needs to be covered (200 to 400kHz in each band) and this can be easily achieved using a VXO and multipliers as the local oscillator.

From the technology point of view it is certainly conven-

ient to use up to date components. High performance and inexpensive microwave semiconductors were developed first for satellite TV receivers and then for mobile communications like GSM or DECT telephones. These new devices provide up to 25dB of gain per stage up to 2.3GHz and up to 14dB of gain per stage up to 10GHz. Many other functions, like schottky mixer diodes or antenna switching PIN diodes are also available.

Using obsolete components makes designs complicated. For example, the well known transistors BFR34A and BFR91 were introduced almost 25 years ago. At that time they were great devices providing almost 5dB of gain at 2.3GHz. Today it makes more sense to use an INA-03184 MMIC to get 25dB of gain at 2.3GHz or in other words replace a chain of 5 (five) amplifier stages with the obsolete transistors.

The availability of active components also influences the selection of passive components. Many years ago, all microwave circuits were built in waveguide technology. Waveguides allow very low circuit losses and high Q resonators. Semiconductor microwave devices introduced microstrip circuits built on low loss substrates like alumina (Al2O3) ceramic or glassfibre-teflon laminates. Conventional glassfibre-epoxy laminates like FR4 were not used above 2GHz due to the high losses and poor Q of microstrip resonators.

However, a zero IF SSB transceiver design does not require a very high selectivity in the RF section. If the circuit losses can be compensated by high gain semiconductor devices, cheaper substrates like the conventional glassfibre-epoxy FR4 can be used at frequencies up to at least 10GHz. The FR4 laminate has excellent mechanical properties. Unlike soft teflon laminates, cutting, drilling and hole plating in FR4 is well known. Even more important, most SMD component packages are designed for installation on a FR4 substrate and may break or develop intermittent contacts if installed on a soft teflon board.

Therefore, losses in FR4 microstrip transmission lines and filters were investigated. Surprisingly, the losses were found inversely proportional to board thickness and rather slowly increasing with frequency. This simply means that the FR4 RF losses are mainly copper losses, while dielectric losses are still rather low. FR4 RF copper losses are high since the copper surface is made very rough to ensure good mechanical bonding to the dielectric substrate.

In fact, if the copper foil is peeled off a piece of FR4 laminate, the lower foil surface is rather dark. On the other hand, if the copper foil is peeled off a piece of microwave teflon laminate, the colours of both foil surfaces are similar. Since different manufacturers use different methods for bonding the copper foil, RF losses are different in different FR4 laminates. On the other hand, the dielectric constant of FR4 was found quite stable. Finally, silver or gold plating of microstrip lines etched on FR4 laminate really makes no sense, since most of the RF losses are caused by the (inaccessible) rough foil surface bonded to the dielectric.

A practical FR4 laminate thickness for microwave circuits with SMD components is probably 0.8mm. A 50ohm microstrip line has a width of about 1.5mm and about 0.2dB/cm of loss at 5.76GHz. Therefore microstrip lines have to be kept short if etched on FR4 laminate. For comparison, the FR4 microstrip losses are about three

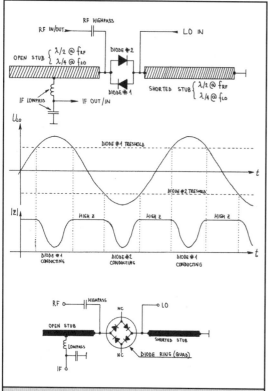

Fig 9.9 : Subharmonic mixer design.

times larger than the microstrip losses of a glassfibre-teflon board and about ten times larger than the losses of teflon semi rigid coax cables.

Although FR4 laminate losses are high, resonators and filters can still be implemented as microstrip circuits. Considering PCB etching tolerances and especially under etching, both transmission lines and gaps in between them should not be made too narrow. A practical lower limit is 0.4mm width for the transmission lines and 0.3mm for the gaps.

As already mentioned, modern semiconductor devices are really easy to use even at microwave frequencies. Silicon MMIC amplifiers provide 25dB of gain (limited by package parasitics) up to 2.3GHz. If less gain is required, conventional silicon bipolar transistors can be used, since their input and output impedances are also close to 50ohms.

GaAs semiconductors are more practical above about 5GHz. In particular, high performance devices like HEMTs have become inexpensive since they have been mass produced for satellite TV receivers. HEMTs operate at lower voltages and higher currents than conventional GaAsFETs, so their input and output impedance are very close to 50ohms at frequencies above 5GHz.

Serious microwave engineers are afraid of using HEMTs since these devices have enough gain to oscillate at frequencies above 50GHz or even 100GHz. In this case it is actually an advantage to build the circuit on a lossy laminate like FR4, since the latter will efficiently suppress any oscillations in the millimetre frequency range. Having the ability to control the loss in a circuit therefore may

represent an advantage!

The availability of inexpensive power GaAsFETs greatly simplifies the construction of transmitter output stages. In particular, the high gain of power GaAsFETs in the 23cm and 13cm bands greatly reduces the number of stages when compared to silicon bipolar solutions.

Zero IF and direct conversion transceivers have some additional requirements for mixers. Mixer balancing is very important, both to suppress the unwanted residual carrier in the transmitter and to suppress the unwanted AM detection in the receiver. At microwave frequencies, the simplest way of achieving good mixer balancing is to use a sub harmonic mixer with two anti parallel diodes as shown on Fig 9.9.

Such a mixer requires a local oscillator at half frequency. Frequency doubling is achieved internally in the mixer circuit. A disadvantage of this mixer is a higher noise figure in the range 10 to 15dB and sensitivity to the LO signal level. Both a too low LO drive or a too high LO drive will further increase the mixer insertion loss and noise figure.

On the other hand, the sub harmonic mixer only requires two non critical microstrip resonators that do not influence the balancing of the mixer. The best performances were obtained using schottky quads with the four diodes internally connected in a ring. The schottky quad BAT14-099R provides about -35dB of carrier suppression at 1296MHz and about -25dB of carrier suppression at 5760MHz with no tuning.

A very important advantage of the sub harmonic mixer is that the local oscillator operates at half of the RF frequency. This reduces the RF LO crosstalk and therefore the shielding requirements in zero IF or direct conversion transceivers. A side advantage is that the half frequency LO chain requires less multiplier stages.

The zero IF SSB transceivers described in this handbook have many parts in common. In particular, the AF and IF sections are identical in all transceivers. The RF sections are similar, however the microstrip filters are necessarily different as well as the low noise and power devices used in each frequency band. Finally, the same VCXO module is used, with small modifications, in all transceivers.

The common modules are described in this chapter. The modules for different frequency ranges are described in the relevant band chapters. Finally, an overview of the construction techniques is given in this chapter as well as shielding of the modules and integration of the complete transceivers.

VCXO and multipliers

Since a relatively narrow frequency range needs to be covered, a VXO followed by multiplier stages is an efficient solution for the local oscillator. The VXO is built as a varactor tuned VCXO with a fundamental resonance crystal, since the frequency pulling range of overtone crystals is not sufficient for this application. A fundamental resonance crystal has a lower Q and is less stable than overtone crystals, but for this application the performance is sufficient.

Fundamental resonance crystals can be manufactured for frequencies up to about 25MHz. Therefore the output of the VCXO needs to be multiplied to obtain microwave frequencies. Frequency multiplication can be obtained by a chain of conventional multipliers including class C

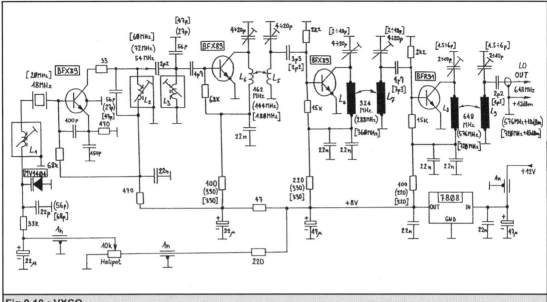

Fig 9.10 : VXCO.

amplifiers and bandpass filters or by a phase locked loop.

Although the PLL requires almost no tuning and is easily reproducible, the PLL solution was discarded for other reasons. A SSB transceiver requires a very clean LO signal, therefore the PLL requires buffer stages to avoid pulling the VCXO and/or the microwave VCO. Shielding and power supply regulation are also critical, making the whole PLL multiplier more complicated than a conventional multiplier chain.

The circuit diagram of the VCXO and multiplier stages is shown on Fig 9.10. The VCXO is operating around 18MHz in the transceivers for 1296MHz and 2304MHz and around 20MHz in the transceiver for 5760MHz. All multiplier stages use silicon bipolar transistors BFX89

(BFY90) except the last stage with a BFR91. The module already supplies the required frequency of 648MHz for the 1296MHz version of the transceiver.

In the 2304MHz version, the module supplies 576MHz by using different multiplication factors. The latter frequency is doubled to 1152MHz inside the transmit and receive mixer modules. In the 5760MHz version, the module supplies 720MHz and this frequency is further multiplied to 2880MHz in an additional multiplier module. Of course, the values of a few components need to be adjusted according to the exact operating frequency, shown in ( ) brackets for 2304MHz and in [ ] brackets for 5760MHz.

The VCXO and multiplier chain are built on a single sided FR4 board with the dimensions of 40mmX120mm as

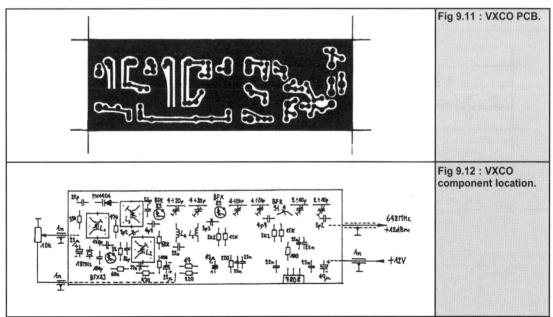

Fig 9.11 : VXCO PCB.

Fig 9.12 : VXCO component location.

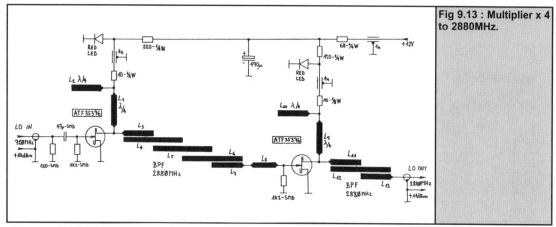

Fig 9.13 : Multiplier x 4 to 2880MHz.

shown on Fig 9.11. The corresponding component location (for the 648MHz version) is shown on Fig 9.12. The exact value of L1 depends on the crystal used. Some parallel resonance crystals may even require replacing L1 with a capacitor. L2 and L3 have about 150nH each or 4 turns of 0.25mm copper enamelled wire on a 10X10mm IF transformer coil former. L4 and L5 are self-supporting coils of 4 turns of 1mm copper enamelled wire each, wound on an internal diameter of 4mm. L6, L7, L8 and L9 are etched on the PCB.

The VCXO module is the only part of the whole transceiver that requires tuning. L2, L3 and the capacitors in parallel with L4, L5, L6, L7, L8 and L9 should simply be tuned for the maximum output at the desired frequencies. In a multiplier chain, RF signal levels can easily be checked by measuring the DC voltages over the base emitter junctions of the multiplier transistors.

When the multiplier chain is providing the specified output power, L1 and the capacitor in parallel with the MV1404 varactor should be set for the desired frequency coverage of the VCXO. If standard computer grade 18.000MHz or 20.000MHz crystals are used, it is recommended to select the crystal with the smallest temperature coefficient.

Unfortunately not all amateurs are allowed to use the international segment around 2304MHz on 13cm. It is a little bit more difficult to find a crystal for 18.125MHz for the German segment around 2320MHz.

The 5760MHz transceiver requires an additional multiplier from 720MHz to 2880MHz as shown on Fig 9.13. The first HEMT ATF35376 operates as a quadrupler while the second HEMT ATF35376 operates as a selective amplifier for the output frequency of 2880MHz. The additional multiplier for 2880MHz is built on a double sided microstrip FR4 board with the dimensions of 20mmX120mm as shown on Fig 9.14. The corresponding component location is shown on Fig 9.15.

The 2880MHz multiplier should provide the rated output power of +11dBm without any tuning. On the other hand, the tuning of L8 and L9 to 720MHz in the VCXO module can be optimised for the minimum DC drain current (max DC voltage) of the first HEMT. The two red LEDs are used as 2V zeners. LEDs are in fact better than real zeners, since they have a sharper knee and do not produce any avalanche noise.

SSB/CW quadrature modulator

The main purpose of the SSB/CW quadrature modulator is to convert the input audio frequency band from 200Hz

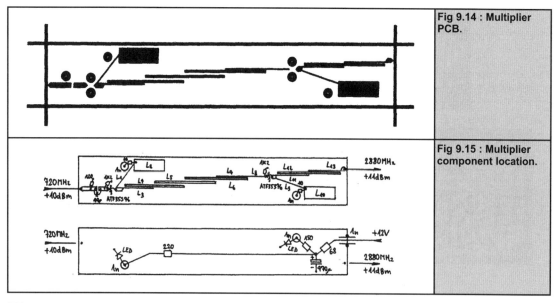

Fig 9.14 : Multiplier PCB.

Fig 9.15 : Multiplier component location.

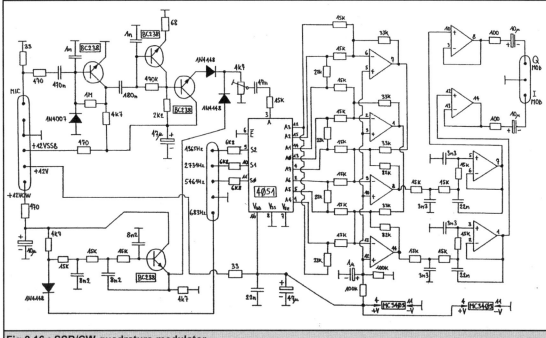

Fig 9.16 : SSB/CW quadrature modulator.

to 2600Hz into two bands 0 to 1200Hz to drive the quadrature transmit mixer. Additionally the module includes a microphone amplifier and a circuit to generate the CW signal. The circuit diagram of the modulator module is shown on Fig 9.16.

The microphone amplifier includes two stages with the transistors BC238. The input is matched to a low impedance dynamic mike with the 33 ohm resistor. The 1N4007 diode protects the input because the microphone input is simply connected in parallel to the loudspeaker output as described later. Finally the output drives an emitter follower with another BC238.

The CW carrier is generated in the same way as the SSB transmission. The 683Hz square wave, coming from the demodulator module, is first cleaned in a low pass audio filter and then processed in the same way as a SSB signal. Both AF modulation sources are simply switched by 1N4148 diodes.

The main component of the modulator is the 4051 CMOS analogue switch. The switch is rotated with the 1365Hz,

2731Hz and 5461Hz clocks coming from the demodulator. The input audio signal is alternatively fed to the I and Q chains. The I and Q signals are obtained with a resistor network and the first four opamps (first MC3403). Then both I and Q signals go through low pass filters to remove unwanted mixing products. Finally there are two voltage followers to drive the quadrature transmit mixer.

The SSB/CW quadrature modulator is built on a single sided FR4 board with the dimensions of 40mmX120mm as shown on Fig 9.17. The corresponding component location is shown on Fig 9.18. Most components are installed vertically to save board space. The SSB/CW quadrature modulator does not require any alignment. The 4.7k ohm trimmer is provided to check the overall transmitter. Full power (in CW mode) should be obtained with the trimmer cursor in central position.

Quadrature transmit mixers

All three transmit mixer modules for 1296MHz, 2304MHz and 5760MHz include similar stages: an LO signal switching, an in phase LO divider, two balanced sub

Fig 9.17 : Modulator PCB.

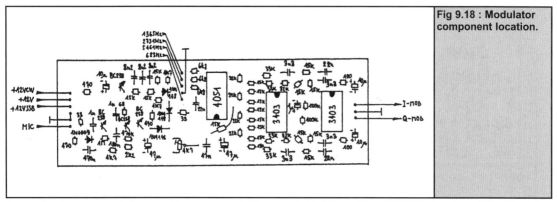

Fig 9.18 : Modulator component location.

harmonic mixers, a quadrature combiner and a selective RF amplifier. LO signal switching between the transmit and receive mixers is performed in the following way :

Most of the LO signal is always fed to the receive mixer. A small fraction of the LO signal is obtained from a coupler and amplified to drive the transmit mixer. During reception the power supply of the LO amplifier stage is simply turned off. This solution may look complicated, but in practice it allows an excellent isolation between the transmit and receive mixers. The practical circuit is simple and the component count is low as well.

The circuit diagram of the quadrature transmit mixer for 1296MHz is shown on Fig 9.19. The 648MHz LO signal is taken from a -20dB coupler and the LO signal level is restored by the BFP183 amplifier stage, feeding two subharmonic mixers equipped with BAT14-099R schottky quads. The 648MHz lowpass filter attenuates the second harmonic at 1296MHz to avoid corrupting the symmetry of the mixers.

The two 1296MHz signals are combined in a quadrature hybrid, followed by a 1296MHz bandpass filter. The latter removes the 648MHz LO as well as other unwanted mixing products. After filtering the 1296MHz SSB signal level is rather low (around -10dBm), so an INA-10386 MMIC is used to boost the output signal level to about +15dBm.

The quadrature transmit mixer for 1296MHz is built on a double-sided microstrip FR4 board with the dimensions of 40mmX120mm as shown on Fig 9.20. The corresponding component location is shown on Fig 9.21. The circuit does not require any tuning for operation at 1296MHz or 1270MHz.

RF front ends

The RF front ends include the transmitter power amplifiers, the receiver low noise amplifiers and the antenna switching circuits. Of course there are major differences among different power amplifier designs, depending not just on the frequency, but also on the technology used and the output power desired. It no longer makes sense to use expensive coaxial relays, since PIN diodes can

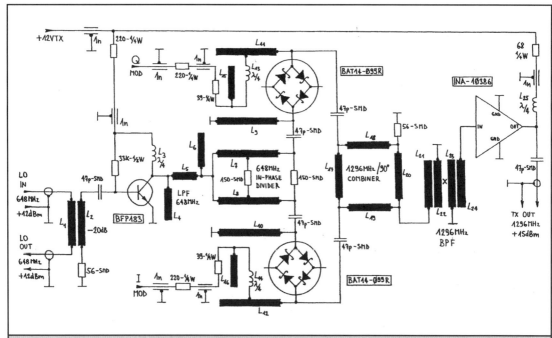

Fig 9.19 : Quadrature transmit mixer for 1296MHz.

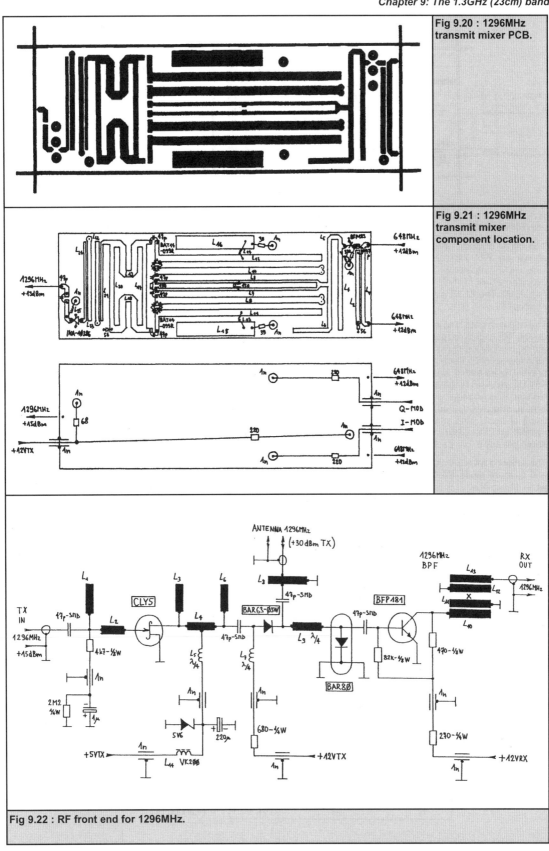

Fig 9.20 : 1296MHz transmit mixer PCB.

Fig 9.21 : 1296MHz transmit mixer component location.

Fig 9.22 : RF front end for 1296MHz.

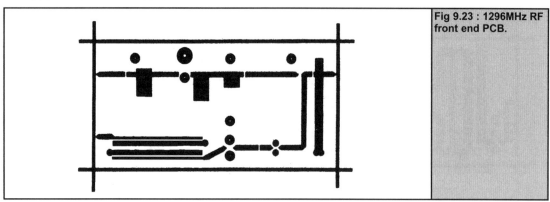

Fig 9.23 : 1296MHz RF front end PCB.

provide the same insertion loss and isolation at lower cost with better reliability and much shorter switching times.

The circuit diagram of the RF front end for 1296MHz is shown on Fig 9.22. The transmitter power amplifier includes a single stage with a CLY5 power GaAsFET, providing a gain of 15dB and an output power of about 1W (+30dBm). The CLY5 is a low-voltage transistor operating at about 5V.

The negative gate bias is generated by rectification of the driving RF signal in the GS junction inside the CLY5 during modulation peaks. The gate is then held negative for a few seconds thanks to the 1uF storage capacitor. To prevent overheating and destruction of the CLY5, the +5VTX voltage is obtained through a current-limiting resistor. This arrangement may look strange, but it is very simple, requires no adjustments, allows a reasonably linear operation and most important of all, it proved very reliable in PSK packet-radio transceivers operating 24 hours per day in our packet-radio network.

The antenna switch includes a series diode BAR63-03W and a shunt diode BAR80. Both diodes are turned on while transmitting. L9 is a quarter-wavelength line that transforms the BAR80 short circuit into an open for the transmitter. The receiving preamplifier includes a single BFP181 transistor (15dB gain) followed by a 1296MHz bandpass filter (-3dB loss). In the 1296MHz RF front-end,

the LNA gain should be limited to avoid interference from powerful non-amateur users of this band (radars and other radionavigation aids).

The RF front-end for 1296MHz is built on a double-sided microstrip FR4 board with the dimensions of 40mmX80mm as shown on Fig 9.23. The corresponding component location is shown on Fig 9.24. The RF front end for 1296MHz requires no tuning. However, since the output impedance of the INA-10386 inside the transmit mixer is not exactly 50ohms, the cable length between the transmit mixer and the RF front-end is critical. Therefore L1 may need adjustments if the teflon-dielectric cable length is different from 12.5cm.

Quadrature receive mixers

All receiving mixer modules include similar stages, an additional RF signal amplifier, a quadrature hybrid divider, two sub harmonic mixers, an in phase LO divider and two IF preamplifiers. The mixers, in phase and quadrature dividers and RF bandpass filters are very similar to those used in the transmitting mixer modules.

The circuit diagram of the quadrature receiving mixer for 1296MHz is shown on Fig 9.25. The incoming RF signal is first fed through a microstrip bandpass filter, then amplified with an INA-03184 MMIC and further filtered by another, identical microstrip bandpass. The total gain of

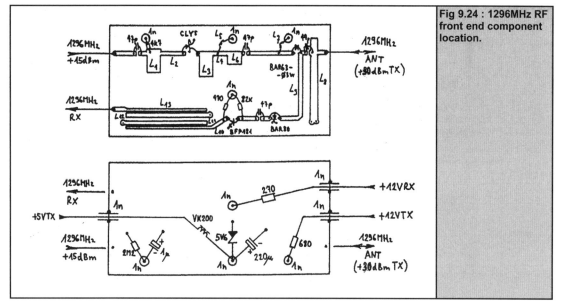

Fig 9.24 : 1296MHz RF front end component location.

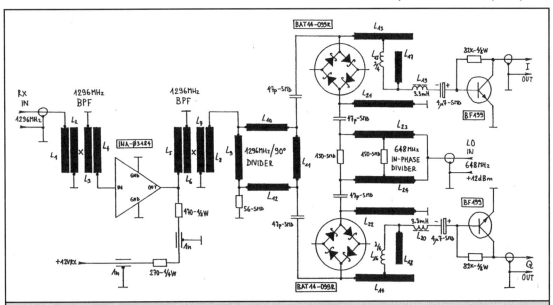

Fig 9.25 : Quadrature receive mixer for 1296MHz.

Fig 9.26 : 1296MHz receive mixer PCB.

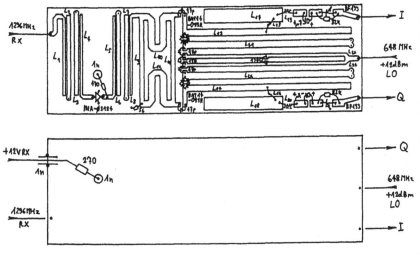

Fig 9.27 : 1296MHz receive mixer component location.

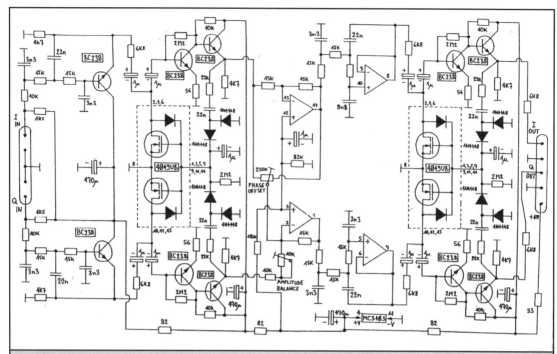

Fig 9.28 : Quadrature ssb IF amplifier.

the chain of the two filters and the MMIC is about 20dB.

A high gain in the RF section is required to cover the relatively high noise figure of the two subharmonic mixers and the additional losses in the quadrature hybrid. The two receiving subharmonic mixers are also using BAT14-099R schottky quads. The mixer outputs are fed through lowpass filters to the IF preamplifiers.

The IF preamplifiers are using HF transistors BF199. These were found to perform better than their BC... counterparts in spite of the very low frequencies involved (less than 1200Hz). HF transistors have a smaller current gain, their input impedance is therefore smaller and better matches the output impedance of the mixers. Both IF preamplifiers receive their supply voltages from the IF amplifier module.

The quadrature receiving mixer for 1296MHz is built on a double-sided microstrip FR4 board with the dimensions of 40mmX120mm as shown on Fig 9.26. The corresponding component location as shown on Fig 9.27. The receiving mixer for 1296MHz requires no tuning.

SSB zero IF amplifier with AGC

The basic feature of direct conversion and zero IF receivers is to achieve most of the signal gain with simple and inexpensive AF amplifiers. Further, the selectivity is achieved with simple RC low pass filters that require no tuning. The circuit diagram of such an IF amplifier equipped with AGC is therefore necessarily different from conventional high IF amplifiers.

A zero IF receiver requires a two channel IF amplifier, since both I and Q channels need to be amplified independently before demodulation. The two IF channels should be as much identical as possible to preserve the amplitude ratio and phase offset between the I and Q

signals. Therefore both channels should have a common AGC so that the amplitude ratio remains unchanged.

The circuit diagram of the quadrature SSB IF amplifier with AGC is shown on Fig 9.28. The IF amplifier module includes two identical low pass filters on the input, followed by a dual amplifier stage with a common AGC. An amplitude/phase correction is performed after the first amplifier stage, followed by another pair of low pass filters and another dual amplifier stage with a common AGC.

The two input low pass filters are active RC filters using BC238 emitter followers. Discrete bipolar transistors are used because much less noisy than operational amplifiers. The input circuit also provides the supply voltage to the IF preamplifiers inside the receiving mixer module through the 1.5k ohm resistors.

The dual amplifier stages are also built with discrete BC238 bipolar transistors. Each amplifier stage includes a voltage amplifier (first BC238) followed by an emitter follower (second BC238) essentially to avoid mutual interactions when the amplifiers are chained with other circuits in the IF strip.

The AGC is using MOS transistors as variable resistors on the inputs of the dual amplifier stages. To keep the gain of both I and Q channels identical, both MOS transistors are part of a single integrated circuit 4049UB. The digital CMOS integrated circuit 4049UB is being used in a rather uncommon way, however the remaining components inside the 4049UB act just as diodes and do not disturb the operation of the AGC.

The IF amplifier module includes two trimmers for small corrections of the amplitude balance (10k ohm) and phase offset (250k ohm) between the two channels. The correction stage is followed by two active RC low pass filters using operational amplifiers (MC3403), since the

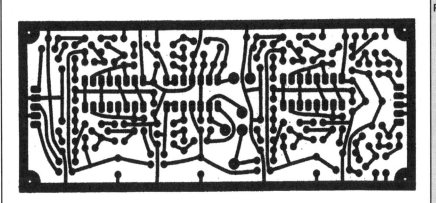

Fig 9.29 : IF amplifier PCB.

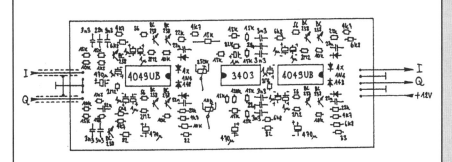

Fig 9.30 : IF amplifier component location.

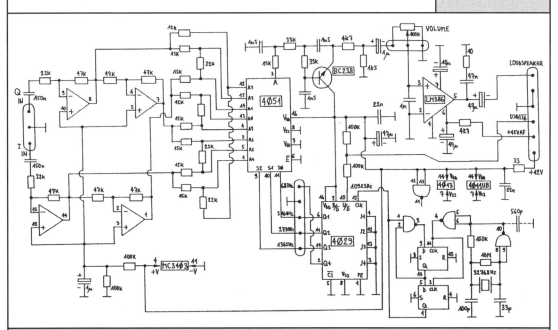

Fig 9.31 : Quadrature ssb demodulator and AF amplifier.

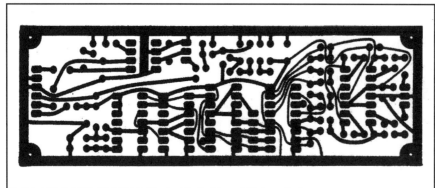

Fig 9.32 : Demodulator PCB.

signals are already large enough and the operational amplifier noise is no longer a problem. Finally there is another, identical dual amplifier stage with its own AGC.

The quadrature SSB IF amplifier is built on a single sided FR4 board with the dimensions of 50mmX120mm as shown on Fig 9.29. The corresponding component location is shown on Fig 9.30. In order to keep the differences between the I and Q channels small, good quality components should be used in the IF amplifier. Using 5% resistors, 10% foil type capacitors and conventional BC238B transistors should keep the differences between the two channels small enough for normal operation. Most components are installed vertically to save board space.

The amplitude balance (10k ohm) and phase offset (250k ohm) trimmers are initially set to their neutral (central) position. These trimmers are only used while testing the complete receiver to obtain the minimum distortion of the reproduced audio signal.

Quadrature SSB demodulator and AF amplifier

The main function of the quadrature SSB demodulator is the conversion of both I and Q IF signals (frequency range 0 to 1200Hz) back to the original audio frequency range from 200Hz to 2600Hz. The same module includes a power AF amplifier and a clock generator for both the phasor rotation in the transmitter and the phasor counter rotation in the receiver. The circuit diagram of the module is shown on Fig 9.31.

The quadrature SSB demodulator includes four operational amplifiers (MC3403) to produce an 8-phase system from the I and Q signals, using a resistor network similar to that used in the modulator. The signal demodulation or phasor counter rotation is performed by the CMOS

analogue switch 4051, rotating with a frequency of 1365Hz. The I and Q signals are alternatively fed to the output or in other words the circuit performs exactly the opposite operation of the modulator.

Unwanted mixing products of the phasor counter rotation are removed by an active RC low pass (BC238). The demodulated audio signal is fed to the 100k ohm volume control. A LM386 is used as the audio power amplifier due to its low current drain and small external component count.

The three clocks required to rotate both 4051 switches in the modulator and in the demodulator are supplied by a binary counter 4029. The 4029 includes an up/down input that allows the generation/demodulation of USB or LSB in this application. The up/down input has a 100k ohm pull up resistor for USB operation. LSB is obtained when the up/down input is grounded through a front panel switch.

USB/LSB switching is usually not required for terrestrial microwave work. USB/LSB switching is only required when operating through satellites or terrestrial linear transponders or when using inverting converters or transverters for other frequency bands. Finally, USB/LSB switching may be useful to attenuate interference during CW reception. An alternative way of USB/LSB switching is interchanging the I and Q channels. When assembling together the modules of the described transceivers one should therefore check the wiring so that the transmitter and the receiver operate on the same sideband at the same time.

The 4029 counter requires an input clock around 11kHz. This clock does not need to be particularly stable and a RC oscillator could be sufficient. In the described transceivers a crystal source was preferred to avoid any tuning. In addition, if all transceivers use the same rotation

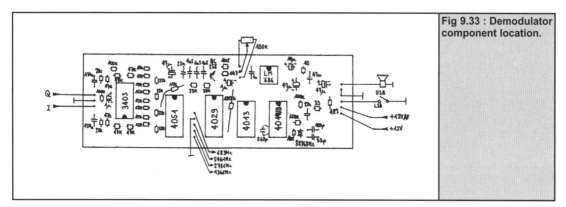

Fig 9.33 : Demodulator component location.

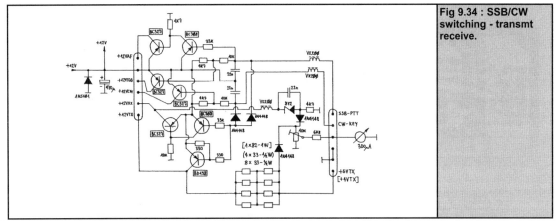

Fig 9.34 : SSB/CW switching - transmt receive.

or counter rotation frequency, the mutual interference is reduced.

The crystal oscillator is using a clock crystal operating on a relatively low frequency of 32768Hz. The dual D flip flop 4013 divides this frequency by 3 to obtain a 10923Hz clock for the 4029 binary counter. The resulting rotation frequency for the 4051 switches is 1365Hz. The latter figure almost perfectly matches the hole in the frequency spectrum of human voice. The same 4029 counter also supplies the CW tone 683Hz to reduce the unwanted mixing products in the transmitter.

The quadrature SSB demodulator and AF amplifier are built on a single sided FR4 board with the dimensions of 40mmX120mm as shown in Fig 9.32. The corresponding component location is shown in Fig 9.33. Most of the components are installed vertically to save board space.

The 32768Hz crystal oscillator will only operate reliably with a 4011UB (or 4001UB) integrated circuit. The commonly available B series CMOS integrated circuits (4011B in this case) have a too high gain for this application. In the latter case a 560pF capacitor may help to stabilise the oscillator. On the other hand, the oscillator circuit usually works reliably with old 4011 or 4001 circuits with an A suffix or no suffix letter at all.

SSB/CW switching RX/TX

A SSB/CW transceiver requires different switching functions. Fortunately both SSB and CW modes of operation require the same functions in the receiver. Of course two different operating modes are required for the transmitter:

SSB voice and CW keying.

The RX/TX changeover is controlled by the PTT switch on the microphone in the SSB mode of operation. In the CW mode of operation, most transceivers use an automatic delay circuit to keep the transmitter enabled during CW keying. This delay circuit was perhaps required in old radios using several mechanical relays. In modern transceivers with all electronic switching it makes no sense, since the RX/TX switching can be performed in less than one millisecond.

It therefore makes sense that SSB transmission is enabled by simply pressing the PTT switch while CW transmission is enabled by pressing the CW key so that no special (and useless) controls are required on the front panel of the transceiver. In the CW mode of operation, no delays are required and the receiver is enabled immediately after the CW key is released (BK mode of opera-

tion).

The circuit diagram of the SSB/CW switching RX/TX is shown on Fig 9.34. In the described SSB/CW transceivers, most modules are enabled at all times with a continuous +12V supply: VCXO and multipliers, receiving mixer, IF amplifier, demodulator and modulator. When enabling the transmitter either by pressing the PTT or CW keys, the RX LNA is turned off (+12VRX) and the TX PA is turned on (+12VTX and +5VTX or +4VTX).

During SSB transmission the RX AF amplifier is turned off (+12VAF), to avoid disturbing the microphone amplifier (+12VSSB). On the other hand, during CW transmission the AF amplifier remains on as well as most receiver stages, so that CW keying can be monitored in the loudspeaker or phones. The +12VCW supply connects the 683Hz signal to the modulator input.

The supply voltages +12VAF, +12VSSB, +12VCW and +12VRX are switched by BC327 PNP transistors. Due to the higher current drain, the +12VTX supply voltage requires a more powerful PNP transistor BD138. The TX PA receives its supply voltage through a current limiting resistor from the +12VTX line. Since the latter dissipates a considerable amount of power, it is built from several smaller resistors and located in the switching unit to prevent heating the PA transistor(s).

The value of the current limiting resistor depends on the version of the transceiver. The 1296MHz PA with a CLY5 requires 8 x 33ohm 1/2W resistors for a total value of 16.5ohms. The 2304MHz PA with a CLY2 requires 4 x 33ohm 1/2W resistors for a total value of 33ohms. Finally, the 5760MHz PA with two ATF35376s requires a single 82ohm 1W resistor.

The switching module also includes the circuits to drive the front panel meter. The latter is a moving coil type with a full scale sensitivity of about 300uA. The meter has two functions. During reception it is used to check the battery voltage. The 8V2 zener extends the full scale of the meter to the interesting range from about 9V to about 15V.

During transmission the meter is used to check the supply voltage of the PA transistor(s). Due to the self biasing operation, the PA voltage will be only 0.5-1V without modulation and will rise to its full value, limited by the zener diode inside the PA, only when full drive is applied. The operation of the PA and the output RF power level can therefore be simply estimated from the PA voltage.

A S-meter is probably totally useless in small portable transceivers as those described in this article. If desired,

245

Fig 9.35 : SSB/CW switching - transmit receive PCB.

the AGC voltage can be amplified and brought to a front panel meter. However, one should not forget that LED indicators are not visible in full sunshine on a mountain-top, so the choice is limited to moving-coil and LCD meters.

Most components of the SSB/CW switching RX/TX are installed on a single sided FR4 board with the dimensions of 30mmX80mm as shown on Fig 9.35. Their location (1296MHz version) is shown on Fig 9.36. Only the reverse polarity protection diode 1N5401 and the 470uF electrolytic capacitor are installed directly on the 12V supply connector. The 10k ohm trimmer is used to adjust the meter sensitivity.

Construction of zero IF SSB transceivers

The described zero IF SSB transceivers are using many SMD parts in the RF section. SMD resistors usually do not cause any problems, since they have low parasitics up to at least 10GHz. On the other hand, there are big differences among SMD capacitors. For this reason a single value (47pF) was used everywhere. The 47pF capacitors used in the prototypes are NPO type, rather large (size 1206), have a self resonance around 10GHz and introduce an insertion loss of about 0.5dB at 5.76GHz. Finally, the 4.7uF SMD tantalum capacitors can be replaced by the more popular tantalum drops.

Quarter wavelength chokes are used elsewhere in the RF circuits. In the 5760MHz transceiver all quarter wave-length chokes are made as high impedance microstrips. On the other hand, to save board space in the 1296MHz and 2304MHz versions, the quarter wavelength chokes are made as small coils of 0.25mm thick copper enam-elled wire of the correct length, chosen according to the frequency: 12cm for 648MHz, 9cm for 23cm mixers (648/1296MHz), 7cm for 1296MHz and for L3, 5.5cm for 13cm mixers (1152/2304MHz) and 4cm for 2304MHz. The wire is tinned for about 5mm on each end and the remaining length is wound on an internal diameter of about 1mm.

The SMD semiconductor packages and pin outs are

shown on Fig 9.37. Please note that due to lack of space, the SMD semiconductor markings are different from their type names. Only the relatively large CLY5 transistor in a SOT-223 package has enough space to carry the full marking CLY5. The remaining components only carry one, two or three letter marking codes.

All of the microstrip circuits are built on double sided 0.8mm thick FR4 glassfibre epoxy laminate. Only the top side is shown in this article, since the bottom side is left un etched to act as a ground plane for the microstrips. The copper surface should not be tinned nor silver or gold plated. The copper foil thickness should be preferably 35µm.

Since the microstrip boards are not designed for plated through holes, care should be taken to ground all compo-nents properly. Microstrip lines are grounded using 0.6mm thick silver plated wire (RG214 central conductors) inserted in 1mm diameter holes at the marked positions and soldered on both sides to the copper foil.

Resistors and semiconductors are grounded through 2mm, 3.2mm and 5mm diameter holes at the marked positions. These holes are first covered on the ground plane side with pieces of thin copper foil (0.1mm). Then the holes are filled with solder and finally the SMD component is soldered in the circuit.

Feed through capacitors are also installed in 3.2mm diameter holes in the microstrip boards and soldered to the ground plane. Feed through capacitors are used for supply voltages and low frequency signals. Some compo-nents like bias resistors, zeners and electrolytic capaci-tors are installed on the bottom side (as shown on the component location drawings) and connected to the feed through capacitors.

The VCXO/multiplier module and all of the microstrip circuits are installed in shielded enclosures as shown on Fig 9.38. Both the frame and the cover are made of 0.4mm thick brass sheet. The printed circuit board is soldered in the frame at a height of about 10mm from the bottom. Additional feed through capacitors are required in the brass walls. RF signal connections are made using

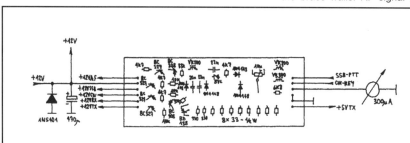

Fig 9.36 : SSB/CW switching - transmit receive component location.

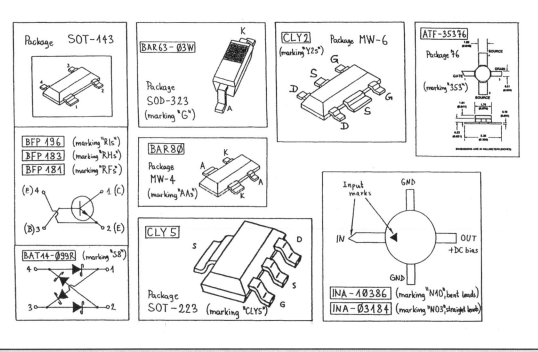

Fig 9.37 : SMD semiconductor packages and pinouts.

thin teflon dielectric coax like RG-188. The coax braid should be well soldered to the brass frame all around the entrance hole. Finally the frame is screwed on the chassis using sheet metal screws.

The covers are kept in place thanks to the elastic brass sheet, so they need not be soldered. An inspection of the content is therefore possible at any time. The VCXO/ multiplier module is built on a single sided board and therefore requires both a top and a bottom cover. The remaining modules are all built as microstrip circuits, so the microstrip ground plane acts as the bottom cover and only the top cover is required.

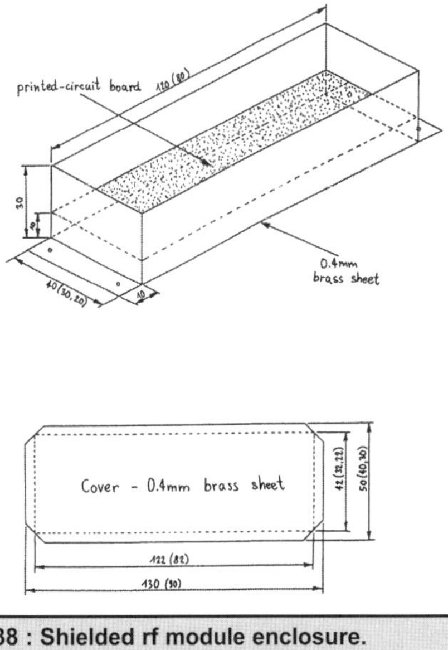

The sizes and shapes of the microstrip circuit boards are selected so that no resonances occur up to and including 2880MHz. Microwave absorber foam is therefore only required in the three modules operating at 5760MHz. To avoid disturbing the microstrip circuit, the microwave absorber foam is installed just below the top cover.

The modules of a zero IF SSB transceiver are installed in a custom made enclosure as shown on Fig 9.39. The most important component is the chassis. The latter must be made of a single piece of 1mm thick aluminium sheet to provide a common ground for all modules. If a common ground is not available, the receiver will probably self oscillate in the from of ringing or whistling in the loud-speaker, especially at higher volume settings.

The chassis carries both the front and back panels as well as the top and bottom covers. All of the connectors and commands are available on the front panel. The later is screwed to the chassis using the components installed, CW pushbutton, SMA connector, meter and tuning helipot. The top and bottom cover are made of 0.5mm thick aluminium sheet to save weight and are screwed to the chassis using sheet metal screws.

The shielded RF modules are installed on the top side of the chassis where a height up to 32mm is allowed. The audio frequency bare printed circuit boards are installed on the bottom side of the chassis. The interconnections between both sides of the chassis are made through five large diameter holes.

Fig 9.38 : Shielded rf module enclosure.

The location of the modules as well as the location of the connectors and commands on the front panel are shown

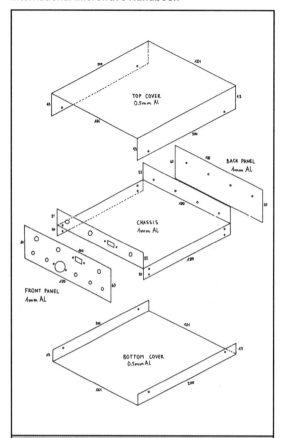

Fig 9.39 : Zero IF ssb/cw transceiver enclosure.

on Fig 9.40 for both sides of the chassis.

The loudspeaker should not be installed inside the transceiver, since the RF receiving modules are quite sensitive to vibrations. On the other hand, the same loudspeaker may also be used as a dynamic microphone for the transmitter. The circuit is designed so that it allows a simple parallel connection of the loudspeaker output and the microphone input. The PTT and CW keys are simple switches to ground.

Checkout of zero IF SSB transceivers

The described SSB/CW transceivers do not require much tuning. The only module that really requires tuning is the VCXO/multiplier. The latter is simply tuned for the maximum output on the desired frequency. Of course, the desired coverage of the VCXO has to be set with a frequency counter.

After the VCXO/multiplier module is adjusted, the remaining parts of the transceiver require a checkout to locate defective components, soldering errors or insufficient shielding. The receiver should already work and some noise should be heard in the loudspeaker. The noise intensity should drop when the power supply to the LNA is removed. The noise should completely disappear when the receiving mixer module is disconnected from the IF amplifier. A similar noise should be heard if only one (I or Q) IF channel is connected.

Next the receiver is connected to an outdoor antenna far away from the receiver and tuned to a weak unmodulated carrier (a beacon transmitter or another VCXO/multiplier module at a distance of a few ten meters). Tuning the receiver around the unmodulated signal one should hear both the desired tone and its much weaker mirror image tone changing its frequency in the opposite direction. The two trimmers in the IF amplifier should be set so that the

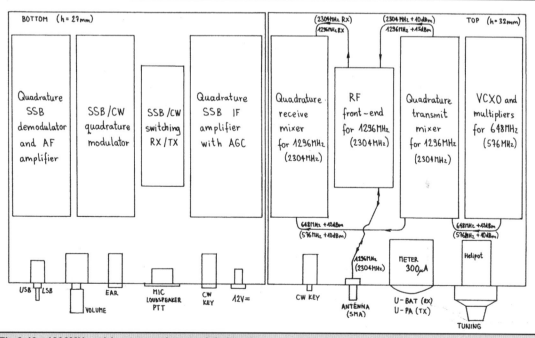

Fig 9.40 : 1296MHz ssb/cw transceiver module location.

mirror tone disappears. The correct function of the USB/LSB switch can also be checked.

Finally, the shielding of the receiver should be checked. A small, handheld antenna (10-15dBi) is connected to the receiver and the main beam of the antenna is directed into the transceiver. If the noise coming from the loudspeaker changes, the shielding of the local oscillator multiplier chain is insufficient.

Next a mains operated fluorescent tube (20W or 40W) is turned on in the same room. A weak mains hum should only be heard when the handheld antenna is pointed towards the tube at 2-3m distance. If a clean hum without noise is heard regardless of the antenna direction, the shielding of the local oscillator multiplier chain is insufficient.

The transmitter should be essentially checked for the output power. The full output power should be achieved with the trimmer in the modulator in a middle position in CW mode. The DC voltage across the PA transistor should rise to the full value allowed by the 5.6V or 4.7V zener diode. The output power should drop by an equal amount if only I or only Q modulation is connected to the transmit mixer.

Finally the SSB modulation has to be checked with another receiver for the same frequency band or best in a contact with another amateur station at a distance of a few km. This is the simplest way to find out the correct sideband, USB or LSB, of the transmitter, since the I and Q channels can be easily interchanged by mistake in the wiring.

The residual carrier level of the transmit mixer should also be checked. Due to the conversion principle this carrier results in a 1365Hz tone in a correctly-tuned SSB receiver. The carrier suppression may range from -35dB in the 1296MHz transceiver down to only -20dB in the 5760MHz transceiver. A poor carrier suppression may be caused by a too high LO signal level or by a careless installation of the BAT14-099R mixer diodes. Note that the residual carrier can not be monitored on another

correctly tuned zero IF receiver, since it falls in the AF response hole of the zero-IF receiver.

The current drain of the described transceivers should be as follows :

Receive

- 1296MHz:105mA

- 2304MHz:175mA

- 5760MHz:300mA.

The current drain of the transmitters is inversely proportional to the output power due to the self biasing of the PA. The minimum current drain corresponds to SSB modulation peaks or CW transmission. Transmitters:

- 1296MHz:650-870mA

- 2304MHz:490-640mA

- 5760MHz:410-440mA.

All figures are given for a typical sample at a supply voltage of 12.6V.

At the end, one should understand that zero IF transceivers also have some limitations. In particular, the dynamic range of the receiver is limited by the direct AM detection in the receiving mixer. If very strong signals are expected, the LNA gain has to be reduced to avoid the above problem. This is already done in the 1296MHz receiver, since strong radar signals are quite common in the 23cm band. The sensitivity to radar interference of the described zero IF 1296MHz transceiver was found comparable to the conventional transverter + 2m transceiver combination.

On the other hand, the 2304MHz and 5760MHz transceivers have a higher gain LNA. If the dynamic range needs to be improved, the second LNA stage can simply be replaced with a wire bridge in both transceivers. Of course, the internal LNA gain has to be reduced or the LNA has to be completely eliminated if an external LNA is used.

Fig 9.41 : Picture of 1.3GHz transverter.

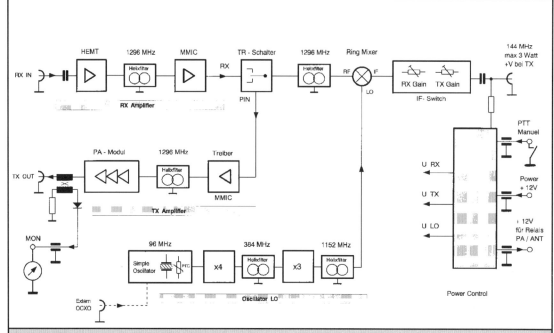

Fig 9.42 Block diagram of DB6NT 1.3GHz transverter.

Transverters

The most popular method to get onto the microwave bands is to use a transverter to translate the output of a commercial vhf transceiver to the required band. The transverter describe here is from Michael Kuhne, DB6NT [2]

The new version of the 23cm transverter is an improved circuit of the transverter described in 1991 [3] By use of modern semiconductors and refinement concerning cooling and the easy replication of the circuit a further optimisation of the transverter had been possible.

The current transverter is a single board construction on R04003 substrate (Fig 9.41). The receiver has a noise figure of typical 0.6dB at more than 20dB gain.

The transmitter achieves an output power of more than 1.5W in a frequency band of 1296 ... 1298MHz at an IF of 144MHz. The spurious rejection is better than 60dB. Harmonic rejection is better than 40dB.

Everything -TX, RX, LO, IF-Switch and T/R control is on a single board housed in a 55x74x30mm large box from tin plate, see block diagram Fig 9.42.

For tuning only a simple detector is necessary. All filters are helical filters with restricted tuning range. The restricted tuning range of the helical filters doesn't allow tuning on 'false' resonances.

Circuit description

The proven 'simple' crystal oscillator (XO) uses the FET SST310 in a grounded gate circuit. The crystal frequency for a 144MHz IF is 96MHz. The coil is tuned by the usual ferrite tuning screw. An extra 40C PTC heater improves the drift characteristic of the XO. Extra pads are provided for fitting additional capacitors that can be selected for temperature compensation. For normal use in a restricted temperature change environment the stability is sufficient. But for more serious work a special outboard solution like the OCXO from DF9LN is required. This can be fed in at the source of the SST310, as indicated in the circuit diagram (Fig 9.43). The crystal and the heater have to be removed in this case.

The XO is followed by a quadrupler to 384MHz which utilises a BFR92P transistor. The fourth harmonic is filtered by a helix band pass filter and drives a tripler using a BFG93A. The output filter selects the harmonic at t 1152MHz. The power at this point is around 5mW (7dBm).

T/R Switch

The IF port of the mixer is terminated by selectable attenuators for transmit and receive. These are switched by PIN Diodes BAR64-03W to a common IF connector. A voltage of at least +9V, which can be supplied by a FT290 for example, activates the T/R switching. Other brands of 2m transceivers have to be modified accordingly.

Whilst this method of T/R switching via the IF coaxial cable is quite elegant, a separate method, via the PTT MAN input, is provided.

An extra output is fitted for TX+, which can be used for external coaxial relays or PAs. This output must be protected by a 0.63A fuse because it is not short circuit protected.

Receiver

The RX chain uses a HEMT Amplifiers (NE32584C) and a second stage with a INA03184 MMIC from HP. The gain of >30dB makes an extra IF amplifier obsolete. The stages are coupled with a helical filter F4. The second stage is coupled to the mixer via another PIN switch and a second helical Filter F3.

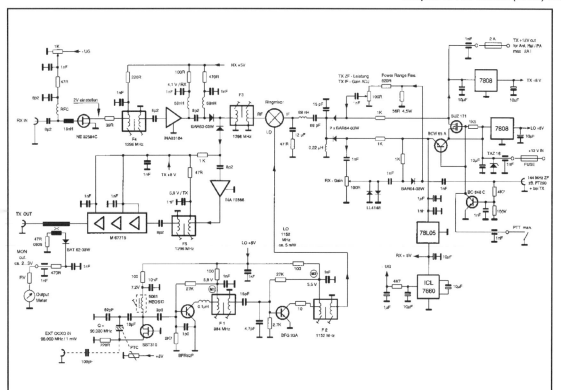

Fig 9.43 : Circuit diagram of DB6NT 1.3GHz transverter.

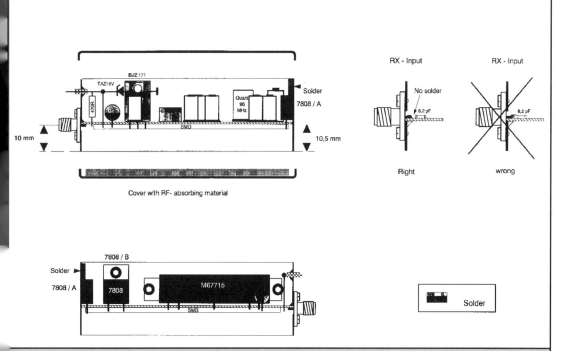

Fig 9.44 : Construction hints for DB6NT 1.3GHz transverter.

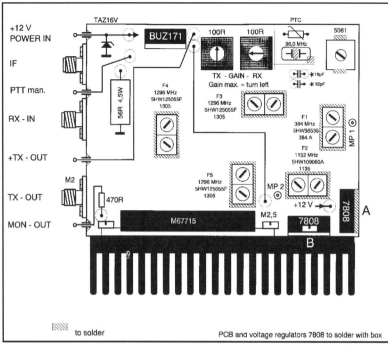

Fig 9.45 Component layout for DB6NT 1.3GHz transverter.

Transmitter

After the PIN switch and the helical Filter F3, which is used both for receive and transmit, a INA1 0386 MMIC from HP follows and drives the Mitsubishi hybrid M67715 via a second helical filter F5.

The Mitsubishi hybrid can deliver around 1.5W output. A directional coupler with a BAT15-03W Schottky diode provides a monitor voltage of the RF output power.

Construction

To achieve successful construction of this transverter the builder has to have experience in the use and handling of SMD parts. Furthermore experience with smaller projects in microwave circuits are valuable. In any case the construction of this transverter is not a beginner's project.

The usual ESD protection measures should be obeyed. Construction Steps

- Solder the walls of the tinplate box and trim the PCB for fitting into the tinplate box.

- Mark the holes for the SMA connectors

- Drill holes for SMA connectors and feed through capacitors

- Solder the PCB into the box (Fig. 9.44). Use a 10.2mm high piece of wood as a ruler to find the right height adjustment.

- Insert the LM7808 (B) regulator into the PCB (Remove middle pin of the regulator!). Drill two holes for the heatsink and one hole for the regulator into the side wall of the box. The heatsink should be in the middle of the PCB.

- Mount the parts onto the PCB (Fig. 9.45). Mount the feed through capacitors. Solder the helix filters. Solder the regulator LM7808 (A) with its heatsinks to the wall of the tinplate box. The FET BUZ171 should be fitted to

the PCB by holding it tightly down when soldered. Clean the finished PCB with alcohol. The tuning screws of the resonators should be removed. Dry the module in a stove (1 h at 80°C) or over night lying on a central heating radiator.

- Mount the LM7808 (B) and the PA hybrid using some heat sink compound to ensure good thermal contact with the heatsinks.

Alignment

- The following steps are necessary for the alignment of the transverter:

- Apply 12V. Use a current limited (< 0.6A) power supply. Check the voltage at the output of the fixed voltage regulators.

- Measure the collector voltage at the BFR92P (Test point MI). Turn the tuning screw of the oscillator coil until a decrease of the collector voltage indicates the proper oscillation. The measurement should read around 5.8V.

- Measure the voltage at M2 (Fig. 9.43). Tune band pass filter FI (384MHz) to minimum voltage (about... 5.5V) at M2.

- Connect dummy load or antenna to the input connector of the receiver.

- Adjust 10k pot for a reading of 2V at the drain of the receiver FET NE32584C.

- Connect a 2m receiver to the IF connector. Turn the receiver gain and transmitter gain pots fully CCW. You will observe an increase in noise level. By tuning the helical filters F3 and F4 you can maximise the noise output. If there is an indication of more than S1 at the 144MHz transceiver you should adjust the receiver gain pot accordingly.

- Connect a 50Ω dummy load to the transmitter output. Switch transverter to transmit by grounding the PTT

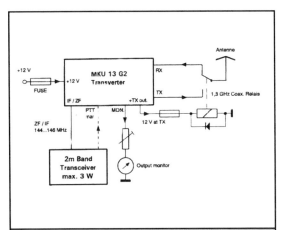

Fig 9.46 : Connection of DB6NT 1.3GHz transverter to a transceiver.

input. Drive the transverter with 1 - 3W on 144MHz. Measure the monitor voltage at MON OUT. It should read 2.3V. Adjust the transmitter gain pot to a reading of about 1 V. Now the helical filter F5 and the LO filter F2 can be readjusted to maximum output.

• Reduce the transmitter gain by clockwise rotation of the transmitter gain pot until the output starts to decrease. A value of 80% of the maximum assures linear operation.

• Connect an antenna to the receiver input. Adjust the XO until a known beacon reads the correct frequency. If the correct frequency cannot be adjusted solder a 220nH choke in parallel to the crystal.

Take some low resistance carbonised foam and glue it into the bottom cover. This damps the resonances of the box. The heatsink should be mounted onto a chassis plate to further reduce the thermal resistance.

Fig 9.46 shows the connection of the unit to a transceiver and antenna. Fig 9.47 shows the output spectrum of the completed unit.

Amplifiers

Most transverters have a limited power output, it is therefore necessary to use a power amplifier. Valve amplifiers are still used but semiconductor amplifiers are in use by many amateurs.

A solid state broadband 80 Watt amplifier for 23cm [4]

This circuit produces 80 Watts of RF on the 23cm band, with 4 Watts of drive. It is made with four coupled hybrid amplifiers M 57762 from Mitsubishi.

Amplifier Description

One of the most interesting points about the unit is that it can be used for mobile or portable use, because it only needs a 12 volt power supply. It is rather compact and can be fitted into a standard 19" rack. It weighs about 10kg with its switching power supply, so it can advantageously replace a 2C39 valve amplifier. For the owners of QRO valve amplifiers (F 6007, TH328, 338, etc.), this circuit can be used as a driver. Unlike valve amplifiers, this amplifier is suitable for all modes. It is broadband, which allows for ATV use, and does not drift, which means that no retuning is required during operation. The broadband qualities of the individual amplifiers are retained in the overall circuit, because broadband power splitters are used.

Multi brick amplifiers have been described before, using splitters made up of four coax stubs. Although much cheaper, such an approach requires extreme care to be used in construction to avoid unbalances. Unbalance produces losses, thus decreasing both gain and power output capabilities.

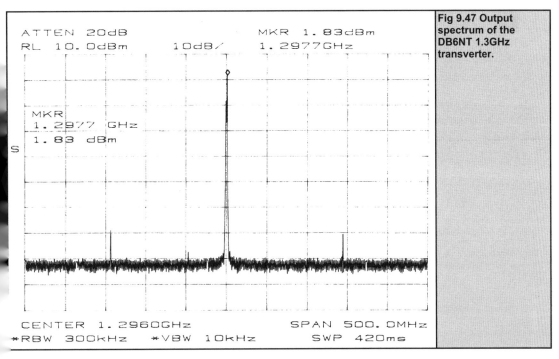

Fig 9.47 Output spectrum of the DB6NT 1.3GHz transverter.

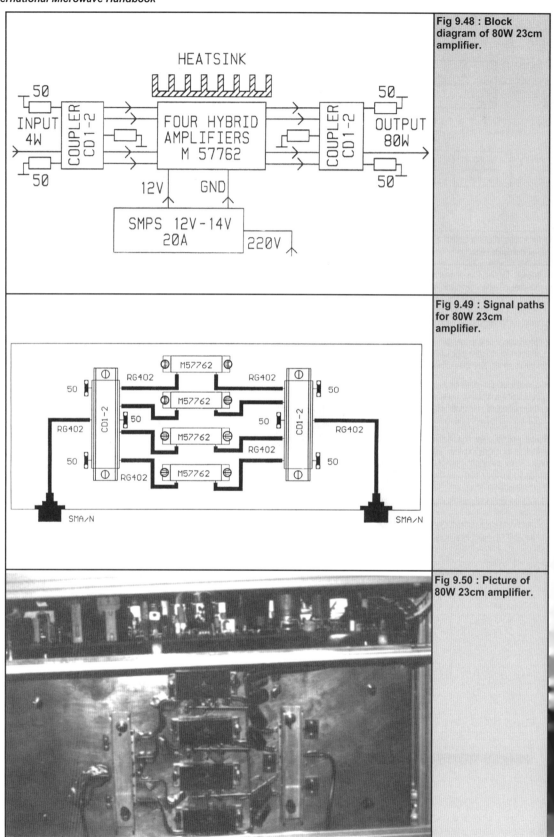

Fig 9.48 : Block diagram of 80W 23cm amplifier.

Fig 9.49 : Signal paths for 80W 23cm amplifier.

Fig 9.50 : Picture of 80W 23cm amplifier.

Fig 9.51 : Close up picture of an hybrid module in the 80W 23cm amplifier.

Furthermore, a coupler made from coax has a much narrower bandwidth than with hybrid couplers, which means that the broadband capability of the hybrid amplifiers is lost. This can be a disadvantage for FM ATV operation for example.

The circuit connections are shown in Fig 9.48 and the RF paths can be seen in more detail in Fig 9.49 and in photographs, Fig 9.50 and Fig 9.51. Table 9.3 shows the component list.

The input RF power reaches the four way power splitter through low loss (semi rigid) coaxial cable. The power splitter is broadband and needs only 50Ω low reactance resistors, which can be seen in Fig 9.50, as external components. The latter are required to absorb any unbalance at the splitters outputs. With 4 Watts at the amplifiers input, allowing for a slight insertion loss in the splitter and the coax stubs (0.3dB overall) about 1 Watt comes out of each of the four splitter outputs, and is brought through four identical lengths of semi rigid coax to the four hybrid amplifiers inputs.

Power supply decoupling is very important and must be carried out using good quality capacitors and short leads at each hybrid amplifiers power input.

No printed circuit board is required for the amplifiers to avoid losses. RF input and output leads are soldered directly to the coax inner conductors. The same technique is used for the power supply leads. This can be seen in Fig 9.52.

The outputs of the four amplifiers are connected with semi rigid coax to a second hybrid power splitter, which is used the other way around to merge the outputs from the four hybrid amplifiers. Three balancing resistors are also used here, somewhat larger than at the input.

Semi rigid type RG 402 coax cable is used throughout the RF paths. It has low losses, can be readily soldered and its power handling capability is quite high at these frequencies.

The whole circuit is mounted on a large copper plate, which acts as a ground and is bolted onto an aluminium

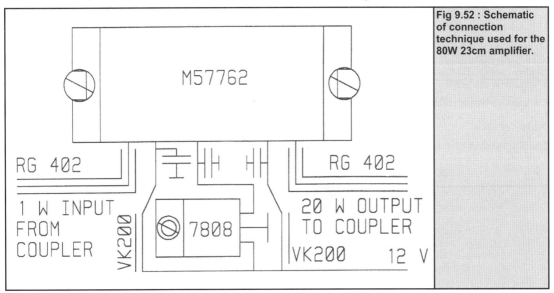

Fig 9.52 : Schematic of connection technique used for the 80W 23cm amplifier.

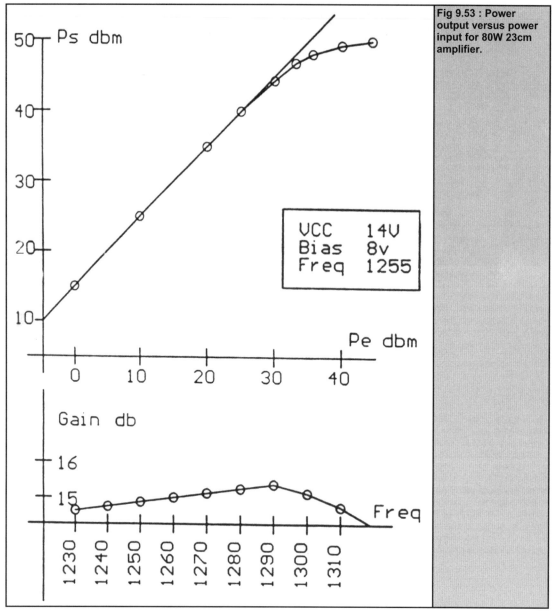

Fig 9.53 : Power output versus power input for 80W 23cm amplifier.

heatsink. About 200 Watts will have to be dissipated and a fan can be useful here.

The coax stubs are soldered directly to the copper plate, which makes a very stable assembly. This is essential in this kind of circuit.

The 50Ω resistors and the hybrid amplifiers are bolted to

the heatsink. The copper plate and the heatsink are drilled together. The holes in the copper plate are then slightly enlarged and only the heatsink is tapped.

The 6dB hybrid couplers are pressed against the copper plate by a U-shaped metal profile, visible in Figs 9.50 and Fig 9.51.

The hybrid couplers must make good contact with the copper plate over their whole length. This requires small holes to be drilled in the copper plate to make space for the rivet heads in the coupler's packages.

Thermal compound should be used under the hybrid amplifiers and between the copper plate and the heatsink. It should be stressed that the final quality of this circuit depends mainly upon the mechanical stability of the assembly and the quality of the RF connections.

Table 9.3 : Component list for 80W 23cm amplifier.	
1 mtr	RG402 semi rigid coaxial cable, RS Components
6	Resistive load, Elhyte Sarl type T-250-500-10 [5]
2	6db coupler, Nucletudes SCD type CD 1-2 [6]
4	Hybrid amplifier, Mitsubishi type M57762

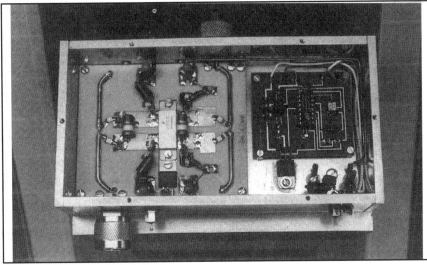

Fig 9.54 : Photograph of 90W 23cm amplifier.

Testing the amplifier

This is the easiest part, because there is nothing to be tweaked! Connect a 50Ω 100Watt 1.2 GHz specified dummy load at the output through a power meter and another dummy load at the input. Turn the power supply on and cheek that the 12 volt and 8 volt rails on the amplifiers. Check that NO RF is present at the output.

Turn the power off and replace the dummy load at the input with an RF generator at the correct frequency.

Increase the power progressively at the input and check the output power as well as the balancing resistors temperature, which should not rise at all unless there is an imbalance somewhere! The gain must be about 15dB. The test equipment used was:

- HP 435A Power Meter
- HP 8481A Power Sensor
- Narda 769 30dB Attenuator
- HP 8558B Spectrum Analyser
- HP 8444A Tracking Generator
- HP 5386 Frequency counter

On the prototype the impedance match was better than

25dB from 1240 to 1310MHz. Saturated output power was 80 Watts with a 12 volts power supply, 90 Watts with 13 volts power supply and 100 Watts with 14 volts.

Power supply current is of the order of 18 amps at 13 volts with output power saturated and 20 amps at 14 volts. Fig 9.53 Shows power output versus power input. Thanks to Marc, F3XY for confirming the results.

A 90W transistor PA for 1.3GHz [7]

In Dubus 4/96, Konrad Hupfer, DJ1EE, introduced a PA for 1.3GHz based on Push Pull transistors from MA/COM and Motorola [8]. These transistors are either very expensive or not available. In the meantime a similar transistor from Ericcson, the PTB-20174 is available and moderate in pricing. Moderate in this case means that the transistors costs about DM 530, (US$ 300.00). This price, which may not be ham-like may give some headaches if this transistor is killed during operation or initial tuning. Table 9.4 shows, that the manufacturers data on 1490MHz can be verified on 1270MHz too. Output power is in excess of 90W with quite good linearity and an excellent efficiency of 57% when driven with a drive power of about 14W can be achieved (Fig 9.54). This value is considerable better than with hybrid modules (<30%) and even better than can be achieved with valves like the 2C39.

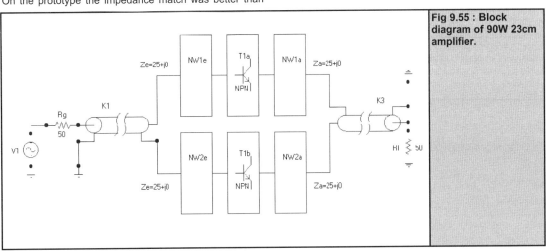

Fig 9.55 : Block diagram of 90W 23cm amplifier.

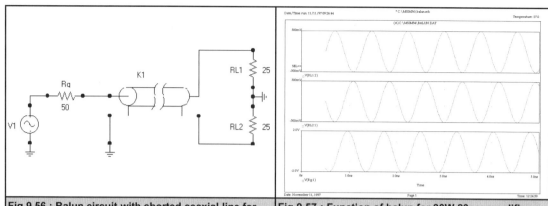

Fig 9.56 : Balun circuit with shorted coaxial line for the 90W 23cm amplifier.

Fig 9.57 : Function of balun for 90W 23cm amplifier.

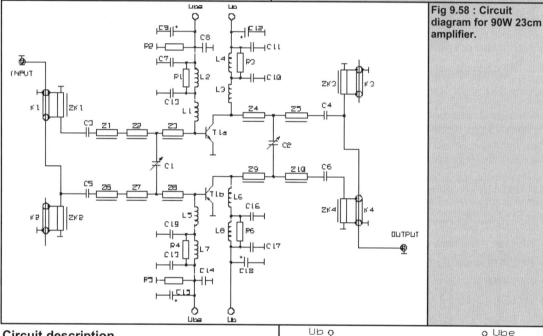

Fig 9.58 : Circuit diagram for 90W 23cm amplifier.

Circuit description

The transformation networks (NW1e, NW2e and NW1a, NW2a) have been calculated to accommodate the transistor PTB20174 and the substrate Rogers 4003 with an εr=3.38. The block diagram can be seen from Fig. 9.55. Short pieces of semi rigid function as baluns to drive the transistors in push pull and combine the output. Fig. 9.56

Table 9.4 : Comparison of manufacturers and measured data for the PTB-20174.

Data	Power Gain (dB)	Output Power (W)	Efficiency (%)	Bias Current (mA)
Manuf. @1490MHz	8.3 (typ)	>90	56	250
Measured @1270MHz	8.1	>90	57	300

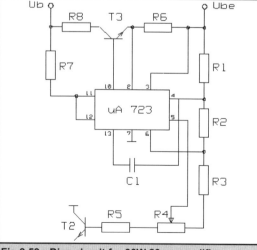

Fig 9.59 : Bias circuit for 90W 23cm amplifier.

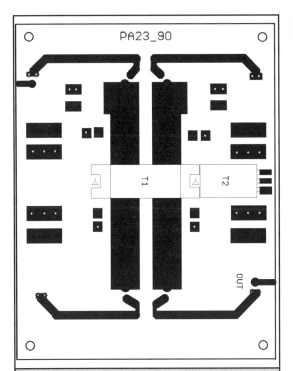

Fig 9.60 : PCB layout for 90W 23cm amplifier.

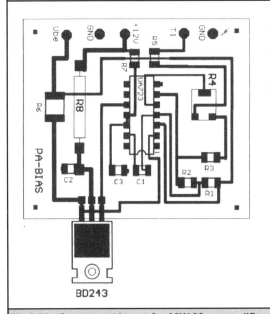

Fig 9.61 : Component layout for 90W 23cm amplifier.

and Fig 9.57 show the function of these baluns. Voltage V1 is divided in half and fed in antiphase to RL1 and RL2.

The circuit diagram of the PA is shown in Fig. 9.58. The baluns are soldered to quarter wave microstriplines ZK1 and ZK2. To preserve symmetry additional lines ZK2 and ZK3 together with additional coaxial lines K2 and K3 have to be introduced. This scheme provides a fully balanced

Table 9.5 Component list for 90W 23cm amplifier.

R1,4	10Ω 0.25W
R2,5	18Ω 0.25W
R3,6	12Ω
C1,2	11pF Tronser / JO 60-0715-10011
C3,5	18pF ATC 100B
C4,6	22pF ATC 100B
C7,13	1.5nF Ker
C8,14,11,17	0.1µF MKL
C10,16,13,19	33pF ATC 100B
C9,12,15,18	10µF 35v
L1,5	3t, 0.6mm CuEl, D=4mm
L2,7	12t, 0.6mm CuEl on R1,4
L3,6	2t, 0.6mm CuEl, D=4mm
L4,8	VK200, Valvo
K1,2,3,4	UT086 (D=2.2mm), Z=50Ω L=35.3mm
Z1,6	Z=13Ω, L=9.6mm
Z2,7	Z=15Ω, L=13.4mm
Z3,8	Z=15Ω, L=4mm
Z4,9	Z=15Ω, L=18mm
Z5,10	Z=15Ω, L=8.2mm
T1	PTB20174
T2,3	BD243 or BD201
PA90	PCB, Rogers 4003, 0.81mm 0.035mm Cu

Parts for bias circuit

T2,3	BD243
R1,4,7	1k
R2	18k
R3	8k2
R5	150Ω
R6	2Ω 0.5W
R8	47Ω 5W
C1	1nF 100v
C2,3	100nF 50v

configuration that works even if the load impedances of the transistor are not the same. The quarter wave microstriplines have a length of 35.3 mm and so do the coaxial lines.

The bias circuit is shown in Fig. 9.59. A temperature

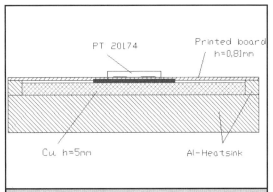

Fig 9.62 : Mounting details for heat spreader on 90W 23cm amplifier.

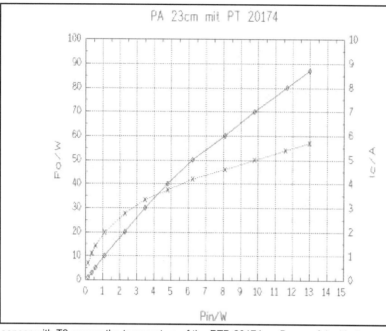

Fig 7.63 : Output power and current versus drive power for 90W 23cm amplifier.

sensor with T2 senses the temperature of the PTB-20174 and serves as a reference voltage for the bias regulator using a uA723. This circuit provides a temperature compensated and low impedance source for the base current. You should operate the bias circuit with 12 -14V, this voltage can be used as a T/R switching.

Construction

Fig. 9.60 Shows the pcb layout, Fig 9.61 Shows the component layout and Table 9.5 shows the components list. Variable trimmer C1 is soldered directly to the bases. C2 has a distance of about 18mm to the collector. C1 is initially tuned with the slug 1 mm inside the body and C2 is on minimum.

Be careful with the semi rigid lines. For initial operation connect a current limited power supply with an output voltage of 26V. Adjust for a bias current of 300mA. Adjust C1 for maximum return loss and C2 for maximum output with moderate drive power of about 5W. Check with an oscilloscope (at the collector circuit), that there is no oscillation on HF (3 - 50MHz).

A well designed heat sink has to be used for this PA. With full output a thermal power of about 80W has to be sinked. The heat sink should have a thermal resistance of less than 0.5°C/W. For example a Fischer type SK418 with a length of 200mm provides a thermal resistance of 0.3°C/W. For effective heat distribution a heat spreader made from a 6mm thick piece of copper is bolted to the transistor Fig 9.62.

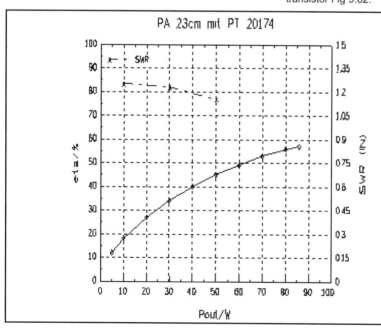

Fig 9.64 : Efficiency and input vswr versus output power for 90W 23cm amplifier,

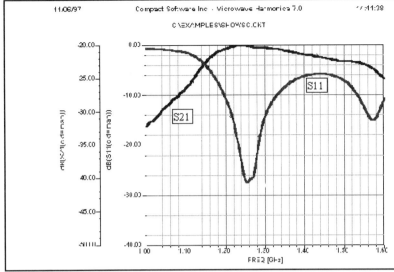

Fig 9.65 : Measuremt of S21 and S11 for low power (<1mW) for 90W 23cm amplifier.

A blower may be worthwhile since the temperature of the heatsink can reach a temperature of nearly 80°C with full output and 100% duty cycle.

Measurement results

The PA has been measured at 1270MHz (Phase II uplink frequency on 23cm). Operation on 1300MHz requires only minor retuning of C1.

Fig. 9.63 shows the dependency of output power PO and collector current IC and drive power Pin. At a drive level of 13.8W an output power of 90W at a collector current of 6A can be achieved.

Fig. 9.64 shows an efficiency of 57% at this power level.

The input VSWR is less than 1.15. Fig. 9.65 shows the return loss and gain at low power.

Antennas

Yagi antennas are now used in most 1.3GHz stations. The main reason for this is that they offer relatively high gain with minimum weight and windage. Dishes less than about 1.5m in diameter are rather inefficient at this frequency and even at this size, may be inconvenient in many home installations. However, larger dishes are becoming more popular with contest and expedition stations since often they do not need to be mounted very high and normal masts can be used to support them.

A wide-band 23cm beam [9]

Noel Hunkeler, F5JIO, designed an array of six half wave dipoles with a reflector plane for a 23cm packet radio link It provides 10dBd gain and an SWR not exceeding 1.1:1 throughout the band.

Curtains of phased dipoles have been used by amateurs since before WWII, examples are shown in the 1937 Jones antenna handbook.

F5JIO consulted Rothammel, the German antenna bible, which gives the following guidelines for the reflector plane: for best F/B ratio, it should extend at least a half wave beyond the perimeter of the curtain on all sides. If made of wire mesh instead of solid sheet to reduce windage, the wire pitch should be λ/20 or less. A reflector

plane spaced 5/8λ behind the radiator adds maximum gain, up to 7dB, but a spacing of 0. 1 - 0.3λ gives better F/B ratio. If spaced at least 0.2λ behind the curtain, the reflector plane does not affect the feed point impedance of the array.

Matching

With the dimensions given in Fig 9.66, the feed point impedance of each dipole pair is approximately 600Ω balanced. Three pairs in parallel give 200Ω balanced. The 4:1 re-entrant line balun transforms this to 50Ω, unbalanced. Note that each dipole is supported at its voltage node, hence the standoffs need not be high quality insulators.

Construction

For 23cm this antenna is small, so a solid aluminium reflector is practical, this plate supports all other components. Slightly bend the phasing rods so they do not touch at the crossovers. For weather protection, a plastic food container serves as a radome. Its RF absorption seems negligible for our application and it is much cheaper than Teflon. Though precision is required, construction of this antenna is not difficult.

Table 9.6 Component list for wideband 23cm beam.

Reflector	400 x 400 (340 min) 2.5 thick aluminium sheet (Qty 1)
Stand-off	Teflon (or PVC) 60L x 20D (Qty 6)
Dipole	Brass, silvered, 108L x 6D (Qty 6)
Rod, phasing	Wire, silvered, 2D (Qty 4)
N connector	(Qty 1)
Feedline	Semi rigid coax, 50Ω, approximately 4D (Qty 1)
Balun	as above, 92.5L (Qty 1)
Bolt	M3 x 8, SS (Qty 4)
Cover	Plastic food container (Qty 1)
Mast clamp	From TV antenna (Qty 1)

All dimensions in mm

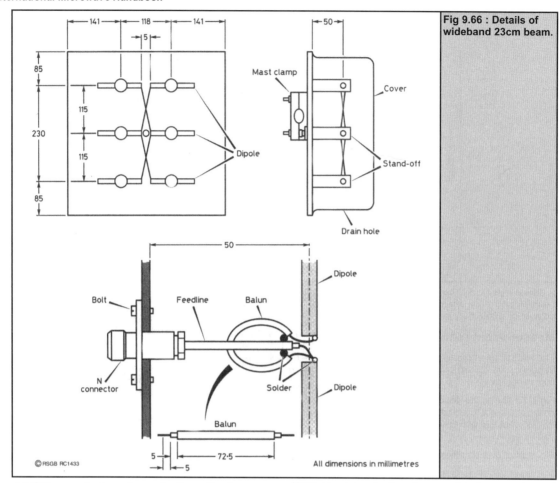

Fig 9.66 : Details of wideband 23cm beam.

1.3GHz conical and parabolic reflectors [10]

For the UHF and microwave operator the ultimate in high gain antenna has for long been the parabolic dish reflector, the larger the better. But for the home constructor a true paraboloid is not an easy shape to come to terms with. The main difficulty is that the paraboloid is a doubly curved surface. If constructed from flat sheets of material, part of the material needs to be either stretched or compressed or both.

An effective alternative was brought to attention by Fred Brown, W6HPH (then also G5AWI). He showed in a 1966 issue of the VHF'er that conical reflectors, constructed very easily from flat sheets of material (hardware cloth in his case) could provide an extremely useful substitute. A conical reflector can give considerably more gain than would a corner reflector antenna yet it is just as easy to build.

A shallow cone has the advantage of being a singly curved surface and can be made from flat circular sheet by removing or overlapping a segment of the material. Performance depends on the fact it is not really vital for the surface of a parabolic reflector to be a true paraboloid. W6HPH pointed out that it is usually accepted that there can be departures of up to one sixteenth of the wavelength at any point on the surface without suffering any significant deterioration of gain and directivity.

While one sixteenth of a micro wavelength may not sound much, in practice it permits quite drastic changes in the overall shape. Up to a certain size, in terms of wavelength, it can be two shallow cones, one inside the other: Fig 9.67. A two cone reflector can be satisfactory up to a diameter of 13.86 wavelengths in certain conditions. This means that for 1296MHz one could build a reflector of some 10ft diameter, or even 31.5ft at 432MHz without having to worry about double curved surfaces, and yet

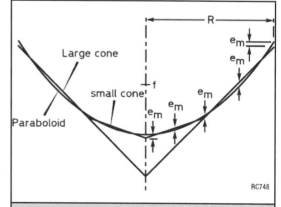

Fig 9.67 : Principle of the polyconic reflector showing the small departures (m) from the true paraboloid.

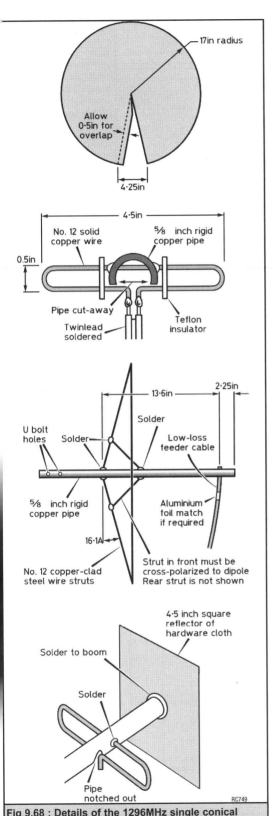

obtain virtually as much gain and directivity as with a true paraboloid.

A single conical reflector is satisfactory up to 3.46 wavelengths. W6HPH provided details of a 1296MHz single conical reflector antennas with a gain of about 16dB: Fig 9.68. He claimed this could be constructed in a few hours for a cost (in 1966) of about $1.50. The cone has a radius of 17in. Both the cone and the small 4.5in square reflector for the folded dipole element were made from 0.5in mesh hardware cloth.

I was reminded of W6HPH's suggestions by an article "High Performance Antenna for 24cm" by John Cronk, GW3MEO (Electronics World & Wireless World, August 1995, pp 699-700). He introduced his design as having fewer critical dimensions than most other configurations, and capable of being easily customised. It comprises a lightweight parabolic shaped plane reflector which is illuminated with a horizontally polarised dipole and reflector feed.

Gain is given as about 15dBd (i.e. about 17dBi) representing a power increase of almost 32 times. Bandwidth is more than ample even for amateur television. Wind resistance is lower than for expanded aluminium mesh.

The mesh reflector was fashioned from a 3ft by 2ft sheet of wire mesh called Handy mesh obtained from a local hardware store. The mesh consists of half inch squares formed by 22swg tinned wire. The sheet was cut in half to form a strip 60in long with the overlapped section strengthened by binding some of the coincident wires with 22swg tinned copper wire and then soldering. The mesh is then fixed to an aluminium square former that is shaped as a parabola.

For full details on building GW3MEO's antenna see the original article in EW & WW. The design includes a balun feed etc and is considerably more complex than the W6HPH design. But it seems worthwhile to show how he obtains a parabola profile, without resorting to mathematics, by using a pin and string: Fig 9.69.

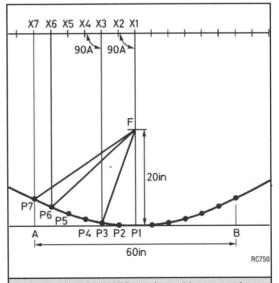

Fig 9.68 : Details of the 1296MHz single conical reflector antenna as designed by W6HPH providing about 16dB gain.

Fig 9.69 : How GW3MEO produces his antenna's parabolic profile drawn without resorting to mathematics by using a pin and string.

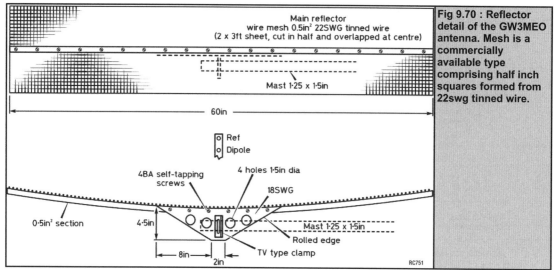

Fig 9.70 : Reflector detail of the GW3MEO antenna. Mesh is a commercially available type comprising half inch squares formed from 22swg tinned wire.

GW3MEO explains this as follows: "First draw line AB, and then at its centre draw line PFX, at right angles to AB. Next either draw line Xy parallel to AB, or use a long rule or tape parallel to AB, this must be marked off with regular divisions - X1, X2 etc. Now fix one end of a piece of string at point F using a pin. Take the string around another pin at P1, and then up to the point X1, and mark this length with a knot. Now plot the curve by moving the knot to X2, and keeping the string at right angles to XY, prick a mark at P2, and so on, until half the curve is marked out. Repeat for the other half of the curve. Draw a line smoothly through the pin pricks to show the shape of the reflector surface."

The curve is later transferred to a home made bending device using hardwood blocks and a vice to avoid crumpling the tubular reflector former which comprises 60in of 0.5in square aluminium tube stock, as commonly used for TV yagi antenna booms: Fig 9.70.

Since this is a planar reflector with a parabolic profile, the problem of the double curve is avoided. GW3MEO gives the acceptable tolerance of one tenth of a wavelength (2.5cm at 1.3GHz), a looser tolerance than W6HPH's one sixteenth wavelength. It would be interesting to compare the two designs in terms of performance, cost and time of construction.

Filters

A simple general purpose filter

A simple filter for low power applications is shown in Fig 9.71. It can be tuned to operate at 1,152 MHz and 1,296MHz. The filter consists of a shortened quarter wave line, tuned by a 2BA screw, with capacitive input/output coupling using two BNC sockets. The outer body can be easily fabricated from copper sheet. The filter is tuned to the desired frequency by means of the 2BA screw and the insertion loss is minimised by screwing the BNC sockets in or out.

An interdigital bandpass filter for 1.3GHz

Most methods of generating RF at 1.3GHz give rise to some unwanted or spurious outputs, e.g. harmonics and mixer products. Varactors and ssb mixers will produce the highest spurious levels and it is often necessary to connect an effective filter between such transmitters and the antenna.

A suitable three pole filter is shown in Fig 9.72. It consists of three shortened, tuned quarter wave lines, with the input and output connections tapped on to the outer lines. Construction should be evident from the figure, the only critical point being that for minimum insertion loss all surfaces that are screwed together should be a good fit. Ideally, they should be hard soldered together, with the top and bottom plates screwed to the frame by four or five screws along each side to ensure good RF contact.

Adjustment of the filter may be carried out by connecting it to the input of the receiver tuned to a small signal source

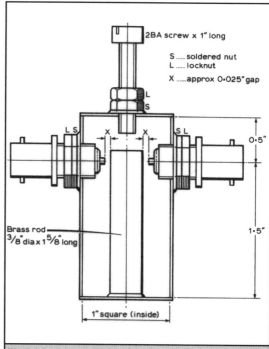

Fig 9.71 : A simple general purpose filter for 1.3GHz.

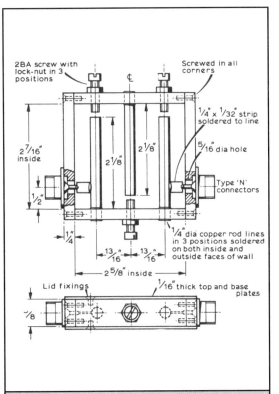

Fig 9.72 : An interdigital filter for 1.3GHz.

Labels within the figure:

- 2BA screw with lock-nut in 3 positions
- Screwed in all corners
- ¢
- ¼" x ¹/₃₂" strip soldered to line
- 2 ⁷/₁₆" inside
- 2 ⅛"
- 2 ⅛"
- ⁵/₁₆" dia hole
- Type 'N' connectors
- ½"
- ¼"
- ¹³/₁₆"
- ¹³/₁₆"
- ¼" dia copper rod lines in 3 positions soldered on both inside and outside faces of wall
- 2 ⅝" inside
- Lid fixings
- ¹/₁₆" thick top and base plates
- ⁷/₈"

set up at a suitable distance from the antenna. Once the filter has been tuned up for maximum signal strength with the weakest signal available, further fine tuning may be carried out by inserting a suitable attenuator between the filter and the receiver. A long piece, say 5 or 10m long, of thin (lossy) coaxial cable is suitable for this purpose.

The insertion loss is small (approximately 0.5dB) and so virtually no loss of output power should be seen when the filter is put directly in the output of the transmitter.

A final return of the transmitter output load and tuning adjustments is recommended after the filter in installed to recover any power lost from small loading changes. The filter could also be used as an image rejection filter in a receiver.

References

[1] From article by Matjaz Vidmar, S53MV, CQ ZRS

[2] From article by Michael Kuhne, DB6NT, Dubus 2/2000 pp 7 - 20. Kits available from Kuhne Electronics, Birkenweg 15, D-95119, Naila, Germany, email , http://www.db-6nt.com.

[3] Transverter for 1.3GHz by DB6NT, Dubus 3/91

[4] From article by Angel Vilaseca, HB9SLV and Serge Riviere, F1JSR, VHF Communications 2/94 pp 85 - 92

[5] Elhyte Sarl, B.P. 34.91620 La Ville Du Bois, France

[6] Nucletudes SCD, Av Du Hoggar, Z.A. du Courtabeuf, B.P. 117, 91944 Les Ulis 2, France

[7] From article by Dieter Briggmann, DC6GC, Dubus Technik V pp 342 - 352

[8] Konrad Hupfer, DJ1 EE, 100W Transistor Linear on 1.3GHz, DUBUS 4/96, pp. 5 - 12

[9] From article by Noel Hunkeler, F5JIO, RadCom July 1997 pp 72

[10] From in Technical Topics, RadCom October 1995, pp 75 - 76

Chapter 10

The 2.3GHz (13cm) band

In this chapter :

- Preamplifier
- Receive converter
- Tranceiver
- Amplifiers
- Antennas
- Filters

T he use of the 2.3GHz band can largely be attributed to the ready availability of commercial equipment on the amateur market. This has allowed the testing of a greater number of paths, due to the ownership of advanced equipment. As the equipment is highly transportable this has allowed easier operation from elevated portable sites.

A number of interesting facets of operation on the band have increased spectrum usage and improved equipment performance. These include:

Television

The 2.3GHz band is widely used for amateur television this is largely due to the availability of commercial TV sender equipment that can easily be used in the amateur band allocation.

Satellites

The provision of equipment operating around 2.4GHz on recent amateur satellites has encouraged development of receiving equipment and techniques for the investigation of space to earth propagation at this frequency.

Moonbounce

The 13cm band is used by amateurs for moonbounce but is not as popular as 23cm and 3cm.

Band planning

A majority of countries have an allocation in the 2.3Ghz to 2.45GHz range. In the UK the amateur service is allocated the range 2,310MHz to 2,450MHz on a secondary basis to the fixed service. The secondary status is shared with mobile and radiolocation services. The amateur satellite service has the allocation of 2,400MHz to 2,450MHz. Both services must accept interference from the industrial, scientific and medical allocations within their spectrum. Fig 10.1 shows the UK band plan and Fig 10.2 Shows the USA band plan.

Propagation and equipment performance

Most of the propagation modes encountered on the VHF/UHF bands are also present on the 2.3GHz band, with the exception of those that require ionised media such as aurora, meteor scatter and sporadic E. The most commonly encountered propagation modes on 2.3GHz are line of sight, diffraction, tropospheric scatter, ducting and aircraft scatter. Moonbounce (eme) is also a possibility at this frequency.

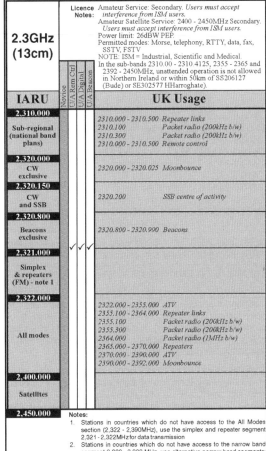

Fig 10.1 : UK Bandplan for 13cm.

Using the techniques of chapter 2, System analysis and propagation, it is possible to predict the possibilities for communication over a given path, when the performance parameters of the equipment are known. Fig 10.3 shows the path loss capability of five different sets of equipment ranging from a basic system of varactor doubler, interdigital converter and single Yagi antenna to a system with

Frequency USA Usage	
2300.0-2303.0	High rate data
2303.0-2303.5	Packet
2303.5-2303.8	TTY packet
2303.9-2303.9	Packet, TTY, CW, EME
2303.9-2304.1	CW, EME
2304.1	Calling frequency
2304.1-2304.2	CW, EME, SSB
2304.2-2304.3	SSB, SSTV, FAX, Packet AM, Amtor
2304.30-2304.32	Propagation beacon network
2304.32-2304.40	General propagation beacons
2304.4-2304.5	SSB, SSTV, ACSSB, FAX, Packet AM, Amtor experimental
2304.5-2304.7	Crossband linear translator input
2304.7-2304.9	Crossband linear translator output
2304.9-2305.0	Experimental beacons
2305.0-2305.2	FM simplex (25 kHz spacing)
2305.20	FM simplex calling frequency
2305.2-2306.0	FM simplex (25 kHz spacing)
2306.0-2309.0	FM Repeaters (25 kHz) input
2309.0-2310.0	Control and auxiliary links
2390.0-2396.0	Fast-scan TV
2396.0-2399.0	High rate data
2399.0-2399.5	Packet
2399.5-2400.0	Control and auxiliary links
2400.0-2403.0	Satellite
2403.0-2408.0	Satellite high-rate data
2408.0-2410.0	Satellite
2410.0-2413.0	FM repeaters (25 kHz) output
2413.0-2418.0	High-rate data
2418.0-2430.0	Fast-scan TV
2430.0-2433.0	Satellite
2433.0-2438.0	Satellite high-rate data
2438.0-2450.0	WB FM, FSTV, FMTV, SS experimental

Note: The 2300 MHz band plan was adopted by the ARRL Board of Directors in January 1991

Fig 10.2 : USA bandplan for 13cm.

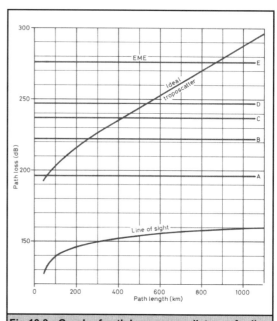

**Fig 10.3 : Graph of path loss versus distance for line of sight, tropospheric scatter and eme propagation, related to the performance of five sets of equipment
(A) 20dBi antenna, 1W tx, 15dB nf, 2.5kHz bw
(B) 1.2m dish, 20W tx, 15dB nf, 2.5kHz bw
(C) 1.2m dish, 20W tx, 2dB nf, 2.5kHz bw
(D) 1.2m dish, 20W tx, 2dB nf, 250kHz bw
(E) 4.0m dish, 25W tx, 2dB nf, 250Hz bw**

moonbounce capability. Also shown on the graph are the path losses associated with three different modes of propagation, line of sight (the least lossy mode of propagation), tropospheric scatter (a reliable, commercially used mode of propagation) and moonbounce.

It can be seen that relatively simple equipment is capable of working very long line of sight paths easily. However, very few paths are purely line of sight and propagation losses are noticeably higher. The tropospheric scatter loss is a reasonable upper limit for a terrestrial path and can be used to estimate the maximum range of equipment under normal propagation conditions.

From the graph it can be seen that the simplest equipment (A) might be expected to work paths of up to 70km long. By adding an amplifier and changing the antenna to a 1.2m dish (B) the coverage distance increases to 260km. By adding a receive preamplifier (E) the coverage distance increases to 425km. Line (D) shows the effect of the equipment of (C) but using a narrow bandwidth IF filter, as one could when receiving cw, the coverage distance increasing to 535km.

At distances over about 860km, it can be seen that the terrestrial path loss is normally greater than the route via the moon. Line (E) shows the parameters of equipment

capable of moonbounce on 2.3GHz. Under anomalous propagation conditions the path losses can be considerably reduced, often approaching the line of sight loss and even simple equipment is then capable of working very long paths.

Preamplifier [1]

A preamp equipped with a PHEMT provides a top notch performance in noise figure and gain as well as unconditional stability for the 13cm amateur band. Noise figure is 0.35dB at a gain of 15dB. It utilises the C band PHEMT NEC NE42484A and provides a facility for an optional second stage on board. The second stage with the HP GaAs MMIC MGA86576 can boost the gain to about 40dB in one enclosure. The preamplifier is rather broadband and usable from 2300 to 2450MHz.

Circuit description

The construction of this LNA follows the proven design of the 23cm HEMT LNA [2] by using a wire loop with an open stub as an input circuit (Fig 10.4). The FETs grounded source requires a bias circuit to provide the negative voltage for the gate. A special active bias circuit (see Fig 10.5) is integrated into the RF board which provides regulation of voltage and current for the FET.

Stub ST and inductance L1 provide a match for optimum source impedance for minimum noise figure. L1 functions as a dielectric transmission line above a PTFE board and has somewhat lower loss than a microstripline. L3 and L4 provide inductive feedback to increase the stability factor and input return loss. R1, R2, L9 and R3 increase the stability factor. The system of C2/L5/C6/L7 and L8 is

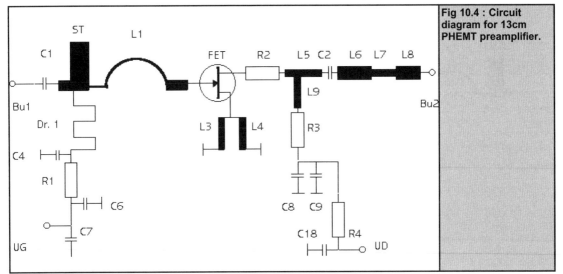

Fig 10.4 : Circuit diagram for 13cm PHEMT preamplifier.

specially designed to match the output of the single stage version to 50Ω and to allow easy insertion of the GaAs MMIC for the two stage version.

In the two stage version it provides the appropriate input and output match to the MMIC (Fig 10.6). This solution was found by doing some hours of design work with Microwave Harmonica. It allows the two versions to have the same PCB.

C4 provides a short on 2.3GHz, because it is in series resonance at this frequency. On all frequencies outside the operating band the gate structure is terminated by R1. Dr.1 is a printed λ/4 choke to decouple the gate bias supply.

The two stage version utilises a HP GaAs MMIC MGA86576 in the second stage. It provides about 2dB noise figure and 24dB gain. Input is matched by a wire loop for optimum noise figure. Output is terminated by a resistor R5 and a short transmission line L10. Together with L7/L8 and C3 a good output return loss is measured.

The source pads have to provide a very low inductance path to the ground plane, to preserve the MMIC's inherent unconditional stability. To achieve unconditional stability four ground connections are needed on each source. Appropriate source pads are provided on the PCB. Simulation indicates a minimum K factor of 1.2 in this arrangement on a 0.79mm thick substrate. A thicker substrate is prohibitive.

The MMIC typically adds 0.07dB to the noise figure of the first stage. This is somewhat difficult to measure, because most converters will exhibit gain compression, when the noise power of the source amplified by more than 40dB will enter the converter.

Construction

The construction uses microstripline techniques on glass PTFE substrate Taconix TLX with 0.79mm thickness with the PCB having dimension is 34x72mm. An active bias circuit, which provides constant voltage and current is integrated into the PCB (Fig 10,8). Fig 10.7 shows a top view of FET's and Fig 10.9 Shows the component layout.

The construction process is:

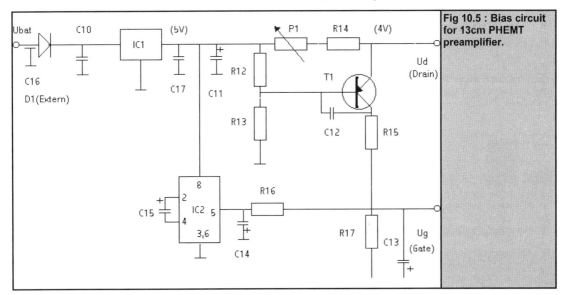

Fig 10.5 : Bias circuit for 13cm PHEMT preamplifier.

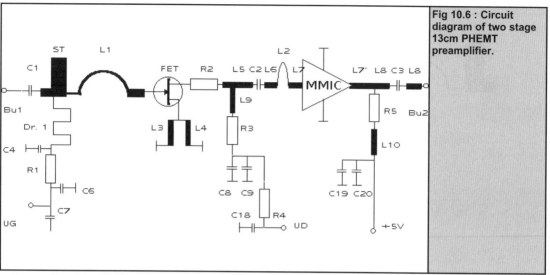

Fig 10.6 : Circuit diagram of two stage 13cm PHEMT preamplifier.

- Prepare tinned box (solder side walls).

- Prepare PCB to fit into box.

- Prepare holes for N connectors. Note, Input and output connector are asymmetrical. Use PCB to do the markings.

- Drill holes for through connections (0.9mm dia.) in the PCB and use 0.8mm gold plated copper wire to make through connections at the positions indicated.

- Solder all resistors onto PCB.

- Solder all capacitors onto PCB.

- For L1 cut a 17mm length of gold plated copper 0.5mm diameter wire. Bend down the ends at 1mm length to 45°. Form wire into a half circle loop as shown in Fig 10.9. Solder into the circuit with 1mm clearance from the PCB. The wire loop has to be flush with the end of the gate stripline and should be soldered in a right angles to it. The wire loop has to be oriented flat and parallel to the PCB.

- Verify the open circuit function of bias circuit. Adjust P1 to 45Ω resistance. Solder 100Ω test resistor from the drain terminal on the PCB to ground. Apply +12V to IC1 and measure +5V at output of IC1, -5V at IC2 Pin5, -2.5V> at collector of T1, +3.6V at emitter of T1, +3V at base of T1, -2.5V at R17 and +2.0V across the 100Ω. If OK, remove 100Ω test resistor.

- Solder PHEMT onto PCB. Use only an insulated soldering iron (Weller), ground the PCB, your body and the power supply of the soldering iron. Never touch the PHEMT on the gate, only on the source or the drain, when applying it to the PCB and solder fast (<5sec).

- Solder N Connectors into sides of the box.

- Solder the finished PCB into box, solder from both sides at the side walls and solder centre pins of the connectors to the microstriplines.

- Solder feed through capacitors into box.

- Connect D1 between feed through capacitor and PCB.

- Connect 12V and adjust P1 for 16mA drain current (measure 160mV across R4 on RF Board). Voltages

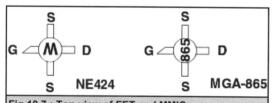

Fig 10.7 : Top view of FET and MMIC.

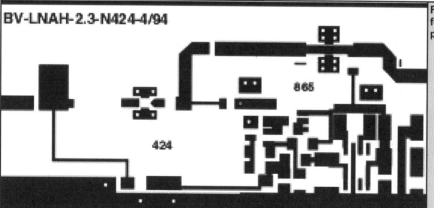

Fig 10.8 : PCB layout for 13 cm PHEMT preamplifier.

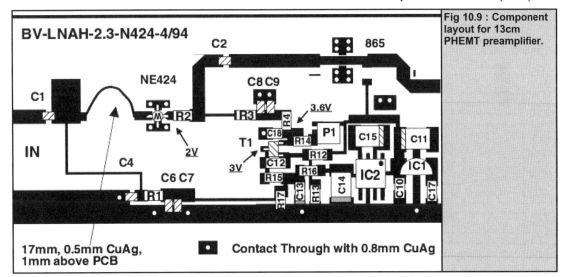

Fig 10.9 : Component layout for 13cm PHEMT preamplifier.

should be around +2.0V at the drain terminal, -0.4V at the gate, +3.6V at emitter of T1.

- Connect LNA to noise figure meter, if you have one, and adjust input wire loop, adjust the clearance to PCB as well as drain current by adjusting P1 for minimum noise figure. Even without tuning the noise figure should be within 0.1dB of minimum because of the limited tuning range of the wire loop.

- Glue conducting foam into inside of top cover and slip into the top of the box.

To add the MMIC amplifier, refer to Fig 10.10 for construction.

- Prepare PCB by cutting slits into the microstriplines around the MGA865. These are a 2mm slit for L2, a 1.8mm slit for the MMIC and a 0.8mm slit for C3.

- For L2 cut a 8mm length of gold plated copper 0.5mm diameter wire. Form wire into a half circle loop as shown in Fig 10.10. Solder wire loop into the circuit. The wire loop has to lie flat on the PCB. The wire loop has to be flush with the end of the gate stripline and should be soldered in a right angle to it.

- Follow other instructions given above

Measurement Results

Noise Figure and Gain

Measurements were taken by a HP8510 network analyser and HP8970B/HP346A noise figure analyser, transferred to a PC and plotted. Fig.10.11 and 10.12 show the

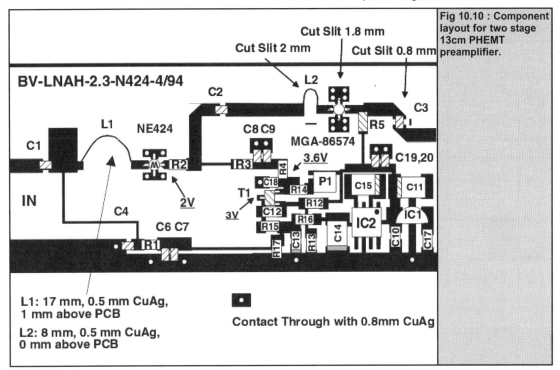

Fig 10.10 : Component layout for two stage 13cm PHEMT preamplifier.

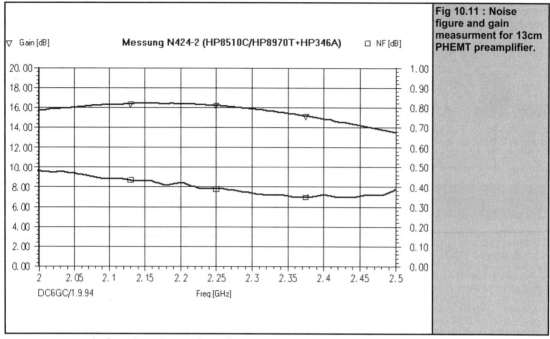

Fig 10.11 : Noise figure and gain measurment for 13cm PHEMT preamplifier.

measurement results for gain and noise figure for the 1 stage and 2 stage version respectively. Using a special PHEMT NEC NE42484A optimised for C Band, a typical noise figure of 0.35dB at a gain of 15 dB can be measured on 2.32GHz. An optional second stage on the same PCB using the GaAs MMIC MGA86576 from HP will boost the gain from 15db to 41dB. The two stage version measures with a noise figure of 0.45dB. This version can be used for satellite operation. For EME, where lowest noise figure is at premium, a cascade of two identical one stage LNAs may be more appropriate.

Both versions are rather broadband. They can cover the various portions of the 13cm amateur allocation from 2300 to 2450MHz without re tuning.

The real surprise is the performance of the C band PHEMT NE424. It performs better than several other HEMTs (FHX35, FHX06, NE324, NE326) tried in this circuit and it measures 0.15dB better than its published noise figure. In fact the Microwave Harmonica simulation predicts a 0.5dB noise figure based on the data sheet value. The lower noise figure measured seems to be due to a special bias current and the lower value of gamma at approximately 0.75 which is due to the gate length of 0.35μm and a gate. This provides optimum properties for application in 2 - 4GHz LNAs.

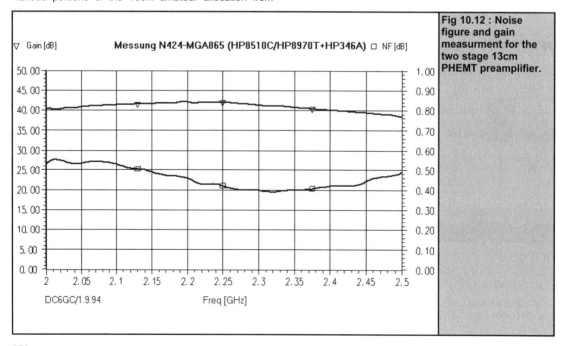

Fig 10.12 : Noise figure and gain measurment for the two stage 13cm PHEMT preamplifier.

Table 10.1 : Component list for 13cm PHEMT preamplifier.

C1	4.7pF Chip-C 50mil (500 CHA 4R7 JG)
C2,3,6	100pF SMD-C, size 0805
C4	5.6pF SMD-C, size 0805
C7,8,19	1000pF SMD-C, size 0805
C9,18,20	10nF SMD-C, slze 0805
C10,12,17	0.1µF SMD-C, size 1206
C11,14,15	10µF SMD-Electro, size 1210
C13	1µF SMD-Electro, size 1206
C16	1000pF Feedthrough
R1,3	47Ω SMD-R size 1206
R2,14	39Ω SMD-R, size 1206
R4,5	10Ω SMD-R, size 1206
R12	6.8kΩ SMD-R, size 1206
P1	100Ω SMD-Pot, Murata 4310
Dr.1	Printed λ/4
L1	Wire loop, 0.5mm gold plated copper wire 18mm long, 1mm above board
L2	Wire loop, 0.5mm gold plated copper wire 8mm long, on PCB
D1	1N4007
FET	NE42484A, NEC
MMIC	MGA-86576, HP
T1	PNP e.g. BC807, BC856, BC857, BC858 BC859, SOT-23
IC1	µA7805A, TO-92
IC2	LTC1044SN8
Bul1,2	N small flange or SMA
PCB	Taconix TLX, 35 x 72mm, 0.79 mm εr=2.55
Box	Tinplate 35 x 74 x 30mm

Stability

Stability is excellent. This has been achieved by a carefully controlled combination of inductive source feedback, resistive loading in the drain and non resonant DC feed structures for drain and gate. A broadband sweep from 0.2 to 20GHz showed a stability factor K of not less than 1.2 and the B1 measure was always greater than zero. These two properties indicate unconditional stability. At the operating frequency of 2.3GHz stability factor is about 1.6. The 2 stage version with the MGA865 measures with K>4 at all frequencies.

Conclusions

The preamplifier provides quantum leap towards the perfect noiseless preamplifier. It uses a low cost and rugged C band PHEMT instead of relying on expensive X band HEMTs. An improvement of about 0.2dB in noise figure has been achieved in comparison to the no tune HEMT preamplifier described in [3]. This improvement provides roughly 1.5dB more S/N in EME or satellite operation but is not noticeable in terrestrial links. However, the new preamplifier has to be tuned. This requires a noise figure meter for alignment. For those, who like a no tune device, the HEMT preamplifier in [3] provides adequate performance. The preamplifier described there measures with 0.55dB NF typically.

Receive converter

13 cm receive converter [10]

This project began with preamplifier for 13cm. That worked well with no problems, all nice and stable. It was then obvious, if one were to place a buffer and a mixer after the RF stage and continue construction along the same lines that the RF stage was built, then it would be possible to construct a really good, sensitive 13cm, stable converter, with a powerful output signal to the backpack receiver, which could be 2m or 70cm. The RF stage comes from Dubus engineering book 4/95, in which DJ9BV is the author of a series of very impressive RF stages for the 23cm, 13cm, 9cm and 6cm amateur bands. The oscillator/multiplier comes from VHF Communications 3/86, albeit with some modifications.

The 13cm converter has been made with a double oscillator/multiplier, enabling it to receive two ranges on 13cm, 2320-2322MHz and the satellite range 2400-2402 MHz, where there is something new and exciting to use it for - AO 40 Phase 3D microwave satellite, where there is a downlink frequency on 13cms and the uplink frequency of 1269 MHz, S/L mode (S = 13cm, L = 23cm). Fig 10.13 shows a block diagram of the receive converter

The converter is constructed on three double sided, 1.6 mm thick fibre glass PCBs, which are highly suitable up to 3 GHz. In order to test the difference between fibre glass and teflon PCBs on 2.3 GHz, a converter has also been made on double sided, 0.79 mm thick teflon PCB. The result was that there was no difference to speak of between it and the converter on a fibre glass PCB.

The measuring equipment used for making adjustments etc. was:

- A sensitive 1GHz frequency counter

- An ordinary standard diode probe with an RF amplifier in front end of approximately 30 dB gain (three MAR2s)

- A signal generator, a crystal controlled oscillator whose 24th harmonic ends up at 2321/2400 MHz.

Construction and circuits

Printed circuit board No. 1 is an oscillator/multiplier, Fig 10.14 shows the circuit diagram. This PCB was made in an A and a B version, see Fig 10.16. The A PCB has an output frequency of 544MHz, it is used if ones backpack receiver is 144MHz. PCB B has an output frequency of 944MHz and is used if the receiver is on 432MHz. The two PCBs are mounted so that all the ceramic capacitors and the two BFR96Ss are located on the bottom of the PCB, which is also the circuit side, see Fig 10.17 for the component layout. One thing is very important: Remember to make approximately 10 wire feed throughs between the top and bottom of the PCB.

PCB No. 2 is two doublers if your receiver is 144MHz, or buffer and doubler if the receiver is 432MHz, Fig 10.18 shows the circuit diagram. The +/-5 volt supply is mounted on this PCB, Fig 10.19 shows the circuit diagram. In the RF circuit, all capacitors are of the SMD type. Resistors etc. are ordinary components. Both circuits are installed on the upper side of the PCB, Fig 10.20 shows the PCB. The entire underside is a ground plane. Remember to connect the four PCB coils at their cold end.

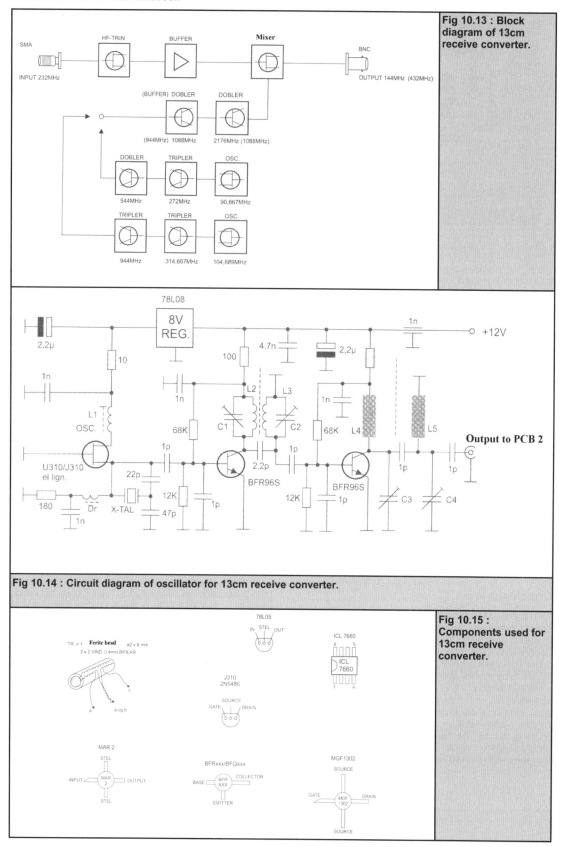

Fig 10.13 : Block diagram of 13cm receive converter.

Fig 10.14 : Circuit diagram of oscillator for 13cm receive converter.

Fig 10.15 : Components used for 13cm receive converter.

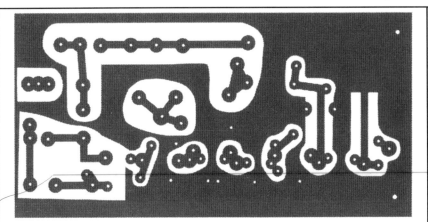

Fig 10.16 : Oscillator/ multiplier PCB for 13cm receive converter.

Type A - 544MHz output, crystal 90.667MHz, L1 3.5 turns 0.4mm wire on 5mm former, L2 & L3 3 turns 1mm wire 5mm diameter, 4 x 10pf trimer

Type B - 944MHz output, crystal 104.889MHz, L1 3.5 turns 0.4mm wire on 5mm former, L2 & L3 3 turns 1mm wire 5mm diameter, 4 x 5pf trimer

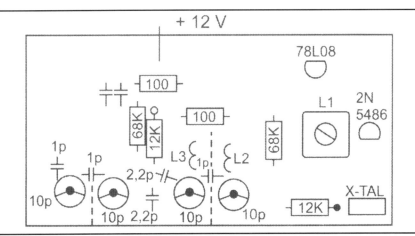

Fig 10.17 : Component layout of oscillator/ multiplier PCB for 13 cms receive converter.

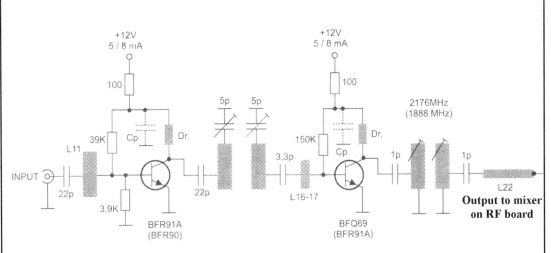

Fig 10.18 : Circuit diagram of dual doubler or buffer doubler for 13 cm receive converter.

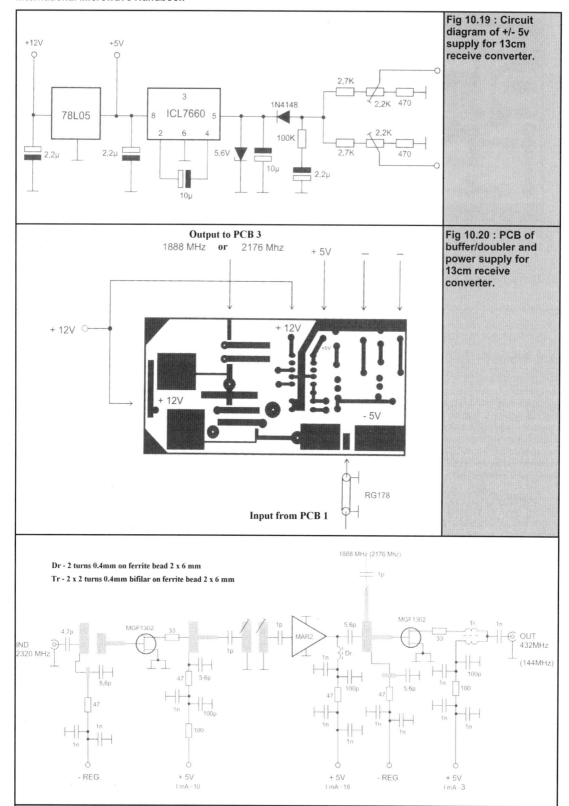

Fig 10.19 : Circuit diagram of +/- 5v supply for 13cm receive converter.

Fig 10.20 : PCB of buffer/doubler and power supply for 13cm receive converter.

Fig 10.21 : Circuit diagram of RF stage for 13cm receive converter.

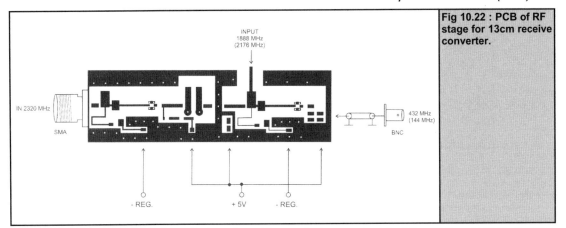

Fig 10.22 : PCB of RF stage for 13cm receive converter.

PCB No. 3 is the RF stage, buffer and mixer stage, Fig 10.21 shows the circuit diagram. All capacitors and resistors here are of the SMD type. In this PCB many wire feed throughs need to be made between the upper and lower surface of PCBs. The MAR2 amplifier must be soldered in carefully, as it is highly sensitive to heat (soldering iron).

In this PCB too, remember to connect the two PCB coils at their cold end, Fig 10.22 shows the PCB layout.

Adjustment

For the oscillator PCBS, place the oscillator on the right frequency. If there are problems with precise frequency, a choke coil can be placed in parallel over the crystal, approximately 20 turns of 0.1 mm wire wound on a 1M resistor. Adjust the two multiplier stages of circuit A to a resonance with an output frequency of 544MHz and approximately 10 mW/50 ohms. Circuit B with an output frequency of 944MHz will produce substantially less

output, approximately 1-3 mW (two triplers). In these stages, be extra careful to tune to the right frequencies.

For PCB No. 2, adjust the BP filter of the input stage using a frequency counter loosely hooked up to the output coil of the BP filter with a link coil (snooper). The next stage, a doubler stage with an output frequency of 1888MHz or 2176MHz, is best trimmed in resonance by listening to an antenna signal fed to the converter. The two PCB coils on the doubler stage are adjusted by either shortening or lengthening them approximately +/-2 to 5 mm. When lengthening the PCB coil, solder a piece of wire approximately 0.4 mm thick onto the hot end of the coil, noting now that there will be mixing noise in the RX, with the noise rising by approximately 3 to 4 dB. Assuming the converter works, you can easily stop one or two of the stages in the multiplier chain, and the overall sensitivity of the converter will only drop about 15 to 20 dB, so you can be misled when tuning up. So, remember mixing noise in the RX.

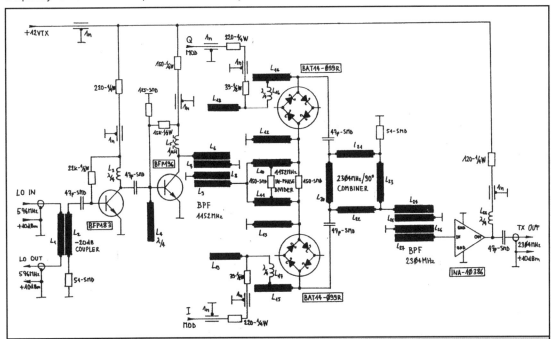

Fig 10.23 : Circuit diagram of 2304MHz quadrature transmit mixer.

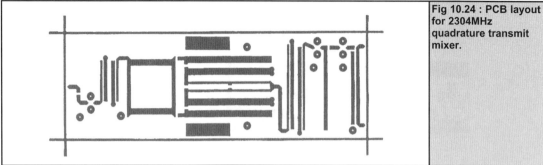

Fig 10.24 : PCB layout for 2304MHz quadrature transmit mixer.

For PCB No. 3, the 2320MHz RF unit of the converter only the two PCB coils in the BP filter need to be adjusted. Use the same procedure as for PCB No. 2. Listen to a weak signal and adjust by a couple of mm.

Transceiver [11]

The design for a zero IF transceiver is shown in chapter 9. This design has RF sections for 2.3GHz which are described here.

Quadrature transmit mixer

The circuit diagram of the quadrature transmit mixer for 2304MHz is shown in Fig 10.23. The 576MHz LO signal is taken from a -20dB coupler, amplified by the BFP183 transistor and then doubled to 1152MHz by the BFP196 transistor. The doubler output goes through a microstrip bandpass filter to feed the two sub harmonic mixers with BAT14-099R schottky quads.

The two 2304MHz signals are combined in a quadrature hybrid, followed by a 2304MHz bandpass filter. The latter removes the 1152MHz LO as well as other unwanted mixing products. After filtering the 2304MHz SSB signal level is rather low (around -11dBm), so an INA-10386 MMIC is used to boost the output signal level to about +10dBm.

The quadrature transmit mixer for 2304MHz is built on a double sided microstrip FR4 board with the dimensions of 40mmX120mm as shown on Fig 10.24. The corresponding component location is shown on Fig 10.25. The circuit

does not require any tuning for operation at 2304MHz or 2320MHz. For operation in the satellite band above 2400MHz, the LO bandpass should be readjusted to 1200MHz by shortening L7 and L8 at their hot ends.

The quadrature transmit mixer does not supply any output signal when the modulation input is absent. For transmitter testing purposes it is possible to obtain an output signal by feeding a DC current of 2-10mA into one or both mixers.

RF front end

The circuit diagram of the RF front end for 2304MHz is shown on Fig 10.26. The transmitter power amplifier includes two stages: a BFP183 driver and a CLY2 final amplifier. The additional BFP183 driver is required since the gain and output power of the INA-10386 in the transmit mixer are smaller at 2304MHz than at 1296MHz. Further the additional bandpass filter at the input of the power amplifier adds some insertion loss.

The BFP183 operates as a class A amplifier while the CLY2 is used in a similar self biasing arrangement like the CLY5 in the 1296MHz front end. Of course the drain current of the CLY2 is smaller, the +5V transmit current limiting resistor must be higher and the RF output power on the antenna connector amounts to about 0.5W (+27dBm). The PIN diode antenna switch is identical to that used in the 1296MHz RF front end with a series diode BAR63-03W and a shunt diode BAR80.

Since there are no powerful users of the 2.3GHz band,

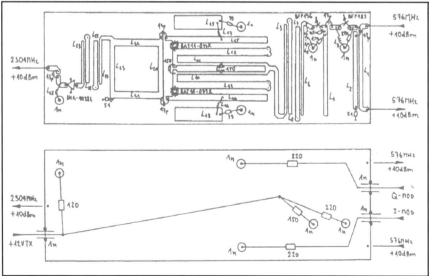

Fig 10.25 : Component layout for 2304MHz quadrature transmit mixer.

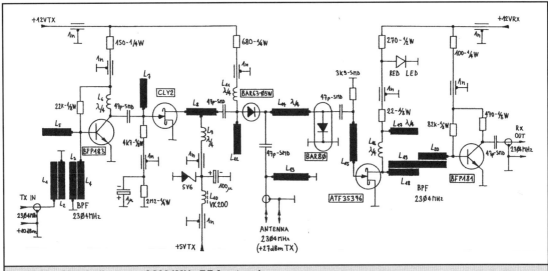

Fig 10.26 : Circuit diagram of 2304MHz RF front end.

the RF front end for 2304MHz includes a two stage LNA: a HEMT ATF35376 in the first stage and a BFP181 in the second stage. The overall gain of the LNA is around 23dB. Since the I_{dss} of the ATF35376 is usually around 30mA, no negative voltage needs to be applied to the gate.

The RF front end for 2304MHz is built on a double sided microstrip FR4 board with the dimensions of 40mm x 80mm as shown on Fig 10.27. The corresponding component location is shown on Fig 10.28. The RF front end for 2304MHz should require no tuning, since both the transmit and the receive chains have a few dB of gain margin. However, in order to squeeze the last milliwatt out of the

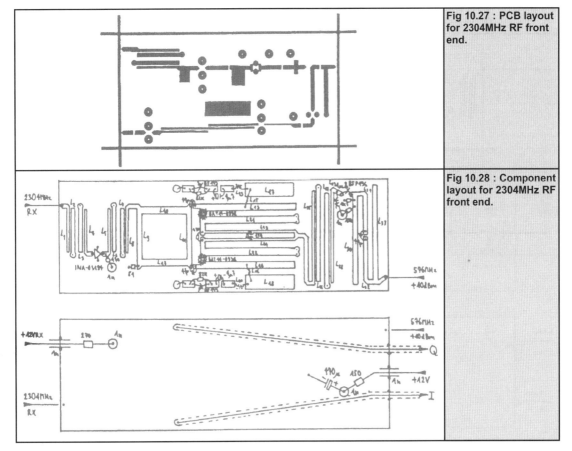

Fig 10.27 : PCB layout for 2304MHz RF front end.

Fig 10.28 : Component layout for 2304MHz RF front end.

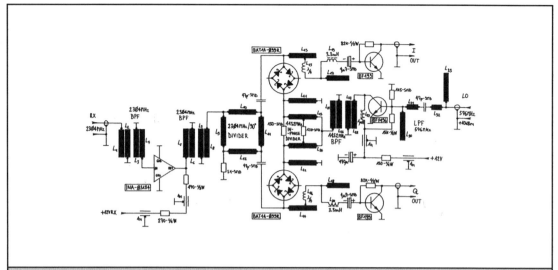

Fig 10.29 : Circuit diagram of 2304MHz quadrature receive mixer.

CLY2 (is it really necessary?) some tuning may be attempted on the output.

Quadrature receive mixer

The circuit diagram of the quadrature receiving mixer for 2340MHz is shown on Fig 10.29. Since the same components, INA-03184 MMIC and BAT14-099R Schottky quads, have similar performances in the 1296MHz and 2304MHz frequency bands, the circuit diagram of the 2304MHz mixer is almost identical to the 1296MHz mixer.

The only difference is the additional frequency doubler to 1152MHz with the transistor BFP196. The multiplier includes a low pass on the input and a bandpass filter on the output. The input low pass filter should prevent unwanted interactions with other circuits operating with the same 576MHz LO signal. The 1μH choke should have a ferrite core for the same reason.

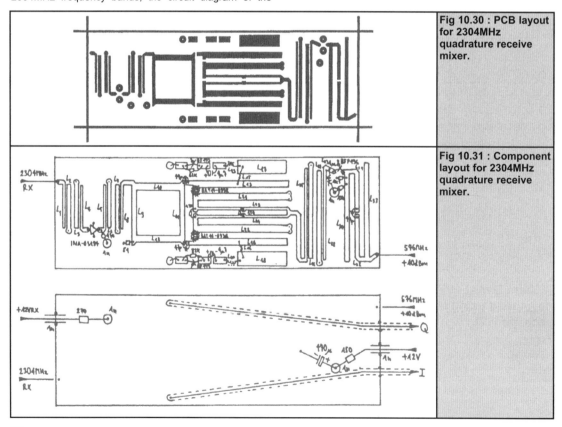

Fig 10.30 : PCB layout for 2304MHz quadrature receive mixer.

Fig 10.31 : Component layout for 2304MHz quadrature receive mixer.

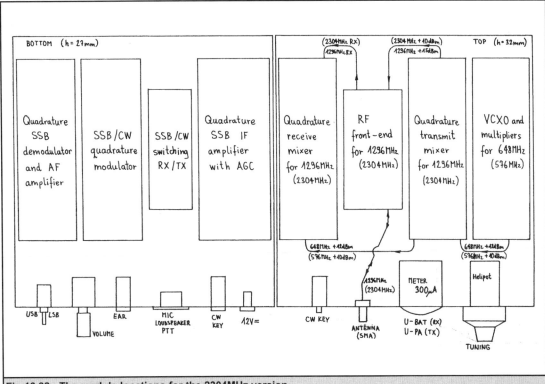

Fig 10.32 : The module locations for the 2304MHz version.

The quadrature receive mixer for 2304MHz is built on a double sided microstrip FR4 board with the dimensions of 40mmX120mm as shown on Fig 10.30. The corresponding component location is shown on Fig 10.31. The receiving mixer does not require any tuning for operation on 2304MHz or 2320MHz. For operation in the satellite band above 2400MHz, the LO bandpass should be readjusted to 1200MHz by shortening L26 and L27 at their hot ends.

Construction

The main construction details are shown in chapter 9, the items specific to the 2304MHz version are as follows.

The location of the modules of the 1296MHz or 2304MHz transceiver as well as the location of the connectors and controls on the front panel are shown on Fig 10.32 for both sides of the chassis.

Amplifiers

Semiconductor amplifier devices are readily available for 2.3GHz, two designs are shown here.

A 13cm GaAsFet Power amplifier developed using the PUFF CAD software package [12]

The 2 stage power amplifier described here supplies an initial output of 5 Watts at 23 dB amplification in the 13cm band. The circuit was developed using the PUFF CAD software package, which makes it amazingly simple to calculate and simulate even relatively complicated micro-wave circuits. Several publications [4], [5], together with

our own research, have already put the capability of the low cost software used to the test, so that very positive results were to be expected.

The goal was to develop several amplifiers using the software, build them, and compare the results with the simulated values. Three different types of amplifier were involved, with different performance figures varying from 4 to 12 Watts in the given frequency range.

Selection of semiconductors, the simulation/analysis of the amplifier circuit using the CAD software, and construction of the 5 Watt amplifier with the results obtained are described.

Selecting semi conductors

The transistors used in the amplifier were Mitsubishi types, from the 0900 range for UFW power amplifiers. They are, in actual fact, N channel Schottky GaAs power FET'S, which had already been successfully used in the construction of several circuits [6] [7], and which could be obtained at relatively low cost. Their power spectrum stretched from 0.6W (the 0904 type) right up to 10W (the 0907 type) for amplification levels of between 8 and 13dB, depending on type and frequency.

The performance figures targeted by the development:

Amplification > 20 dB at k > 1

Output   min 5 W at max 1dB compression

Band width   100 MHz

Zin = Zout = 50Ω at return loss $\geq$ 20dB

These can be attained with a 2 stage amplifier.

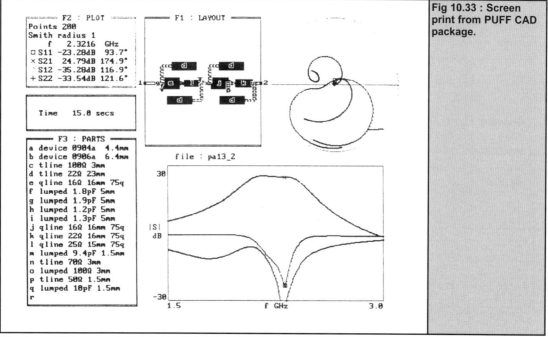

F2 : PLOT
Points 200
Smith radius 1
  f   2.3216 GHz
□ S11 -23.28dB 93.7°
× S21 24.79dB 174.9°
  S12 -35.28dB 116.9°
+ S22 -33.54dB 121.6°

Time   15.0 secs

F3 : PARTS
a device 0904a 4.4mm
b device 0906a 6.4mm
c tline 100Ω 3mm
d tline 22Ω 23mm
e qline 16Ω 16mm 75q
f lumped 1.8pF 5mm
g lumped 1.9pF 5mm
h lumped 1.2pF 5mm
i lumped 1.3pF 5mm
j qline 16Ω 16mm 75q
k qline 22Ω 16mm 75q
l qline 25Ω 15mm 75q
m lumped 9.4pF 1.5mm
n tline 70Ω 3mm
o lumped 100Ω 3mm
p tline 50Ω 1.5mm
q lumped 10pF 1.5mm
r

F1 : LAYOUT

file : pa13_2

|S|
dB

30

-30
1.5   f GHz   3.0

Fig 10.33 : Screen print from PUFF CAD package.

The type 0906 seemed a suitable high level stage transistor. It displayed particularly high operational thermal stability because of its large ceramic housing and, in contrast to the 0905, which was usually running under strain, easily supplied 37dBm = 5W at 1dB compression, thus guaranteeing stable operation with permanent output e.g. for ATV transmitters. The type 0904 was a suitable driving transistor, because it displayed a high level of amplification (13dB) for a compression free output amounting to almost 28dBm = 630mW. The S parameters of the selected transistors required for the development of the circuit came from the Mitsubishi data bank, and applied under the following DC conditions:

- MGF0904: UDS = 9 V at ID = 0.2 A

- MGF0906: UDS = 1 0 V at ID = 1. 1 A

The efficiency of these transistors is normally about 40%, so that a DC input power of more than 12 Watts is required in operation, the resulting power loss has to be dissipated through a heat sink of sufficiently large dimensions.

Simulation and analysis of amplifier circuit using cad software

The operation of the PUFF CAD software is comprehensively described in [4], [5] and [8]. Therefore only the results obtained will be described.

Fig.10.33 shows the screen dump from PUFF with the draft layout of the circuit, the associated Smith diagram, the parts and the paths of the scatter parameters over the frequency range selected (1.5 to 3.0 GHz). The plot window (top left) also shows the size and phase of the scatter parameters for the selected operating frequency (2.3216 GHz) in the order:

- input impedance (S11) with return loss value

- amplification (S21)

- feedback (S12)

- output impedance (S22) with return loss value.

The stability factor of the amplifier at the operating frequency can be determined from the calculated scatter parameters. The theoretical relationships required for this can be found in [9]. Determining the absolute stability (K > 1) using this factor has been tried and tested as best for normal RF amplifiers, so that from knowledge of the scatter parameters the frequency range over which the circuit will be stable can easily be estimated. A quadripole (amplifier) is absolutely stable if it always remains stable whatever the load at the input and output.

The gain slope obtained (S21) as a function of the frequency showed a marked resemblance to that of a coupled band filter. This characteristic was obtained, firstly through the lengthwise layout of the transmission lines (qlines/tlines) for each stage and secondly, through the 50Ω coupling of the two stages.

The figures shown in Fig 10.33 gave the following output values for the test circuit:

- Return loss input: -23dB

- Return loss output: -33dB

- Amplification at 2,320 MHz: 24.8dB

- Feedback: -35dB

- K factor at 2,320MHz: 5

- Bandwidth (-3dB): -280/+120 MHz

Fig 10.34 shows the PCB layout generated by the CAD software as a printout from a laser printer for Teflon based material with a substrate thickness of 0.79mm. A correction factor is required to generate a precise photographic image.

The tracks on the edge of the board are earth surfaces, which are through hole plated to the earth side of the PCB.

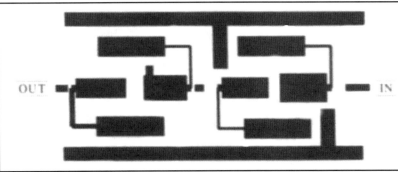

Fig 10.34 : PCB layout for the 13cm GaAsFet power amplifier, generated by PUFF.

The parts listed in the PUFF output as "lumped" (Fig 10.33) shows the discrete components required. These are capacitors and resistors that are required for the circuit to operate. Fig 10.35 shows the circuit diagram for the amplifier and Table 10.2 the parts list.

Table 10.2 : Component list for 13cm GaAsFet power amplifier.	
C1,2,3,4	2.5pf Teflon/Ceramic trimmer
C5	10pF Tekelec chip
C6,7	4.7pF ATC chip
J1,2	N connector
L1,2,3	Zs=100/24mm stripline
L4	Zs=70/24mm stripline
T1	0904 Mitsubishi N GaAsFet
T2	0906 Mitsubishi N GaAsFet
Zs	Zs=22/23mm stripline
Z1,3	Z=16/16mm stripline
Z2	Z=25/15mm stripline
Z4	Z=22/1=63mm stripline

Assembling the amplifier

The amplifier circuit was built on a Teflon board (εr = 2.33) with dimensions of 109 x 54 x 0.79 (mm). It was screwed to an aluminium block (110 x 100 x 10mm), which was used for fastening and as a heat sink for the power transistors and voltage controllers (Fig 10.36).

The DC power supply system was assembled on double sided epoxy board (91 x 20 x 1.6mm), which was soldered to the long side of the housing (Fig 10.36). The circuit used was from the article [6].

Figs 10.37, 10.38 and Table 10.3 show the circuit, PCB layout and parts list for this power supply. The components are mounted on the copper side, so that the earth surfaces have to be through hole plated.

Grooves were milled in the heat sink so that the drain and gate connections of the transistors could be soldered flush to the board. For this purpose, the Teflon board had recesses measuring 4.4 x 17 (mm) and 6.4 x 22 (mm), into which the transistors were inserted and then screwed to the heat sink (see Fig 10.36). A copper foil was used between the board and the heat sink (115 x 57 x 0.08mm), which was then soldered to the tinplate housing. It provided a very good earth connection between the transistors, the board, the housing and the heat sink.

Before the board was mounted, the earth surfaces had to be through hole plated with 2mm copper (hollow) rivets. At least 4 rivets per long side and earth connector are required for this (see Fig 10.36).

The board was fastened to the heat sink at 6 points, using M2 screws. The transistors each required 2 threaded holes in the baseplate for the source connection.

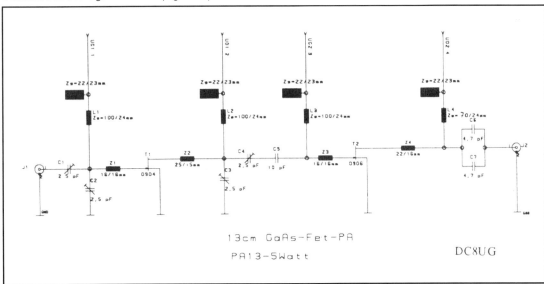

Fig 10.35 : Circuit diagram of the 13cm GaAsFet power amplifier.

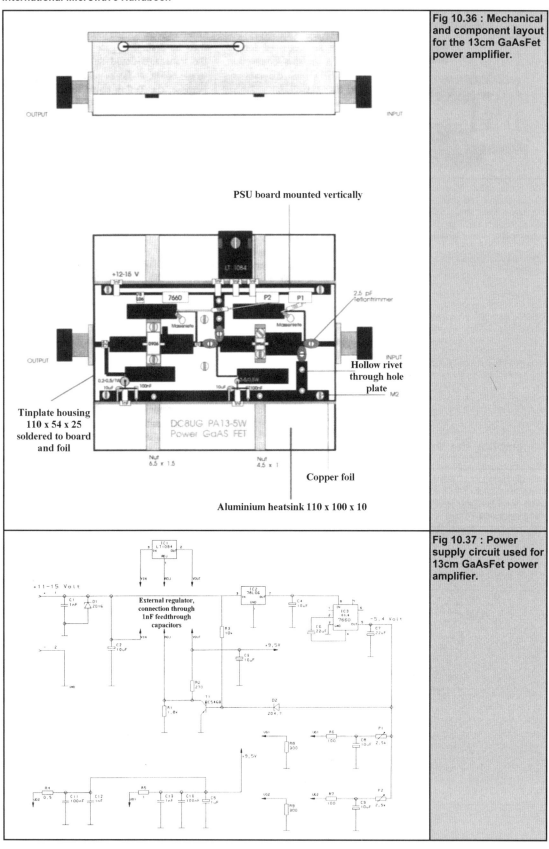

Fig 10.36 : Mechanical and component layout for the 13cm GaAsFet power amplifier.

PSU board mounted vertically

+12-15 V

7660   P2   P1

2.5 pF Teflontrimmer

OUTPUT   INPUT

Hollow rivet through hole plate

Tinplate housing 110 x 54 x 25 soldered to board and foil

DC8UG PA13-5W Power GaAS FET

Nut 6.5 x 1.5   Nut 4.5 x 1

Copper foil

Aluminium heatsink 110 x 100 x 10

Fig 10.37 : Power supply circuit used for 13cm GaAsFet power amplifier.

+11-15 Volt

External regulator, connection through 1nF feedthrough capacitors

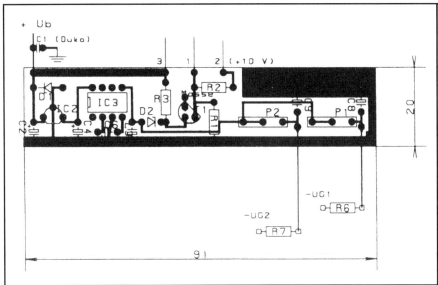

Fig 10.38 : PCB layout for power supply used on 13cm GaAsFet power amplifier.

The dimensions of the tinplate housing were 110 x 55 x 28mm and before assembly it was drilled with the necessary holes for the feed through capacitors and recesses for the N sockets. The housing itself was then soldered together and soldered to the long sides of the board. The soldered on sockets could then be screwed to the heat sinks on the front faces as well.

The best construction and commissioning procedure is as follows:

- Assemble and mount power supply board

- Gate resistances (R6, R7) should already be soldered onto power supply for better mounting (see Fig 10.38)!

- Assemble and wire up the 6 feed through capacitors (1nF) and the blocking capacitors C3, C5, C10, C11

- Fasten and connect up voltage controller by means of feed through capacitors

- Mount and connect up resistances (R4, R5, R8, R9) on RF board

- Mount trimmers (C1, C2, C3, C4)

- Mount chip capacitors (C5, C6, C7)

- The power supply can now be tested.

- Mount GaAsFET's

The static current levels can now be set:

- 0904 - ID = 0.2A

- 0906 - ID = 1.1A

Note: for continuous operation in unfavourable conditions, it is advisable to mount the amplifier on an additional heat sink to ensure stable operation.

Test results

The prototype amplifier was constructed so that a 5 Watt output could be achieved with an input of 25mW at 2,320 MHz. The measurement was carried out using a type BP 432 Wattmeter and a 30dB attenuator from Narda.

Fig 10.39 shows the transfer characteristic of the amplifier.

Table 10.3 : Component list of power supply use for 13cm GaAsFet power amplifier.

C1,12,13	1nF Feedthrough
C2.3.4,8,9	10µF tantalum 16v
C5	1µF tantalum 16v
C6,7	22µF tantalum 10v
C10,11	100nF Sibatit
D1	ZD16 zener diode
D2	ZD4.7 zener diode
IC1	LT1084
IC2	78L06
IC3	ICL7660
P1,2	2.5k potentiometer
R1	2k metal film
R2	270Ω metal film
R3	10k metal film
R4	0.2 - 0.5Ω 1W metal film
R5	3 - 5Ω 0.5W metal film
R6,7	100Ω metal film
R8,9	300Ω metal film
T1	BC546B

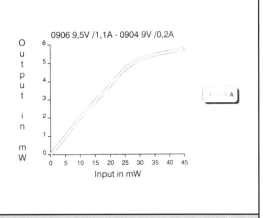

Fig 10.39 : Transfer characteristics of 13cm GaAsFet power amplifier.

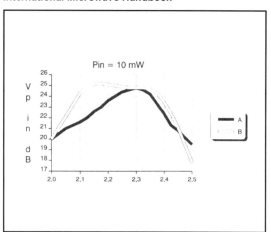

Fig 10.40 : Comparison of measured and simulated power amplification for 13cm GaAsFet power amplifier.

At 5 Watts output, the compression range begins, i.e. a further increase in power leads to a considerable worsening of the inter modulation interval; (1dB compression -33dBin).

Fig 10.40 shows the power amplification with an input of 10mW over the frequency range.

Curve A shows the measured gradient with the amplifier tuned to 2,320MHz. Curve B shows the gradient obtained through simulation, in accordance with Fig 10.33

The amplifier has a band width of 300MHz. Its amplification reduction at the band limits is of course somewhat less than in curve B. The reasons for this are the additional losses of the circuit that can not be covered completely by the simulation.

The linear amplification obtained is only slightly different from the calculated value. If the amplifier is broad band tuned, so that its response corresponds to curve B, the amplification falls by about 1dB (20%) as the band width increases.

To sum up, we can say that using PUFF low cost software to develop simple integrated high frequency circuits can be highly recommended. True, the efficiency is very much reduced by comparison with high end products such as, for example, Super Compact, but the results obtainable are more than adequate for the amateur sector.

40W GaAsFet power on 13cm [13]

Two TPM2026-14 Fets are combined via branch line couplers. An output power of 40W can be achieved at a gain of 13dB, Fig 10.41 shows the circuit diagram. The Power FETs used should preferably be from the same lot.

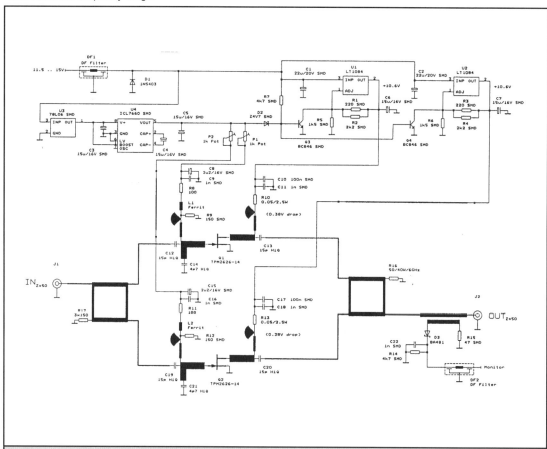

Fig 10.41 : Circuit diagram of 40W 13cm GaAsFet power amplifier.

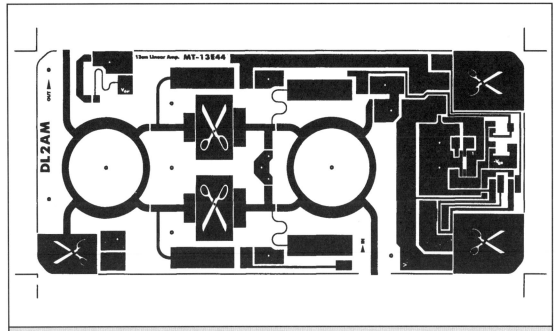

Fig 10.42 : PCB for 40W 13cm GaAsFet power amplifier.

Construction

First cut holes for the Fets and the regulator into the PTFE board with a scalpel, Fig 10.42 shows the PCB layout. Put board into the machined box and fasten the 16 brass screws. Be careful that holes in the PCB fit to the nuts in the box and have a larger size, at least 0.3mm. This is to avoid short circuits. Mount N connectors. Mount regulators with an insulating sheet. Solder all resistors, capacitors and the feed throughs. Tantalum capacitors have the positive polarity on the side with the bar!!! Mount the box on a heat sink with maximum thermal resistance of 0.4°C/W.

Check everything visually and with an ohmmeter. Connect the power and measure 10.6V at Pin 2 of the regulators. Set bias pots to maximum negative voltage of 3.2V at the gate terminals.

Put a 39Ω resistor from Pin 5/7660 to ground. This disables the negative supply. Check that the safety circuit causes the output voltage to drop from 10.6V to 1.2V.

If everything works ok, mount the power Fets carefully. The legs are cut to half their length. Use as little solder as possible.

Inspect everything again and clean the whole PCB.

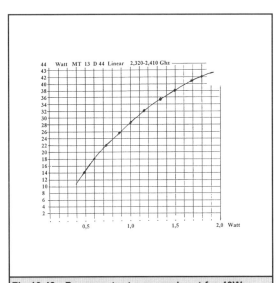

Fig 10.43 : Power output v power input for 40W 13cm GaAsFet power amplifier.

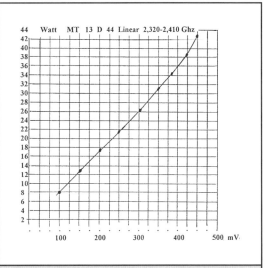

Fig 10.44 : Monitor voltage output v power output for 40W 13cm power amplifier.

Alignment

Connect a dummy load of at least 50W power capacity and a power meter to the output. Connect to a current limiting power supply - overload trip should be set to 12A - and adjust bias current of the FETs to 3.7A each. Across R10 and R13 you can measure a voltage drop of 0.17V.

Then drive the unit with about 500mW from a transverter. Check output power at about 14dB gain. It should read about 12W. Fig 10.43 and 10.44 show the performance of the amplifier.

Then increase drive while observing the output power. An output power of 40W should be reached with 1.6 to 2W of drive. Don't use more drive power than 2W.

If you are brave enough and have the money for spare FETS, you can even try to tune. This is accomplished by the attachment of small tuning stubs made from 0.1 mm thick copper. These stubs can be attached at gate and drain of the power FETs and possibly at the branch line couplers. If you have already more than 36W of output power without tuning you probably should forget about tuning.

Antennas

The most commonly used antenna at 2.3Ghz is a Yagi, there is one design shown in chapter 3 with another shown below.

If a wider beamwidth is wanted the omni directional Alford slot antenna may be used.

A suitably fed parabolic dish is also popular mainly due to the high gain achievable with a relatively small sized antenna. A graph of dish gain against dish diameter is shown in Fig 10.45. The parabolic surface of the dish may be solid, but in some circumstances this may lead to unacceptable wind loading on the mast. As an alternative, the reflecting surface may be made from wire mesh as long as the holes are not greater than $\lambda/10$, i.e. 13mm) in the plane of polarisation. Two suitable types of dish feed

are to be described here, the dipole/splashplate feed and the feed horn. This latter feed can be used on its own as a low gain, broad beamwidth radiator that can be useful for local testing.

DL design of a 13cm Yagi antenna [`14]

It is nice when you can say I really got my money's worth there. This is a warming experience I had when I tried my hand at manufacturing some 13cm yagi antennas, a 10 element and a 22 element yagi. Having seen and read some German design articles on 13cm antennae, I naturally had to try some of them. One of them was a long Yagi, admittedly a little too long for my purposes, but if that was going to work in a quarter or half size model of the original design, it looked as if it would be just the job for my, mostly portable, purposes.

The design for the 13cm antennas comes from a popular antenna designer, DL6WU, who has been much copied by many radio hams. While very popular, his writings are also courtesy of Mr H. Yagi from Japan, dating right back to 1928. I have previously made 70cm and 23cm antennas on the basis of his articles, and they are my best antennae for the two bands to date.

The Danish version of this yagi (Fig 10.46) is made of standard dimension brass materials, worth mentioning, as you will often see measurements for the materials used that are not in stock at your local metal goods dealer. Common standard dimensions like 2.0, 2.5, 3.0mm etc. are off the peg goods, so once the shopping has been done and the brass has been delivered, you're well on your way towards your 13cm antenna project.

Making the antenna follows the customary recipe for brass work. use good tools. Mark out the antenna boom and check the marking an extra time before positioning the 20 centre punch marks and then drilling 20 two millimetre diameter holes for the directors, which first need to be soldered into the boom. Solder the dipole on top of the boom. Fig 10.47 shows the dimensions of the elements and Fig 10.48 shows details of the dipole.

Drill two 10mm diameter holes in the 80 x 80 mm, 1 mm brass plate reflector plate, soldering it firmly to the antenna boom as well. Behind the reflector, drill two 4mm diameter holes spaced 75 mm apart. On top of the holes I have soldered two 4 mm diameter nuts so that the antenna can be secured to a mast fixture or something else.

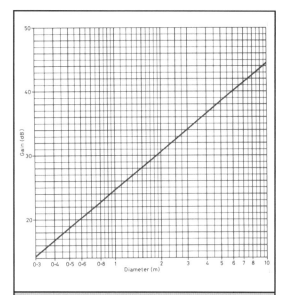

Fig 10.45 : Gain v diameter for a parabolic dish at 2.3GHz, assuming 50% feed efficiency.

Fig 10.46 : Picture of 13cm 22 element Yagi.

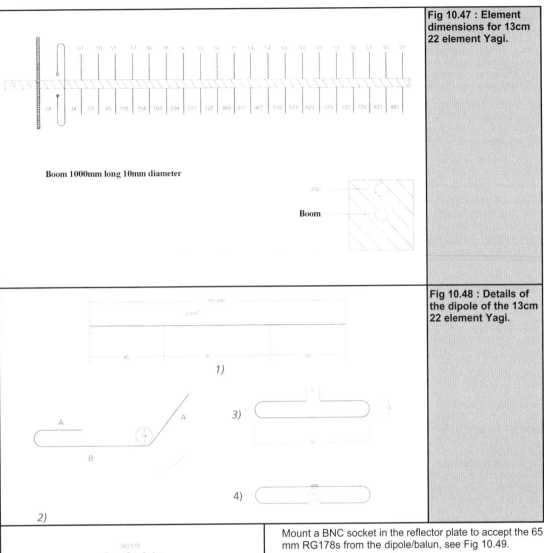

Fig 10.47 : Element dimensions for 13cm 22 element Yagi.

Boom 1000mm long 10mm diameter

Fig 10.48 : Details of the dipole of the 13cm 22 element Yagi.

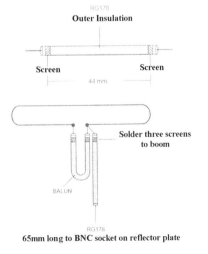

Fig 10.49 : Details of Balun for 13cm 22 element Yagi.

Mount a BNC socket in the reflector plate to accept the 65 mm RG178s from the dipole/balun, see Fig 10.49.

The really important thing is to solder the screen directly onto the antenna socket and allow the screen to go right up to the inside pin on the BNC socket, where the inner conductor in the cable will be soldered.

If the antenna is to be sited outdoors, Araldite can be used to make 50ohm cabling etc. watertight.

The finished antenna fares well in practical use, it is what is known as the real McCoy. In terms of gain, I have made the following observations in order to familiarise myself with the antenna's gain. I found an absolutely stable test signal that was received on a half wave dipole connected to a 13cm converter that mixes down the 2.3GHz to 432MHz, using my Yaesu FT790 as a receiver. With the dipole antenna, an approximately S1 signal is received, with plenty of noise on the loudspeaker. A 10 element Yagi was then substituted for the dipole and a nice, approximately S3 signal was received. Then the 22 element antenna was connected. It's then that you get your money's worth, an impressive S7 signal was received with a clear, pure tone heard over the loudspeaker.

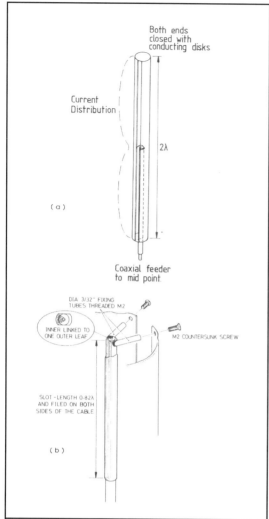

Fig 10.50 : Construction of the 2.3GHz Alford slot antenna using a dual slotted cylinder. The impedance feedpoint is 200Ω. (a) Dimensions for 2,320MHz are, slot length 280mm, slot width 3mm, tube diameter 19mm by 18swg. (b) Construction of a suitable balun. The balun slots are 1mm wide and 26mm long.

According to the original designer of the antenna, a boom length of 7λ and approximately 20 elements makes for something over 15dBd.

Alford slot antenna

Developed by M Walters, G3JVL, this antenna fulfils the need for an omni directional horizontally polarised antenna. This makes it particularly useful for beacons, fixed station monitoring purposes and mobile operation. Mechanical details of this antenna are shown in Fig 10.50. The prototype was made from 22mm outside diameter copper water pipe. Material is removed from one part of the tubing to produce a slotted tube with an outside diameter of 18.5mm and a slot width of 2.6mm. To ensure circularity, the tube is best formed around a suitable diameter mandrel. Small tabs are soldered at the top and bottom of the tube to define the slot length of 229mm. A

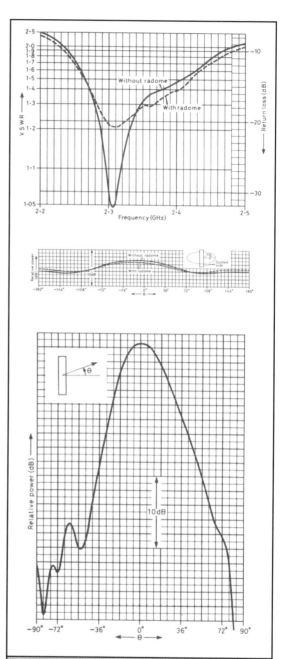

Fig 10.51 : Performance of a 2.3GHz Alford slot antenna. Top - impedance. Middle - horizontal polar diagram. Bottom - Vertical polar diagram.

plate is soldered across the bottom of the tube to strengthen the structure.

The RF is fed via a length of 0.141 (3.6mm) RG141 semi rigid coaxial cable up the centre of the tube to the centre of the slot via a 4:1 balun constructed at the end of the cable. The detailed construction of the balun is shown in Fig 10.50b. The two diametrically opposite slots are cut carefully using a small hacksaw with a new blade. The inner and outer of the cable are shorted using the shortest possible connection and balun is attached to the slot using two thin copper foil tabs.

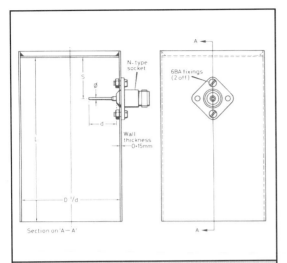

Fig 10.52 : Construction of a 2.3GHz feed horn. Dimensions are, D=80.5mm, L=310mm, s=76mm, d=27mm and φ=3.5mm.

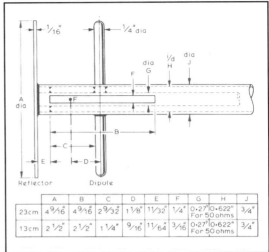

	A	B	C	D	E	F	G	H	J
23cm	4⁹⁄₁₆"	4⁹⁄₁₆"	2⁹⁄₃₂"	1¹⁄₈"	1¹⁄₃₂"	¼"	0·27" For 50 ohms	0·622"	¾"
13cm	2½"	2½"	1¼"	⁹⁄₁₆"	1¹⁄₆₄"	³⁄₁₆"	0·27" For 50 ohms	0·622"	¾"

Fig 10.54 : Construction of a 2.3GHz disc and dipole feed.

If suitable test gear is available, the match of the antenna can be optimised by carefully adjusting the width of the slot by squeezing he tube in a vice, or by prising the slot apart with a small screwdriver.

Typical antenna characteristics are shown in Fig 10.51. The gain of the antenna has been measured to 6.4dBi.

Feed horn for 2.3GHz

The feed horn is ideally suited to feed a dish having an f/D of between 0.2 and 0.4. It consists of a short length of circular waveguide with a coaxial to waveguide transition made from an N type socket and a short probe. Constructional details are shown in Fig 10.52. The circular waveguide section is made from a cylinder closed with a plate one end (e.g. a large coffee tin) cut to the specified length. All dimensions are fairly critical (especially the length of the probe) and hence care should be taken during construction if a good match is to be obtained without adjustment. If a different can is used, the length and position of the probe, and the length of the tin may need altering for minimum vswr. The N type socket should be fixed to the tin by using four fixings, using washers as spacers to accommodate the curved surface of the tin. A completed feed, made from a coffee tin, is shown in Fig 10.53.

Dipole/splashplate feed

The version of this feed to be described here is suitable for dishes having an f/d of between 0.25 and 0.45. The feed is built around a length of fabricated coaxial line. The ratio of inner conductor diameter to internal outer diameter is chosen to produce the desired impedance, in this case 50Ω. Details of the feed are shown in Fig 10.54. It is constructed from a piece of 15mm outer diameter copper water pipe, the length being chosen to allow point x to be positioned at the focus of the dish. The 43mm diameter disc is turned or filed from a 1.6mm thick sheet of brass or copper. The two slots may either be milled or hacksawed

Fig 10.53 : Photograph of G4DKK's feed horn constructed from a coffee tin.

Fig 10.55 : Photograph of the 2.3GHz feed in the author's 1.2m mesh dish.

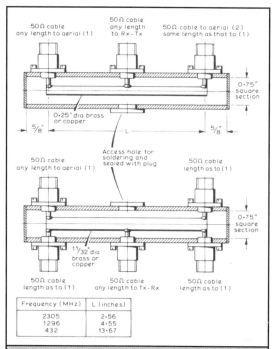

Frequency (MHz)	L (inches)
2305	2·56
1296	4·55
432	13·67

Fig 10.56 : Two way and four way antenna combiners for 2.3GHz.

and filed carefully. The ptfe spacers marked Sp are 13.5mm o.d. by 5mm thick spacers with a 6.14mm hole drilled in the centre; they are used to ensure line concentricity.

The feed can be held in position in the dish by using a modified 15mm cold water tank compression fitting. The internal step in the fitting is filed away to leave a constant 15.2mm bore. The fitting mounted in the dish in a 35mm hole using the supplied nut. The position of point x can be altered by sliding the feed in the fitting. When the position of the feed has been optimised by monitoring a remote signal, the feed can be locked in place by tightening the nut over the olive. This type of fitting must be fitted before attaching the RF connector.

The N type female line connector is fixed to the line by counter boring its collar 15mm diameter by 4mm deep and soldering the collar to the line. The centre pin is soldered to the line inner and the connector assembled with its supplied ptfe insulators. The completed feed in the author's 1.2m dish is shown in Fig 10.55.

Multiway combiners for 2.3GHz antennas

The power splitters shown in Fig 10.56 enable either two or four 50Ω antennas to be fed from a single 50Ω cable. The units consist of a length of fabricated coaxial line that performs the appropriate impedance transformations. In this case the inner is made λ/2 long between the centres of the outer connectors, and the outer is made approximately 32mm longer. In the original design, the outer was made from square section aluminium tubing, the ends of which (and the access hole for soldering the centre conductor) were sealed with aluminium plates bonded with adhesive. Alternatively, copper or brass tubing may be used and the plates soldered. Any other size of inner or outer within reason may be used provided that the ratio of the inside dimension of the outer to the diameter of the

inner conductor is suitably chosen.

The cables connecting each antenna to the splitter must be 50Ω and can be of any length provided they are the same in all cases. Preferably, all the cable is taken from the same batch. Note that the two way combiner may be used to combine a broad beamwidth antenna of moderate gain with a high gain narrow beamwidth antenna providing both antennas are 50Ω impedance.

Filters

13cm bandpass filters

A three stage and a single stage bandpass filter are described here [15].

Manufacturing and adjustment of the 3 circuit filter

The filter is made from 1mm semi rigid brass plate. It is very important to have some decent handheld tools to work with. Use new drill bits, put a new blade in the hacksaw etc. The work of cutting out various small side pieces, baseplate and top piece will be a straightforward job. Fig 10.57 shows the dimensions of the 3 circuit filter.

The holes for the three 4mm diameter brass rods (1/4 wave coils) must be made so that the brass rods are pressed firmly into the holes, after which they must be soldered firmly into the side piece. The nuts for the three 4mm diameter screws must be soldered home on the exterior of the side piece. The ready mounted side pieces can then be soldered onto the baseplate, after which the work should be polished and cleaned with some wire wool. Fit the SMA sockets and solder the inner pin 5mm from the frame to the 1/4 wave coils on either side of the filter. Solder the top plate firmly all the way around the filter.

Adjustment is perfectly simple. Turn the three 4mm diameter screws right in until they stop at the coils, followed by a 1/4 turn back, and lock with the nut specially used for the purpose. The locking nut has not been sketched on the working drawing of the filter.

The single circuit bandpass filter

The small filter is made following the same procedure as the 3 circuit filter. A brass tube measuring 24 mm in diameter internally and 30 mm high is not exactly an off the shelf item, it can be made from a piece of 76 x 30 mm brass plate, bent over a pipe firmly secured in a vice. Solder the joint produced by bending to give you a round brass tube. At the bottom, file off a little material so that the inner pin from the SMA socket will go into the 24mm diameter tube. Figs 10.58 and 10.59 show the dimensions of the single circuit filter When ready to assemble the filter, start by soldering the 5mm diameter tube firmly into the baseplate. Then the 24mm diameter tube firmly into the baseplate, and solder the SMA sockets firmly to the baseplate as well. The two small link coils can now be fitted, finally soldering on the 4mm diameter round top plate.

The single circuit filter is then ready for adjusting to the frequency. At 2.3GHz, the 4mm diameter screw has to be rotated in until it stops at the 5mm diameter tube (the 1/4 wave coil), then turn back about 4 revolutions and you will be very close to the frequency. Fine tuning can be performed using a transmitter or receiver.

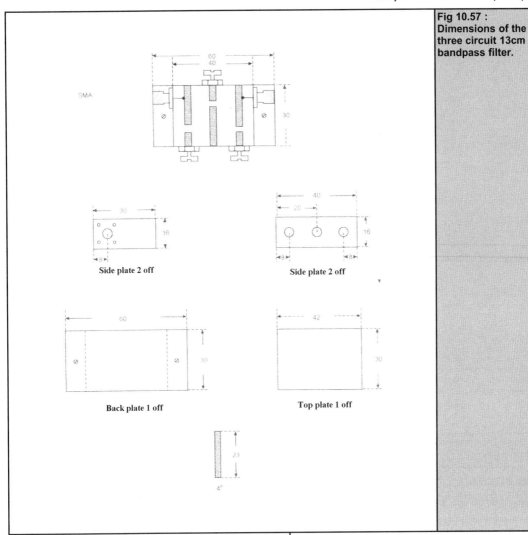

Fig 10.57 : Dimensions of the three circuit 13cm bandpass filter.

Application

The bandpass filters can be used for different purposes on 13cm:

- After the mixer stage or PA stage outlet to the antenna to ensure that output is actually on the right frequency, because at 2.3GHz an amplifier stage can easily be 500 MHz wide, so there are many good reasons for using filters on 13 cm.

- On the receiver side I use a filter between my signal generator and the 13cm converter. The signal generator is a crystal controlled 96,708330 MHz signal whose 24th harmonic will be precisely 2321,000 MHz, and in a reasonably good receiver this signal will be heard approximately 20 - 30dB above the background noise of the receiver.

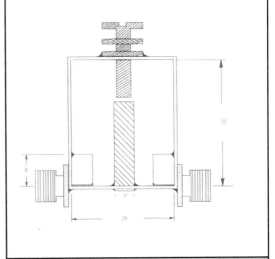

Fig 10.58 : Cross sectional view of the single circuit 13 cm bandpass filter.

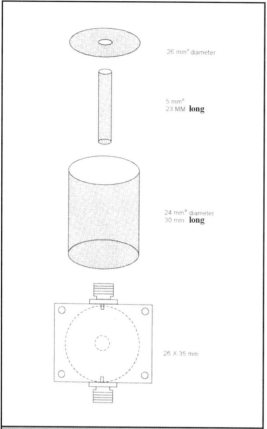

26 mm° diameter

5 mm°
23 MM **long**

24 mm° diameter
30 mm **long**

26 X 35 mm

Fig 10.59 : Dimensions of the single circuit 13cm bandpass filter.

References

[1] From an article by Rainer Bertelsmeier, DJ9BV,Dubus

[2] R. Bertelsmeier, DJ9BV, "HEMT LNAs for 23cm", DUBUS TECHNIK IV, pp 177 - 190

[3] R. Bertelsmeier, DJ9BV, "No Tune HEMT Preamp for 13 cm", DUBUS TECHNIK IV, pp 191 - 197

[4] Bertelsmeier, R.: PUFF Design Software, Dubus-Info, vol. 18 (1989), no. 4, pp. 30 - 33

[5] Lentz, R.E.:PUFF - CAD Software for Microwave-Stripline Circuits VHF Communications, 2/91, p. 66

[6] Kuhne, M.: High-Power GaAsFET Amplifier for 9cm, Dubus-Info, vol. 20 (1991), no. 2, pp. 7 - 16

[7] Schmitt, G.: 13cm Power Amplifier with GaAs, Dubus-Info, vol. 20 (1991), no. 4, pp. 55 - 56

[8] Wedge, S.W., Compton, R. & Rutledge, D.:PUFF Computer Aided Design for Microwave Integrated Circuits, Puff Distribution, California Institute of Technology, Pasadena

[9] Unger/Harth:High-Frequency Semi-Conductor Electronics Hirzel-Verlag, Stuttgart ISBN 37776 02353

[10] From an article by Leo Lorentzen, OZ3TZ, OZ 11/2001 pp 633 - 639

[11] From an article by Matjaz Vidmar, S53MV, CQ ZRS

[12] From an article by Harald Fleckner, DC8UG, VHF Communications 3/1994 pp 130 - 141

[13] From an article by Phillipp Prinz, DL2AM, Dubus Technik V pp 353 - 359

[14] From an article by Leo Lorentzen, OZ3TZ, OZ 1/2001 pp 18 - 20

[15] From an article by Leo Lorentzen, OZ3TZ, OZ 11/2000 pp 633 - 635

Chapter 11

The 3.4GHz (9cm) band

In this chapter :

- Preamplifier
- Transceiver
- Amplifier

- Antenns
- Filters

The 3.4GHz band may be considered as a transitional band between the lumped circuit techniques of the lower microwave bands and the dominant waveguide techniques of the higher microwave bands. The relatively short wavelength has a big influence on the antenna designs used on the band in that relatively small dish antennas will have a very usable gain. The band also represents the highest frequency at which the Yagi may be considered a reasonable antenna prospect.

Band planning

A majority of countries, worldwide, have an allocation in the 3,300 to 3,500MHZ range. In the UK, the amateur service is allocated the range 3,400 to 3,475MHZ on a secondary basis to the fixed, fixed-satellite and radiolocation services Fig 11.1 shows UK band plan and Fig 11.2 shows th USA band plan .

Propagation and equipment performance

Most of the propagation modes encountered on the VHF/UHF bands are also present at 3.4GHz, with the exception of those which require ionised media such as aurora, meteor scatter and Sporadic E. The most commonly encountered propagation modes at 3.4GHz are line of sight, diffraction, tropospheric scatter, ducting and aircraft scatter. Moonbounce is also a possibility at this

Frequency (MHz) USA usage	
3300 -3500	Band allocated
3456.3-3456.4	Propagation beacons

Fig 11.2 : USA band plan for 9cm.

frequency. Further information on some of these modes can be found in chapters 1 and 2.

Using the techniques of chapters 2 and 3 it is possible to predict whether communication is possible over a given path, when the performance parameters of the equipment are known. Fig 11.3 shows the path loss capability of five different sets of equipment ranging from a basic system of

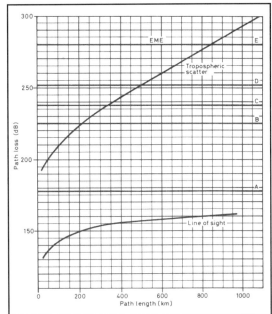

Fig 11.3 : Graph of path loss versus distance for line of sight, tropospheric and eme propagation, related to the performance of five sets of equipment.
A: 20dBi antenna, 1W tx, 15db nf, 25kHz bw
B: 1.2m dish, 10W tx 15db nf, 2.5kHz bw
C: 1.2m dish, 10W tx, 4db nf, 2.5kHz bw
D: 1.2m dish, 10W tx, 4dB nf, 100Hz bw
E: 4.0m dish, 15W tx, 1dB nf, 100Hz bw

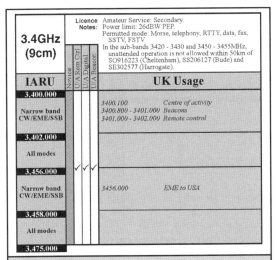

Fig 11.1 : UK band plan for 9cm.

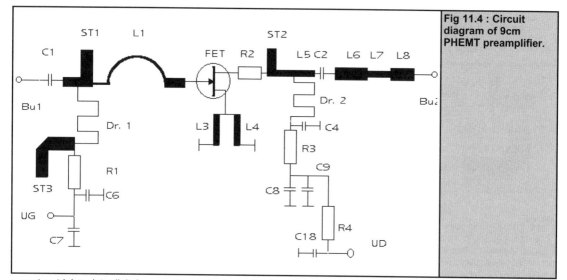

Fig 11.4 : Circuit diagram of 9cm PHEMT preamplifier.

varactor tripler, interdigital converter and single Yagi antenna to a system with moonbounce capability. Also shown on the graph are the path losses associated with three different modes of propagation; line of sight (the least lossy mode of propagation), tropospheric scatter (a reliable, commercially used mode of propagation) and moonbounce.

It can be seen that relatively simple equipment such as varactor multiplier transmitter generating 1W of nbfm, an interdigital converter and a single loop quad Yagi (A) is capable of working very long line of sight paths easily. However, very few paths are purely line of sight and propagation losses are usually noticeably higher. The tropospheric scatter loss is a reasonable upper limit for a terrestrial path and can be used to estimate the maximum range of equipment under normal propagation conditions. By adding a single stage valve amplifier producing 10W, changing the antenna to a 1.2m dish and changing to ssb or cw (B), the equipment is capable of covering a troposcatter path of up to 210km. By adding a 3.5dB noise figure bipolar receive preamplifier, (C), the coverage distance increases to 330km. Line (D) shows the effect of the equipment of (C) but using a narrow bandwidth i.f.

filter, as one could when receiving cw, the coverage distance increasing to 510km. At distances over about 840km it can be seen that the terrestrial path loss is greater than the route via the moon. Line (E) shows the parameters of equipment capable of moonbounce on 3.4GHz.

Under anomalous propagation, conditions path losses are considerably reduced, often approaching the line of sight loss and under these conditions even simple equipment is then capable of working very long paths.

Preamplifier

The preamplifier equipped with a PHEMT provides top notch performance in noise figure and gain as well as unconditional stability for the 9cm amateur band. The noise figure is 0.45dB at a gain of 14 dB. It utilises the C band PHEMT NEC NE42484A and provides a facility for an optional second stage on board. The second stage with the new HP GaAs MMIC MGA86576 can boost the gain to about 38dB in one enclosure. The preamplifier is broadband and usable from 3400 to 3700MHz.

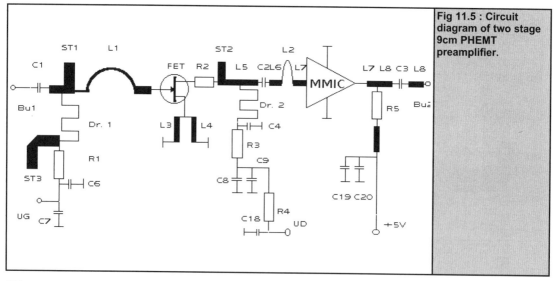

Fig 11.5 : Circuit diagram of two stage 9cm PHEMT preamplifier.

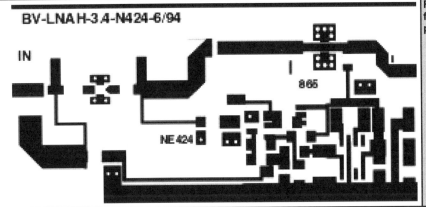

Fig 11.6 : PCB layout for 9cm PHEMT preamplifier.

Circuit description

The construction of this LNA follows the proven design of the 23cm HEMT LNA [1] by using a wire loop with an open stub as an input circuit, Fig 11.4 shows the circuit diagram. The FETs grounded source requires a bias circuit to provide the negative voltage for the gate. A special active bias circuit is integrated into the RF board that provides regulation of voltage and current for the FET, this circuit is common with the 13cm PHEMT preamplifier described in chapter 10 and shown in Fig 10.5.

The design is basically the same as the13cm PHEMT preamplifier described in chapter 10 [2]. There is only an additional stub ST2 in the output for matching. Quarter wave Stub ST3 provides a short on 3.4GHz. On all frequencies outside the operating range the gate structure is terminated by R1. Dr.1 is a printed $\lambda/4$ choke to decouple the gate bias supply. C4 provides a short on 3.4GHz, because its in series resonance. On all frequencies outside the operating range the drain structure is terminated by R3. Dr.2 is a printed $\lambda/4$ choke to decouple the drain bias supply.

The two stage version uses an HP GaAs MMIC MGA86576 in the second stage, Fig 11.5 shows the circuit diagram. It provides about 2dB noise figure and 23dB gain. Its input is matched by a wire loop for optimum noise figure. Its output is terminated by a resistor R5 and a short transmission line L10. Together with L7/L8 and C3

a good output return loss is achieved.

The MMIC typically adds 0.07dB to the noise figure of the first stage. This is somewhat difficult to measure, because most converters will exhibit gain compression, when the noise power of the source amplified by more than 40dB will enter the converter.

Construction

The construction uses microstripline techniques on a glass PTFE substrate, Taconix TLX 0.79mm thick. An active bias circuit, which provides constant voltage and current is integrated onto the PCB which measures 34 x 72mm, see Fig 11.6 for the PCB layout and Fig 11.7 for the component layout. Construction should be carried out as follows :-

• Prepare the PCB to fit into aluminium box.

• Prepare holes for the SMA connectors. Note, input and output connector are asymmetrical, use the PCB to do the marking.

• Drill holes for through contacts (0.9mm diameter) in the PCB and use 0.8mm gold plated copper wire to make connections at the positions indicated.

• Solder all resistors onto PCB.

• Solder all capacitors onto PCB.

• For L1 cut a 9mm length of 0.25mm diameter gold

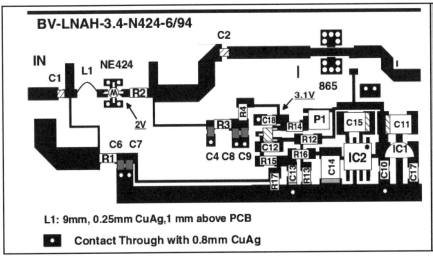

Fig 11.7 : Component layout for 9cm PHEMT preamplifier.

L1: 9mm, 0.25mm CuAg,1 mm above PCB

■ Contact Through with 0.8mm CuAg

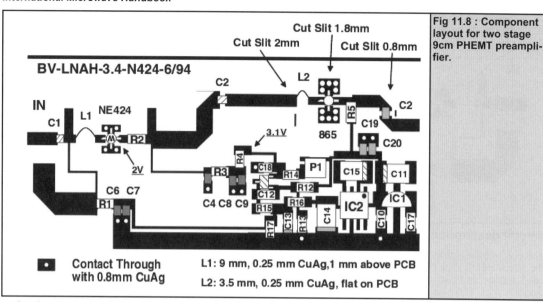

Fig 11.8 : Component layout for two stage 9cm PHEMT preamplifier.

plated copper wire. Bend down the ends at 0.5mm length to 45°. Form the wire into a half circle loop according to Figs 11.6 and 11.7. Solder into the circuit with 1mm clearance from the PCB. The wire loop has to be flush with the end of the gate stripline and should be soldered in at right angles to it. The wire loop should be flat and parallel to the PCB.

- Verify open circuit function of bias circuit. Adjust P1 to 50Ω resistance. Solder 100Ω test resistor from the drain terminal on PCB to ground. Apply +12V to IC1 and measure +5V at output of IC1, -5V at IC2/Pin5, -2.5V at collector of T1, +3.1V at emitter of T1, +2.5V at base of T1, -2.5V at R17 and +2.0V across the 100Ω. If O.K. remove 100Ω test resistor.

- Solder PHEMT NE424 onto PCB. Use only an insulated soldering iron (Weller), ground the PCB, your body and the power supply of the soldering iron. Never touch the PHEMT at the gate, only at the source or the drain, when applying it to the PCB and solder fast (<5sec).

- Mount the finished PCB into box. Use M2 or M2.5 screws.

- Mount the SMA connectors. Cut the inner conductor to 1mm length. Solder to striplines with the least amount of solder possible.

- Mount feed through capacitors into box.

- Connect D1 between feed through capacitor and PCB.

- Connect 12V and adjust P1 for 16mA drain current (measure 160mV across R4 on RF board). Voltages should be around +2.0V at the drain terminal, -0.4V at the gate, +3.1V at emitter of T1.

- Connect the LNA to a noise figure meter, if you have one, and adjust the input wire loop, adjust the clearance to PCB as well as drain current by adjusting P1 for minimum noise figure. Even without tuning the noise figure should be within 0.1dB of minimum because of the limited tuning range of the wire loop.

- Glue conducting foam into inside of top cover and and slip to the top of the box.

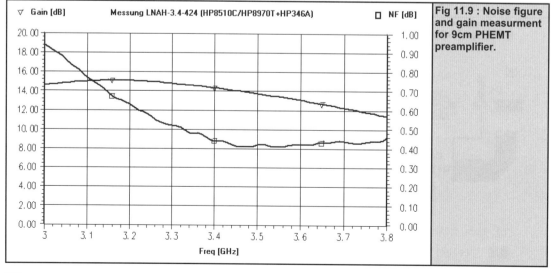

Fig 11.9 : Noise figure and gain measurment for 9cm PHEMT preamplifier.

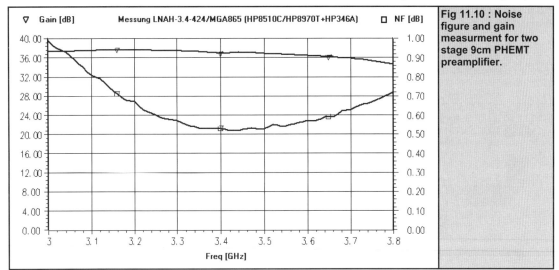

Fig 11.10 : Noise figure and gain measurment for two stage 9cm PHEMT preamplifier.

To add the MMIC amplifier, refer to Fig 11.8 for construction.

- Prepare PCB by cutting slits into the micro striplines around the MGA865. These are a 2mm slit for L2, a 1.8mm slit for the MMIC and a 0.8mm slit for C3.

- For L2 cut a 3.5mm length of gold plated copper 0.25mm diameter wire. Form wire into a half circle loop as shown in Fig 11.8. Solder the wire loop into the circuit. The wire loop has to lie flat on the PCB. The wire loop has to be flush with the end of the gate stripline and should be soldered at right angles to it.

- Follow other instructions given above

Measurement Results

Noise Figure and Gain

Measurements were taken with a HP8510 network analyser and HP8970B/HP346A noise figure analyser, transferred to a PC and then plotted.

Fig 11.9 and 11.10 show the measurement results for gain and noise figure for the 1 stage and 2 stage version respectively. Using a special PHEMT NEC NE42484A optimised for C Band, a typical noise figure of 0.45dB at a gain of 14 dB can be measured on 3.4GHz. An optional second stage on the same PCB with the GaAs MMIC MGA86576 from HP will boost the gain from 14db to 37dB. The two stage version measures with a noise figure of 0.55dB. This version can be used for satellite operation. For EME, where lowest noise figure is at premium, a cascade of two identical one stage LNAs may be more appropriate.

Both versions are broadband, they can cover the various portions of the 9cm amateur allocation from 3400 to 3700MHz without retuning.

Stability

A broadband sweep from 0.2 to 20 GHz shows a stability factor K of not less than 1.6 and the B1 measure was always greater than zero (Fig 11.11). These two properties indicate unconditional stability.

The 2 stage version with the MGA865 measures with K>4 at all frequencies

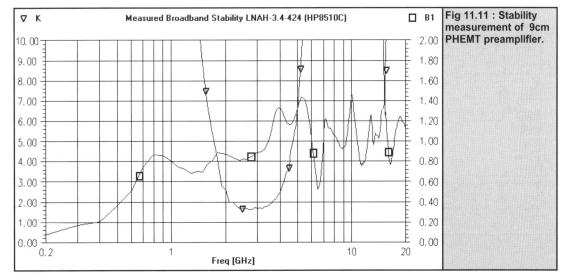

Fig 11.11 : Stability measurement of 9cm PHEMT preamplifier.

Table 11.1 : Component list for 9cm PHEMT preamplifier.

C1	1pF Chip-C 50mil (500 1R0 JG) Tekelec
C2,3	1.5pF SMD-C, size 0805
C4	1.8pF SMD-C, size 0805
C6	100pF SMD-C, size 0805
C7,8,19	1000pF SMD-C, size 0805
C9,18,20	10nF SMD-C, size 0805
C10,12,17	0.1µF SMD-C, size 1206
C11,14	10µF SMC-Electro, size 1210
C13	1µF SMD-Electro, size 1206
C16	1000pF Feedthrough
R1	47Ω SMD-R, size 1206
R2	33Ω SMD-R, size 1206
R3	22Ω SMD-R, size 1206
R14	68Ω SMD-R, size 1206
R16,17	22kΩ SMD-R, size 1206
P1	100Ω SMD-Pot, Murata 4310
L1	Wire loop, 0.25mm gold plated copper wire 9mm long, 1mm above board
L2	Wire loop, 0.25mm gold plated copper wire 3.5mm long, on PCB
D1	1N4007
FET	NE42484A NEC
MMIC	MGA-86576 HP
T1	PNP e.g. BC807, BC856, BC857
IC1	µA7805A TO-92
IC2	LTC1044SN8
Bu1,2	SMA
PCB	Taconix TLX, 34 x 72mm, 0.79mm, εr=2.55
Box	

Transceiver [3]

Since June 1998 amateur radio operations on the 3.4GHz frequency band were authorised for the first time in Slovenia (S5). Of course we all wanted to activate the new band as soon as possible. The easiest way was to modify the successful and simple zero IF microwave SSB/CW transceiver described in chapter 9.

The 3.4GHz (9cm) band is not very popular among microwave amateurs, at least not in Europe. In ITU region II the amateur allocation extends from 3300MHz to 3500MHz. On the other hand from a few years ago there were only a few European nations that allowed amateur activity in the 3400 - 3475MHz segment on a secondary basis. Long distance narrow band amateur operations were mainly concentrated around 3456MHz both in Europe and in North America, since 3456MHz is a convenient multiple of several common frequencies: 144MHz (x24), 432MHz (x8) and 1152MHz (x3) (local oscillator in transverters).

According to new common European frequency plans, a small part of the 9cm band, usually 3400 - 3410MHz, was allocated to the amateur radio service in many European nations. The same frequency band 3400 - 3410MHz was also allocated to the amateur radio service in Slovenia on a secondary basis. Accordingly narrow band activity in Europe was shifted down to 3400MHz.

Of course the 3.4GHz zero IF SSB transceiver presented in this article makes use of the proven designs for 2.3GHz and 5.7GHz shown in chapters 10 and 12. The low frequency and audio stages are the same, as well as the sub harmonic mixers with BAT14-099R Schottky quads, these are described in chapter 9. Besides GaAs HEMTs and INA MMIC amplifiers, the 3.4GHz zero IF transceiver uses new, high performance and inexpensive silicon bipolar transistors like the BFP420.

The zero IF SSB transceiver for 3400MHz uses a single PIN diode (BAR81) antenna switch just like the 5.7GHz and 10GHz versions. Therefore the transmitter output stage (CLY2) makes an integral part of the antenna switch. Although the CLY2 was not designed for operation at 3.4GHz, it can still provide more than 0.5W of output power, although the power gain (8-9dB) is smaller than in the 13cm band. All of the RF stages are built as microstrip circuits on 0.8mm thick glass fibre epoxy laminate FR4.

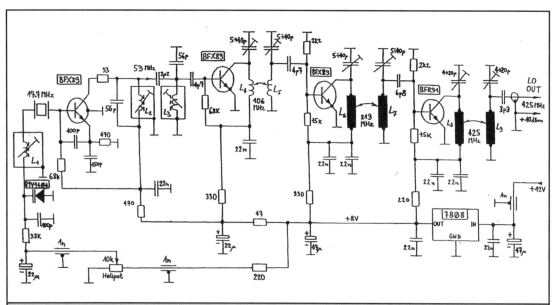

Fig 11.12 : Circuit diagram of modified VCXO and multiplier stages for 3400MHz transceiver.

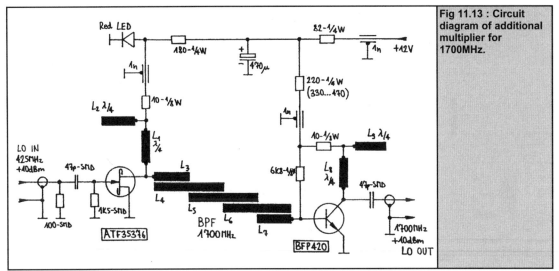

Fig 11.13 : Circuit diagram of additional multiplier for 1700MHz.

Modified VCXO and multiplier stages

A transceiver for 3456MHz could use the same VCXO and multiplier chain up to 576MHz as in the 2304MHz version, since the two frequencies are in an exact ratio 3:2. The 576MHz signal is doubled to 1152MHz to drive the sub harmonic mixers in the 2304MHz transceiver. The same 576MHz signal could be tripled to 1728MHz to drive the sub harmonic mixers in the 3456MHz transceiver.

Since the narrow band segment in the 9cm band was shifted down to 3400MHz, a slightly different crystal has to be used in the VCXO. This frequency multiplication could be used, however it does not represent an efficient technical solution. Above 1GHz silicon bipolar transistors become quite inefficient as frequency multipliers, while GaAs FETs and HEMTs mainly produce even harmonics due to their quadratic transfer function.

The 3.4GHz transceiver therefore includes a slightly modified multiplier chain. The VCXO is using a 17.7MHz

(fundamental resonance) crystal. The VCXO output is first tripled to 53MHz, then doubled to 106MHz, again doubled to 213MHz and finally doubled to 425MHz in the same VCXO module. The VCXO module is followed by an additional HEMT multiplier used to multiply 425MHz by four to obtain 1700MHz to drive the sub harmonic mixers.

The modified VCXO and multiplier stages are shown on Fig 11.12. For operation on 3400MHz, a fundamental resonance crystal for 17.708MHz is required in the VCXO. Unfortunately this frequency is not standard and a special crystal has to be ordered. The closest standard crystal is the PAL subcarrier multiple at 17.734MHz (used in many TV sets in Europe) that multiplies up to 3405MHz in the described circuit.

Although narrow band operation is permitted at 3405MHz in all countries where amateurs have access to the 9cm band, it is probably very difficult to find any other stations on this frequency. On the other hand, if a standard PAL crystal for 17.734MHz is pulled down by 26kHz with a

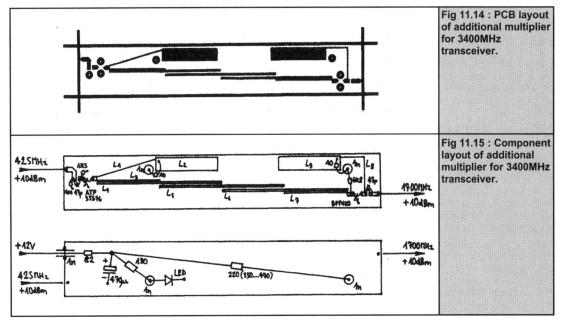

Fig 11.14 : PCB layout of additional multiplier for 3400MHz transceiver.

Fig 11.15 : Component layout of additional multiplier for 3400MHz transceiver.

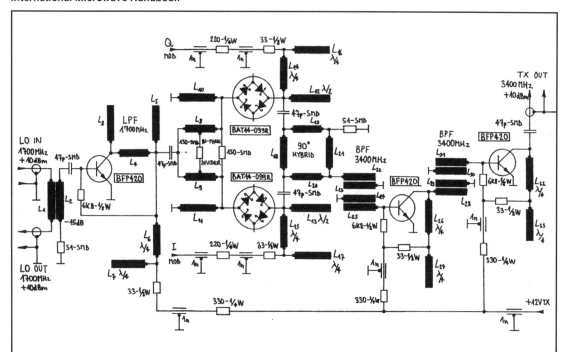

Fig 11.16 : Circuit diagram of quadrature transmit mixer for 3400MHz transceiver.

series inductor, the resulting VCXO will not be very stable. In the latter case, just the power on warm up frequency drift may reach 50kHz at the final frequency of 3.4GHz.

In order to pull a standard PAL subcarrier crystal down to 17.708kHz, a relatively high series inductor L1 is required in the range of 5...8µH, depending on the crystal used and other parasitic capacitances in the circuit. The stability can be improved by grounding the crystal case. The suggested components (varicap MV1404 and 100pF parallel capacitor) provide a coverage of about 600kHz around 3.4GHz that should be sufficient for narrow band modes including the correction of the VCXO drift.

The multiplier chain is tuned to lower frequencies by increasing the values of the resonant and coupling capacitors, while the inductors are left unchanged. Capacitive trimmers (Philips foil type, 7.5mm diameter with three leads) in the 106MHz and 213MHz bandpass filters are increased to 5-40pF (violet body), while the trimmers in the output stage at 425MHz are increased to 4-20pF (green body).

The printed circuit board of the VCXO and multiplier stages is of course identical as those for the 23cm, 13cm, 5cm and 3cm versions of the zero IF SSB transceiver, see Fig 9.10, 9.11 and 9.12 in chapter 9. There are only a few changes of capacitor values in the component location. If a standard 17.734MHz crystal is pulled down to 17.708MHz in the described circuit, special care has to be taken that the oscillator operates in a stable way and does not jump on another resonant mode of the crystal.

The circuit diagram of the additional multiplier for 1700MHz is shown on Fig 11.13. The multiplier includes an ATF35376 HEMT, overdriven with the 425MHz (+10dBm) input signal. The following microstrip bandpass filter selects the fourth harmonic at 1700MHz. The 1.7GHz bandpass filter is followed by an amplifier

(BFP420) to boost the 1700MHz LO signal level to about 10mW (+10dBm).

The additional multiplier for 1700MHz is built as a microstrip circuit on a double sided, 0.8mm thick FR4 glass fibre epoxy printed circuit board with the dimensions of 20mm x 120mm. The upper side is shown on Fig 11.14 the lower side is not etched to act as a ground plane for the microstrip circuits. The component location on both sides of the printed circuit board is shown on Fig 11.15.

A correctly assembled multiplier for 1700MHz should not require any tuning. One should however check the I_{dss} of the ATF35376 HEMT. When checking the overall transceiver, the LO power level may be too high, degrading both the noise figure of the receive mixer and the carrier suppression of the transmit mixer. If that is the case, increase the supply resistor of the 1700MHz amplifier BFP420 from the original 220Ω up to 330Ω or even 470Ω.

Quadrature transmit mixer for 3400MHz

The circuit diagram of the transmit mixer for 3400MHz is shown on Fig 11.16. There are two sub harmonic mixers, an in phase LO divider and a quadrature hybrid coupler for 3.4GHz The module also includes a coupler and buffer amplifier for the 1700MHz LO and two output RF amplifier stages, to boost the 3.4GHz output signal level to about 10mW (+10dBm). All bandpass filters and other passive RF components are built as microstrip circuits on double sided, 0.8mm thick FR4 glass fibre epoxy laminate.

The receiver and transmitter of the zero IF SSB transceiver for 3400MHz both require the same local oscillator signal at 1700MHz. The switching of the latter is performed in the same way as in the zero IF SSB transceivers for 1.3GHz, 5.7GHz and 10GHz. Most of the LO signal power is fed to the receive mixer all of the time. A small part of the LO signal is taken by a -15dB coupler and

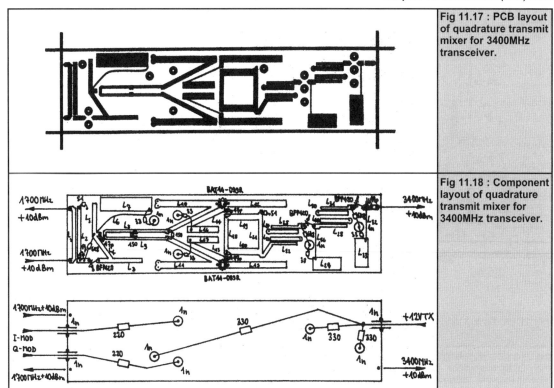

Fig 11.17 : PCB layout of quadrature transmit mixer for 3400MHz transceiver.

Fig 11.18 : Component layout of quadrature transmit mixer for 3400MHz transceiver.

amplified by a BFP420 transistor to drive the transmit mixer. The 1700MHz LO signal is filtered in a low pass filter (L3, L4 and L5) before being applied to the two sub harmonic transmit mixers through an in phase divider L8/L9.

The BAT14-099R schottky quads provide good mixer symmetry at relatively low frequencies, so no special measures are required to improve the unwanted carrier suppression at 3400MHz. Of course the mixer symmetry can be further improved by orienting both BAT14-099R schottky quads in the same direction in the circuit. In practice this means that one of the two BAT14-099R has to be soldered upside down in the circuit.

Both mixer outputs are combined in a quadrature hybrid to obtain the desired SSB signal at 3400MHz. The hybrid is followed by a 3400MHz bandpass (L22, L23, L24 and L25) to remove the LO residuals at 1700MHz and other unwanted mixing products far away from the 3.4GHz frequency band. Two amplifier stages with BFP420 transistors boost the relatively low signal level (about -12dBm) up to about 10mW (+10dBm). Another 3400MHz bandpass filter (L28, L29, L30 and L31) is built between the two amplifier stages.

The module of the quadrature transmit mixer for 3400MHz is built as a microstrip circuit on a double sided, 0.8mm thick FR4 glass fibre epoxy printed circuit board with the dimensions of 30mm x 120mm. The upper side is shown on Fig 11.17 the lower side is not etched to act as a ground plane for the microstrip circuits. The component location on both sides of the printed circuit board is shown on Fig 11.18.

A correctly assembled quadrature transmit mixer for 3400MHz should not require any tuning. One should

however check the output power (+10dBm) and the mixer balance (at least 25dB of unwanted carrier suppression). If the carrier suppression is not sufficient, check the LO power level and the installation of the mixer diodes.

RF front end for 3400MHz

The circuit diagram of the RF front end of the zero IF SSB transceiver for 3400MHz is shown on Fig 11.19. The RF front end includes the transmitter power amplifier, the receiver RF preamplifier with bandpass filter and the antenna switch with a PIN diode. The RF front end is built as a microstrip circuit on a double sided, 0.8mm thick FR4 glass fibre epoxy printed circuit board like in the transceivers for 1.3GHz, 2.3GHz and 5.7GHz.

The transmitter power amplifier is designed with the inexpensive GaAs transistor CLY2 in a plastic package. The CLY2 can produce more than 0.5W of RF power on 3.4GHz, however the available gain is much smaller than on 2.3GHz. The gain of the CLY2 amounts to only 8-9dB at 3400MHz and requires a drive power of 60-80mW. An ATF35076 HEMT (high I_{dss}) is used in the driver stage. Since the gain of the ATF35076 is far too high for this application, it is partially reduced by the 100Ω resistor from gate to ground. The ATF35076 still provides a few dB of gain margin for the overall transmitter chain.

The antenna switch uses a single PIN diode BAR81 in shunt configuration on the receiver input. Due to the low parasitic capacitance of the BAR81, no negative bias voltage is required to bring the diode to the OFF state in the 3400MHz band. No special circuit is therefore required to drive the PIN diode. A simple resistor to +12VTX is sufficient.

While transmitting, the short circuit from the BAR81 diode is transformed in an open circuit at the summing node by

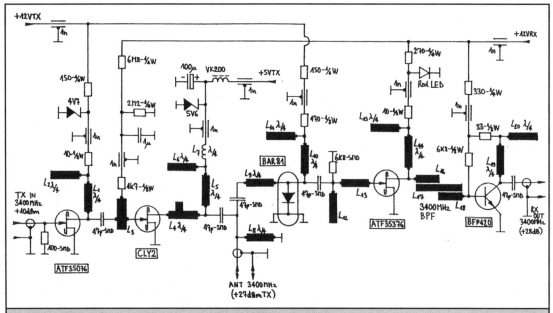

Fig 11.19 : Circuit diagram of RF front end for 3400MHz transceiver.

the L9 quarter wavelength line. The BAR81 provides more than 20dB attenuation to protect the receiver. On the other hand, the transmitter output stage remains connected to the antenna at all times. While receiving the CLY2 drain voltage is removed and a positive bias is applied to the gate. The CLY2 then behaves as a short circuit, further transformed by L4 in an open circuit at the summing node.

The RF front end includes a two stage receiving preamplifier including a simple bandpass filter for 3400MHz. The first stage uses an ATF35376 HEMT, while a BFP420 silicon transistor is used in the second stage. The overall gain amounts to about 28dB including the losses in the inter stage bandpass filter. Since there are no strong

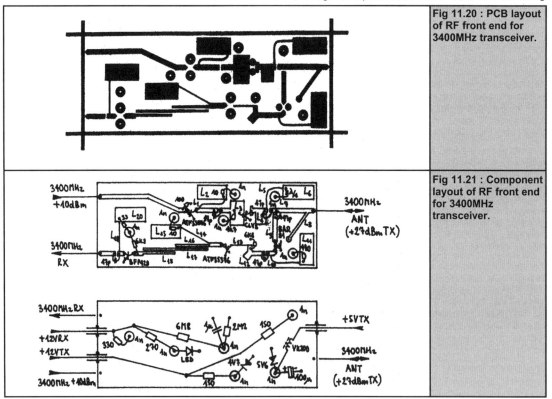

Fig 11.20 : PCB layout of RF front end for 3400MHz transceiver.

Fig 11.21 : Component layout of RF front end for 3400MHz transceiver.

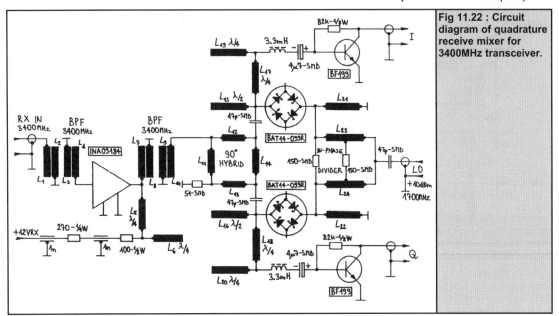

Fig 11.22 : Circuit diagram of quadrature receive mixer for 3400MHz transceiver.

signals expected in the 3400MHz band, the preamplifier gain can be kept high to improve the noise figure.

The RF front end of the zero IF transceiver for 3400MHz is built as a microstrip circuit on a double sided, 0.8mm thick FR4 glass fibre epoxy printed circuit board with the dimensions of 30mm x 80mm. The upper side is shown on Fig 11.20 the lower side is not etched to act as a ground plane for the microstrip circuits. The component location on both sides of the printed circuit board is shown on Fig 11.21. L7 is the only quarter wavelength choke that is not printed on the circuit board, but is a small self supporting coil made of a 25mm long piece of 0.15mm thick copper wire.

A correctly assembled RF front end for 3400MHz should not require any tuning. Before installing the printed circuit board in the brass frame it makes sense to check the I_{dss} of all of the FETs. Small corrections to the output circuit may bring a few milliwatts more output power, but any improvement is usually less than 1dB.

Quadrature receive mixer for 3400MHz

The circuit diagram of the quadrature receive mixer for 3400MHz is shown on Fig 11.22. The circuit of the quadrature receive mixer for 3400MHz is very similar to the equivalent mixer for 1296MHz except that all band-pass filters, hybrids and chokes are designed for a higher operating frequency.

Although the available gain from the INA03184 MMIC drops quickly above 2.5GHz, this device still provides the highest gain on 3400MHz with a modest DC current drain. Besides the INA03184 MMIC amplifier, the input signal is fed through two bandpass filters for 3400MHz before reaching the quadrature hybrid. Both sub harmonic mixers use BAT14-099R Schottky quads, fed in phase with the 1700MHz local oscillator. Finally, the quadrature receive mixer module also includes two IF preamplifiers with two BF199 transistors.

The module of the quadrature receive mixer for 3400MHz is built as a microstrip circuit on a double sided, 0.8mm thick FR4 glass fibre epoxy printed circuit board with the dimensions of 30mm x 120mm.
The upper side is shown on Fig 11.23 the lower side is not etched to act as a ground plane for the microstrip circuits. The component location on both sides of the printed circuit board is shown on Fig 11.24.

A correctly assembled quadrature receive mixer for 3400MHz should not require any tuning. Some BF199 transistors may cause excessive popcorn noise and may have to be replaced. The mixer noise figure may also increase if the LO signal level is too high.

Modification of other modules

The zero IF SSB/CW transceiver for 3400MHz is using the same quadrature modulator, IF amplifier and demodulator as the similar transceivers for 1296MHz, 2304MHz, 5760MHz or 10368MHz. The only differences are in the supply circuit for the transmitter power amplifier and in the driver of the PIN diode antenna switch.

The SSB/CW switching RX/TX module includes a current limiting resistor for the transmitter power amplifier. The value of the latter should match the RF device used in the particular transmitter output stage. In the 3400MHz transceiver the CLY2 is operated at higher currents than in the 2304MHz version, so the current limiting resistor has to be somewhat smaller. Recommended values are around 25-27Ω. The exact value is chosen so that the operating voltage of the output stage may rise up to the full value of 5.6V limited by the zener diode in the RF front end.

The SSB transceiver for 3400MHz does not require a special PIN diode driver like the 5.7GHz and 10GHz versions, since the capacitance of the BAR81 PIN diode is small enough for operation at 3400MHz without reverse bias. However, a PIN driver may be required when using higher capacitance diodes like the BAR80.

Assembly and checkout of the 3400MHz SSB transceiver

Although the assembly of the SSB transceiver for 3400MHz is not as demanding as the assembly of the 5.7GHz and 10GHz versions, it is recommended to

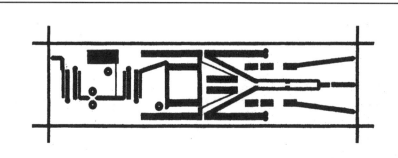

Fig 11.23 : PCB layout of quadrature receive mixer for 3400MHz transceiver.

carefully select the components to be used in the RF section. In particular it is recommended to use new SMD capacitors of the size 0805 or smaller, since the older, larger 1206 types may have strange responses in the microwave frequency range. One should absolutely avoid using capacitors made of brown ceramic, since the latter has a high temperature coefficient and high dielectric losses at microwave frequencies.

The 3400MHz version uses a new component, the BFP420 silicon transistor (SMD marking AMs). The latter is packaged in the very small SOT-343 package. Also the pin out is changed, since the BFP420 has the emitter grounded internally to improve its RF performance. The wide pin is one of the emitters, the other emitter being located diagonally on the opposite side of the package. The base is on the same side as the wide pin while the collector is on the opposite side of the wide pin. The BFP420 transistors operate at low voltages 2-3V just like HEMTs, but they do not tend to oscillate at very high frequencies like HEMTs.

The RF modules of the SSB transceiver for 3400MHz have the same dimensions as the modules of the 5.7GHz and 10GHz versions. Therefore the same brass boxes, overall transceiver enclosure (60mm x 180mm x 180mm) and module location as in the 5.7GHz or 10GHz versions can be used. The only difference is in the position of some connections, requiring different holes on the brass boxes. All of the microstrip circuits of the 3400MHz SSB transceiver are designed so that no microwave absorber is necessary in any of the RF modules. Also the VCXO module does not require a bottom cover.

The checkout of the transceiver should start with the alignment of the VCXO and multiplier stages. The VCXO should be tuned to cover the desired frequency range. The multiplier stages are adjusted for the maximum output signal, measured as a voltage drop on the base of the following stage. The maximum on 425MHz is found by measuring the drain current (voltage) of the HEMT multiplier for 1700MHz. The additional multiplier for 1700MHz does not require any tuning except for the adjustment of the output power (around +10dBm).

Both the receiver and the transmitter should already be operational, but they still need several checks. In the receiver one should first check the overall gain. There should be a clear drop in the noise when the supply to the LNA is removed and the noise should drop almost to zero when the IF preamplifiers are disconnected. If the drop of the noise power is more pronounced at low supply voltages (10V or less), the LO signal level is probably too high.

Next the receiver is tuned to a weak signal (beacon or another VCXO with multiplier chain) and the received signal is carefully monitored. Besides the main response a mirror signal moving in the opposite direction while tuning should also be heard. The mirror response should be suppressed with the two trimmers in the IF amplifier. Any frequency instabilities are probably caused by the VCXO, especially when using a cheap PAL subcarrier crystal pulled down to 17.708MHz.

The transmitter output power should be checked while adjusting the trimmer in the modulator. In CW mode, the full power should be achieved at about 1/3 of the full resistance. The supply voltage of the output stage should then rise to 5.6V limited by the zener diode. In SSB mode the output power and supply voltage of the output stage should both drop to almost zero without modulation. The

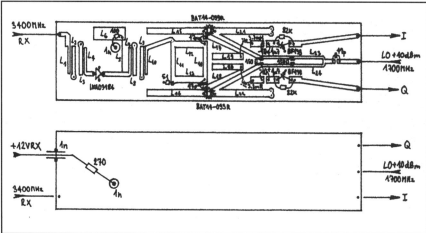

Fig 11.24 : Component layout of quadrature receive mixer for 3400MHz transceiver.

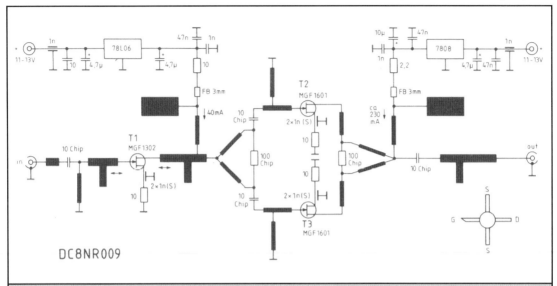

Fig 11.25 : Circuit diagram of 9cm GaAsFet power amplifier.

unwanted carrier suppression should be at least 25dB. An insufficient carrier suppression is probably also caused by a too high LO signal level.

The SSB modulation as well as the carrier suppression should also be checked with another station operating in the 3400MHz band. Both LSB and USB have to be checked, since it is relatively easy to swap I and Q lines in the internal wiring of the transceiver. The mode of operation (USB or LSB) of the receiver can also be checked by tuning the receiver to an unmodulated carrier (beacon).

Finally the overall shielding of the transceiver has to be checked. Waving with a hand usually produces a 1.4kHz whistle in the loudspeaker. This effect is common to all zero IF radios and is caused by the leakage of the LO signal because of insufficient shielding. The frequency of the reflected signal from a moving hand is shifted by the Doppler effect. Finally this signal is collected by the antenna and fed to the receiver front end.

The current drain of the described SSB/CW transceiver for 3400MHz is about 240mA during quiet reception at a nominal supply voltage of 12.6V. The current drain rises to about 590mA during CW transmission at an output power of 500mW. During SSB transmission the current drain is even higher, since it is inversely proportional to the modulation level, and it may reach 750mA during speech pauses.

Amplifier [4]

The following article describes a two stage amplifier module which is able to deliver an output of 0.5 W and has a gain of at least 23 dB. It is therefore very suitable for portable use after an output digital filter [5] has been added.

Circuit description

There is not a good supply of efficient and reasonably priced silicon semiconductor devices available for the 9 cm band. It would therefore seem worth while to place

several small signal transistors in parallel as proposed in [6] in order to achieve the required output power. This power amplifier has all its transistors working at the limits of their ratings (BFG 91 A: Icav = 35 mA!) and despite all the trouble taken, it still worked in the saturation area at full output power. The small stage gain of 3 to 4 dB with four parallel transistors is a sure indication of this.

In the proposed amplifier, GaAs-FETs are used (Fig 11.25), the driver stage being the well known MGF 1302. The final stage uses two power FETS, MGF 1601, working in parallel but isolated inputs and outputs using a Wilkinson divider [7]. Experiments have shown that this method is very efficient due to the well defined impedance ratios that can be achieved.

In the interests of linearity, all stages work in class A. Thermal runaway is not possible with FETs as they possess a negative temperature coefficient. This is not the case with bipolar devices, however.

Automatic bias is used for the FETs in order to place the gate at a negative potential with respect to the source. The source's decoupling must be carried out carefully, using plate capacitors, to ensure a good RF ground with low inductance. This will pay off in terms of amplifier stability and gain. As the test data indicates, this method of biasing does not involve any real disadvantages at these frequencies compared to direct source grounding.

The MGF 1601 has been available for some time now, it is quite cheap and the manufacturers claim an output power, at 8GHz, of 150mW (1 dB compression) and at 6 to 8 dB gain (V_{DS} = 6 V, I_D=100mA). At about an octave lower, the performance will be even better.

Construction

The circuit is constructed on a 0.79 mm thick, double sided, copper coated printed circuit board (DC 8 NR 009) with a PTFE substrate. It is 72 x 72 mm and fits into a proprietary tin plate box 74 x 74 x 30 mm.

The placement of the few components on the board (Fig 11.26) is, with the exception of the source decoupling

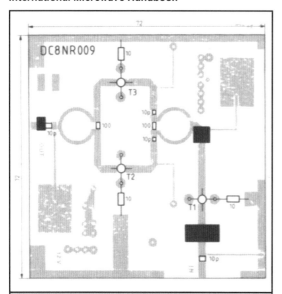

Fig 11.26 : RF component locations on the track side of the 9cm GaAsFet power amplifier PCB.

Fig 11.28 : Photograph of completed 9cm GaAsFet power amplifier.

capacitors, not critical. More details are given in Figs 11.27 and 11.28.

The distance between the two holes for each transistor is determined by the diameter of the plate capacitors used. These capacitors should be as small and as thin as possible (lightly coloured ceramic material). The holes should be drilled with a small drill and then filed out so that there is 2 mm separation between the circumference of the holes for the MGF 1302, and 2.8 mm for the MGF 1601. The capacitors should be a push fit, therefore care is required with the filing process! In addition, both final transistors have a 2.8 mm wide slot filed out between these holes. This will accept the transistor body at a later stage. This slot is now covered on the track side with a piece of copper foil 0.5 to 0.8 mm thick. To do this, the

track is tinned at the appropriate places and the tinned foil is then pressed down onto the board with the hot soldering iron. Using a soldering iron temperature of 450°C, the high heat conductivity of copper will ensure that this operation will be completed quickly enough to avoid damage to the copper track side of the board.

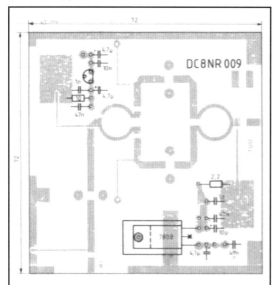

Fig 11.27 : DC component locations on the ground plane side of 9cm GaAsFet power amplifier PCB.

Fig 11.29 : Photograph of completed 9cm GaAsFet power amplifier and PCB cut outs for power FETs.

Finally, the plate capacitors are placed in the appropriate holes and soldered in using the same technique as with the copper foil.

The printed circuit board may become a little buckled and should now be carefully straightened out before it is soldered into its tin plate housing. Care should be taken that the socket pins line up exactly with the appropriate pads in order that they can be soldered directly to the track side of the board.

The inner conductors of the connectors should be shortened before fitting and the PTFE collar removed as far as possible. If N connectors are used, it is better to use the precision version these have a smaller flange than the standard version and require only a 4mm mounting hole to be drilled. They also have a better match to the board transmission lines. The coupling capacitors used were all high Q types with a light coloured ceramic substrate.

Setting up

Before soldering in the semiconductors, the voltage at the output of the voltage regulators should be checked.

Because of the relatively high dissipation of heat from the output devices, they must be cooled sufficiently. This is achieved by pressing in the transistors, with the markings underneath, into the slots provided in the board. The transistor leads are then soldered on to the tracks. A little heat paste will ensure better heat conductivity.

The gate and drain leads are cut off so that only 3mm remains and the source connection is cut off so that it does not project beyond the plate decoupling capacitor. During the mounting process the usual precautions must be followed to prevent high static electrical charges from damaging these expensive transistors. Also the inputs and outputs must be directly terminated during the tuning process in order to prevent a mismatch occurring due to a high VSWR at the end of the test cables. Pads of 3 dB or more should be suitable for this purpose.

Following the application of the supply voltage, the quiescent current given in the circuit diagram should be set. This value should only slightly increase when drive is applied. The symmetry of the output stage is then checked. This is achieved by measuring the voltages across the source resistors. These voltages must only differ from one another by only a few millivolts, whether drive is applied or not.

If the devices come from the same batch, as was the case with the author, then there will be no difficulty with the symmetry. If balancing is necessary, however, it can be achieved either by changing source resistors or a small capacitive tuning foil soldered to the source lead.

The tuning procedure consists of holding a small piece of copper foil with a bent matchstick, sliding it up and down and in contact with the tracks, and watching for the largest peak in output power. The foil is then soldered to that position using as little solder as possible. This procedure is carried out three times. The dimensions of the foil strips and their distances from the device connections are: -

• Gate: 6 x 13 mm, distance 7 mm

• Drain: 5.5 x 7 mm, distance 10 mm

• Output: 4 x 4mm

These dimensions should be regarded as guidance only, a little experimenting with different foil sizes and distances may result in a few more tenths of a dB output power.

Test results

This two stage module has an astoundingly high gain. It amounted to some 28 dB for an output power of 0.5 W. The 1 dB compression point is somewhat under this power, higher drive then results in saturation. Nevertheless, 0.7W with a gain of 24dB was achieved. Input drive powers of more than 2mW should be inhibited by means of input attenuator pads.

The final transistors are stable, even when working under changing load conditions and putting on the cover caused little change. To ensure a good heat exchange, it is perhaps advisable not to put a cover on the component side. An alternative approach is to perforate the cover over the voltage regulators.

Antennas

The most commonly used antenna at 3.4GHz is a suitably fed parabolic dish. Their popularity is mainly due to the high gain achievable with a relatively small sized dish. A graph of dish gain against dish diameter is shown in Fig 11.30. The parabolic surface of the dish may be solid, but in some cases this may lead to excessive wind loading on the mast. As an alternative, the reflecting surface may be fabricated from wire mesh, as long as the holes are not greater than λ /10 (9mm) in the plane of polarisation.

Two suitable types of dish feed are described here, the dipole/splashplate feed and the feed horn. This latter design may be used independently of the dish as a low gain, broad beamwidth radiator that can be useful for local testing.

Where circumstances do not permit the use of a dish, a quad loop Yagi can still offer a useful gain when constructed carefully.

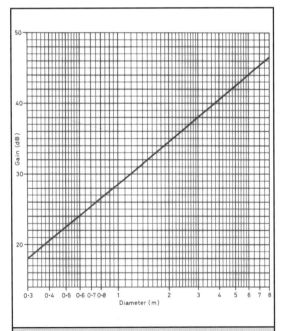

Fig 11.30 : Gain vs diameter for a parabolic dish at 3.4GHz, assuming a 50% feed efficiency.

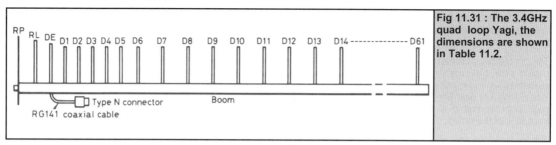

Fig 11.31 : The 3.4GHz quad loop Yagi, the dimensions are shown in Table 11.2.

Table 11.2 : Dimensions of 3.4GHz quad loop.

Boom:

Boom diameter	12.5mm o.d.
Boom length	2.0m
Boom material	aluminium alloy

Elements:

Driven element	1.6mm diameter welding rod
All other elements	1.6mm diameter welding rod

Reflector size:

Reflector plate	52.4 x 42.9mm

Element lengths:

Length L(mm)(see Fig 11.32a):

Reflector loop	99.2
Driven element	92.7
Directors 1-12	83.9
Directors 13-20	81.2
Directors 21-30	78.2
Directors 31-40	76.1
Directors 41-50	75.1
Directors 51-60	74.6

Cumulative element spacings (mm):

RP	0.0	RL	29.5
DE	38.6	D1	49.2
D2	57.2	D3	74.1
D4	91.1	D5	103.0
D6	125.0	D7	158.9
D8	192.8	D9	226.7
D10	260.6	D11	294.5
D12	328.4	D13	362.3
D14	396.2	D15	430.1
D16	464.1	D17	498.0
D18	531.9	D19	565.8
D20	599.7	D21	633.6
D22	667.5	D23	701.4
D24	735.3	D25	769.2
D26	803.1	D27	837.0
D28	871.0	D29	904.9
D30	938.8	D31	972.7
D32	1006.6	D33	1040.5
D34	1074.4	D35	1108.3
D36	1141.2	D37	1176.1
D38	1210.0	D39	1244.0
D40	1277.9	D41	1311.8
D41	1345.7	D43	1379.6
D44	1413.5	D45	1447.4
D46	1481.3	D47	1515.2
D48	1549.1	D49	1583.1
D50	1617.0	D51	1650.9
D52	1684.8	D53	1718.7
D54	1752.6	D55	1786.5
D56	1820.4	D57	1854.3
D58	1888.2	D59	1922.1
D60	1956.1	D61	1990.0

The quad loop antenna

This antenna was originally designed by M Walters, G3JVL. The design of the antenna is shown in fig 11.31 with the critical dimensions being shown in Table 11.2. The design is a scaled version of the 1,296MHZ design adapted to the narrow band segment at 3,456MHZ. The number of directors is 61 thus the boom length is 2m. The construction of the antenna is quite straightforward providing care is taken in the marking out process. Measurements should be made from a single point or datum. In marking the boom for instance, measurements of the position of the elements should be made from a single point rather than marking out individual spacings.

All elements are made from 1.6mm diameter welding rod cut to lengths shown in the table then formed into a loop as shown in Fig 11.32a. The driven element is brazed to an M6 x 25 countersunk screw drilled 3.6mm to accept the semi rigid coaxial cable. All other elements are brazed onto the heads of M4 x 25 countersunk screws.

All elements, screws and joints should be protected with a coat of polyurethane varnish after assembly. If inadequate attention is paid to weatherproofing the antenna, then the performance of the antenna will gradually deteriorate as a result of corrosion. Provided the antenna is carefully constructed, its feed impedance will be close to 50Ω. If a suitably rated power meter or impedance bridge is available, the match may be optimised by carefully bending the

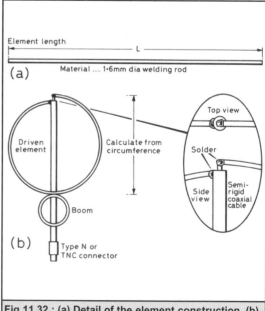

Fig 11.32 : (a) Detail of the element construction. (b) Assembly of the driven element.

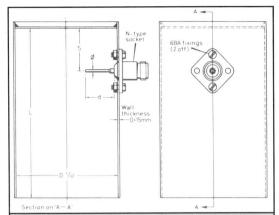

Fig 11.33 : Constructional detail of a 3.4GHz feed horn. Dimensions are, D=56mm, L=208mm, s=51mm, d=18mm and φ=2.1mm.

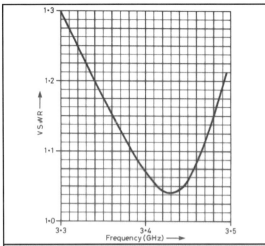

Fig 11.35 : VSWR response of the 3.4GHz feed horn.

reflector loop toward or away from the driven element. The antenna can be mounted using a suitable antenna clamp. It is essential that the antenna be mounted on a vertical support, as horizontal metalwork in the vicinity of the antenna can cause severe degradation in its performance.

Feed horn for 3.4GHz

The feed horn described here is ideally suited to feed a dish having an f/D ratio of between 0.2 and 0.4.

The feed horn consists of a short length of circular waveguide with a coaxial to waveguide transition made from an N type socket and a short probe. Constructional details are shown in Fig 11.33. the circular waveguide section can be made from an empty food tin closed with a plate at one end. All the dimensions are fairly critical (especially the length of the probe) hence, care should be taken during construction if a low vswr is to be obtained without adjustment. If a different diameter can is used the length and position of the probe and the length may need

altering for best match. The N type socket should be fixed to the horn using four fixings, using washers as spacers to accommodate the curved surface of the tin. Alternatively an SMA socket may be used. Fig 11.34 is a photograph of the finished feed horn. The vswr response of the feed horn is shown in Fig 11.35, from which it can be seen that the feed is an excellent match at 9cm.

Dipole/splashplate feed

The version of the feed to be described here was developed by G Coleman, G3ZEZ. It is suitable for use in a dish having an f/D ratio between 0.25 and 0.45.

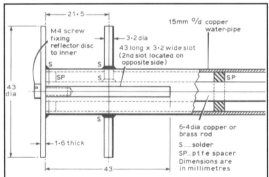

Fig 11.36 : Construction of a 3.4GHz disc and dipole feed.

Fig 11.34 : Photograph of complete 3.4GHz feed horn. Note that this version is constructed from brass tube and plate and uses an SMA socket.

Fig 11.37 : Photograph of the modified plumbing fitting for mounting disc and dipole feed.

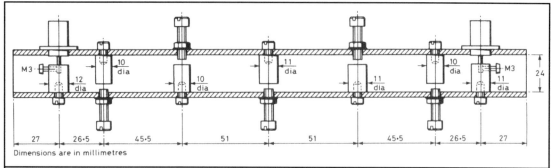

Fig 11.38 : Dimensions of a 3.4GHz interdigital filter.

The feed is constructed around a length of fabricated coaxial line, the ratio of the inner diameter of the outer conductor to the outer diameter of the inner conductor being chosen to produce the desired impedance in this case 50Ω.

Details of the feed are shown in Fig 11.36. It is constructed from a length of 15mm diameter copper water pipe, the length of which is chosen to enable point X to be positioned at the focus of the dish. The 43mm diameter disc is turned or filed from a 1.6mm thick sheet of brass or copper. The two slots may be either milled to shape or carefully hacksawed and filed. The ptfe spacers marked SP are 13.5mm o.d. by 5mm thick spacers with a 6.35mm hole drilled centrally. They are used to maintain line concentricity.

The feed is held in position with a modified 15mm cold water tank compression fitting (Fig 11.37), preferably one with a mounting nut on either side of a central flange. The internal step in the fitting is filed away to leave a constant 15.2mm bore. The fitting is mounted in the dish in a 35mm diameter hole using one of the supplied nuts. The position of point X can be altered by sliding the feed in the fitting. When the position of the feed has been optimised by monitoring a remote signal, the feed can be locked in place by tightening the nut over the olive. This type of fitting must be fitted on the line *before* attaching the N type connector.

The N type female connector is fixed to the line by counterboring the collar 15mm diameter by 2mm deep and soldering the collar to the line. The centre pin is soldered to the line inner and the connector assembled with its supplied ptfe insulators.

Filters

A 3.4GHz interdigital bandpass filter

The narrow band filter to be described here was originally published in [8]. It has an insertion loss of around 1dB and a 3dB bandwidth of 48MHz. The rejection of signals 144MHz away from the centre frequency is greater than 90dB.

The filter is constructed from brass, although it could, with advantage, be made from copper. The sidewalls are each made from pieces of 300 x 24 x 4mm bar drilled as shown in Fig 11.38. the base and coverplate are identical and made from 300 x 32mm pieces of 1mm sheet. The covers are attached to the sidewalls using a total of 28 M3 x 6 brass screws to ensure a good RF seal. The input and output coupling elements are made as shown in Fig 11.39. the remaining coupling elements are shown in Fig 11.40. Each have an 8mm deep tapped M4 hole in one end. They are attached to the sidewalls using M4 x 8 screws and tuned at the opposite end with M4 x 12 screws, each fitted with an M4 locknut. The input and output connectors are four hole fixing N types. The size of the central hole drilled for mounting these connectors may need altering, depending on the size of the boss of the connector.

The filter can be aligned by connecting it between 3,456MHz transmitter and receiver, including suitable

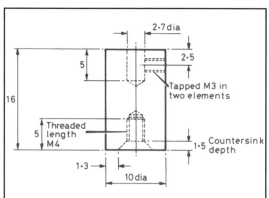

Fig 11.39 : The construction of the interdigital filter input and output coupling rods. Diameter D is 17mm.

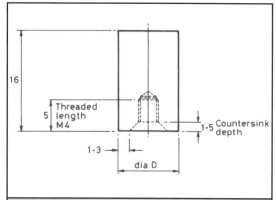

Fig 11.40 : The construction of the remaining interdigital coupling rods. Dimension D is 15.3mm.

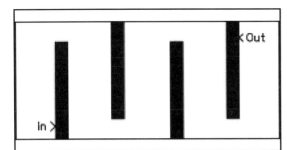

Fig 11.41 : The basic structure of an interdigital filter.

attenuators on the filter input and output to provide both a 50Ω termination for the filter, but also to avoid damaging the receiver. The filter should then be tuned for maximum received signal. The filter may need slight retuning in its final configuration to accommodate any slight source and termination impedances.

Waveguide interdigital filters [9]

Most microwave transverters, especially the no tune variety, need some additional filtering to operate in locations with RF pollution. Accessible mountain tops are notoriously bad environments. Waveguide post filters provide superior performance at 10 GHz [10] and 5760 MHz [11] but become large and heavy at lower frequencies. Interdigital filters are excellent performers at the lower microwave frequencies [12] but the usual construction techniques require some machining, mostly tedious tapping of threads in many holes. One of the beauties of waveguide filters is that the basic structure is accurately defined by the waveguide, so construction requires only drilling and soldering. Since surplus waveguide is reasonably plentiful, I wondered if it could be used to build interdigital filters for the lower microwave bands. As we shall see, my experiments were quite successful.

Interdigital filters

The basic structure of an interdigital filter, shown in Fig 11.41, is a group of coupled resonators in a metal housing. Each resonator is an electrical λ/4 long, but physically shortened by capacitance at the open end. The resonators are interdigitated, with the position of the open ends of the resonators alternating as depicted in Fig 11.41. (A similar filter with all resonators aligned in the same direction is called a comb line filter.) The coupling between resonators is controlled by their separation. Several methods are commonly used to make input and output connections, but a simple one is to use taps on the input and output resonators.

The starting dimension for an interdigital filter is the width of the housing, which should be λ/4 at the operating frequency. All of the other dimensions are interrelated, a change in one affects others so that empirical design of a filter would be difficult and frustrating. Fortunately, computer programs are available to design interdigital filters. A BASIC program [13] by N6JH appeared in Ham Radio Magazine. I translated this into PASCAL and compiled it. My version, INTFIL.EXE, is available for downloading at http://www.qsl.net/n1bwt/intfil.zip. One 1296 MHz filter that I built using this program was carefully measured using an automatic network analyser and found to match the predicted performance almost perfectly with no tuning. This gave me confidence in the accuracy of the program.

Table 11.3 : Waveguide dimensions for interdigital filters.

Waveguide	Wide Dimension (inches)	Frequency (λ/4, MHz)
WR-340	3.4	868
WR-284	2.84	1039
WR-229	2.29	1289
WR-187	1.872	1577
WR-159	1.59	1857
WR-137	1.372	2152
WR-112	1.12	2636
WR-90	0.90	3280
WR-75	0.75	3937
WR-62	0.622	4747
WR-50	0.51	5789

Filter Design

The first part of filter design is the same for all types of filters, calculation of coupling coefficients and other parameters to achieve the desired performance. These are tabulated in The ARRL Handbook [14] and other reference books [15] for the most common types of filters. The Butterworth (maximally flat) and the Chebyshev response, which trades some pass band ripple (amplitude variation) for somewhat steeper skirts at the pass band edges. The tabulated parameters, g_{mn}, are for a normalised prototype filter, so that further calculations are required to find actual component values for a desired frequency and impedance.

The second part of filter design is to convert the normalised parameters into component values or physical dimensions. The calculations are quite tedious, so graphical solutions were often published [16] before computers were commonly available. Now these calculations are easily performed on a PC, allowing us to evaluate multiple filter designs before choosing one to build.

Design of an interdigital filter begins with the choice of a required bandwidth. Simple filter programs such as INTFIL are only reliable for bandwidths between about 1% and 10% of the centre frequency, and very narrow bandwidth filters are lossy and require tight tolerances in construction. Therefore, a 3% to 5% bandwidth is recommended as a good starting point. The next step is to decide how steeply the skirts roll off at the pass band edges. For example, steeper skirts are required to reject an image close to the operating frequency. Generally, filters with steeper skirts require more resonators and have more loss, so a compromise may be in order. It is possible to calculate the number of resonators required, but a few trial designs on the computer should yield the same result and provide some insight as well.

Designing waveguide interdigital filters

Now let's design a filter in an available waveguide. As mentioned above, we should start with the width of the enclosure at λ/4 at the centre frequency. This would be the large inside dimension of a waveguide used as the enclosure. Table 11.3 lists the inside dimensions for some commonly available waveguides and the frequencies for which the wide dimensions are λ/4. The large dimension for WR-229 is λ/4 at 1289 MHz, so it is a logical material for a 1296MHz filter. Simply designing a filter for the 1289MHz centre frequency with enough bandwidth to include 1296MHz does the trick. I chose a 50MHz

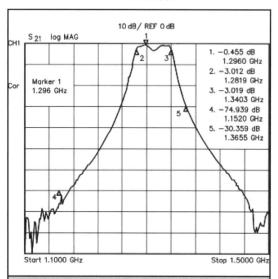

Fig 11.42 : Pass band of 1296MHz waveguide interdigital filter.

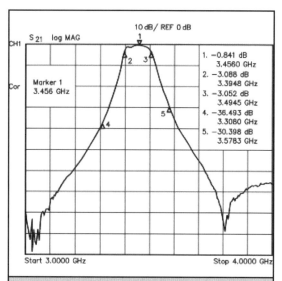

Fig 11.44 : Pass band of 3456MHz waveguide interdigital filter.

bandwidth and used INTFIL to calculate the rest of the dimensions for a resonator diameter of 3/8 inch. Since WR-229 is being scrapped as 4GHz telephone microwave links are decommissioned, I was able to find all I could carry at a flea market for $5.

Once the filter was assembled, I found that the bandpass was slightly above 1296MHz, a little higher than the design. This was easy to fix, however: I drilled and tapped holes for tuning screws opposite the open ends of the resonators and inserted screws to add capacitance and lower the frequency. On the other hand if the frequency ended up a bit low, then it would be necessary to shorten the resonators slightly.

I first adjusted the screws for minimum insertion loss at 1296MHz, then readjusted them for minimum SWR at both ends. The second adjustment is a bit more involved since each adjustment affects both ends and a few

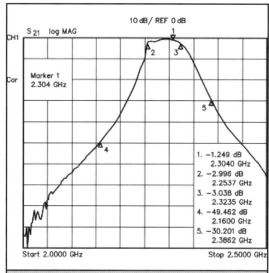

Fig 11.43 : Pass band of 2304MHz waveguide interdigital filter.

reversals were needed. The final pass band, shown in Fig 11.42, is about 58MHz wide with an insertion loss less than 0.5 dB at 1296MHz.

Tuning is straightforward for these filters, with moderate bandwidth and a reasonable number of resonators. However, filters with very narrow or wide bandwidths, or with many resonators, require a more complex tuning procedure. Dishal's procedure [17], [18] allows the tuning of one resonator at a time.

For other bands, no waveguide exactly matches λ/4, but there are some good candidates that fall within about 10% of the desired frequency. The ubiquitous X band waveguide, WR-90, is close to λ/4 at 3456 MHz, while WR-137 (used in 6GHz microwave links) is close to 2304 MHz. For these two, simply making a wide filter is not good enough. The bandpass would include the commonly used LO frequencies for a 144MHz IF. Thus, we need a design procedure that can move the centre frequency slightly.

The design procedure that I use makes two similar designs, one at the desired frequency and one at the λ/4 frequency waveguide, using the same percentage bandwidth (bandwidth centre frequency) for both designs. Using the same percentage bandwidth results in two designs differing only in the resonator lengths and tap positions, and the difference is small because the frequencies are close together. Since the higher frequency design also calls for the λ/4 distance to be shorter, making the resonators this short would result in less capacitance and an actual resonant frequency higher than desired. My compromise is to split the difference between the two design lengths and make the resonator lengths halfway between the two designs.

I followed the above design procedure for two more filters, one for 3456MHz in WR-90 waveguide with 108MHz bandwidth and the other for 2304MHz in WR-137 waveguide with 75MHz bandwidth. Each design uses four resonators with a Butterworth response. After fabrication, both filters had passbands that included the design frequency, as shown in Figs 11.43 And 11.44. Thus, they are usable with no further tuning. Any elective tuning

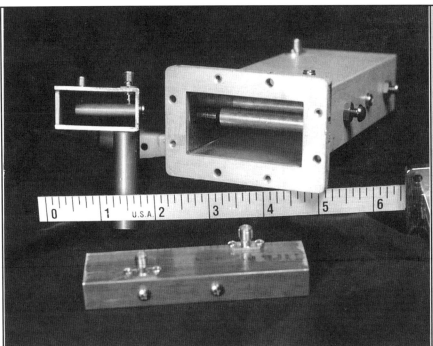

would optimise the input and output SWR at the desired frequency.

To simplify testing, the effects of the end walls are minimised by locating them relatively far from the end resonators. The result is that leaving the end walls off during testing makes little difference in performance. I located the end walls one inch from the end resonators in the 1296MHz filter, and could find only a slight perform-ance difference with the end walls in place. There was no detectable difference for the higher frequency filters. Of course, the end walls should be installed for operation, stray leakage could otherwise negate the effect of the filter.

Construction

The three completed filters are shown in Fig 11.45. Each resonator is attached by a screw through a narrow wall of the waveguide and the coaxial connectors are mounted in a wide wall of the waveguide with short leads to the tap points on the end resonators.

Resonator lengths and spacings are fairly critical, so accurate measurement is needed. The holes are best made with a drill press, see Fig 11.46 for details of tools for interdigital filter construction. Start with a centre drill or small drill bit to spot the hole, then follow with a drill bit of the desired diameter. The mounting holes in the end of the resonators should be tapped and countersunk slightly,

---

I use a small metal lathe to trim the interdigital filter resonators to length. A lathe is the ideal tool for this work, but is a luxury for most hams. Some other tools are great for homebrewing and inexpensive imports have made them quite affordable. The two that I find almost indispensable are a drill press and a dial calliper.

A drill press is a sturdy drill on a stand, with an adjustable table to hold the work. For the interdigital filters, it drills the holes square and true. The mounting holes in the resonator rods can be first drilled and then threaded, by putting a screw tap in the drill press and feeding it by hand (power off!), with the rod held in a vice. Imported tabletop drill presses are available for less than $60. [20], [21] I used and abused one of these constantly for 16 years before buying a larger floor mounted model.

A dial calliper is a measuring instrument capable of resolving dimensions to a precision of 0.001 inches on a dial. Most of the dimensions in the interdigital filters, and for much microwave work, must be more precise than I can measure with a ruler, so my dial calliper also sees constant use. I even scribe dimensions directly with the calliper tip, a gross abuse of the tool that I justify by its low replacement cost. Imported 6inch dial callipers are available for less than $20, [22], [23] a very modest investment compared to the alternative: eyestrain and frustration.

Everything else can be done with common hand tools plus patience. A hacksaw and file can cut the waveguide to length and trim the resonators, measuring frequently with the dial callipers. The holes are carefully marked, centre punched, and drilled to size with a drill press. A set of number sized drills provides many more choices of hole size than ordinary fractional sizes, and is available at reasonable cost from any of the suppliers already mentioned.

A modest investment in tools, inexpensive, but not cheap, can add to the pleasures of homebrewing and improve the results.

Fig 11.46 : Description of tools for interdigital filter construction

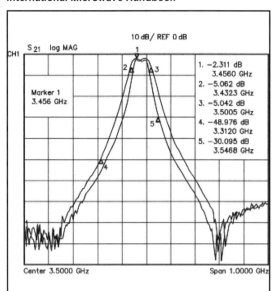

10 dB/ REF 0 dB

CH1 S 21 log MAG

Marker 1
3.456 GHz

Center 3.5000 GHz   Span 1.0000 GHz

1. −2.311 dB
   3.4560 GHz
2. −5.062 dB
   3.4323 GHz
3. −5.042 dB
   3.5005 GHz
4. −48.976 dB
   3.3120 GHz
5. −30.095 dB
   3.5468 GHz

Fig 11.47 : Pass band of simplified 3456MHz waveguide interdigital filter, before and after tuning.

so contact is made around the resonator perimeter. For initial testing, I don't solder the input and output connections, but rather make them slightly long with a sharp point contacting the tap point on the resonators.

Using waveguide as the housing makes the filters easy to build, and results in a robust, stable filter, suitable for rover operations. Some of my previous experiments in filter construction using hobby brass and PC board were less successful due to mechanical instability: Vibrations or the weight of connecting coax cables affected their performance. One notably bad filter was so unstable that the frequency response would vary during measurement.

After building and testing the filters in Fig 11.45, I wondered if there was an even simpler way to make these filters. Since the resonator length for the 3456 MHz filter is just a hair's breadth over 3/4inch, perhaps an ordinary 3/4inch long, 1/4inch diameter threaded stand off could be used as a resonator. The resonator spacings and the tap point dimensions are the critical ones, so I calculated a filter with 75MHz bandwidth using 1/4inch diameter resonators, then built it with threaded stand offs. It took less than an hour to complete. A quick measurement showed nearly 3dB loss at 3456 MHz, so I took it apart to add tuning screws to see if I could improve it. Careful tuning only reduced the loss to 2.4dB, versus 0.8 dB for the filter with machined brass resonators. Fig 11.47 shows the response before and after tuning. The response is slightly higher in frequency before tuning, but otherwise there is little difference. I attribute the higher loss to three factors:

- I arbitrarily designed this version for a 75MHz (2%) bandwidth, compared to a 108MHz (3%) bandwidth for the other 3456MHz filter. As previously mentioned, filters with narrow bandwidths tend to be more lossy.

- The threaded stand offs are plated with nickel, a lossy metal.

- The stand offs are chamfered at the ends, so the contact area is smaller at the shorted end where currents are highest.

Table 11.4 : Waveguide interdigital filter examples.

	WR-229	WR-137	WR-90	WR-90
Waveguide				
Target Frequency (MHz)	1289	2304	3456	3456
Bandwidth (MHz)	50	75	108	79

Resonator (Designed for waveguide λ/4)

	WR-229	WR-137	WR-90	WR-90
Diameter inches	0.375	0.25	0.1875	0.25
End Length inches	1.983	1.099	0.732	0.727
Interior Length inches	1.971	1.095	0.73	0.728
Tap Point inches	0.23	0.127	0.089	0.089
λ/4 Frequency MHz	1289	2155	3280	3280
Bandwidth MHz	same	70	100	75

Resonator (Designed for target frequency)

	WR-229	WR-137	WR-90	WR-90
Diameter inches	0.375	0.25	0.1875	0.25
End Length inches	1.983	1.187	0.777	0.773
Interior Length inches	1.971	1.183	0.775	0.772
Tap Point inches	0.23	0.136	0.093	0.095
Spacing 1_2,3_4 inches	1.47	0.864	0.58	0.653
Spacing 2_3 inches	1.632	0.951	0.636	0.709

Compromise Dimensions

	WR-229	WR-137	WR-90	WR-90
Resonator Diameter inches	same	0.25	0.1875	0.25*
End Length inches		1.144	0.756	0.75*
Interior Length inches		1.14	0.752	0.75*
Tap Point inches		0.132	0.09	0.092
Loss, Calculated dB	0.2	0.3	0.4	0.6
Loss, Measured dB	0.45	1.25	0.8	2.3
B/W, Measured MHz	58	70	100	68
LO Frequency MHz	1152	2160	3312	3312
LO Rejection, calc dB	−59	−47	−34	−45
LO Rejection, Mea dB	−75	−49	−36	−49

* = threaded standoff

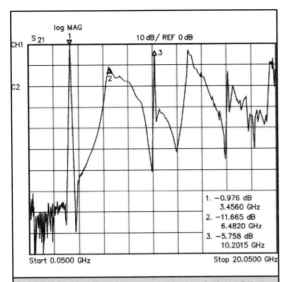

log MAG
CH1 S 21   1   10 dB/ REF 0 dB
C2

1. −0.976 dB
   3.4560 GHz
2. −11.665 dB
   6.4820 GHz
3. −5.758 dB
   10.2015 GHz

Start 0.0500 GHz   Stop 20.0500 GHz

Fig 11.48 : The transmission characteristics of the 3456MHz filter from 50MHz to 20GHz.

Performance

The waveguide interdigital filters exhibit excellent performance with low insertion loss in the pass band and high rejection of undesired frequencies. Steep skirts provide good rejection of possible spurious signals, such as LO leakage only 144MHz away from the operating frequency. LO rejection is much greater for the 1296MHz filter than for the others because the relative LO separation is much greater at 1296MHz, being 9% of the operating frequency at 1296MHz, versus 6% at 2304MHz and 4% at 3456MHz.

Table 11.4 lists the filter dimensions and compares the measured performance with the design values, as calculated by INTFIL. The measured performance is quite close to the design values. The dimensions shown are for the two designs for each filter, plus the compromise values that I fabricated, to illustrate the design procedure.

The only performance flaw for these filters is poor harmonic rejection. At frequencies much higher than the operating frequency, the waveguide enclosure behaves as a waveguide rather than just an enclosure. This behaviour is not unique to the waveguide interdigital filters all conductive enclosures will propagate waveguide modes at frequencies above the cutoff frequencies for the interior dimensions. Fig 11.48 shows the transmission characteristics of the 3456MHz filter from 50MHz to 20GHz. Out of band rejection is excellent (> 70 dB) below the pass band and good above the pass band up to about 6GHz. Above 6GHz, various waveguide modes are propagated and limit the attenuation.

Conclusion

Good filters are important for operation in locations with RF pollution, and are recommended when no tune transverters are followed by broadband power amplifiers. Interdigital filters offer excellent performance for the lower microwave bands. The filters are easily constructed in a waveguide housing without extensive machining and require little tuning, resulting in a robust, high performance filter.

References

[1] R. Bertelsmeier, DJ9BV, HEMT LNAs for 23cm, DUBUS TECHNIK IV

[2] R. Bertelsmeier, DJ9BV, No-Tune HEMT Preamp for 13cm, DUBUS TECHNIK IV

[3] From an article by Matjaz Vidmar, S53MV, CQ ZRS

[4] From an article by Werner Rahe, DC8NR, VHF Communications 4/1989 pp 249 - 254

[5] Burfeindt, H., DC9XG, A SSB Transmit Mixer and Linear Amplifier for 3456 MHz VHF Communications, 1/1984, p. 13 - 22

[6] Team of authors, Ein 9 cm Transvertersystem moderner Konzeption, CQ-DL 9 - 11/88

[7] Hupfer, K., DJ1EE, Wideband Power-Divider/Combiner for the 2 m and 70 cm Bands, VHF Communications, Vol. 20 1/1988, p. 2 - 7

[8] Narowband filters for the 23cm, 13cm and 9cm band, D Vollhardt, DL3NQ, VHF Communications, January 1978, pp 2 - 11

[9] From an article by Paul Wade, W1GHZ, QEX January 1999

[10] G. Elmore, N6GN, A Simple and Effective Filter for the 10GHz Band, QEX, July 1987, pp 3-5

[11] P. Wade, N1BWT, A Dual Mixer for 5760 MHz with Filter and Amplifier, QEX, August 1995, pp 9-13

[12] R. Fisher, W2CQH, Interdigital Bandpass Filters for Amateur VHF/UHF Applications, QST, March 1968, p 32

[13] J. Hinshaw, N6JH, and S. Monemzadeh, Computer-Aided Interdigital Bandpass Filter Design, Ham Radio, January 1985, pp 12-26

[14] R. Dean Straw, N6BV, Editor, The ARRL Handbook for Radio Amateurs, ARRL, 1997. See pp 30.22 onwards

[15] G. Matthei, L. Young, and E. M. T. Jones, Microwave Filters, Impedance Matching Networks, and Coupling Structures, McGraw-Hill, 1968

[16] W.S. Metcalf, Graphs Speed Design of Interdigital Filters, Microwaves, February 1967, pp 91-95

[17] M. Dishal, Alignment and Adjustment of Synchronously Tuned Multiple-Resonator-Circuit Filters, Proc. IRE, November 1951, pp 1448-1455

[18] G. Matthei, L. Young, and E. M. T. Jones, Microwave Filters, Impedance Matching Networks, and Coupling Structures, McGraw-Hill, 1968, pp 668-673

[19] Harbor Freight Tools, 3491 Mission Oaks Blvd, Camarillo, CA 93011; tel 800-423-2567

[20] Grizzly Industrial, Inc, 1821 Valencia St, Bellingham, WA 98226; tel 800-523-4777

[22] MSC Industrial Supply Co, 151 Sunnyside Blvd, Plainview, NY 11803, tel 800-645-7270. The street address is for the corporate offices. Call for a location near you

[23] Eastern Tool & Supply, 149 Grand St, New York, NY 10013; tel 800-221-2679

Chapter 12

The 5.7GHz (6cm) band

In this chapter :

- Preamplifier
- Transceiver
- Transverter

T he 5.7GHz band is the lowest amateur microwave band to be dominated by waveguide techniques. It is also the first band on which dish antennas are extensively used in preference to Yagi antennas.

Over the past few years the activity has increased on this band due its use for amateur TV and Satellite working. There are a number of kit or ready built tranceivers for these activities. This chapter includes some representative examples of the equipment available and the techniques being used.

Band planning

The current UK band plan is shown in Fig 12.1 and the USA band plan is shown in Fig 12.2.

Propagation and equipment performance

Most of the propagation modes encountered on the VHF/UHF bands are also present at 5.7GHz, with the exception of those which require ionised media, such as aurora, meteor scatter and sporadic E. The most commonly encountered propagation modes at 5.7GHz are line of sight, diffraction, tropospheric scatter, ducting, aircraft scatter. Moonbounce is also possible at this frequency.

Futher information on some of these modes can be found in chapters 1 and 2. Using the techniques described in these chapters, it is possible to predict whether communication is possible over a given path, when the performance parameters of the equipment are known. Fig 12.3 shows the path loss capability of three different sets of equipment ranging from a basic system of low power transverter to a system with moonbounce capability. Also shown are the path losses of three types of propagation, line of sight (the lowest loss mode of propagation), tropospheric scatter (a reliable terrestrial mode of propagation) and moonbounce. It can be seen that even relatively simple narrow band equipment is capable of working very long line of sight paths easily. However very few paths are purely line of sight and propagation losses are usually noticeably higher.

The tropospheric scatter loss is a reasonable upper limit for a terrestrial path and can be used to estimate the maximum range of equipment under normal propagation conditions. From the graph it can be seen that simplest equipment (A) might be expected to work paths of up to 30km long. By employing a larger dish and a slightly more powerful transmitter (B) the coverage distance increases to 120km. By adding a receive preamplifier the coverage distance increases to 200km. Equipment (C) is more powerful again and the coverage distance is increased to 425km. At distances over 830km it can be seen that the

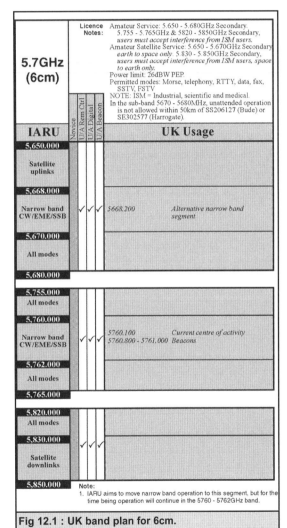

Fig 12.1 : UK band plan for 6cm.

terrestrial path loss is greater than the route via the moon. Line (E) shows the parameters of equipment capable of moonbounce at 5.7GHz. Under anomalous propagation path losses are considerably reduced, often approaching the line of sight loss and even simple equipment is then capable of working very long paths.

Frequency (MHz)	USA Usage
5650-5925	Band allocated
5760.3-5760.4	Propagation beacons

Fig 12.2 : USA bandplan for 6cm.

Preamplifier [1]

A preamp equipped with a PHEMT provides top notch performance in noise figure and gain as well as unconditional stability for the 6cm amateur band. The noise figure is 0.65dB at a gain of 13dB. It utilizes the X band PHEMT NEC NE32684 and provides a facility for an optional second stage on board.

Circuit description

This LNA utilizes a specially designed quarter wave stripline coupler in the input. This innovative technique provides about 0.1dB less loss than a high Q ATC100A coupling capacitor, hence 0.1dB less noise figure, see Fig 12.4) LNAs and, last but not least, it is very inexpensive. Using the NEC32684 PHEMT a noise figure of 0.65dB can be measured at 5.760GHz with a gain of 13dB.

Fig 12.5 shows the circuit diagram and Fig 12.6 the PCB with the dimensions 34 x 47mm. It is printed on RT Duroid 5870 with 0.508mm thickness. The quarter wave coupler at the input has a length of 9.3mm and a slit width of 100µm. The substrate thickness makes the coupler low loss and provides low inductance plated through holes necessary for stability.

The grounded source requires a bias circuit to provide the negative voltage for the gate. A special active bias circuit (Fig 12.7) was developed which provides regulation of voltage and current for the FET and allows for independ-

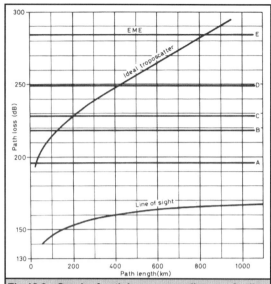

Fig 12.3 : Graph of path loss versus distance for line of sight, tropospheric scatter and eme propagation, related to the performance of 5 sets of equipment :
A: 0.7m dish, 4mW tx, 10dB nf, 2.5kHz bw
B: 1.2m dish, 100mW tx, 10dB nf, 2.5kHz bw
C: 1.2m dish, 100mW tx, 3dB nf, 2.5kHz bw
D: 1.2m dish, 1W tx, 3dB nf, 250Hz bw
E: 4.0m dish, 25W tx, 3dB nf, 250Hz bw

ent adjustment of these parameters. The PNP transistor T1 operates in a feedback loop with the FET and maintains a constant current through the FET by using D2,D3 as a reference voltage, sensing the drain current of the FET at the emitter of T1 and controlling the gate voltage of the FET. P2 provides current adjustment (5 - 20mA). IC1 is a variable voltage regulator (2.2 - 3.4V).

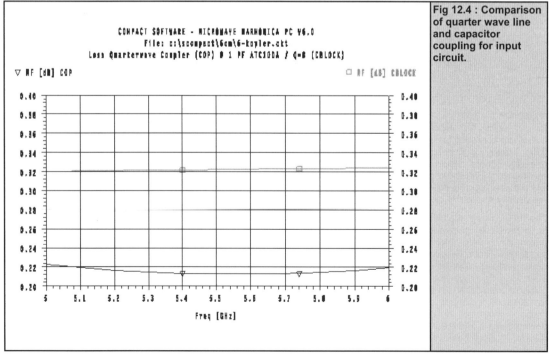

Fig 12.4 : Comparison of quarter wave line and capacitor coupling for input circuit.

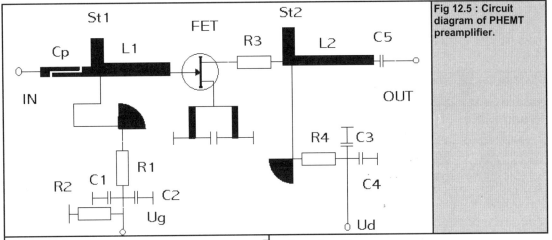

Fig 12.5 : Circuit diagram of PHEMT preamplifier.

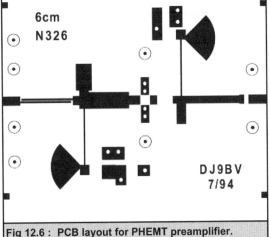

Fig 12.6 : PCB layout for PHEMT preamplifier.

Construction

The construction uses micro stripline techniques on glass PTFE substrate RT-Duroid 5870 0.508mm thick. See Fig 12.8 for construction of the RF PCB and Table 12.1 for the component list.

- Mount SMD parts on bias board and check function of bias board, see Table 12.2 for component list.

- Prepare PCB to fit into aluminium box.

- Prepare holes for SMA connectors into box. Note: Input and output connectors are asymmetrical. Use PCB to do the marking.

- Drill holes for through board contacts (0.9mm dia.) and connect with 0.8mm gold plated copper wire at indicated positions.

- Solder all resistors onto PCB.

- Solder all caps onto PCB.

- Solder PHEMT NE326 onto PCB. Use only an insulated soldering tool (Weller), ground the PCB, your body and the power supply of the soldering iron. Never touch the PHEMT by the gate connection, only at the sources or the drain when placing on the PCB and solder quickly (<5sec).

- Mount PCB into box with M2 screws.

IC2 generates the negative voltage. The circuit provides reverse bias protection for the gate and cannot provide more than 4V on the drain even if P1 (Drain Voltage Adjustment) fails (open circuit). A small (25 x 35mm) PCB (Fig 12.9) can be soldered vertically onto the RF Board. The DC connections of the RF Board mechanically fit on the output pins of the bias board. Table 12.2 shows the parts list of the bias circuit.

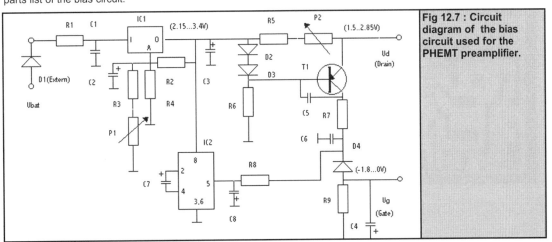

Fig 12.7 : Circuit diagram of the bias circuit used for the PHEMT preamplifier.

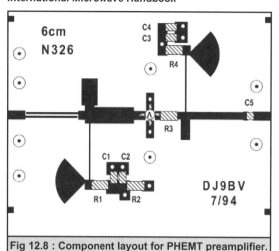

Fig 12.8 : Component layout for PHEMT preamplifier.

- Mount SMA connectors and solder inner line to strip-line.

- Mount bias board into box.

- Connect D1 between feed through cap and bias board. Connect gate terminal and drain terminal of RF PCB to bias board.

- Connect +12V and adjust P1 for 12mA drain current (measure 260mV across R4 on RF Board). The drain voltage should be around +2.0V.

- Glue conducting foam into inside of top cover and mount top plate.

The active bias circuit, which provides constant voltage and current is on a small (25x35mm) PCB (Fig 12.9). It can be soldered vertically onto the RF Board. The DC connections of the RF Board mechanically fit on , see fig 12.10 for the component layout.

Measurement Results

Noise Figure and Gain

Measurements were taken by a HP8510 network analyser and HP8970B/HP346A noise figure analyser, transferred

Table 12.1 Component list for PHEMT preamplifier.

C1,3	SMD-C	100pF	
C2,4	SMD-C	1000pF	Size 0805
C5	Chip-C	1pF	500 CHA 1R0 JG
R1	SMD-R	47Ω	Size 1206
R2	SMD-R	100kΩ	Size 1206
R3	SMD-R	4.7Ω	Size 1206
R4	SMD-R	22Ω	Size 1206
FET		NE32684	NEC
PCB	Duroid 58 x 70mm, 0.508, εr=2.3		
Box	Aluminium 35 x 74 x 30mm		

Table 12.2 Componet list for bias circuit used with PHEMT preamplifier.

C1,6	SMD-C	0.1µF	Size, 1206
C2,4	SMD-Elko	1µF/16V	Size, 1206
C3,7,8	SMD-Elko	10µF/16V	Size, 1210
C5	SMD-C	0.01µF	Size, 0805
T1	Si PNP	BC807	
IC1	Regulator	LM317	Na, LM317LZ
IC2	Inverter	LTC1044	LT, LTC1044SN8 (ICL7660)
D1	Si Diode	1N4007	Mo
D2,3,4	Si Diode	1N4148	Mo
P1	SMD Pot	1k	Vitrohm, 4310
P2	SMD Pot	100Ω	Vitrohm, 4310
R1	SMD R	10Ω	Size, 1206
R2,3	SMD R	220Ω	Size, 1206
R4	SMD R	680Ω	Size, 1206
R5	SMD R	33Ω	Size, 1206
R6	SMD R	1.2k	Size, 1206
R7	SMD R	4.7k	Size, 1206
R8	SMD R	10k	Size, 1206
R9	SMD R	22k	Size, 1206
PCB	Board,	Epoxy 25x35mm DC3XY, Bias Board	

to a PC and then plotted. Fig 12.11 shows the measurement results for gain and noise figure for the one stage amplifier. A typical noise figure of 0.65dB at a gain of 13dB can be measured on 5.76GHz. The preamp is rather broadband, a low noise figure is maintained from 5.4 to 6GHz.

Fig 12.9 : Bias circuit PCB for PHEMT preamplifier.

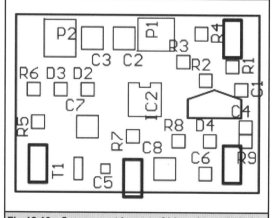

Fig 12.10 : Component layout of bias circuit PCB for PHEMT preamplifier.

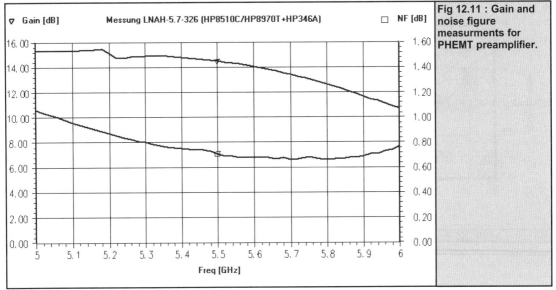

Fig 12.11 : Gain and noise figure measurments for PHEMT preamplifier.

Stability

A broadband sweep from 0.2 to 20 GHz shows a stability factor K of not less than 1.1 and the B1 measure was always greater than zero. These two properties indicate unconditional stability.

Conclusions

The preamp is a no tune design, unconditionally stable and has a very low noise figure.

Transverter

The most popular method to get onto the microwave bands is to use a transverter to translate the output of a commercial vhf transceiver to the required band. The transverter described here is from Paul Wade, W1GHZ.

A single board transverter for 5760MHz and Phase 3D [3]

Single board transverters account for most of the activity on the microwave bands below 5760 MHz, but there has been a dearth of high performance designs for higher bands. This transverter provides the compact convenience of a single board unit without compromising performance, making it easy to get on 5760 MHz or the Phase 3D satellite without requiring scarce surplus components.

The main reasons for the scarcity and complexity of previous transverter designs for this band are a lack of suitable inexpensive gain devices and the difficulty in

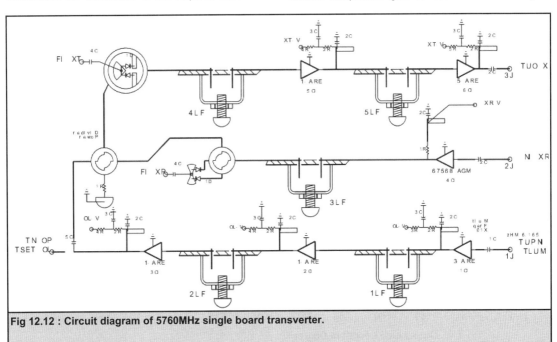

Fig 12.12 : Circuit diagram of 5760MHz single board transverter.

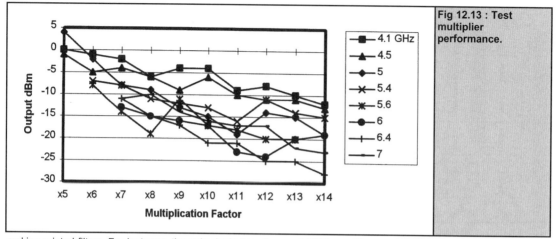

Fig 12.13 : Test multiplier performance.

making printed filters. For instance, the only single board transverter article previously, by KK7B [4], was simply a bilateral mixer and printed filter; a high IF, 1296MHz, was required because of poor filter selectivity.

Recently, the availability of inexpensive GaAs and heterojunction bipolar silicon (ERA series) MMIC devices has solved the gain problem, and pipe cap filters [5] offer a good alternative to printed filters. In a previous 5760MHz article [6], I suggested that the next improvement would be to add a transmit and receive amplifier stage to that dual mixer board. This article goes a step further, by including a multiplier chain for the local oscillator and enough amplifier stages to make a complete transverter station.

Description

The heart of this transverter is the printed circuit dual mixer [6], a design that has worked well for several years with a number in use. The receive mixer is augmented with a GaAs MMIC preamplifier preceding the pipe cap filter. This preamp provides good noise figure with adequate gain. The transmit side has two stages of gain with two pipe cap filters. One follows the mixer and there is one between the MMIC devices, to adequately filter out the LO and image signals. The LO chain consists of three active stages separated by filters, an MMIC working as a times 10 frequency multiplier, followed by a pipe cap filter,

an MMIC amplifier, a second pipe cap filter, and another MMIC amplifier to raise the power to the level required for the mixers. The MMIC devices keep the schematic diagram in Fig 12.12 straightforward enough so that a separate block diagram is not necessary.

The local oscillator input, at 561.6MHz (or 552.4MHz for Phase 3D), comes from the same KK7B local oscillator board [7] used in lower frequency single board transverters. These boards are readily available and work so well that no alternative was seriously considered. When I built the LO board, I calculated the resistor values for operation at 9V, so that it would operate at the same regulated voltage as the rest of the transverter. I also carved out a small block of Styrofoam to fit around the crystal for better thermal isolation, in the hope of better frequency stability.

Multiplier

The new ERA series of MMIC devices from Minicircuits [8] offer usable gain up to 10GHz at low cost, so they were obvious choices for the amplifier stages. However, an article by N0UGH [9], which described using the ERA-3 as a frequency multiplier for 10GHz, showed additional possibilities. Using a hobby knife, I made a breadboard using an ERA-3 followed by a pipe cap filter on a scrap of Teflon PC board to test the multiplier performance. A 3/4 inch pipe cap filter can be tuned roughly from 4 to 7GHz, so I tuned it to several different frequencies and varied the

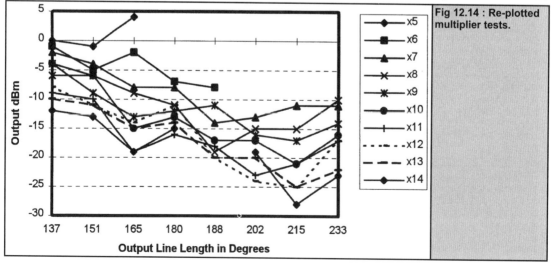

Fig 12.14 : Re-plotted multiplier tests.

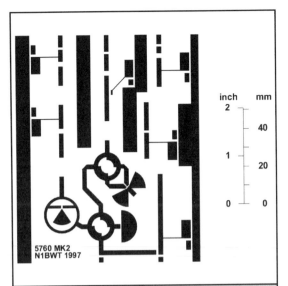

inch | mm
2
40
1
20
0 | 0

5760 MK2
N1BWT 1997

Fig 12.15 : PCB layout for 5760MHz single board transverter.

applies to MMIC frequency multipliers, I calculated electrical line lengths for my breadboard and replotted the data as shown in Fig 12.14. Now we can see that some multiplication factors are more affected by line length than others, and we can design to optimize for the desired multiplication factor. Fig 12.14 also shows the x10 multiplication used in this transverter to be relatively insensitive to line length; the output power change with output line length can be attributed to increasing multiplication factors.

Filters

Printed circuit filters are not suitable for two reasons:

- Dimensions become very critical at bandwidths less than 10%

- Radiation from them increases at higher frequencies (where the board thickness is a significant fraction of a wavelength). In a recent 3456MHz design [12] KH6CP (now W1VT) used thinner Teflon PC material to reduce radiation from the filters, however, this requires that dimensional tolerance also be reduced proportionally, to the point where printed patterns would not be reproducible. If high Q structures like printed filters are eliminated from the board, radiation is reduced, and then the thicker dielectric material is usable.

The pipe cap filters [5] are made with readily available copper plumbing fixtures and offer good performance but require tuning. My previous experience [7] showed that a single pipe cap filter did not provide adequate LO rejection at 5760MHz, so multiple filters were required. I was uncertain whether it would be possible to tune up multiple filters without sophisticated test equipment, so I tested the tuning on the multiplier breadboard. I found that the tuning screw varied the frequency by 300 to 400MHz per revolution, or about 1MHz per degree of rotation. Also, the frequency could be set repeatably by measuring the height of the tuning screw. So, it is possible to preset the

input frequency and power with a signal generator to try various multiplication factors, from x4 to x15. The results, plotted in Fig 12.13, show pretty good multiplier performance, but the curves have too many ups and downs for predictable performance.

While I was trying to understand the data in Fig 12.13, Steve, N2CEI, suggested that an article [10] on MMIC frequency multipliers by WA8NLC might offer some insight. The article referred me to Hewlett Packard application note AN-983 [11], which describes how diode frequency multiplier performance is affected by the phase shift of the transmission line length between the diode and filter. Since WA8LNC showed that the same phenomenon

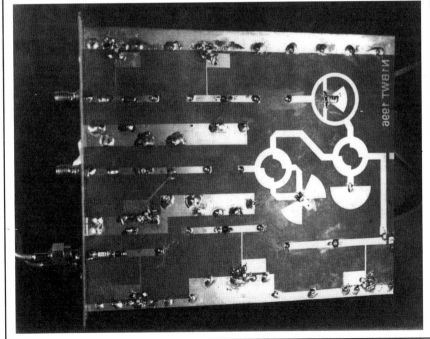

Fig 12.16 : Picture of components mounted on 5760MHz single board transverter PCB.

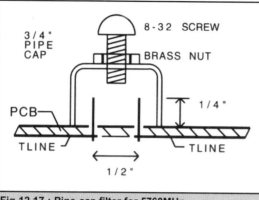

Fig 12.17 : Pipe cap filter for 5760MHz.

Fig 12.19 : Plot of simple detector sensitivity.

tuning screws close to the desired frequency, or to easily retune from 5760MHz to 5668MHz for Phase 3D. The difference in tuning should only be about one quarter of a turn, but the filters are sharp enough that retuning is required. Finally, since the filters are separated by amplifier stages, it is possible to tune them individually with minimum interaction.

Construction

Layout of the printed circuit board is shown in Fig 12.15. All components except the pipe cap filters go on the top surface, as shown in the photograph, Fig 12.16. The pipe caps are soldered on the ground plane side. Since a torch is used to solder the pipe caps, it is a good idea to install them first, but not until all holes are drilled and a clearance area is cut in the ground plane around the probe pins for the filters.

A pipe cap filter sketch is shown in Fig 12.17. This is the procedure I use to install the pipe cap filters:

- In preparation, I drill tight fitting holes for the probes and make clearance holes in the ground plane around the probe holes.

- For each pipe cap, I measure from the holes and scribe a square on the ground plane that the pipe cap just fits inside.

- Next I prepare each pipe cap by drilling and tapping (use lots of oil) the hole for an 8-32 tuning screw, then flattening the open end by sanding on a flat surface.

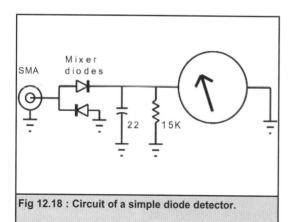

Fig 12.18 : Circuit of a simple diode detector.

- Then I apply resin paste flux lightly to the open end, and to the area around the screw hole for a brass nut to extend the thread length as shown in Fig 12.17. The nut is not necessary, but it makes tuning smoother. The nut is held in place for soldering with a temporary stainless steel screw. (Solder won't stick to it.)

- Next I center the open end of each cap in a scribed square on the PC board. The flux holds it in place.

- Finally, I fit a circle of thin wire solder around the base of each pipe cap and nut. The caps are soldered one at a time, starting with the center one. I hold each cap down with gentle pressure and heat it for a few seconds with a propane torch until the solder melts and flows into the joint, then let the solder harden before releasing pressure. Don't be shy with the torch, melt the solder quickly and remove the heat. Keep the flame on the top of the pipe cap to avoid damage to the PC board.

For the coupling probes, I use brass escutcheon pins that are 1/32 inch in diameter. The desired probe length inside the pipe cap filter is 1/4 inch, so I cut them to a length of 9/32 inch, not counting the head, to compensate for the PC board thickness. The probes are not installed yet, but later, as part of the tune up procedure. To install them, I put a small amount of flux under the head of each pin, then insert it into the tinned hole and solder.

The 5760MHz end of the board, with J1, J2 and J3, needs a robust way to attach the connectors. The PC board was planned to fit into an extruded aluminum box (made by Rose Enclosure Systems) that is available from Down East Microwave [13]. The box is supplied with aluminum endplates, but I believe that the PC board ground plane must be soldered to the endplate around the connectors to provide a proper microwave ground. Aluminum does not solder well, so I cut out a brass endplate to match the aluminum ones supplied with the enclosure, drill holes for the connectors and solder the ground plane of the board (top and bottom) to the brass plate. The connector mounting screws only provide mechanical strength. Connector J1 need not be brought outside the box; I only did so for convenient testing.

Once the heavy soldering is complete, the other components may be installed, starting with the MMICs. On my hand made boards, there are no plated through holes, so I ensure short connections for the ground leads by mounting the MMICs on the bottom (ground plane) side of the board and bending the input and output leads up through the board to reach the printed transmission lines. It takes

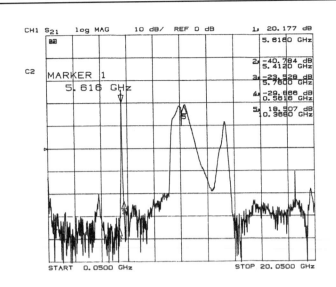

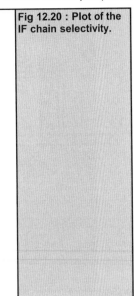

Fig 12.20 : Plot of the IF chain selectivity.

a bit of trimming at each hole (with a hobby knife) for lead clearance, but the sides of the holes should fit tightly to keep the ground contact as close as possible. Alternatively, Down East Microwave sells boards with plated through grounding holes, so all devices can be mounted on top of the board with short lead length.

Bias resistors and decoupling capacitors are also soldered on top of the board, but the dc connections are brought out through small clearance holes so that the dc wiring is on the far side of the ground plane and does not cause unwanted feedback. Resistor values shown in the schematic diagram, Fig 12.12, are for operation from a 9 V source, provided from a three terminal voltage regulator IC. The IC should maintain a stable voltage for the transverter and LO board, even with a partially discharged 12V battery, to keep frequency and power output stable.

Next, solder the dc blocking capacitors (in series with the transmission lines) in place. The capacitor connecting the LO to the mixers, C5, is initially connected to the test point for tune up, rather than to the mixers. Then the probe pins are inserted into filter FL1 only. FL2 is bypassed by soldering short pieces of enameled wire to each of the transmission line ends connecting to FL2, with the two wires parallel and closely coupled to form a small capacitor. Now we are ready to begin tune up. The other filters will be added and tuned one at a time.

Tune up

The tune up procedure will tune one filter at a time, adding additional sections of the circuit sequentially. Since the filters are separated by amplifier stages, interaction between them is minimal and repeaking previous adjustments is unnecessary.

The LO section is tuned first. An SMA connector is temporarily attached to the LO test point so that we can monitor output power here. I tune the transverters using only a power meter, and check the LO frequency with a surplus wavemeter. If you do not have a power meter, a diode detector is usable. Fig 12.18 shows one that can be quickly assembled "dead bug" style on the flange of an SMA connector using a mixer diode pair and chip components, the values are not critical. Fig 12.19 is a plot

of the detector sensitivity I measured using a Simpson 260 analog VOM. analogue meters make tuning much easier. The plot shows that the expected power levels should provide reasonable output voltage, but should not be taken as a calibration curve, since sensitivity may change with frequency, temperature and different components.

Now install the tuning screw into filter FL1. I use 3/4 inch long flathead brass screws rather than the round head screws shown in Fig 12.17. When tuned to 5616 MHz, the head extends 0.29 inches above the top of the pipe cap. If you begin with the screw at approximately this setting, tuning should go smoothly. (*Note: A 3/4 inch flathead screw has an overall length of 1/4inch, while a roundhead screw measures 3/4 inch to the bottom of the head. So a round head screw could be used for tuning by measuring the extension to the bottom of the head. However, in the enclosure I used, the extra height of the screw head prevented the cover from fitting properly. A 5/8 inch long screw would also solve this problem, but was not available at the local hardware store where I found the flathead screws.*)

Next the LO source board at 561.6MHz is connected to J1, and power is applied to the source and to IC1, IC2 and IC3. (If you can measure power, the LO power input to J1 should be +6 to +10dBm.) Look for output power at the test point while adjusting FL1, there should be a peak within about a half turn of the screw from the initial setting. If much more tuning is required, its possible that you have found the wrong harmonic. If you have any way of measuring frequency, check it now.

Once you are confident that filter FL1 is tuned to the correct harmonic for 5616MHz output, peak the output and tighten the locknut on the tuning screw. Then proceed to filter FL2. Remove the bypass wires, insert and solder the probe pins and adjust the tuning screw to the same depth as FL1. Apply power again, peak the tuning screw of FL2, and tighten the locknut. If you are able to measure power, the output at the test point should be +5 to +10 dBm. If the power is slightly low, trimming excess metal from the ends of the transmission lines around the filter probe pins will help.

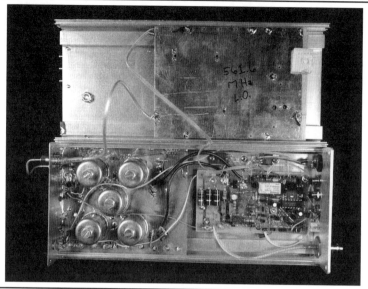

Fig 12.21 : Photograph of a transverter board packaged as a compact rover system in a Rose extruded box.

Now that the local oscillator is complete, the transmitter is next. Move blocking capacitor C5 to the transmission line connecting to the mixers and remove the temporary SMA connector from the LO test point. Insert probe pins into FL4, and bypass FL5 with a wire capacitor just as FL2 was earlier. Initially insert the tuning screw into FL4 the same depth as FL1. Move the output power indicator to J3, then apply power to IC5, IC6 and the whole LO chain. Turn the FL4 tuning screw counterclockwise, decreasing the screw depth to raise the frequency. You should find a peak within about one half turn of the screw, since filter FL4 should be tuned 144 MHz higher than FL1. Peak the tuning and tighten the locknut.

Proceeding to FL5, remove the bypass wires, insert and solder the probe pins. Set the tuning screw to the same depth as FL4. Apply power again, peak the tuning screw and tighten the locknut. The transmitter is now complete, with a typical output power of +10dBm.

The final step is to tune the receiver section. Insert and solder the probe pins in FL3, and apply power to the LO chain and to IC4 through the 3 terminal regulator, IC8 (Fig 12.22, IC8 and surrounding components are attached to the back of the board, dead bug style). Peak FL3 while listening to another station or to a weak signal source, [14] and tighten the locknut. If you can measure noise figure, adjust the voltage to IC4 for best noise figure by varying the 910Ω resistor attached to IC8. Otherwise, leave the voltage at approximately 6V, since the voltage for best noise figure is between 5V and 7V and not very critical.

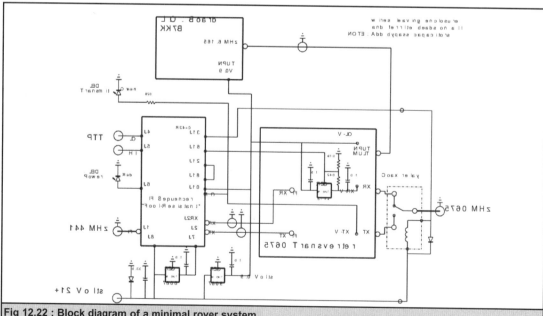

Fig 12.22 : Block diagram of a minimal rover system.

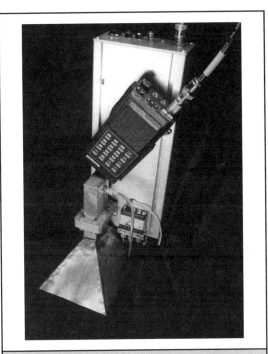

Fig 12.23 : Photograph of minimal rover system.

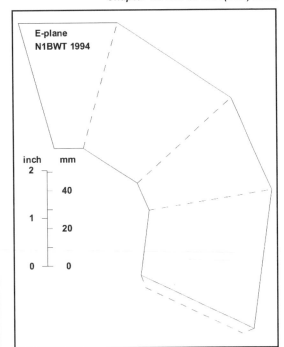

Fig 12.25 : Template of feedhorn for RCA DSS offset dish.

Performance

As a barefoot transverter, this unit makes an excellent rover rig, with about 10 mW transmitter output and about 3.5dB receiver noise figure. Both the LO and the transmitter outputs are pretty clean due to the double filtering. Fig 12.20 is a plot of the IF chain selectivity (when operated as a straight through amplifier rather than a frequency multiplier). All LO outputs at the test point are more than 60dB down, except for the third harmonic of the LO sources at 1684.8MHz, which is only 50dB down. The

transmitter output has some LO leakage, about 35dB down from peak carrier output, and the image signal is about 45dB down. Since the LO leakage is about 45dB down when the transmit section is not powered, more shielding would be required for significant improvement. All other outputs are more than 60dB down, except the

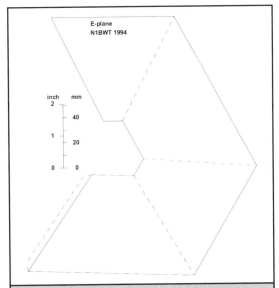

Fig 12.24 : Template of 15dBi horn for 5760MHz.

Fig 12.26 : Rover system with RCA DSS ofset dish.

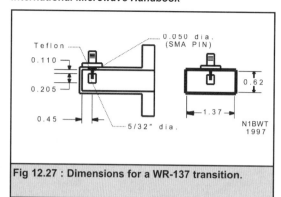

Fig 12.27 : Dimensions for a WR-137 transition.

second harmonic, which is only 30dB down; as Fig 12.20 shows, the pipe cap filters have little rejection above 10GHz.

On the receive side, a single filter provides adequate image rejection to maintain a good noise figure [15].

This performance is probably superior to lower frequency "no tune" transverters and should be an excellent foundation for a high performance system, with the addition of power amplifiers and a low noise preamp.

System

The transverter board does the RF work, but does not make a complete system without help. Fig 12.21 is a photograph of a transverter board packaged as a compact rover system in a Rose extruded box. One half of the extrusion contains the LO source board, mounted face down, while the other half contains the transverter board, also mounted face down, and an IF sequencer interface board [16]. The two RF boards are mounted face down so that the active circuitry is sandwiched between the PC board ground plane and the grounded wall of the case, thus maximizing isolation between different sections. The block diagram of this basic system is shown in Fig 12.22. The minimal rover system in Fig 12.23 uses the RF sensing function of the IF board to operate with only a simple HT and a horn antenna.

Unless you are willing to move the antenna cable to switch from transmit to receive or to use separate antennas, an antenna switch is required. Down East Microwave stocks some affordable used SMA coax relays that require 24V to operate. I disassembled one of these and rewound the coil to operate at 12V, so the whole system in Fig 12.23 can be powered by an automobile battery.

The IF sequencer board also provides all the circuitry and sequencing needed to safely control external preamplifiers and power amplifiers for a high performance system.

Antennas

Only a simple horn antenna [17] is needed for a basic rover system, I used my HDLANT [18] computer program (download from http://www.arrl.org /qexfiles/) to create the template in Fig 12.24. Use the template to build a horn that has 15 dBi gain, like the one in Fig 12.23, from a bit of flashing copper. Just tape a full size copy of the template to a sheet of copper or brass, cut it out, fold on the dotted lines, and solder the metal horn together on the end of a piece of waveguide.

For a rover system with better performance, feed a small

DSS dish [19] with an offset feed horn [20] (see template in Fig 12.25). This and a multimode transceiver, for CW and SSB capability,make up the excellent rover system shown in Fig 12.26. The system is mounted on top of a 10 GHz transverter. With a quick change feed mounting arrangement I have a two band rover station sharing the same dish. Of course, larger dishes can provide even better performance [21].

Both of the antenna templates are for horns that mate with WR-137 waveguide (1.37" × 0.62" internal dimensions), but the larger WR-162 and WR-187 sizes also work fine at 5760MHz. Surplus waveguide is available, so you should be able to find one of the three sizes or make a reasonable imitation from copper or brass sheet. Waveguide to coax transitions are more difficult to find, but easy to build. Fig 12.27 shows dimensions for a WR-137 transition. The coax probe is a section of 5/32 inch diameter brass rod with a hole drilled in it to fit over an SMA connector pin, which is soldered in position. Trim the SMAs Teflon insulation flush with the inside of the waveguide. This unit has a low SWR from 5.2 to 7.5GHz and outperforms several commercial units I have tested. If yours needs adjustment, vary the length of exposed SMA pin (nominal 0.110 inches) by moving the 5/32" diameter section and re soldering it at the new location; repeat until good SWR is achieved.

Potential

Even the basic rover system in Fig 12.23 is capable of communications over *any* line of sight terrestrial path. I can make this claim because we have frequently demonstrated that a simple 10GHz WBFM Gunnplexer system is capable of distances more than 100 km. This simple 5760MHz system using NBFM has comparable power output and antenna aperture, but has a 13dB advantage in receive bandwidth (approximately 10 kHz vs 200 kHz) and a noise figure at least 5dB better. The improvement of 18dB translates, using the inverse square law, to eight times better range capability (more than 800 km) enough for all line of sight paths.

The improved rover system shown in Fig 12.26 adds another 10dB of antenna gain and 5dB better receiver bandwidth (even more for CW) to provide the potential for over the horizon or partially obstructed paths.

From there, we can move up to larger antennas, a receive preamplifier and transmit power amplifiers, all the way up to EME capability.

Phase 3D

The Phase 3D satellite requires only a transmit capability at 5668MHz. If there is another interested station not too far away, however, receive capability makes it possible to check out both systems before tackling the complexities of satellite tracking.

Published estimates [22] of Phase 3D path loss at this frequency are 181.5dB at perigee and 201.4dB at apogee. With the estimated 22dBic satellite antenna gain, the system in Fig 12.25 should be able to provide an uplink signal that is weak at perigee, and lost in the noise at apogee. One solution would be a larger dish, but that would have a beam width narrower than the manageable 8° beam width of the DSS dish. A better solution would be a modest power amplifier.

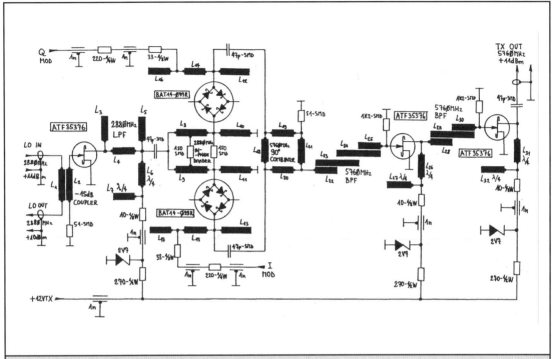

Fig 12.28 : Circuit diagram of 5760MHz quadrature transmit mixer.

Conclusion

This transverter provides the compact convenience of a single board unit without compromising performance. While it is not a "no tune" design, tune up is straightforward and systematic. Now it is possible to get on 5760 MHz without any hard-to-find surplus components.

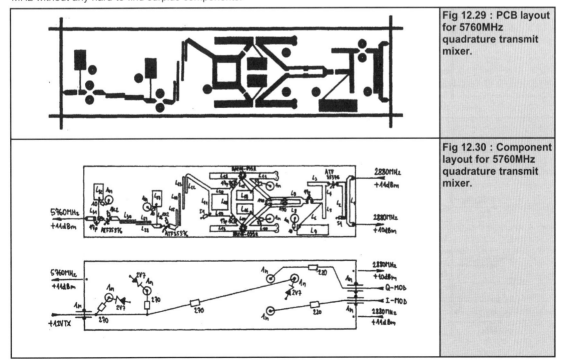

Fig 12.29 : PCB layout for 5760MHz quadrature transmit mixer.

Fig 12.30 : Component layout for 5760MHz quadrature transmit mixer.

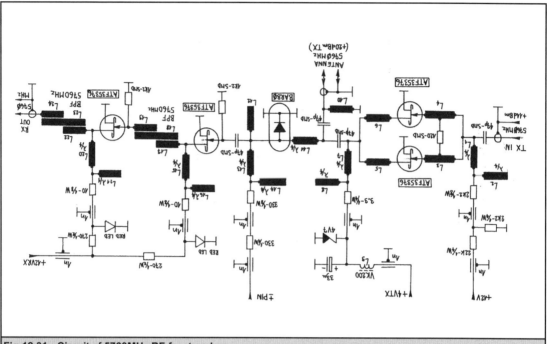

Fig 12.31 : Circuit of 5760MHz RF front end.

Transceiver

The design for a zero IF transceiver is show in chapter 9. This design has RF sections for 5.7GHz which are described here.

Quadrature transmit mixer

The transmit mixer module for 5760MHz includes the following stages: an LO signal switching, an in phase LO divider, two balanced sub harmonic mixers, a quadrature

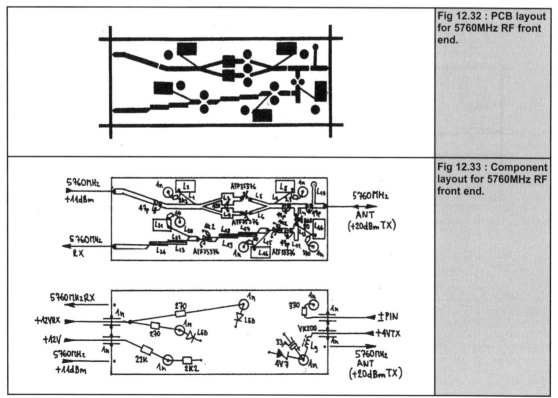

Fig 12.32 : PCB layout for 5760MHz RF front end.

Fig 12.33 : Component layout for 5760MHz RF front end.

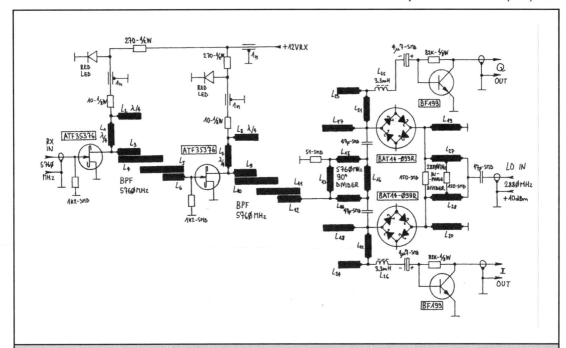

Fig 12.34 : Circuit diagram for 5760MHz quadrature receive mixer.

combiner and a selective RF amplifier. LO signal switching between the transmit and receive mixers is performed in the following way :

Most of the LO signal is always fed to the receive mixer. A small fraction of the LO signal is obtained from a coupler and amplified to drive the transmit mixer. During reception the power supply of the LO amplifier stage is simply turned off. This solution may look complicated, but in practice it allows an excellent isolation between the

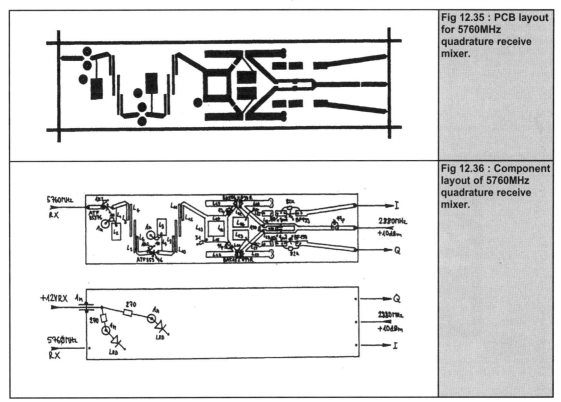

Fig 12.35 : PCB layout for 5760MHz quadrature receive mixer.

Fig 12.36 : Component layout of 5760MHz quadrature receive mixer.

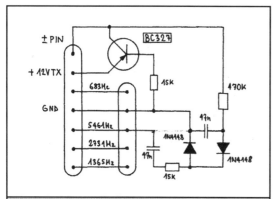

Fig 12.37 : Pin driver for 5760MHz transceiver.

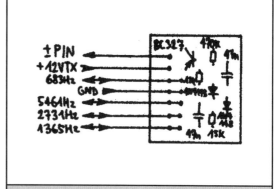

Fig 12.39 : Component layout for 5760MHz Pin driver.

transmit and receive mixers. The practical circuit is simple and the component count is low as well.

The circuit diagram of the quadrature transmit mixer for 5760MHz is shown on Fig 12.28. The 2880MHz LO signal is taken from a -15dB coupler and the LO signal level is restored by the ATF35376 amplifier stage, feeding two sub harmonic mixers equipped with BAT14-099R Schottky quads. The 2880MHz low pass attenuates the second harmonic at 5760MHz to avoid corrupting the symmetry of the mixers.

The two 5760MHz signals are combined in a quadrature hybrid, followed by a 5760MHz band pass filter. The latter removes the 2880MHz LO as well as other unwanted mixing products. After filtering the 5760MHz SSB signal level is rather low (around -14dBm), so two amplifier stages with ATF35376 HEMTs are used to boost the output signal level to about +11dBm.

The quadrature transmit mixer for 5760MHz is built on a double-sided microstrip FR4 board with the dimensions of 30mmX120mm as shown on Fig 12.29. The corresponding component location is shown on Fig 12.30. The circuit does not require any tuning for operation at 5760MHz.

RF front end

The RF front end includes the transmitter power amplifier, the receiver low noise amplifier and the antenna switching circuit. It no longer makes sense to use expensive coaxial

Fig 12.38 : PCB layout for 5760MHz Pin driver.

relays, since PIN diodes can provide the same insertion loss and isolation at lower cost with better reliability and much shorter switching times.

The circuit diagram of the RF front end for 5760MHz is shown on Fig 12.31. The transmit power amplifier uses two HEMTs ATF35376 in parallel to obtain about 100mW (+20dBm) on the antenna connector. The gain of the HEMTs is around 13dB, however circuit losses both in the input matching network and in the antenna switch on the output amount to about 3dB, so that about +10dBm of drive power is required.

The two PA HEMTs receive a positive bias on the gates both while transmitting and while receiving. In transmission the PA HEMTs generate a self bias just like the CLY5 and CLY2 power GaAsFETs. Of course the +4VTX supply line requires a current limiting resistor.

The antenna switch includes a single shunt diode BAR80 to protect the receiver input during transmission. During reception, the two PA HEMTs act as short circuits thanks to the positive gate bias. The short circuit is transformed through package parasitics, L5, L6 and interconnecting lines (total electrical length ¾ lambda) into an open circuit at the summing node.

Since the shunt diode BAR80 was not designed for operation above 3GHz, its capacitance introduces additional insertion loss in the receiving path at 5.76GHz. This additional loss can be substantially reduced if a reverse bias is applied to the BAR80 diode. Therefore a negative bias voltage is applied to the BAR80 during reception and a positive current is applied during transmission through the command line +-PIN.

Since there are no strong signals expected in the 5.7GHz band, the LNA for 5760MHz includes two stages with ATF35376 HEMTs. The total insertion gain including the losses in the antenna switching network and the two 5760MHz band pass filters amounts to about 23dB.

The RF front end for 5760MHz is built on a double sided microstrip FR4 board with the dimensions of 30mm x 80mm as shown on Fig 12.32. The corresponding component location is shown on Fig 12.33. The RF front end for 5760MHz requires no tuning. It is however recommended to select the ATF35376 HEMTs according to the I_{dss}. The highest I_{dss} devices should be used in the transmitter PA while the lowest I_{dss} devices should be used in the receiver LNA.

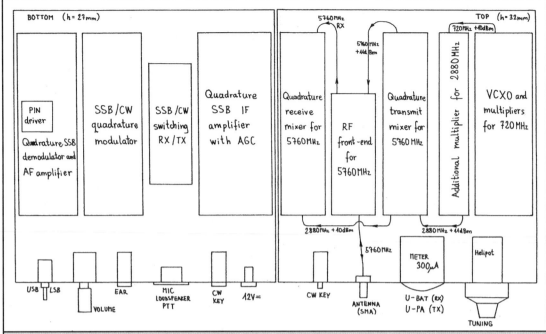

Fig 12.40 : The module location for the 5760MHz version.

Quadrature receive mixer

The receiving mixer module includes the following stages, an additional RF signal amplifier, a quadrature hybrid divider, two sub harmonic mixers, an in-phase LO divider and two IF preamplifiers. The mixers, in-phase and quadrature dividers and RF band pass filters are very similar to those used in the transmitting mixer modules.

The circuit diagram of the quadrature receiving mixer for 5760MHz is shown on Fig 12.34. Since a component with the gain comparable to the INA-03184 is not available for 5.7GHz, two RF amplifier stages are required to obtain about 20dB of gain. ATF35376 HEMTs are used in both RF amplifier stages. Otherwise the circuit is almost identical to the quadrature receiving mixer for 1296MHz.

The quadrature receiving mixer for 5760MHz is built on a double sided microstrip FR4 board with the dimensions of 30mm x 120mm as shown on Fig 12.35. The corresponding component location is shown on Fig 12.36. The receiving mixer for 5760MHz requires no tuning.

Construction

The main construction details are shown in chapter 9, the items specific to the 5760MHz version are as follows.

The 5760MHz version of the SSB/CW transceiver requires an additional PIN diode driver to provide a negative bias to the BAR80 PIN diode during reception. The corresponding circuit diagram is shown on Fig 12.37. The negative voltage is obtained from the 5461Hz clock, while the BC327 PNP transistor applies a positive voltage during transmission.

The PIN driver is built on a small FR4 board with the dimensions of 23mm x 20mm as shown on Fig 12.38. The corresponding component location is shown on Fig 12.39. The whole PIN driver module is then installed piggy-back

on the demodulator board, using as support the 5-pole connector carrying the clocks to the modulator.

The module location for the 5760MHz version is shown on Fig 12.40. The location of the connectors and controls is the same for all three transceivers (see chapter 9).

References

[1] From article by Rainer Bertelsmeier, DJ9BV, Dubus Technik IV

[2] R. Bertelsmeier, DJ9BV, "HEMT LNA for 13cm", DUBUS TECHNIK IV, pp. 191 - 197

[3] From article by Paul Wade, W1GHZ, QEX Nov 1997

[4] R. Campbell, KK7B, "A Single-Board Bilateral 5760-MHz Transverter," *QST*, Oct 1990, pp 27-31

[5] K. Britain,WA5VJB, "Cheap Microwave Filters," Proceedings of Microwave Update *88* (Newington: ARRL, 1988) pp 159-163. Also appears in ARRL UHF/Microwave Project Book (Newington: ARRL, 1992) pp 6-6 to 6-7

[6] P. C. Wade, N1BWT, "A Dual Mixer for 5760 MHz with Filter and Amplifier," *QEX*, Aug 1995, pp 9-13

[7] R. Campbell, KK7B, "A Clean, Low-Cost Microwave Oscillator, QST, Jul 1989, p 15. Also appears in ARRL UHF/Microwave Project Book, pp 5-1 to 5-9

[8] Mini-Circuits Labs, PO Box 350166, Brooklyn, NY 11235-0003; tel 718-934-4500

[9] D. Nelson, NØUGH, "MMIC Multiplier for 10.8 GHz Local Oscillator," Feedpoint (North Texas Microwave Society), March/April 1996

[10] J. Davey, WA8NLC, "Frequency Multipliers Using Silicon MMICs," ARRL UHF/Microwave Project Book, pp 5-13 to 5-15

[11] Comb Generator Simplifies Multiplier Design, Application Note 983, Hewlett-Packard

[12] Z. Lau, KH6CP/1 (now W1VT), "3456 MHz Transverter," QEX, Sep 1996, pp14-20

[13] Down East Microwave Inc, 954 Rte 519, Frenchtown, NJ 08825; tel 908-996-3584

[14] P. C. Wade, N1BWT, "Weak Signal Sources for the Microwave Bands," ARRL UHF/Microwave Project Book, pp 5-16 to 5-18

[15] P. C. Wade, N1BWT, "Noise Measurement and Generation," QEX, Nov 1996, pp 3-12

[16] P. C. Wade, N1BWT, "A Fool-Resistant Sequenced Controller and IF Switch for Microwave Transverters," QEX, May 1996, pp 14-22

[17] P. C. Wade, N1BWT, "Practical Microwave Antennas, Part 1," QEX, Sep 1994, pp 3-11

[18] P. C. Wade, N1BWT, "Practical Microwave Antennas, Part 2," QEX, Oct 1994, pp 13-22

[19] MCM Electronics, 650 Congress Park Dr, Centerville, OH 45459; tel 800-543-4330

[20] P. C. Wade, N1BWT, "More on Parabolic Dish Antennas," QEX, Dec 1995, pp14-22

[21] P. C. Wade, N1BWT, "High-Performance Antennas for 5760 MHz," QEX, Jan 1995, pp 18-21

[22] W. A. Tynan, W3XO, "Phase 3D, A New Era for Amateur Satellites," Proceedings of the 30th Conference of the Central States VHF Society (Newington: ARRL, 1996) pp 47-61

The 10GHz (3cm) band

T he 10GHz band has been the subject of probably one of the longest development histories of any of the amateur microwave bands, extending to some 60 years or more in the UK and, therefore, must occupy a special place in the annals of UK amateur radio. The learning and experimental phase continues to this day, for the 10GHz band is still, in many respects, an ideal beginners band.

References to the theory and use of waveguides and magnetron oscillators to generate centimetre waves, then referred to as micro-rays, started to appear in the early 1930s and klystron oscillators were described in 1939, just before the second world war. Most of the microwave tubes developed in the 30s and 40s are still, with some refinements, in use today.

The *RSGB Bulletin* carried a series of articles during 1943 entitled Communication on centimetre waves, although the earliest recorded amateur 10GHz experiments, over a distance of two miles (a little over three km), took place in 1946 between W2RJM and W2JN. In October 1947, the RSGB published a 54 page booklet entitled M*icrowave Technique* which sold for the princely sum of two shillings and three pence (including postage).

UK 10GHz experimentation appears to have begun in earnest some two years later, in 1949, with experiments between G3BAK and G3LZ who used the 430MHz band to provide talkback, no mean feat in its own right at that time. The culmination of these experiments was the first recorded UK amateur 10GHz contact in January, 1950.

Band planning

The band available in the UK on a shared, secondary basis is 500MHz wide, extending from 10.0GHz to 10.5GHz. The first 450MHz is allocated to the amateur service and the top 50MHz to the amateur satellite/space service. Common band usage is illustrated in Fig 13.1 and this bandplan has come about, over the years, by evolutionary growth rather than active band planning as such. The band plan for USA is shown in Fig 13.2

History

Amateur exploitation started with the availability of low powered ex-radar klystrons such as the 723 A/B and its derivatives. Such devices were originally designed to work in the region of 8.0 to 9.5GHz (i.e. the lower part of X band). With ingenuity and mechanical modification some klystrons could be persuaded to yield a few milliwatts up to about 10.10GHz, but more commonly no higher in the band than 10.05GHz. Valuable archival material can be found in [1], a then definitive three part article by G3BAK (VK5ZO and still active, in 1991, on 10GHz). It was natural that the earliest established practice was to operate in this part of the band using modified klystrons as receiver, local oscillator and transmitter. The stability left much to be desired and the design of power supplies, particularly for portable use, to provide the several operating voltages needed had quite a bearing on the use of klystrons. Problems arose from both thermal drift of the inter electrode spacings of the device and the frequency pushing effects of short term voltage drift on those electrodes. Supplies of between 450V and 1000V were needed to operate them and this was potentially dangerous in portable operation. Nevertheless, working systems were established and used by a handful of pioneering amateurs. Those who wish to continue to use klystrons, for historical or sentimental reasons, should read [1], [2], [3] and [4]. Wideband beacons were established at around 10.1GHz once activity demanded them, although this was much later than the early work mentioned.

In the mid to late 1970s, Gunn effect devices started to appear on the surplus market at attractive prices in the form of Gunn Diodes, complete oscillators capable of acting as self oscillating mixers, dual cavity oscillator/ mixers, all designed for doppler intruder alarms or speed measuring modules. Most of these modules operated at around 10.67GHz and it was soon found that optimum efficiency could be obtained by re tuning the modules into the area around 10.35GHz to 10.40GHz. This part of the band was thus adopted as preferred for wideband operation, being adjacent to both the narrow band segment and to beacons at 10.40GHz.

Gunn oscillators offered much improved stability compared with klystrons plus the added benefit of very simple, low voltage power supplies. The ready and inexpensive availability of such modules has made the 10GHz band one of the easiest and most accessible of the amateur microwave bands for the beginner in the field of low power (QRP) microwave communication.

Concurrent with the availability of Gunn devices, a few dedicated microwave operators started to develop narrow band equipment and most of the work in the UK centred around the design work of Mike Walters, G3JVL. He pioneered construction and use of waveguide 16 iris coupled multi cavity filters for transmitting, receiving and transverting. In the early 1980s, amateur radio saw the arrival of relatively inexpensive GaAs fets usable at 10GHz (and beyond) and GaAs fet dielectric resonator stabilised oscillators (dro's). These devices, designed for use in satellite tvro (tv receive only) systems at 11 to

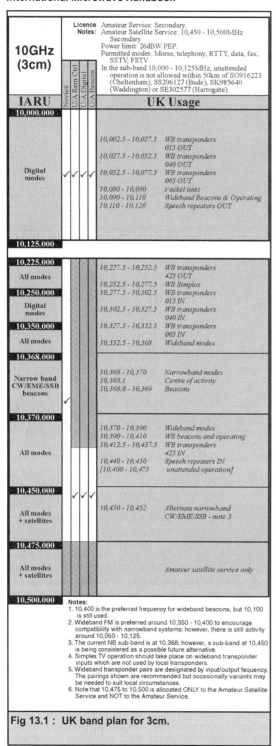

Fig 13.1 : UK band plan for 3cm.

Frequency (MHz)	USA Usage
10000-10500	Band allocated
10368	Narrow band calling frequency
10368.3-10368.4	Propagation beacons
10364	Calling frequency

Fig 13.2 USA band plan for 3cm.

less power dissipation, they were considerably more stable (allowing the use of narrower receiver bandwidths) and had a cleaner spectrum than the Gunn which, in its turn, was better than the klystron. Developments in the 90's centred around the use of pcb technology and active mixers, resulting in several "no tune" designs for transverters.

These features, coupled with the techniques becoming more accessible to amateurs, point towards the 10GHz band growing in importance as an amateur allocation, particularly from the point of view of the beginner in microwaves. It should be pointed out, if it is not immediately obvious to the reader, that the acquisition of skills at 10GHz will also give a firm foundation to those amateurs who may subsequently become interested in satellite tv reception. The RF techniques are very similar and much amateur built equipment for the 10GHz band can be stretched to cover the 11 to 12GHz band.

Nature of the band

All the tropospheric modes of propagation described in chapters 1, 2 and the individual band chapters can be expected to manifest themselves at 10GHz. As with the other microwave bands, there is no possibility of ionosphere modes operating.

Atmospheric phenomena, such as heavy rain, can be expected to play a significant part in propagation effects although less so than in the higher bands where water vapour or oxygen absorption are of high significance. Contacts via reflection and scattering from heavy rainstorms are well known to amateurs, as are the effects of frontal systems, inversions and ducting. Under normal weather conditions, forward troposcatter is always present, as described in [5].

The wavelength is such that significant diffraction effects can be observed over and around sharply defined objects and usable reflections can occur from buildings, gas holders, cooling towers, islands and other similar objects. Significant refraction effects can be observed under ducting conditions and particularly over water by the formation of a super refractive duct of limited depth which readily propagates these frequencies much in the manner of a waveguide. Much amateur communication takes place over line of sight paths but even here atmospheric effects may be significant.

Path losses for various propagation modes are given in Fig 13.3, together with an indication of the capability of different types of equipment in terms of dB of path loss capability. If sufficient power can be generated by amateur equipment and coupled with very high antenna gains, even the eme path is surmountable within the limits allowed by the amateur licence. This was shown during 1989 by WA5VJB and WA7CJO who made an 888.5 mile (1,421km) contact from Texas to Arizona, via the moon.

12GHz, could be modified to operate at the high end of the band. They proved to be intermediate between the then accepted amateur standards of wideband and narrow band operation. Power levels comparable with low powered Gunn oscillators could be achieved with similarly simple power supplies, but with the advantage that the GaAs fet dro is more efficient. They gave more output for

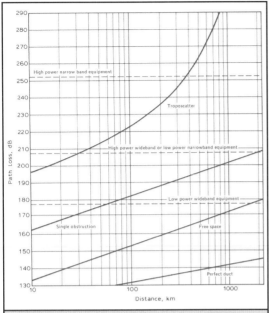

Fig 13.3 : Path loss and equipment capability.

Table 13.1 : Receiver sensitivity for different modes.

Mode	Detector Threshold	Bandwidth (kHz)	System Gain	Typical Sensitivity	Notes
wbfm	+10	250		-110	(2)(3)
wbfm	+10	25	10	-120	(2)(3)(4)
nbfm	+10	5	7	-127	
ssb	0	2.5	16	-143	(5)
cw	0	0.5	7	-150	
cw	0	0.1	7	-157	

Notes
(1) Detector thresholds assumed, ssb/cw 0dB, fm +10dB
(2) Basic optimised in line Gunn system with no antenna gain and negligable feeder loss, as might be measured using a professional signal generator/attenuator straight into a microwave head.
(3) Assumes no image rejection for wbfm equipment.
(4) To utilise the gain from a narrower receiver bandwidth, it may be necessary to employ effective afc.
(5) Assumes 3dB additional gain from image recovery in the change from wbfm to ndfm. Image recovery assumed for ssb and cw modes.

Modes

All the modes used on the lower bands are now in use at 10GHz. Both wideband modes (wbfm, fast scan atv and data) and narrow band modes (nbfm, fsk, afsk, ssb and cw) are in use.

Wideband (fm) modes, using bandwidths of between about 50kHz and many MHz, are most easily and directly generated in simple equipment using the voltage frequency characteristics of Gunn devices. Similar characteristics exist with klystrons, but these have now been totally superseded in amateur usage. The use of narrow band modes implies a much more stable source so that the signal generated will remain in the passband of the receiver tuned to it. Thus, the source is almost invariably crystal controlled and derived from a lower frequency by multiplication, phase locking or by a combination of multiplication, mixing and filtering. The architecture of the narrow band system is much more complex than that of a wideband system. It is now possible to use dros to provide a signal stable enough to class as pseudo narrow band whilst retaining most of the simplicity of the Gunn transceiver. With careful design and application of the dro it should be possible to generate nbfm of stability adequate to support the use of receiver bandwidths of 25kHz or less.

Equipment designed to take advantage of the simplicity of wbfm systems seldom, if ever, has any protection against image reception. To give an example, if the receiver local oscillator is set to 10.380GHz and the receiver intermediate frequency is 30MHz, then the receiver will hear two signals at F(osc) + F(i.f) and F (osc) - F(I.F); in this case at 10.410GHz (upper image or sum product) and 10.350GHz (lower image or difference product). If the wanted signal is, in fact, on 10.350GHz then what the receiver hears is the sum of the wanted signal plus the noise in the image channel. In theory, removing this image noise will result in a system sensitivity gain of 3dB.

A further penalty to be paid in using fm is that the detector threshold is 10dB above that of a product detector such as that used to detect ssb or cw. This means that the audio signal produced tends to be present at many dB above this threshold, or completely inaudible in the noise. This is in contrast to cw where the operator with a trained ear can hear signals quite distinctly even when almost down in the noise. Some of this disadvantage can be recovered by employing higher oscillator stability, narrower deviation and therefore narrower receiver bandwidths.

Narrow band equipment set up to receive ssb or cw will normally reject the unwanted image and employ a product detector, giving the receiving system an immediate additional gain of about 13dB. The more stable signal will also allow the use of narrower bandwidth receiver filters, thus enabling further system sensitivity to be realised. The cumulative effects of all these changes are summarised in Table 13.1. The figures given are not absolute but serve as a fair representation of the *order* of system improvement that is likely to be achieved in moving from a simple wideband system to a narrow band system. Once this level of sophistication has been reached, it is well worth the operators efforts to squeeze the last dB out of the system. This might be achieved by optimising antenna feeds, tuning out all mismatches, optimising the mixer and so forth.

Waveguide components

The theory of many of both the passive waveguide components and other, active, components commonly used at 10GHz has already been described in chapters 4 and 5. The construction of many components for 10GHz which are of particular relevance to test equipment were described in chapter 9. Other generally useful components for 10GHz are given in this chapter, often in some detail (where this is appropriate). Waveguide loads, calibrated variable attenuators, directional couplers, detectors and wavemeters when assembled together as a test bench and used properly enable many measurements to be made at home with sufficient accuracy to be

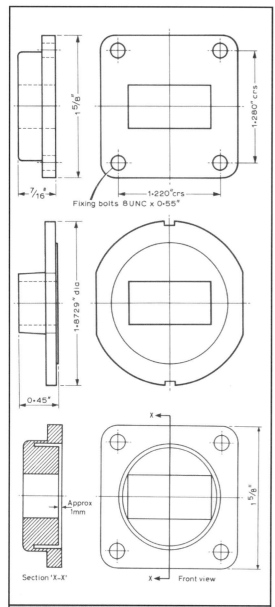

Fig 13.4 : Standard WG16 flanges.

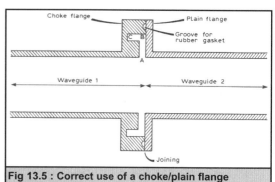

Fig 13.5 : Correct use of a choke/plain flange combination.

for amateur purposes. Good results can be obtained from components constructed by hand, almost literally, on the kitchen table!

The most suitable rectangular guide for use at 10GHz is WG16 which has external dimensions of 1in (25.4mm) by 0.5in (12.7mm). The internal dimensions are 0.9in (22.9mm) by 0.4in (10.2mm). It is available most commonly in copper, brass and aluminium. It can be obtained from a number of sources either new or second hand on the surplus market. Copper guide should always be used where high Q is required, for instance in the construction of filters, or where high thermal conductivity is required to assist in the heat sinking of higher powered Gunn diodes than those commonly used in simple amateur equipment. For most other purposes brass, which is more rigid, easier to work and less expensive, is quite adequate. Aluminium or aluminium alloy guide should be avoided as it is very difficult for the amateur to solder parts together, whereas this is very easy with either copper or brass, provided that the metal is clean.

Two shapes of standard flange are available, with the dimensions of each given in Fig 13.4. Rectangular flanges are less expensive than the round type, and require only four bolts to fasten them together. The constructor can make substitute flanges of this type from thick brass plate suitably cut, drilled and filed to fit the waveguide. The boss on the rear of commercially produced flanges assists accurate fitting but is not essential to their function. The round types require a pair of special screw coupling/locating rings which make them quick release and are particularly useful where the operator may need to repeatedly attach or detach different pieces of equipment to or from a common antenna feed, for example.

Both types of flange are available in plain or choke format. Plain flanges normally have a shallow groove machined into them which takes an O ring gasket to provide a weather seal. The plain type is designed to fit flush with the end of the waveguide and, when using this type, the constructor should try to make the ends of the waveguide flat and square with the waveguide neither protruding from, nor recessed in, the flange. The consequences of improper fitting might be either some RF leakage or mismatch at the joins.

A choke flange has a recessed face, relative to the mating surfaces, and another groove (other than that for the O ring seal) machined into it. This groove and the recessed face forms a quarter wave choke which reduces leakage from the joint and does not rely on metal to metal contact for its operation. It is designed to be matched with a plain flange to provide a leak-proof RF connection which should

meaningful without continual recourse to the use of expensive professional test gear. It will also become apparent that some components that cannot be readily made or obtained from surplus sources (such as circulators and isolators) are also valuable adjuncts to testing as well as operating.

Waveguides and flanges

10GHz equipment with components built in waveguide16 (WG16) is most familiar. There is nothing magic about this particular size of guide and it is possible to use other similar sizes. Other materials can be used (such as copper water pipe) with quite wide tolerances, because the dimensions are often not very critical and the precision matching of professional components is not essential

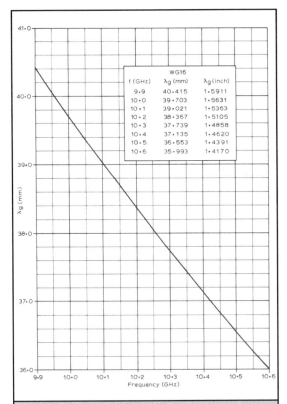

	WG16	
f (GHz)	λg (mm)	λg (inch)
9·9	40·415	1·5911
10·0	39·703	1·5631
10·1	39·021	1·5363
10·2	38·367	1·5105
10·3	37·739	1·4858
10·4	37·135	1·4620
10·5	36·553	1·4391
10·6	35·993	1·4170

Fig 13.6 : Relationship of waveguide wavelength (λg) to frequency.

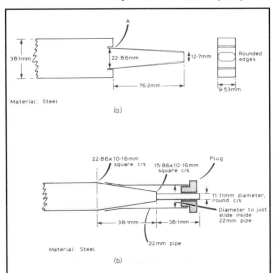

Fig 13.7 : Two types of tool for making rectangular to circular waveguide transitions (G8AGN). Tool (b) has a tight plug which prevents the round end of the transition distorting whilst forming the transition.

be better matched than possibly misaligned plain flange joints. Fig 13.5 shows such a pair of flanges diagramatically in cross section; the distance CBA is λ/2, the short circuit at C being reflected as a short circuit at A, effectively joining the ends of the waveguide electrically, even if there is no close physical connection between the two. A sound mechanical connection between the flanges is thus not essential. It should be obvious that a choke flange should *always* be matched to a plain flange to be effective in this way.

Brass WG16 will exhibit a loss of about 5.5dB per 100ft (approximately 31 metres) whilst copper guide should be about 40% of this, i.e. about 2.2dB for a similar run.

Fig 13.6 gives the relationship between the waveguide wavelength (λg) and the frequency in GHz for WG16 to enable calculation of critical dimensions.

With the ever increasing cost of new rectangular

Table 13.2 : Standard copper water pipes used as waveguide.

OD (mm)	Wall Thickness (mm)	Cutoff frequency (MHz)	Next mode cutoff (MHz)
15	0.5	12557	16404
22	0.6	8452	11041
28	0.6	6560	8569
35	0.7	5232	6835
42	0.8	4351	5684
54	0.9	3368	4400

waveguide it seemed useful to investigate a cheaper alternative copper water pipe for long runs such as might be used in a beacon or fixed home station installation.

For minimal losses the wave needs launching from conventional rectangular guide into the pipe (or vice versa), since the propagation modes are different in the two forms. The fabrication of waveguide components such as oscillators and mixers in circular guide is much more difficult than with rectangular guide and this justifies the time spent in making transitions.

Water pipe is available in a range of standard sizes as shown in Table 13.2, the first few smaller sizes being most readily available at diy supermarkets, the larger sizes from builders merchants.

It can be seen that 22mm pipe is usable at 10GHz, 35mm pipe at 5.7GHz, and possibly 54mm pipe at 3.4GHz, though the latter would be operating rather close to cutoff. Since this chapter covers 10GHz, attention was concentrated on assessing the suitability of 22mm pipe, which costs about 50p per foot compared with at least ten times this for new WG16. Before any attenuation measurements could be carried out, some transitions from normal rectangular waveguide had to be built. This was achieved by using short pieces of pipe, approximately 3in (75mm) long and deforming one end of each pipe into an approximately rectangular shape. Initial attempts to do this simply by squeezing the pipe in a vice were not very successful. Better results were obtained by gently hammering a tapered plug, as shown in Fig 13.7a, into the pipe and then gently squeezing the rectangular/elliptical cross section of the pipe right up to the plug at point A. The resulting cross section is then roughly the same as standard WG16. Although the plug tool shown in Fig 13.7a is adequate if used with care, a second design was conceived which has specific provision to ensure that during the deformation of one end of the pipe from a circular to a rectangular cross section, the other end of the pipe is kept circular by means of a separate plug which is a tight fit into the pipe. The general arrangement is as shown in Fig 13.7b. This second plug tool has the minor

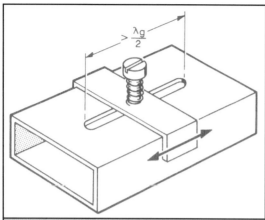

Fig 13.8 : A single. critically placed screw may be used for tuning out mismatches. Alternatively, the sliding version shown here will give a wider range of matching.

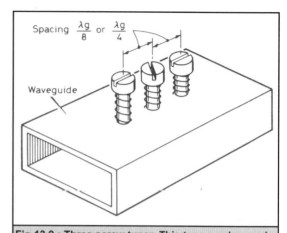

Fig 13.9 : Three screw tuner. This type can be used instead of the adjustable single screw tuner in Fig 13.8.

disadvantage of being able only to be used on short lengths of pipe, it does, however, enable a better (i.e. more smoothly varying) cross section transition to be made.

Normal square waveguide flanges were adapted to fit onto both circular and rectangular ends of the transitions Standard self soldering Yorkshire fittings can be used to join sections of permanently fixed pipe together for instance in a home station feeder run. Such a run should preferably, be kept as straight as possible. Bends are permissible provided that they can be made without the wall of the tube collapsing or becoming corrugated and that then take place over a number of wavelengths, although there is still a risk of mode changes. It is stressed that transition should be made in copper, since this is malleable and ductile, allowing the metal to be stretched without fractures provided the operation is carefully carried out.

The measured insertion loss of a pair of such transitions connected back to back was 0.8dB at 10.368GHz.

Sections of flexible or flexible and twistable waveguide are occasionally available from surplus sources and could be used to correct guide misalignments in an otherwise rigid system. However, such waveguide can introduce significantly higher losses into the system together with mismatches which may need to be subsequently tuned out.

Tuning and matching waveguide components

One of the attractive features of working in waveguide is the ease with which certain operations can be carried out compared with the lower (non-waveguide) frequencies. An example of this is the tuning and matching of waveguide components. Two methods are used, the (sliding) single screw tuner or the electrically similar three screw tuner. The three screw tuner is simpler to construct and therefore is the one most often used in amateur built equipment. A single screw, if fitted in the correct position along the centre line of the broad face of the waveguide, can be made to tune out almost any mismatch in the system simply by adjusting the depth of penetration of the screw. This adjustment will hold over a range of frequency that, although narrow in absolute terms (in the region of

1%), is usually more than adequate by amateur standards.

The matching screw is preferably fitted close to the component that requires matching. The further the screw is from the mismatch, the narrower the bandwidth of the matching, so that the screw should ideally be placed within a wavelength or so of the mismatched component. Its position is fairly critical within the range $\pm\lambda g/4$ and a method of optimising this is to mount the screw on a carriage that slides along the waveguide as shown in Fig 13.8. Both the carriage and the screw should be spring loaded to eliminate mechanical play. This design is not easy to make and use, so the constructor will normally use the three screw tuner. One possibility is the use of a short length of OE waveguide as the carriage. The $\lambda_g/2$ slot is best milled into the waveguide face. It is possible to create a sufficiently accurate slot by drilling a line of small holes along the centre line and opening this into a slot by means of a hacksaw blade followed by a fine file and considerable patience.

The three screw tuner is illustrated in Fig 13.9 and consists, simply, of three screws mounted centrally along the broad face of the waveguide. During experimental work the screws can be spring loaded to eliminate play, but after adjustment they should be secured in place by lock nuts. As with the sliding screw tuner, the screws may be mounted anywhere along the waveguide but preferably near to the component to be matched. Additional support for the screws can be provided by soldering a piece of thick brass plate to the face of the waveguide before drilling and tapping the screw holes.

Setting up either the single screw or the three screw tuner can be done using a vswr indicator, if one is available. Alternatively, they can be set during final testing of the equipment. For example, if fitted to a mismatched antenna, they are tuned to maximise signals received or transmitted and, if fitted to a mixer, to optimise the signal/noise ratio on a received signal.

It will usually be found that only one, or perhaps two, of the three screws will have a significant effect. The screws should be inserted the minimum amount necessary to achieve a good match. If the screws have to be inserted well into the waveguide then the component is probably a bad match, if none of the screws have any beneficial

Fig 13.10 : Photograph of a home made slotted line vswr detector (G3PFR). Photo by G6WWM.

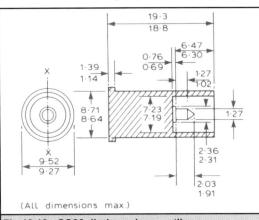

(All dimensions max.)

Fig 13.12 : SO26 diode package outline.

effect, then the component is probably already well matched.

Waveguide slotted line for vswr measurement

A rudimentary but satisfactory vswr indicator can be made by modifying a sliding screw tuner design to take a diode mount, the diode being coupled into the waveguide field by means of a small probe. The probe should be made from thin, stiff wire so that it causes minimum field disturbance whilst making measurements and it will be necessary to arrange for the probe penetration to be adjustable for the same reason. Minimum penetration consistent with an adequate indication should always be used. The coupling slot should be made at least one guide wavelength long in order to ensure that several maxima and minima of the standing wave can be detected. An example of a simple amateur built vswr slotted waveguide and probe is given in Fig 13.10. In this instance the sliding carriage is made from a split, milled block of brass, the two parts of which are held together by four countersunk screws. These can be tightened to make the carriage a smooth sliding fit on the waveguide. Probe penetration can be adjusted by sliding the diode mount up or down in the block and clamping in place by means of a knurled screw. To minimise the effects of variable metal to metal contact whilst sliding the block along the waveguide it was found necessary to glue a thin ptfe sheet to the carriage to isolate it from the slotted top surface of the

waveguide. This completely eliminated the effects of contact problems, which manifested themselves as erratic readings on the indicating meter, and allowed smooth detection of maxima and minima.

For quick indications of vswr a directional coupler and diode detector would be more useful, as it would give continuous readout.

Waveguide diode mounts, mixers/ detectors

Mixers and detectors can be constructed in many ways, some complex, some simple. Three main diode packages commonly used at these frequencies and the construction of the mixer or detector may be determined by the outline of the diode to be used.

The most familiar package is the DO22/SOD47, illustrated in Fig 13.11, which is used for the 1N23 type point contact diode and for some of the recent Schottky barrier diodes such as the BAV92 series. Older versions had a fixed anode collet whilst more modern versions, such as the 1N415 series, are supplied with a removable collet which will fit either the anode or cathode spigot of the device. This package is designed to fit within the narrow dimension of WG16. The cartridge package, SO26, shown in Fig 13.12, is typical of the CV2154/55 and SIM2/3 series and is designed to be mounted external to the waveguide

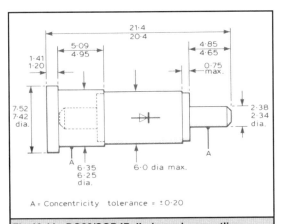

A = Concentricity tolerance = ±0·20

Fig 13.11 : DO22/SOD47 diode package outline.

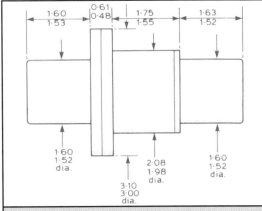

Fig 13.13 : SOD31 diode package outline.

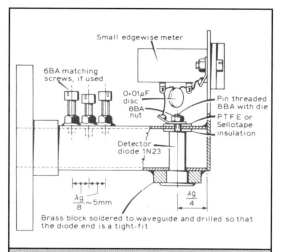

Fig 13.14 : Simple mixer/detector in WG16.

and coupled into it by means of a probe or cross bar coupling.

The third package is the SOD31 outline illustrated in Fig 13.13 and is familiar as the package used for low to medium power Gunn diodes. Like the Gunn diode, it is designed for post mounting within the waveguide.

A simple detector for 10GHz

The design of a simple detector is shown in Fig 13.14 and consists of a 1N23 type diode mounted centrally across the waveguide. The live end of the diode is decoupled by the small capacitor formed between the end of the diode package and the inside wall of the waveguide and between the nut/solder tag and the outside surface of the guide wall.

The diode is matched into the waveguide by adjusting the three matching screws for maximum current output with RF input at the chosen operating frequency.

Offset mixer using choke decoupling and a 1N415 type diode

The construction of this mixer is shown in Fig 13.15. Note that the diode is offset from the centre line of the broad face of the waveguide in order to improve matching.

- Drill a 7/64in (2.8mm) hole in the choke block, in the correct position, and also through the two broad faces of the waveguide in the position shown. Drill and tap 6BA (or 2.5mm) holes for the matching screws.

- Deburr and remove swarf.

- Ensure all metal parts are bright, clean and grease free.

- Jig the bearing nuts in place with non solderable screws and locate the choke block by using the shank of the drill bit (7/64in or 2.8mm, whichever has been used) passed through it and the waveguide. The end plate can be held in position by a small G clamp, whilst the flange should be self jigging.

- Solder all the joints using a small, hot gas flame and a minimum of solder.

- Allow the assembly to cool thoroughly before disturbing it.

- Trim off any excess from the end plate, open the hole in the choke block and upper waveguide wall to 3/8in (9.5mm) diameter and then remove all traces of flux and any excess solder, especially from inside surfaces.

The RF choke is obviously best turned in a lathe but can also be fabricated in one piece using hand tools or a stand drill as indicated in the constructional techniques chapter. A suitable choke can sometimes be salvaged from ex doppler units of a type not really suited to amateur use (for example the side by side dual cavity oscillators and mixers), in which case the choke block should be drilled out to suit the diameter of the salvaged choke. Such chokes are usually already insulated, either with a coating of some durable epoxide or by anodising, and are typically 1/4in (6.3mm) in diameter. When installing such chokes, care should be taken that the insulation is not

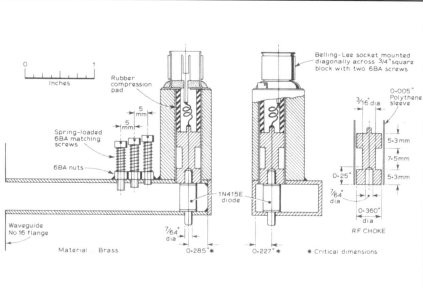

Fig 13.15 : Improved mixer using ofset diode mounting and choke decoupling. Note the three screw tuner for matching, immediately in front of the diode mount. A simple type of 75Ω coaxial connector is used at the IF output and is quite adequate up to UHF.

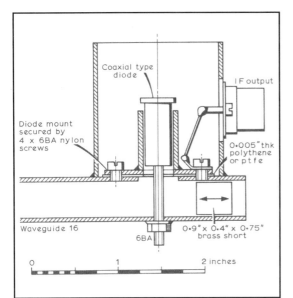

Fig 13.16 : Mixer using a cartridge diode. The mixer may be tuned by sliding the diode up or down in its mount and by adjustment of the sliding short circuit.

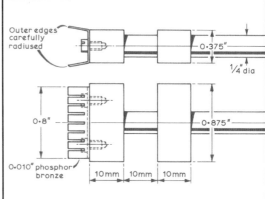

Fig 13.17 : Adjustable waveguide short circuit using contacting fingering, backed by two quarter wave spaced, quarter wave long choke blocks.

punctured by roughness inside the choke block or by too tight a fit into the block. If damaged, Sellotape, ptfe tape or Mylar tape can be used to re insulate the choke before use.

A mixer using a cartridge diode

A mixer using the CV2154/55 or SIM2/3 type diode is shown in fig 13.16. It is tuned by an adjustable RF short and by raising the diode within its mount. Construction can be simplified by replacing the sliding short by a fixed short as in the preceding designs and the diode centre line to shorting plate distance can be made any *odd* number of $\lambda_g/4$ if this is convenient to the form of construction adopted. A three screw tuner can then replace the sliding short.

Matched loads and attenuators

Professional matched loads are usually made from a lossy ferrite material, the form of which is designed to absorb virtually all the incident power. This is usually achieved by step matching or by tapering the ferrite over several wavelengths in both the narrow and the broad dimensions of the waveguide until the load occupies the whole cross sectional area of the guide. Occasionally external cooling fins may be attached to the waveguide in order to increase the power dissipation capability of the load. However, power dissipation is seldom a problem with most amateur 10GHz equipment. Low power matched loads can be made using everyday materials, such as wood, without resort to the more exotic materials used professionally. Professional loads are often obtainable from surplus sources. Fixed attenuators can be constructed in a similar manner, although even more useful variable attenuators are more difficult to construct. The attenuating material consists of a resistive card tapered at both ends in order to provide a good match in either direction.

Directional couplers

Directional couplers are widely used at lower frequencies

and most amateurs will be familiar with them as so called swr meters where they are used to measure the forward and reflected power sampled from the transmission line in an antenna system.

The theory of waveguide directional couplers has been covered in chapter 4. Transmission lines and components and so will not be discussed further here, except to say many designs exist, that they are relatively simple to make and, again, are very useful components.

Adjustable waveguide short circuits

It is sometimes useful to terminate a waveguide component with an adjustable short circuit as distinct from a fixed $\lambda_g/4$ short circuit, e.g. when matching a detector diode. The short circuit is more versatile, if adjustable, when experimenting with waveguide component design and optimisation.

An adjustable short circuit can also be used as a wavemeter as described in chapter 9 where the design used a combination of a cross coupler and adjustable short as a simple form of self calibrating wavemeter. Although such a wavemeter does not offer *very* good frequency resolution, it represents one of the most fundamental measurements which can be made, the *direct* measurement of guide wavelength.

The simplest form of adjustable waveguide short consists of a block which is a firm press fit in the waveguide and which can be clamped in place after adjustment. The rear face of the block can be tapped to take a screw that can be used to adjust the position of the block and subsequently removed. If the block is a good fit in the waveguide, then its length is uncritical. If not, then it should be about $\lambda_g/4$ in length so that it becomes, in effect, a choke. A slightly smaller block, $\lambda_g/4$ long, insulated with Sellotape (or self adhesive ptfe or polyester tape) will provide more reproducible adjustment. One form of choke block uses fingering to ensure contact, as shown in Fig 13.17. The ends of the fingers should be radiused and smoothed using the fine emery paper so that adequate contact pressure can be used without scoring the walls of the waveguide during repeated adjustment. The length of each of the choke sections is made $\lambda_g/4$, i.e. approximately 10mm at 10GHz. This would *not* be suitable for tuning an oscillator as intermittent contact of the fingering would cause frequency jumps.

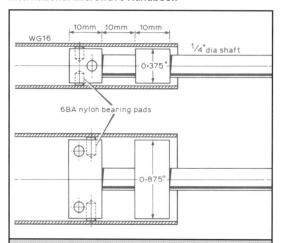

Fig 13.18 : Alternative form of adjustable waveguide short circuit using non contacting choke blocks.

A highly recommended form of adjustable short is shown in fig 13.18. The choke is floated on pips of an insulator which protrude about 0.01in (0.254mm). A method for making these inserts is to drill 1/8in (3.174mm) diameter holes to about the same depth in the appropriate positions on the choke and to press a suitable piece of insulator into each hole. The ptfe body of a standard feed through insulator is ideal. The ptfe is cut off flush with the surface of the choke and then removed with a needle. The bottom of the hole is then packed with material of thickness equal to half the difference between the size of the choke and the dimensions of the waveguide to be fitted and the insert replaced. A second method is to tap a hole in the choke of just sufficient length to take a screw and then to fit a nylon screw. Using a micrometer to monitor the process, the nylon screw is cut back so that the height of the pip is equal to half the difference between the width of the choke and the inside dimensions of the waveguide.

The insert on the opposite face is fitted in the same way and trimmed so that the total width over the inserts is equal to the internal dimension of the waveguide. With a little care and patience it is possible to eliminate play almost completely and obtain very smooth, jump-free tuning.

A third alternative is to make a block 0.88in (22.35mm) wide by 0.38in (9.65mm) high by about 0.75in (19mm) long and to create a v-shaped groove about 0.1in (2.5mm) from one end of it. This groove, which should be about 2mm wide and the same depth at its centre, should be cut into all four faces of the block by marking and filing. The groove does not need to be particularly accurate in profile as irregularities are not too important when the groove is subsequently filled with the contacting medium. A short length of subminiature coaxial cable of 1.5mm o.d. has its outer sheath removed leaving the bare braid, dielectric and centre conductor, the overall diameter of which is about 1.3mm. This is formed and cut to fit the groove. Insertion of the block into the waveguide will compress the cable somewhat and the braid will act as a multi contact wiper against all four walls of the guide, much like the fingering of the earlier design.

Wavemeters

Most simple equipment in current use on 10GHz employs free running oscillators based on Gunn devices and whilst

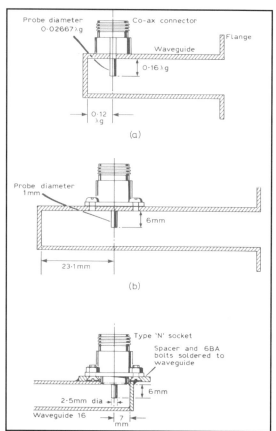

Fig 13.19 (a) General form of WG to coaxial transition. (b) WG16 to N type transition (schematic, non optimised). (c) Optimised WG16 to N type transition.

their stability is remarkably good, it is poor by comparison with crystal controlled standards. It is very desirable that the constructor (and more particularly the operator) of such equipment has a ready means of frequency calibration available, either as part of a test bench or, better still, built into the system. Better frequency setting in the field will considerably improve the operators chances of making successful contacts.

Coaxial to waveguide transitions

Coaxial cables are not used much at 10GHz, other than in very short lengths, because of the high losses involved compared to waveguide. Heliax can be useful, e.g. Andrews FHJ4-50 exhibiting a loss of about 4dB per 20ft. Short lengths of conventional flexible cable terminated in N type connectors may sometimes be found, particularly in older X band equipment, but it is more likely that short 0.141 in (3.58mm) semi rigid coaxial leads terminated in SMA connectors will be found on more modern equipment. It is useful, therefore, to have coaxial to waveguide transitions available when experimenting with pcb mounted circuits or some of the more recent professional equipment appearing on the surplus market. Two types are described.

General design dimensions

The general form of a coaxial to waveguide transition is shown in Fig 13.19a, where D is the probe diameter, L is the probe length and S is the distance between the probe centreline and the waveguide short.

Work carried out by Mike Walters, G3JVL, at 10GHz gave the optimum dimensions for a transition from waveguide to 50Ω coaxial line as being:

- D = 0.027λ_g
- L = 0.160λ_g
- S = 0.120g (theoretically λ_g/4)
- or S = 0.620g(theoretically 3λ_g/4)

These relationships can be scaled for other frequencies. For instance, the successful construction of such transitions for the 3.4GHz band in WG11 was reported by G4KNZ [6].

For 10.369GHz λ_g is 37.32mm and, from this, the following dimensions are derived:

- D = 0.995mm (1mm)
- L = 5.970mm (6mm)
- S = 4.478mm (5mm))
- or S = 23.14mm (23mm)

It is quite in order to use the rounded figures given in brackets, no really noticeable change in performance results from this. An exact match can be obtained by fitting a single screw tuner between the probe and the waveguide short to alter the centre frequency and, if necessary, a three screw tuner between the probe and the flange.

A coaxial (N type) to WG16 transition

N type connectors are specified for use only up to about 10GHz and can be used successfully although there is a slight risk of over moding in the connectors or the cable, typically a short length of FHJ4-50 used with them. The use of conventional cables is not advised because the losses involved are high and the flexible nature of the cable can lead to unpredictable impedance changes. However, some of the older surplus X band equipment is fitted with these connectors.

A square flange N type socket should be selected. The dimensions of the flange are 18 by 18mm with 13mm fixing centres for 3mm screws. The solder spill is usually 3mm in diameter and hollow. The dielectric insulation carrying the centre pin is normally held in place by a spun collar or rim of metal that protrudes below the flange and goes through a clearance hole in the panel on which the socket is mounted. This collar is normally deeper than the thickness of the waveguide wall (0.05in, 1.27mm) and would protrude into the waveguide if mounted with the flange flush on the waveguide surface. The diameter of the solder spill is also larger than the optimum diameter discussed above. Thus, the socket needs modification before it can be used easily. Occasionally, surplus sockets can be found which do not have this rim, they may also have an elongated solid metal spill. This type of socket is very easy to modify, as all that is required is that the spill is turned down to the required diameter and then cut to length.

First the rim around the insulation on the back of the socket is removed, by careful filing or sawing, to enable the socket to fit flush with the waveguide face. This may cause the dielectric bearing the pin to become loose in the socket, so care is needed. In the final assembly, the dielectric will be held in place by compression against the outer surface of the waveguide to which the socket is attached. The dielectric is cut back flush with the flange surface, using a sharp knife or scalpel.

Next the spill diameter is modified by cutting it almost flush with the dielectric and soldering a 1mm diameter probe of stiff copper wire of the required length in place of the original spill.

In order to accommodate the socket mounting flange on the waveguide without complicating construction, the probe spacing from the waveguide short has been increased by λ_g/2, from 0.12λ_g to 0.62λ_g, i.e. to 23.1mm. This allows the use of a simple, soldered end plate as the waveguide short. The probe enters the waveguide through a hole a little smaller in diameter than the measured diameter of the dielectric, so that the insulation carrying the centre pin of the socket is firmly held in place between the socket body and the waveguide wall. The modified socket can be fixed to the waveguide by soldering or, better, by means of 3mm screws inserted into tapped holes in the waveguide wall. If this method of fixing is used, then the fixing screws should be trimmed in length so that they end up flush with the inner face of the guide wall. This has the advantage that the socket is easily removable for trimming the length of the probe during testing.

In general the use of a larger than optimum diameter probe will increase the effective bandwidth of the transition, which is of little consequence in amateur installations, but may lead to the need to experiment with the length of the probe for best match and signal transfer.

The two alternative forms of construction of an N type transition are given in fig 13.19b, optimised as described above, and 13.19c, not optimised, but of acceptable performance.

A coaxial (SMA) to WG16 transition

SMA connectors are specified for use up to 18GHz and are used with 0.141in (3.58mm) semi rigid coaxial cables. This series of connectors should be the first choice for this band and special versions are available, with a carefully dimensioned spill, to act as a microstrip launcher. However, for the purposes of a WG16 transition, the ordinary SMA socket should be used. This normally has a solder spill that is 1.3mm diameter and 6.5mm long.

There is no need to make the diameter of the spill smaller as this diameter will still provide a reasonable match. The dielectric does not protrude behind the mounting flange and needs no modification. The hole necessary to match the dielectric of the socket is 4mm diameter.

It is possible to mount the socket on the waveguide with a 4.25mm spacing from the end short, provided that it is soldered in place. If four fixing screws are used then this distance must be increased to about 5.5mm to accommodate the rear two screws. Although this is slightly greater than the optimum, it has been found to make little difference in practice. If screws are used they should be 1.5mm, cut so that they do not protrude into the waveguide. Dimensions and a photograph for 10.380GHz are given in Fig 13.20. None are unduly critical and slight mismatching can be corrected by adjustment of the three screw tuner. A good match with wider bandwidth was still obtained with the probe diameter increased to 3mm by means of a soldered sleeve.

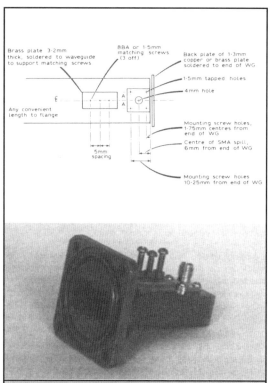

Fig 13.20 : Top - diagram of WG16 to SMA transition. Bottom - WG16 to SMA transition (G3PFR). Photo by G6WWM.

Waveguide switches

Waveguide switches fall into two categories, mechanical and electronic. The former consist of some mechanism which effectively moves one section of waveguide relative to a fixed section. This could be as simple as a plug and socket type of arrangement, a flapper across a waveguide junction or a finely engineered rotary switch, such as those produced commercially and occasionally found on the surplus market. The change over might be effected manually or by means of a remotely switched solenoid.

The electronic switch usually consists of strategically placed, biased pin diodes. Although simple in principle, this type of switch has been little used by amateurs, largely because it offers performance inferior to mechanical switches, both in terms of isolation and insertion losses

The most common application is to switch an antenna between a transmitter and receiver. Ideally, operation of the switch should not require the movement of any of the guides attached to it and for this reason the flapper or professional mechanical types are preferred.

Simple plug and socket switching

Manual switching can be accomplished by the use of round flanges and locking rings, albeit a slow and cumbersome process. Change over can be made easier if the male locating ring is fixed in position on its flange by the use of glue, for it is often difficult, in the field with cold fingers, to manipulate both the rings and the equipment simultaneously!

A waveguide three port flapper switch

This type of switch was described in outline in [7] and in more detail [8]. It consists of two sections of WG16 suitably joined to form a Y junction. At the apex of this junction a hinged flapper is introduced which can lie across either switched waveguide port and actuated by a lever, as shown in Fig 13.21. A coupling hole is cut in the flapper plate so that oscillator injection to the mixer is optimised with the switch in the receive position. This will need to be adjusted on test, to give the optimum mixer current for the particular diode in use. Thus the size of the hole should start small and be progressively increased on test until the optimum is reached. If separate receive and transmit oscillators are to be used, then this coupling hole can be omitted.

An interesting variant of the flapper three port switch was briefly described in [9], in turn derived from [10]. It consisted of a Y shaped assembly of OE waveguide as

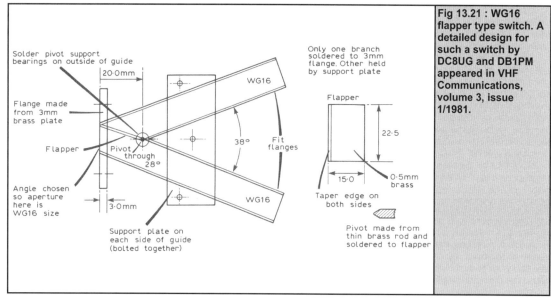

Fig 13.21 : WG16 flapper type switch. A detailed design for such a switch by DC8UG and DB1PM appeared in VHF Communications, volume 3, issue 1/1981.

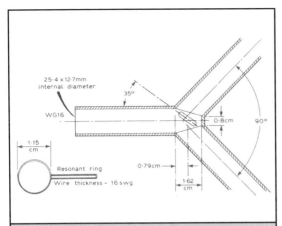

Fig 13.22 : Resonant loop switch in Old English waveguide.

shown in fig 13.22. The original design was for a frequency of 9.375GHz but the dimensions of the loop should scale linearly with the free space wavelength (λ_0), i.e. 1.15 x 9,375/10,400 at 10.400GHz. Unfortunately this is just too large to fit into WG16, but an oval ring should work just as well as a circular one. No work has yet been done on this design but the original switch was claimed to have an input vswr better than 1.06, a bandwidth of ±3% and an isolation between switched arms of at least 25dB.

Two alternative simple switches

The first, shown diagramatically in Fig 13.23, consists of two sectors of a disc hinged together, one being fixed and the other movable. Into the movable section is cut a

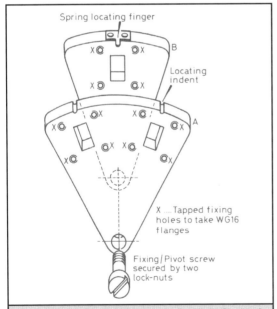

Fig 13.23 : Schematic of a sectoral switch. Sector A is fixed and sector B moves relative to it. The antenna is connected to sector B by a short length of flexible waveguide and the transmitter and receiver to sector A. The only critical dimensions are the windows and flange mounting bolt holes which must match WG16.

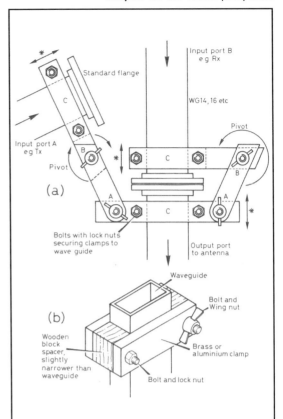

Fig 13.24 : (a) Simple mechanical arrangement allowing changeover by moving sections of waveguide. Ports A and B are pivoted around the wing nuts on the swing arms. The swinging arms are attached to clamps [C] secured to the waveguides as shown. Losses, mismatch and leakage are minimised by use of choke/plain flange pairs. The clamps and wing nuts are adjusted for correct alignment and close fit of the flanges. Then tighten wing nuts A and B as a pivot. Wooden spacers are not needed in the A-B section of the clamps. (b) Detail of the clamp blocks.

rectangular hole whose dimensions correspond to the inside dimensions of WG16. The antenna port waveguide is attached to this hole and is a short length of flexible, twistable WG16 held in place by screws fitted through its flange and into tapped holes in the sector. The fixed sector has two holes cut in it as shown in the diagram. Some kind of locating mechanism will need to be incorporated in the switch and could consist of two v-shaped grooves cut into the periphery of the fixed sector. A suitably profiled spring could then be attached to the moving sector by means of tapped screws and positioned so that it clicks the movable sector into position when a changeover is made. Coupling of the fixed sector to the equipment can be made with conveniently shaped lengths of WG16.

Another mechanical arrangement is shown in Fig 13.24, a design used by ZL2AQE on 5.7GHz. There is no reason why this concept should not work at 10 or 24GHz. Provided there is reasonable alignment of the flanges there should be little insertion loss. Leakage from the

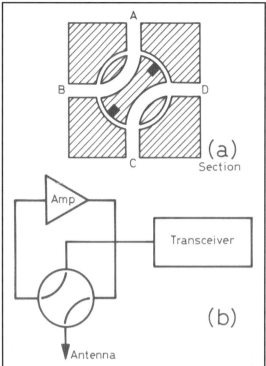

Fig 13.25 (a) Simplified diagram of a professional four port rotary changeover switch. Waveguide flanges bolt on to the body of the switch at A, B, C and D. The central barrel rotates in 90° steps, alternately connecting A-B and C-D (as shown) or A-D and B-C. The black bars are ferrite chokes between cavities to improve isolation. (b) Using a four port switch to enable use of an amplifier on both receive and transmit. As shown, the switch is in the transmit position.

flanges and losses by slight misalignment can be minimised by using plain/choke flange combinations as described earlier.

Professional mechanical waveguide switches

These usually consist of a cylindrical barrel into which are formed two WG16 sized 90° radiussed bends, as shown in Fig 13.25a. The barrel is mounted, together with a locating mechanism, within an outer cylinder machined to fine tolerances. The barrel can rotate within the cylinder and therefore acts as a four-port changeover switch. Isolation between ports is often improved by slabs of ferrite incorporated in chokes in the barrel, between the adjacent cavities. Insertion losses in a well designed professional switch will be fractions of a dB and the isolation between ports may well be in excess of 60dB.

One advantage to the four port switch is that the extra port can be used to take a waveguide load. On receive, a separate transmit oscillator is best switched off since there may be enough leakage of the transmit source to partially de-sensitise the receiver. However, only trial and error will reveal whether this is necessary. Using a commercial, ferrite isolated switch it was found possible to allow a 20mW transmit Gunn to run continuously into the load without undue problems whilst receiving. The re-

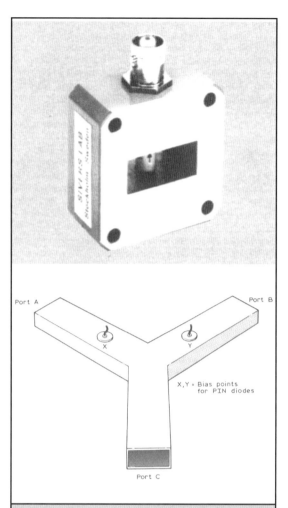

Fig 13.26 : Top - A pin switch/modulator (Sivers Lab data sheet). Insertion loss typically 2dB, isolation minimum 10dB, switching time 10 μS, maximum RF power 1W average, bias -1V at 20mA (insertion) and 0V (isolation). Bottom - A three port changeover pin switch.

ceiver oscillator should, ideally, be left running continuously (for stability) and when in transmit mode its escaping power will be absorbed into the load. Another advantage is that it can allow an amplifier to be used in front of a low power transceiver, on receive as a preamplifier and on transmit as a power amplifier, by reversing its direction as in Fig 13.25b.

PIN switches and modulators

A pin switch consists of a pin diode, usually in a 1N23 type package mounted centrally across a short length of waveguide, as shown in figure 13.26a. When reverse biased (or to zero voltage), the diode presents a very high impedance and when forward biased by a dc current, a very low impedance. A single diode mounted in this way can be used as a modulator, if a modulating signal is superimposed upon the dc bias. Provided that the Gunn source is suitably isolated from the pin diode, the modulation so produced is am rather than the customary fm and this method of modulation is often used professionally in test benches to produce square wave modulated signals.

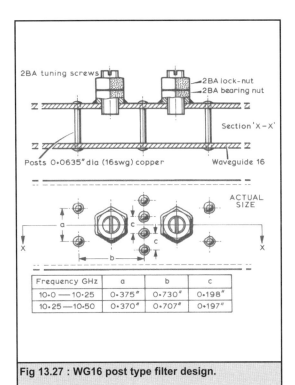

Fig 13.27 : WG16 post type filter design.

Frequency GHz	a	b	c
10·0 — 10·25	0·375"	0·730"	0·198"
10·25 — 10·50	0·370"	0·707"	0·197"

Typical performance of a single pin switch/modulator might be:

- Insertion loss 2dB

- Isolation (min) 10dB

- Switching time 10 microseconds

- Power handling 1W average

- Switch off voltage 0V

- Switch on voltage 1V

In order to act as a changeover switch, it is necessary to employ pin diodes mounted in two of the legs of a Y junction (Fig 13.26b), much in the manner of the mechanical switch discussed above. The third leg, without the diode, is the antenna port. In use forward bias is applied to, say, diode A and a zero voltage or reverse bias to diode B. Under these conditions, diode A will be low impedance and because it is a quarter wave from the junction, the A leg will present a low impedance and will be off. The reverse will be true for leg B which will be on as the diode is a high impedance across the guide. It is possible that increased switching isolation could be obtained, at the expense of increased insertion loss, by placing additional pin diodes, spaced at n x $\lambda_g/4$ (where n is an odd number) apart in each switched leg, although this has not been tried. Decoupled mountings suitable for use with pin diodes are described in the earlier sections on waveguide diode mounts.

Waveguide filters

10GHz filters usually take one of two forms. These are the post type and the iris coupled type. Both are comparatively easy to make, giving acceptable performance, with a minimum of tools and precision metal working.

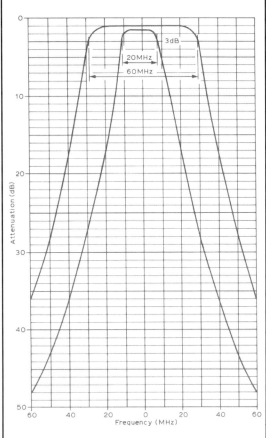

Fig 13.28 : Performance of G3JVL multi cavity iris coupled filters.

Waveguide post filters

The design of two filters (due to G3WJG) which, between them, cover the whole of the 10GHz allocation is shown in Fig 13.27. Eight posts are fitted across the guide in fairly precise positions and two tuning screws are used to set the frequency. The filters have an 80MHz equal ripple bandwidth, a 3dB bandwidth of 120MHz and an insertion loss of 0.3dB. Frequencies 1GHz away are attenuated by about 40dB. Compared with some other filter designs, this type is fairly tolerant in terms of construction errors. Perfectly satisfactory filters have been made by marking out using a magnifying glass and an accurate rule and drilling instead of the more precise use of a milling machine. A recommended method of assembly is as follows:

- Straighten the copper wire by stretching it slightly. Cut oversized lengths and squeeze one end sufficiently to stop the wires passing through the holes in the waveguide.

- Fit the wires and flanges and jig the 2BA bearing nuts in place with non-solderable screws.

- Using solder with a water soluble flux, solder the flanges in position using a gas blow lamp. Solder the post joints on the *underside* of the waveguide only, which will prevent solder running into the inside of the guide.

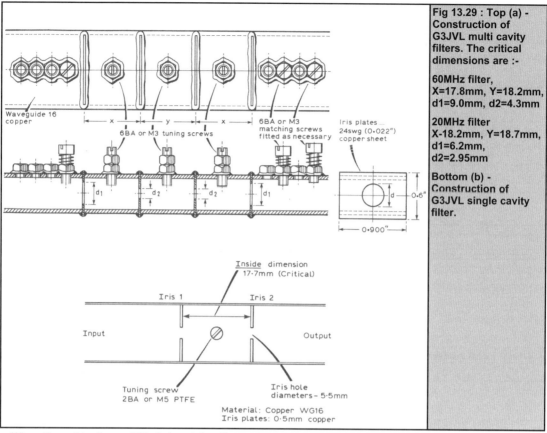

Fig 13.29 : Top (a) - Construction of G3JVL multi cavity filters. The critical dimensions are :-

60MHz filter, X=17.8mm, Y=18.2mm, d1=9.0mm, d2=4.3mm

20MHz filter X-18.2mm, Y=18.7mm, d1=6.2mm, d2=2.95mm

Bottom (b) - Construction of G3JVL single cavity filter.

- When the solder has solidified, invert the filter and solder the two bearing nuts and the remaining post joints again from the underside, allow to cool and then wash well in running water.

The filter can be aligned using a Gunn source, a directional coupler, variable attenuator and a mixer used as a detector.

Waveguide iris coupled cavity filters

The response of some higher performance bandpass filters (due to G3JVL) having a slightly different method of construction is shown in Fig 13.28. The filters have 3dB bandwidths of either 20 or 60MHz depending on their dimensions as indicated in the caption to Fig 13.29a. They are particularly suitable for use in the local oscillator chain of a receiver to attenuate the noise, generated at signal frequency, which would otherwise degrade the receiver performance. Even when a relatively low IF is used, significant rejection of noise can be obtained; for example, at 30MHz the rejection is 27dB. The 60MHz filter can be used to effectively eliminate the image (second channel) response of the receiver. Either type of filter could also be used to remove the unwanted harmonics from the output of an oscillator/multiplier chain to be used as a beacon.

As shown in Fig 13.29a, the filter consists of a length of WG16 in which iris plates are used to define three resonant cavities which are coupled by centrally placed holes of specified dimensions. The basic design frequency is 10.5GHz but the tuning screws fitted may be used to tune the filter down to 9.5GHz at least. Matching screws are fitted at each end of the filter as necessary to

tune out any mismatch with the external circuitry.

In order to maintain the Q of the filter, it should be made entirely from copper. Brass should only be used if it can be copper or silver plated after assembly and cleaning. The tuning and matching screws can be of brass only if little of each projects into the cavities, otherwise threaded copper rods should be used. The filter is constructed, in outline, as follows

- Scribe grooves deeply into the top and bottom broad faces of the guide, corresponding to the iris positions.

- Using a fine saw such as a junior hack-saw blade, first cut slots at each corner and then use these as a guide to extend the cuts across the full width of the top and bottom walls of the waveguide.

- Fit the iris plates in their correct positions and fix them in place by bending the corners of the iris plates. Jig the tuning and matching screw support nuts into place using non-solderable screws. Fit any flanges required and solder *all* joints at one operation using a gas flame. Alternatively, use the bottom only method described for the assembly of post filters.

Note that it is not necessary for the iris plates to make good contact with the sidewalls of the waveguide. Alignment of such filters is dealt with in a later section.

If it is necessary to *raise* the filter frequency, then large screws (0BA or 6mm) can be fitted centrally through the sidewalls of each cavity, although it would be better to make another filter specifically for the higher frequency.

The design of whole band single cavity iris coupled filters

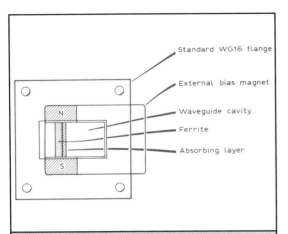

Fig 13.30 : Construction (cross section) of a field displacement isolator. Not to scale.

is given in Fig 13.29b. The main features are:

- No matching screws are needed or intended.

- Design centre frequency is 10.5GHz, bandwidth 10MHz, ripple 1dB.

- Calculated insertion loss (copper) 1.9dB.

- Tuned to 10.368GHz, adjacent products 1,152MHz away are attenuated by 40dB.

- The filters were not designed for highest possible performance, but as a reliable, easily made piece of test equipment suitable for aligning multipliers.

Miscellaneous components

A number of components that are difficult or impossible for the amateur to construct are extremely useful and may, with experience, be almost indispensable. They can occasionally be found on the surplus market and should be obtained if available.

Non reciprocal and other devices

Several forms of non reciprocal or uni directional device exist. Non reciprocal isolators and circulators rely upon the special properties of ferrites for their operation. The hybrid T, which *is* a reciprocal but uni directional device, relies solely on waveguide symmetry.

Isolators

Isolators are two port devices exhibiting low insertion loss in the forward direction (typically 0.5dB) and very high attenuation in the reverse direction (typically 30dB or more).

They are particularly useful when interposed between a Gunn oscillator and its load, particularly if the load can exhibit a varying impedance which would pull the oscillator frequency. Isolators will commonly provide between 20 and 30dB isolation, as measured on surplus isolators of unknown design frequency but constructed in WG16. They are usually quite broad band devices covering about ±500MHz around their design centre frequency in X band. Thus isolators for 10.6GHz still offer acceptable performance at 10.3GHz. It may prove possible to improve performance by fitting matching screws at the input side of the isolator and tuning these for minimum insertion loss. This has not been found necessary on several

Fig 13.31 : Photograph of WG16 field displacement isolator (G6WWM).

isolators of different designs which were used in both experimental transceiver and prototype wb (Gunn based) beacons.

Since the performance of the isolator depends on the degree of magnetisation of the ferrite within the guide, care should be taken not to damage the external magnet by either mechanical shock (dropping the device), overheating, exposure to strong external magnetic fields (loudspeakers and the like) or by proximity to ferrous objects that could cause either temporary magnetic distortion or, at worst, more permanent damage.

The field displacement isolator consists of a short length of waveguide with an externally mounted permanent magnet, looking something like Fig 13.30. Inside the waveguide there will be a slab of ferrite, coated with what looks like plastic or ceramic. This is an absorbing layer and it is this layer which limits the power which can be handled by the isolator. Fig 13.31 is a photograph of such an isolator.

The magnetic field is arranged so that the ferrite is magnetised transversely well below resonance and, because of its high permittivity, the RF field within the waveguide is distorted. The permeability further distorts the field in a non reciprocal manner and the absorbing layer is placed at a point where the electric (E) filed is zero for the forward wave and maximum for the reverse wave. Thus sited, the device offers minimum insertion loss and maximum reverse attenuation. The insertion sense will usually be indicated by an arrow somewhere on the external surface of the component, but is easy to determine by trial and error.

Circulators

The circulator is similar in concept and function to the isolator except that it is a three port junction, usually arranged in the form of an equi angular (120°) Y, although sometimes in a T arrangement. At the junction of the three legs there is a magnetised post or disc of ferrite and, usually, three screw tuners are present in each leg. The

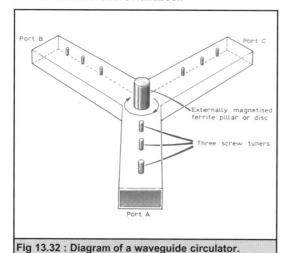

Fig 13.32 : Diagram of a waveguide circulator.

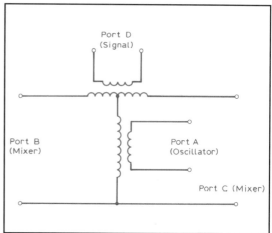

Fig 13.34 : Equivalent electrical circuit of a Hybrid T.

magnetised ferrite functions in such a way that RF energy is transferred uni directionally to one adjacent port, much in the manner of a traffic roundabout on a road.

Fit 13.32 shows this in diagrammatic form, power applied to port a will be coupled mainly into port B. A small amount of leakage into port C might occur, to some degree dependent on the match of whatever load is attached to port B. Isolation between adjacent ports in the reverse direction to that indicated as forward (for example from A to C in the direction opposite to the arrow indicating the forward direction) may lie between about 15 and 30dB, again measured on surplus circulators of unknown frequency specification.

Circulators are, again, quite wideband devices. One

circulator designed for 10.68GHz, measured 30dB at its design frequency and was still about 20dB at 10.3GHz. It is thus often unnecessary to try to alter the degree of magnetisation as was described in [11]. Isolation can be improved, at the expense of bandwidth (seldom important in the amateur context) by adjusting the three screw tuners normally present in the device. By this means the reverse isolation in the circulator mentioned above was improved to nearly 28dB with an insertion loss between ports of about 0.3dB. Again similar care should be taken against damaging the magnetic field when using such a device.

The hybrid T

This is a four port device suited to making balanced mixers. Unlike the previous two devices, it is reciprocal and does not rely upon ferrites for its function, being constructed entirely from waveguide. It is shown diagrammatically in Fig 13.33.

In theory, it is a combination of a shunt T network and a series T network and is analogous to the balanced bridge circuit, used commonly in line telephony, which produces the sum and difference of two signals fed into it. The equivalent circuit is shown in Fig 13.34. When a bridge circuit of this configuration is properly terminated by matching impedances, the signal introduced into port A divides equally between ports B and C and not at all into port D. Similarly if the signal is applied into port D, it will be equally divided between ports B and C, but not into port A.

The same happens in the waveguide hybrid. A signal introduced into port A will produce equal amplitude, in phase signals in ports B and C, but no (or very little, if properly matched loads are attached to each port) signal at port D. Any output appearing in port D is the result of asymmetry in the B/C arms or reflections from the loads attached to them.

If the signal is introduced into port D, there is little or no coupling into port A, but the signals produced in ports B and C will be equal in amplitude and 180° out of phase. These effects are due solely to the device symmetry and are independent of frequency within the pass band of the dominant (TE10) mode in the waveguide.

If signals are applied to both the B and C ports, the vectorial sum of the signals will appear at port A and the

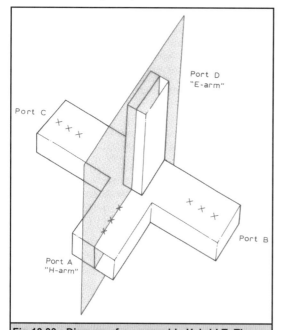

Fig 13.33 : Diagram of a waveguide Hybrid T. The shaded area shows the plane of symmetry on which the device depends. Three screw tuners may be fitted at the points marked xxx.

HEMT devices are more easily damaged by static than 'ordinary' GaAsFETs. The following precautions are recommended by the designer, G3WDG, and should be applicable not only to these latest designs

A simple static free workstation should be made and used. This can be as simple as a sheet of metal, such as thin aluminium, to which a few flexible 'fly leads' are attached. One fly lead should be connected preferably to a proprietary anti static wristband (these are available from, for instance, Tandy) or to a thin, flexible wire attached to a finger ring. If using the latter, put a high resistance (say 100k) in series between the wire and the metal work plate. The body earthing device should be worn all the time whilst working on the FETS.

The device may now be unpacked and placed on the work surface. FET leads may be safely cut to length using un-insulated tools which, when not in use, are kept on the work surface. If you use insulated side cutters, touch the metal of the cutters to the work surface *before* allowing them to touch the FET leads.

Once these operations have been completed, the equipment into which the device is to be soldered should be placed on the work surface. Another fly lead (possibly using a small crocodile clip) should be connected between the box in which the PCB is housed (or the ground plane of the PCB, if that is more convenient) and the work surface.

Next, arrange another fly lead between the work surface and the soldering iron bit, separate from the mains earth. Check that even when hot there is a low resistance path from the soldering iron tip to the work surface. It is also worth checking that there is no leakage in the iron by measuring the resistance between the tip and the heater connections when both hot and cold, but disconnect the mains first!

Solder the source leads first, making certain that the iron tip does not touch either the gate or drain lead during the soldering operation. Before soldering the gate or drain, disconnect the mains supply from the iron whilst maintaining the tip to work surface connection. Once in place, the device is safe! If at any time you have to do any further soldering operations, make sure you take the same precautions again.

Quoting G3WDG directly, "Some irons claim to have an ESD connection, I would not trust this! Sometimes after zapping, devices do not fail catastrophically, the noise figure rises by a few tenths of a dB but otherwise the amplifier seems to behave normally. Poor final performance may therefore be related to device damage". Whilst these remarks apply primarily to HEMT devices, they are equally applicable to the more expensive power devices, but for a different reason.

Fig 13.35 : Electrostatic handling techniques by G3WDG.

vectorial difference at port D. When used in this way the output signals are 0° (in phase) or 180° (out of phase) relative to the input. This is then known as a 180° hybrid.

One disadvantage of the hybrid T is that it presents a considerable vswr even if all ports are correctly terminated. To avoid this, three screw tuners may be placed in the H plane and E plane arms of the hybrid and used to match out the vswr without destroying the inherent symmetry. This will limit the bandwidth of the device although this will be of little consequence to amateurs. A hybrid T matched in this way is known as a Magic T.

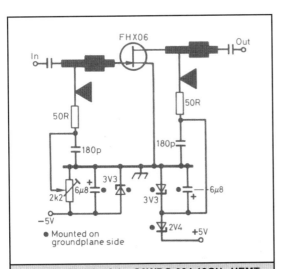

Fig 13.36 : Circuit of the G3WDG-004 10GHz HEMT LNA.

Preamplifier

Much of the equipment used by amateurs in the UK comes from designs by G3WDG. The preamplifier design is one of the full range of designs for 10GHz

The G3WDG-004 LNA

Two versions of the -004 amplifier are available, one with SMA input and one with WG16 input. All prototypes averaged 1dB nf or slightly under. Where the noise figure was fractionally above 1dB there was some evidence of degradation during construction, possibly by static, see Fig 13.35 for G3WDG's recommendations for electrostatic handling.

The circuit for either amplifier is identical and is shown in Fig 13.36. A short kit is available from the RSGB Microwave Components Service [12], it contains both the HEMT and a detailed construction sheets. The advantage of the WG input version is that a probe inputs signal directly from the waveguide to the HEMT. The SMA version will almost certainly require a least one more SMA connector and a short length of semirigid "cable" to the amplifier input. Thus the noise figure will be degraded by a few tenths of a dB, it is vital to win that little loss back if you are searching for the ultimate performance, for instance for moonbounce!

The layout details of the board of either version is given in Fig 13.37 to give you some idea of the work involved. There is also a universal regulated power supply that is used on many of the microwave designs available, the circuit diagram is shown in Fig 13.38 with details of the components for this LNA are given in Table 13.3. The preamplifier gives a gain of about 10dB with a noise figure of 1dB or a little less.

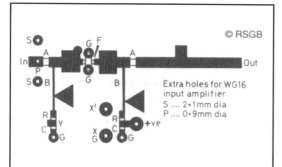

**Fig 13.37 : PCB and component of the G3WDG-004 HEMT LNA. Key to components :-
A - 2.2pF ATC capacitor, B - bias wires, C - 180pF chip capacitors, F- HEMT, G - grounding vero pins, R - 47Ω chip resistors, X Y positions of connection to bias pot X is -ve input.**

The additional holes shown on the layout are needed for the waveguide 16 (WG 16) version of the amplifier which is more difficult to construct than the SMA input version. This is because more 'mechanical' work is needed for this version in order to gain those few valuable tenths of a dB noise figure where ultimate performance is needed. Fig 13.39a shows the engineering work needed to adapt a Type 7750 tin plate box (Piper Communications) to take both the short length of WG16 (input), the PCB and output SMA connector. Construction should follow the order suggested by G3WDG:

- Solder the tin plate box together in the usual fashion. Then cut out the hole needed for the WG16 by drilling a lot of small holes inside the outline of the large hole and then join them and file out to be a tight fit on the WG16

- Cut the piece of WG16 to length, square off the ends and drill/tap all the holes shown in Fig 13.39b

- Cut a short circuit plate (thin copper sheet or double sided FR4/G10 PCB material can be used) slightly larger than the waveguide and solder it to the end of the waveguide where shown.

- File off any excess, especially where the PCB abuts the WG.

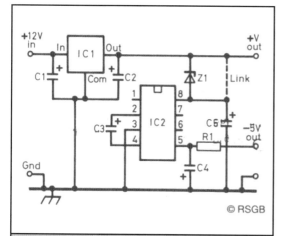

Fig 13.38 : Basic circuit of the G4FRE-023 dual output power regulator module.

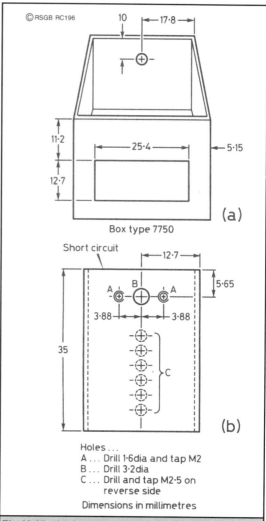

Holes ...
A ... Drill 1·6dia and tap M2
B ... Drill 3·2dia
C ... Drill and tap M2·5 on reverse side

Dimensions in millimetres

Fig 13.39 : (a) Preparing a tin plate box for the G3WDG-004 HEMT LNA (WG version). (b) Waveguide 16 dimensions and drilling sizes.

Now refer to Fig 13.40 and prepare the PCB by very carefully cutting away the copper around the hole P with a Vero cutter, to about 3mm diameter. Trim the PCB to be a neat fit into the box, be sure to leave enough on the input end of the board to allow 2 x 4mm screws to fit inside the box! These screws serve to clamp the PCB to the waveguide (holes S on the PCB to holes A on the WG). Cut off part of the input line, leaving 1mm to the left of the centre line of hole P. Fit and solder all the grounding pins. Fit the Veropin probe and solder it in place in hole P. Cut and fit the PTFE washer, referring to Fig 13.40b On to the

Table 13.3 Component list for dual output power regulator use on G3WDG-004 HEMT LNA.

R1	2k2
C1,2	1µF SMD tantalum
C3,4	22µF SMD tantalum
C5	10µF SMD tantalum
Z1	Not fitted - replace with link
IC1	78L05
IC2	ICL7660SCPA

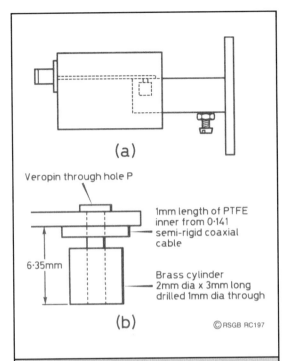

(a)

Veropin through hole P

1mm length of PTFE
inner from 0·141
semi-rigid coaxial
cable

6·35mm

Brass cylinder
2mm dia x 3mm long
drilled 1mm dia through

(b)

©RSGB RC197

Fig 13.40 : (a) General side view of G3WDG-004 WG16 HEMT amplifier assembly. (b) Detail of WG to PCB transition probe.

pin on the ground plane side of the PCB, pushing it firmly down to the PCB, but taking care not to damage the pin or board.

Fit the piece of WG into the box, through the hole, offer up the PCB, check alignment and loosely fit the 4mm screws. If the board does not align with the probe pin in the centre of the WG probe hole, Fig 13.40b, then file the hole in the box as required for proper alignment. Mark the centre of the hole for the SMA output connector (the 17.8mm dimension shown in Fig 13.39a may need to be altered!). Remove the screws and dismantle PCB and WG, drill the holes in the box for the SMA connector and open the centre clearance hole to 3.3mm. Refit the WG, PCB and 4mm screws. Then cut the spill of the SMA connector to 1.5mm and fit it to the box. Fit the PCB into place so that the output track touches the spill and solder it to the spill.

Solder three sides of the WG to the tinplate box, but not the side to which the PCB mates. Remove the SMA fixing screws, unsolder the spill and remove the connector. Remove the 4mm screws and the PCB. Carefully solder the last face of the waveguide, but do not melt the solder on the WG shorting plate or you'll have to start again! After soldering, make sure the mating surface of both the waveguide and the PCB are clean, bright and free from oxide and flux. Fit the brass collar, as shown in Fig 13.40b to the probe pin, the length shown is critical to 0.05mm. Complete the WG assembly by soldering a WG 16 flange to the input end, again being careful not to melt the solder on the rest of the assembly. Apply a small amount of conductive epoxy to the waveguide where the PCB mates, especially around the probe hole, but not so much that a short circuit occurs when the PCB and WG are squeezed together! Reassemble the PCB and SMA connector. Solder the output track to the SMA spill and

tighten the 4mm screws. Check there is not a short between the WG and input probe caused by 'squeezing' of excess epoxy. Cure the epoxy by heating at about 150C for an hour or so. After curing, recheck for absence of short circuit on the input probe - if there is a short, remove the PCB and try again! This may damage the PCB, so do it with care and try to get it right the first time!

Finally, solder all around the ground plane/ box junction. Finish off the amplifier by mounting all the components as for the SMA version. Incidentally, the holes C on the WG 16 are used for matching screws to literally "screw the last tenth of a dB" out of the amplifier!

Transceiver [13]

Advantages and drawbacks of zero IF transceivers

Zero IF transceivers have both advantages and drawbacks when compared to conventional SSB transceivers with crystal filters and many frequency conversions. Considering the current state of technology, zero IF transceivers are probably most suitable for the low amateur microwave bands: 1296MHz, 2304/2320MHz and 5760MHz. Therefore working radios for these frequency bands were developed first [14].

Although the zero IF transceivers for 1296MHz, 2304/2320MHz and 5760MHz allow many modifications and improvements of the original design, one would like to extend the zero IF design to other frequency bands as well. However, the transfer of a transceiver design to another frequency band may not be straightforward.

Amateur radio SSB transceivers for 432MHz and especially 144MHz require a very high dynamic range. A 144MHz SSB receiver should both withstand local stations with kilowatt transmitters as well as achieve a low noise figure. The dynamic range requirement for a 144MHz SSB receiver is probably even more demanding than for an HF (3-30MHz) receiver.

Extending the 1296MHz SSB transceiver design, described in chapter 9, to 432MHz or even to 144MHz does not make much sense. The dynamic range of the sub harmonic mixers used in the 1296MHz transceiver is certainly not sufficient for the current usage of the 144MHz amateur band. Due to several high power stations, even direct AM demodulation in the simple sub harmonic mixers would become a problem. A zero IF SSB transceiver for 144MHz requires much better mixers operating at the fundamental LO frequency.

On the other hand it is not easy to find suitable components for very high frequencies. Components for 10GHz (transistors, diodes etc) can be readily found on the market and it takes much effort to make them work on 24GHz. There are few parts suitable for 47GHz or higher frequencies. In addition, frequency accuracy, mixer balance and quadrature, mechanical stability and efficient shielding are more difficult to handle at higher frequencies.

A zero IF SSB transceiver for 10368MHz is described here. The design is based on the same components as the 5760MHz transceiver, described in chapter 12. HEMTs are used as amplifiers and BAT14-099R Schottky quads are used as sub harmonic mixers. Although 10GHz does not represent a problem for HEMTs, this seems to be the upper limit for the BAT14-099R quads. The

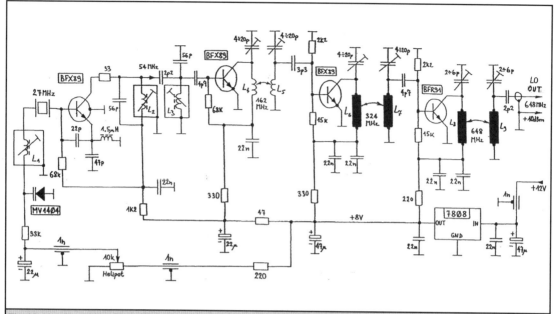

Fig 13.41 : Circuit diagram of VXCO and multipliersfor 648MHz used in 10GHz transceiver.

BAT14-099R quads are packaged in the relatively large and asymmetric SOT-143 package causing severe mixer imbalance problems.

The 10GHz SSB transceiver also includes PIN diode antenna switching. The RF front end is built on teflon laminate, while all other microstrip circuits are built on conventional FR4 glass fibre epoxy, including several bandpass filters for 10GHz. The IF and AF sections are of course identical to those used in the transceivers for 1296, 2304 and 5760MHz, described in chapters 9, 10 and 12 respectively.

Modified VCXO and multiplier stages

Although for the radio amateur 10GHz frequency band allocation extends from 10000MHz to 10500MHz, most narrow band operation is concentrated slightly above 10368MHz. In the near future one may also expect narrow band activity in the satellite segment (10450-1050 0MHz), probably concentrated around the lower end around 10450MHz.

10368MHz is an integer multiple of many popular frequencies. For example, 10368MHz is the ninth harmonic of 1152MHz, a reference frequency that also generates 2304MHz, 3456MHz, 5760MHz and 24192MHz. 10368MHz is also the eighth harmonic of 1296MHz suggesting the use of the same VCXO and multiplier chain for 648MHz with some additional multiplier stages (x8) to obtain 5184MHz for the sub harmonic mixers.

However, it makes sense to modify the VCXO itself for operation in the 10GHz transceiver. The relative crystal frequency pulling should be eight times smaller at 10GHz due to the additional multiplier stages. On the other hand, frequency stability is much more critical at 10368MHz than it is at 1296MHz.

Both requirements can be met by replacing the original VCXO using a 18MHz fundamental resonance crystal with a modified VCXO using a 27MHz third overtone

crystal. Overtone crystals have a higher Q than fundamental crystals therefore providing better frequency stability. On the other hand, the frequency pulling range of an overtone crystal is very restricted and it is barely sufficient for a 10GHz transceiver.

Fortunately the circuit diagrams of both VCXOs are similar and the same printed circuit board can be used for both of them, including the multiplier chain up to 648MHz. Of course the 10GHz transceiver requires an additional multiplier for 5184MHz built in microstrip technology just like in the 5760MHz transceiver. The 648MHz input frequency is first multiplied by four to 2592MHz and then doubled to 5184MHz.

The modified VCXO and multiplier chain up to 648MHz are shown on Fig 13.41. A 27.000MHz crystal is required for operation at 10368MHz. The frequency pulling range is very small in spite of the large capacitance ratio of the MV1404 varactor. The frequency coverage amounts to only 150-200kHz at the final frequency centred around 10368.100MHz.

A coverage of 150kHz is however sufficient for normal 10GHz narrow band work, provided that the frequency stability is reasonable. Fortunately the 27.000MHz crystal is used in teletext decoders inside many TV sets. Thanks to very high volume production these inexpensive crystals have an excellent frequency stability.

The overtone oscillator itself (BFX89) is designed to reduce the loading and therefore the heating of the crystal, to further improve the frequency stability. The VCXO is followed by two resonant circuits (L2 and L3) tuned to 54MHz. The following multiplier stages are identical to those used in the 1296MHz transceiver and are tuned to 162MHz (X3), 324MHz (X2) and 648MHz (X2).

Since the VCXO module is followed by an additional multiplier for 5184MHz in the 10368MHz transceiver, an output power of 10mW (+10dBm) is sufficient at 648MHz.

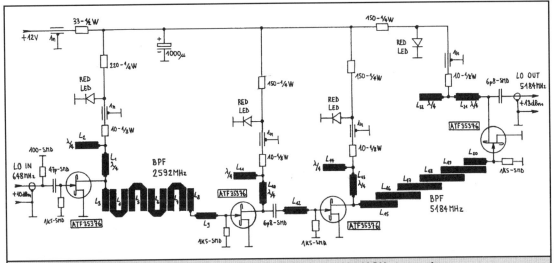

Fig 13.42 : Circuit diagram of additional multiplier for 5184MHz used in 10GHz transceiver.

Therefore the power supply resistors of the multiplier stages inside the VCXO module are increased to 330Ω, 330Ω and 220Ω. The printed circuit board is the same as in the 1296MHz version (see Fig 9.10 - 9.12) and there are just a few minor variations in the component locations

The circuit diagram of the additional multiplier for 5184MHz is shown on Fig 13.42. The circuit includes four HEMTs ATF35376. The first HEMT is overdriven by the 648MHz signal to produce many harmonics. The following microstrip bandpass (L3, L4, L5, L6, L7 and L8) selects the fourth harmonic at 2592MHz. The second HEMT amplifies the 2592MHz signal to drive the third HEMT operating as a frequency doubler. The doubler is followed by a 5184MHz bandpass (L15, L16, L17, L18, L19 and L20) and the 5184MHz signal is finally amplified by the

last HEMT to obtain 20mW (+13dBm).

The additional multiplier for 5184MHz is built on a double sided microstrip 0.8mm FR4 board with the dimensions of 20mm x 120mm as shown on Fig 13.43. The corresponding component location is shown on Fig 13.44. The 5184MHz multiplier should provide the rated output power (+13dBm) without any tuning, provided that all of the components are installed and grounded correctly.

Quadrature transmit mixer for 10368MHz

The circuit diagram of the quadrature transmit mixer for 10368MHz is shown on Fig 13.45. The 5184MHz LO signal is taken from a -15dB coupler and the LO signal level is restored by the ATF35376 amplifier stage, feeding

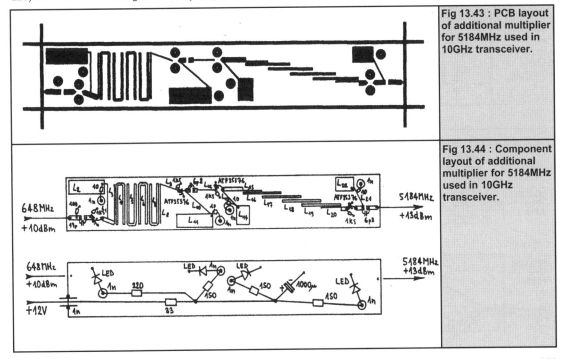

Fig 13.43 : PCB layout of additional multiplier for 5184MHz used in 10GHz transceiver.

Fig 13.44 : Component layout of additional multiplier for 5184MHz used in 10GHz transceiver.

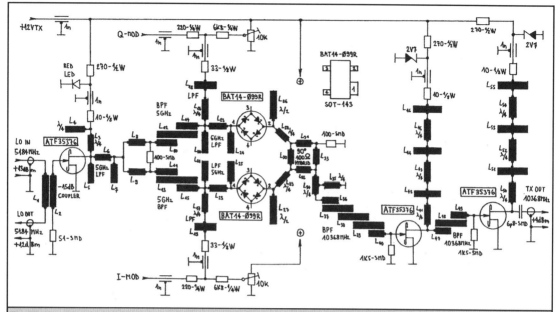

Fig 13.45 : Circuit diagram of quadrature transmit mixer for 10GHz transceiver.

two sub harmonic mixers equipped with BAT14-099R Schottky quads. The 5GHz low pass filter attenuates the second harmonic at 10GHz to avoid corrupting the symmetry of the mixers.

The unwanted carrier rejection of the BAT14-099R sub harmonic mixers is only 10-15dB at 10368MHz. The main reason for the rather poor carrier rejection is the large and asymmetrical SOT-143 package. Pin 1 of this SMD package is wider than the remaining three pins. Unfortunately, microwave |Schottky quads in suitable symmetric

packages are hardly available and much more expensive than the BAT14-099R.

On the other hand, the sub harmonic mixer balance can be corrected by a DC bias applied to the diodes. The quadrature transmit mixer for 10368MHz therefore includes two 10kohm trimmers to adjust the mixer balance. In this way the unwanted carrier rejection can be improved to better than 30dB.

The two 10GHz signals are combined in a quadrature hybrid. The hybrid is built as a 100Ω circuit to save board

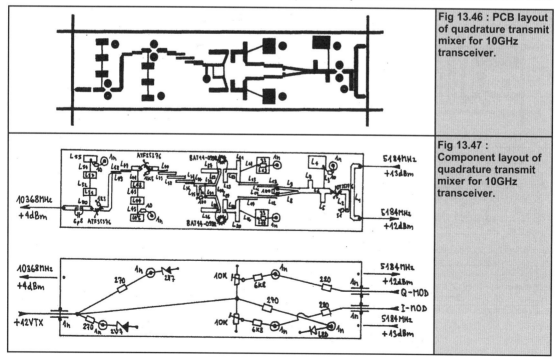

Fig 13.46 : PCB layout of quadrature transmit mixer for 10GHz transceiver.

Fig 13.47 : Component layout of quadrature transmit mixer for 10GHz transceiver.

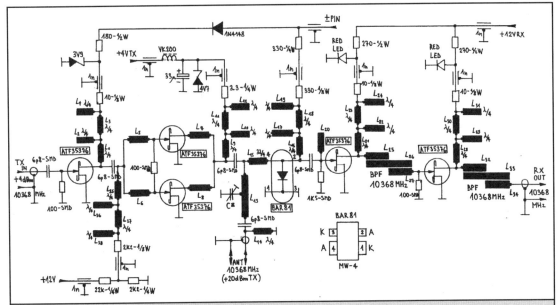

Fig 13.48 : Circuit diagram of RF front end for 10GHz transceiver.

space. Quarter wave transformers are used to restore the impedances to 50Ω. The circuit diagrams of both sub harmonic mixers are slightly different from those used in the 1296, 2304 or 5760MHz versions for the same reason.

The hybrid is followed by a 10368MHz bandpass filter (L36, L37, L38, L39 and L40). The latter removes the 5184MHz LO as well as other unwanted mixing products. After filtering the 10368MHz SSB signal level is rather low (around 30µW or -15dBm), so two amplifier stages with ATF35376 HEMTs are used to boost the output signal level to about 2.5mW (+4dBm).

The quadrature transmit mixer for 10368MHz is built on a double sided microstrip 0.8mm FR4 board with the dimensions of 30mm x 120mm as shown on Fig 13.46. The corresponding component location is shown on Fig 13.47. Since only positive bias is available through the 10kohm trimmers for mixer balancing, the BAT14-099R packages should be oriented correctly in the circuit. In particular, one of the two packages has to be installed upside down.

Except for the balancing trimmers, the quadrature transmit mixer for 10368MHz should not require any tuning. Both balancing trimmers are simply adjusted for the minimum output power when no modulation is present. This adjustment is best performed when the whole transceiver is assembled.

RF front end for 10368MHz

The circuit diagram of the RF front end for 10368MHz is shown on Fig 13.48. The RF front end includes a transmit power amplifier, a receive low noise amplifier and a PIN diode antenna switch. Unlike the RF front ends for 1296, 2304 or 5760MHz, all built on FR4 laminate, the RF front end for 10368MHz is built as a microstrip circuit on 0.5mm thick teflon laminate to reduce RF losses.

Building the microstrip circuit on glass fibre teflon laminate (thickness 0.5mm or 0.020, dielectric constant εr=2.55)

allows an 1-2dB increase of the transmitter output power and an 1-2dB improvement of the receiver noise figure. The output power of the 10368MHz transmitter is in fact even slightly higher than the output power of the 5760MHz transmitter. Although both transmitters use the same semiconductor devices (HEMTs), the latter is built on lossy FR4 laminate.

The transmitter power amplifier is designed with inexpensive HEMTs, so the output power is limited to 100mW (+20dBm) at the antenna connector. The amplifier includes an ATF35376 HEMT driver stage followed by two ATF35376 HEMTs in parallel in the output stage. The 100ohm resistor between L5 and L6 improves the power dividing and prevents push pull parasitic oscillations of the output stage.

The two output HEMTs receive a positive bias on the gates both while transmitting and while receiving. During transmission the transistors generate a self bias as in the 5760MHz transmitter. Of course the +4VTX supply line requires a current limiting resistor.

The antenna switch is using a single shunt PIN diode just like the 5760MHz counterpart with the PIN diode BAR80. However, the parasitic capacitance of the BAR80 is far too high for operation at 10368MHz, so the new diode BAR81 (SMD component marking ABs or BBs, same MW-4 package) has to be used. The parasitic capacitance of the new BAR81 PIN diode is less than half that of the old BAR80. The insertion loss of the BAR81 is reduced by applying a negative bias (±PIN) while receiving.

During transmission, the BAR81 is turned on and the short circuit is transformed by L15 into an open circuit at the summing node. The insertion loss of the BAR81 in the receive path exceeds 20dB and this is sufficient to protect the receiver. During reception, the two transmitter output HEMTs act as short circuits thanks to the positive gate bias. The short circuit is transformed through package parasitics, L7, L8 and interconnecting lines (total electrical length 3λ/4) into an open circuit at the summing node.

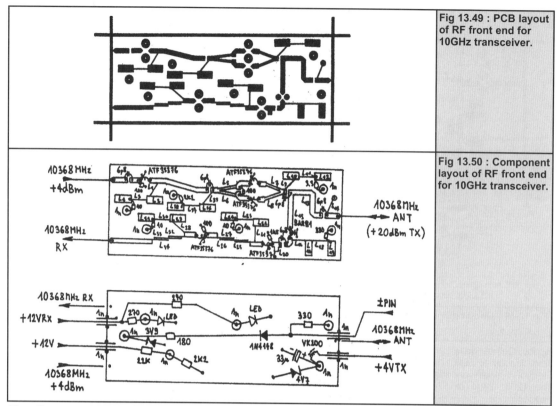

Fig 13.49 : PCB layout of RF front end for 10GHz transceiver.

Fig 13.50 : Component layout of RF front end for 10GHz transceiver.

Since there are no strong signals expected in the 10GHz band, the LNA for 10368MHz includes two stages with ATF35376 HEMTs. The total insertion gain including the losses in the antenna switching network and the two 10368MHz bandpass filters amounts to about 23dB.

The RF front end for 10368MHz is built on a double sided microstrip 0.5mm glass fibre teflon board with the dimen-

sions of 30mm x 80mm as shown on Fig 13.49. The corresponding component location is shown on Fig 13.50. A low loss teflon laminate allows a higher output power and a better noise figure. On the other hand, a low loss teflon laminate does not suppress the parasitic oscillations of HEMTs in the millimeter frequency range. These oscillations have to be controlled by damping resistors, usually 100Ω connected between gate and source.

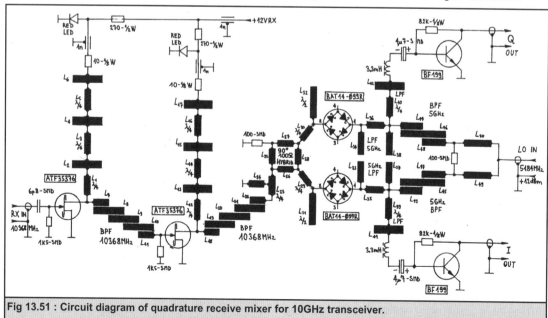

Fig 13.51 : Circuit diagram of quadrature receive mixer for 10GHz transceiver.

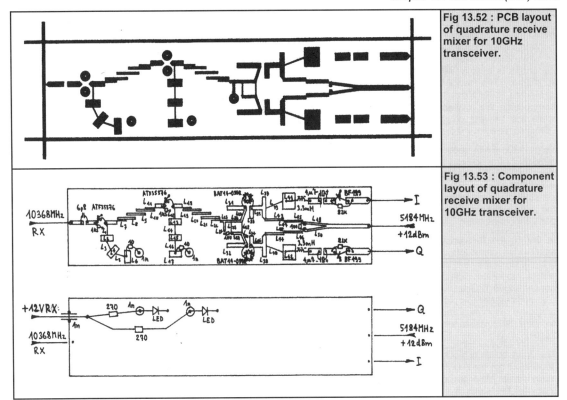

Fig 13.52 : PCB layout of quadrature receive mixer for 10GHz transceiver.

Fig 13.53 : Component layout of quadrature receive mixer for 10GHz transceiver.

The RF front end for 10368MHz includes a single tuning point. A capacitive tuning stub (a 2mm x 3mm piece of copper foil) is added to L13 to improve the antenna matching, which in turn depends on the installation of the antenna cable and connector. This matching stub will improve the output power by about 1dB or in other words, the output power may be as low as 80mW without any tuning.

Quadrature receive mixer for 10368MHz

The circuit diagram of the quadrature receive mixer for 10368MHz is shown on Fig 13.51. It differs from the similar 5760MHz mixer in the design of the quadrature hybrid and sub harmonic mixers with BAT14-099R diodes. The module includes two RF amplifiers with ATF35376 HEMTs, two 10368MHz bandpass filters, two sub harmonic mixers operating in quadrature and two identical IF preamplifiers with BF199 transistors.

The sub harmonic mixers are identical to those in the transmitter using BAT14-099R schottky quads. Both mixers are supplied in phase (L49 and L50) with the LO signal. The RF input signal is split by a 100ohm quadrature hybrid (L25, L26, L27 and L28). Impedance matching is provided by quarter wavelength lines L23, L29 and L30.

The mixers are followed by two IF preamplifiers with BF199 transistors identical to those used in the 1296, 2304 and 5760MHz receivers. The 3.3mH chokes can be replaced by lower values, since powerful signals (both amateur and out of band) are not expected in the 10GHz band. Lower value chokes are less sensitive to disturbing low frequency magnetic fields, including disturbing fields generated in the transceiver itself.

The quadrature receiving mixer for 10368MHz is built on a double sided microstrip 0.8mm FR4 board with the

dimensions of 30mm x 120mm as shown on Fig 13.52. The corresponding component location is shown on Fig 13.53. In the 10GHz band, a quarter wavelength is only 4mm long on FR4 or 5mm long on teflon boards, so more complicated microstrip circuits can be used. For example, the supply/bias chokes are built as two section low pass filters on the RF front end teflon board and as three section low pass filters on both mixer FR4 boards. These improved RF chokes introduce less insertion loss and allow a lower crosstalk.

The receiving mixer for 10368MHz requires no tuning. However, the BF199 IF preamplifier transistors should be selected for the lowest noise. It seems that these transistors do not have a guaranteed l/f noise specification. In all of the prototypes built the Philips BF199 transistors produced the least amount of noise.

Construction of the zero IF SSB transceiver for 10368MHz

The SSB/CW transceiver for 10368MHz is using the same quadrature modulator, IF amplifier and demodulator as the similar transceivers for 1296, 2304/2320 and 5760MHz. Since the same semiconductors are used as in the 5760MHz version, the current limiting resistor in the SSB/CW switching RX/TX module has to be set to 82Ω, 1W.

The new PIN diode BAR81 also requires the small PIN driver module also used in the 5760MHz transceiver. Since the new BAR81 is much improved with respect to the old BAR80, the resulting receiver sensitivity loss amounts only to a few dB if no negative bias is provided (no PIN driver used) and the ±PIN line is simply connected to +12VTX. It therefore makes sense to use the new BAR81 also in the 5760MHz transceiver, while the old BAR80 is good enough for 1296 or 2304/2320MHz.

In the 10GHz frequency range, even small SMD components are rather large when compared to the wavelength of only 29mm. The size of the SMD resistors and capacitors is usually indicated in hundredths of an inch (0.254mm). The first SMD resistors and capacitors were of the size 1206 (about 3mm x 1.5mm). Today, most SMD components are available in the sizes 0805 and 0603, while the newest components are of the size 0402 (dimensions 1mm x 0.5mm).

Large 1206 SMD capacitors should not be used in the 10368MHz transceiver, since they have parasitic internal resonances in the 10GHz frequency range. The resonant frequencies are further decreased by the high dielectric constant of the ceramic used to build these capacitors. In the 10368MHz transceiver only 0805 or smaller SMD parts should be used.

In the 10GHz frequency range it makes sense to use small value capacitors (mainly 6.8pF in the circuit diagrams), since they are built from low loss WHITE ceramic with a moderate dielectric constant and internal resonances above 18GHz. Higher value capacitors made from coloured ceramic (purple or brown) have higher RF losses and lower resonance frequencies. Finally, the newest and smallest 0402 resistors and capacitors are useful even at 24GHz.

The main building block of the RF circuits operating at 10368MHz are the ATF35376 HEMTs, although there are many similar devices produced by other manufacturers that offer the same S parameters at similar bias conditions. When selecting these devices one should take care of the I_{dss}, since most of these transistors operate at zero bias for circuit simplicity.

An I_{dss} of about 30mA is desirable. Devices with higher I_{dss} are only useful in the transmitter output amplifier. If devices with a sufficiently low I_{dss} can not be obtained, then the 270Ω 1/2W (or similar) resistors should be reduced to allow proper operation of the zener or LED shunt regulator. Lower NF selections of the same device, like the ATF35176 or the ATF35076, usually have a higher I_{dss}.

The intermediate/audio frequency section of zero IF or direct conversion transceivers also requires a careful selection of active devices with low l/f or popcorn noise. Experience accumulated by building many transceivers for 1296, 2304/2320, 5760 and 10368MHz shows that Philips BF199 transistors perform best in the IF preamplifiers. However, several very noisy samples of the BC238 transistors had to be replaced in the quadrature IF amplifiers as well. It seems that the factory rejects exceeding the l/f noise requirements are sent to hobbyist shops. On the other hand, industrial leftovers found at flea markets perform best, since these devices had to go through severe input quality controls.

The RF modules of the 10368MHz SSB transceiver have the same dimensions as the corresponding units of the 5760MHz counterpart, but the exact locations of the electrical connections are slightly different. Their shielded enclosures should be carefully manufactured out of thin brass sheet, since the 10368MHz SSB transceiver is even more sensitive to microphonics and RF leakage. Both transmit and receive mixers and the RF front end require 1cm thick microwave absorber foam (anti static foam) installed under the covers of the shielded enclosures.

The VCXO module does not require a bottom cover like in the 1296MHz version, since 10368MHz is only the sixteenth harmonic of 648MHz. On the other hand, the additional multiplier for 5184MHz and the RF modules should be accurately shielded. Efficient feed through capacitors should be used elsewhere. All internal RF connections should be made with double braid flexible teflon coax or UT085 semirigid.

The complete 10368MHz SSB transceiver can be installed in the same enclosure as the 1296, 2304 or 5760MHz versions with a central chassis and internal dimensions of 60mm (height) x 180mm (width) x 180mm (depth). The module location within the enclosure is the same as in the 5760MHz version, including a similar wiring among the modules.

The loudspeaker should not be installed in the same case to avoid microphonics. Microphonics could perhaps be reduced by using machined enclosures for the RF modules. Mechanical vibrations can also be controlled by inserting pieces of plastic foam between the modules to act as an acoustic absorber.

Testing of the zero IF SSB transceiver for 10368MHz

Testing of the transceiver should start with the alignment of the VCXO and multiplier stages. The VCXO should be adjusted for the desired frequency coverage. The multiplier stages are simply adjusted for the maximum output at the desired frequency. The maximum is observed as the rectified voltage drop on the base of the following transistor, measured through a suitable RF choke. The maximum on 648MHz is measured as the drain current dip of the 2592MHz multiplier. Although the additional multiplier does not require any adjustments, the output signal level (+13dBm at 5184MHz) should be checked.

Since the receiver does not require any tuning, it should already work. First, the overall amplification should be checked. The output noise should drop when the supply to the LNA is removed. When the IF preamplifiers are disconnected, the noise should drop almost to zero.

Next the receiver is connected to an antenna and tuned to a weak unmodulated carrier (distant beacon etc). Besides the desired signal its rather weak image should also be heard. The image can be detected since its frequency changes in the opposite direction when tuning the receiver. The image is then attenuated by adjusting the two trimmers, phase quadrature and amplitude balance, in the IF amplifier.

The transmitter should be first checked for the output power. The full output power should be achieved with the trimmer in the modulator at about 1/3 resistance in CW mode. The DC voltage across the PA transistors should rise to the full voltage allowed by the 4V7 zener. Finally, the output power is optimised with the tuning stub on L13.

The transmitter is then switched to SSB to adjust the balance of both transmit mixers. The two 10kohm trimmers are simply adjusted for the minimum output power with no modulation. Unfortunately this setting is sensitive to the mixer diode temperature and 5184MHz LO drive level, so the carrier suppression may not stay as good as it was when adjusted during the testing of the transceiver.

The SSB modulation should be checked by a radio contact with another amateur station on 10368MHz. In particular the correct sideband, USB or LSB, should be

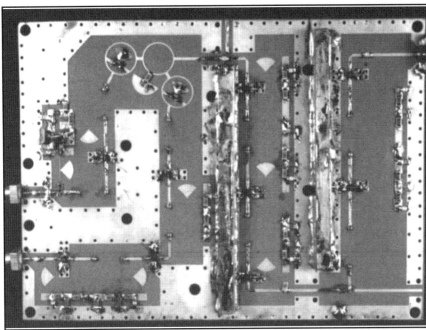

Fig 13.54 : Picture of first prototype 10GHz transverter.

checked, since the I and Q modulation lines are easily interchanged by mistake. The other station should also check the carrier leakage or transmit mixer imbalance, heard as a 1365Hz tone added to the modulation.

Finally, the shielding of the transceiver should be checked. Waving your hand in front of the antenna usually causes a 1365Hz whistle in the loudspeaker of the receiver. The latter is caused by local oscillator leakage, frequency shifted by the Doppler effect of the moving hand and finally collected by the antenna. While this effect can not be eliminated completely with the suggested mechanical construction, it should be small enough to allow normal use of the transceiver. Of course, the receiver sensitivity and shielding can also be checked with a mains operated fluorescent tube as described for the 1296, 2304 and 5760MHz transceivers.

The current drain of the described 10368MHz transceiver should be around 390mA during quiet reception at a nominal supply voltage of 12.6V. The current drain of the transmitter is inversely proportional to the output power and ranges from 550mA (CW or SSB peak power) up to 580mA (SSB no modulation). The current drain could be reduced substantially if a more efficient regulator were used to power the several HEMT stages with an operating voltage of only 2V.

Transverter [15]

A single board transverter for 10GHz

The ARRL 10GHz cumulative contest has spawned a lot of interest in 10GHz operation. In recent contests there have been over 40 stations active in New England alone, with many others having an interest but no equipment to enable them to participate. In other VHF contests, 10GHz is the second most active band above 432MHz, only 1296MHz sees more activity.

Many of us started on 10GHz with simple wideband FM transceivers [16]. Some of these were as simple [17] as a converted police radar detector or supermarket door

opener, with a modified FM broadcast receiver as an IF radio. WBFM operation is fun and capable of impressive distances, over 200km for a clear path, with quite basic equipment and simple horn antennas.

For real DX, narrow band modes of operation are needed, usually SSB and CW. Just narrowing the receive bandwidth from 200kHz for WBFM to hundreds of Hertz for CW provides roughly 30dB improvement, and equipment improvements can provide further increases. By the inverse square law, doubling the path length requires 6dB more signal, so +30dB is a major increase in capability.

The missing piece needed for affordable 10GHz narrow band operation is a transverter which can be readily reproduced. The goal of this single board transverter design is to provide reproducibly good performance.

Background

A simple WBFM transceiver relies on a Gunn diode oscillator to provide all the microwave magic. Narrow band equipment is much more complex, usually a transverter from some lower band. The narrow bandwidth requires a very stable oscillator, and linear amplification is needed for SSB operation. Amplification at microwave frequencies has historically been difficult and expensive, but improvements in semiconductor technology have helped significantly in recent years.

The most useful semiconductor device is the microwave GaAsFET. Although intended mainly for low noise amplifiers, hams discovered that they could be pushed to provide modest amounts of power. In 1993, I built a 10GHz transverter [18] using a number of modular building blocks containing GaAsFets. Each of the modules was hand built and hand tuned, but a number of hams have used copies of some of the modules as well as surplus components to assemble working transverters. Around the same time, KH6CP [19] (now W1VT) and G3WDG [20], in England, described a modular transverter.

The only commercial equipment for the 10GHz band

Fig 13.55 : Picture of first prototype 10GHz transverter showing the pipe cap filters.

comes from Germany. The initial cost seems quite high, but many of us spent a good deal more on parts and test equipment before we had a working system.

At the lower microwave frequencies, single board no tune transverters are very popular. They provide good performance at low cost, using MMIC amplifiers and stripline filters printed on the circuit board. Above about 4GHz, printed filters are hard to make consistently due to tolerances in both dimensions and dielectric constant. They also suffer from radiation as the board thickness becomes a significant fraction of a wavelength. Making the board thinner to reduce radiation loss only makes the problem worse.

In 1997, I described a single board transverter for 5760 MHz [21] which uses tuned filters made from common copper plumbing pipe caps. Similar filters [22] were used in the modular 10GHz designs cited above, but the modular construction allowed individual tuning. Although the 5760 MHz single board unit uses five tuned filters, the tuning procedure is simple and straightforward, demonstrating that a no tune design is not necessary, hams are capable of tuning.

Design Considerations

The success of the 5760 MHz single board transverter design brought a frequent question: What about a 10GHz version? However, there were still several problems to solve for a design that can be reproduced in quantity, without requiring an undue amount of expertise:

1 Without excellent grounding, gain and stability of the MMIC amplifiers is degraded.

2 Radiation from the microstrip transmission lines becomes significant at higher frequencies, resulting in losses and unwanted coupling.

3 The pipe cap filters have more loss at 10GHz.

4 Filtering for the popular 144MHz IF is more difficult at 10GHz.

5 The transverter circuit board is large enough to flex

during handling, resulting in cracked solder joints on some of the surface mount resistors.

6 The LO multiplication factor is too large for a single frequency multiplier. Experience suggests that the frequency multiplication from 568MHz to 10GHz would work better if done in two steps.

As a first step, I cut out the LO multiplier section from a 5760 transverter board and assembled it with 1/2 inch pipe caps rather than the normal 3/4inch ones. NØUGH [23] had reported good results at 10GHz from a x3 frequency multiplier using an ERA-3 MMIC, so I tested the assembly with a x3 frequency multiplication ratio. The output power at 10.224GHz was more than adequate for a dual mixer, solving one problem, inadequate LO power to the mixer, a weakness in many microwave designs.

Now I needed a way to get from 568MHz to 3408MHz to drive the x3 multiplier. I was confident that another MMIC frequency multiplier would do the job, but a compact filter at 3408MHz was still a problem, since a 1inch pipe cap would be too high for the aluminum enclosure used for the 5760MHz transverter. N2CEI tried longer screws in the 3/4 inch pipe cap filters and found that they could be tuned low enough in frequency, completing the solution to problem number 6 above.

Problems 3 and 4 are solved together, by adding an additional pipe cap filter in both transmit and receive chains for filtering, and an additional MMIC stage in each chain to compensate for the losses of all the filters. The trick is that separating the filters with an amplifier stage makes the tuning of each filter independent from its neighbors, and the loss of the filters isolates the amplifier stages from each other, improving stability of the amplifiers. KK7B once explained to me that this is the secret that makes no tune transverters possible.

However, some of the isolation is lost if there is significant radiation from the transmission lines. Most of us have built preamplifiers that work well until we put a cover on the box!

Problems 1 and 2 are made more difficult by the desire for

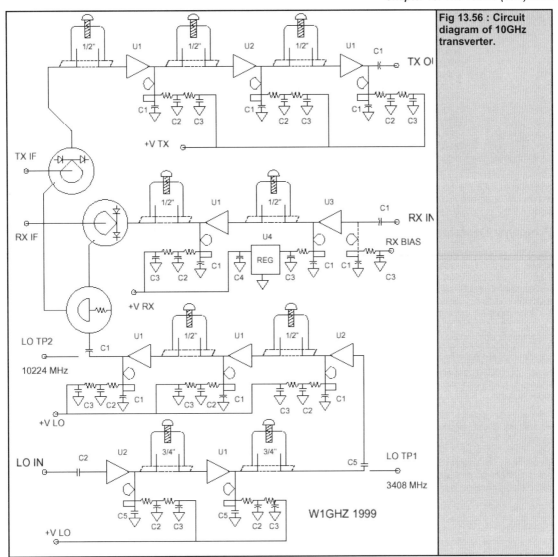

Fig 13.56 : Circuit diagram of 10GHz transverter.

W1GHZ 1999

reproducibility. As an example, AF1T and K2CBA have each made a single board 10GHz transverter with similarities to this one. Each had transmission lines cut out by hand with a hobby knife, and good grounding was accomplished by carefully bending the MMIC leads and fitting them into small slots cut in the board. Radiation was controlled by adding shielding strips soldered vertically to the board. The process is tedious and painstaking, performed by skilled and experienced microwavers, but the end result works well.

The reproducible alternative calls for a professionally made printed circuit board with plated through holes for grounding. In the past, Down East Microwave [24] has produced affordable kits for hams based on some of my articles, so this time I approached Steve, N2CEI, before starting the design so that he could help make it reproducible. For instance, the best way to reduce radiation from the microstrip lines is to make the dielectric thinner. Steve confirmed that his PC board supplier could supply very thin Teflon boards with adequate line tolerance, so we chose a 0.015 inch board thickness, half as thick as the previous 5760MHz design, to control radiation.

An additional advantage to the thin board is that the length of the plated through ground holes is also halved, so that the ground inductance is also cut in half. Problems 1 and 2 are solved together. The disadvantage of the thin board is that it is much more difficult to produce a prototype in the basement, we chose to have a small number made professionally, knowing that we would find small problems and end up discarding them.

A real disadvantage of the thin Telfon board is that it is *very* flexible. When I built the first prototype, shown in Fig 13.54 and 13.55, I added several stiffeners of brass angle stock. Even so, one of the solder joints for a chip resistor cracked during tune up, making the output intermittent. A better solution to problem 5 was needed. Steve and I discussed several possibilities. One we considered was backing up the thin Teflon board with a much thicker backup printed circuit board of conventional fiberglass material. Finally, Steve came up with an elegant solution. A mounting plate of thick aluminum with cutouts for the pipe caps that also serves to align them during soldering. The aluminum is thick enough to allow threaded holes for connector mounting as well. The aluminum mounting

Table 13.4 Component list for 10GHz transverter.

C1	1pF ATC chip
C2	100 pF chip
C3	0.1µf chip
C4	0.33µf
C5	5 to 10pF ATC chip
R(all)	51Ω chip
U1	ERA-1 MMIC
U2	ERA-2 MMIC
U3	MGA-86576 MMIC
Mixer diode pair	HSMS-8202

plate would take me hours to machine but modern CNC machinery apparently works a lot faster and cheaper than I do!

Prototype Description

Once the problems were worked out, we could start the actual design. The schematic diagram, shown in Fig 13.56, is an extension of the 5760MHz single board transverter. The LO section is split into two frequency multiplier sections, and both the transmit and receive sections have an extra amplifier stage and filter. For the layout, I made some sketches, calculated the important dimensions, and turned it over to Steve to have the CAD design done for the printed circuit. I know, this doesn't sound much like homebrewing, but I have already built a bunch of 10GHz transverters by hand and have three spares to lend out, so it is time to find another way to get more activity.

After several cycles of review and correction, we thought we had the data base ready for production of a trial lot of boards. While they were being made, I prepared some pipe caps so that I was ready to solder them on immediately, as shown in Fig 13.55. Then I added the brass stiffeners and started adding components, the completed prototype is shown in Fig 13.54

Tune up went quite well, since there are test points allowing each major section to be tested independently. I started with the LO, tuning each frequency multiplier section separately. When I connected them together, the output was a bit low, so I poked around and added a small tuning flag near the input to the second frequency multiplier stage, just after test point TP1. This brought the LO output at 10.224 MHz up to the level needed to properly drive a dual mixer, +7 to +10 dBm.

Next I connected the LO to the mixers and applied power to the transmit section. After tuning the pipe cap filters, the output power was about +5 dBm, with very little LO leakage. The three stages of pipe caps provide adequate filtering.

Finally, I applied power to the receive section and peaked the receive filters on a strong signal from another transverter in the backyard. Then I used a noise figure meter for final tuning, which yielded a noise figure around 6 dB.

Once everything was tuned up, I checked the frequency stability. This transverter, like the 5760MHz version and the no tune units for lower frequencies, uses the KK7B oscillator board [25] for the 568MHz LO source. At 10GHz, the crystal frequency is multiplied roughly 100 times, so any instability is greatly magnified. For 5760MHz, I found the stability to be marginal, with about a half hour warm up; at 10 GHz, it would be even worse. To improve the stability, we investigated small crystal heaters. I obtained some clip on heaters from the RSGB component service [26], but, for some reason, the Japanese manufacturer is unable to find them in the USA. As an alternative, Steve tried some simple thermistors. These devices have a resistance that increases dramatically with temperature; when a voltage is applied, they draw current, which produces heating, which increases the resistance, reducing the current. With the right combination of thermistor and voltage, a stable temperature is quickly reached. Steve found that soldering the thermistor to the crystal was a good way to control the crystal temperature and improve stability. With the thermistor in operation, the frequency stays within the receiver SSB passband after about one minute, and stability is good enough for SSB operation after less than 5 minutes. However, stray air currents can move the frequency a few kHz, so enclosing the heated crystal in a small block of foam insulation is probably a good idea for windy mountaintops.

The basic transverter performance is adequate for a modest rover station. It is comparable to the WBFM Gunnplexer system cited above, with a 30dB potential advantage in signal bandwidth. Since the design provides sufficient filtering to provide a clean signal, it can also be the basis for a more advanced station, with the addition of power amplifiers, preamplifier, and large antennas.

Another advantage is low power consumption. The whole transverter may be powered from a 12volt battery, and draws less total current than the surplus phase locked oscillator bricks I used in previous transverters. A power amplifier would certainly increase power consumption, but only while transmitting, and most operators aren't calling CQ on 10GHz.

The Final Version

Once the prototype was working, I went over it carefully and made a list of eleven changes. Steve added some improvements to aid producibility, plus provisions for some hotter MMICs that may become available, and sent it off to the CAD designer. The final layout is shown in Fig 13.57. Eventually, some corrected boards were produced, as well as the aluminum support plates and housings. These, plus the LO source board, an IF sequencer board, and voltage regulators for stable operation from 12volts, make up a complete transverter unit. Fig 13.58 shows a unit during assembly in an extruded aluminum box made by Rose Enclosure Systems. All of the pieces are available from Down East Microwave.

Construction

The aluminum mounting plate is designed to aid in assembly as well as providing rigidity. The PC board is bolted to the mounting plate, and the pipe cap filters are aligned during soldering by the square holes in the mounting plate. Then the components are soldered on the surface of the board (some form of optical aid is useful for those of us who wear bifocals). A temperature controlled soldering iron with a small tip is ideal for the chip components. However, the excellent grounding provided by the plated through holes requires more heat for ground connections, so a larger soldering iron may be needed for some connections. The blocking capacitors between sections should be placed initially so that the output of each section goes to a test point rather than to the next section.

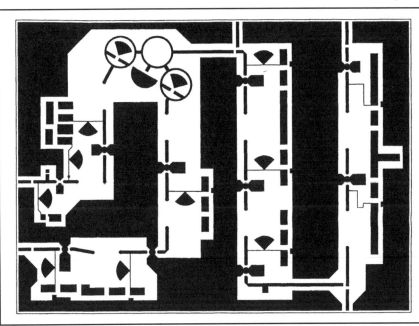

Fig 13.57 : The final layout for the 10GHz transverter.

This is how I prepare the pipe caps:

- In each pipe cap I drill and tap (use lots of oil) a hole in the center for a tuning screw: 4-40 thread for the 1/2 inch caps and 8-32 thread for the larger ones.

- Next, I flatten the open end by sanding on a flat surface. Then I apply resin paste flux lightly to the open end and place it in position on the PC board, the flux holds it in place.

- Finally, I put some thin wire solder around the base of each pipe cap. The caps are soldered one at a time, starting with the centre one. I hold each cap down with gentle pressure and heat it for few seconds with a propane torch until the solder melts and flows into the joints, then let the solder harden before releasing pressure. Don't be shy with the torch, melt the solder quickly and remove the heat. Keep the flame on the top of the pipe cap to avoid damage to the PC board.

- The input and output connections for the pipe cap filters are plated through holes in the microstrip lines which are sized for a tight fit around the 0.7 mm diameter brass nails used as probes, so that the probes stay aligned vertically inside the pipe cap. I measured the response of a single pipe cap with different probe lengths; a shorter probe narrows the filter bandwidth but increases the loss. The recommended probe length is 5/32 inch to 3/16 inch for the 1/2 inch pipe caps, where the length is measured to the head of the nail, so that the length inside the pipe cap is shorter by thickness of the circuit board, 0.015 inch. For the 3/4 inch pipe caps, Lengths of 5/16 inch for the first stage and 1/4 inch for the second stage provided best performance.

Finally, the SMA connectors for 10GHz transmit and receive are bolted through the end panel of the aluminum enclosure. Take care to line up the connector pins squarely on the microstrip lines, tighten the screws, and then solder the pins to the lines.

All other connections, both DC and RF, pass through clearance holes in the aluminum mounting plate. The RF connections should be made with small coax, properly grounded at each end.

Tune up

The aluminum mounting plate is also designed to aid in tune up, with extra tapped holes for mounting SMA connectors at the test points. The rigidity it provides also eliminates many of the intermittent problems encountered during tuning.

The KK7B LO source can provide a significant amount of output power, since it must drive a diode multiplier in some of the no tune transverters. The MMIC multipliers need less drive, typically +5dBm at 568MHz. Steve has been developing a new LO source that uses a higher frequency crystal and provides a higher frequency output, at 1132MHz for this transverter. At the higher frequency, even less drive is required; around 0dBm (one milliwatt) is enough.

I recommend testing and adjusting one section at a time, starting with the LO. Monitor the output power at test point TP1 and power only the first two stages. The 3/4 inch pipe cap filters require one inch long screws to reach 3408 MHz; screw them all the way in. Connect the LO source oscillator input and start tuning. Adjust the two screws out a half turn at a time until some output is detected, which should occur within a few turns. Then peak for maximum output, and check the power level and frequency, if possible. The tuning screws ended up with the flat part of the head about 1/4 inch from the top of the pipe cap. The output power should be +4 to +9dBm (or more) at 3408MHz. Finally, check to see that the output level from the LO source board is right for the frequency multiplier and not critical. Vary the output level from the LO source board by adjusting its supply voltage; the 3408MHz output should vary smoothly and not too rapidly as the voltage to the LO source board changes.

Now it is time to tune the rest of the LO. Move the blocking capacitor so that the 3408MHz output is driving the next stage, and monitor the output level at test point TP2. The tuning screws on the 1/2 inch pipe caps should be 1/2 inch long with a 4-40 thread. Preset the two screws

Fig 13.58 : Picture of final 10GHz transverter during assembly showing aluminium housing.

in the LO section so that the flat part of the head is 0.27 inches from the top of the pipe cap. Apply voltage to all the LO stages and start tuning in small increments until some output is detected. Then peak for maximum output, and check the power level and frequency, if possible. The output power should be +5 to +10dBm (or more) at 10224MHz. A LO power level in this range is fine for the dual mixer, but it will still work with somewhat less. Once again, vary the output level from the LO source board (by adjusting its supply voltage) to be sure that power levels are right for the frequency multipliers. If they are, the 10224MHz output should vary smoothly and not too rapidly as the voltage to the LO source board changes.

The next section is the transmitter, which is tuned up in two steps. First, move the blocking capacitor so that the 10224MHz output is feeding the dual mixer, and monitor the output power level at the transmit output connector. Preset the three tuning screws to match the final setting of the two in the LO section. Apply power to all LO and transmit stages, and start tuning in small increments until some output is detected. At this point, we are peaking the transmit filters on the LO leakage through the mixer, rather than filtering out the leakage. Peak the filters for maximum output. I typically see around one milliwatt.

Now it is time to add some 144MHz input to the transmit mixer, from a signal generator or an IF rig. The input level should be about 0dBm. We know that the filters are set at 10224MHz, so the tuning screws must move out to raise the frequency. The required adjustment is less than one quarter turn of each screw. Move them all out together (in sequence) by the same amount. The output will drop as we tune away from 10224MHz and then increase as we reach 10368MHz. Peak the output turning, favouring the high side to maximize LO rejection. Key the IF signal on and off to be sure that the output is due to mixing and not LO feed through. With proper tuning, the LO leakage should be more than 40dB down. Finally, adjust the IF level to find the maximum linear output; if possible, vary the LO level as well to find the best operating point. After all, we dont want a lot of intermod on 10GHz when it gets crowded!

To tune up the receiver, we need a reasonably strong signal at 10.368GHz. It could be another 10GHz rig, a signal generator or weak signal source, or a harmonic from a 1296MHz rig, but it must be far enough away so that the signal is coming in through the antenna. Connect a receiver with an S meter to the IF receive output, and some sort of small antenna to the receiver 10GHz input connector. Preset the filter tuning screws to match the transmit filter setting as a starting point, then apply power to the LO and receive stages. Tune in the signal and peak the tuning screws for maximum S meter reading. Do the usual checks to make sure the signal is coming in the antenna and is not a birdie. If you have access to a noise figure meter, the tuning screws and voltages on the front end may be tweaked for best noise figure.

Performance Summary

After completing the adjustment procedure, the transmit power is about +10dBm, or 10 milliwatts, and the receiver noise figure is around 6dB. As a basic system with an attached horn antenna, these numbers are comparable to a Gunnplexer system in a similar sized package, except for the 30dB bandwidth advantage for the narrow band system.

The output spectrum is reasonably clean, with the only outputs being leakage from the LO multiplier chain: -45dBc at 10224MHz, -30dB at 6816MHz, and more than 50dB down at 3408MHz. Since the transmit section has 3 filters, these are probably due to leakage inside the box, a result of the single board construction.

Other than leakage, the LO chain is quite clean and delivers adequate power to the dual mixer.

The entire transverter board operates from a voltage of +8volts. A 7808 three terminal regulator IC, which guarantees stable operation with a battery voltage as low as 11volts, easily supplies this power. Don't forget the recommended capacitors (0.1µF) on the input and output of the IC, and adequate heat sinking; bolting it to the aluminum enclosure is probably enough. The complete LO current drain is about 330mA, plus another 70mA for the receive stages or 100mA for the transmit stages. Thus, a complete transverter with coax relay and IF switch should not draw much more than ½ amp, less than most VHF transceivers.

10GHz Systems

A basic transverter with a simple horn antenna is a good way to start out and is sufficient for any line of sight path. For higher performance, the single board transverter can be the heart of a range of 10GHz systems, from simple backpack rover system to a high performance EME system. Obviously, the different requirements of these systems affect the choice of other components needed to get on the air. Weight and power consumption are critical in backpack systems, while performance is paramount for EME. Other systems, for home systems or vehicles, have intermediate requirements, but cost is a common constraint.

The pieces common to all systems include a source board for the LO, an IF interface, voltage regulators, some sort of enclosure, and an antenna. An IF transceiver is also needed to drive the transverter, typically at 144MHz. The new MFJ-9402 SSB transceiver for 2 meters looks like a worthy successor to the venerable IC-202. I've been using one to tune up my latest transverter.

For the LO source, either the KK7B board with 568MHz output or the new DEMI LO board with 1132MHz output work very well. Another alternative would be a phase locked source. The older bricks could drive the dual mixer directly from test point TP2, while some newer synthesized oscillators have output at 2556MHz which could drive the second frequency multiplier from test point TP1. For stability, some sort of crystal heater or oven is recommended; for an EME station, a good oscillator that runs continuously is a possibility.

The IF interface handles the transmit receive switchover and switches power to the appropriate part of the transverter. It can be very simple, but slightly more complex designs like my [27] Fool resistant IF sequencer can help prevent untimely catastrophes. Dont try to get by with multiple switches that you throw in sequence, this is a recipe for disaster!

The transverter and LO boards are designed to operate from a regulated +8volts, easily generated from a 12 volt battery with a simple three terminal regulator like a 7808, with enough margin to keep going with when the battery gets low. Dont forget capacitors on both the input and output of the three terminal regulator, to prevent oscillations and to keep out stray RF.

The enclosure has several crucial functions: it keeps internal frequencies inside, keeps stray RF out, keeps rain out of the electronics, and keeps out stray fingers. For portable work, a good enclosure is rugged enough to protect the contents when the wind blows everything over onto rocks. Metals are the best materials to meet all these requirements. However, the price of a new enclosure can be breathtaking. My secret is to look for useless old electronics at flea markets and surplus outlets that happen to be in a convenient sized enclosure that I can reuse.

I haven't seen any solid state switches that provide good 10GHz performance at reasonable cost, so a TR relay is needed for most systems. The alternative is to use separate antennas. Small coax relays with SMA connectors are fairly common surplus items, and most of them will work at 10GHz. Some of them have worn out contacts, but a check for DC continuity while switching a few times will usually sort out the bad ones

The antenna is the real key to system performance, since an antenna of modest size can provide very high gain. The small horn antennas on Gunnplexer systems are ideal for beginners, they are rugged and easy to point and still provide 17 dB of gain. We later added a lens in front of the horn for additional gain. A larger horn can also provide higher gain, and horns almost always work as designed. Both horns and lenses are easy to design and build. My HDL_ANT computer program [28] does the calculations and even prints a template for horn construction.

Parabolic dish antennas can provide even higher gain. My favorite is the now ubiquitous DSS satellite dish. These little 18inch offset fed dishes will provide more than 30dB of gain with a simple horn feed [29] and cost only a few dollars [30]. Another advantage is that the transverter can be mounted just below the feedhorn without affecting performance, so that feeder loss can be minimized. Coax cable loss adds up quickly at 10GHz, and there aren't any good ones

For better receive performance, a good preamplifier will help, if it is near the feed, otherwise feeder loss will limit performance. Many of us use a converted LNB [31] from a DSS system, the older ones (1.4 dB NF) are being replaced by lower noise ones, and brand new units are getting pretty reasonable. For even lower noise figure, you can homebrew a HEMT preamplifier. My favorite is a KH6CP design [32], which seems to work with almost any good transistor and provide a noise figure of 1 dB or less. The PC board is available from Down East Microwave.

A really good receiver should be balanced by increased transmit power, so that the other station can hear you as well. There are some surplus solid state amplifiers available that can be retuned to 10GHz if you have patience, but other alternatives are needed. We are working on a small companion power amplifier for this transverter.

What you *don't* need is a TWT amplifier. I frequently see messages on the internet looking for a 10watt TWT, but there arent many available. I've tried to use a TWT amplifier for rover operation, but it is big, HEAVY, and needs a generator to provide 600watts of AC for 10watts of RF. Then I get to the mountain top and find that the fog is so thick that I dare not turn on the high voltage, so I end up operating with a much smaller solid state amplifier. My best DX, 549km, was done with less than 100milliwatts and a small dish. Save the TWT for EME work where you need the power.

Whatever you decide on for a system, don't spend a year engineering it the first time, you will always want to make changes. Put it together, take it out and make some contacts, and you'll figure out how you really want to put things together, and have fun doing it.

Finally, mountains arent necessary. Height is not what matters, just getting the antenna higher than nearby obstructions. What counts is giving the signal a good start without any obstructions for a few miles. If you can find a good path over water, standing on the beach makes you higher than anything in the path.

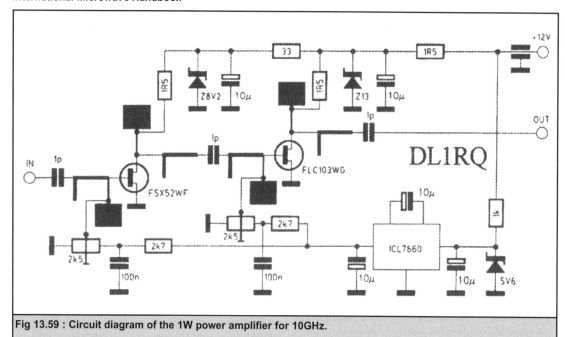

Fig 13.59 : Circuit diagram of the 1W power amplifier for 10GHz.

Conclusion

Our goal has been to make a 10GHz transverter reproducible and affordable enough that everyone can get on the band. There are no design breakthroughs, just a lot of learning from previous efforts. It offers good performance in a small package and can be the heart of many different 10GHz systems.

Amplifier [33]

The following article introduces two power amplifiers for 10 GHz. Both modules were developed by the author with a view to ease of copying and reliable long term operation. Numerous examples of the 1W power amplifier have been running for some years (some of them installed on masts) and none has yet given rise to any problems.

The last few years have seen the emergence onto the market of output GaAs FET's in the 5W range, which certainly do drain the "research budget" of a radio amateur, right down to the last penny. But "better 5 Watts in the GHz rucksack than a designer suit in the wardrobe". Thus the tried and tested existing circuit was expanded using a TIM 0910-4 from Toshiba, which was internally tuned to 50Ω. Five examples of the 5W version have been assembled so far. Long term operation in the laboratory, and use in several competitions, have indicated that this power amplifier can also offer the operational reliability of the 1W version. In one example, at least, a faulty SMA relay triggered a ten minute crash test with full power drive but open output. The transistor survived the experience, though the author was forced to part with the relay.

In spite of all efforts at reproducibility, the cumulative total of the small tolerances in the component values and the assembly can eventually lead to significant individual deviations (-3 dB is normal) in the amplification and output power. But there is some comfort in the fact that, with patience, experience and good measurement facilities, a power amplifier can be trimmed to the rated values with a fine calibration using the "small disc method".

The following descriptions are intended to provide a stimulus to the interested radio amateur.

1W power amplifier for 10GHz

The circuit diagram (Fig 13.59) is comparable in structure to the author's two stage 5.7GHz power amplifier, published in [34] and [35], including the additional voltage inverter for the negative gate voltage. Good experiences with the reliable FSX52WF transistors (drive) and FLC103WG transistors (high level stage) from Fitjitsu led to trials at 10 GHz, which were immediately successful, although the FLC103WG is only specified for use up to 8GHz by the manufacturer. 0805 model 1pF SMD capacitors were used as high frequency coupling elements (2.0 mm. x 1.25 mm.). Research carried out only recently as part of a specialist project [36] showed that the SMD capacitors used by the author, with a series inductance of $L_{series} = 0.66 \pm 0.01$nH, differ from those components available by normal mail order, with $L_{series} = 0.72 \pm 0.02$nH. Unfortunately, the author was not in a position to identify the individual manufacturer. In any case, a rough calculation quickly shows that the series resonance frequency of these SMD capacitors lies around 6.0GHz. So, at 10GHz the coupling "capacitors" should be considered more as DC disconnecting components with inductive behaviour. Naturally, this inductance clearly has an affect on the matching by the striplines (at the cost of narrowing the band).

The power supply was deliberately made simple. Stabilisation was provided through 1.3W zener diodes, which simultaneously provided protection against over voltage and false polarity. The 1.5Ω/0.25W axial carbon film resistors act as both isolation resistances and safety resistances. A circuit for protection if the negative power supply failed was dispensed with following an involuntary 24 hour test without any negative supply voltage, which did not damage the semiconductors.

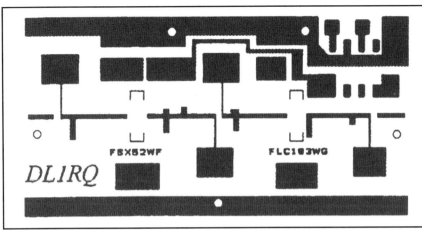

Fig 13.60 : PCB layout of the 1W power amplifier for 10GHz.

Assembly

The assembly and board layout of an amplifier for microwaves are determined by two essential requirements:

- The high frequency transition from the earth surface of the board to the source flange of the transistor must be as close to ideal and as smooth as possible

- The extraction of the transistor's lost heat must be as close to ideal as possible

A board with a sandwich construction has proved to be a way of being able to fulfil both requirements simultaneously. The layout (Fig 13.60) is etched onto an RT/Ditroid D-5870 board measuring 68.5mm. x 34mm. x 0.25mm. After pre tinning of the earth surface, the board is soldered onto a 1.0mm. thick copper plate under high pressure. Next, two oval grooves 0.75 mm. deep are milled, using a 2.5 mm. diameter bore groove milling cutter, for the source flanges of the transistors. When the five 2.1mm. diameter holes have been made for the board to be screwed into the housing and for the contacts to be connected up, and when the tracks have been tin plated (or silver plated), the board is assembled as far as the two 10μF electrical capacitors (on the drain side) and the transistors (Fig 13.61). In order to guarantee good heat transfer between the copper plate and the milled aluminium housing (Fig 13.62), some heat conducting paste is smeared over the aluminium base in the vicinity of the transistors. To ensure good connection between the high frequency section and the earth in the input and output areas, silver conducting lacquer can be smeared there (very sparingly, of course). The partly assembled board is now fitted into the suitably prepared aluminium housing and screwed down by five M2 brass screws. When the connections to the feed through capacitor have been completed, it is possible to check the DC function. For this purpose, the two trimmers are pre-set to a gate voltage of about -1.5V.

The trickiest stage in the procedure is the soldering of the GaAsFET into the milled grooves. To this end, the aluminium housing is first heated, with the board inside, to precisely 150°C. Each milled groove is then pre tinned, using low temperature solder with a melting temperature of 140C. Excess tin is then removed using a de soldering pump. The transistors are then placed in the grooves, all the relevant safety measures known must be taken. Normally, the tin binds very well with the gold plated flanged base, something that can easily be tested by a visual check of the flanged bores. Naturally, this soldering process should be carried out as rapidly as possible. The housing is then immediately placed on a cold copper block or a large cooling body, so that the temperature quickly falls.

Drain and gate connections are soldered onto the stripling, all the relevant safety measures must be taken. The two 10μF capacitors on the drain side are fitted and the SMA flanged bushes are screwed on. The power amplifier is ready for tuning (Fig 13.63).

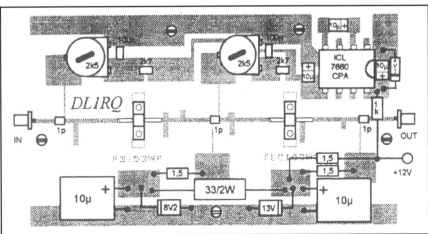

Fig 13.61 : Component layout of the 1W power amplifier for 10GHz.

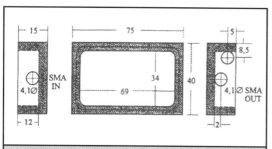

Fig 13.62 : Dimensions of the milled aluminium housing for the 1W power amplifier.

Tuning

First, the no signal currents are set as follows:

* For the FSX52WF at approximately 70mA, this corresponds to a voltage drop of 105 mV, across a 1.5Ω protective resistor. For the FLC103WG, at approximately 240mA, it corresponds to a voltage drop of 360mV, across a 1.5Ω protective resistor.

* With 30mW drive at the desired frequency, an output of approximately 400mW (in the worst case) and of 1 Watt (in the ideal case) should be measurable. As already

mentioned initially, the "small disc method" is normally of assistance. Small discs, measuring about 2-4 mm^2, a few toothpicks to press down and push, a lot of patience and, above all, the greatest care in watching out for short circuits will (hopefully) soon lead you to achieve full output power. After tuning, an aluminium cover plate 1mm thick can be fitted. In the overwhelming majority of the power amplifiers measured, almost no influence from the cover could be detected. Of course, there were just a few cases in which minimal self excitation was detected when the cover was put on. This is caused by astonishingly stable housing resonance, slightly above the calibration frequency, with a few milliwatts of power at the output. Even this undesirable oscillation disappeared with a low powered drive. A strip of absorbent material about 5mm wide and about 10mm long, glued to the inside of the cover in the area above the FSX52WF, provided a reliable remedy here.

Measurements

A comparison of the output data from the semiconductors (Fig 13.64 and Fig 13.65) with the readings from a typical power amplifier (Fig 13.66 - Fig 13.68) makes clear how successful the project is in practice.

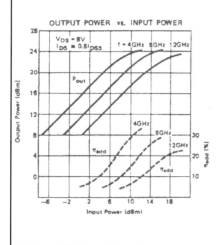

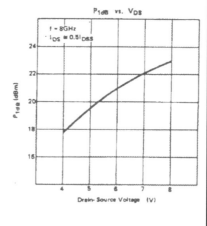

Fig 13.63 : Example of a 1W power amplifier for 10GHz.

Fig 13.64 : Power data for the FSX52WF taken from the Fujitsu data sheet [38].

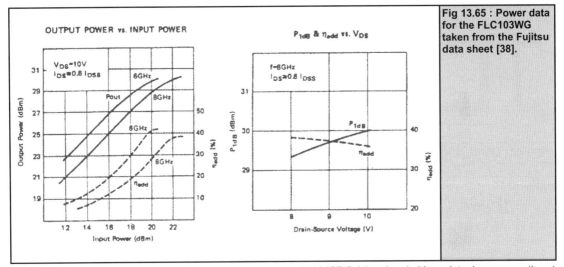

Fig 13.65 : Power data for the FLC103WG taken from the Fujitsu data sheet [38].

5W power amplifier for 10GHz

The circuit diagram for the 5 Watt power amplifier (Fig 13.69) differs from the circuit diagram for the 1 Watt power amplifier due to the inclusion of the additional power device, the Toshiba TIM0910-4 This is an internally matched power GaAsFET and needs a more expensive power supply circuit, provided by DB6NT [37]. The

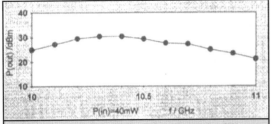

Fig 13.66 : Typical power bandwidth for the 1W 10GHz power amplifier.

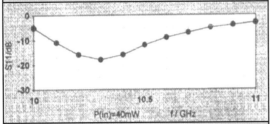

Fig 13.67 : Typical input matching for the 1W 10GHz power amplifier.

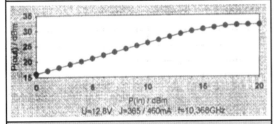

Fig 13.68 : Typical linearity for the 1W 10GHz power amplifier.

LT1084CP 5-A low threshold regulator is permanently set to a drain voltage of 9.5V, and is switched off through the BC848 transistor if the -5V supply fails. The relatively high resistance gate circuitry of the TIM0910-4 corresponds to the recommendations from the manufacturer. This explains reduction in the power consumption from approximately 2.5A without drive to approximately 2.1A with full drive. The first appearance of gate current can be seen in the linearity graph (Fig 13.76) as a slight bend in the characteristic line at an output of approximately 3dBm, 2W. Wire resistors of 10Ω or 0.1Ω are used as drain resistances. They have adequate inductance and the voltage drop across them can be used to measure the drain current. The ferrite beads threaded on them could be described as "hope beads", as they are use in the hope that they contribute to the suppression of undesirable housing resonances. A Siemens directional coupler with a BAT 15-098 low barrier Schottky diode is provided for the monitoring of the output.

Assembly

There are some differences between the assembly of the 5W power amplifier and that of the 1W power amplifier. The layout (Fig 13.70) is etched onto an RT/Duroid D-5870 board measuring 101.6mm. x 50.8mm. x 0.25mm. (4 x 2 x 0.01 in.). These dimensions arc used because of the availability of a ready-made milled nickel plated aluminium housing from Telemeter Electronic, type ZG42-N [39]. Following the pre-tinning of the earth surface, the board is soldered to a 1.5mm. thick copper plate under high pressure. Two 0.75mm. deep oval grooves are then cut using a 2.5 mm. diameter bore groove milling cutter for the source flanges of the two drive transistors. Countersinking is carried out for the LTI084CP voltage controller, with a precisely fitting opening for the TIM0910-4 transistor, in accordance with the dimension specified in the layout. The two components are then screwed directly onto the 8mm thick housing base. This provides exactly the height required by the stripline for the TIM0910-4. After drilling the seven 2.1mm diameter securing and through plating holes and tin plating (or silver plating) of the tracks, the board is assembled, as far as the 10µF electrical capacitor (on the power supply line), the LTI0S4CP and the three GaAs transistors (Fig 13.71). In order to guarantee good heat transfer between the copper plate and the housing, some heat conducting paste is smeared onto the aluminium base in the area of the two

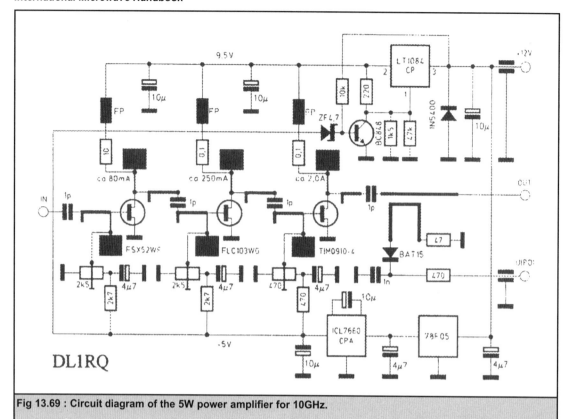

Fig 13.69 : Circuit diagram of the 5W power amplifier for 10GHz.

drive transistors. Silver conductive lacquer can be smeared on to produce a good transition from the high frequency section to earth in the vicinity of the input, the output, the gate and the drain of the TIM0910-4. Of course, silver conductive lacquer must be applied extremely sparingly. Even a small leak while the board is being screwed down can lead to considerable problems. (The author's prototype was tested without conductive lacquer. No difference could be detected from later versions with conductive lacquer).

The partly assembled board is now put into the housing, which has all the necessary holes and is screwed down using seven M2 brass screws. When the LT1084CP has been added (don't forget the little insulating discs!) and

the connection to the feed through capacitor has been completed, the DC function can be checked. All trimmers are pre-set to -1.5V for the gate voltage. The two drive transistors are soldered in just as described for the 1W high level stage. When the housing has cooled, the remaining components and the SMA bushes are mounted (with suitable Teflon collars). The final step is the incorporation of the TIM0910-4, into the opening, using two M2 screws. Since a certain amount of lever action can not be prevented when the screws are tightened, the gate and drain lugs should not be soldered to the stripline until afterwards.

It is vital to bear in mind that until sometime early in 1993, Toshiba identified the drain lug by a chamfer. Since then,

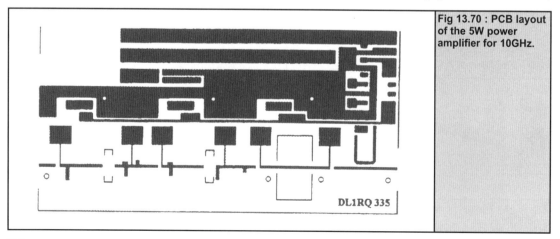

Fig 13.70 : PCB layout of the 5W power amplifier for 10GHz.

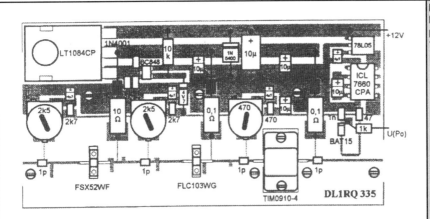

Fig 13.71 : Component layout of the 5W power amplifier for 10GHz.

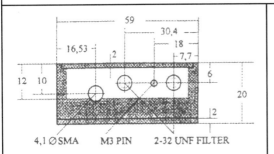

Fig 13.72 : Dimensions of the Telemeter ZG4-2-N housing. Dimensions in mm.

the gate lug has been chamfered. To verify the connections, a suitable Ohmmeter can be used for measurements. Battery operated equipment with a low measuring voltage (< 0.5V) is suitable. The drain to source resistance is usually markedly less than an Ohm. The gate to source resistance, by contrast, will be very high resistance and this is independent of the polarity of the measurement voltage. Naturally, measurements must be carried out very carefully as static charges may be present. The author had to decide whether to give preference to the best heat contact (by an abundant use of heat conducting paste) or to the best high frequency contact (by doing without the paste), eventually just a smear of paste was used.

Tuning

First, the no signal currents are set:

- FSX52WF to approximately 80mA this corresponds to a voltage drop of 800mV with a drain resistance of 10Ω
- FLC103WG to approximately 250mA this corresponds to a voltage drop of 25mV with a drain resistance of 0.1Ω
- TIM0910-4 to approximately 2.0A this corresponds to a voltage drop of 200mV with a drain resistance of 0.1Ω

Should the readings not match the above values, fine tuning should be carried out using the "small disc method".

Cooling

Estimating the cooling of the power amplifier we can calculate as follows:

Approximate power loss for the TIM0910-4:

$$P_v = 9.5V \times 2.1A = 20W$$

The RF output is not taken into account

Temperature differential between channel and flange:

$$\Delta\theta C°C = 3.5 \text{ K/W} \times 20W \approx 70°C$$

i.e. a flange temperature of, for example, 60°C. gives a

Fig 13.73 : Example of a 5W power amplifier for 10GHz.

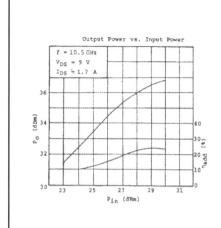

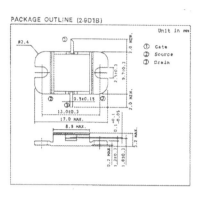

Fig 13.74 : Power data and case details for the Toshiba TIM0910-4 [40].

channel temperature of 130°C. The maximum permissible channel temperature is 175°C. If you want to "keep on the safe side", the power amplifier housing should be cooled in such a way that, even when the ambient temperature is at its most unfavourable a flange temperature of 60°C will not be greatly exceeded at TIM0910-4.

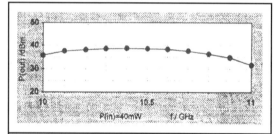

Fig 13.75 : Typical power bandwidth for the 5W 10GHz power amplifier.

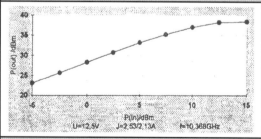

Fig 13.76 : Typical linearity for the 5W 10GHz power amplifier.

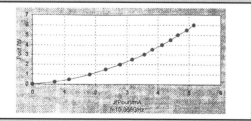

Fig 13.77 : Relative power measurement using a BAT15.

I would like to thank all those who have contributed to this project. Special thanks go to my fellow radio enthusiast Manfred Deutsch, DC4UI, who was always ready to help and advise me in selecting and testing components.

Antennas and feeds

The theory and general design considerations have already been discussed. What is given here, therefore, is a number of different practical designs for 10GHz, together with some ideas to aid their construction. Where appropriate, a little of the theory is restated to clarify the design objectives.

Horn antennas

Large pyramidal horns can be an attractive form of antenna for use at 10GHz and above. They are fundamentally broadband devices showing virtually perfect match over a wide range of frequencies, certainly over the amateur band. They are simple to design, tolerant of dimensional inaccuracies during construction and need no adjustment. Horns are particularly suitable for use with transmitters and receivers employing free running oscillators, the frequency of which can be very dependent on the match of their load (antenna). Another advantage is that their gain can be predicted within a dB or so (by simple measurement of the size of the aperture and length) which makes them useful for both the initial checking of the performance of systems and as references against which other antennas can be judged. Their main disadvantage is that they are bulky compared with other antennas having the same gain.

Large (long) horns, such as that illustrated in Fig 13.78, result in an emerging wave which is nearly planar and the gain of the horn is close to the theoretical value of $2\pi AB/\lambda^2$, where A and B are the dimensions of the aperture. For horns which are shorter than optimum for a given aperture, the field near the edge lags in relation to the field along the centre line of the horn and causes a loss in gain.

For very short horns, this leads to the production of large minor lobes in the radiation pattern. Such short horns can, however, be used quite effectively as feeds for a dish.

The dimensions for an optimum horn for 10GHz can be calculated from the information given in Fig 13.79 and, for

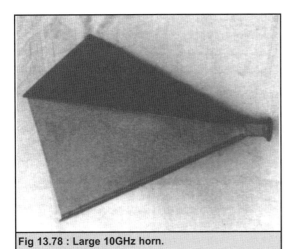

Fig 13.78 : Large 10GHz horn.

a 20dB horn, are typically:

A = 5.19in (132mm)

B = 4.25in (108mm)

L = 7.67in (195mm)

There is, inevitably, a trade off between gain and physical size of the horn. At 10GHz this is in the region of 20dB or perhaps slightly higher. Beyond this point it is better to use a small dish. For instance a 27dB horn at 10GHz

would have an aperture of 11.8in (300mm) by 8.3in (210mm) and a length of 40.1in (1,019mm) compared to a focal plane dish which would be 12 in (305mm) in diameter and have a length of 3in (76mm) for the same gain.

Construction of 10GHz horns

Horns are usually fabricated from solid sheet metal such as brass, copper or tinplate. There is no reason why they should not be made from perforated or expanded metal mesh, provided that the size and spacing of the holes is kept below about $\lambda/10$. Construction is simplified if the thickness of the sheet metal is close to the wall thickness of WG16, i.e. 0.05in or approximately 1.3mm. This simplifies construction of the transition from the waveguide into the horn. The geometry of the horn is not quite as simple as appears at first sight since it involves a taper from an aspect ratio of about 1:0.8 at the aperture to approximately 2:1 at the waveguide transition. For a superficially rectangular object, a horn contains few right angles, as shown in Fig 13.80 that is an approximately quarter scale template for a nominal 20dB horn at 10.4GHz. If the constructor opts to use the one piece cut and fold method suggested by this figure, then it is strongly recommended that a full sized template be drafted on stiff card that can be lightly scored to facilitate bending to final form. This will give the opportunity to correct errors in measurement before transfer onto sheet metal and to prove to the constructor that, on folding, a pyramidal horn is formed!

The sheet is best sawn (or guillotined) rather than cut with

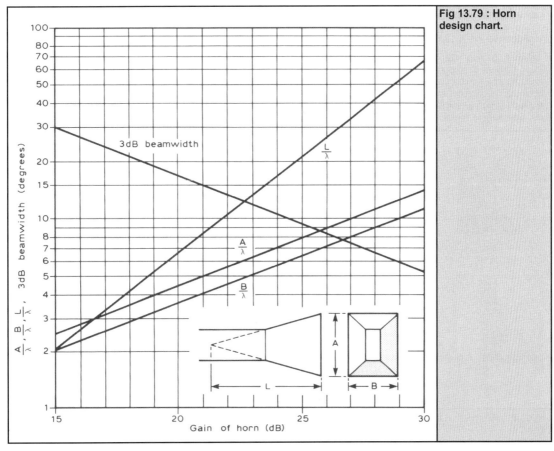

Fig 13.79 : Horn design chart.

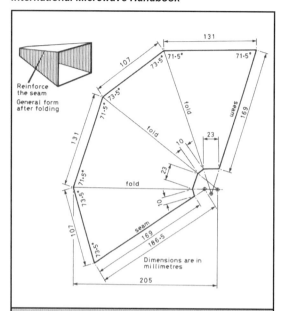

Fig 13.80 : Dimensioned template for single piece construction of a 20dB horn.

tin snips, so that the metal remains flat and undistorted. If the constructor has difficulty in folding sheet metal, then the horn can be made in two or more pieces, although this will introduce more soldered seams which may need jigging during assembly and also strengthening by means of externally soldered angle pieces running along the length of each seam. Alternative methods of construction are suggested in Fig 13.81.

It is worth paying attention to the transition point that should present a smooth, stepless profile. The junction should also be mechanically strong, since this is the point where the mechanical stresses are greatest. For all but the smallest horns, some form of strengthening is necessary. One simple method of mounting is to take a short

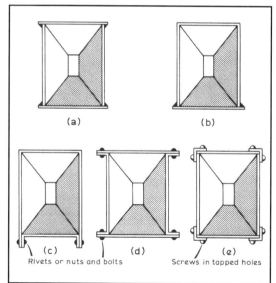

Fig 13.81 : Alternative construction methods for horns.

length of Old English (OE) waveguide, which has internal dimensions matching the external dimensions of WG16, and slitting each corner for about half the length of the piece. The sides can then be bent (flared) out to suit the angles of the horn and soldered in place after carefully positioning the OE guide over the WG16 and inserting the horn in the flares. One single soldering operation will then fix both in place. After soldering, any excess of solder appearing inside the waveguide or throat of the horn should be carefully removed by filing or scraping. The whole assembly can be given a protective coat of paint.

An alternative method would be to omit the WG16 section and to mount the horn directly into a modified WG16 flange. In this case the thickness of the horn material should be a close match with that of WG16 wall thickness and the flange modified by filing a taper of suitable profile into the flange.

Whichever method of fabrication and assembly is used, good metallic contact at the corners is essential. Soldered joints are very satisfactory provided that the amount of solder in the horn is minimised. If sections of the horn are bolted or riveted together, then it is essential that many, close spaced bolts or rivets are used to ensure such contact. Spacing between adjacent fixing points should be less than a wavelength, i.e. less than 30mm.

10GHz horns from oil cans and other materials

Empty 5 litre oil cans are an excellent source of tinplate for the purposes of building horn antennas, if you can still find a suitable can! Select cans with no rust visible either inside or outside. Cut off the top and bottom of the can with tin snips or a hacksaw, then cut along the length of the four corners and flatten out the four pieces of tinplate thus obtained.

Although a horn of any gain could be built, two values suggest themselves from the sizes of tinplate available. A 130 by 160mm horn of 22dBi gain requiring two cans (it uses four of the larger sheets), or an 80 by 80mm horn of 16dBi gain made by cutting down two of the larger sheets (or using four smaller sheets) which need only be about half the length.

Remove any sharp edges or slivers of metal, then wash and degrease the sheets. It is not necessary to remove most of the paint, only where you want to solder. This can be done with wire wool or paint stripper. Mark out the centre lines along each piece, then the dimensions of the open end of the horn, and the appropriate outside dimensions of the waveguide, separated by the length of the horn. Join up the ends of these lines and extend them about 20mm past the waveguide end line. See Fig 13.82(a).

Cut out one pair of opposite sides *exactly* along the marked lines; on the other pair, leave about 3mm extra along the sides, see Fig 13.82(b). Then fold the tabs (a) back about 10° to 15°. Hold two adjacent sides in position and make several spot solder joins along the seam. Repeat this process until all four sides are joined up. Make sure that the narrow end is a snug fit over the waveguide by holding the waveguide in place while doing the spot soldering. Then extend the solder joints all along the four seams, both inside and outside, for rigidity. All this soldering can be done with a medium powered electric soldering iron.

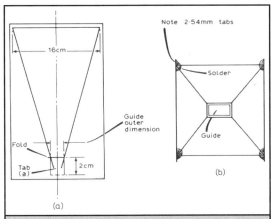

Fig 13.82 : Horns made from oil can tinplate.

Then insert the waveguide into the tabs, making sure that they grip it firmly. A gas flame will probably be needed to help solder the waveguide in place. A suggestion is to prop the horn horizontally on a gas stove, with the guide a few centimetres above the flame. The flame should only be used to keep the metal just below the melting point of the solder, and the soldering iron used to melt the solder locally where required. This avoids the whole structure falling apart when all the solder melts at once! Also solder the end of the waveguide to the inside of the horn, making a nice smooth joint at the throat of the horn. This process is made much easier if all the mating surfaces around the guide and tabs are tinned before assembly; if all the surfaces are very clean and a separate flux is available, this precaution may not be necessary. While the

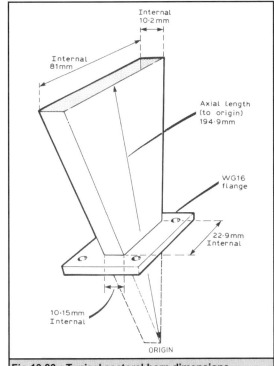

Fig 13.83 : Typical sectoral horn dimensions.

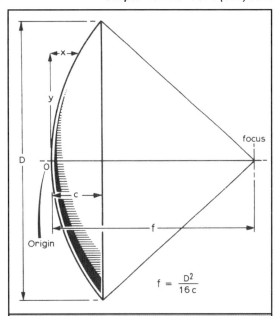

$$f = \frac{D^2}{16c}$$

Fig 13.84 : Basic parabolic dish geometry.

waveguide is hot, the flange can also be soldered into place.

When the horn has cooled down, scrape off any remaining flux and give it several coats of paint, both inside and outside, to prevent it rusting. A red brown epoxy based paint made for car repairs is recommended for this and other outdoor items, as it seems to adhere to surfaces very tenaciously.

Having said earlier that excess solder should be removed from the inside of the horn, it is apparent that this precaution has not been (and could not be) taken with this particular design. The outcome is simply some loss of gain that proved acceptable for the purposes to which the horn was to be put. If the horn is to be constructed as a reference antenna, then such precautions *must* be taken.

Double clad fibreglass pcb material has also been used successfully to make light weight horns of the type described here. These are eminently suitable for occasional light use and can be easily and very cheaply made by adapting the methods outlined above. Soldering with a very small soldering iron is possible, although some ingenuity is needed in making the horn to waveguide transition!

Sectoral horns

The most useful type of sectoral horn is that with an H plane flare. With the broad faces of the guide vertical a horizontally polarised field is produced which has an azimuth pattern of nearly 180° but a vertical pattern which is compressed into a few degrees (depending on the gain). This makes it useful for a beacon where semi omni directional coverage is needed. Construction is similar to that described above and the dimensions of a nominal 10dBi horn are given in Fig 13.83.

Dish antennas and feeds at 10GHz

The geometry and general points concerning paraboloidal reflectors has been fully discussed elsewhere. Their main

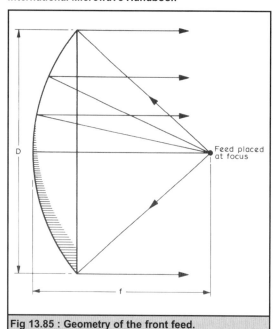

Fig 13.85 : Geometry of the front feed.

advantage is that they offer high gain for their size, can be designed for any gain and can operate at any frequency. Their disadvantages are that they are not easy to make accurately, they can be difficult to mount and feed and may have a high windage. Nevertheless, they are widely used by amateurs and some practical dishes and feeds are discussed next. The basic geometry of a dish antenna is given in Fig 13.84, together with the relationship between diameter, depth and focus. From a practical point of view the important relationship is the ratio of f to D (focus to diameter), for this will determine the optimum feed design.

A dish can be illuminated by :-

- A front feed, where the radiating element is fixed in front of the reflector (dish) by means of a tripod arrangement

- A rear feed, where the radiating element is mounted at the end of a length of feeder (waveguide) which goes through the centre of the dish and acts as the support for the radiator. The rear feed is usually more convenient and can take several forms, according to the f/D ratio of the dish.

Direct feed dishes

The geometry of the front feed is illustrated in Fig 13.85. The phase centre of the feed is placed at the focus of the dish and power radiated by the feed as a spherical wave is converted by reflection at the paraboloid into a plane wave. Some years ago it was found that an ordinary galvanised dustbin lid of the smoothly rounded type was a sufficiently accurate paraboloid to be usable as a dish reflector at 10GHz. Most of the lids examined have a relatively long focus, with an f/D ratio in the range 0.7 to 0.9. A lid with the smallest ratio will make a more compact antenna. Lids which have dents or wrinkles greater in depth than about λ/10 at the design frequency should be avoided.

It is well worth spending some time in measuring the lid to make sure that it is a reasonable parabola. The mathematical methods for this are described elsewhere. Suffice it to say that the 19in (483mm) lid chosen for the purpose only deviated from a true parabola by λ/17 at the centre and by λ/50 (at 10GHz) elsewhere.

The front feed most suitable is a small horn and the means of supporting the feed is illustrated in Figs 13.86 and 13.87. Fig 13.88 is a photograph of the completed antenna. Fuller details of the construction of such an antenna will be found in [41].

Fig 13.89 gives the dimensions of a small feed horn as a function of the f/D ratio of the dish. The horn used should

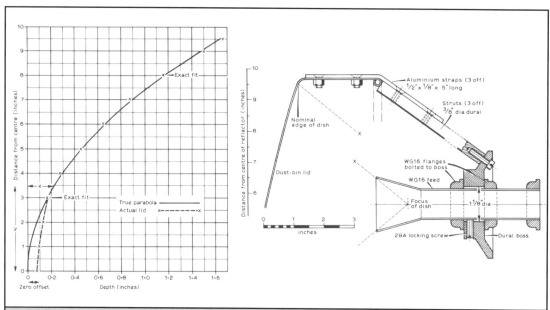

Fig 13.86 : Dustbin lid antenna. (a) The profile of the lid, as measured, compared to a true parabola. (b) The feed method (G3RPE).

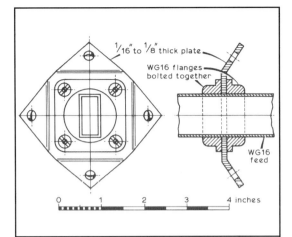

Fig 13.87 : Dust bin lid feed mount (G3RPE).

be designed using this data once the f/D ratio of the lid is determined. As an example, if a paraboloid with an f/D ratio around 0.53 is available, the dimensions of a suitable feed horn are shown in fig 13.90. A dish of this type is shown, complete with flexible waveguide feed, in Fig 13.91.

A simple modified Cutler feed, which has come to be known as the penny feed because an old (pre decimalisation) penny was just about the right size for the end disc, is shown in fig 13.92. This was originally described by G4ALN as being suitable for dishes with an f/D ratio in the range 0.25 to 0.3. Whilst it will work reasonably well with a focal plane dish (f/D 0.25), there is a significant degree of under illumination and consequential loss of gain. However, the constructor may be prepared to trade this loss of efficiency for the simplicity of construction and relative lack of criticality of the feed dimensions.

Fig 13.88 : Photograph of a complete dust bin lid antenna (G3RPE).

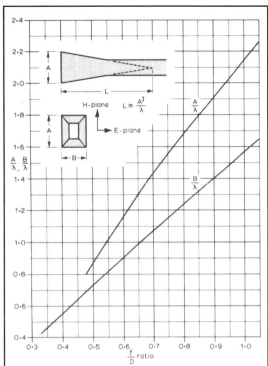

Fig 13.89 : Small feed horn design chart.

The feed is made by cutting two appropriately dimensioned grooves in the end of a length of WG16 and soldering on a circular end disc (the penny). The length of the slot formed and the diameter of the disc are thought to be not critical within a few percent, and the width of the slots even less so. Signals with standard horizontal polarisation are produced with the broad walls of the waveguide vertical. The feed can be used without any attempt to improve the match, the vswr is typically about 1.5:1. The match may be improved by conventional

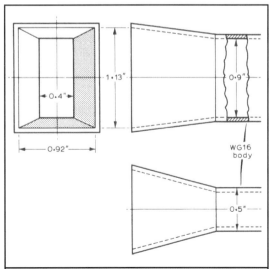

Fig 13.90 : Dimensions of a feed horn for a dish with 0.53 f/D.

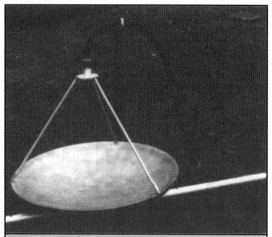

Fig 13.91 : Photograph of a complete front fed dish (G3RPE).

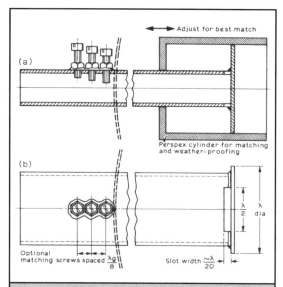

Fig 13.92 : Modified Cutler (Penny) feed (G4ALN).

matching screws fitted behind the dish as shown in the drawing or by means of the Perspex protective and matching sleeve also illustrated there. A professional 610mm (24in) dish complete with WG16 feed, designed for 11.1 to 11.2GHz, was acquired. The feed was a version of the feed just described. In the knowledge that this had been designed for use at (nominal) 11,150MHz and on the assumption that a professional designer would not have used this kind of feed if it had not been acceptably efficient. Also in light of reports that this type of feed was not very efficient with certain types of dish, it was decided to make some careful physical measurements of both the dish and feed. The results of these measurements, when compared with the original version, are quite interesting and may help to shed some light on the apparent inefficiency of the modified Cutler feed.

First the dish was measured and it weighed in with the following characteristics:

  Diameter (D) = 610mm

  Depth (C) = 115mm

  Focus (f) = 202.23mm

  F/D ratio = 0.332

The measurements of C and calculations for f were confirmed by measuring the distance between the detachable feed mounting plate and the *centre of the slots*. Next the feed was examined and a number of interesting points

emerged. The feed is illustrated in the dimensioned diagrams (Fig 13.93) and summarised and compared with the original G4ALN dimensions in Table 13.5. Suggested (calculated) dimensions for 10.380GHz are also given (all in mm) in Table 13.5. The most significant differences between the professional feed and G4ALNs version are:

Table 13.5 : The Penny or modified Cutler feed.

Measurment :	Professional	G4ALN	Suggested Dimensions
Frequency :	(11.15GHz)		(10.38GHz)
Disc Diameter	1.25λ	1.0λ	1.25λ
	(33.36)	(28.9)	(35.83)
Slot length	0.5λ	0.5λ	0.5λ
	(13.35)	(14.45)	(14.34)
Slot width	0.085λ	0.05λ	0.085λ
	(2.25)	(1.45)	(2.42)
Scatter Pins:			
Diameter	0.08λ		0.08λ
	(2.1)		(2.3)
Length	0.1λ		0.1λ
	(2.8)		(3.0)

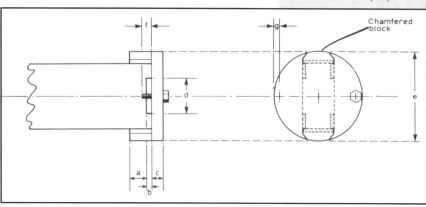

Fig 13.93 : Modified Penny feed (G3PFR). Dimensions are given in Table 13.5.

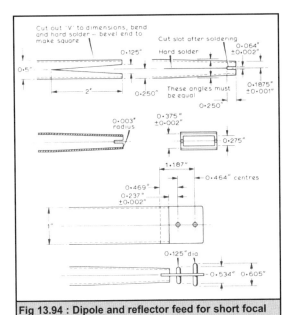

Fig 13.94 : Dipole and reflector feed for short focal length dishes.

- The disc is much thicker, about 0.185λ (4.9mm) and this may be significant. Scaled for 10.380GHz, this would be 5.26mm.

- The disc is backed by two chamfered blocks, as shown in the diagrams. It is probable that these, too, have a significant effect on dish illumination.

- The width of the slots is greater in the professional feed.

- The presence of two scatter pins mounted near the rim of the disc and lying above the midline of the broad face of the guide. Since they were pieces of studding fitted into tapped holes and locked with a lock nut, they were presumably tuneable at some stage of assembly.

- The f/D ratio of the dish is significantly greater than that described in the original text.

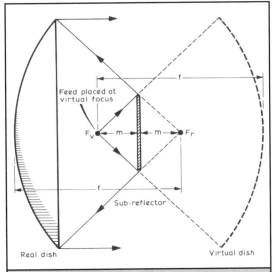

Fig 13.95 : Geometry of the plane sub reflector.

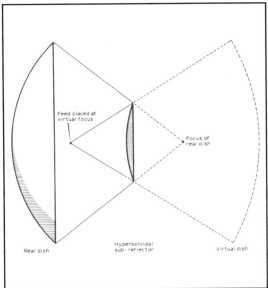

Fig 13.96 : Geometry of the Cassegrain sub reflector feed.

A dipole and reflector feed

The problem of feeding short focal length dishes, and particularly focal plane dishes which are often used by amateurs simply because they are available, is highlighted by the last feed design.

A form of feed known to be very effective with a focal plane dish is the dipole and reflector shown in Fig 13.94. It is not an easy feed to make but, constructed properly and accurately (particularly with regard to dipole dimensions), it is capable of very good results with dishes of f/D = 025. The method of construction is self evident, except that the dipole and reflector are best made longer than specified and trimmed to length after assembly. Care must be taken to remove excess solder at the base of the dipole and at the seams between the web and waveguide.

The E plane and H plane phase centres are not coincident, but their separation can be neglected for most purposes. The mean phase centre is located approximately one third of the distance between the dipole and reflector, nearer to the dipole. This point should lie at the focus of the dish and the exact position determined by experiment. If the feed waveguide passes through a sliding bush in the dish, then the whole assembly can be easily adjusted and may be locked in position after optimisation, either by set screws or by soldering. Again, a professional focal plane dish and dipole/reflector feed was obtained and measurements taken. The figures confirmed the dimensions given in the diagram, the only difference being that the reflector was a disc 29mm (1.14in) in diameter, rather than the rod reflector shown. This disc was also used to support a thin plastic (polythene) sleeve which, fitted over the entire length of the feed and clamped to the mounting boss inside the dish, provided weather protection. As in the previous design, it is possible to improve the feed match by fitting conventional matching screws immediately behind the dish.

Indirect feeds

The basic geometry of the sub reflector feed is shown in

Fig 13.97 : Photograph of a Cassegrain fed dish showing the sub reflector mounting and feed horn (G3RPE).

Fig 13.95, where the sub reflector is a plane disc. The disc area should not exceed 30% of the dish area if losses, due to aperture blockage, are not to exceed 1dB. Once the size of the disc is chosen, its position is automatically fixed, it must *just* intercept lines drawn from the real focus of the dish (Fr) to its rim. The position of the feed is also fixed and its distance from the sub reflector must be equal to the distance of the disc from the real focus (Fr), the distance "m" in the figure.

With this type of feed it is usual to adopt a Cassegrain configuration in which the plane reflector is replaced by a hyperboloid and the geometry of this system is illustrated In Fig 13.96. The main result of this change is that the virtual dish seen by the feed has a longer focal length than the real dish and thus a dish of short focal length, which can be difficult to illuminate efficiently, can be converted into one of longer apparent focal length

Making the sub reflector is not easy for amateurs without a lathe. However it is possible, using either annealed copper or soft alloy, to carefully beat a disc into shape with repeated *light* hammer blows, with the disc supported on a firm but resilient backing (for example stiff foam plastic) or on a hard wood profile block. If the former method is used, then progress is best judged by using a template cut from stiff card or another piece of metal. Gentle beating should continue until the profile is reached. The convex surface of the hyperboloid can then be smoothed with fine emery paper to remove the hammer blemishes (which should be few, if the hammering has been sufficiently light!). A good enough profile for amateur purposes can be achieved with patience.

The sub reflector, when complete, is supported in the dish

as shown in Fig 13.97 and the feed horn (designed from the data given in Fig 13.79) fitted to complete the assembly.

An alternative method of construction, for those possessing the necessary equipment and skills, might be to make a profile from two hardwood blocks, one male, one female, and to use a press to form the metal sheet. However, this is probably beyond the means of most amateur constructors and the slow, patient manual approach may be followed.

Periscopes for microwaves, 10GHz without feeder loss [42]

Microwave operation from mountain tops is very effective and great fun in summer. In winter, however, most of our mountain tops are inaccessible, and all are inhospitable. More importantly, we are rarely on a mountain during short lived propagation enhancements. We would all like to be able to operate from home, but many of us dont not have super locations, my house is surrounded by trees, and my tower barely reaches the treetops.

Some tests had shown that 10GHz operation is possible at my QTH. I have made rain scatter and snow scatter contacts pointing straight up through a skylight, and other contacts by aiming a dish at the tops of trees to scatter off them. If I could get the 10GHz signal to the top of the tower, better results should be possible.

Traditionally, we use a transmission line feed to get signals to the antenna. Feeder loss is a difficult problem for microwave antennas. Low loss feeder is expensive, and tends to be large in diameter. At 10GHz, coaxial feeder larger than about a half inch diameter will support waveguide modes that increase loss, so coax is not a good choice. Waveguide has lower loss than coax, but not good enough for a decent tower, the loss approaches 10dB per 100 feet of WR-90 waveguide at 10GHz.

One alternative is to mount parts of the system on the tower. Many hams have been using this approach successfully, but there are problems with weatherproofing and stability over temperature extremes. Tales of climbing a tower during a New England winter for repairs make this approach sound less attractive.

For several years now, Dick, K2RIW, has been talking about the merits of a periscope antenna system for microwaves. He convinced me to do some reading [43]. Then I made some performance estimates for a reasonable size system. The numbers looked good, so I decided to put a periscope system together and found that it really worked. Then, in order to improve the system, a better understanding was desirable.

Description

A periscope antenna system consists of a ground mounted antenna pointed up at an elevated reflector that redirects the beam in a desired direction. A simple version, with a dish on the ground directly under a flat 45° reflector, is shown in Fig 13.98. The flat reflector is often referred to as a flyswatter, and we will use that terminology. The lower antenna does not have to be under the flyswatter reflector; the reflector tilt angle can compensate for offset configurations, as shown in Fig 13.99. The geometry is a bit more complicated, but a personal computer could easily do the calculations and accurately control the flyswatter pointing.

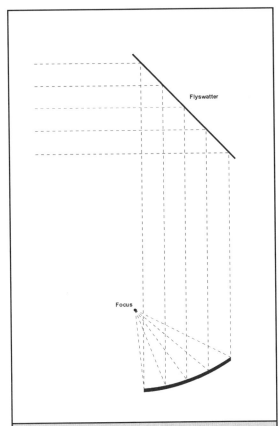

Fig 13.98 : A simple version of periscope antenna.

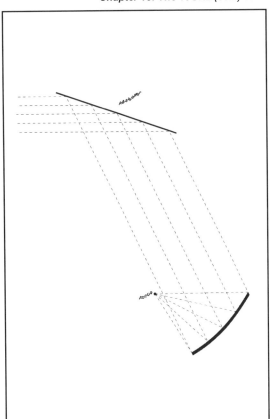

Fig 13.99 : Compensating for periscope antenna system which is not vertical by using reflector tilt.

Periscope antenna systems have been used for fixed microwave links with good results, but are no longer allowed in the USA by the FCC for new commercial installations. The reason seems to be that most good sites are so crowded with antennas that low sidelobes are required, and stray reflections from edges and supports of the flyswatter reflector make it difficult to meet the requirements. However, for amateurs, antenna selection is a matter of individual choice.

The only recent publications found in major databases on periscope antennas are in Russian. In English, there is a description in the Antenna Engineering Handbook [43], [44] which relies on a confusing graph for explanation. After some study and working out a few examples, it dawned on me that the graph is attempting to display an equation with four variables in a two dimensional media.

The only amateur reference [45] I have seen for periscope antennas is by G3RPE, and W1JOT was kind enough to provide a copy. G3RPE limits his analysis to 10GHz, thus eliminating one variable, and provides six additional graphs to illustrate some of the possible combinations. This is an improvement but still doesn't provide much intuition.

I used the G3RPE graphs and the ones in the Antenna Engineering Handbook to work out estimates for a number of possibilities. It appeared that the gain of a periscope antenna system can be within a few dB of the gain of the ground antenna, given the right combination of dimensions. Usually, the flyswatter has a larger diameter

than the ground antenna. Thus, reasonable antenna gain is possible with a dish on the ground, so that feeder loss is low. The feeder loss in the flyswatter is power radiated from the ground antenna that misses the flyswatter and continues into space. This loss is obviously dependent on the geometry of the periscope components. Some combinations of dish, flyswatter, and spacing can even provide more gain than the dish alone, like a feeder with gain!

The beamwidth of the periscope system is similar to the beamwidth of a dish with the same diameter as the flyswatter.

For some combinations, this can result in a beamwidth significantly narrower than a dish with equivalent gain, a minor disadvantage for the periscope antenna.

Another minor problem with periscope antennas concerns polarisation. If the flyswatter rotation is independent of the ground antenna, then the polarisation changes with rotation. There are several ways to compensate:

- Rotate the ground antenna as well as the flyswatter.

- Rotate the feed polarisation.

- Use circular polarisation and accept 3dB loss in all directions.

For rain and snow scatter, my observation is that polarisation does not seem to be particularly critical.

Since the flat flyswatter is not frequency sensitive, it can be used on other microwave bands as well. The ground

Fig 13.100 : Photograph of flyswatter mount.

antenna could be a dish with feeds for multiple bands, or separate ground antennas could be used, adjusting the flyswatter angle for each band.

Construction

The first step in constructing a periscope was to find a good sized piece of aluminium for the flyswatter. I located a 30inch octagon left over from one of my daughter's high school adventures (I have been assured that it came from scrap somewhere and not from a road). To stiffen the mounting area, I attached a heavy aluminium frying pan with a flat rim to the centre of the octagon, and bought a new pan for the kitchen.

I have seen commercial periscope installations with a fixed flyswatter and they look pretty straightforward, the flyswatter is attached to the side of the tower and bolted down after adjustment. However, there are no examples of a flyswatter that rotates and tilts, so the difficult part was figuring out the mechanics.

One approach for rotation would be to mount the flyswatter on the central mast below other antennas. The mast could pass through a central hole in the flyswatter. For a larger flyswatter, the area blocked by the tower would be small, but I wanted to do initial testing with a reasonable size reflector. An alternative, shown in the G3RPE article, is to mount the flyswatter on one side of the mast with the ground antenna following it around the tower as it rotates.

A better approach is to mount the flyswatter on the side of the tower, with rotator and support above it and out of the RF path. Since I didn't trust an ordinary rotator in tension, with the antenna weight pulling it apart, I chose the style that has the mast passing through the body of the rotator. This is attached to a very solid side bracket, available from IIX Equipment [46]. The tower causes some blockage in one direction, but not enough to prevent successful contacts.

Tilting the flyswatter is a little more difficult, but is important for scatter propagation. For a pivot, I drilled a hole through the mast for a stainless steel rod that passes through sleeve bearing on each side. Two pieces of angle aluminium are mounted to the frying pan with a bearing fit in each, the mast is sandwiched between them. Fig 13.100 is a photograph of the assembly which should make this clearer. To power the tilt mechanism, K2CBA

Fig 13.101 : Photograph of the trial flyswatter assembly.

provided an old TVRO dish actuator. The trial flyswatter assembly is shown in Fig 13.101.

When we tested the flyswatter before raising it, we found that the tilt actuator would bind up. The actuator needed to pivot as the flyswatter tilted. Another bearing was needed in place of the U bolts around the actuator in Fig 13. 100. A quick trip to the hardware store located a large swivel caster, normally used for moving heavy machinery, with mounting holes that matched the U bolt holes. When the caster wheel was removed, the actuator fit in the space and was clamped in by the axle bolt, as shown in Fig 13.102. Now the tilt operated smoothly, and we could move the assembly into position near the top of the tower, as shown in Fig 13. 103.

On the air performance

For initial testing, I set up a 10GHz rover system with an 18inch DSS dish directly under the flyswatter. Separation between dish and flyswatter was about 10 metres, so the estimated gain from the curves was about 5dB down from the 18inch dish, but with the narrow beamwidth of a 30inch dish. Also, only crude azimuth and elevation indication were available. Clearly, this was not the optimum configuration, but adequate for initial testing.

The first 10GHz tests were disappointing. Without any rain or other propagation enhancement, stations at a moderate distance were extremely weak, and closer ones were audible in all directions with no discernible peak. With trees in all directions, wet foliage was scattering and absorbing signals. After a couple of days without rain, signals were not much better. The flyswatter does not clear the treetops, and foliage has significant attenuation at 10GHz.

Fig 13.102 : Photograph of modified flyswatter mounting.

Rainscatter performance is much better. With rain predicted for the 1998 June VHF QSO party, I made a radome over the 10GHz rover system using clear plastic garbage bags. We had several inches of heavy rain during the contest, which produced strong rainscatter signals. I was able to work stations in four grids, with best DX of 131 miles, and probably could have worked more if the rain had not stopped on Sunday morning. Most of the contacts were on CW, but AF1T, 41 miles away, was so loud that we switched to SSB. If you'd like to hear Dale's signal, there are some sound clips at http://www.qsl.net/n1bwt .

Rainscatter signals typically have fairly broad headings, so my crude azimuth indication was adequate. For elevation, a local beacon peaked broadly somewhat above horizontal, so I left the tilt at that setting. For normal propagation, beamwidth of the periscope antenna should be quite narrow, like any high gain antenna, so a better readout system is needed.

Since these initial tests, I have added a larger fixed dish at the base of the tower and a digital tilt indicator. With these improvements, I am able to make local 10GHz contacts without enhancement, plus more distant home stations whenever there is any precipitation, typically in three grid squares. Fortunately, I am also able to work mountaintop stations without precipitation enhancement, since the precipitation tends to discourage the less hardy ones. To date, I have worked a total of six grid squares, three without precipitation enhancement, on 10GHz with the periscope antenna system.

We have noticed that rain scatter has an auroral quality, probably from random doppler shift from raindrops falling at different speeds. Snow scatter can provide outstanding signals, if there is no wind and the snowflakes are large, the flakes fall slowly and good SSB quality is possible. Unfortunately, the northeast part of the USA had a severe drought for most of 1999 and another in 2001, so good conditions have been rare.

One last point concerns wind loading: I have never seen the flyswatter move or waver in any wind condition, even though the rotator is a cheap TV model with plastic gears. A dish of the same diameter has much more wind load than the flat reflector and would probably have long ago stripped the gears.

For a very large flyswatter, it might be prudent to tilt it flat when not in use to further reduce wind load.

Fig 13.103 : Photograph of flyswatter mounted on a tower.

Analysis

While my periscope antenna appeared to work quite well, I really wanted to understand it better. There are two papers on periscope antenna systems referenced in the Antenna Engineering Handbook [43], [44], but I was not able to locate copies until several months later. The first paper, by Jakes [47], used an analogue computer for the analysis which produced the confusing graph in Antenna Engineering Handbook. The second paper, by Greenquist and Orlando [48], proved more promising, since it included not only a more detailed analysis of periscope gain, but also some measured results from actual antennas. The gain at 4GHz is presented as a series of graphs similar to Fig 13.104. These curves are for a flyswatter with a square aperture, but a circle or other shape would only change the gain by a dB or so. Also, the true aperture is the projected aperture, viewed from the 45° angle, so that a rectangular flyswatter is needed to provide a square aperture and an ellipse is required for a round aperture. For simplicity, we will refer to the length of one side or diameter as the flyswatter aperture dimension, and not quibble about that last dB or so.

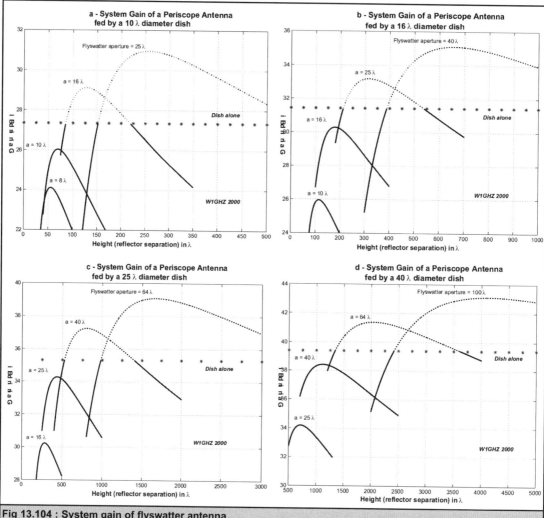

Fig 13.104 : System gain of flyswatter antenna.

A simple view of a periscope antenna is to consider it as a reflector antenna, just like a dish. The flat flyswatter reflector can be considered as a parabola with infinite focal length, so it must be fed from infinitely far away for the illuminating energy to be a true plane wave. Far field radiation from an antenna is approximately a plane wave, so the feed dish can provide a plane wave if it is far enough away so that the flyswatter is in the far field, or Fraunhofer region, of the dish that is, beyond the Rayleigh distance. The gain of the periscope would be the gain of the flyswatter aperture area if the dish were able to illuminate it efficiently; a larger flyswatter would provide more gain.

However, as we shall see, the periscope antenna provides much better performance if the reflector spacings are smaller, less than the Rayleigh distance of one of the reflectors. As a result, a more complicated analysis is required to calculate the gain. We must account for not only the imperfect illumination of the flyswatter, but also the path loss, or space attenuation, between the dish and flyswatter. Since path loss is defined between two isotropic antennas, we must also include the gain of the dish, and flyswatter; both must be compensated for operation in

the near field, or Fresnel region. In the Fresnel region, a large flyswatter may be illuminated with more than one Fresnel zone, and the second zone is out of phase with the first, causing losses. In total, five terms are necessary for the periscope gain calculation:

- G1, the gain of the dish, with Fresnel correction factor

- Space attenuation, the path loss between dish and flyswatter

- G2, the aperture gain of the flyswatter when intercepting power from the dish, including Fresnel correction factor

- Edge effect, loss of efficiency at the edge of the flyswatter due to diffraction

- G3, the aperture gain of the flyswatter radiating into free space, corrected for illumination taper

The system gain of the periscope antenna is the sum total of these gains and losses.

The periscope gain calculations involve a couple of difficult functions, so it took some work before I was able

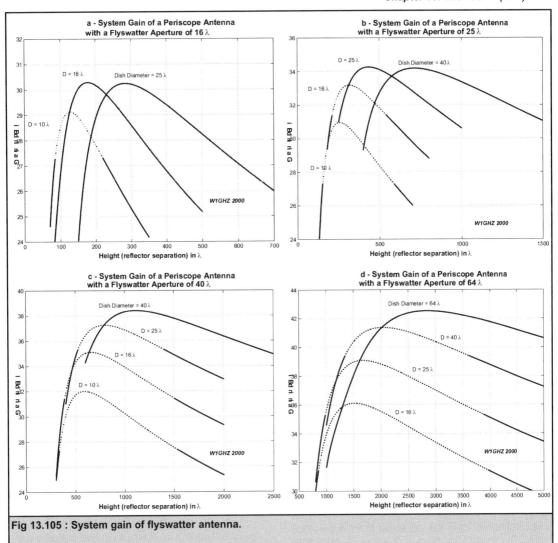

Fig 13.105 : System gain of flyswatter antenna.

to do the calculations. One function, the fresnel sine and cosine, we had needed previously [49] to calculate the phase centre of horn antennas. The code for this was written by Matt, KB1VC. To correct for the illumination taper, a spherical Bessel function $\Lambda_p(v)$ is needed. The paper says this function proved more difficult. "Spherical Bessel functions of the form $\Lambda_p(v)$ are tabulated". For these functions, Silver [50] refers to Tables of Functions [51] by Jahnke and Emde. Fortunately, Byron, N1EKV, had a copy and lent it to me, so that I could make and verify the calculation. Today, of course, we have personal computers, so no one uses books of tabulated functions. (Note: the Spherical Bessel functions $\Lambda_p(v)$ are not the same as Bessel functions for spherical co-ordinates $j_n(x)$ found in modern references [52], [53], [54].)

Once I was able to calculate periscope gain and reproduce the results of Greenquist & Orlando, I wanted to understand the complex relationship between the different dimensions. The equations are far too difficult to offer any insight, so my approach was to graph the results to try and visualise the relationships.

The first step was to replot the curves from Greenquist &

Orlando in terms of wavelengths, so they are usable at any frequency. Each curve in Fig 13.104 shows the periscope gain for a specific dish size as a function of flyswatter aperture and height (reflector spacing). We can see that it is possible with some combinations to achieve a system gain several dB higher than the gain of the dish alone (the curves are dotted lines where the system gain is higher than the dish alone). Best gain occurs when the flyswatter is larger than the dish, and there appears to be an optimum height, but we can't see what produces the optimum. However, we know that path loss follows an inverse square law, doubling the distance increases path loss by 6dB. To increase reflector gain by 6dB requires a doubling of aperture diameter. Thus we can understand why large gains require a large dish and larger flyswatter.

We can also see that large gains also require extremely large reflector spacing. Rearranging the curves for specific flyswatter sizes, in Fig 13.105, shows the same trend but does not really add any insight. The dotted portions of the curves again represent combinations where the system gain is higher than the gain of the dish alone.

The problem is that we are trying to display a four

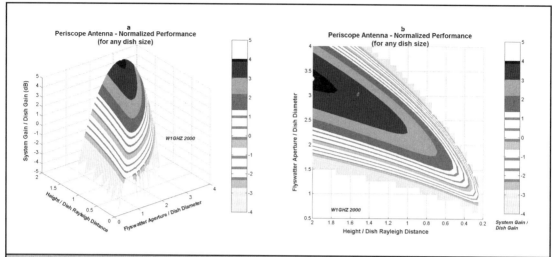

Fig 13.106 : Periscope antenna normalised gain.

dimensional problem in a two dimensional medium. A three dimensional graph might help, if we could reduce the problem to three dimensions. My approach is to normalise the other quantities in relation to the dish diameter, so that one axis is the ratio of flyswatter aperture to dish diameter, and the gain is the effective gain of the periscope, the ratio of system gain to dish gain. The height, or reflector spacing, is normalised to the Rayleigh distance of the dish diameter D:

$$\text{Rayleigh distance} = \frac{2D^2}{\lambda}$$

In Fig 13.106, we can see the effective gain, or increased gain provided by the periscope system over the dish alone, increasing as the relative flyswatter aperture increases. The 3D plot in Fig 13.106a also shows that the range of optimum combinations is narrow and gain falls off quickly if we miss. The maximum effective gain shown is about 4dB, with gain still increasing at the edge of the graph. The graph from Jakes [47] shows an asymptotic value of 6dB as the flyswatter becomes infinitely large. Other numerical values are difficult to discern from the 3D plot, so a 2D version is shown in Fig 13.106b, looking

down from the top. We must rely on shading and the gain bar to find effective gain values.

In Fig 13.106, it is apparent that the gain is not limited by the Rayleigh distance of the dish. However, if we instead normalise the height to the flyswatter Rayleigh distance, using the flyswatter diameter or square side for D in the calculation, the optimum combination becomes apparent in Fig 13.107. The 3D plot illustrates that the range is narrow, and the contour lines below the plot indicate the values. The contour lines for high effective gain are all in the range of 0.2 to 0.3 on the horizontal axis, the ratio of height to flyswatter Rayleigh distance. Thus, the height for best effective gain is roughly 1/4 of the flyswatter Rayleigh distance, regardless of the flyswatter size. Usually, we already have a tower of height h and would like to find the optimum size flyswatter aperture A:

$$A \cong \sqrt{2h\lambda}$$

A flyswatter larger than this optimum size suffers excessive losses due to the Fresnel and illumination taper effects. In the plot, this is the area to the right of the gain peak. To the left of the peak is the area where the distance is too large or the flyswatter too small, and the gain decrease is inverse square, due to space attenuation, the path loss between the two reflectors.

Since the Rayleigh distance is a function of the square of the aperture, the relationship between the dimensions is still not obvious. A realistic example should help. My periscope installation is about 20 metres high, or roughly 700λ at 10GHz. The effective gain for this height is plotted in Fig 13.108 as a function of dish and flyswatter size. The optimum flyswatter size is roughly 40λ, about what we would calculate using the formula above. However, the effective gain increases as dish diameter decreases (note that this axis is reversed for better visibility). Yet we know that dish gain increases with diameter, so what is happening to the system gain? Look at Fig 13.109, the system gain has a broad peak with a dish diameter around 30λ and a flyswatter aperture around 40λ. If we can't find these exact sizes, any combination inside the contour circle below the 3D plot will be within a couple of dB of optimum.

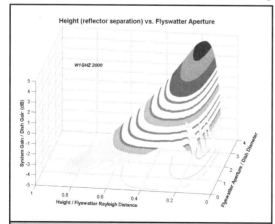

Fig 13.107 : Normalised periscope antenna performance.

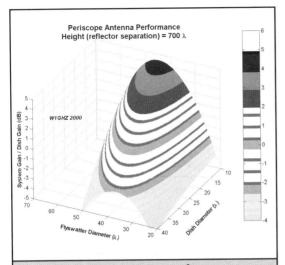

Fig 13.108 : Periscope antenna performance.

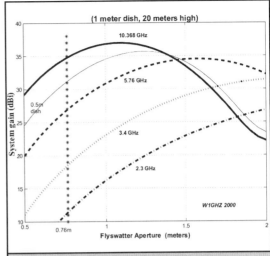

Fig 13.110 : System gain of W1GHZ periscope antenna.

What we learn from Fig 13.108 and Fig 13.109 is that a smaller dish contributes only a small gain to the system, while a larger dish is too large to illuminate the flyswatter effectively at this distance. Best system performance is a compromise between the effective gain of the periscope and the gain of the dish alone, and must be determined for a particular height.

Reaching an intelligent compromise requires that we be able to estimate performance. The graphs included here should be adequate for rough estimates, they were created using MATLAB [55] software, which is powerful but a bit expensive for amateur use. However, I was able to create a Microsoft Excel spreadsheet that performs the periscope performance calculations, so that you may make accurate estimates for any dimensions. You may download periscopegain.xls from http://www.w1ghz.org. Colour versions of these graphs are also available on this web site.

Enhancements

The original periscope trial system described above was rather small, both the dish and the flyswatter are under-size for 20 metres separation, resulting in the estimated 5dB loss. One solution would be to reduce the separation, but this would elevate the bottom dish and require a lossy feeder. A better solution would be to increase the size of the dish and flyswatter.

I recently increased the dish size and made the installation more permanent by adding a fixed, one metre diameter, offset fed dish at the base of the tower. The flyswatter mounting structure was designed to accommodate a flyswatter at least 3 feet wide and 4 feet high, so there is room for improvement here. Also, since dishes and flat reflectors are not frequency sensitive, it would be great to use the periscope system on more than one band. The calculated periscope system gain in Fig 13.110 shows roughly 3dB improvement at 10GHz with the one metre dish over the line for the previous 18inch (~0.5M) dish.

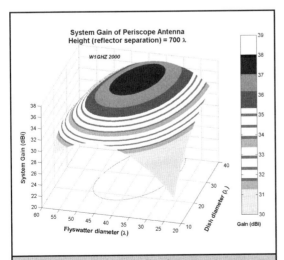

Fig 13.109 : System gain of periscope antenna.

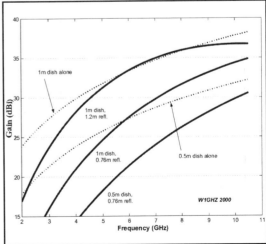

Fig 13.111 : Gain of W1GHZ periscope antenna vs frequency.

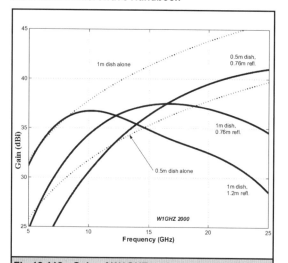

Fig 13.112 : Gain of W1GHZ periscope antenna vs frequency.

Fig 13.110 also shows the system gain for several microwave ham bands as a function of flyswatter aperture. Clearly, a larger flyswatter than the current 0.76-metre aperture is desirable, particularly at lower frequencies. The optimum flyswatter aperture for 10GHz is about 1.2 metres, which is not unmanageably large. Fig 13.111 plots the periscope system gain vs frequency for the one metre dish with a 1.2 metre flyswatter as well as the current 0.76 metre one with both the one metre dish and the original 18inch dish. The dotted curves show the gains of the dishes alone. The current system is comparable to a 24inch dish at 10GHz and falls off at lower frequencies, but the largest combination provides performance close to a one metre dish, not only at 10.368GHz, but also at 5760 and 3456MHz. This looks like a winner! Even at 2304MHz, the mediocre gain is comparable to a medium sized loop Yagi.

These additional bands are almost free; only a feed horn is required. The offset dish has a huge advantage for multiband operation, as the feed horn and support are out of the beam. I envision a carousel of feed horns next to the dish, changing bands by rotating the appropriate horn into position at the focal point of the dish. Two band operation is much simpler, a dual band feed horn [56] will do the job.

Will the periscope work on higher bands? Of course it will, if the dish and flyswatter reflectors have surfaces that are good enough, parabolic or flat within about 1/16λ. As an example, I made some estimates for my system at 24GHz. For a height of 20 metres, our formula estimates an optimum flyswatter aperture A of 0.7 metres, close to my current system, so I extended the curves of Fig 13.111 to include 24GHz in Fig 13.112. The curve for my current combination, a one metre dish with a 0.76 metre flyswatter, has maximum gain around 15GHz and falls off by 24GHz to provide less gain than at 10GHz. My proposed larger flyswatter is even worse at 24GHz, with low gain and very narrow beamwidth due to the large flyswatter aperture, a bad combination. However, the smaller flyswatter with a small 0.5 metre dish looks good, the gain is higher than the dish alone. Feeder without loss at 24GHz is a real miracle!

As usual, we cannot have everything, and must compro-

mise. With the larger dish and flyswatter, I could get good performance on 10GHz as well as 5760 and 3456MHz. With the smaller dish and flyswatter, I could have 10GHz and 24GHz but not much gain on the lower bands. Since my biggest obstacle is foliage, I would probably do better at lower frequencies. However, a more complicated alternative might be to use the smaller flyswatter with both dishes, moving and tilting the flyswatter to change bands.

Conclusion

The periscope antenna system has enabled me to achieve better microwave results from the home QTH than I could before. Without it, I can't get out of my own backyard if the weather is good. The system works well on rain and snow scatter; in many areas, rain is easier to find than altitude.

The periscope is worth considering as an alternative to high feeder losses or tower mounted systems.

I don't believe I have explored the full potential of this antenna, so I urge you to try it and report the results. The analysis and graphs shown here should enable you to understand periscope antenna operation and allow you to design one with confidence.

References

[1] D Clift, G3BAK, RSGB Bulletin , March, April and May 1953

[2] VHF/UHF Manual, 2nd Edition (1971), pp5.31 to 5.34, RSGB

[3] VHF/UHF Manual, 3rd Edition (1976), pp8.23 to 8.26 and 8.33 to 8.35, RSGB

[4] VHF/UHF Manual, 4th Edition (1983), pp9.39 to 9.41, RSGB

[5] J N Gannaway, G3YGF, Radio Communication, August 1981, pp710 to 714 and 717

[6] S J Davies, G4KNZ, RSGB Microwave Newsletter, 09/82, October 1982

[7] C Scrase, G8SHF, RSGB Microwave Newsletter, 10/81, November 1981

[8] H Fleckner, DC8UG and G Hors, DB1PM, VHF Communications, 1/81, Verlag UKW Berichte

[9] B Chambers, G8AGN, RSGB Microwave Newsletter, 06/85, July 1985

[10] Microwave transmission circuits, G L Ragan, Vol.9, MIT Radiation Labs

[11] VHF/UHF Manual, 4th Edition (1983), p9.27, RSGB

[12] Microwave committee component service, c/o Mrs P Suckling, G4KGC, 314A Newton Road, Rushden, Northants NN10 0SY

[13] From an article by Matjaz Vidmar, SM53MV, CQ ZRS

[14] Matjaz Vidmar: NO-TUNE SSB TRANSCEIVERS FOR 1296, 2304 & 5760MHz

[15] From article by Paul Wade, W1GHZ, w1ghz@arrl.net

[16] 10GHz Gunnplexer Communications, The Radio Amateurs Handbook, ARRL, 1992 edition

[17] L.L. Filby, K1LPs, More on the Flavoradio Receiver Conversion , Proceedings of Microwave Update '91, ARRL, 1991

[18] P. Wade, N1BWT, Building Blocks for a 10GHz Transverter, Proceedings of the 1993 (19th) Eastern VHF/UHF Conference, ARRL, 1993

[19] Z. Lau, KH6CP, Home-Brewing a 10GHZ SSB/CW Transverter, QST, May 1993, pp. 21-28, and June 1993, pp. 29-31

[20] C. Suckling, G3WDG & P. Suckling, G4KGC, Modern 10GHz Transverter System, DUBUS Technik IV, DUBUS Verlag, 1995, pp. 276-339

[21] P. Wade, N1BWT, A Single-Board Transverter for 5760MHz and Phase 3D, QEX, November, 1997, pp. 3-14

[22] K. Britain,WA5VJB, Cheap Microwave Filters, Proceedings of Microwave Update '88, ARRL, 1988, pp. 159-163. Also appears in ARRL UHF/Microwave Project Book, ARRL, 1992, pp. 6-6 to 6-7

[23] D. Nelson, NUGH, MMIC Multiplier for 10.8GHz Local Oscillator, Feedpoint (North Texas Microwave Society), March/April 1996

[24] R. Campbell, KK7B, A Clean, Low-Cost Microwave Oscillator, QST, July 1989, p.15. Also appears in ARRL UHF/Microwave Project Book, ARRL, 1992, pp. 5-1 to 5-9

[25] Down East Microwave Inc., 954 Rt. 519 Frenchtown, NJ 08825, 908-996-3584

[26] P.C. Wade, N1BWT, A Fool-resistant Sequenced Controller and IF Switch for Microwave Transverters, QEX, May 1996, pp. 14-22

[23] D. Nelson, NUGH, MMIC Multiplier for 10.8GHz Local Oscillator, Feedpoint (North Texas Microwave Society), March/April 1996

[24] R. Campbell, KK7B, A Clean, Low-Cost Microwave Oscillator, QST, July 1989, p.15. Also appears in ARRL UHF/Microwave Project Book, ARRL, 1992, pp. 5-1 to 5-9

[25] Down East Microwave Inc., 954 Rt. 519 Frenchtown, NJ 08825, 908-996-3584

[26] P.C. Wade, N1BWT, A Fool-resistant Sequenced Controller and IF Switch for Microwave Transverters, QEX, May 1996, pp. 14-22

[27] P.C. Wade, N1BWT, More on Parabolic Dish Antennas, QEX, December1995, pp.14-22

[28] MCM Electronics, 650 Congress Park Drive, Centerville, OH 45459, 800-543-4330

[29] P. Wade, N1 BWT, and D. Twombley, WB1FKF, Modification of TVRO LNBs for 10GHz, QEX, April 1995, pp. 3-5. (reprinted in Proceedings of the 1995 (21st) Eastern VHF/UHF Conference)

[30] Z. Lau, KH6CP, The Quest for 1 dB NF on 10 GHz , QEX, December 1992, pp. 16-19. (reprinted in Proceedings of the 1995 (21st) Eastern VHF/UHF Conference)

[31] P. Wade, N1 BWT, and D. Twombley, WB1FKF, "Modification of TVRO LNBs for 10 GHz," QEX, April 1995, pp. 3-5. (reprinted in Proceedings of the 1995 (21st) Eastern VHF/UHF Conference)

[32] Z. Lau, KH6CP, "The Quest for 1 dB NF on 10 GHz ," QEX, December 1992, pp. 16-19. (reprinted in Proceedings of the 1995 (21st) Eastern VHF/UHF Conference)

[33] From article by Peter Vogl, DLIRQ, VHF Communications 1/1995 pp 52 - 63

[34] Peter Vogl: 6-cm. Transverters in Modem Stripline Technology; VHF-UHF Munich 1990, pp. 49-66

[35] Peter Vogl: 6-cm. Transverters in Stripline Technology, Part 2 (Conclusion); VBF Communications, 2/1991, pp. 69 - 73

[36] Roland Richter: Determination of Parasitic Inductances on Capacitors, Special Project at Dominicus von Linprun Grammar School Viechtach, 1994

[37] Michael Kuhne: Power FET Amplifier for the 9-cm. Band; VIIF-UBF Munich 1992, pp. 50 - 58

[38] Fujitsu Distributor; Melatronic Nachrichtentechnik GmbH, Unterschleiisheim

39] ZG4-2-N Aluminium Housing; Telemeter Electronic GmbH, Donauwbrth

[40] Toshiba Distributor; Tricom Mikrowellen GmbH, Freising

[41] VHF/UHF Manual, 4th Edition (1983), pp9.68 to 9.70, RSGB

[42] From article by Paul Wade, W1GHz

[43] H. Jasik, Antenna Engineering Handbook, First Edition, McGraw-Hill, 1961, pp. 13-5 to 13-8

[44] R.C. Johnson, Antenna Engineering Handbook, Third Edition, McGraw-Hill, 1993, pp. 17-24, 25

[45] D. Evans, G3RPE, Observations on the flyswatter antenna, Radio Communication, August 1977, pp. 596-599

[46] IIX Equipment Ltd., 4421 W. 87th ST., Hometown, IL 60456. 708-423-0605

[47] W.C. Jakes, Jr., A Theoretical Study of an Antenna-Reflector Problem, Proceedings of the I.R.E., February 1953, pp. 272-274

[48] R.E. Greenquist and A.J. Orlando, An Analysis of Passive Reflector Antenna Systems, Proceedings of the I.R.E., July 1954, pp. 1173-1178

[49] P. Wade, N1BWT, and M. Reilly, KB1VC, Metal Lens Antennas for 10GHz, Proceedings of the 18th Eastern VHF/UHF Conference of the Eastern VHF/UHF Society, ARRL, 1992, pp. 71-78

[50] S. Silver, Microwave Antenna Theory and Design, McGraw-Hill, 1949, p. 194

[51] E. Hahnke and F. Emde, Tables of Functions: with Formulae and Curves, Dover, 1945, p. 128, 180-189

[52] C.D. Hodgman, S.M. Selby, R.C. Weast, Mathematical Tables from Handbook of Chemistry and Physics, Eleventh Edition, Chemical Rubber Publishing, 1959, p. 287

[53] M. Abromowitz and I.A. Stegun, Handbook of Mathematical Functions, Dover, 1972

[54] W.H. Press, S.A. Teukolsky, W.T. Vetterling, B.P. Flannery, Numerical Recipes in C: The Art of Scientific Computing, Second Edition, Cambridge University Press, 1992, pp. 251-252

[55] The Mathworks, Inc, Natick, MA

[56] P. Wade, N1BWT, Dual-band Feedhorn for the DSS Offset Dish, Proceedings of the 23rd Eastern VHF/UHF Conference of the Eastern VHF/UHF Society, ARRL, 1997, pp. 107-110

Chapter 14

The 24GHz (12mm) band

In this chapter :

- Low noise oscillator source
- Modular designs, preamplifier, power amplifier, doubler and mixers

- Amplifiers
- Antennas

T he techniques used at 24GHz are similar to those used at 10GHz and the equipment is similar, though physically only about half the size. The devices are usually mounted in waveguide, and other components such as directional couplers, attenuators, loads, etc. are also constructed in waveguide.

A feature of the band distinguishing it from 10GHz and lower frequencies is the marked dependence of propagation on the weather. The frequency is close to 22GHz, where signals are significantly absorbed by water vapour, and so high humidity and rain can greatly increase propagation losses. For this reason, this part of the spectrum has been little used professionally in the past and, as a consequence, there is little surplus equipment available. Indeed, the range 18 to 26GHz, known as K band, was often omitted in microwave test equipment and in a manufacturers range of devices.

The band is now being increasingly used for doppler radar applications such as intruder alarms and speed measurement systems. This type of equipment has to be low cost to be competitive and so Gunn diodes and mixer diodes for 24GHz are now available at reasonable prices. Complete Gunn oscillator and in line oscillator/mixer assemblies are also available which considerably simplify getting started on the band.

Propagation

At 24GHz there is a significant loss due to the presence of water in the atmosphere. This attenuation is due to absorption by water vapour Molecules and scattering by raindrops. At higher frequencies (e.g. 48GHz) oxygen also causes absorption, but the effect is small at 24GHz just over 0.01dB/km, compared to water.

The attenuation due to water vapour molecules depends upon the humidity and typically varies from around 0.05 to 0.20dB/km. Fig 14.1 shows the attenuation (in dB/km) for different temperatures and humidities, a figure of 0.1dB/km is typical for the United Kingdom.

The attenuation due to rain scattering depends upon the size and shape of the raindrops, and the number intercepted along the path. Fig 14.2 shows the expected loss, in dB/km, for rainfall rates between 0.1 and 100mm/hour. The graph strictly applies to steady, widespread rain and horizontally polarised signals. Thus, for a line of sight path, the attenuation comprises the sum of several components:

- Free space loss
- Water vapour absorption loss

- Rain scattering loss (if present)

If it is only raining over part of the path, a sufficiently accurate estimate of the rain scattering loss can be obtained by multiplying the loss per kilometre (from Fig 14.2) by the length of the path over which it is raining.

The droplets in dense fog also cause scattering and give an attenuation comparable with rainfall of 2.5 to 5.0mm/hour. Snow and hail also cause a significant loss but little data is available on these effects. Fig 14.3 shows the path losses for various types of path: basic free space, free space plus water vapour, light drizzle and heavy rain. Apart from the free space line, water vapour loss at 0.1dB/km has been included. The light drizzle path loss assumes an additional loss of 0.1dB/km due to 1 mm/hour rainfall; the heavy rain path loss assumes a loss of 1.3dB/km due to 10mm/hour rain.

It seems likely that the best time to try a long path may be in winter on a dry, cool, clear day, when the humidity is very low. However, do not discount operation in showery weather; heavy rain is unlikely to occur completely over a long path and it may be possible to take advantage of dense rain clouds by using them as passive reflectors.

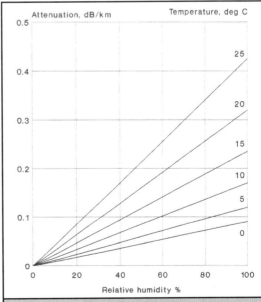

Fig 14.1 : Loss due to water vapour at 24GHz.

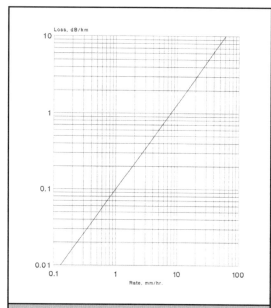

Fig 14.2 : Loss due to rainfall at 24GHz.

Ducting over water should not be discounted either, this phenomenon is well known on 10GHz and should occur on 24GHz. The difference is that the water vapour needed to form ducting conditions will also absorb the signals. Thus, for an enhancement of signals to occur, the gain due to ducting must outweigh the additional absorption.

Band planning

The current UK band plan is shown in Fig 14.4. In the USA 24.0 - 24.25GHz is licenced for all modes except for novices.

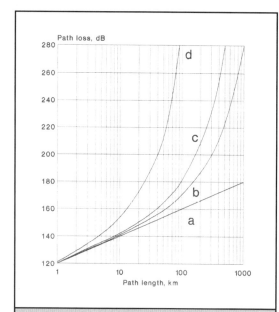

Fig 14.3 : Path loss at 24GHz. (a) Free space loss. (b) Free space plus water vapour. (c) Light drizzle. (d) Heavy rain.

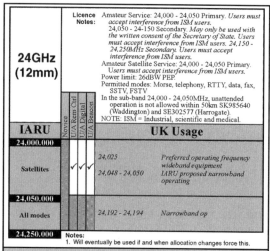

Fig 14.4 : UK band plan for 12mm.

Equipment capability

A simple wideband station might consist of a 7mW Gunn oscillator, a 1N26 receive mixer with an overall noise figure of, say, 15dB and a 20dB horn antenna. An advanced wideband station might use an 12inch diameter dish having a gain of 35dBi, an output power of 10mW, and achieve a lower noise figure of 10dB. A bandwidth of 200kHz is used in each case. The path loss capability of these two types of station (in each case, working to a similar station) is shown in Table 14.1. For comparison, a high performance narrow band station is also included, using a 3kHz bandwidth IF.

It must be stressed that the ranges shown in Table 14.1 assume an average loss of 0.1 dB/km due to water vapour, while in practice this will vary considerably. The results suggest that reliable operation in temperate zones such as the UK over paths greater than 150km can be achieved with relatively modest equipment by using narrower bandwidths than commonly used on this band, and coupling this with taking advantage of cold, dry weather conditions. Judging by the success of Italian amateurs in operating over a 450km path on 24GHz in the Mediterranean, it seems likely that long distance contacts may also be assisted by anomalous propagation conditions, e.g. ducting, but this has yet to be investigated properly.

Table 14.1 : Path loss capability at 24GHz.

Transmitter	Receiver	Path loss capability	Approximate line of sight range
7mW WB 20dB horn	15dB NF 20db horn	142dB	10km
10mW WB 35dBi dish	10db NF 35dBi dish	179db	150km
10mW NB 35dBi dish	10dB NF 35dBi dish	207dB	350km

Low noise local oscillator [1]

Amateurs are using narrow band modulation including CW, SSB and NBFM on ever higher frequencies. In the US, SSB is commonplace on all microwave bands through 10 GHz and is spreading to the 24 and 47GHz millimetre wave bands. In Europe, narrow band operation has taken place as high as 411GHz

The local oscillator (LO) used at these higher, millimetre wave frequencies must be much more stable than at lower, microwave frequencies. On SSB, the indicated frequency should be within 500Hz at both the transmitter and receiver, or you may not hear the station calling you. During a microwave contest, you don't want to adjust both the antenna and the frequency while trying to make a contact. I wasted many hours during the last 10GHz and up contest because the LO in my transverter was 85kHz off frequency at 24.192 GHz.

A lesser known problem is phase noise. When extended to millimetre wave bands, most published LO designs for amateur microwave transverters have excessive phase noise that limits the dynamic range of the transverter. Many contesters on mountain tops have experienced this desensitisation from commercial equipment on nearby frequencies or amateur beacons operating at the same site.

I decided to replace the LO in my 24GHz transverter and solve both the phase noise and stability problems. This article describes the crystal oscillator and multiplier designs that resulted. They work nicely with existing transverters using the KK7B LO design [2] with a few modifications and can be adapted to others. The KK7B LO was originally designed for a 2304MHz transverter and was extended by N1BWT (now W1GHZ) with an additional x5 multiplier for his 10GHz transverter [3] Replacing the original LO with the circuit described here and modifying the first multiplier board make this technology much more usable on the millimetre wave bands at 24 GHz and above.

The Phase Noise Problem Explained

The LO for a microwave or millimetre wave transverter is usually a crystal oscillator operating around 100MHz, which is then multiplied up to the amateur band in use. It is critical that this oscillator have a very low noise floor because each stage of multiplication adds noise to the signal at a rate of 6 dB each time the frequency doubles. In addition, the frequency multiplication process is lossy; signal levels can decrease rapidly with high order multiplication, and the noise inherent in low level amplifiers can degrade the noise floor even faster. Multiplying the crystal frequency by 100 to reach 10GHz increases the noise floor by at least 40 dB. At 250GHz, the added noise would be 68 dB, or more.

Many VHF crystal oscillator circuits have a noise floor no better than 155 dBc/Hz. This means that noise 155 decibels below the carrier power will appear in each hertz of bandwidth at the oscillator output. Multiplying to 245GHz increases this to at least 87dBc/Hz. This is the best case, and the degradation will always be several decibels worse due to the noise figure of components within the multiplier chain.

The problem with a high LO noise floor is that signals outside the receiver pass band mix with this noise and appear as increased noise within the pass band, reducing

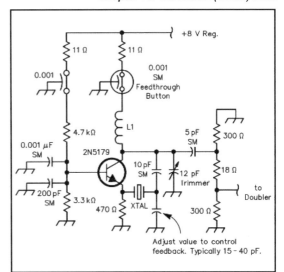

Fig 14.5 : The traditional common base Butler oscillator.

sensitivity. If NBFM is being used, the noise that appears in the 16kHz bandwidth will be -87 + 42dB, only 45dB below the level of the interfering signal, the receiver dynamic range is 45dB. Therefore, any signal appearing within the pass band of the RF circuitry (several gigahertz wide) that is 45dB (about 7½ S units) stronger than the desired signal will mask it completely. When transmitting, broadband noise will be radiated that is only 45dB below your signal. The problem is smaller at lower frequencies but still results in a receiver that is easily overloaded. The same example at 10GHz results in a 71dB dynamic range, which is more acceptable, but still 15dB worse than most VHF/UHF transceivers and more than 30dB worse than many HF transceivers.

Improving Stability

The first issue is frequency stability. Quartz crystals have inherent temperature sensitivity that can vary the resonant frequency by ±10 ppm from 0 to +70°C, which amounts to ±100kHz at 10GHz. Temperature compensated crystal oscillators and those in ovens do better at ±0.3 ppm or ±3kHz at 10GHz. This is totally unacceptable at millimetre wave frequencies, where the drift is multiplied to ±7kHz at 24GHz or ±75kHz at 250GHz.

To provide a rock stable LO, the only solution is to phase lock the crystal oscillator to something more stable. Small rubidium frequency standards are now available at moderate cost on the surplus market. Typically, they are removed from obsolete radio navigation equipment [4]. The long term accuracy over temperature is 1 part in 10^9 or 0.001 ppm. This results in an accuracy of ±250Hz after multiplication to the highest amateur band at 241 - 250GHz.

Improving Phase Noise

The phase noise problem is solved by building a crystal oscillator with a lower noise floor. The ratio of the noise at the input of the transistor to the signal arriving from the crystal ultimately determines this noise floor level.

The traditional common base Butler oscillator described in The ARRL UHF/Microwave Experimenters Manual (see Fig 14.5) shows the problem. The input resistance of the

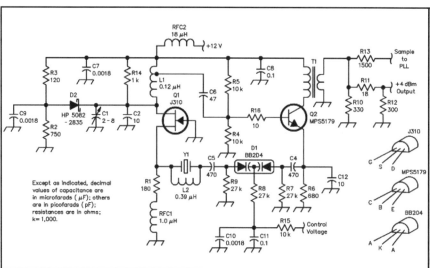

Fig 14.6 : The VXCO of the low noise, phase locked crystal oscillator. See table 14.4 for additional component details.

2N5179 emitter is very low. With the transistor biased for 5mA emitter current, as shown, the input impedance is approximately 5Ω. At resonance, the fifth overtone crystals used in this circuit have a series resistance of about 60Ω. In this type of oscillator, the peak current through the crystal is approximately equal to the standing current; the RMS current through the crystal is 3.5 mA, and the power dissipated in the crystal is 0.735 mW.

The trouble is that the amount of power appearing at the 2N5179 emitter is only 0.061 mW or 12dBm, and the noise figure of the 2N5179 is probably about 10dB in this configuration. The noise floor can be calculated by taking the noise power caused by circuit resistance, -177 dBm, adding the noise figure and subtracting input power level. This yields -177 + 10 - (-12) = -155 dBc. This is only a first order estimate, and the actual oscillator could be several decibels worse.

The power level in the crystal cannot be raised, as it would cause excessive ageing and instability. We must provide more input to the transistor by increasing its input impedance to provide a better match to the crystal. With a bipolar transistor amplifier, this means reducing the emitter current; but this would also reduce the power available, exacerbating the problem.

The input impedance of a common gate FET is the inverse of the transconductance and is independent of the standing current. The transconductance for a J310 FET is about 8,000-18,000mS, its input impedance in a common gate configuration is therefore between 56Ω and 125Ω. It also has a noise figure of less than 2dB at 100MHz. Replacing the 2N5179 with a J310 and keeping the same crystal dissipation results in an input power of 0.686 mW or -2dBm in the worst case (lowest input impedance). We can also assume that the J310 noise figure may be degraded somewhat, to 3dB, by noise in later stages. The noise floor is lowered to -177 + 3 - (-2) = -172dBc, an improvement of 17dB.

The increased input impedance does have one drawback: It is in series with the crystal, so the loaded Q of the crystal is lower than with a low input impedance circuit. In this application, it is not a problem. The notional capacitance of a 90 to 100MHz crystal is about 2.4fF (femtofarads), which results in a reactance of about 1.6 M. The unloaded Q is about 27,000 (1,600,000/60) and the

loaded Q is about 8700 in the worst case. This results in a bandwidth of about 11kHz, which is fine for SSB and CW operation. To ensure that there is no additional degradation in Q, the crystal must be driven from a low impedance source.

Once we have a low phase noise oscillator, we need to make sure that it is not degraded by succeeding stages in the LO chain. The amplifier(s) immediately following the VCXO must contribute very little noise, and the initial frequency multiplication must be done in small increments to minimize reduction of the LO level at any intermediate point. As the multiplication process proceeds, the LO noise floor rises. As it does, we can be less stringent in our requirements by using higher order multipliers and having less constraint on noise figure.

The VCXO Circuit

The basis of the low noise, phase locked crystal oscillator (LNPLXO) is the voltage controlled crystal oscillator (VCXO) circuit shown in Fig 14.6.

A J310 JFET (Q1) is a common gate amplifier providing a high impedance input for the signal from the crystal (Y1). A 2N5179 bipolar transistor (Q2) is an emitter follower providing low impedance drive to the crystal. Y1 is a fifth overtone, AT cut crystal ground for operation in the series resonant mode with a load capacitance of 30pF. The feedback path is completed through a resonant circuit consisting of C1, C2 and L1 that selects the desired overtone. R14 loads the drain of Q1 to ensure linear operation. It also sets the loop gain of the oscillator. The dual varactor diode, D1, in series with the crystal, provides for pulling of the crystal frequency by ±500 Hz. L2 cancels out the parallel capacitance of the crystal to enable a wider pulling range.

The circuit composed of D2, R2, R3 and C9 controls the amplitude of the oscillator. The voltage across R3 is about 1.6V. When the RF voltage on Q1s drain reaches -2V, D2 conducts, and the signal is limited. This is done without affecting the operating point of Q1, so it remains in a linear, low noise mode. R2 can be adjusted to change the output level of the oscillator, while making sure that the power dissipated in Y1 never exceeds 1mW.

To ensure minimum effect on loop gain by external load variations, the output of the oscillator is taken from the

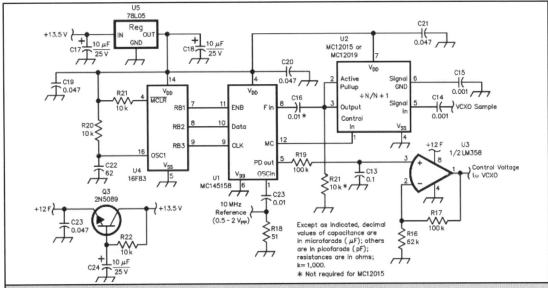

Fig 14.7 : PLL circuit diagram. See table 14.4 for additional component details.

collector of Q2. T1, a 9:1 broadband transformer, provides impedance matching, 6 to 7dBm is available at its secondary. A 3dB attenuator provides more isolation and reduces the signal level to prevent overdrive of the next stage in the LO chain. The 1500Ω resistor is used to couple some of the output to the PLL circuitry.

The PLL Circuit

Fig 14.7 shows the PLL components required to lock the VCXO frequency to the 10MHz output of a rubidium frequency standard. A sample of the VCXO output is applied to a dual modulus pre scaler chip, U2. This can be a Motorola MC12019 for division by 20/21, or a Motorola MC12015 for division by 32/33. The pre scaler divides the VCXO frequency so that it will not exceed the 20MHz maximum clock frequency of the PLL chip, U1.

The Motorola MC145158 PLL chip at U1 is the heart of the circuit. It accepts the 10MHz reference frequency at pin 1 and divides it to a user settable internal reference frequency using the R counter. The pre scaled VCXO output is applied to pin 8, the input of the A and N counters. The A counter determines when the pre scaler will be switched from the N to N+1 mode. The N counter divides the pre scaler output before application to the phase detector. The PLL will lock at the frequency determined by all of these counters as shown in Table 14.2.

The phase detector output at pin 5 of U1 is filtered by R19 and C13, then amplified by U3. U3 is required to increase

Table 14.2 : Frequency vs Divider Values R, A, and N.

MC12019 prescaler
Frequency = (10 MHz / R) x (A + N x 20)

MC12015 prescaler
Frequency = (10 MHz / R) x (A + N x 32)

Note that N cannot be less than the prescaler division ratio.

the phase detector output, from 5 to 13.5V. Note that R19 and C13 are not the values predicted by the equations Motorola supplies for the PLL. The crystal is a high Q device and there is a time delay when changing its frequency. The low pass filter time constant had to be determined experimentally. The loop has been verified to be adequately damped with three crystals of different manufacture, so the time constant should not need to be changed. To ensure stability, the PLL reference frequency should not be set below 50kHz. It should also not go above 200kHz to ensure that the phase detector output is accurate. This frequency range should be adequate for all amateur microwave and millimetre wave LO requirements. Choose your division ratios carefully.

The PLL (U1) is configured by data entered serially on the clock, data and enable pins. A PIC16F83 micro controller unit (U4, MCU) provides these data. Bits 1, 2 and 3 of U4 port B are used to drive these lines. Since the MCU does not require an accurate clock, the RC oscillator configuration is used. The use of an MCU may seem like overkill, but a PIC16F83-04 with 1 KB of program memory and 68 bytes of RAM is only $6.25 in single unit quantities. The single 18pin DIP package is also smaller than any other solution.

LO Chain

My 24.192GHz transverter uses a harmonic mixer with an anti parallel diode pair that requires injection at half the LO frequency. With a 1296MHz IF, this works out to 11.448GHz. Thus, the output of the VCXO at 95.4MHz needs to be multiplied by 120. I planned to use the same frequency multiplication chain that N1BWT used in his 10GHz transverter [3]. This multiplied the crystal frequency by six in the first stage, followed by two stages of multiplication by four and five. Each was a single PC board. However, measurements showed that it degraded the phase noise significantly. The first multiplication in this LO chain reduced the LO signal level below -23dBm. Coupled with a noise figure of 6dB, this could not support an LO to noise ratio of more than -148dB, a degradation of 7dB on the theoretical minimum of 155dBc.

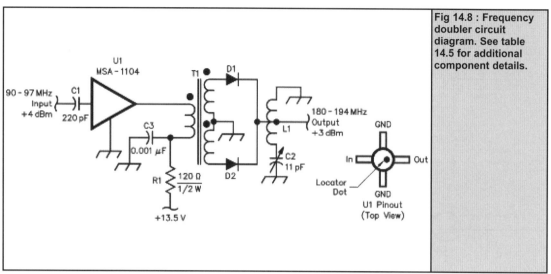

Fig 14.8 : Frequency doubler circuit diagram. See table 14.5 for additional component details.

The LO chain was redesigned to add an additional stage, resulting in successive multiplication by 2, 3, 4 and 5. The KK7B LO board was modified to act as a tripler and a new doubler stage was designed to precede it.

Frequency Doubler

The doubler stage in Fig 14.8 is essentially a full wave rectifier. This circuit suppresses the fundamental and odd harmonics [7], which results in much cleaner output from the following tripler. U1, an MSA-1104, provides the low noise figure and high output level that are critical to maintaining a low noise floor. D1 and D2 are driven with about +16dBm and produce about +3dBm of output, which is then filtered by L1 and C2. Minimal filtering is required here, given the six poles of filtering in the tripler that follows.

Frequency Tripler

Modifying a KK7B x 6 multiplier board to perform the tripler function turned out to be fairly easy. Remove U1, Q2 and R7 from the board (see Fig 14.9) to disable the original crystal oscillator. Connect the doubler output to the free end of C10. Replace U2 with an MSA-0485-less gain is needed, as the input level is larger. It amplifies the +3dBm from the doubler to +11dBm, which is enough to drive the following diode tripler. Also, replace L3 with a 0.22mH RFC.

The original diode sextupler is converted to a tripler by removing L4 and C12 and changing C11 to 8.2pF. This creates a low Q series resonant circuit at about 190 MHz to isolate the tripler output from U2. D1 is also removed and replaced with two PIN diodes in anti parallel. PIN

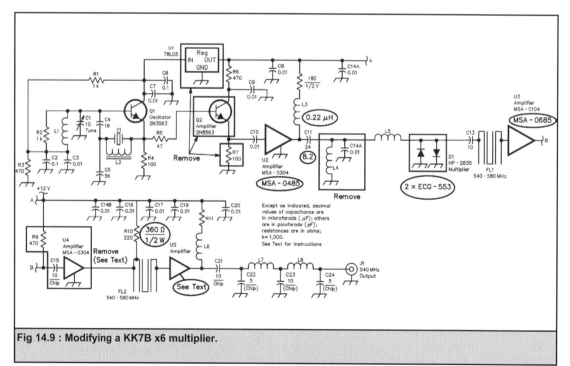

Fig 14.9 : Modifying a KK7B x6 multiplier.

Table 14.3 : Output Spectrum.

Harmonic	Frequency (MHz)	Level (dBc)
1	92	< -75
2	184	-62
3	276	< -75
4	368	-56
5	460	-70
6	552	0
7	644	< -75
8	736	-67
9	828	< -75
10	920	-72
11	1012	< -75
12	1104	-43

diodes produce about 3dB more output in this application than Schottky diodes [6]. The anti parallel circuit also results in 6dB more output than a single diode and good suppression of even harmonics, resulting in a cleaner LO. I used two ECG-553 diodes because they were easily available, but an HP HSMP-3821 or two 5082-3188s should work just as well. I thought about designing a more complex impedance matching circuit for the PIN diodes, but tripler output was better than predicted by the HP application note, so I discarded the idea in favor of simplicity.

U3 is replaced with an MSA-0685 to reduce the noise figure. Since the signal levels are much higher, U4 is removed and replaced with a copper strap to the filter. Also remove C15 and R9. Replace C15 with a copper strap and change R10 to a 360Ω, 1/2W resistor. U5 can be an MSA-0485 for +12 dBm output or an MSA-1104 for +16 dBm output. The MSA-0485 is adequate to drive the KK7B x 4 multiplier board.

Construction

I built the VCXO and PLL on copper clad Vector board. Make sure to shield the VCXO from the PLL, otherwise leakage from the PLL digital circuitry will show up as spurs at -50 to -60dBc after multiplication to 11GHz. One

Table 14.4 : SSB Phase Noise at 552 MHz.

Offset (kHz)	Noise (dBc)
2	-133
5	-143
10	-146
25	-148
50	-150
500	-154

Table 14.5 : Parts lists for low noise oscillator source.

VXCO

C1	2 - 8pF NP0 ceramic trimmer
C2	10pF NP0 ceramic disc
L1	0.12 mH, 8 turns #22 AWG enamelled wire on T30-12 with a tap 2 turns from cold end
L2	0.39 mH, 15 turns #26 AWG enamelled wire on a T37-12 powdered-iron core.
T1	Primary 6 turns #30 AWG enamelled wire, secondary 2 turns #28 AWG enamelled wire on a BN-61-2402 ferrite core.
Y1	Fifth overtone crystal, series resonant, 30 pF load capacitance, series resistance less than 60Ω.

PLL

C16 and R21 are not required if U2 is a MC12015.	
C22	62 pF ceramic, ±10% tolerance
C14,15,16,23	ceramic, ±20% tolerance
C13	Monolithic ceramic, ±20%
C19,20,21	Monolithic ceramic, +80%/-20%
C17,18,24	Tantalum Electrolytic
U1	MC145158 PLL
U2	MC12019 (,20/21) or MC12015 (,32/33) prescaler
U3	LM358
U4	PIC16F83 microcontroller
U5	78L05 5-V regulator

Frequency doubler

C2	11 pF trimmer
D1,2	HP5082-2835
L1	6 turns #24 AWG enameled wire 1/4 inch long on a 3/16-inch-diameter form. Tap at 11/2 turns from grounded end.

Unless otherwise specified, use 1/4 W, 5%-tolerance carbon composition or film resistors.

doubler was constructed in a minibox for testing and the other on the KK7B board by chopping up the tracks in the old crystal oscillator area. It could probably be added to the VCXO board.

Circuit construction on copper clad perfboard was described in amateur literature in the late 70s, but it is less well known today. It requires the use of old fashioned leaded components, but avoids the need to etch PC boards. I've found that this method can be used up to 150 MHz if care is taken to provide large interconnected areas of copper as a ground plane.

The board I use is made by Vector Electronics [5]. It is an epoxy glass material with 0.042 inch diameter holes punched on a 0.1 inch grid and is coated with copper on one side. Components are mounted on the epoxy glass side and the grounded ends of components are just soldered to the copper. Where ungrounded connections are required, a pad cutting tool is used to remove the copper around the holes. With good layout, most interconnections in analogue circuits can be accommodated by mounting the components in adjacent holes and soldering the leads together on the non component side of the board. Occasionally, you will need to run a wire between holes to make connections.

MCU Programming

The PIC16F83 MCU must be programmed to load the appropriate values into the MC145158 PLL chip. A listing of the program used may be downloaded from the ARRL Web page [9].

The program is very simple. On power up, it initializes the port used to communicate with the PLL, executes a delay routine to ensure that the PLL chip has powered up, programs the PLL registers then shuts off. The PLL counter division values are sent one bit at a time, with the most significant bit first. In this example, the PLL phase detector runs at 100kHz, so the R counter is set for division by 100. The N and A counters are set for division by 49 and 15, respectively, to achieve division by 995 when used with a MC12015 pre scaler.

The program uses common subroutines to send ones and zeros to the PLL chip and enable latching of the values sent. The program may be easily altered by changing the calls to the subroutines ONE and ZERO to reflect the binary values to be loaded into the counters

Adjustment

Adjustment of the LNPLXO is straightforward. Apply power and adjust C1 for maximum output on a power meter. This can be as simple as a 50Ω resistor, a Schottky diode rectifier and a voltmeter. Then connect the 10MHz reference oscillator and adjust C1 for approximately 5V on D2. This detuning pulls the crystal frequency slightly to centre it in the PLL lock in range. Check the power output again to be sure that it has not dropped by more than 0.5dB. If it has, add a capacitor in series with the crystal to raise the frequency, or an inductor to lower the frequency and repeat the adjustment. Crystals ordered with a 10 ppm tolerance and 30pF load should not require any circuit modifications. If you have crystals on hand that are calibrated for series resonance without a reactive load (so called 0pF load), insert a 100nH inductor in series with the crystal.

Results

The output at 552MHz was examined on a spectrum analyzer and all spurious outputs except the second harmonic were below -55dBc, as shown in Table 14.3.

Phase noise measurement was done by building two identical 92MHz LNPLXOs and multiplier chains then connecting them to a common 10MHz crystal oscillator. The outputs (approximately +17dBm) were then applied to a double balanced mixer (DBM) through 10 and 20dB attenuators, resulting in a +7dBm LO and -3 dBm at the RF port. The output of the DBM was ac coupled to a low noise amplifier. This arrangement converts the carrier to dc and the phase noise sidebands on each side of the carrier to frequencies between 2kHz and 500kHz, where they were measured on a HP 8553 spectrum analyser. See [8] for more information on this technique.

To measure the phase noise in the absence of the multiplier chain, the LNPLXOs were connected to low noise isolation amplifiers and a Level-23 DBM. This DBM did not support an IF below 100kHz, so measurements were limited to offsets between 100 and 500kHz.

The noise floor of the basic LNPLXO was measured to be -171dBc. After multiplication by six to 552MHz, the noise floor at a 500kHz offset was measured at -154dBc. This is very close to the theoretical minimum, considering that my

measurements are only accurate within ±2dB. The phase noise versus frequency for these two cases is shown in Table 14.4. The SSB phase noise rises slowly from -150dBc at 50kHz to -143dBc at 5kHz, and then jumps to -133dBc at 2kHz. The slow rise is probably caused by flicker noise in the J310 FET and AM PM conversion in the multipliers. The more rapid rise below 5kHz occurs within the pass band of the crystal itself, and is as expected.

This performance should be adequate for my 24GHz transverter, which has a noise figure of 1.9dB and a calculated third order input intercept of -28dBm. This would result in a two tone dynamic range of 76dB with a perfect local oscillator. I estimate that the LO phase noise floor will degrade by 34dB (32dB for multiplication by 40, plus an extra 2dB for circuit losses) to -120dBc after multiplication to 22.896GHz. This would limit dynamic range to 78dB in a 16kHz bandwidth, so LO noise is not the limiting factor for NBFM, SSB and CW operation on the 12mm band.

Low LO phase noise will be even more important for the 6mm band, where LO phase noise will degrade by 7dB, reducing the dynamic range to 71dB. This results in a dynamic range at 47GHz that is similar to that of 10GHz transverters using the traditional Butler oscillator circuit. In addition, the frequency will stay within 50Hz even if it is 115°C in the shade.

Modular designs [10]

Recent, low cost, GaAs devices have proven to be usable at 24GHz for the construction of LNA (low noise amplifiers), power amplifiers and frequency multipliers. This article presents a low cost solution to build 24GHz modules, using a common design for the mechanics. Thus enabling experimentation with several devices and configurations without the cost and time associated with the manufacturing of a milled case for each design. Also presented here are three basic modules for 24GHz, an amplifier, a frequency doubler and a fundamental mixer.

Introduction

Packaged GaAs Fets, used in the TVRO industry, prove to be usable at 24GHz. Although they are usually specified up to 18 (and sometimes to 20GHz) they can be put to work at 24GHz with careful tuning. Low noise amplifiers were developed [11], [12] and also frequency multipliers operating at 24GHz [13], [14] and even some attempts at 47GHz have been made by the author, by DB6NT, JE1AAH [15], [16] and certainly by others. One major limitation on the experimentation at these frequencies is the fact that each circuit needs to be properly mechanically installed in a shielded box (usually an aluminium milled box) and provided with the necessary coaxial or waveguide input and output. Every time a new circuit is to be experimented one must build a new box to accommodate the PCB. Therefore (excepting a few audacious people) only certified or quasi certified designs are worth the effort and money.

The designs

To make the experimentation easier I decided to attack the problem the other way around. Only one mechanical design is necessary (or two if you want, explained later) and the several modules will all fit inside. This requires that all circuits must have the connections placed at the same position, which is not a restriction for the kind of

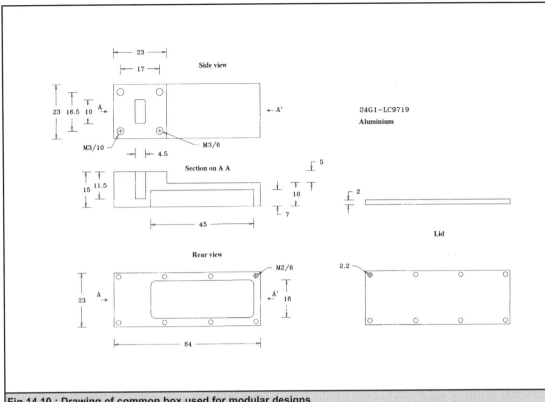

Fig 14.10 : Drawing of common box used for modular designs.

modules envisaged. The maximum number of FETs is likely to be 3, as in the DB6NT amplifier [11], all the other circuits are expected to be smaller than a 3 FET amplifier. I used this assumption to define the maximum room needed for a module, resulting in 45 x 16 mm internal dimensions. All power supply components will be external to save on the amount of PCB needed and also to have the smallest overall size possible. A configuration where a WG input and coaxial output (or vice versa) (fig 14.10) gives the maximum possibilities for interconnection. So a low noise amplifier will have WG input and coax output, while a power amplifier will have coaxial input and waveguide output and a frequency doubler would be coaxial input (at 12GHz) and WG output (at 24GHz). As input and output are at the same relative position, any of the modules can be placed inside the box either way

Fig 14.11 : Circuit of the low noise amplifier.

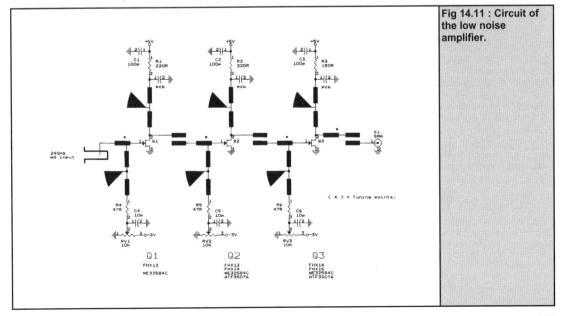

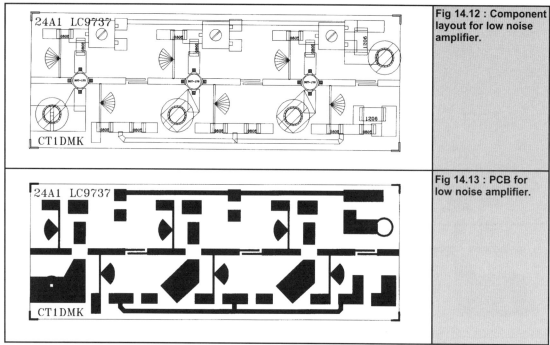

Fig 14.12 : Component layout for low noise amplifier.

Fig 14.13 : PCB for low noise amplifier.

round. If you want to have the same modules but only SMA connectors and no WG port a second milled case can be built. The WG input (or output) transition central pin is made of semi rigid UT085 cable, using only the teflon inner conductor, that fits inside a hole with the same diameter as the cables shield. The transition from the stripline to the waveguide is done in the same plane as the WG flange (that is, parallel to the bottom of the box, see drawing and photos for clarification). This arrangement resulted in little return loss and proved to be better than the 90° solution commonly used on 10GHz and 24GHz.

Low noise amplifier

The LNA consists of three HEMT stages connected via λ/4 couplers on 50Ω lines,similar to DB6NTs [11] but rearranged to fit inside the box and without negative bias voltage generation (fig 14.11). The first stage connects directly to the waveguide for the lowest possible loss, while the output is DC isolated with a λ/4 coupler. The prototype PCB uses only straight 50Ω lines (fig 14.12 and 14.13) and requires tuning on each FET (at gate and drain) by placing small stubs along the lines while monitoring the relevant characteristics (noise figure or gain). In this way some different types of transistors could be tested and their possible use on 24GHz evaluated.

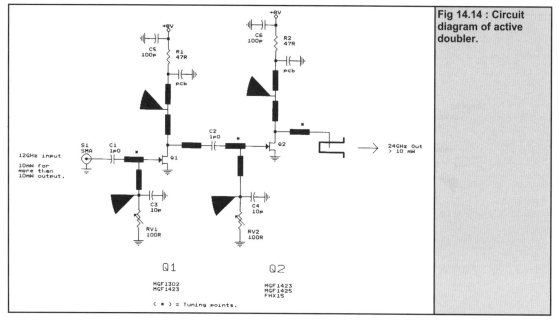

Fig 14.14 : Circuit diagram of active doubler.

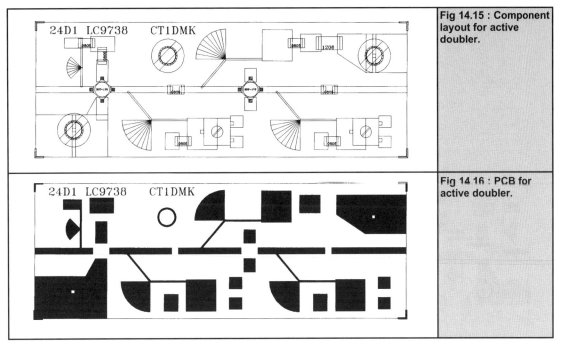

Fig 14.15 : Component layout for active doubler.

Fig 14.16 : PCB for active doubler.

The sets FHX13, FHX13, FHX14 and NE32463, NE32463, NE32563, were the ones that produced the best results. A gain better than 24dB, with noise figures below 2.0dB, should always be possible to obtain. The best I could get was from the NEC transistor set and was 26dB gain and 1.4dB (+/- 0.2dB) noise figure. Noise figure measurements where done with the Hot/Cold method using a load at different temperatures and also the sky of several clear days, as suggested by DK8CI in [17], for people without a good noise source and no liquid nitrogen at hand. To operate, the amplifier requires +5Volt and -5Volt from a external supply. The FETs bias should be adjusted for maximum gain or lowest noise figure on the first FET (also on the second if desired). Mitsubishi HEMT transistors were not available at my supplier, therefore I

could not test them, but they should work as well.

Power amplifier

A 10 milliwatt power amplifier can be built using the same strategy as the LNA. Basically the PCB design is the same, only with the $\lambda/4$ coupler now on the input and direct stripline on the output to the waveguide port. Also the FET biasing resistors are modified in order to make them draw more current from the 5V, especially on the last stage (reduce the R1, R2 and R3 to 47Ω). The HEMT transistors used here may be the ones with higher NF (lower cost) like the NE32863 or the FHX15, since in principle, they have all the same characteristics other than noise figure. Attempts with a MGF1303 and MGF1425 were done with good results but lower gain was obtained

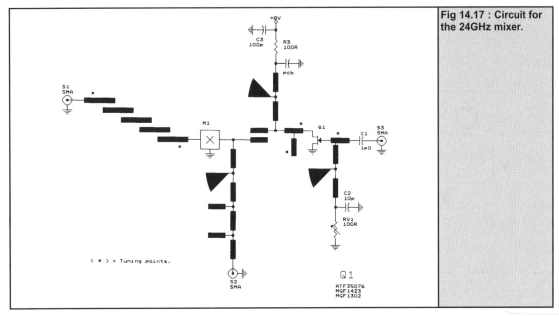

Fig 14.17 : Circuit for the 24GHz mixer.

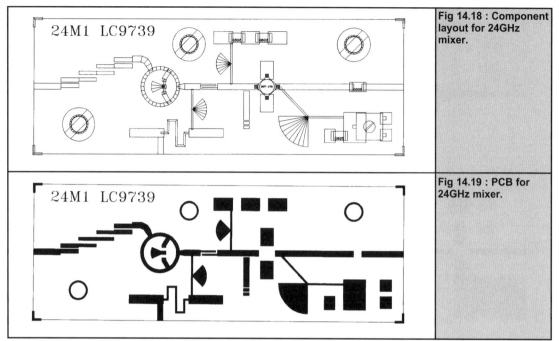

Fig 14.18 : Component layout for 24GHz mixer.

Fig 14.19 : PCB for 24GHz mixer.

(15dB) comparing with the HEMT devices (23dB). Output power, at 1dB compression, vary from device to device and are in the range of 8 to 12dBm. A 10dBm output is easy to obtain.

Frequency doubler

To use fundamental mixers on 24GHz (sometimes, good WG units are found at flea markets) a good source for local oscillator is mandatory. Also for those experimenting at 47GHz a good source of 24GHz is of interest. This design uses two transistors, the first as a buffer for the 12GHz, and the second as the multiplying device (fig 14.14). Again, to allow the test of several types of

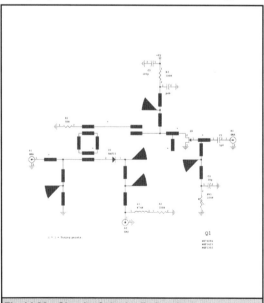

Fig 14.20 : Circuit of single diode mixer.

transistors, all signal lines are 50Ω and tuning is required to match every transistor. Tests were made using a MGF1303 plus a MGF1425 and more than 10dBm were obtained after quick tuning (for 10dBm at input). Levels as high as 15dBm may be reached by increasing the current and exhaustive tuning, but also the transistor may show its limits during the tests. If you don't have many transistors or dont like GaAs smoke, dont push much above 10 dBm. The PCB layout is shown in Fig 14.16 and the component layout in fig 14.15.

24GHz Mixer

This is the last module type presented here is a fundamental one to make a transverter. Different versions are presented, one with integral microstrip band pass filter and another without filter plus a PHEMT and single diode mixer. For transverter designs using 144MHz as intermediate frequency the filter required is too narrow to be feasible in a microstrip design. In this case a mixer module without filter should be used and installed with the input to the WG port, allowing direct connection to the waveguide filter. The module with integral band pass filter is suitable only for configurations having an IF greater than 1GHz since the filter has about 1.5GHz band pass at 3dB (fig 14.17). The filter design resulted in a compromise between insertion loss and bandwidth. The present design already has 3 to 4dB of insertion loss making the realisation of narrower filters impractical. Although the 1.5GHz bandwidth can be tweaked in such a way that it starts to cut just below the frequency of interest, making the image suppression greater than 30dB for an IF of 1296MHz. This module requires a local oscillator in the 12GHz range and it employs a GaAs FET as the local oscillator frequency doubler to get about 7dBm to drive a rat race mixer. The diodes should be the best that one can get. The BAT15 works reasonably well while the BAT30 (beam lead) should be the way to go for best performance. The PCB layout for the mixer is shown in fig 14.19 and the component layout in fig 14.18.

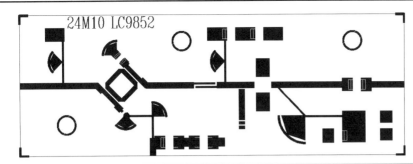

Fig 14.21 : PCB for single diode mixer.

Single diode mixer and PHEMT mixer

Balanced mixers using two diodes [12] will certainly require a little bit more power than a single device mixer. Also good balancing is desirable for good operation and therefore diodes are required to be as good a match as possible. The performance of the rat race mixers at 24GHz, using SMD packaged devices such as the BAT15 may not be as good as we would like. Conversion losses of 8 to 10dB can be obtained, but without much tuning effort figures like 15dB are most probable. These difficulties led me to try a single diode mixer design and a PHEMT mixer design, to take advantage of recent devices. Both modules share the same configuration the mixing element changes from one design to the other.

The PHEMT mixer resulted in a receive only mixer but with largely improved specification compared to any other types of mixers that I have tried so far.

Single diode mixer

The local oscillator doubler section remains unchanged, and it is expected to deliver as much as 10dBm at 24GHz to the input of the mixer. The mixer combines the RF and LO signals in a ring coupler that also provides the RF/LO isolation (Fig 14.20). The ring coupler design can be found in some detail in the literature [18], [19] and will not be described here. Several diodes from Siemens and NEC were tested and the results were in general 2dB better than the balanced design. One of the best encap-sulated devices in this configuration was the BAT15 in a plastic package. While the best results, 7dB conversion loss, were obtained from a single BAT30 in beam lead package.

Tuning the mixer for best performance is not too difficult, however the ring coupler has a sharp tuning point that is obtained by placing a small stub in the lower right corner of the ring (Fig 14.24). This adjustment must be done before any other optimisation otherwise the diode would not receive enough LO power. The PCB for the single diode mixer is shown in Fig 14.21.

PHEMT receive Mixer

The design using a PHEMT device is in all respects identical to the one using a single diode but uses a PHEMT transistor as the mixing element (Fig 14.22). Additionally this circuit has to include the transistor bias elements. Several transistors were tested, and most of them performed excellently. However the lowest noise figure devices were not the best for this application, as my tests showed. The best device tested was the NE42484 in which I could obtain the impressive figures of 3dB conversion gain and 4dB noise figure (as far as I could measure).

Tuning is similar to the single diode mixer design. There is also the gate bias to adjust this is somewhat critical to set. As the device is being used as a mixer, drain voltages and currents do not match the recommended values for the

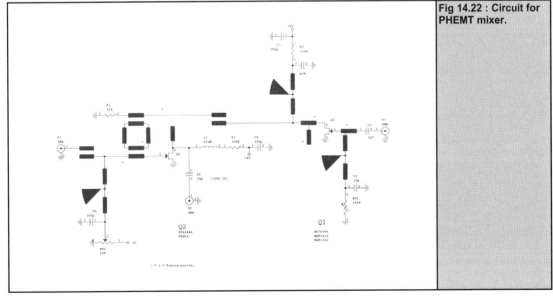

Fig 14.22 : Circuit for PHEMT mixer.

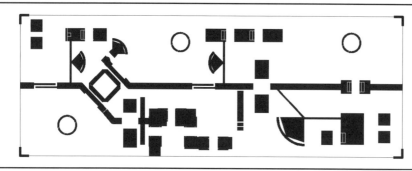

Fig 14.23 PCB for PHEMT mixer.

linear operation so experimenting is necessary to find the right values for gate voltage and for the drain resistor. The PCB layout for the PHEMT mixer is shown in Fig 14.23

Benefits of single diode and PHEMT mixers

These two mixers make it possible to reduce the amount of gain necessary in the receive chain, namely the LNA. Also, the PHEMT design, can reduce significantly the impact of mixer noise figure in the overall system noise. Single diode mixers as good as 7dB conversion loss are now easy to construct and PHEMT mixer can have the impressive 3dB conversion gain and 4dB NF that can even enable it to be used as the first element in simple transverter setups, or receive only stations Thanks to all that helped with info and components.

Construction

All PCBs are made of Rogers RT5780 10mils. After etching, 1.5mm holes are drilled at the grounding positions. Then solder them to a 0.8mm thick cooper sheet cut to fit inside the milled box. Some filing may be necessary for them to fit correctly. After cleaning, to remove the solder flux they are ready to be populated. The modules can be placed in and changed as many times as necessary. Assembly of all components should be made with the modules out of the aluminium box. The ground connection of the FETs sources should be done at the exact position where the ceramic case ends (to make the connection as short as possible). Apply all general rules for GaAs FETs and SMD described in so many other articles. Hardcopies of some PCBs at 4:1 scale can be found at my web page: http ://escriba.cfn.ist.utl.pt cupido

Conclusion

I believe that one or two more dB may be obtainable on any of the developed modules, but it is also my opinion that the tuning of such circuits becomes limited by hand movement precision and patience (that last no longer than one hour), rather than by the devices themselves. There is plenty of room for your skills here.

Amplifiers

Generating power at 24GHz is still an expensive business. The two semiconductor designs shown here represent what is currently achievable.

Super low noise HEMT amplifier [20]

Usually expensive and hard to get single chip capacitors have been used in 24GHz amplifiers. One solution to the problem of DC separation was the construction of home made capacitors made from PTFE copper clad substrate. The amplifiers which used this type of coupling were prone to oscillations on 8GHz, because of the box waveguide resonances.

The new amplifier uses coupled lines, which act as a band pass filter. The use of the NEC HEMT NE32584C proved to be another advantage. Even without tuning, the gain was more than 15dB. After tuning, a gain of 30dB is possible. Noise figure of the waveguide version measured as 1.5dB and of the coaxial version as 1.8dB. Fig 14.26

Fig 14.24 : Tuning the mixer.

Fig 14.25 : Picture of completed mixer.

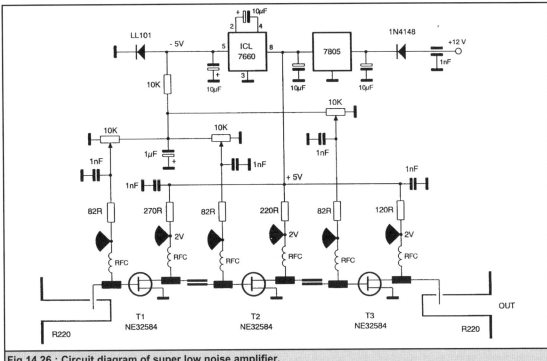

Fig 14.26 : Circuit diagram of super low noise amplifier.

shows the circuit diagram and Fig 14.27 the component layout.

Construction

The amplifier can be constructed as a coaxial version, Fig 14.28, or a waveguide version Fig 14.29.

Cut the PCB to fit into the box. Use the component layout to find the position for through board contacts, these are made using 0.1mm copper foil.

Clean the PCB and screw the PCB into the box. Epotek H20E silver epoxy cement should be used for a solid contact.

The waveguide transitions are made from UT85 semi rigid coax, use the mechanical drawing for details.

For the coaxial version microstrip launchers with 2.16mm thick PTFE (SMA) connectors are preferred.

Alignment

Apply supply voltage and adjust pots for 10mA drain current on each device. Drain voltage is about 2V. Drive the amplifier with 50µW and optimize the output power by applying small copper foils on the microstriplines. A saturating output power of 15mW should be reached. If you can lay hand on a noise figure meter, a tuning for minimum noise figure can be provided at the input.

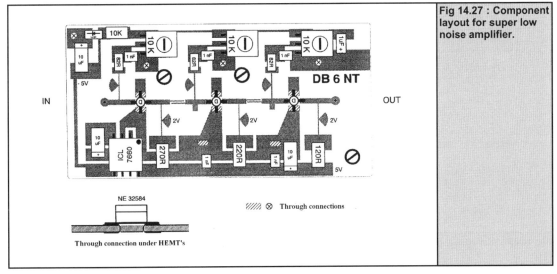

Fig 14.27 : Component layout for super low noise amplifier.

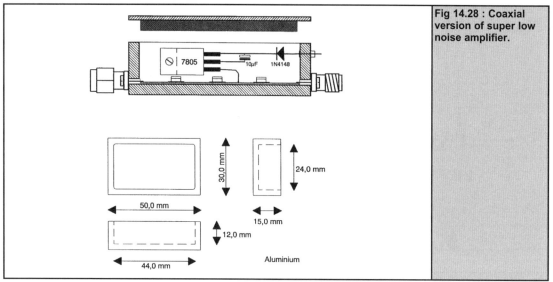

Fig 14.28 : Coaxial version of super low noise amplifier.

30,0 mm
50,0 mm
24,0 mm
15,0 mm
12,0 mm
44,0 mm
Aluminium

200mW GaAsFET Amplifier

Many stations have been constructed for the 24GHz band in recent years that can generate transmit power levels ranging from a few 100µW to a maximum of 100mW. These values could be obtained using low cost transistors (selected MGF 1303 parallel). The power gain obtained from a 100mW PA with MGF 1303 was approximately 13 dB.

A 200mW amplifier will be described with an amplification of 20 dB. FLRO16FH and FLRO26FH K Band Power FET's are used, from the Fujitsu company.

Building 24GHz amplifiers calls for great experience in the SHF range and for a lot of patience. You have to work for every dB or mW when you're calibrating, using "little flags" and putting MOS foam and copper strips into the housing!

The amplifier circuit

Fig 14.30 shows the 24GHz amplifier circuit. The amplifier is built on a 0.25 mm. thick Teflon printed circuit board made from RT/duroid 5870. Stages TI and T2 are connected to one another through 50Ω striplines and coupling capacitors, and are capacitively tuned by means of soldered on "little flags".

The parallel high level stages with T3 and T4 are connected to one another through four λ /4 couplers. This

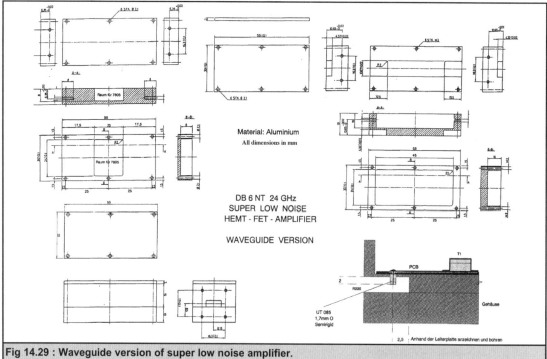

Material: Aluminium
All dimensions in mm

DB 6 NT 24 GHz
SUPER LOW NOISE
HEMT - FET - AMPLIFIER

WAVEGUIDE VERSION

Fig 14.29 : Waveguide version of super low noise amplifier.

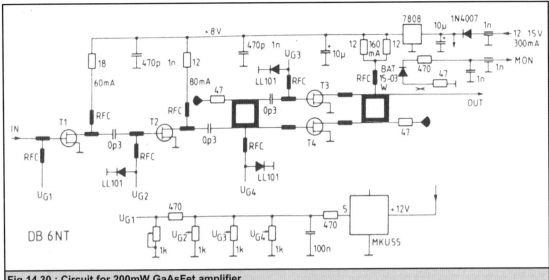

Fig 14.30 : Circuit for 200mW GaAsFet amplifier.

circuit option makes it possible to obtain good decoupling without feedback between the two end transistors, and makes a considerable contribution to the stability of the circuit. Should any asymmetry arise in the power stages, the radio frequency output this generates is absorbed in the resistor connected to port 4. These 47 - 50Ω resistors on the couplers are small SIU, types 1.4 x 1.4 mm. format.

Table 14.6 : Power output measurement for 3 test amplifiers.

P in (mW)	P out test set 1 coax (mW)	P out test set 2 WG (mW)	P out test set 3 coax (mW)
0.2	21		20
1.0	105		100
2.0	195		190
5.0	275		255
10.0	300	230	255
20.0	310	230	255

Di-Cap capacitors are used as coupling capacitors between the stages.

The negative voltage for the FET's is generated through an MKU 55 hybrid module. This is fastened to the internal wall of the tinplate housing. At the output, the circuit has a directional coupler with a BAT 15-03W SMD Schottky diode from Siemens. The power delivered can be monitored at any time at this test output using a moving coil instrument.

The LL 101 diodes are SMD Schottky diodes and help to delay a breakdown of the FET's in the event of any short circuiting of the coupling capacitors. SMA 3.5 mm. microstrip connectors should be used as coaxial jacks.

Assembly

The amplifier can be assembled in either a waveguide or a coaxial version. Different housings and printed circuit boards are used in each case.

First the 0.25mm thick Teflon printed circuit board (Fig 14.31, coaxial format) is cut to dimension, drilled and through hole plated. 0.1 mm, copper foil should be used for through hole plating.

The printed circuit board is now inserted into the aluminium housing as a drilling jig and the appropriate holes are drilled in the housing. The components shown in the layout diagram (Fig 14.32) are now placed on the track side (top) of the printed circuit board.

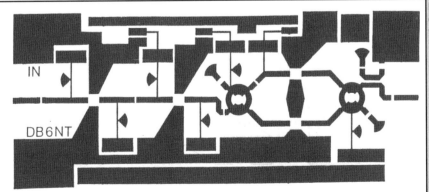

Fig 14.31 : PCB of coaxial format 200mW GaAsFet amplifier.

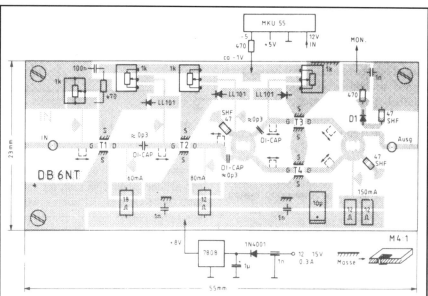

14.32 : Component layout for 200mW GaAsFet amplifier.

The BAT 15-03W diode is soldered "overhead" to obtain lower housing inductivity levels.

Before being put into the housing, the printed board assembly should be washed down with spirit and then aligned straight to guarantee that it is installed flat. The underside of the board is coated with heat conducting paste in the centre to improve the heat dissipation from the FET'S. The underside of the printed circuit board is coated with a second adhesive. Optimum mounting is obtained by pressing on the printed circuit board while the adhesive dries. This procedure must be done quickly and precisely, since the adhesive dries very quickly and no changes can be made to the printed circuit board's position thereafter. The printed circuit board is now fastened using 5 x M2 screws.

As an example, Fig 14.33 shows a milled two part aluminium housing.

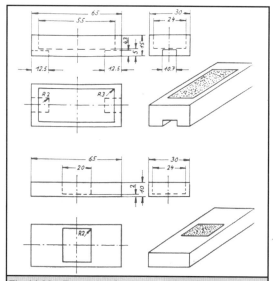

Fig 14.33 : Example of milled box for 200mW GaAsFet amplifier.

Calibration

A power supply with fine control current limit should be used for calibration. A 24GHz signal source with an adjustable output and a suitable power meter are required for calibration. The 1k trimmers are used to set UG1, UG2, UG3 and UG4 to give the drain curents shown in the circuit diagram (Fig 14.30)

Table 14.7 : Parts list for 200mW GaAsFet amplifier.

TI	FLR 016 Fujitsu
T2	FLR 026 Fujitsu
T3	FLR 026 Fujitsu
T4	FLR 026 Fujitsu
DI	BAT 15-03W Siemens
3 x	LL 101 SMD Schottky diode
1 x	7808
1 x	IN4001
1 x	MKU 55
4 x	1kΩ Minitrimmer
3 x	47Ω SHF type, Rohrer, Munich
2 x	470Ω 0204
3 x	12Ω SMD
1 x	18Ω SMD
1 x	10μF tantalum SMD
1 x	1nF SMD
1 x	100nF SMD
3 x	Di-Cap 0.2 - 0.5pF
2 x	1μF-DF, threaded

Measurements were carried out at various input powers, the results are shown in Table 14.6 An output of 200mW was reliably obtained from all of the test rigs.

Measuring equipment:

HP 435 Power Meter with BP 8485A Power Sensor, together with Midwest 550-20 damping unit and Rosenberger R 220 3.5mm wave guide transition.

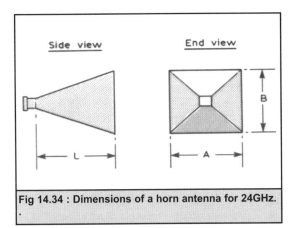

Fig 14.34 : Dimensions of a horn antenna for 24GHz.

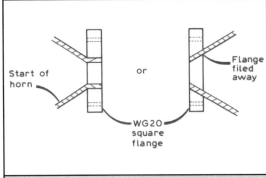

Fig 14.36 : Joining a horn directly to a flange.

Antennas

At 24GHz, very high gains can be achieved with quite compact antennas, and either horns or dishes will give useful gains. A reasonable sized horn will give 30dB gain (5° beamwidth) and will be fairly uncritical to build. Dish profiles will need to be accurate to within about 1mm but very high gains can be achieved. A suitable feed is described later. These very high gains will be associated with very small beamwidths, so some careful thought must be given to the construction of a rigid mount so that the antenna can be pointed accurately.

Suitable materials are pcb material, brass or copper sheet or tin plate. If pcb material is used, the doubled sided type will enable the joins to be soldered both inside and out for extra strength. An ideal source of tin plate is an empty oil can.

The transition from the guide to the horn should be smooth. Use either a butt joint or file the waveguide walls to a sharp edge, see Fig 14.35. The material should be cut to size and then soldered together and onto the end of a piece of waveguide. Alternatively the horn may be coupled directly to the inside of a flange. In this case the material should be the same thickness as the waveguide would be and it is bent as it enters the flange, or the flange may be filed to be part of the taper, see Fig 14.36.

Table 14.8 : Gain of 24GHz horns.

Gain	Beamwidth (degrees)	Length (mm)	A (mm)	B (mm)
15dB	30	26	31	25
20dB	17	81	55	45
25dB	9	270	98	79
30dB	5	810	174	141

Table 14.9 : Gain of 24GHz dishes.

Diameter	Gain	Beamwidth (degrees)
12 in	35dB	3
18 in	38dB	2
24 in	41dB	1.5

Horns

A horn is a very easy antenna to construct, as its dimensions are not critical and it will provide a good match without any tuning. The dimensions, as per Fig 14.34, for optimum gain horns of various gains (at 24GHz) are shown in Table 14.8. Above 25dB the horn becomes very long and unwieldy and it becomes more practical to use a dish.

Dish gains

Very high gains can be achieved at 24GHz with relatively compact dishes. However, the resulting narrow beamwidth will make accurate pointing difficult. The most useful size for portable working is probably 12in or 18in. The gains and beamwidths for various sizes of dish are shown in Table 14.9.

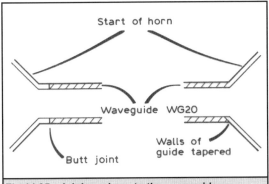

Fig 14.35 : Joining a horn to the waveguide.

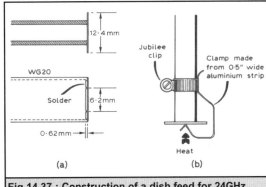

Fig 14.37 : Construction of a dish feed for 24GHz.

As at 10GHz, only solid dishes are suitable and the accuracy of the profile needs to be within λ/10, i.e. within about 1 mm.

Dish feeds

A convenient feed for a dish is the so called "penny feed" due to G4ALN. The dimensions of a version for 24GHz constructed in WG20 are shown in Fig 14.37 (a). The method used to construct the feed is as follows.

A piece of WG20 of sufficient length to reach the focus of the dish is cut, and its ends squared off by filing. The positions of the slots are marked out using vernier calipers and a right angle, and the slots filed out with a needle file. Repeated checking of the dimensions of the slots during filing will ensure accuracy, paying most attention to the length of the slots.

The end disc is made from 0.036inch (0.9mm) thick brass sheet. A 0.5inch (12.7mm) square piece of this is cut out and soldered to a 0BA brass washer. Using the washer as a guide, the corners are filed off until the brass is the same size as the 0BA washer. A small amount of further filing is then sufficient to reach the final size. The brass disc is then unsoldered from the washer, deburred, and the solder filed off.

The assembly of the disc on the waveguide requires special care to ensure accurate alignment. The clamping arrangement shown in Fig 14.37(b) is used to hold the disc firmly against the end of the waveguide. The disc is then moved around until it was centrally located, as indicated by measurement with vernier calipers. With the clamp still in place, the disc is soldered to the waveguide above a small gas flame, with the waveguide held vertically. Even though a minimum of solder is used, some solder may flow into the slots and this can be removed after soldering by cutting it away with a scalpel blade, followed by the insertion of the end of a junior hacksaw blade into the slots (after removing one of the pins from the hacksaw blade).

The assembly is completed by sliding a 1/8inch (3.175mm) thick brass plate, with a 0.25 by 0.5inch (6.35 x 12.7mm) slot filed in its center, on to the waveguide. This plate is for bolting to the dish to hold the feed in place. A homemade WG20 flange is then soldered on to the end of the waveguide. The assembly is held in the dish, and the feed slid backwards and forwards to find the point of maximum gain by listening to a remote signal source. The brass plate is then soldered in position, using a setsquare to ensure that the plate is perpendicular to the waveguide in both planes.

References

[1] From an article by John Stephenson, KD6OZH, QEX Nov 1999

[2] R. Campbell, KK7B, "A Clean, Low-Cost Microwave Local Oscillator," QST, July 1989, pp 15-21.

[3] P. Wade, N1BWT, "Building Blocks for a 10GHz Transverter," ARRL UHF/Microwave Projects Manual, Vol 2, pp 3-33 to 3-40.

[4] I use a surplus Efratom FRK-L rubidium oscillator obtained from Lehman Scientific, 85 Surrey Dr, Wrightsville, PA 17368. They will also perform calibration of oscillators, traceable to NIST. Dont try to calibrate these using WWV or WWVH. Received WWV and WWVH signals are only accurate to 0.1ppm because of propagation variations caused by short term instability of the ionosphere. These used oscillators sell for about $500.

[5] Vector Electronics products are available from most distributors and some local electronic parts stores in the US. Vector part number 169P84WEC1 is a 17 x 8.5 inch board. Vector part number P138 is a pad cutting tool.

[6] Low Cost Frequency Multipliers Using Surface Mount PIN Diodes, Application Note 1054, Hewlett Packard. (A PDF file of this note is available at http://ftp.hp.com/ pub/accesshp /HP-COMP /rf/4_downld /lit/diodelit /an105 4.pdf -Ed.)

[7] W. Hayward, W7ZOI and D. DeMaw, W1FB, Solid-State Design for the Radio Amateur, p 42

[8] V. Manassewitsch, Frequency Synthesizers-Theory and Design (New York: John Wiley and Sons, 1987)

[9] You can download this package from the ARRL Web http://www.arrl.org/files/qex/. Look for STEP1199.ZIP

[10] From an article by Luis Cupido, CT1DMK, Dubus 3/1998 pp 12 - 21 and 4/1999 pp 5 - 11

[11] Michael Kuhne, DB6NT, Low noise 24GHz amplifier, DUBUS 3/96

[12] Erich Zimmermann, HB9MIN, 24GHz Rat Race mixer, DUBUS, 2/93

[13] Erich Zimmermann, HB9MIN, Rauscharmer 24GHz Vorvestrker, DUBUS 1/94

[14] Michael Kuhne, DB6NT, 47GHz Transverter, DUBUS, 1/94

[15] T. Takamizawa, JE1AAH, 47GHz High performance components, MW update 96

[16] T. Takamizawa, JE1AAH,47GHz Active doubler, DUBUS, 4/94

[17] Hermann Hagen, DK8CI, in CT1WW Memorial 1997, Oliv.Azmeis/Portugal.

[18] Artech House, "Microstrip circuit design"

[19] E.H.Fooks, "Microwave engineering using microstrip circuits", Prentice Hall

[20] From an article by Michael Kuhne, DB6NT, Dubus Technik V pp 196 - 200

[21] From an article by Michael Kuhne, DB6NT, VHF Communications 3/1995 pp 147 - 152 pp 196 - 200

Chapter 15

The bands above 24GHz

In this chapter :

- Transmission lines
- Solid state devices
- Antennas

- Experimental equipment for 47GHz
- Laser transceiver

U ntil 1979 the highest frequency microwave band available to amateurs was 24GHz. The World Administrative Radio Conference 1979 (WARC 79) allocated a number of new bands above 24GHz to the Amateur and Amateur Satellite Services. They are listed in Table 15.1 and in the text they are referred to as the 47, 76, 120, 142 and 241GHz bands.

Since that time there has been comparatively little work done on these new higher bands, largely because of the non availability of suitable components to amateurs. Components such as mixer, detector, Gunn and multiplier diodes, GaAs fets and the like have been available to professional engineers, but their price has precluded amateur use. As a result, the development of amateur techniques and practical designs have lagged behind developments on other microwave bands. Therefore much of what follows in this chapter is tentative and based on the limited experiences of a few amateurs. There is some activity on these bands now (2002) with the world record QSO standing a 286km on 47GHz, 145km on 76GHz, 62km on 145GHz, 2.1km on 241GHz and 0.05km on 411GHz.

This chapter also introduces another area of the spectrum for amateur experimentation, the light wavelength. It is a relatively new field that has lots of potential for the keen constructors.

The Bands

The actual bands available to amateurs at present vary from country to country, but it is to be expected that at least the exclusive bands will be eventually allocated world wide. Parts of these bands are harmonically related to presently active lower frequency microwave bands and the International Amateur Radio Union (IARU), part of the International Telecommunications Union (ITU), has recommended that initial operation in the new bands should use these harmonically related frequencies. It is thus possible to verify the frequency of operation by listening for harmonics from transmitters in the bands below and, eventually, when the necessary techniques have been developed, it will be possible to generate RF at the higher frequencies by frequency multiplication.

Fig 15.1 illustrates some of these harmonic relationships together with some other useful relationships *not* related to amateur frequency allocations. Fig 15.2 gives some simple relationships based upon harmonics from readily available Gunn sources in the 23 to 25GHz range. Some of the successful European work, described later, has

Table 15.1 : Amateur millimetre bands.

Band	Status
47.0 - 47.2GHz	Exclusive
75.5 - 76.0GHz	Exclusive
76.0 - 81.0GHz	Secondary to radio location
119.98 - 120.02GHz	See footnote
142.0 - 144.0GHz	Exclusive
144.0 - 149.0GHz	Secondary to radio location
241.0 - 248.0GHz	Exclusive
248.0 - 250.0GHz	Secondary to radio location

These bands with the exception of 120GHz are allocated to the amateur satellite services.

been based upon multipliers from frequencies in this range. The awkward multiplication factor of x7 from 144MHz is best avoided, for instance by generating 1,008MHz by some other route and providing amplification to a satisfactory power level at 1,008MHz, before further multiplying up into some of the millimetre bands.

It is also practical to start at higher frequencies in the tables; thus an oscillator at 8.064, 16.128GHz or in the range 23 to 25GHz would be a prolific source for the millimetre bands.

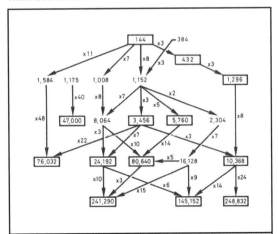

Fig 15.1 : Harmonic relationships of the millimetre amateur bands. Amateur band frequencies in MHz are shown in boxes.

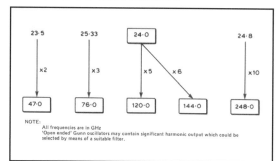

Fig 15.2 : Simple harmonic relationships derived from tuneable Gunn oscillators. Frequency in GHz.

Propagation

All of the propagation phenomena known on the lower microwave bands can be expected to manifest themselves in the millimetre bands, but the scale of the phenomena will be smaller. Effective ducts may be only a few metres thick. Smaller objects are effective reflectors and local weather may be of more consequence than is the case on the lower frequency bands.

One phenomenon appears on the millimetric bands, which is of much less importance on bands below 20GHz, atmospheric absorption. We are used to referring to the atmosphere as transparent to light, but a moments thought will remind us that we can see further on a clear day than a misty one. The same effect manifests itself with radio frequencies above about 20GHz. Depending on the humidity, there is a variable loss due to water vapour, there is a fixed loss due mainly to oxygen and finally there is a considerable occasional loss due to rain (or other hydrometeors).

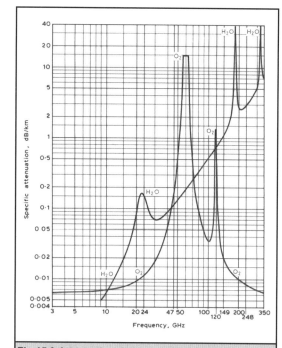

Fig 15.3 Attenuation due to oxygen and water vapour at pressure of 1 atmosphere, temperature 20°C, water vapour 7.5g/m³ (source CCIR)

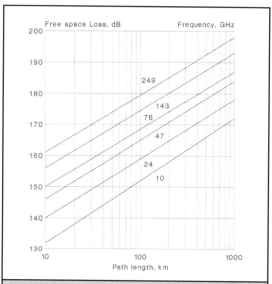

Fig 15.4 : Free space path loss versus frequency for the bands from 10 to 249GHz (except 120GHz).

Loss due to absorption follows a totally different law to that associated with radiation, the loss is a *linear* function of distance and is expressed in dB/km. If there is a loss of, say 5dB in the first km, then there is the same loss in each subsequent km. This causes the loss to build up much more rapidly than is due to the inverse square law of radiation and, sooner or later, the loss due to absorption will determine the maximum range that can be worked.

Figure 15.3, taken from CCIR reports, shows the attenuation per km on a horizontal path due to oxygen and water vapour (at 7.5g/m³), the two gases, which cause the greater part of the losses. *This is in addition to the normal free space loss.*

At a Microwave Round Table [1], G8AGN gave a valuable assessment paper which looked in more detail at the effects of atmospheric absorption and rain attenuation and considered the potential of the amateur millimetre bands. Much reference was made to the 24GHz band as a model, because it may be considered as a longer mm waveband and also lies close to the principal atmospheric water absorption band. The remainder of this section is based largely on that paper and it should become apparent that the mm bands above offer more potential than at first sight.

Free space loss

For line of sight paths, the free space path loss (in dB) may be calculated from the classical formula:

$$Loss = 92.45 + 20\log(f) + 20\log(d)$$

where f is the frequency in GHz and d is the path length in km. Fig 15.4 shows this relationship in graphical form for all the bands above 10GHz, with the exception of 120GHz.

Water vapour and oxygen attenuation

Fig 15.3 has already given a general overview of the additional path losses to be expected due to water vapour and oxygen absorption. The loss due to oxygen is fixed and there is nothing that can be done about it. In

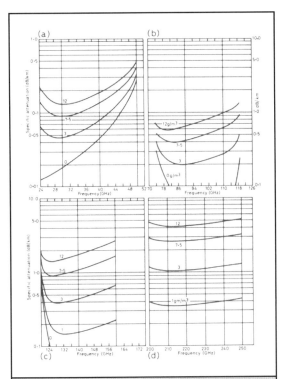

Fig 15.5 : Specific attenuation due to water vapour. (a) 24 to 52GHz. (b) 70 to 126GHz. (c) 102 to 176GHz. (d) 200 to 250GHz. The figures on the curves indicate the water content of the atmosphere, expressed in g/m³ (Source JPL model).

Table 15.2 : Estimated atmospheric absorption loss.

Band	Loss
10GHz	$\alpha = 0.0066 + 0.0011W$
24GHz	$\alpha = 0.012 + 0.0185W$
47GHz	$\alpha = 0.13 + 0.016W$
76GHz	$\alpha = 0.2 + 0.034W$
142GHz	$\alpha = 0.152W$
241GHz	$\alpha = 0.417W$

Loss (α) in dB/km, W in gm water/cubic metre

amount of water held in the air and the absolute humidity. This may be quite high in the summer but very low on a cold night in the winter.

A more useful picture is therefore given by the data derived from a Jet Propulsion Laboratory model, which is shown in Fig 15.5. Here absorption predictions for several values of atmospheric water vapour content are shown.

From these curves, a set of empirical relationships between absorption loss (in dB/km) and atmospheric water content (in gm per cubic metre) have been derived for the amateur bands above 10GHz and these are shown in Table 15.2. These relationships consist of two terms, a constant term, due largely to oxygen, and a term proportional directly to the water vapour content.

For these to be of use, a knowledge of the atmospheric water content over the path of interest is thus required but such information is not directly available and hence must be sought. Fortunately it is at hand in the form of the relative humidity (RH). This may be defined as:

Actual water content of air in gm/m³ at temperature T
Water content in gm/m³ if air saturated at temperature T

Note that RH is thus expressed as a percentage.

It is important to note that RH is mainly dependent on the temperature and, although not obvious from the formula given above, it is also slightly dependent on pressure, but this is ignored here. The weight of water, W, held in saturated air at different temperatures is given by the curve shown in Fig 15.6. An approximate relationship between W and T has been derived from this. Hence, if the air temperature and RH can be measured, then the actual water content and, thus, the additional absorption term in the path loss can be determined.

The most satisfactory way for the radio amateur to measure RH is using a pair of mercury thermometers, one of which has its bulb cooled by evaporation from a water wetted wick surrounding it. This records the so called wet bulb temperature. Provided that the air flowing over the bulbs is moving at a velocity of at least 1m/sec, then the RH can be determined from simultaneous reading of both thermometers, as outlined in Table 15.3. Such measurements are facilitated by using an instrument known as a whirling psychrometer - two identical thermometers, one dry and one wet, mounted side by side in what looks like an old fashioned football supporters rattle. This is whirled rapidly round and the wet and dry bulb temperatures read off the two thermometers without delay. Calculation is eliminated by use of a psychrometric table supplied with the instrument. It becomes merely a matter of reading off the dry bulb temperature against the difference dry minus wet (depression) to obtain RH directly.

particular, at 120GHz there is an absorption band due to oxygen. The maximum loss occurs just below the amateur band at 118.75GHz but, even so, a loss of around 0.5dB/km is to be expected. This is enough to make long distance contacts unlikely. The loss due to water vapour, however, is not fixed, it is dependent upon the total

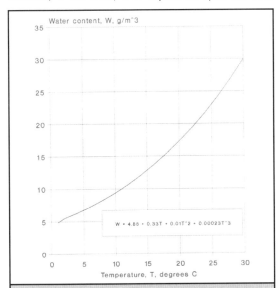

Fig 15.6 : Graph of weight of water, W, in saturated air plotted as a function of temperature, T.

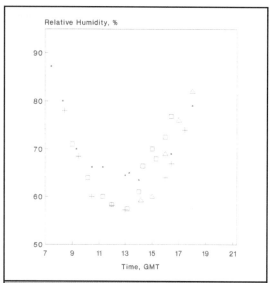

Fig 15.7 : Typical diurnal variation of humidity in northern England as measured over several days by G8AGN.

Fig 15.7 shows the typical variation of RH throughout several days, as measured by G8AGN in the northern UK. At first sight it would appear that the most favourable time for operation on the mm bands would be in the early afternoon. This figure does not, however, tell the whole story since it is the *actual* water content of the air, which matters, and this is dependent on temperature. Thus, even though the RH may be quite low around the early afternoon, the air temperature will be high, as will be the amount of water that the air could hold if saturated. In practice, therefore, the optimum time for operation may be determined by measuring both RH and air temperature at both ends of the path and estimating the increased path losses due to water absorption.

Table 15.4 gives the maximum water content in the atmosphere at various temperatures and the corresponding attenuation for the four principal millimetre bands. Below 0°C, the absolute humidity falls rapidly so that, at least in the 47 and 76GHz bands, the loss due to water

Table 15.3 : Determination of relative humidity.

$$RH = \frac{100\left(e_w - AP(T_d - T_w)\right)}{e_d}$$

P = atmospheric pressure in mB (milliBar)
T_d = dry bulb temperature in °C
T_w = wet bulb temperature in °C
e_d = Saturated water vapour pressure at T_d
e_w = Saturated water vapour pressure at T_w
A = A constant dependant on the velocity of the air flowing over the thermometer bulbs

For an air velocity of 1 to 1.5m/sec and T_d 0°C

A = 0.000799

e_d and e_w are related to T_d and T_w by:

e = $t^{4.9283}$ + $10^{(23.5518 - (2937.4/t))}$

where t = T + 273 (i.e. degrees K)

Table 15.4 : Effect of water vapour at various temperatures.

Temp (°C)	Max water Content (gm/m^3)	Attenuation (dB/km) 47GHz	76GHz	142GHz	241GHz
0	4.8	0.20	0.36	0.73	2.0
5	6.8	0.24	0.43	1.03	2.8
10	9.4	0.28	0.52	1.43	3.9
15	12.8	0.33	0.64	1.95	5.3
20	17.3	0.41	0.79	2.63	7.2
25	23.1	0.50	0.99	3.51	9.6
30	30.4	0.62	1.23	4.62	12.7
35	39.6	0.76	1.55	6.02	16.5
40	51.0	0.95	1.93	7.75	21.3

vapour can be ignored. The figures are maxima for the bands when the air is saturated with water, i.e. 100% relative humidity. The actual loss can be taken as proportional to the relative humidity at any temperature. In temperate climates, 100% relative humidity over a large area is uncommon, but 90% occurs sufficiently frequently to limit the range of regular contacts. In arid regions the humidity may be less than 10% for much of the time. It follows that the best way of working very long ranges on the millimetre bands is either to operate in an arid climate or on a freezing winter's night!

Rain attenuation

The other main factor limiting long distance contacts is the increased path loss (a) due to rain attenuation. This may be estimated by using the relationship:

$$\alpha \text{ (dB/km)} = A x R^B$$

where R is the rainfall rate in mm/hour and the constants A and B are given in Table 15.5. Actual values of rain attenuation may be estimated using the curves of Fig 15.8, given that typical rainfall rates are 0.25, 1, 4 and 16mm/hour for drizzle, light, moderate and heavy rain respectively. In practice, such rainfall rates do not apply uniformly along the whole path, since rain is often concentrated in localised cells which are typically 1 to 10km in extent.

Band Capabilities and Uses

The 47 and 76GHz bands

The 47 and 76GHz bands can be used for point to point operation in the same way as the 10 and 24GHz bands, with the proviso that, at least in Europe and the coastal

Table 15.5 : Estimated attenuation due to rain.

Attenuation $\alpha = AR^B$ (dB/km)		
R rainfall mm/hour		
A = cf^d (frequency in GHz)		
B = ef^g (frequency in GHz)		
For f<2.9GHz	c=6.39 x 10^{-5}	d=2.03
f>2.9 <54GHz	c=4.21 x 10^{-5}	d=2.42
f>54 <180GHz	c=4.09 x 10^{-2}	d=0.699
f>180GHz	c=3.38	d=-0.151
For f<8.5GHz	e=0.851	g=0.158
f>8.5 <25GHz	e=1.41	g=-0.0779
f>25 <164GHz	e=2.63	g=-0.272
f>164GHz	e=0.616	g=0.0126

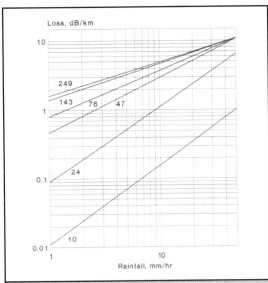

Fig 15.8 : Attenuation in the amateur millimetre bands due to rain plotted as a function of the rate of rainfall.

regions of America, water vapour will limit ranges to a few tens of kilometres for most of the time. Fig 15.9 shows the expected path losses at 47GHz with the capability of simple wideband equipment and state of the art low power narrow band equipment marked for reference. Fig 15.10 shows the expected path losses at 76GHz.

Fig 15.11 shows the path lengths possible at 47GHz for wideband equipment based, perhaps, on generating some second harmonic from a 23.5GHz oscillator and using a harmonic mixer. For comparison, the first world record for this band, a little over 50km, was made using just such wideband equipment. Fig 15.12 shows that paths of 10 to 20km should be possible at 76GHz using very modest equipment, again based on harmonic transmitters and mixers.

It can be seen that, even with very modest equipment, the 47 and 76GHz bands offer some considerable scope for experimentation and dx working. As pointed out later, it is expected that mm wave devices will be used more widely in both the professional and consumer areas and so will become available to amateurs on the surplus market. When this happens, then the prospect of narrow band operation at these frequencies will transform the situation and allow longer distance contacts than those predicted here, certainly on the lower mm wavebands.

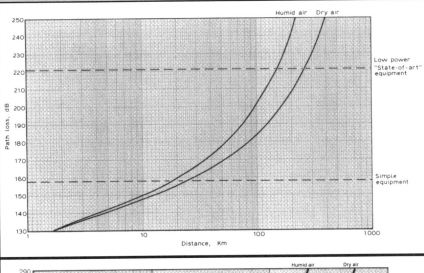

Fig 15.9 : Path loss versus frequency for the 47GHz band, for both dry and humid conditions.

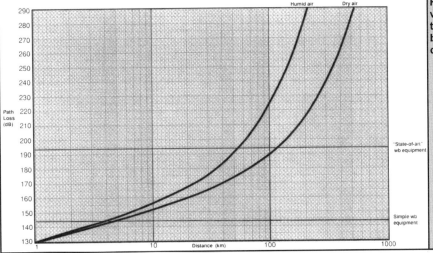

Fig 15.10 : Path loss versus frequency for the 76GHz band, for both dry and humid conditions.

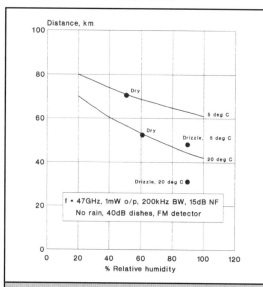

Fig 15.11 : Predicted performance of simple wideband equipment at 47GHz.

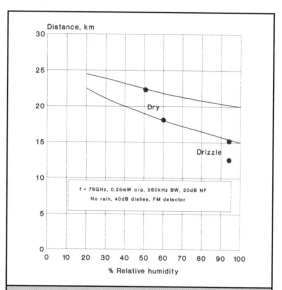

Fig 15.12 : Predicted performance of simple wideband equipment at 76GHz.

The higher mm bands

The 120GHz band is on the edge of an absorption band due to oxygen. The loss from this gas of about 0.4dB to 0.5dB per kilometre is more than that to be expected from water vapour and is such that quite potent equipment would be required for communication over any great distance. What this band does offer is privacy, because long distance transmission is practically impossible, so

are interference and eavesdropping.

Possible uses of the 142 and 241GHz bands are more problematical. Consolation can be found in the fact that, apart from the inevitable move to higher frequencies as the lower bands fill, no clear consensus has appeared among professional users how best to exploit the special characteristics of the higher millimetre bands.

Table 15.6 : Waveguide for the millimtre bands.

Type No	Outside dimension (mm)	Inside dimension (mm)	Cut off (GHz)	Band (GHz)	λ_0 (mm)	λ_g (mm)	λ_g/λ_0	Loss (dB/m)
WR42 WG20	12.7/6.35	10.67/4.32	14.047	24.0	12.49	14.41	1.234	0.45
K&S 268	9.53/4.76	8.74/4.00	17.1	24.0	12.49	18.18	1.454	0.60
WR28 WG22	9.14/5.59	7.11/3.56	21.07	24.0	12.49	26.11	2.090	0.75
K&S 266	7.93/3.97	7.14/3.18	21.0	24.0	2.49	25.72	2.060	0.75
WR22 WG24	7.72/4.88	4.78/2.39	26.24	47.0	6.38	7.70	1.208	0.78
K&S 264	6.35/3.18	5.55/2.38	26.98	47.0	6.38	7.79	1.221	0.90
K&S 262	4.76/2.38	3.96/1.60	37.8	47.0	6.38	18.06	2.830	1.8
WR12 WG26	5.13/3.58	3.10/1.55	48.35	76.0 81.0	3.94 3.70	5.11 4.61	1.296 1.247	2.5
WR8 WG28	3.96 dia	2.03/1.02	73.84	120.0	2.50	3.17	1.268	5.0
WR7 WG29	3.96 dia	1.65/0.825	90.84	142.0 149.0	2.11 2.01	2.75 2.54	1.301 1.261	6.0
WR5 WG30	3.96 dia	1.30/0.648	115.75	142.0 149.0	2.11 2.01	3.64 3.19	1.725 1.587	9.0
WR4 WG31	3.96 dia	1.09/0.546	137.52	241.0 250.0	1.24 1.20	1.51 1.44	1.215 1.196	12.0

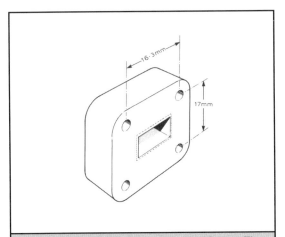

Fig 15.13 : General purpose waveguide flange, The holes are 0.144 in. for either 4BA or M3 bolts. The waveguide is central with respect to bolt holes. Overall dimensions are not critical.

Transmission Lines

Waveguide and flanges

Table 15.6 lists the standard waveguide sizes suitable for the amateur millimetre bands together with a range of rectangular brass tubing available from model shops and which can be used for the 47 and 76GHz bands. The quite high losses to be expected from standard rectangular waveguide should be noted. The values given in Table 15.6 are for copper; brass waveguide will give figures some 30% worse and silver about the same amount better.

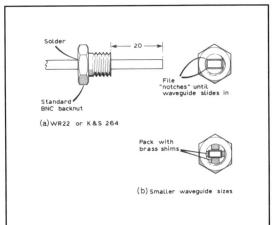

Fig 15.14 : Use of BNC or TNC connector instead of a flange.

Two flange types are available for millimetre waveguide, square flanges with four bolt holes and round flanges with removable, screwed rings. The latter are more popular among professional users but are more expensive. Fig 15.13 gives the dimensions of a general purpose waveguide flange much used professionally for test gear. The same sized flange is used with a range of waveguide sizes and it can therefore be used as a standard connector between non standard waveguides.

As an alternative to the use of flanges, it is possible to adapt popular coaxial connectors for use with waveguide. Fig 15.14 shows the use of a BNC or TNC connector in this way. Since these two connectors have the same basic dimensions differing only in the use of a bayonet or

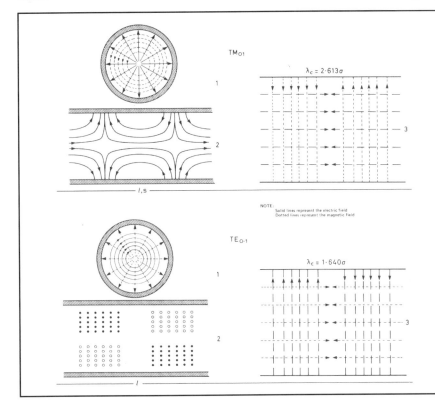

Fig 15.15 : TM and TE modes in circular waveguide. Solid lines represent E field and dotted lines the H field.

© Artech House Inc. from Theodore S Saad, Microwave Engineers Handbook Vol2 1971. Reprinted by permission.

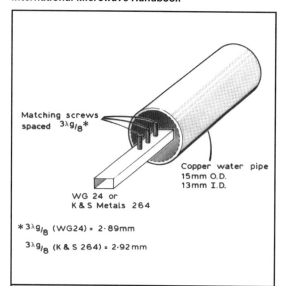

Fig 15.16 : Mode transition for rectangular to circular waveguide.

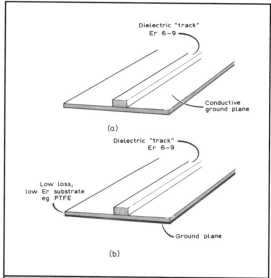

Fig 15.18 : Dielectric waveguide.

screwed clamp, it is possible to use the same arrangement with BNC fittings for easy connection and disconnection in the workshop and TNC fittings for a more reliable connection in the field. The internal coaxial components of either type of plug are discarded and for the larger waveguides the connector will need to be drilled through to allow the waveguide to fit.

Overmoded waveguide

As can be seen from Table 15.6, the loss of conventional waveguide is prohibitive, except for very short runs inside the equipment. If we wish to operate with an elevated antenna, which involves a long transmission line, then the losses associated with a waveguide run are too high and other techniques are required. If a waveguide is larger than necessary to propagate the fundamental waveguide mode then a variety of other modes are possible in the guide. If not controlled, this overmoding is undesirable but, if controlled, particular modes may propagate with much lower loss than the fundamental TE10 mode used in conventional rectangular waveguide.

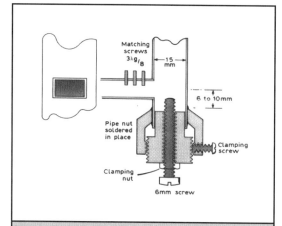

Fig 15.17 : Launching the TM01 mode, rectangular to circular guide transition.

The most useful low loss modes occur in a circular tube, and are the TE01 and TM01 modes. They are symmetrical, the TM01 mode has a radial electric field and a circumferential magnetic field, while the TE01 mode has a radial magnetic field and a circumferential electric field. These are shown in Fig 15.15. In each case the fields are low at the circumference of the metal guide; this means that the currents flowing in the waveguide wall are minimal, so the losses are correspondingly small.

The TE01 mode shows the lowest loss, a 15mm diameter copper tube operating at 47 or 76GHz can be expected to have a loss of not more than 0.01dB/m. The mode can be launched from a rectangular guide by the arrangement shown in Fig 15.16. The transition is quite badly mismatched and matching screws are required in the rectangular guide as shown. The TM01 mode has a higher loss, a 15mm copper tube as above would have a loss of 0.1 dB/m. This mode is much easier to launch by the arrangement shown in Fig 15.17 and, for the short feeder runs is likely to be used by amateur stations, the additional loss is not important. Since these waveguide modes are radially symmetrical, the connections at the two ends need not be in the same plane. If the circular guide is run up a mast, the antenna can be arranged to rotate about the centre line of the circular guide and a simple choke joint is all that is required to provide for rotation.

It is important that the overmoded waveguide runs are kept straight, the effect of bends is to couple between modes, and energy is transferred into other higher loss modes. These modes are also mismatched at the terminations, so that very high standing waves are set up and a variety of unacceptable resonances will be produced in the guide, leading to further losses.

Other transmission lines

Whilst amateurs are most likely to use waveguide, it has several disadvantages for the professional user, including, the need for close tolerances in building components. It is difficult to use mass production techniques and waveguide does not lend itself to integration techniques.

Table 15.7 : Comparison of transmission line losses.

Transmission line	Loss at 47GHz (dB/cm)	Loss at 76GHz (dB/cm)
Rectangular guide (copper)	0.011	0.022
50Ω microstrip (on quartz)	0.112	0.225
Dielectric guide (alumina)	0.039	0.078

There are alternative transmission lines that can be used to overcome these problems, one is microstrip, which is popular at centimetre wavelengths, but at millimetre wavelengths the losses start to become excessive. An alternative is dielectric waveguide.

Dielectric waveguide

This consists of a rectangular strip of dielectric and is normally used spaced from a ground plane by a separate, thick dielectric layer, as shown in Fig 15.18. The electro magnetic field is confined in the rectangular dielectric due to its refractive properties, much as optical fibres behave with light. It can provide fairly good performance, much better than that of microstrip, as shown in Table 15.7. Most of the familiar waveguide components can be made from dielectric waveguide, directional couplers, attenuators, phase shifters, filters and isolators are all used professionally.

Coaxial lines

Coaxial cables are almost impractical, in the 47GHz band and above, because of overmoding, which occurs when the circumference of the cable approaches one wavelength in the dielectric. For example, at 47GHz the largest coaxial line with polythene dielectric (dielectric constant of 2) that can be used would have a circumference of:

$$\frac{300}{47 \times \sqrt{2}} = 4.5mm$$

giving a diameter of 1.4mm. At 76GHz, the maximum safe diameter is 0.9mm. Cables of such diameter are available (small diameter semi rigid cable), but are generally too lossy to be useful.

Optical transmission lines

In the higher frequency bands it is possible to dispense with the metal walls of the waveguide by focusing radiation into a sufficiently tight beam and catching all of the radiated power in a similar collector. By using a high gain antenna which produces an almost parallel beam for a short distance and collecting the power in another antenna of similar size, it is possible to transmit power over tens of metres with relatively low loss. The losses due to side lobes or other defects of the antennas are likely to be less than those from the same length of conventional waveguide. The antennas are necessarily located within each other's near fields and best results will be obtained with identical antennas at the two ends of the link.

Solid State Devices

Devices for the millimetre bands are not commonplace. There are only a few manufacturers of both oscillator and

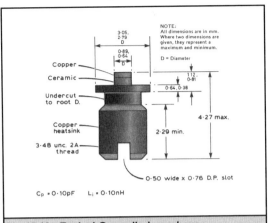

Fig 15.19 : Typical Gunn diode package.

detector/mixer diodes up to about 100GHz, notably Alpha Industries and M/A-COM (Microwave Associates). Devices at these frequencies are often listed in a catalogue by band (as in X band for 10GHz), but confusion can arise since there are several letter designation systems and some frequency bands are different in each system. Common designations are:

- Ka band, covering 18 to 26.5GHz. This stands for K above, since K band is 12.4 to 26.5GHz, and Ku (for K under) is 12.4 to 18GHz.

- Q band, covering 40 to 75GHz. However, the range 40 to 60GHz is sometimes referred to as U band (e.g. by Alpha Industries).

- W band, covering 75 to 110GHz.

Oscillators

Klystrons may be found and semiconductor oscillators, Gunn devices and impatt diodes are available for the lower frequency millimetre bands. Low power Gunns will give around 10mW of output power, while high power devices giving some 100mW are available but are very expensive. They are usually operated with 4.5 to 5.5V bias, similar to 24GHz devices. A typical package for a device operating in the 40 to 75GHz range is shown in Fig 15.19.

Mixers and detectors

Probably most diodes in the pill package (SOD31, SOD45, SOD46 and similar outlines) will function as useful rectifiers even if they are not too good as mixers. In general, the chips used in modern microwave diodes are usable up to at least 100GHz, it is the mount which sets the limit to the frequency of use and the problem is getting the RF into the chip. Techniques to overcome this are described later.

Quite low noise figure diodes are available using Schottky baffler construction. Noise figures as low as 8dB at 60GHz are attainable. Like high power Gunn diodes, these better performance devices are very expensive.

Multipliers

At millimetre wave frequencies, multiplication usually means high order multiplication and step recovery diodes are available, suitable for output frequencies of up to

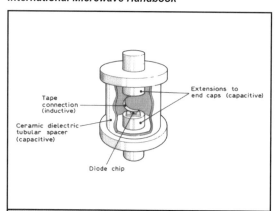

Fig 15.20 : Internal construction of a pill mixer diode.

100GHz. They, too, are expensive and the amateur is probably better off trying to use lower frequency devices. These will function with reduced efficiency, but should give enough output power for a receiver local oscillator or a harmonic generator/calibrator.

Techniques

Mounting components

Most of the readily available pre packaged devices are physically large compared with a wavelength and it is quite impracticable to arrange connections to terminals in the sense appropriate to lower frequencies. Usually the actual semiconductor chip is physically small and, in devices designed to operate efficiently at frequencies of 20 or 30GHz, it is usual for the chip to perform quite well at several times that frequency. The problem is, then, to get the RF into or out of the chip.

Fig 15.20 illustrates the internal construction of a typical mixer diode in a pill (SOD31) package, showing the device chip mounted on one cap of the package and tape connection made from the other cap. Gunn devices, varactors and pin diodes usually have similar arrangements. Since the physical size of the device package is an appreciable part of a wavelength even at 47GHz, we need to find some way to bypass the terminals. The technique used in waveguide mixers is, therefore, to immerse the diode package in the RF field and to compensate as best

we can for the reactance introduced by the packaging.

The ceramic package is equivalent to bulk capacitive susceptance across the waveguide and is tuned out by a sliding short behind the diode. A series of tuning screws then match the resistive term and tune out the internal inductance of the mount. Since we are interested only in relatively narrow bandwidths, the whole 47GHz band is only 0.43% wide, the restricted bandwidth introduced by these matching methods is of no consequence. The method is illustrated in the description of a mixer mount for 47GHz, which follows later.

Self oscillating mixers

A receiver can be built using a Gunn diode as both the local oscillator and a mixer - it is referred to as a self oscillating mixer. This technique means that a complete receiver (and transmitter, of course) can be built which requires only one expensive semiconductor device. Construction is also greatly simplified , only one diode mount need be built and, of course, there is no need for a cross coupler.

Fig 15.21 shows the cross section of a 47GHz self oscillating mixer assembly, with separate dc and IF feeds. The oscillator uses a plunger tuning arrangement. One problem to watch out for is the occurrence of low frequency parasitic oscillations, which are normally suppressed in Gunn oscillators by a shunt capacitor across the choke. A capacitor cannot be used here because it would shunt the IF signal to ground, in the example given, the Gunn was operated at the highest permissible voltage and this cured the problem. At these frequencies the receive performance can approach that of a separate oscillator and mixer and a system built in this way is an easy way of getting started on the band.

Antennas

Horns

The simplest antenna to make and use is a horn, quite large gains, e.g. 30dB, can easily be achieved in a compact size. For higher gains, a parabolic dish can be used. There are other alternatives, e.g. lenses and microstrip antennas, but these are really only of practical interest to professional users.

Methods described for 10 and 24GHz horns can be adapted to make horns for these higher frequencies. A satisfactory method is to make a former from hardwood and build up the horn soldering the four sides on a cut and try basis, and finishing off with a wrapping of copper wire or some reinforcing plates at the joint of the horn with the waveguide. The same former can be used for several horns, producing transmit and receive antennas for two or more experimental stations. Some examples of calculated horn dimensions for 47 and 76GHz are given in Fig 15.22.

Dishes

Small dishes are useful as directional antennas but large dishes are hopelessly impracticable. For instance, a gain of 46dB implies a beamwidth of less than one degree in both azimuth and elevation, requiring an accurate knowledge of the location of the station to be worked and a very rigid mounting to enable the antenna to be aligned and kept pointed correctly. Exceptionally precise setting up would be required to point the dish at the other station, probably involving optical sighting tubes or telescopes.

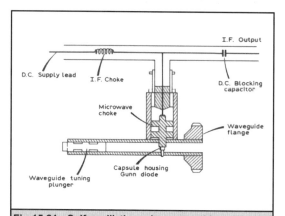

Fig 15.21 : Self oscillating mixer assembly.

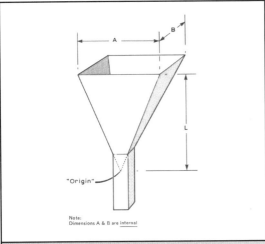

Fig 15.22 : Typical horn dimensions for 47 and 76GHz.
For 47GHz WG24 is used.
20dB horn: A=28.5, B=22.3, L=33.2
30dB horn: A=88.6, B=71.5, L=385.7
For 76GHz WG26 is used.
20dB horn: A=17.6, B=13.8, L=20.4
30dB horn: A=54.8, B=44.2, L=238
All dimensions in mm.

Fig 15.23 extends the range of dish size versus gain to the millimetre bands. Note that the diameter is now given in centimetres, not feet!

Parabolic antennas for the millimetre bands can be cut from solid on a lathe. If a block of metal is used, then the reflector can be used immediately. If a block of hardwood is used then it should be covered in kitchen aluminium foil. The foil should be laid in parallel strips with an overlap of at least $\lambda/4$ at the lowest frequency of operation and parallel to the electric plane of polarisation. Make strips

wide enough so that nowhere are there more than two layers and do not attempt to lay the strips radially as it is too difficult to avoid a muddle at the centre of the dish.

The required accuracy is not very high, $\lambda/10$, which is 0.6mm at 47GHz, falling to 0.1mm at 240GHz. This should present no difficulty on a metal working lathe. If a front feed is used then an f/D ratio of 0.5 is recommended, this is the ratio correctly fed with a sectoral horn that tapers in the narrow face only. Initially, a satisfactory feed can be made for experimental work by just radiating from an open guide. Such dishes for the millimetre bands are so small that it is quite good practice to make them oversized and then underfeed them. This produces a highly efficient design in the sense that no RF is wasted and the best gain is obtained for a given beamwidth. For example, for an f/D ratio of 0.5, the complete dish is turned and the feed is in the plane of the aperture where it is very easy to support.

Lenses

At millimetre wavelengths, techniques similar to optics become practical. Components are much larger than the wavelength and dielectrics exhibit properties, which refract the waves in the same way as light is refracted (bent) in a lens. Thus, not only will a parabolic dish focus microwaves as a concave mirror does light, so a dielectric lens will focus millimetre waves. It might be worthwhile for the experimenter to produce a piano convex lens by turning a disc of a dielectric such as polythene or perspex and trying this as an alternative to a dish or horn!

Examples of experimental equipment for 47GHz

Availability of components

The bands immediately below the four principal amateur bands are assigned to the fixed and mobile services while the upper sections of the bands are shared with radio

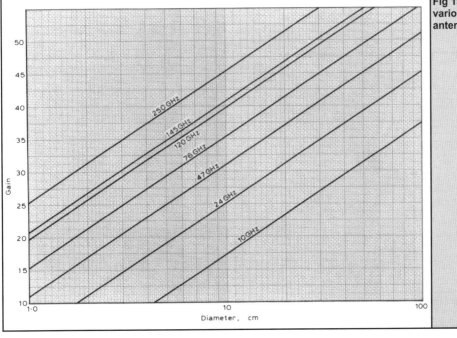

Fig 15.23 Gain of various sized dish antennas.

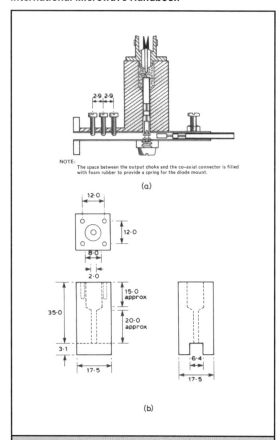

NOTE:
The space between the output choke and the co-axial connector is filled with foam rubber to provide a spring for the diode mount.

(a)

(b)

Fig 15.24 : A mixer/detector for 47GHz.

location. It is to be expected that components will be developed that can be adapted to conventional station to station QSOs in the lower parts of the bands. It will be worth watching the surplus market for components suitable for wideband operation in the higher parts of the bands.

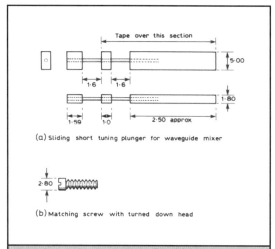

(a) Sliding short tuning plunger for waveguide mixer

(b) Matching screw with turned down head

Fig 15.25 : Detail of tuning plunger for the 47GHz mixer/detector.

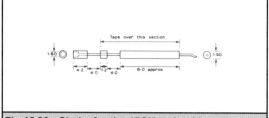

Fig 15.26 : Choke for the 47GHz mixer/detector.

A mixer/detector for 47GHz.

This design is based on old designs for 10GHz, but using a pill diode instead of a 1N23/1N415 type diode. Fig 15.24 shows the general assembly and the loose components are shown in Figs 15.25, 15.26 and 15.27. The waveguide is soldered to the end of a squared up brass block. The drawing shows the waveguide set into a slot cut into the block, but this is not really necessary. The size of the block is not important; the sizes shown are suitable for a BNC connector and can be adjusted to suit a different connector. The minimum height is 25mm to allow for the coaxial choke in the output lead and the spill on the output connector. A piece of brass sheet is also soldered to the top of the waveguide at the input to provide thickness for the threads of the matching screws. A flange, or other connector, should be soldered in place at the same time.

The sliding short comprises two sections of high/low transformer, the first of which has an air dielectric and the second plastic. It is backed up by a larger piece of metal for convenience of handling. The prototype was made of brass. The block is first filed up in the solid to be about 0.1mm (5 thou) smaller than the inner dimension of the waveguide you are using. The dimensions given are for K & S Metals size 264 tubing. The changes for professional waveguide are obvious. Do not be too concerned if these dimensions are not precise; what is important is that the sides of the block are parallel.

Next, the block is drilled lengthways with a number 74 drill (for 22swg wire) or 0.5mm drill (for 0.5mm wire). First select your wire and then drill to suit. It does not matter if the hole is off centre (in fact the block is easier to assemble if the hole is not exactly on the centre line) but do try to make the hole parallel with the long axis. Cut off

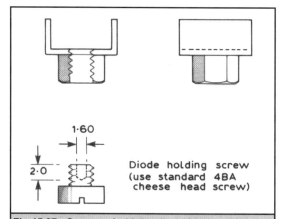

Diode holding screw (use standard 4BA cheese head screw)

Fig 15.27 : Screw to hold the diode for the 47GHz mixer/detector.

a small piece of the block and file it down to 1.59mm long. It is worth taking time and care over this dimension, making sure that the block is not tapered. This is one of the reasons the original block was made too long, you can afford to throw away your failures! Make two of this length. Cut off another small piece and file down to 1mm. This component is less important electrically; its main job is for mechanical support.

Now re-assemble on a piece of wire as shown in Fig 15.25. The best method is to apply a touch of solder paste to the parts to be joined and to apply a hot iron to the remote end of the big block and allow the heat to work its way through. Do not underestimate the time it takes for the block to heat and cool. Clean up the block, e.g. with a nylon burnishing pencil available from model shops.

Next, bend and twist the end sections gently until they look straight and wind on a few layers of thin, self adhesive plastic tape. The tape should preferably be ptfe, but polyester or even Sellotape will do. Here, accuracy of the end section tells, the better it is, the less important is the dielectric used in the back section. Don't use double sided sticky tape. Now peel off the tape, one face of the block at a time, until the main block is a nice smooth fit in the waveguide and make final adjustments to the alignment of the end sections so the whole block moves smoothly. The point here is to test the size using the large block and tweak the smaller blocks to line up. Do not cut extra tape from the small blocks to make them slide. Now cut away the tape from the end block so that this block is left with a 0.05mm air gap between it and the waveguide.

Next, we tackle the output choke. Because a coaxial transmission line too large in diameter will permit a waveguide mode so that RF bypasses the filter, we are limited to a maximum diameter of 2.0mm. We first drill through the whole block and the waveguide with a 2mm drill. Next open out the top of the block with a larger drill to clear the connector you are using and to provide for the compression pad. The choke inner is turned from brass to a diameter of 1.9mm and while the piece is still in the lathe, a hole (drill 71 or 0.5mm as before) is drilled the full length. Next, pieces are cut off and faced off to lengths of 4.18 and 1.0mm. The three pieces are then assembled on to a straight piece of wire, which at this stage can protrude at both ends, and be soldered up. The wire extended at the connector end becomes the output terminal and is cut to length. The piece is then returned to the lathe and the copper wire faced off flush with the brass end and touched with a centre drill. Now drill a 1.6mm (1/16in) hole, 2mm deep, into the end of the piece to receive the end of the diode. As before, wrap tape around the piece until it is a smooth fit into the 2mm hole and then cut away the tape as shown in Fig 15.26 to form the complete choke section. By now you will have noticed that the 2mm hole is too small to permit assembling this mixer by dropping the diode down the output tube, as we do with a 1N23/1N4I5 diode mixer on 10GHz. Thus a detachable diode mount is needed. This is a 4BA screw fitted to a nut soldered to the face of the waveguide. First measure a few nuts and pick half a dozen or so with the same thickness; these will be our stock of such mounts. Cut off a short (10mm) piece of waveguide, drill a 3.1mm hole (size 32) in one face and cut away the other face, taking care not to distort the piece. Tap the hole 4BA and using a chrome plated or rusty steel screw to align the nut, solder this to the outer face of the waveguide. Ease out the end of the nut with a 4.0mm drill (size 26) so that a 4BA screw will screw down hard and now you will have the jig shown in Fig 15.27(a).

This will enable you to file down a 4BA screw to the required length. Do this and put the screw in the lathe and drill the 1.6mm hole, 2mm deep, as shown in Fig 15.27(b). Drill and tap 10BA for the matching screws and fit the nut for the clamping screw for the sliding short, once again using an untinnable screw to locate it.

Finally make some matching screws. The heads of the screws must be turned down so that they can be spaced 2.9mm (3λ/8) apart without fouling each other. 10BA screws are preferred, but 8BA are just possible if the head is turned down to the same diameter as the thread.

Testing and setting up

The instructions for setting up assume you have no source of 47GHz, but do have a source at 24GHz. There is no need, yet, to re-tune it to 23.5GHz. Set the matching screws to be flush on the inside of the guide. Connect the most sensitive meter you have to the output of the diode unit and offer it up to the open end of the source waveguide. Adjust the sliding short for maximum reading, if you are lucky you will find a series of settings about 4mm apart corresponding to half wavelengths in the guide. Choose the setting nearest to the diode and clamp the short in place. Finally set the matching screws for maximum meter reading.

If your source has a high harmonic output, you may be confused by the presence of 72GHz, and the setting of the short may not be too obvious. Move the mount as far from the source as possible while still getting a reasonable meter reading, and draw up a graph of the meter reading against position of the short. You should then see a pattern and be able to select the best working position.

Finally, retune the oscillator to 23.5GHz and repeat all the optimisations described above for the new frequency, which should now lie in the 47GHz band.

Alternative approaches to 47GHz

The following is an abstract of the ideas used by HB9MIN and HB9AMH [2] which led to the establishment of the first world record for amateur communication at 47GHz, of slightly more than 50km, which stood from 1984 until August 1988. One of the main problems for the amateur (and, it seems, many professionals too!) is access to suitable test equipment; some home made test equipment is described in outline in this section.

Three possibilities for the generation of 47GHz signals were considered and examined:

- Use of the second harmonic from a 23.5GHz Gunn oscillator, with the fundamental suppressed by filtering.

- The use of a 23.5GHz Gunn oscillator, followed by a Schottky diode doubler.

- The use of a 47GHz fundamental Gunn oscillator.

The third alternative was immediately dismissed because of the very high cost of Gunn diodes and professional oscillator assemblies. Such oscillators are very temperature sensitive (about 3MHz/°C) and are pulled by at least 100MHz by slight load variation. Exact design frequency is difficult for the amateur to attain because of the unpredictable nature of the effect of the package parasitic elements on the cavity frequency.

The first alternative offers perhaps 1 or 2mW output from a commercial 24GHz oscillator; the temperature effect is typically 1.5MHz/°C whilst frequency pulling will be

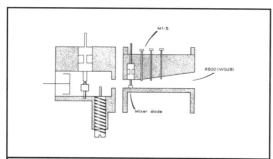

Fig 15.28 : Use of 23.5GHz oscillator followed by a doubler (HB9MIN).

around 10MHz, both figures being much better than with a 47GHz oscillator.

Although the remaining alternative is possible, the increase in power output did not seem to be worth the effort of mounting an expensive diode in a special mount which tapers from WG20 to a reduced height diode section and then back to WG23. Especially when the 23.5GHz oscillator could be used effectively as either a sub harmonic self oscillating mixer or with a low barrier or zero bias Schottky diode in an in line configuration. Ideally the mixer device should be GaAs because silicon diodes are near their limit, although still just usable.

The equipment used consisted of an M/A-COM 24GHz oscillator re tuned for 23.5GHz and a low barrier Schottky diode mounted in WG23 waveguide and used in an in line configuration as shown in Fig 15.28. The antenna used was a 70cm dish (lamp shade) with an f/D ratio of 0.3, measured gain of 40dB and fed with a scaled down

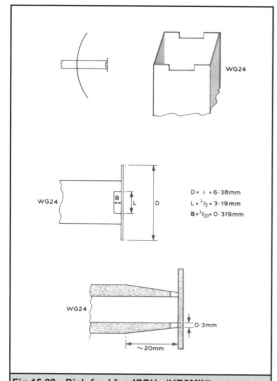

Fig 15.29 : Dish feed for 47GHz (HB9MIN).

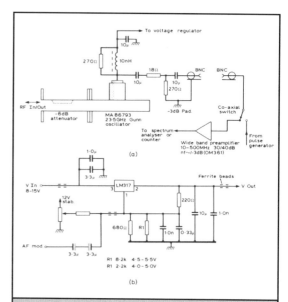

Fig 15.30 : (a) Schematic for an up down converter. (b) Stabilised power supply for the Gunn oscillator (HB9MIN).

version of the penny feed, shown in Fig 15.29 Great accuracy is needed in the construction of the feed and it may be more practical to use a large horn. The pointing accuracy needed for such a dish is better than 0.5°.

To test and align such transceivers, some form of measuring equipment is essential; professional equipment at these frequencies is not common. Several pieces of home made equipment are described briefly. The first is an up/down converter based on the self oscillating mixer principle, the IF output of which is routed to a counter or spectrum analyser at lower frequency. The block diagram of the converter is given in Fig 15.30. The same converter, when fed from a pulse generator at IF, will generate test signals in the band.

A power meter was also constructed using a zero bias Schottky diode type HSCH3206, mounted in a reduced height guide, taper matched into WG23 as shown in Fig 15.31. This is normally used with the test set up shown in Fig 15.31. Fig 15.33 gives some detail of the -23dB cross coupler. This, too, is difficult to make and great accuracy is needed in its construction. It must be pointed out that these ideas, whilst practical, do need a great deal of care and attention in their construction and the degree of inaccuracy which might be tolerable at, say 10GHz, certainly is *not* at 47GHz.

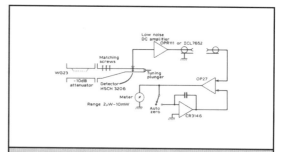

Fig 15.31 : Schematic of a power meter (HB9MIN).

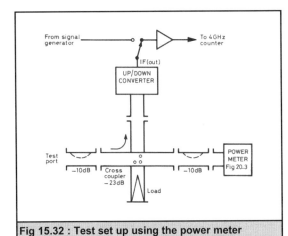

Fig 15.32 : Test set up using the power meter (HB9MIN).

Acknowledgements

The authors would like to acknowledge the assistance of E Zimnermann, HB9MIN, in preparing the content of the preceding part of this chapter.

Versatile Laser transceiver [4]

This article provides a snapshot of the status of development work. Like much DIY equipment, this transceiver will never be quite finished. Hopefully a great many people who will continually make suggestions for improvements and extensions will use it.

In the early days, radio amateurs were pioneers with their work on identifying high frequency engineering relationships. Nowadays laser communication offers them a new opportunity.

Optical communication is amateur radio using light as the information carrier. Only the narrow wave range between approximately 400 and 760nm is visible to the human eye. Short wavelengths produce extremely high frequencies of 394-750THz. In an analogous manner to our amateur bands, we can talk about a 394THz band here. Frequen-

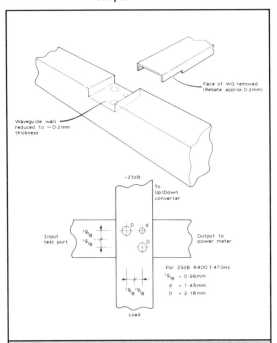

Fig 15.33 : Cross coupler dimensions for 47GHz (HB9MIN).

cies are not usually specified where light is concerned.

Light has to be modulated for communications use. In base band operation, the light is directly amplitude modulated by the information. Alternatively for FM operation the light is amplitude modulated by an auxiliary carrier, which is then frequency modulated. This has the advantage of being independent of intensity variations.

Inside the receiver a detector converts the light into current, which is proportional to the intensity of the light. This current is amplified and demodulated in FM operation. The detector signal follows the envelope of the light wave and does not contain any light frequency fractions.

Fig 15.34 : Photograph of laser transceiver (DH5FFL).

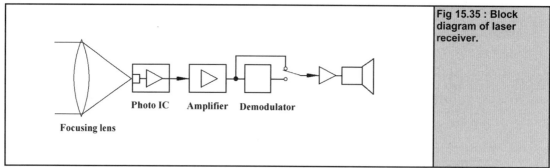

Fig 15.35 : Block diagram of laser receiver.

Photo IC   Amplifier   Demodulator

Focusing lens

This is called incoherent reception, it is simple but not very sensitive. It is considerably lower than for a coherent receiver, which takes the light waveform and mixes it down to an intermediate frequency with an oscillator. Coherent receivers in the optical range are still at the developmental stage today. All amateur radio receivers for lower frequencies operate coherently.

In spite of this handicap, distances of more than 10km can be achieved with relatively low outputs e.g. under 1mW using laser transmitters. This is due to the fact that light can easily be concentrated, which leads to a very high antenna gain. The price that has to be paid is that the transmitter and the receiver must be very precisely aligned with one another, which is familiar to microwave amateurs.

At the same time we have to deal with optical propagation where each leaf on a tree interrupts the beam path. In practice, finding a long propagation path is difficult. This makes it necessary to find an exposed point, and to carry out some careful work there using a map, a compass and a pair of binoculars.

The availability of opto electronic components such as laser modules with integrated, ready adjusted beam concentration (collimation), ultra bright LEDs, photo diodes and optical receiver ICs at reasonable prices has enormously simplified the assembly of equipment for communication in the optical range. The laser transceiver described here (Fig 15.34) uses only standard components which are easy to obtain, and keeps the expenditure as low as possible. To gain experience, a test rig was built using an article by Dipl. Ing. Michael Biller. The experience obtained led to the transceiver described here, which certainly does not represent the final stage in development. Since the development and construction of your own laser transceiver is the very first step. Any DIY enthusiast is naturally completely free to impose his/her own requirements on the project.

Before you begin

Some basic decisions were taken at the start. These were:

- Use of visible laser diodes. If infra red lasers are used, the beam can only be seen with image converters or TV cameras. It is a great simplification if the laser can be seen by the naked eye. The human eye has a lid closing reflex that closes the eye immediately when there is threat of blinding. If you accidentally look into the beam of light you will be dazzled immediately, so you can never expose your eye to the laser beam without being aware of it.

- Use of standard components. Having had some bad experiences with delivery of special components, I now

use only standard components. Suppliers that have the components available are listed.

- Use of FM operation. The very narrow beam from the laser module is particularly sensitive to atmospheric disturbances. Using FM makes it possible to limit the signal in the receiver and thus to suppress amplitude variations. Operation in base band should be possible as an alternative, and simple arrangements for switching over are planned for a later stage.

- Use of an unmodified laser module. The laser module is from a laser pointer that has such a narrow beam (with the integrated lens) that it is necessary to use a telescopic sight for reception. It is virtually impossible to obtain better concentration using inexpensive components and the means available to an amateur. Besides, adjustments would be even trickier.

Transceiver Assembly

The function blocks of the equipment are covered in the sequence of signal flow.

Receiver

Fig 15.35 shows the block diagram. The focusing lens focuses the light onto the receiver surface of the detector. With correct adjustment, all light falling onto the lens is focussed and the sensitivity is increased by the factor of lens area/detector area, which corresponds to an antenna gain. Thus the first important factor is the area or the diameter of the lens. The focal distance of the lens, together with the size of the detector area, determines the aperture angle of the receiver. The optical quality of a standard focusing lens is quite adequate for our purposes. Large diameter lenses are big, heavy, rare and expensive.

Westfalia has an inexpensive glass lens in its range with a diameter of 90mm and focal length 215mm. This is referred to as a large hand held magnifying glass. It can be ordered under no. 458166, costing DM9.95/€5.09, and it is completely adequate for our purposes. I advise you not to use plastic lenses.

For the pre-selection of the receiver, as in the case in high frequency engineering, a filter is included in the optical reception path. Ideally, from the mixture of incident light, this allows only the transmitter's wavelength to pass. This weakens the ambient light and improves the signal/noise ratio. A red filter for photography can be very effective. Check that it lets the laser light pass without attenuation. Commercially available coloured discs with precisely specified transmission are better. Glass such as the RG 630 is of interest, the number is the wavelength in nm above which light can pass through the filter. Select a figure 10-20nm from the wavelength of your laser. This

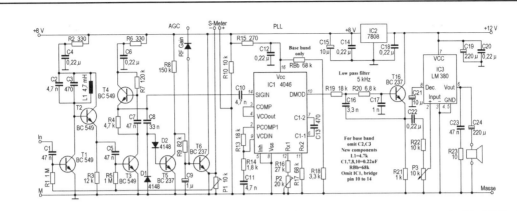

Fig 15.36 : Circuit diagram of laser receiver.

glass is commonly available as samples measuring 50mm. x 50mm.. Unfortunately such a piece of glass is much too big for us and must be cut up. And at approximately DM 60/30€ it is not cheap either, quite apart from the fact that the company supplying you will impose a minimum order charge.

The ultimate filtration is obtained using interference filters with transmission ranges in the range of 20-50nm but they cost at least DM150/€77.

The unit described as front end consists of the detector and an amplifier. The detector converts the incident light that is transformed into a current in the amplifier and is amplified at a low noise level. It is also possible to use photo ICs, in which photo diodes and amplifiers are integrated.

The photo IC that is of interest is the OPT 101P, which involves only a little wiring. It can be obtained from Reichelt as OPT 101P for DM9.95/€5.09. A 2.5mm jack plug is used to connect the front end with the receiver, with a screened cable, e.g. RG-174/U which will make full duplex communication possible later.

In principle, the receiver is a controlled amplifier with a PLL used as a demodulator for FM operation. This is followed by a low frequency, low pass filter with a bandwidth of 5kHz and a small power amplifier for loudspeaker operation.

Special attention was paid to good stability and an S meter, which responds well to low signals. The latter is important for calibration, since the human eye has trouble in distinguishing small differences in brightness. The maximum amplification is approximately 110 dB for a control range of approximately 90 dB in the base band. 1µV is sufficient for almost noise free reception and for FM operation 2µV is sufficient. The amplification is infinitely variably for base band operation.

The circuit of the receiver is shown in (Fig 15.36), the PCB layout is shown in (Fig 15.37) and the component layout in Fig 15.41. This can be used for base band or for FM operation with a 32.768 kHz auxiliary carrier. It contains all the circuits required and fits into a tinplate housing measuring 55.5mm x 74mm x 30mm (e.g. Andys Funkladen, order no. 2101065).

The printed circuit board for the receiver must be assembled differently for base band or FM. A logic level switch over is planned as the next step in development.

Table 15.8 : Technical data for laser transceiver.

Receiver:	
Supply voltage (internal, 8V)	12V
Input voltage for control unit /	
S meter display	1.5µV
S meter display	110dB
Control range	90dB
Low frequency band width	5kHz
Front end (in prototype)	OPT-210
Antenna gain for 9cm. lens of receiver, as against detector alone	Aprox. 35dB
Transmitter:	
Supply voltage (internal, 8V)	12 V
Maximum output current	500mA
FM auxiliary carrier	32.768kHz
FM dispersion	5kHz
Test tone	512Hz
Antenna gain for laser module of DB6NT, as against isotropic antenna	Approx. 44dB

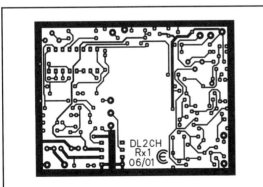

Fig 15.37 : PCB layout for laser receiver.

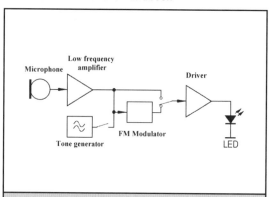

Fig 15.38 : Block diagram of laser transmitter.

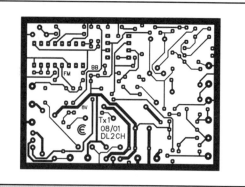

Fig 15.39 : PCB layout for laser transmitter.

Transmitter

The block diagram of the transmitter is shown in Fig 15.38). The microphone feeds an amplifier with an integrated sound generator. It supplies a switchable 512Hz tone that is of assistance for searching and calibrating. The downstream FM modulator modulates the frequency of a voltage controlled oscillator. The driver stage converts the input voltage into an output current of up to 500mA. This flows through the laser LED to produce the light for transmission.

The link between the driver stage and the laser, like that between the front end and the receiver uses a 2.5mm jack plug and a screened cable.

Where Does The Light Come From?

The laser module used has an ultra bright LED, its intensity alters linearly with the current. Looked at physically, this is injection luminescence, in which every electron crossing through the diode generates a light quantum. The laser module already contains a precisely calibrated lens, and emits a beam with a very small aperture angle, which for our purposes can be used without any alteration.

There is a PCB for the transmitter containing all of the circuitry described above. The laser driver follows the principles of John Yurek, K3PGP [3], designed to protect the costly laser from current and from voltage peaks. John's circuit has been modified to give improved linearity, so that future experiments will be possible, for example, with SSB. A Darlington transistor in the circuit has been replaced by a low powered FET for FM operation, since the Darlington was too slow for the rise time of the 32.768kHz square wave signal and the current overshot. The transmitter PCB (Fig 15.39) fits into the same tinplate housing as that for the receiver. The component layout for the transmitter is shown in Fig 15.42.

An FM/base band switchover, controlled by a short circuit to earth, has been incorporated in the transmitter. Fig 15.40 shows the circuit diagram.

Calibration

To calibrate the receiver, first short circuit the input and then set the voltage at the collector of T3 to 2V with P1.

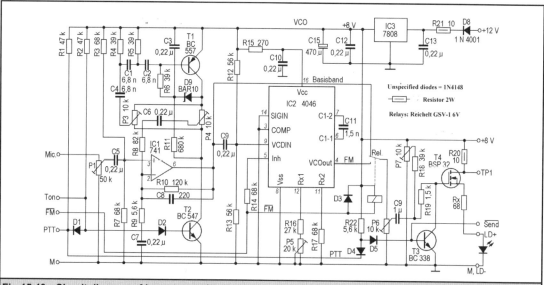

Fig 15.40 : Circuit diagram of laser transmitter.

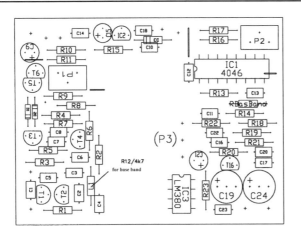

Fig 15.41 : Component layout for laser receiver.

When calibrating the transmitter, P1 controls the microphone gain, P3 the tone pitch of the test tone and P4 the volume of the test tone. The test tone is muted when the PTT is pressed, or there is a Low on Tone. P5 determines the mean FM frequency. P7 controls the mean laser current, the carrier on the base band and P6 the modulation depth. Both the base band signal and the FM square wave should give a value of approximately 8V at the modulator input.

To calibrate the maximum laser current, P6 and P7 are turned down to the lower stop. Rx=100 . Then connect up the voltmeter between TP1 and 8V. A laser current of 100mA produces a reading of 1V on the display. Apply the operating voltage and turn potentiometer P7 slowly, always keeping an eye on the laser power, until a level of 50% has been attained. If this is not attained, Rx must be reduced.

Now the laser current modulation is increased using P6, until the maximum laser power is attained at the modulation peaks. If the maximum power is not attained by rotating P6, once again, reduce Rx. At maximum current there is not much safety but the driver is failsafe, as discussed in the work done by John Yurek, Rx and R20 limit the current. You can be really safe and use a high power 8.2V zener diode from 8V to earth.

Mechanics And Optics

Without a stable mechanical structure, frustration and anger are assured. As a result of this, the quality of the optics and the electronics of the transceiver must be very good. Using normal stands with tapered heads or spherical heads as the base is simply not stable enough for our purpose. The best choice is a proper theodolite stand, as normally used in surveying.

Such stands are on offer in flea markets here and there for about DM100/€51.13. Hopefully the rapidly increasing number of laser stations will soon exhaust this source. Christian Huber, DL2MFB, has pointed out an aluminium stand in the Westfalia catalogue, order no. 878777. The illustration shows the stand as a normally solid theodolite stand made from solid aluminium, just like the one I own. It has height adjustment between 1m. and 1.7m. and priced at DM119/€60.84.

The laser's extremely sharply focusing beam calls for a telescopic sight with cross hairs. The magnification should not be too low x7 makes finding the opposite station and aligning with it precisely much easier. Westfalia is marketing a 3-7 x20 telescopic sight under order no. 106146-02 at a price of DM39.90/€20.40. It can be used for our purpose, especially at this price. Its good colour correction is advantageous, because it causes the red laser dot to be be slightly out of focus, and can not be concealed by

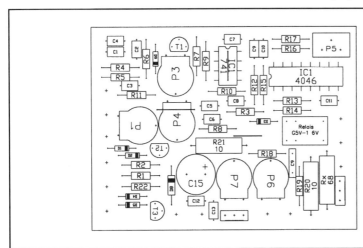

Fig 15.42 : Component layout for laser transmitter.

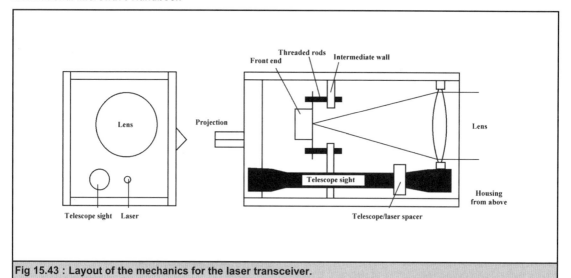

Fig 15.43 : Layout of the mechanics for the laser transceiver.

the cross hairs.

The housing takes all optical and electrical components (Fig 15.43). In the housing, the optical axes of the laser, the telescopic sight and the receiver must be aligned to be precisely parallel to one another and must remain so permanently. Moreover, the housing as a whole must be horizontally and vertically free from movement and it must be possible to adjust it accurately.

The material recommended is 12mm water resistant glued plywood. It is easy to work and when glued and screwed it gives a light rigid structure. It is best if the components can be cut precisely in the DIY superstore. The holes for the lens, the telescopic sight and the laser beam can be cut quickly using a keyhole saw.

The most critical part of the assembly is that between the laser and the telescopic sight. To do this, two holes are bored in an acrylic block with the diameters of the telescopic sight and the laser. The block is then sawn through the centres of the two holes with a thin saw. The block is fitted around the telescopic sight and the laser then pressed together with two securing screws.

The laser can be aligned using cats eye 200m away sighted through the cross hairs of the telescopic sight. Adjust the screws to achieve a good alignment. Then fix the laser with two component adhesive. The cross hairs are precisely aligned onto the laser dot using the adjustment screws on the telescopic sight. This unit is inserted into the housing in such a way that the assembly is not

mechanically stressed. By adjusting the leg length, the stand can be roughly aligned, and the fine calibration is done with the transceiver set screws.

The front end sits on a sheet metal plate with a hole bored through the centre, which is mounted on an intermediate wall in the housing parallel to the lens using three threaded rods. The distance to the lens can be established by moving the intermediate wall and then using the nuts. The front end can be vertically and horizontally moved on the plate itself, at a constant distance to the lens, in order to set the optical axis of the receiver parallel to that of the laser and the telescopic sight.

First the receiver must be set to focus on infinity within the housing. To achieve this a single lens reflex camera is aligned on the lens of the receiver, so that it can see the receiver through the lens. The distance between the lens and the receiver is set in such a way that when the camera is set to focus on infinity the receiver is in sharply focusing, which can be seen precisely on the split image indicator. This somewhat laborious procedure needs to be carried out only once.

For the horizontal and vertical calibration of the receiver, we need a strong light source at some distance, something like a spotlight or a laser. First align the housing on the light source, using the cross hairs. If you now look into the lens, using a little mirror, you can see the position of the light source image spot as a small bright dot.

Now move the receiver in such a way that the dot of light

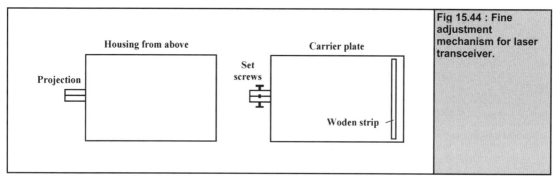

Fig 15.44 : Fine adjustment mechanism for laser transceiver.

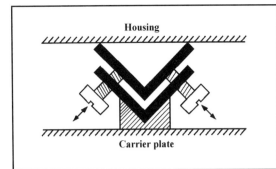

Fig 15.45 : Fine adjustment mechanism for laser transceiver.

lies precisely at the centre of the detector. Then continue to check the precise alignment of the receiver to the light source in the telescopic sight.

Slow Feed Indispensable

The housing is now aligned. It should be possible to pivot it horizontally and vertically, sensitively and free from shaking. A very simple solution for the problem is as follows: A carrier plate, which is larger than the housing, carries a wooden strip mounted crosswise to the direction of the beam. It is rounded at the top and at the front. The housing rests on this at the approximate position of the lens. Both the carrier plate and the housing have a piece of angle mounted like downward pointing "V" projecting at the back (Fig 15.44).

The angle on the carrier plate has two set screws in threaded holes. The housing angle rests on the rounded ends of the two set screws and is held, without play, by gravity. By turning the set screws, you can make very fine adjustments to the position of the housing in both diagonal directions (Fig 15.45).

The simplicity and absolutely play free structure counteracts the slight disadvantage of the diagonal movement when the set screw is turned. But you get the hang of it after a short time. Once you are OK with this arrangement, you can improve the mechanics later at any time.

For a QSO, centre the set screws and align the station by putting the stand down and adjusting the leg length to the opposite station as precisely as possible. Then push the housing along the wooden strip for the horizontal orientation and finally use the set screws for fine setting. The whole stand can be inclined if you press sensitively on the notches of the stand (provided for this purpose) with your foot and press the pointed ends of the legs deeper into the ground. [5]

About The Laser Article

This article provides a snapshot of the status of development work. Like many pieces of DIY equipment, this transceiver will undoubtedly be developed further, since it will hopefully be used by a great many people, who will continually make suggestions for improvements and extensions.

The circuit diagrams were drawn shortly before the final edit and, unfortunately, there are some small errors:

- Transmitter: In the wiring diagram, the laser is switched with a MOS-FET, but on the board it is still switched with a Darlington transistor.

- Receiver: Potentiometer P1 in the receiver is bridged to earth - delete these lines. But the board is OK. The positive connection of the S meter is directly earthed making R10 pointless. There should be a voltage divider there with R11=68k between the connection and the earth, instead of the connection shown.

You can find other current developments on my homepage [4].

References

[1] Propagation above 10GHz - the potential of the amateur mm-wave bands, B Chambers, G8AGN. Proceedings of the Martlesham Microwave Round Table, November 1988, published by the Martlesham Radio Society, with acknowledgements to the British Telecom (BT) Research Laboratory, at whose establishment the meeting was held

[2] 47GHz Praktischer Teil, E Zimmerinann, HB9MIN.

[3] John Yurek's Page with circuits for front end and laser driver

[4] From an article by Hans Hellmuth Cuno, DL2CH, CQDL 10/2001 pp 727 - 730 and 11/2001 pp 810 - 811. His web site has a White Paper on laser communication. Other links are shown as well

[5] Jim Moss, WB9AJZ Link collection and instructions for operating laser QSOs

Even if you are only dealing with light, you should familiarise yourself with the safety rules when using lasers. The eyes are particularly vulnerable to injury. If the laser beam goes into the eye, it is re-focussed by the eyepiece lens. As a consequence of this, a dazzling beam strikes the retina with the full power of the laser. This can damage the photoreceptor cells in the eye. Photoreceptor cells that have been damaged do not regenerate. So never look at the laser beam! Laser classes 1, 2, 3A or 3B are precisely defined in the standard DIN EN 60825-1. They refer to lasers with continually increasing power, going up from 1 to 3B. Each higher class includes all the lower classes, though laser class 2 includes class 1 only in the wavelength range 400-700nm. Here we are certainly still talking about lasers, but the class covers other sources such as, for example, LEDs.

The basis of the formulations in the standard DIN EN 60825-1 runs: In the visible range 400 - 700nm., for continuous operation and for a point outlet, a laser with an initial power of 1mW falls within laser class 2.

Fig 15.47 : Safety and laser radio.

The original text in the appendix for additional Callsign approval in accordance with #16 AfuV (AFuV = Amateurfunkverordnung) of the Amateur Radio Directive:

Frequencies above 500THz can at present be used freely by anyone on the basis of the rules laid down by the TKG. 500THz corresponds to a wavelength of λ = 600nm i.e. orange light. The TKG is the Telecommunications Law. This provision allows everyone to transmit communications in the optical range at wavelengths shorter than 600nm provided the prevailing safety regulations are obeyed, of course. But this is not amateur radio operation, and the use of amateur radio callings is not allowed! There are two options for amateur radio operation in the optical range:

Approval in accordance with #16 of the AFuV, without an additional callsign

The original wording of an original document from: Approval in accordance with #16 AFuV, the Amateur Radio Directive covering experiments in the frequency range above 300GHz with radio apparatus which falls within laser classes 1 or 2 in accordance with DIN EN 60825-1. With this approval, experiments can be carried out in the optical range with radio equipment belonging to laser classes 1 or 2, using the Callsigns allocated. This rule is valid until 31/2/2002 and there is no charge. From October 2001, lasers from class 3A should also be approved. If an application is made, it must begin with the correct wording (word for word!):" I hereby apply for approval in accordance with #16 of the AFuV for experiments in the frequency range above 300GHz using radio equipment which belongs to laser classes 1 or 2, in accordance with DIN EN 60825-1." Following this, you must state your intention to carry out experiments on the propagation, working distance, type of modulation, etc., in accordance with amateur radio regulations.

Additional Callsign approval in accordance with #16 of the AFuV

Original wording of an original document : In view of the law governing amateur radio activities...the radio amateur with the name / class / callsign / address shall be..allocated the additional callsign DA5F.../ approval class 1 / approval number... for special experimental and technical scientific studies, in accordance with #16 of the AFuV.This Callsign may be used to operate radio equipment belonging to laser classes 1, 2, 3A or 3B, in accordance with DIN EN 60825-1. The appendix lists the conditions of allocation and use and can also approve unattended operation. There is an administration charge of DM50 for the issuing of an additional Callsign. The application for an additional Callsign should begin in the following way (word for word!): "I hereby apply for the issuing of an additional Callsign, in accordance with #16 of the AFuV for special experimental and scientific studies in the optical frequency range above 300GHz." As with the approval, you then express your intention to carry out experiments on the propagation, working distance, type of modulation, etc. for the purposes of amateur radio activities. The conditions governing laser class 2 can be found in the approval, which may also list the essential data for laser classes 3A and 3B.

Peter Greil, DL7UHU and Hans-Hellmuth Cuno, DI2CH

Fig 15.46 : Details of licensing conditions in Germany for laser operation.

Data

T he main purpose of the appendix is to present selected data and information supplementary to that already given in the main text of the book or which would not fit logically into the main narrative text. As far as possible, the order in which this information is given corresponds to the chapter order.

- For further data, the following books are recommended:

- Radio Data Reference Book, by G R Jessop, published by the RSGB, 1985.

- Engineering Tables and Data, by Howatson, Lund and Todd, published by Chapman and Hall, 1972.

- Reference Data for Radio Engineers, by ITT, published by H W Sams, 6th edition, 1979.

- Microwave Engineers Handbook, Vols. 1 and 2, edited by Theodore S. Saad, published by Artech House, 1971.

Manufacturer's literature

Almost all electronic component manufacturers publish data sheets for the devices they manufacture. These usually contain very useful data such as maximum ratings, tolerances, performance at different frequencies etc. In addition, many publish applications notes, which may give details of test and evaluation circuits, useful to the experimenter. Devices are often rapidly outdated or updated and improved and for this reason no specific valve and semiconductor data is given.

Most manufacturers now have extremely good web sites

with data sheets available for download. This means that the amount of information available to radio amateurs has increased dramatically. Table A1.1 below shows a short list of web site addresses, but using one of the search engines such as google.com usually produces good results even if you only have a component part number.

Some other useful web sites are :

RSGB Radio Society of Great Britain
www.rsgb.org

ARRL American Radio Relay League
www.arrl.com

Down East Microwave
www.downeastmicrowave.com

European Microwave News
www.emn.org.uk

A useful link site
www.k1dwu.net

A useful link site
www.wa1mba.org/interest.htm

Table A1.1 : Web site addresses of some manufactures.

Manufacturer	Description	Web site address
Agilent (Hewlett Packard)	Manufacturer of a wide range of electronic components	www.agilent.com
Ansoft	Supplier of microwave simulation software	www.ansoft.com
Aplac	Supplier of microwave simulation software	www.aplac.com
Mini Circuits	Manufacturer of microwave components	www.minicircuits.com
National Semiconductors	Manufacturer of a wide range of electronic components	
Philips	Manufacturer of a wide range of electronic components	www.philips.com
Sonnet	Supplier of microwave simulation software	www.sonnetusa.com
Synergy	Supplier of microwave components	www.synergymwave.com
Transtech	Manufacturer of dielectric resonators and ceramic filters	www.trans-technic.com
Vectron	Manufacturer of SAW filters and crystal oscillators	www.vectron.com

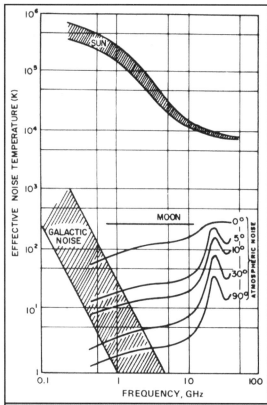

Fig A1.1 : Galactic and solar noise. Degrees on the atmospheric noise curves indicate the angle above the horizon. Minimum noise from these sources is found in the frequency range from 1 to 10GHz. © Artech House Inc, from Theodore S Saad, Microwave Engineers Handbook Vol 2. Reprinted by permission.

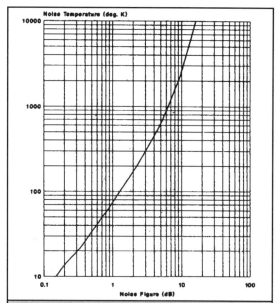

Fig A1.2 : Noise figure vs noise temperature.

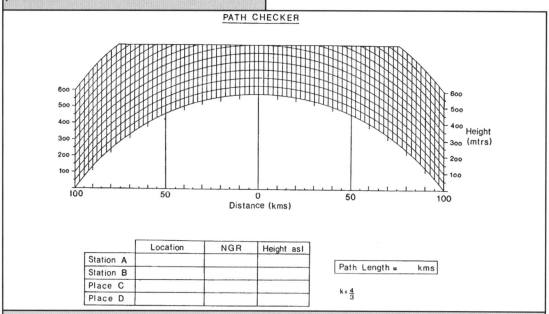

Fig A1.3 : Metric earth profile plotting paper, this can be used instead of the imperial graph given in chapter 1 (Fig 1.1).

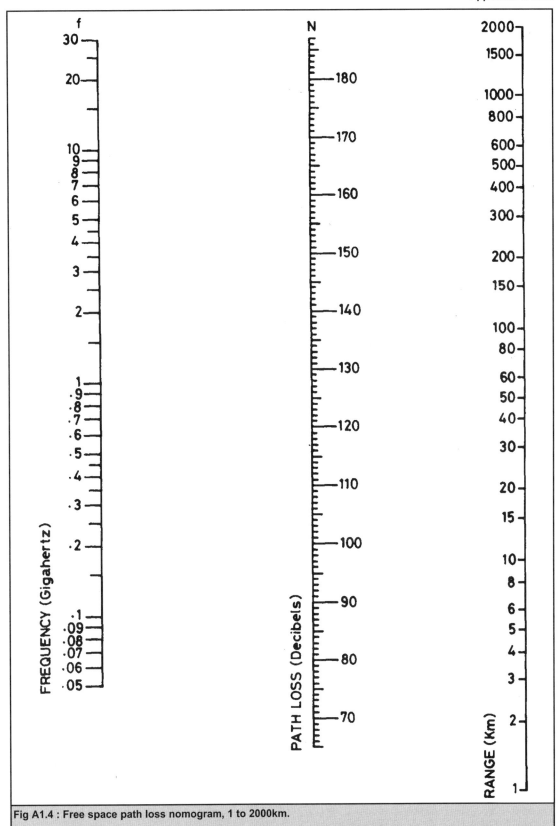

Fig A1.4 : Free space path loss nomogram, 1 to 2000km.

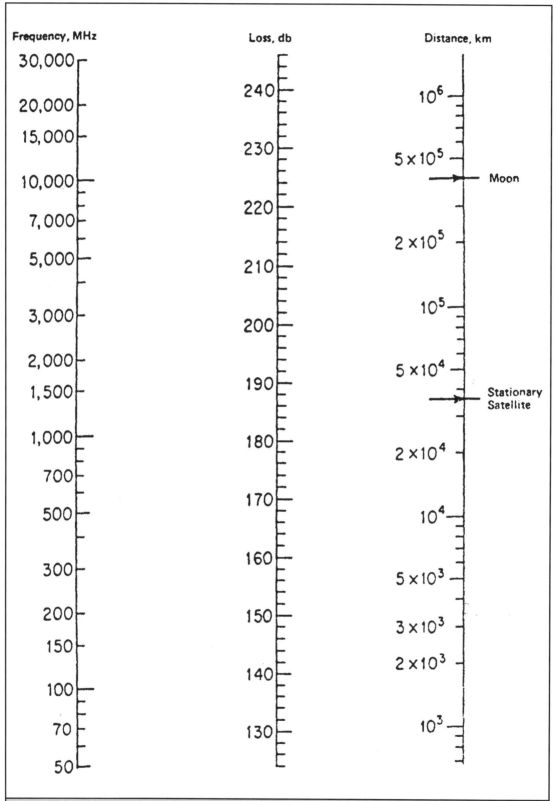

Fig A1.5 : Free space path loss nomogram, 1000 to 10⁶km. From Krassner & Micaels, Introduction to Space Communications, McGraw Hill, © 1964. Reprinted by permission of McGraw Hill Inc.

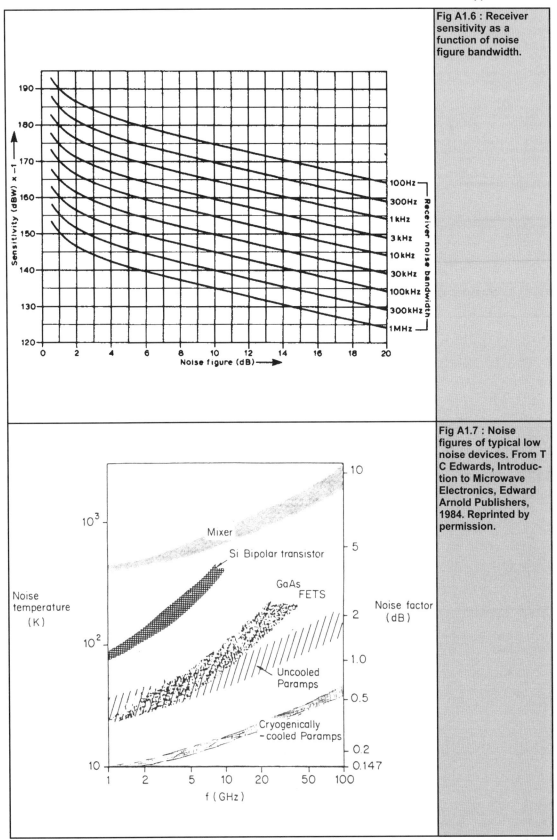

Fig A1.6 : Receiver sensitivity as a function of noise figure bandwidth.

Fig A1.7 : Noise figures of typical low noise devices. From T C Edwards, Introduction to Microwave Electronics, Edward Arnold Publishers, 1984. Reprinted by permission.

Man made fibre ropes

Marina Nylon. The synthetic fibre with excellent shock absorbing properties.

Marina Polyester. Almost as strong as nylon with lower stretch properties. Excellent wear resistance.

Marina Polypropylene. Lower in strength than both nylon and polyester. Extremely light in weight and buoyant. Available in multifilament, Marina blue (monofilament), and Staplespun.

Quality. All ropes in the Marina range are made from the best quality material available and are manufactured to the highest standard.

Marina Hystrain Kevlar. The strongest low stretch fibre rope with a polyester overbraid to give protection against abrasion and ultraviolet degradation.

Tests confirming the weights and breaking loads of the new cordage specified here shall be undertaken according to the methods of test quoted in British Standard 5053.

N.B. KEVLAR HYSTRAIN BASED ON 10mm DIA OVER JACKET

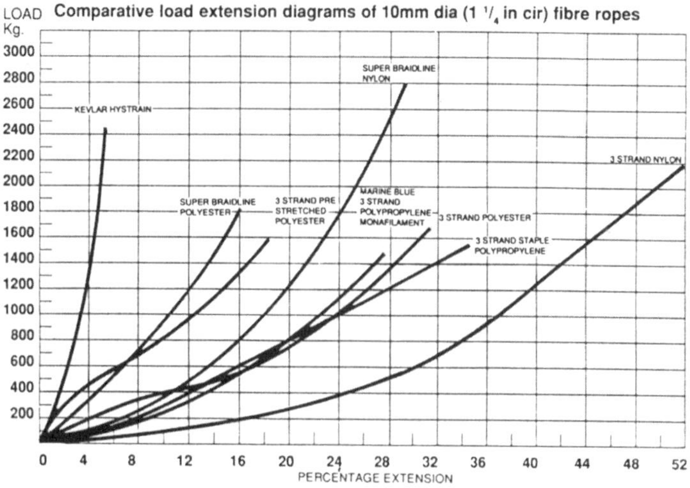
Comparative load extension diagrams of 10mm dia (1 ¹/₄ in cir) fibre ropes
LOAD Kg.
3000
2800
2600
2400
2200
2000
1800
1600
1400
1200
1000
800
600
400
200
0
4
8
12
16
20
24
28
32
36
40
44
48
52
PERCENTAGE EXTENSION
KEVLAR HYSTRAIN
SUPER BRAIDLINE NYLON
3 STRAND NYLON
SUPER BRAIDLINE POLYESTER
3 STRAND PRE STRETCHED POLYESTER
MARINE BLUE 3 STRAND POLYPROPYLENE MONAFILAMENT
3 STRAND POLYESTER
3 STRAND STAPLE POLYPROPYLENE

Fig A1.8 : Load/extension characteristics of man made ropes. Courtesy British Ropes.

	Marina Super Braidline		Marina Braidline	Marina Plaited Dinghy Ropes		
	Super Polyester Braidline	**Super Nylon Braidline**	**Braidline**	**Polyester Matt Finish or Cont. Filament**	**Spun Nylon**	**Multifilament Polypropylene**
dia mm	Min Break Load kg	Min Break Load kg	Min Break Load kg	Min Break Load kg	Min Break Load kg	Min Break Load kg
4	—	—	—	—	—	180
5	—	—	—	225	—	225
6	650	950	350	295	—	295
8	1 175	1 450	675	565	905	565
9	—	—	—	635	1 135	635
10	1 800	2 725	1 125	905	1 405	905
12	2 575	3 400	1 475	1 360	1 815	1. 250
14	3 650	4 300	—	—	—	—
16	4 525	5 450	—	—	—	—
18	5 675	7 700	—	—	—	—
21	7 925	9 525	—	—	—	—
24	10 200	12 700	—	—	—	—

Three Strand Construction

Size	Polyester	Nylon	Polypropylene (Multi-filament, Marina Blue, Fibrefilm, Staple)
dia mm	Min Break Load kg	Min Break Load kg	Min Break Load kg
4	295	320	250
5	400	500	350
6	565	750	550
7	770	1 020	740
8	1 020	1 350	960
9	1 270	1 700	1 150
10	1 590	2 080	1 425
12	2 270	3 000	2 030
14	3 180	4 100	2 790
16	4 060	5 300	3 500
18	5 080	6 700	4 450
20	6 350	8 300	5 370
22	7 620	10 000	6 500

Fig A1.9 : Breaking loads of man made fibre ropes. Courtesy British Ropes.

Stainless steel wire rope & strand

Size		1 x 19 Strand	6 x 7 with steel core Standing Rigging	6 x 19 Running Ropes
dia mm	Approx circ in	Min Break Load tonne	Min Break Load tonne	Min Break Load tonne
2	¼	—	0·24	—
3	⅜	0·72	0·55	0·51
4	½	1·28	0·97	0·91
5	⅝	2·00	1·51	1·42
6	¾	2·88	2·18	2·04
8	1	—	3·87	—

Galvanised wire rope

Size		6 x 7 with steel core Standing Rigging	6 x 19 Running Ropes	6 x 7 with steel core Galvanised P.V.C. (White)
dia mm	Approx circ in	Min Break Load tonne	Min Break Load tonne	Min Break Load tonne
2	¼	0·28	—	0·28
3	⅜	0·63	0·50	0·63
4	½	1·12	0·88	1·12
5	⅝	1·75	1·38	1·75
6	¾	2·52	1·99	—
7	⅞	3·43	2·71	—
8	1	4·14	3·54	
9	1⅛	5·23	4·48	—
10	1¼	6·46	—	—
12	1½	9·30	—	—

Fig A1.10 : Breaking loads of stainless steel and galvanised wire rope. Courtesy British Ropes.

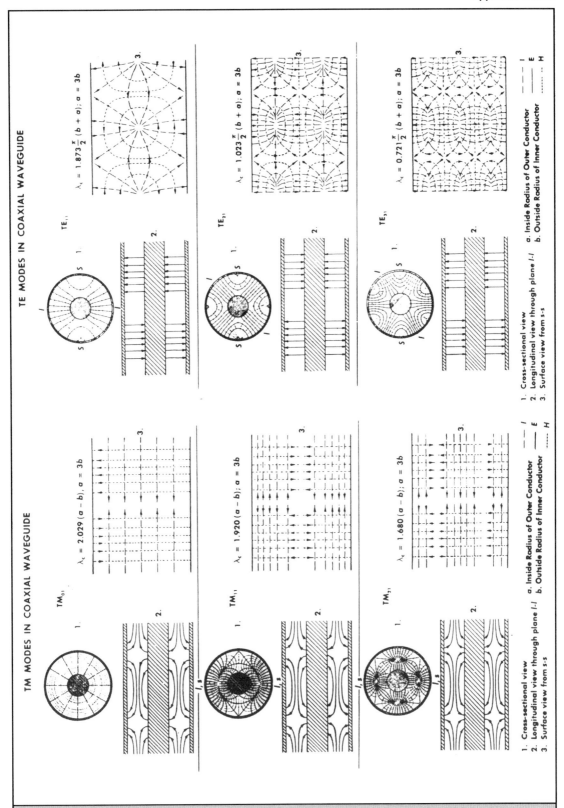

Fig A1.11 : Mode patterns in coaxial lines. From Marcuvitz (ed) Waveguide Handbook, MIT Radiation Laboratory Series Vol 10, McGraw Hill. © Reprinted by permission of McGraw Hill.

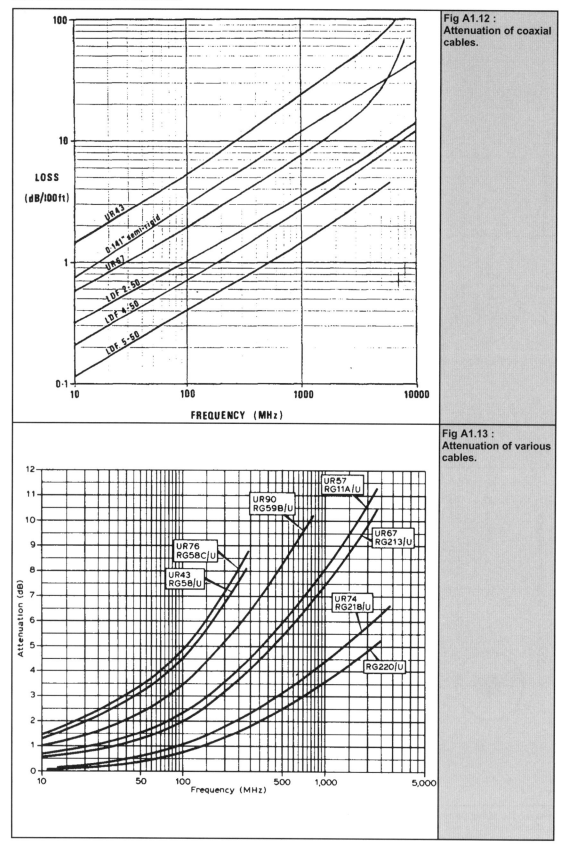

Fig A1.12 : Attenuation of coaxial cables.

Fig A1.13 : Attenuation of various cables.

Characteristics of typical British radio frequency feeder cables

Type of cable	Nominal impedance Z_0 (Ω)	Centre conductor	Dimensions (in) over outer sheath	Dimensions (in) over twincores	Velocity factor	Approximate attenuation (dB per 100ft) 70Mhz	145MHz	430MHz	1250MHz	Remarks
Standard tv feeder	75	7/·0076	0·202	—	0·67	3·5	5·1	9·2	17	—
Low-loss tv feeder (semi-air spaced)	75	0·048	0·290	—	0·86 approx.	2·0	3·0	5·4	10	Semi-air spaced or cellular
Flat twin	150	7/·012	—	0·18 / 0·09	0·71	2·1	3·1	5·7*	11*	*Theoretical figures, likely to be considerably worsened by radiation
Flat twin	300	7/·012	—	0·405 / 0·09	0·85	1·2	1·8	3·4*	6·6*	
Tubular twin	300	7/·012	—	0·446	0·85	1·2	1·8	3·4*	6·6*	

British UR series coaxial cables

UR No	Nominal impedance Z_0 (Ω)	Overall diameter (in)	Inner conductor (in)	Capacitance (pF/ft)	Maximum operating voltage rms	Approximate attenuation (dB per 100ft) 10MHz	100MHz	300MHz	1,000MHz	Approx RG. equivalent
43	52	0·195	0·032	29	2,750	1·3	4·3	8·7	18·1	58/U
57	75	0·405	0·044	20·6	5,000	0·6	1·9	3·5	7·1	11A/U
63*	75	0·853	0·175	14	4,400	0·15	0·5	0·9	1·7	—
67	50	0·405	7/0·029	30	4,800	0·6	2·0	3·7	7·5	213/U
74	51	0·870	0·188	30·7	15,000	0·3	1·0	1·9	4·2	218/U
76	51	0·195	19/0·0066	29	1,800	1·6	5·3	9·6	22·0	58C/U
77	75	0·870	0·104	20·5	12,500	0·3	1·0	1·9	4·2	164/U
79*	50	0·855	0·265	21	6,000	0·16	0·5	0·9	1·8	—
83*	50	0·555	0·168	21	2,600	0·25	0·8	1·5	2·8	—
85*	75	0·555	0·109	14	2,600	0·2	0·7	1·3	2·5	—
90	75	0·242	0·022	0	2,500	1·1	3·5	6·3	12·3	59B/U

All the above cables have solid dielectric with a velocity factor of 0·66 with the exception of those marked with an asterisk which are helical membrane and have a velocity factor of 0·96.

This table is compiled from information kindly supplied by Aerialite Ltd, and BICC Ltd and includes data extracted from Defence Specification, DEF-14-A (HMSO).

Fig A1.14 : Data for common UK flexible cables.

USA RG Series coaxial Cables

Cable No.	Nominal impedance Z_0 (Ω)	Cable outside diameter (iZ_0 (Ω)	Velocity	Approximate attenuation (dB/100ft)					Capac-tance (pF/ft)	Maximum operating voltage (rms)
				1MHz	10MHz	100MHz	1GHz	3GHz		
RG-5/U	52.5	0.332	0.659	0.21	0.77	2.9	11.5	22.0	28.5	3,000
RG-5B/U	50.0	0.332	0.659	0.16	0.66	2.4	8.8	16.7	29.5	3,000
RG-6A/U	75.0	0.332	0.659	0.21	0.78	2.9	11.2	21.0	20.0	2,700
RG-8A/U	50.0	0.405	0.659	0.16	0.55	2.0	8.0	16.5	30.5	4,000
RG-9/U	51.0	0.420	0.659	0.16	0.57	2.0	7.3	15.5	30.0	4,000
RG-9B/U	50.0	0.425	0.659	0.175	0.61	2.1	9.0	18.0	30.5	4,000
RG-10A/U	50.0	0.475	0.659	0.16	0.55	2.0	8.0	16.5	30.5	4,000
RG-11A/U	75.0	0.405	0.66	0.18	0.7	2.3	7.8	16.5	20.5	5,000
RG12A/U	75.0	0.475	0.659	0.18	0.66	2.3	8.0	16.5	20.5	4,000
RG-13A/U	75.0	0.425	0.659	0.18	0.66	2.3	8.0	16.5	20.5	4,000
RG-14A/U	50.0	0.545	0.659	0.12	0.41	1.4	5.5	12.0	30.0	5,500
RG-16/U	52.0	0.630	0.670	0.1	0.4	1.2	6.7	16.0	29.5	6,000
RG-17A/U	50.0	0.870	0.659	0.066	0.225	0.80	3.4	8.5	30.0	11,000
RG18A/U	50.0	0.945	0.659	0.066	0.225	0.80	3.4	8.5	30.5	11,000
RG-19A/U	50.0	1.120	0.659	0.04	0.17	0.68	3.5	7.7	30.5	14,000
RG-20A/U	50.0	1.195	0.659	0.04	0.17	0.68	3.5	7.7	30.5	14,000
RG21A/U	50.0	0.332	0.659	1.4	4.4	13.0	43.0	85.0	30.0	2,700
RG-29/U	53.5	0.184	0.659	0.33	1.2	4.4	16.0	30.0	28.5	1,900
RG-34A/U	75.0	0.630	0.659	0.065	0.29	1.3	6.0	12.5	20.5	5,200
RG-34B/U	75	0.630	0.66		0.3	1.4	5.8		21.5	6,500
RG-35A/U	75.0	0.945	0.659	0.07	0.235	0.85	3.5	8.60	20.5	10,000
RG-54A/U	58.0	0.250	0.659	0.18	0.74	3.1	11.5	21.5	26.5	3,000
RG-55/U	53.5	0.206	0.659	0.36	1.3	4.8	17.0	32.0	28.5	1,900
RG-55A/U	50.0	0.216	0.659	0.36	1.3	4.8	17.0	32.0	29.5	1,900
RG-58/U	53.5	0.195	0.659	0.33	1.25	4.65	17.5	37.5	28.5	1,900
RG-58C/U	50.0	0.195	0.659	0.42	1.4	4.9	24.0	45.0	30.0	1,900
RG-59A/U	75.0	0.242	0.659	0.34	1.10	3.40	12.0	26.0	20.5	2,300
RG-59B/U	75	0.242	0.66		1.1	3.4	12		21	2,300
RG-62A/U	93.0	0.242	0.84	0.25	0.85	2.70	8.6	18.5	13.5	750
RG-74A/U	50.0	0.615	0.659	0.10	0.38	1.5	6.0	11.5	30.0	5,500
RG-83/U	35.0	0.405	0.66	0.23	0.80	2.8	9.6	24.0	44.0	2,000
RG-213/U*	50	0.405	0.66	0.16	0.6	1.9	8.0		29.5	5,000
RG-218/U**	50	0.870	0.66	0.066	0.2	1.0	4.4		29.5	11,000
RG-220/U#	50	1.120	0.66	0.04	0.2	0.7	3.6		29.5	14,000

* Formerly RG8A/U ** Formerly RG17A/U # Formerly RG19A/U

Fig A1.15 : Data for common USA flexible cables.

Diameter (inches)	MIL–C–17E	Impedance Ohms	Capacitance pF/ft(max)	Attenuation, dB/100ft				Max.V (rms)
				0.1GHz	1.0GHz	10GHz	18GHz	
0.047	–	50 ±2.5	32.0	10.7	34.0	112.0	157.0	1000
0.086	RG–405	50 ±1.0	32.0	5.7	20.0	75.0	110.0	1500
0.141	RG–402	50 ±0.5	29.9	3.0	12.0	45.0	68.0	2500
0.250	RG–401	50 ±0.5	29.9	1.95	9.0	30.0	46.0	3000

Fig A1.16 : Characteristics of common semi rigid coaxial cables.

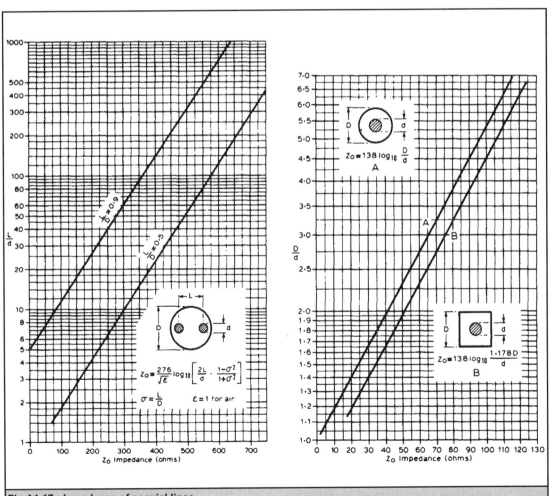

Fig A1.17 : Impedance of coaxial lines.

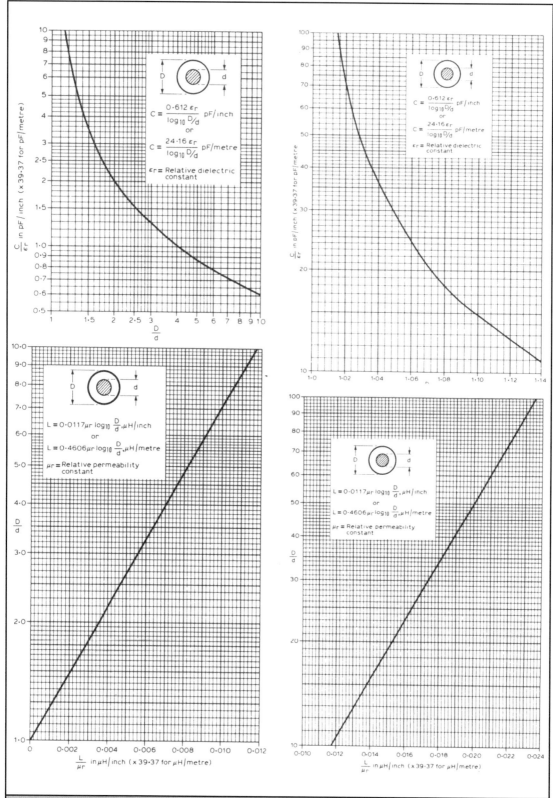

Fig A1.18 : Inductance and capacitance of coaxial lines.

Coaxial plug assembly: BNC type, male contact

- Cut end of cable even and remove 7.9mm (5/l6in) of outer sheath.

- Slide the clamp-nut and pressure sleeve over the cable. Comb out braid.

- Fold the braid back. Insert the ferrule between the braid and dielectric. Trim off excess braid. Remove 5.2mm (13/64in) of the dielectric without damaging the centre conductor. Tin the end of the conductor.

- Slide the rear insulator over the centre conductor and locate the shoulder of the insulator inside recess in ferrule. Slide the contact over the centre conductor until the shoulder of the contact is pressed hard against the rear insulator. Solder the contact to the conductor but avoid over-heating.

- Fit front insulator in body and push sub-assembly into the body as far as possible. Slide pressure sleeve into body and screw in the clamp-nut tightly to clamp cable.

Coaxial plug assembly: BNC type, female contact

- Cut end of cable even and remove 7.9mm (5/16in) of outer sheath.

- Slide the clamp-nut and pressure sleeve over cable. Comb out braid.

- Fold the braid back. Insert the ferrule between the braid and dielectric. Trim off excess braid. Remove 5.2mm (13/64in) of the dielectric without damaging the inner conductor. Tin the end of the inner conductor.

- Slide the rear insulator over the centre conductor and locate the shoulder of insulator inside recess in ferrule. Slide the contact over the centre conductor until the shoulder of the contact is pressed hard against the rear insulator. Solder contact to centre conductor but avoid over-heating.

- Fit front insulator in body and push sub-assembly into the body as far as possible. Slide pressure sleeve into body and screw in clamp-nut tight to clamp cable.

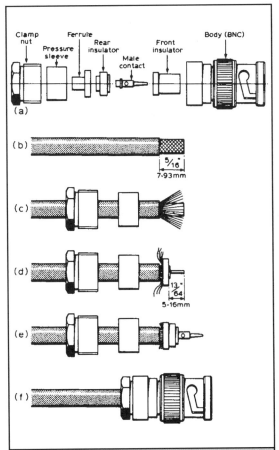

Fig A1.19 : Coaxial plug assembly, BNC type, male contact.

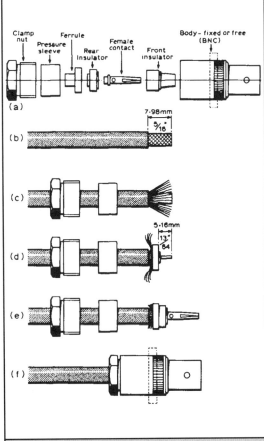

Fig A1.20 : Coaxial plug assembly, BNC type, female contact.

451

Coaxial plug assembly: N type

- Cut end of cable even and remove 8.7mm (11/32in) of outer sheath.

- Slide the clamp-nut and pressure sleeve over cable.

- Fold the braid back. Insert the ferrule between the braid and dielectric. Trim off excess braid. Remove 5.5mm (7/32in) of dielectric without damaging the centre conductor. Tin end of the centre conductor.

- Slide rear insulator over centre conductor and position against end of dielectric. Slide the contact over the centre conductor until the shoulder of the contact is pressed hard against the rear insulator Solder the contact to the conductor but avoid overheating.

- Fit front insulator in the body and push sub-assembly into the body as far as possible. Slide pressure sleeve into body and screw in the clamp-nut tightly to clamp the cable.

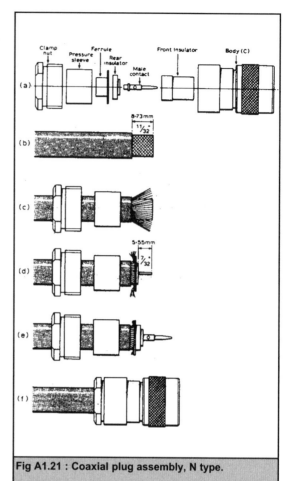

Fig A1.21 : Coaxial plug assembly, N type.

Dielectric	E_r	Loss tangent at 1MHz	Loss tangent at 10GHz	Loss tangent at 25GHz	Manufacturer
Cu Clad 217	2.17	–	0.0009	–	1
Duroid 5880	2.20	0.0004	0,0009	–	2
Cu Clad 250	2.45–2.55	0.0008	0.0018	–	1
Fused Silica Quartz	3.78	0.0001	–	0.00025	–
Duroid 6006	6.00	–	0.0025	–	2
Epsilam 6	6.00	–	0.0018	–	1
Beryllia 97%	6.90	0.0002	0.0003	–	–
Alumina 99.5%	9.70	0.0002	0.0003	–	–
DiClad 810	10.2	–	0.0027 (6.7GHz)	–	3
Epsilam 10	10.2	–	0.002	–	1
Duroid 6010.5	10.5	–	0.0028 (max) 0.0015 (typ)	–	2

Manufacturers: 1. 3M Microwave Products
 2. Rogers Corporation
 3. Keene Corporation

Fig A1.22 : Properties of some microwave dielectric pcb materials.

LDF4-50 CONNECTOR FITMENT

Tools and Materials Required for Assembly

Scale	Spacing gauge (supplied)
Knife	Solder, 63/37. RMA flux core
Pliers	Garnet cloth, 240 grit or finer
Flat file	Three wrenches: two 13/16''; one 1''
Wire brush	Solvent, comothene, vythene, or other non-
Damp cloth	flammable cleaning fluid
Jeweler's saw	Soldering iron, min. 150 W: a resistance type
or fine-toothed	iron is recommended when soldering in
hacksaw	low-temperature environments

Read Instructions Thoroughly Before Assembly

1. Prepare Cable. Straighten the end of the cable for at least 10 in (254 mm) and remove some of the jacket with a knife as shown to expose the outer conductor. Deburr the sharp end of the outer conductor.

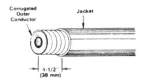

2. Mark Conductor and Remove Jacket. Scribe a cutting line with a knife on the top of the crest of the exposed, corrugated outer conductor. Remove the jacket to the dimension shown, using a straight-edged piece of heavy paper wrapped around the cable to guide the cut.

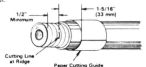

3. Clean Conductor and Add O-ring. Clean the outer conductor with solvent. Add the thick O-ring to second, fully exposed conductor groove from the jacket. Apply a thin coating of silicone grease with your finger tip to the outer surface of the O-ring and to the gasket lead chamfer in the clamping nut. **Note:** Clamping nut threads must be kept free of grease.

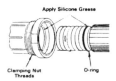

4. Add Clamping Nut and Cut Cable. Push the clamping nut fully onto the cable with a smooth twisting motion so that the spring contacts snap into the conductor groove and the O-ring seats properly against the inside surface of the clamping nut. Check that the conductor cut line is aligned with the edge of the clamping nut.

Tightly grip the clamping nut and carefully cut through the outer conductor with a fine-toothed saw. The cut must be flush with the clamping nut and shallow so that the inner conductor is not damaged. Then pull off the outer conductor with pliers. Carefully clean all foam from the inner conductor with a knife.

Cut the inner conductor to the length shown and deburr the cut end with a file.

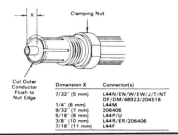

Dimension X	Connector(s)
7/32'' (5 mm)	L44N/EN/W/EW/J/T/NT DF/DM/48923/204516
1/4'' (6 mm)	L44M
9/32'' (7 mm)	206405
5/16'' (8 mm)	L44P/U
3/8'' (10 mm)	L44R/ER/206406
7/16'' (11 mm)	L44F

5. Separate Foam from Outer Conductor. Insert the tip of a knife to a depth of 1/16 inch (1.6 mm) between the foam and the outer conductor of the cable and separate them so that the outer conductor can be flared. Move the knife around the entire circumference of the outer conductor. Scrape away any foam clinging to the outer conductor and remove any burrs from the inside edge. Remove copper particles from the foam with a wire brush.

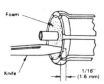

6. Flare Outer Conductor. Thread the connector outer body onto the clamping nut and tighten the connection with wrenches. Hold the clamping nut and turn only the outer body. The flaring surface of the outer body will flatten the outer conductor against the clamping nut ring. Disassemble the connection and inspect the flare to ensure good metal-to-metal contact on final assembly.

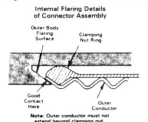

7. Install Inner Connector and Outer Body. Clean the inner conductor with solvent and slide the inner connector onto the conductor. (**Note:** The inner conductor of L44 F is installed after installing the outer body.) Insert the spacing gauge to properly position the inner connector for soldering. (Gauge is not required for L44F/P/U connectors.)

Solder the inner connector in place using the solder hole provided. Cool the connection with a damp cloth and clean the surface with garnet cloth. Make sure the connector is aligned with the axis of the cable.

The inner connector is different for each type of connector assembly as shown in the following illustration. Differences in outer body details have been deleted to simplify the illustration.

Add the large O-ring to the connector clamping nut. Apply a thin coating of silicone grease to the outer surface of the O-ring. Keep all connector threads free of grease. Thread the outer body onto the clamping nut and tighten the connection with wrenches. Hold the clamping nut and turn only the outer body.

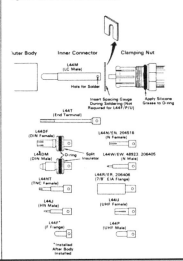

Fig A1.23 : Coaxial plug assembly, Andrews Heliax™ LDF4-50. Courtesy Andrews Antennas.

LDF5-50 CO-AX CONNECTOR ATTACHMENT
Reproduced by kind permission of Andrew Antennas

Description

These connectors are designed for self-flaring of the outer conductor and self-tapping (thread cutting) of the inner conductor of the coaxial cable. A rod (supplied) inserted through the inner connector aids in tapping the cable inner conductor. Connector L45T has screw terminals for external cable connections.

Tools and Materials Required for Assembly

Scale	Hacksaw, fine-toothed blade
Knife	Wrenches: 1-1/4 in (32 mm);
Flat file	L45F also requires 1-5/16 in (33 mm)
Wire brush	Solvent: comothene, vythene, or other non-flammable cleaning fluid

Notice

The installation, maintenance or removal of antenna systems requires qualified, experienced personnel. Andrew installation instructions have been written for such personnel. Antenna systems should be inspected once a year by qualified personnel to verify proper installation, maintenance and condition of equipment.

Andrew disclaims any liability or responsibility for the results of improper or unsafe installation practices.

Read Instructions Thoroughly Before Assembly

1. Prepare Cable. Straighten the end of the cable for at least 10 inches (254 mm) and remove some of the jacket with a knife to expose the outer conductor. Deburr the sharp end of the outer conductor.

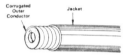

2. Mark Conductor and Remove Jacket. Scribe a cutting line with a knife on the ridge of exposed, corrugated outer conductor. Remove the jacket to the dimension shown, using a straight-edged piece of heavy paper wrapped around the cable to guide the cut.

5. Separate Foam from Outer Conductor. Insert the tip of a knife to a depth of 3/32 inch (2 mm) between the foam and the outer conductor of the cable and separate them so that the outer conductor can be flared. Move the knife around the entire circumference of the outer conductor. Scrap away any foam clinging to the outer conductor. Remove any burrs from the inner edges of both conductors. Remove copper particles from the foam with a wire brush. Remove copper and foam particles from the interior of the inner conductor by holding the assembly downward and sharply tapping the clamping nut.

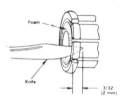

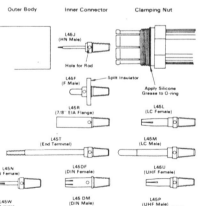

3. Clean Conductor and Add O-ring. Clean the outer conductor with solvent. Add the thick O-ring gasket to the second, fully-exposed conductor groove from the jacket. Apply a thin coating of silicone grease with your finger tip to the outer surface of the O-ring and to the gasket lead chamfer in the clamping nut. **Note:** Clamping nut threads must be kept free of grease.

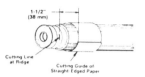

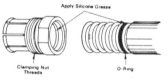

4. Add Clamping Nut and Cut Cable. Push the clamping nut fully onto the cable with a twisting motion so that the spring contacts snap into the conductor groove. Check that the conductor cutting line is aligned with the edge of the clamping nut.

Tightly grip the clamping nut and carefully cut off the cable with a fine-toothed saw so that the cable end is flush with the end of the clamping nut. **Note:** After cutting, verify that the saw cut is flush with the clamping nut. If the cable protrudes, file it flush with the nut.

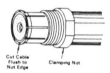

6. Flare Outer Conductor. Thread the connector outer body onto the clamping nut and tighten the connection with wrenches. Hold the clamping nut and turn only the outer body. The flaring surface of the outer body will flatten the outer conductor against the clamping nut ring. Disassemble the connection and inspect the flare to ensure good metal-to-metal contact on final assembly.

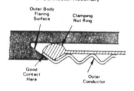

7. Install Inner Connector and Outer Body. Insert the rod supplied through the hole in the inner connector. Twist the threaded end of the inner connector clockwise into the cable inner conductor and continue until the shoulder of the connector touches the inner conductor. Lubricate the threads with a small amount of solvent to aid tapping. Unscrew the connector slightly after every few turns if tapping becomes difficult.

The inner connector is different for each type of connector assembly as shown in the following illustration. Differences in outer body details have been deleted to simplify the illustration.

Add the split insulator to the inner connector of L45F.

Add the large O-ring to the connector clamping nut. Apply a thin coating of silicone grease to the outer surface of the O-ring. Keep all connector threads free of grease. Thread the outer body onto the clamping nut and tighten the connection with wrenches. Hold the clamping nut and turn only the outer body.

Fig A1.24 : Coaxial plug assembly, Andrews Heliax™ LDF5-50. Courtesy Andrews Antennas.

WG no	EIA Desig.	IEC Desig.	Recommended Freq Range (GHz)	Cutoff Freq (GHz)	Internal Dimensions (mm)	External Dimensions (mm)	Wall (mm)	Aspect Ratio
WG00	WR2300	R3	0.32 – 0.49	0.257	584.2 292.1	590.6 298.5	3.175	2.00
WG0	WR2100	R4	0.35 – 0.53	0.281	533.4 266.7	539.8 273.1	3.175	2.00
WG1	WR1800	R5	0.42 – 0.62	0.328	457.2 228.6	463.6 235.0	3.175	2.00
WG2	WR1500	R6	0.49 – 0.75	0.393	381.0 190.5	387.4 196.9	3.175	2.00
WG3	WR1150	R8	0.64 – 0.96	0.513	292.10 146.05	298.45 152.40	3.175	2.00
WG4	WR975	R9	0.75 – 1.12	0.605	247.65 123.82	254.00 130.18	3.175	2.00
WG5	WR770	R12	0.96 – 1.45	0.766	195.58 97.79	201.93 104.14	3.175	2.00
WG6	WR650	R14	1.12 – 1.70	0.908	165.10 82.55	169.16 86.61	2.032	2.00
WG7	WR510	R18	1.45 – 2.20	1.157	129.54 64.77	133.60 68.83	2.032	2.00
WG8	WR430	R22	1.70 – 2.60	1.372	109.22 54.61	113.28 58.67	2.032	2.00
WG9	WR340	R26	2.20 – 3.30	1.686	88.90 44.45	92.96 48.45	2.032	2.00
WG9A	WR284	R32	2.60 – 3.95	1.736	86.36 43.18	90.42 47.24	2.032	2.00
WG10	–	R35		2.078	72.14 34.04	76.20 38.10	2.032	2.12
–	–	–		2.258	66.37 29.50	70.37 33.50	2.000	2.25
WG11	–	–		2.448	60.25 28.50	63.50 31.75	1.626	2.11
WG11A	WR229	R40	3.30 – 4.90	2.577	58.17 29.08	61.42 32.33	1.626	2.00
–	–	R41		2.630	57.00 25.33	61.00 29.33	2.000	2.25
WG12	WR187	R48	3.95 – 5.85	3.152	47.55 22.15	50.80 25.50	1.626	2.15
WG12A	–	–		2.951	50.80 16.92			3.00
WG13	WR159	R58	4.90 – 7.05	3.711	40.39 20.19	43.64 23.44	1.626	2.00
WG14	WR137	R70	5.85 – 8.20	4.301	34.85 15.80	38.10 19.05	1.626	2.21
WG15	WR112	R84	7.05 – 10.0	5.260	28.50 12.62	31.75 15.88	1.626	2.26
Old English				5.902	25.40 12.70			2.00
WG16	WR90	R100	8.20 – 12.4	6.557	22.86 10.16	25.40 12.70	1.270	2.25
WG17	WR75	R120	10.0 – 15.0	7.869	19.05 9.525	21.59 12.07	1.270	2.00
WG18	WR62	R140	12.4 – 18.0	9.488	15.80 7.899	17.83 9.931	1.016	2.00
WG19	WR51	R180	15.0 – 22.0	11.57	12.95 6.477	14.99 8.509	1.016	2.00
WG20	WR42	R220	18.0 – 26.5	14.05	10.67 4.318	12.70 6.350	1.016	2.47
WG21	WR34	R260	22.0 – 33.0	17.36	8.636 4.318	10.67 6.350	1.016	2.00
WG22	WR28	R320	26.5 – 40.0	21.08	7.112 3.556	9.144 5.588	1.016	2.00
WG23	WR22	R400	33.0 – 50.0	26.34	5.690 2.845	7.722 4.877	1.016	2.00
WG24	WR19	R500	40.0 – 60.0	31.39	4.775 2.388	6.807 4.420	1.016	2.00
WG25	WR15	R620	50.0 – 75.0	39.88	3.759 1.880	5.791 3.912	1.016	2.00
WG26	WR12	R740	60.0 – 90.0	48.37	3.099 1.549	5.131 3.581	1.016	2.00
WG27	WR10	R900	75.0 – 110	59.01	2.540 1.270	4.572 3.302	1.016	2.00
WG28	WR8	R1200	90.0 – 140	73.77	2.032 1.016	4.064 3.048	1.016	2.00
WG29	WR7	R1400	110 – 170	90.79	1.651 0.826	3.175 2.350	0.762	2.00
WG30	WR5	R1800	140 – 220	115.75	1.295 0.648	2.819 2.172	0.762	2.00
WG31	WR4	R2200	170 – 260	137.27	1.092 0.546	2.616 2.070	0.762	2.00
WG32	WR3	R2600	220 – 325	173.49	0.864 0.432	2.388 1.956	0.762	2.00

Abbreviations: IEC – International Electrotechnical Commission, EIA – Electronic Industries Association (USA).

Fig A1.25 : Waveguide sizes and characteristics.

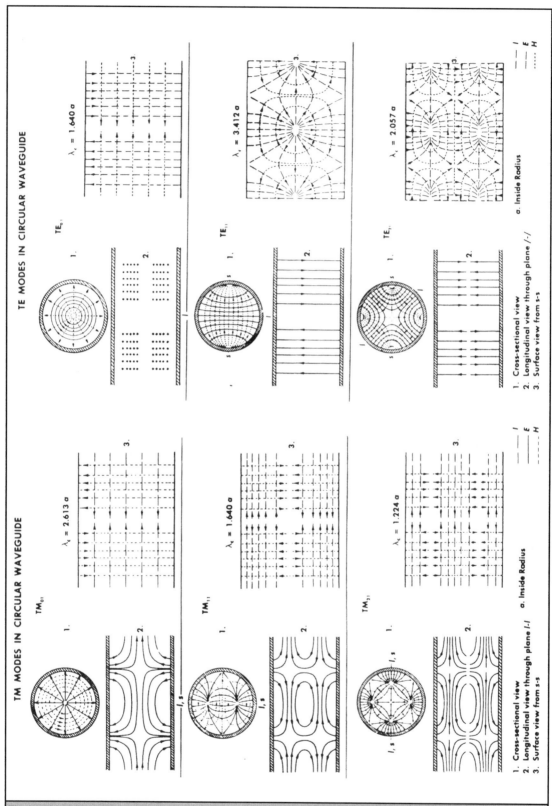

Fig A1.26 : Modepatterns in circular waveguide. From Marcuvitz (ed), Waveguide Handbook, MIT Radiation Laboratory Series Vol 10, McGraw Hill © 1951. Reprinted by permission of McGraw Hill Inc.

MIL		IEEE (std 521)		Revised JCS		Hewlett Packard	
Desc.	Frequency	Desc.	Frequency	Desc.	Frequency	Desc.	Frequency
		HF	3–30MHz	A	0–250MHz		
		VHF	30–300MHz	B	250–500MHz		
		UHF	300MHz–1GHz	C	500MHz–1GHz		
		L	1–2GHz	D	1–2GHz		
		S	2–4GHz	E	2–3GHz	S	2.6–3.95GHz
		C	4–8GHz	F	3–4GHz		
						G	3.95–5.85GHz
		X	8–12GHz	G	4–6GHz		
		Ku (J)	12–18GHz	H	6–8GHz		
						J	5.85–8.2GHz
K	18–26.5GHz	K	18–27GHz	I	8–10GHz		
						X	8.2–12.4GHz
Ka	26.5–40GHz	Ka	27–40GHz	J	10–20GHz	M	10–15GHz
Q	33–50GHz	(Q)				P(Ku)	12.4–18GHz
						N	15–22GHz
U	40–60GHz	mm	40–300GHz	K	20–40GHz	K	18–26.5GHz
V	50–75GHz					R(Ka)	26.5–40GHz
E	60–90GHz			L	40–60GHz		
W	75–110GHz						
F	90–140GHz			M	60–100GHz		
D	110–170GHz						
G	140–220GHz						

Fig A1.27 : Designation of frequency bands, there are many and confusing band designations. Here are some of the more widely used ones.

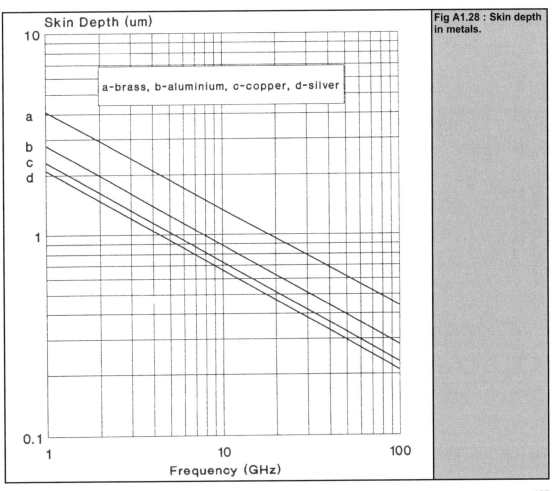

Skin Depth (um)

a-brass, b-aluminium, c-copper, d-silver

Frequency (GHz)

Fig A1.28 : Skin depth in metals.

Summary of Transmission Line Equations

Quantity	General Line Expression	Ideal Line Expression
Propagation constant	$\gamma = \alpha + j\beta = \sqrt{(R + j\omega L)(G + j\omega C)}$	$\gamma = j\omega\sqrt{LC}$
Phase constant β	Imaginary part of γ	$\beta = \omega\sqrt{LC} = \dfrac{2\pi}{\lambda}$
Attenuation constant α	Real part of γ	0
Characteristic impedance	$Z_o = \sqrt{\dfrac{R + j\omega L}{G + j\omega C}}$	$Z_o = \sqrt{\dfrac{L}{C}}$
Input impedance	$Z_{-t} = Z_o \dfrac{Z_r + Z_o \tanh \gamma\ell}{Z_o + Z_r \tanh \gamma\ell}$	$Z_{-t} = Z_o \dfrac{Z_r + j Z_o \tan \beta\ell}{Z_o + j Z_r \tan \beta\ell}$
Impedance of short-circuited line. $Z_r = 0$	$Z_{s.c.} = Z_o \tanh \gamma\ell$	$Z_{s.c.} = j Z_o \tan \beta\ell$
Impedance of open-circuited line. $Z_r = \infty$	$Z_{o.c.} = Z_o \coth \gamma\ell$	$Z_{o.c.} = -j Z_o \cot \beta\ell$
Impedance of line an odd number of quarter wavelengths long	$Z = Z_o \dfrac{Z_r + Z_o \coth \alpha\ell}{Z_o + Z_r \coth \alpha\ell}$	$Z = \dfrac{Z_o^2}{Z_r}$
Impedance of line an integral number of half wavelengths long	$Z = Z_o \dfrac{Z_r + Z_o \tanh \alpha\ell}{Z_o + Z_r \tanh \alpha\ell}$	$Z = Z_r$
Voltage along line	$V_{-t} = V_i(1 + \Gamma_o e^{-2\gamma\ell})$	$V_{-t} = V_i(1 + \Gamma_o e^{-2j\beta\ell})$
Current along line	$I_{-t} = I_i(1 - \Gamma_o e^{-2\gamma\ell})$	$I_{-t} = I_i(1 - \Gamma_o e^{-2j\beta\ell})$
Voltage reflection coefficient	$\Gamma = \dfrac{Z_r - Z_o}{Z_r + Z_o}$	$\Gamma = \dfrac{Z_r - Z_o}{Z_r + Z_o}$

Useful approximations for small VSWR's

If u and v are both small with respect to unity,

(a) $(1 + u)(1 + v) = 1 + u + v$

(b) $(1 - u)(1 - v) = 1 - u - v$

(c) $(1 \pm u)^2 = 1 \pm 2u$

(d) $\dfrac{1 + v}{1 + u} = 1 + v - u$

Examples:

(a) If $r_1 = 1.03$ and $r_2 = 1.08$, then $r_1 r_2 = 1.11$

(b) If $\Gamma_1 = .96$ and $\Gamma_2 = .95$, then $\Gamma_1 \Gamma_2 = .91$

(c) If $r = 1.04$, $r^2 = 1.08$

If $\Gamma = .96$, $\Gamma^2 = .92$

(d) If $r_1 = 1.03$ and $r_2 = 1.08$, then $\dfrac{r_2}{r_1} = 1.05$

Fig A1.29 : Summary of transmission line equations. From Kagan (ed), Microwave Transmission Circuits, MIT Radiation Laboratory Series Vol 9, McGraw Hill © 1951. Reprinted by permission of McGraw Hill.

SWG	mm	SWG	mm	SWG	mm	SWG	mm
7/0	12.7	9	3.66	24	0.599	39	0.132
6/0	11.8	10	3.25	25	0.508	40	0.122
5/0	11.0	11	2.95	26	0.457	41	0.112
4/0	10.2	12	2.64	27	0.417	42	0.102
3/0	9.45	13	2.34	28	0.376	43	0.091
2/0	8.84	14	2.03	29	0.345	44	0.081
0	8.23	15	1.83	30	0.315	45	0.071
1	7.62	16	1.63	31	0.295	46	0.061
2	7.01	17	1.42	32	0.274	47	0.051
3	6.40	18	1.22	33	0.254	48	0.041
4	5.89	19	1.02	34	0.234	49	0.030
5	5.38	20	0.914	35	0.213	50	0.025
6	4.88	21	0.813	36	0.193		
7	4.47	22	0.711	37	0.173		
8	4.06	23	0.610	38	0.152		

Fig A1.30 : Standard wire (and sheet metal) gauge. The dimension columns (mm) refer to wire diameter and sheet metal thickness.

Diameter	Pitch	Clearance	Tapping
1.6	0.35	1.65	1.2
2	0.40	2.05	1.6
2.5	0.45	2.60	2.0
3	0.50	3.10	2.5
4	0.70	4.10	3.3
5	0.80	5.10	4.1
6	1.00	6.10	4.9
8	1.25	8.20	6.7
10	1.50	10.20	8.4
12	1.75	12.20	10.1

Fig A1.31 : Tapping and clearance drill sizes. I.S.O. metric, coarse thread. All dimensions in mm. Note that these screws increase in steps of 0.2mm from 1.2mm to 2.2mm, steps of 0.5mm from 2.5 to 5.0mm and steps of 1mm from 5mm to 11mm. Larger sizes are not given, but can be obtained from any engineer's handbook.

No.	Diameter	Pitch	Clearance	Tapping
0	6.0	1.00	6.10	5.10
1	5.3	0.90	5.40	4.50
2	4.7	0.81	4.80	4.00
3	4.1	0.73	4.20	3.40
4	3.6	0.66	3.70	3.00
5	3.2	0.59	3.30	2.65
6	2.8	0.53	2.90	2.30
7	2.5	0.48	2.60	2.05
8	2.2	0.43	2.25	1.80
9	1.9	0.39	1.95	1.55
10	1.7	0.35	1.75	1.40
11	1.5	0.31	1.60	1.20
12	1.3	0.28	1.40	1.05

Fig A1.32 : Tapping and clearance drill sizes, British Association (BA) thread. Dimensions are given in mm. Odd numbered sizes may occasionally be used on older electrical equipment. Even numbered sizes were, before metrification, "preferred" for electronic and model making use.

Inductance of a single-layer coil

The inductance of a single-layer coil of length at least equal to its radius is given by

$$L = \frac{N^2 r^2}{9r + 10l} \ \mu H$$

where r = radius of coil (in)
  l = length of coil (in)
  N = number of turns

This applies to both close-wound and spaced-turn coils. Correspondingly, the number of turns for a given inductance is

$$N = L \sqrt{\frac{9r + 10l}{r^2}}$$

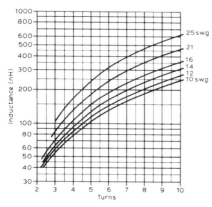

The inductance of 0·375in internal diameter coils with turns spaced one diameter apart

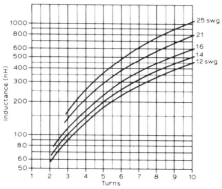

The inductance of 0·5in internal diameter coils, with turns spaced one diameter apart

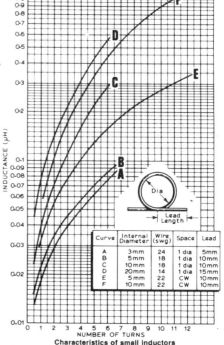

Characteristics of small inductors

Curve	Internal Diameter	Wire (swg)	Space	Lead
A	3mm	24	1 dia	5mm
B	5mm	18	1 dia	10mm
C	10mm	18	1 dia	10mm
D	20mm	14	1 dia	15mm
E	5mm	22	CW	10mm
F	10mm	22	CW	10mm

Self inductance of a straight wire

At radio frequencies, the self inductance of a straight round wire is given by

$$L = 0 \cdot 0021 \ (2 \cdot 303 \log_{10} \frac{4l}{d} - 1) \ \mu H$$

where l = length in centimetres
  d = dia in centimetres

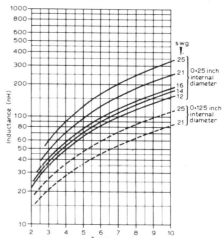

The inductance of 0·25in internal diameter coils with turns spaced one diameter apart

---

Fig A1.33 : Inductance of small coils.

Power and voltage ratios in decibels

Power and voltage ratios are normally expressed in decibels where

$$N(dB) = 10\log_{10}\frac{P_2}{P_1}$$

where P_1 and P_2 are the power ratios being compared. On the assumption of constant impedance, the corresponding voltage ratios V_2 and V_1 may be used,

$$N(dB) = 20\log_{10}\frac{V_2}{V_1}$$

A value in decibels only has absolute meaning if the reference level is stated. The expressions dBm and dBW

are frequently used to express decibels with respect to 1mW and 1W respectively. The table below gives the decibel equivalents of a wide range of voltage and power ratios.

Relationship between dBm and voltage

A power level of 1mW into a 600Ω or 50Ω resistance has become a standard for comparative purposes. It is the datum 0dBm. Signal levels above and below this datum are expressed in ± dBm; they correspond to finite voltage (or current) levels not ratios.

0dBm into a 600Ω resistance corresponds to 0.775V

0dBm into a 50Ω resistance corresponds to 0.225V

Decibel table

Voltage ratio (equal impedance)	Power ratio	dB +	Voltage ratio (equal impedance)	Power ratio
1·000	1·000	0	1·000	1·000
0·989	0·977	0·1	1·012	1·023
0·977	0·955	0·2	1·023	1·047
0·966	0·933	0·3	1·035	1·072
0·955	0·912	0·4	1·047	1·096
0·944	0·891	0·5	1·059	1·122
0·933	0·871	0·6	1·072	1·148
0·923	0·851	0·7	1·084	1·175
0·912	0·832	0·8	1·096	1·202
0·902	0·813	0·9	1·109	1·230
0·891	0·794	1·0	1·122	1·259
0·841	0·708	1·5	1·189	1·413
0·794	0·631	2·0	1·259	1·585
0·750	0·562	2·5	1·334	1·778
0·708	0·501	3·0	1·413	1·995
0·668	0·447	3·5	1·496	2·239
0·631	0·398	4·0	1·585	2·512
0·596	0·355	4·5	1·679	2·818
0·562	0·316	5·0	1·778	3·162
0·531	0·282	5·5	1·884	3·548
0·501	0·251	6·0	1·995	3·981
0·473	0·224	6·5	2·113	4·467
0·447	0·200	7·0	2·239	5·012
0·422	0·178	7·5	2·371	5·623
0·398	0·159	8·0	2·512	6·310
0·376	0·141	8·5	2·661	7·079
0·355	0·126	9·0	2·818	7·943
0·335	0·112	9·5	2·985	8·913
0·316	0·100	10	3·162	10·00
0·282	0·0794	11	3·55	12·6
0·251	0·0631	12	3·98	15·9
0·224	0·0501	13	4·47	20·0
0·200	0·0398	14	5·01	25·1
0·178	0·0316	15	5·62	31·6
0·159	0·0251	16	6·31	39·8
0·141	0·0200	17	7·08	50·1
0·126	0·0159	18	7·94	63·1
0·112	0·0126	19	8·91	79·4
0·100	0·0100	20	10·00	100·0
$3·16 \times 10^{-2}$	10^{-3}	30	$3·16 \times 10$	10^3
10^{-2}	10^{-4}	40	10^2	10^4
$3·16 \times 10^{-3}$	10^{-5}	50	$3·16 \times 10^2$	10^5
10^{-3}	10^{-6}	60	10^3	10^6
$3·16 \times 10^{-4}$	10^{-7}	70	$3·16 \times 10^3$	10^7
10^{-4}	10^{-8}	80	10^4	10^8
$3·16 \times 10^{-5}$	10^{-9}	90	$3·16 \times 10^4$	10^9
10^{-5}	10^{-10}	100	10^5	10^{10}
$3·16 \times 10^{-6}$	10^{-11}	110	$3·16 \times 10^5$	10^{11}
10^{-6}	10^{-12}	120	10^6	10^{12}

Fig A1.34 : Decibel table.

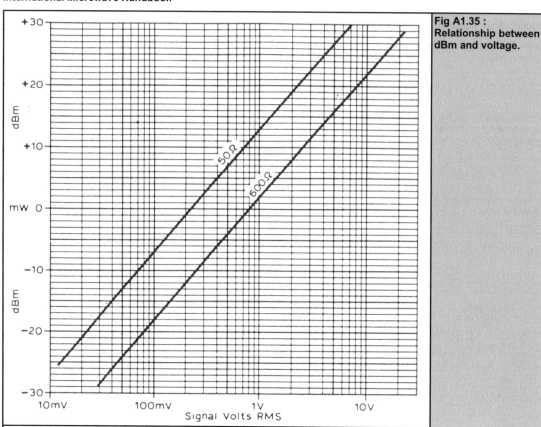

Fig A1.35 :
Relationship between
dBm and voltage.

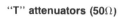

"T" attenuators (50Ω)

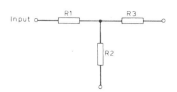

Attenuation (dB)	R1 Resistance (Ω)	R1 Dissipation (mW)	R2 Resistance (Ω)	R2 Dissipation (mW)	R3 Resistance (Ω)	R3 Dissipation (mW)	Total dissipation (mW)
1	2·875	57	433·3	102	2·875	46	205
2	5·731	114	215·2	180	5·731	72	360
3	8·55	171	141·9	242	8·55	87	446
4	11·31	226	104·8	285	11·31	90	520
5	14·01	280	82·24	314	14·01	88	682
6	16·61	332	66·93	332	16·61	83	747
7	19·12	387	55·80	341	19·12	76	804
8	21·53	430	47·31	343	21·53	68	841
10	25·97	519	35·14	328	25·97	52	899
12	29·92	598	26·81	300	29·92	38	936
15	34·9	710	18·36	235	34·9	21	966
20	40·91	818	10·10	168	40·91	10	996
25	44·67	894	5·641	98	44·67	5	997
30	46·9	938	3·165	59	46·9	1	998
35	48·25	970	1·779	29	48·25	<1	999
40	49·01	980	1·000	20	49·01	<1	1,000

Dissipations based on 1W input with a 50Ω termination.

Fig A1.36 : "T" attenuator resistor values.

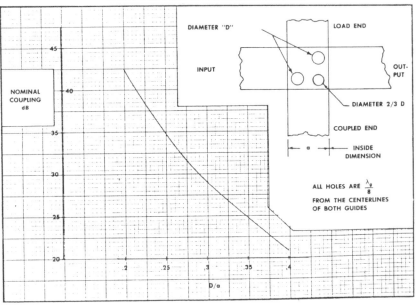

ROUND HOLE CROSS GUIDE DIRECTIONAL COUPLER

Courtesy of Gershon J. Wheeler.

Fig A1.37 : Design of round hole directional couplers. © Artech House Inc, from Theodore S Saad, Microwave Engineers Handbook Vol 2 1971. Reprinted by permission.

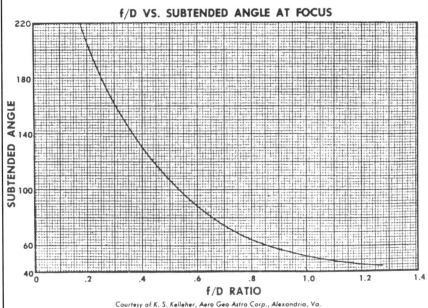

f/D VS. SUBTENDED ANGLE AT FOCUS

Courtesy of K. S. Kelleher, Aero Geo Astro Corp., Alexandria, Va.

Fig A1.38 : Angle subtended at a dish feed. © Artech House Inc, from Theodore S Saad, Microwave Engineers Handbook Vol 2 1971. Reprinted by permission.

Index